Control Systems

(With Matlab Programs)

Control Systems

(With Matlab Programs)

Sonveer Singh
Acting Head
Electronics Engineering
Technical Education Department
Government of Uttar Pradesh

Sanjay Agrawal
Professor
Electrical Engineering
School of Engineering and Technology
IGNOU, New Delhi

TECHSAR™
NEW DELHI

Published and Printed by

Techsar Pvt. Ltd.
4435-36/7, Ansari Road, Daryaganj
New Delhi-110 002 (India)
E-mail: tech_info@techsarworld.com
Website: www.techsarworld.com

ISBN: 978-93-90620-01-2

Reprint 2025

Dedicated to

The Goddess Ma Saraswati

and

Parents

Who are responsible for

What I am today

Foreword

The world is incomplete without engineers. We cannot think of a day's work without being surrounded by marvels designed and executed by engineers. As we embark on this engineering course that would require us to imagine things that don't exist and lay the groundwork for turning the intangible into tangible. we must know that people will know us more for our work than our words.

We must remember to design using science and present it as a perfect art piece. This book has been written to educate us about the various engineering control systems and their essence in our daily lives. Undergraduate engineering students of different branches in different universities have been waiting for a book that deals exclusively with the principles of control theory and effective designing techniques to get the desired behaviour in a controlled manner.

As someone who believes that design and science must co-exist and not replace each other to achieve the best results, we hope that the contents of this book will satisfy our appetite for learning and solve our queries regarding some of the important facets of engineering.

The contents of this book are the culmination of many years of hard work and experience of the authors.

I compliment the authors: Prof Sanjay Agrawal and Dr Sonveer Singh who perceived this book in manner that it presents the most complex subject matters in the simplest form possible.

Studying can be hard, but not when we do it with a devoted mind. I am sure that this book will be appreciated and coveted by aspiring engineers who dream of making this world a more liveable place for all of us.

To students and learners reading this book, we must encourage them to pay close attention to every idea presented in this book. The book focuses on what it says if we want to get the most out of our engineering education. If we want to succeed and excel down the road as an engineering professional, it is imperative for us to learn these things.

New Delhi
20th September, 2022

Prof. Nageshwar Rao
Vice-Chancellor

इन्दिरा गाँधी राष्ट्रीय मुक्त विश्वविद्यालय
Indira Gandhi National Open University
मैदान गढ़ी, नई दिल्ली–110068 भारत | www.hindi.ignou.ac.in
Maidan Garhi, New Delhi–110068 | www.ignou.ac.inwww.

Preface

This book is designed for an undergraduate course in control system which will be very useful for the faculty who is teaching this subject as well as the students who are pursuing undergraduate engineering course and also those students who are appearing in the professional examinations. It will also help the practising engineers to enhance their knowledge on this subject. The authors have made sincere attempts to explain the basic principle of controls systems in simple language, and provide step-by-step mathematical explanation for better understanding.

The problems and their solutions are listed in the order of difficulty and complexity. Most of the examples also help to understand the theoretical topics very clearly. Most of the problems/questions have been collected from the previous year's examination papers of GATE, IES or of equivalent standard and other universities. This will help the students gain more confidence in tackling the examinations.

The authors welcome constructive criticism and helpful suggestions for further improvement of this book.

Sonveer Singh
Sanjay Agrawal

Acknowledgements

The authors would like to acknowledge the contributions of faculty and students, across the India, for their encouragement and support. We would also like to express our sincere thanks to Professor Nageshwar Rao, Vice-Chancellor, Indira Gandhi National Open University, New Delhi. We also take this opportunity to express my gratitude and thanks to Prof. D.K. Chaturvedi and Prof. A.L. Saxena, Department of Electrical Engineering, Dayal Bagh Educational Institute, Deemed University Agra, Prof. D.S. Chauhan, Vice-Chancellor, GLA University, Mathura and Prof. (Rtd.). S. Basu, Electrical Engineering, AMU, Aligarh for their suggestions and encouragement was vital to the completion of this volume.

The patience of our family in bearing with this lengthy project in good humour is greatly appreciated. Last but not the least, we express our deep gratitude to our respected parents and loving family members for their blessings, which helped us to reach our target.

Sonveer Singh
Sanjay Agrawal

Contents

CHAPTER

1 Introduction to Control System

1.1 INTRODUCTION

In modern era, a Control System plays a vital roles in human life. The question arises that: What is a Control System? To answer the question, we can say that in our daily life there are numerous objectives that need to be accomplished. For example, in the domestic purpose, we need to control the temperature and humidity of homes, offices and buildings for comfortable living. Another example for transportation we need to control the automobile vehicle and airplane to go from one place to another place accurately and safely.

Control System is an interconnection of components forming a system configuration in which quantity of interest is maintained or altered in accordance with a desired manner.

Basic control system: The basic part of the control system can be described by:

1. Input or objective of control
2. Control action or control system
3. Response or output.

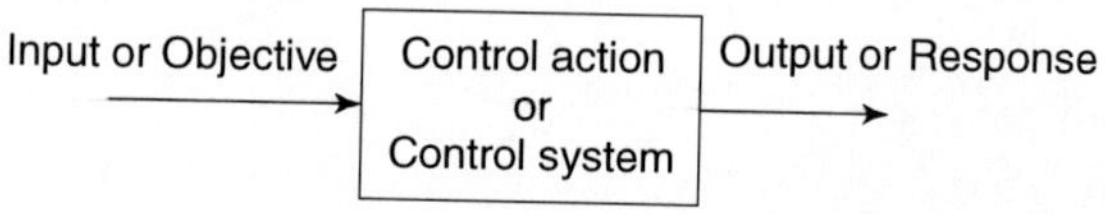

Figure 1.1 Basic control system

Examples of Control System. Simple example of control system, consider the steering control of an automobile vehicle.

Input. Direction of the steering wheel.

Output. Direction of the two front wheels.

Control action. Composed of the steering mechanism and the dynamics of the entire automobile.

The control system can be classified as open loop control system and closed loop control system.

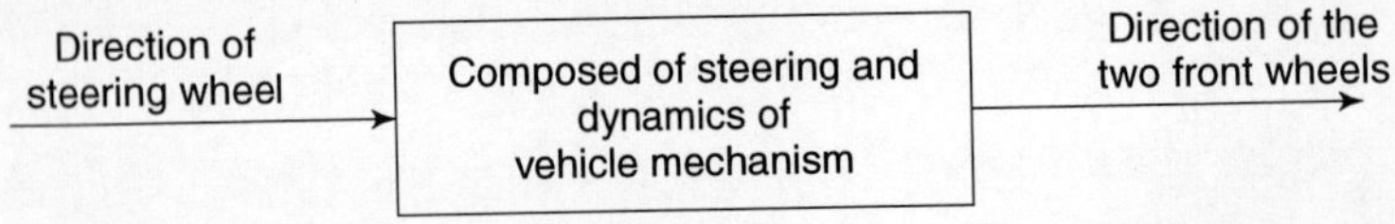

Figure 1.2

1.2 OPEN LOOP CONTROL SYSTEM

Without feedback or non-feedback control system is known as open loop control system.

The element of an open loop control system can usually be divided into two parts: the actuating device and the controlled process. As shown in Fig. 1.3. The actuating device may be filter, amplifier, attenuator, mechanical linkage, etc., depend upon the system.

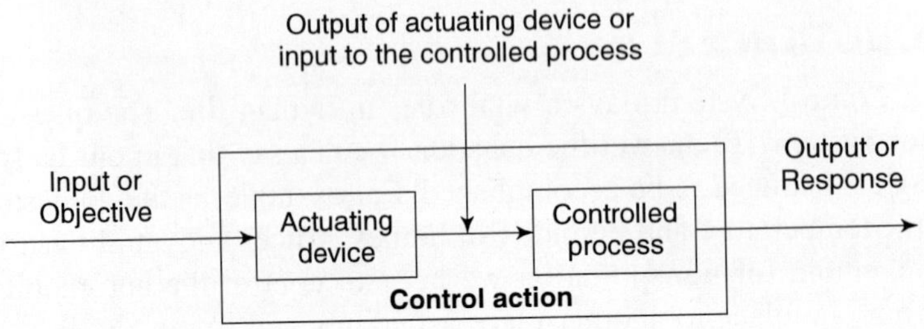

Figure 1.3 Element of an open loop control system

Because of the simplicity and economy of open loop control system, we find this type of system in many non-critical applications. Basically, these systems are economical but usually inaccurate.

Example of Open Loop Control System:

1. ***Bread toaster***

   Input. Raw bread

   Output. Bread toasted

   Control action. Heating the bread

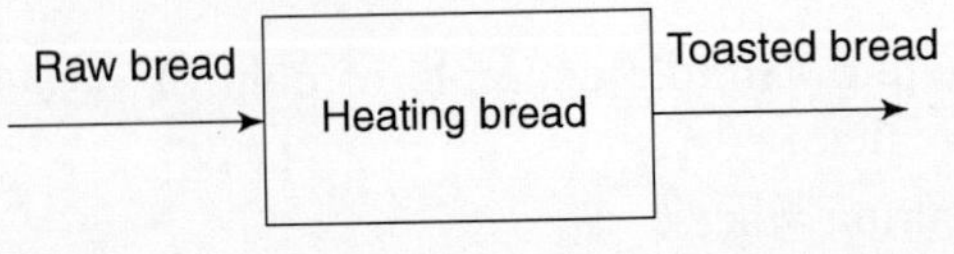

Figure. 1.4 Block diagram of bread toaster

This control action, has nothing to do with the status of toast like no bread in toaster, its burnt, etc., only work with in set time so it is open loop.

2. ***Electric heater***

Input. Current

Output. Heat

Control action. Power loss (i^2 R)

This control action, has nothing to do with the amount of heat and starts and stops after a fixed time, so it is open loop.

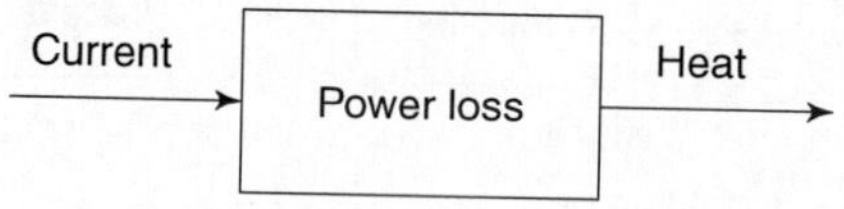

Figure 1.5 Block diagram of electric heater

3. ***Automatic washing machine***

Input. Dirty clothes

Output. Clean clothes

Control action. Washes the clothes for set time

Figure 1.6 Block diagram of washing machine

In case of automatic washing machine, it rotates the clothes with set time, it does not worry about the degree of cleanness and there is no change in mechanism accordingly.

1.2.1 Advantages and Disadvantages of Open Loop Control System

Advantages

1. Open loop control systems are simple in construction.
2. Open loop control system is cheap.
3. Generally, the open loop systems are stable.
4. The maintenance required for open loop system is less.
5. Calibration of open loop control system is easy.

Disadvantages

1. Open loop systems are inaccurate. The accuracy depends upon the calibration of input.
2. The open loop systems are not reliable, the operation of these is affected due to the presence of non-linearties in its element.
3. Optimization of open loop system is not possible.

- *An open loop control system utilizes an actuating device to control the process directly, that means the output has no effect (or control) on the control action.*

1.3 CLOSED LOOP CONTROL SYSTEM

Feeddback control system is known as closed-loop control system. It maintains a prescribed relationship between the output and the reference input by comparing them and using the difference as a means of control (called error signal).

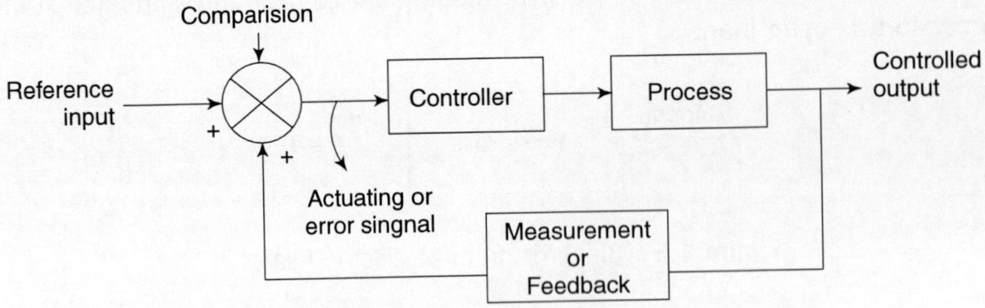

Figure 1.7 Block diagram of closed-loop feedback control system

In this type of system the output is compared with the reference input, the output of comparator is actuating or error signal is input to the controller and the output of controller is input to the process and desired output is obtained.

Examples of Closed-loop Control System

1. *Room temperature control system.* In room temperature control system, it measures the actual room temperature and compares it with the reference temperature, the thermostat turns on or off the heating or cooling equipment.

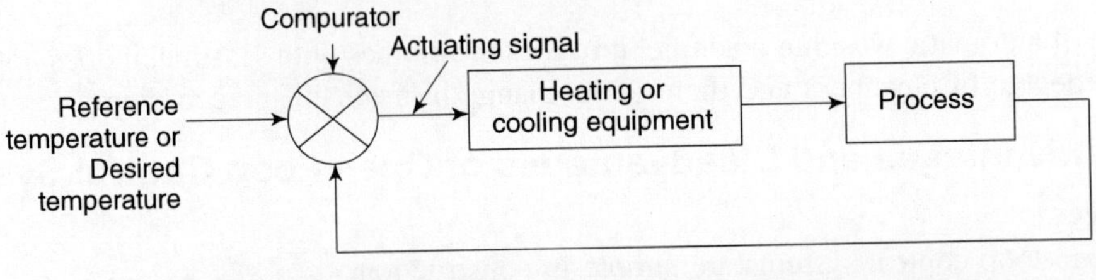

Figure 1.8 Black diagram of room temperature control system

2. *Body temperature and blood pressure control.* Feedback control systems are not limited to engineering. This can be found in various non-engineering fields like human body, for instance, is highly advanced. Body temperature and blood pressure are kept constant by means of physiological feedback.

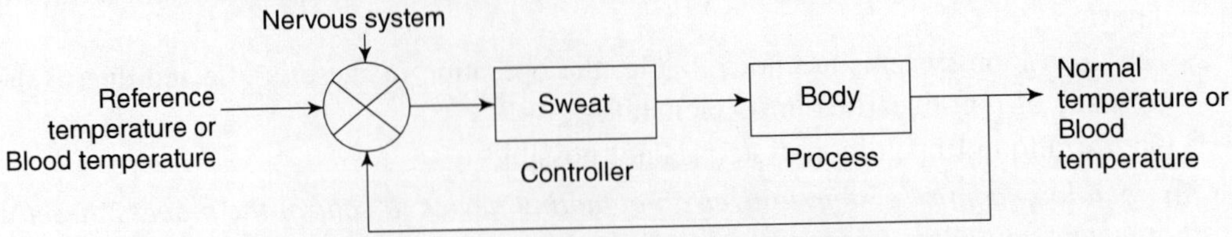

Figure 1.9 Body temperature control

1.3.1 Advantages and Disadvantages of Cloosed-Loop Control System

Advantages

1. The use of feedback makes the system response relatively insensitive to internal disturbances and internal variations in system parameter.
2. Optimization is possible.
3. Closed-loop systems are accurate.
4. Closed-loop systems are reliable because they adjust non-linearities present in its element.

Disadvantages

1. Closed-loop control systems are complex in construction.
2. Closed-loop systems are more expensive.
3. Closed-loop control systems can become unstable under certain conditions.
4. More maintenance is required.

- *A closed-loop control system uses a measurement of the output and feedback of this signal to compare it with the reference input that means the output has an effect on the control action.*

1.4 SERVOMECHANISM

Servomechanism is a process in which the mechanical position is controlled using feedback control. The controlled variable is mechanical position or derivative of position with time like velocity and acceleration.

Servo system, used to position a load shaft, is shown Fig. 1.10.

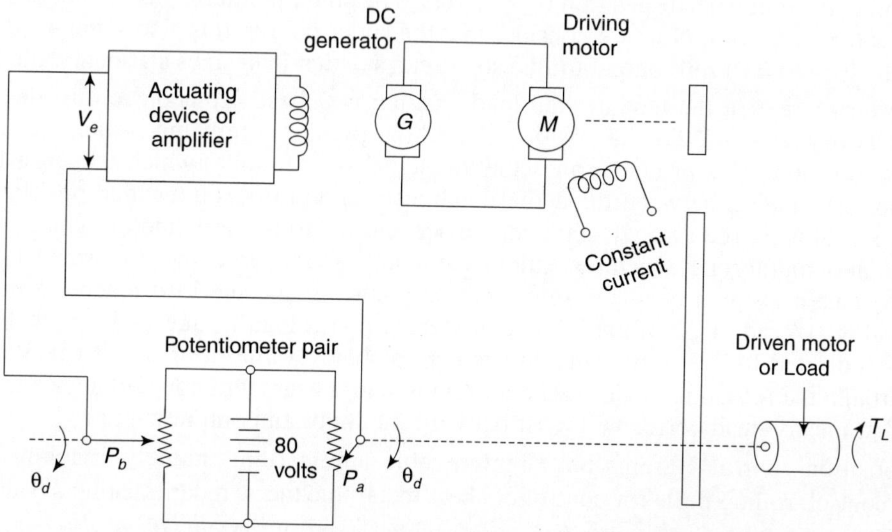

Figure 1.10 Position control system potentiometer pair

Construction

In Fig. 1.10. the driving motor is geared with driven motor (load) to be moved. The output position θ_o and desired position θ_d are measured and compared by potentiometer pair. Potentiometer pair is a combination of two potentiometers like input potentiometer and feedback potentiometer. The output of potentiometer pair is V_e (error voltage) and feed to the actuating device (amplifier). Output of actuating device is feed to the dc generator which supplies the armature voltage to the motor.

$$P_a = \text{Slider pointer}$$

$$P_b = \text{Slider pointer } b$$

$$\theta_o = \text{Output position}$$

$$\theta_d = \text{Desired position}$$

$$\theta_e = \theta_d - \theta_o$$

$$= \text{Error in angular position}$$

$$V_e = \text{Output voltage of potentiometer pair}$$

Output voltage of potentiometer $\propto$ error in angular position

$$V_e \propto \theta_e$$

$$V_e = K_p \theta_e$$

$$K_p = \text{Potentiometer constant volt/rad}$$

Let $$K_p = 80 \text{ volt/rad}$$

$$G = \text{Generator}$$

$$M = \text{Driven motor}$$

Working

For understanding the working of a servomechanism system let us assume $K_p = 80$ volts/rad and assuming the output shaft position $\theta_0 = 0.5$ rad. The slider pointer b has a voltage of + 40 volts ($80 \times 0.5 = 40$ volt). Now we consider that the slider pointer a is also set at + 40 volts. This gives ($40 - 40 = 0$) zero output torque so we can say that load stays stationary at 0.5 rad.

Now we assume that the new desired load position is 0.4 rad. To achieve this, the slider pointer a is placed at + 32 volts position, while the pointer b remains instantaneously at + 50 volts position. This produces an actuating signal of – 8 volts, which is a measure of lead of correspondence between the actual load position and the commanded position. The actuating signal is passed through actuating device and fed to the servomotor, which in turn, generates an output torque which repositions the load. The system comes to a standstill only when the actuating signal becomes zero, i.e., the pointer b and the load reach the position correspond to 0.4 rad (+ 32 volts). Now considering that a load torque T_L is applied at the output as indicated in Fig. 1.10. This will require a steady value of error voltage V_e which acting through the actuating device, generator, motor, and gears will counterbalance the load torque, that means steady error will exist between the input and output angles.

The position control systems have innumerable applications, namely, machine tools, position control, rolling mills for control of sheet metal thickness, radar tracking system, etc.

- *Servomechanism is restricted to use feedback control system, if we disconnect the potentiometer b, then we can easily see that any operator acting as part of feedback loop will find it difficult to adjust qq_0 to a desired value and to be able to maintain it.*

Physical Examples

1. *Automatic aircraft landing system.* The automatic aircraft landing system consists of three basic components: Radar unit, controlling unit and the aircraft. The function of radar unit is to measure the lateral and vertical position of the aircraft. These two positions are fed to the controlling unit. The function of controlling unit is to calculate the bank and pitch commands. These commands are transmitted to the aircraft autopilot by transmitter.

The automatic aircraft landing system is shown in Fig. 1.11 below.

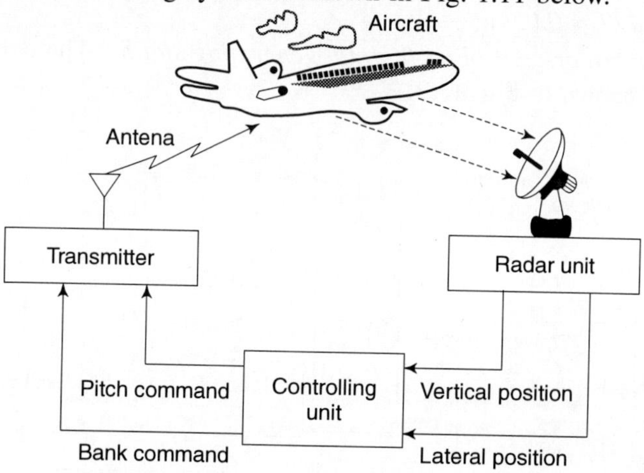

Figure 1.11 Automatic aircraft landing

For the simplicity of the system we are considering that the vertical control and lateral control are independent of each other. So we are considering only the lateral landing control system as shown in Fig. 1.12. The distance of the aircraft from the extended centre line of the landing area on the deck of aircraft is called carrier as lateral distance and denoted by $x(t)$. The aircraft autopilot attempts to force $x(t)$ to zero.

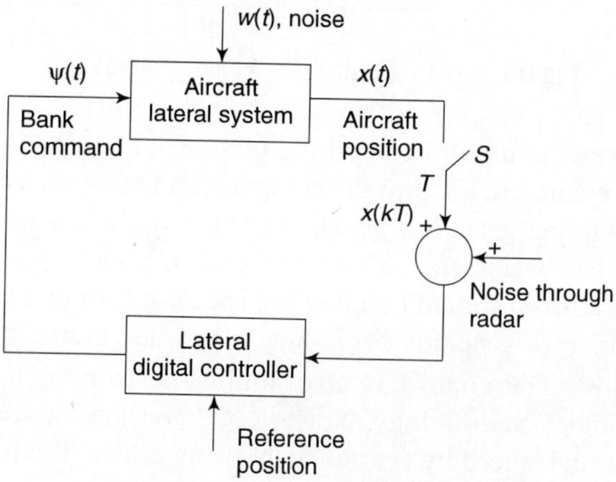

Figure 1.12 Lateral landing system

The function of radar unit to take the sampled value of $x(t)$ as $x\ (kT)$ with $T = 0.05$ sec and $k = 0, 1, 2, 3...$. The function of digital controller is to process the sampled values and produce the discrete bank command which is constant until the next value is received. So we can say that the bank command is updated at every time instant 0.05 sec. This time instant is called the sampling period. The aircraft changes the lateral position according to the bank command.

- *The lateral digital controller must be able to process the control signal within one sampling period ($T = 0.05$s).*

2. *Boiler and generator control on thermal generating station.* The control of boiler and generator are shown in Fig. 1.13.

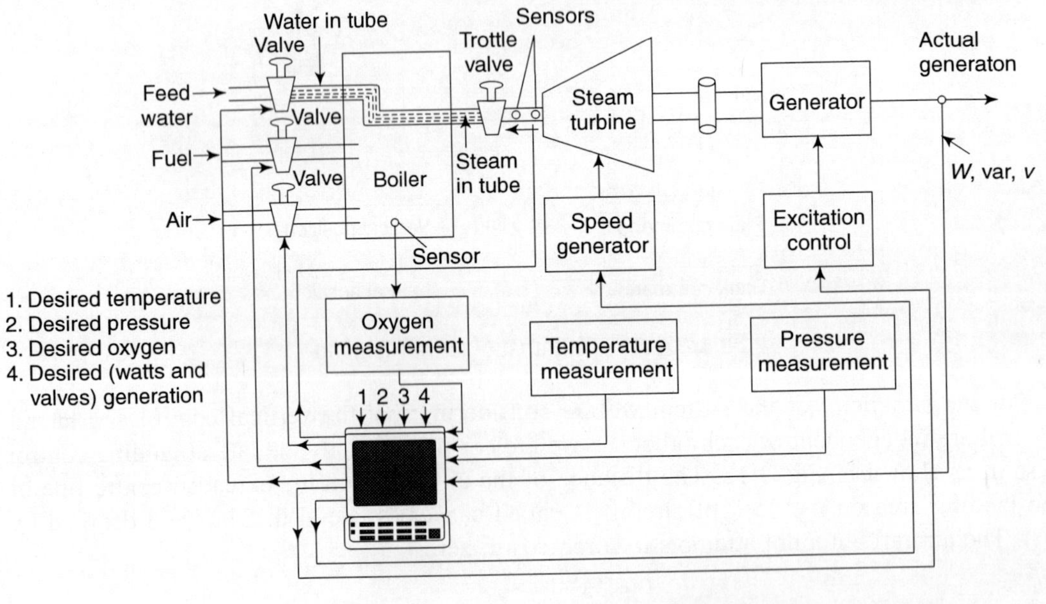

Figure. 1.13 Control for boiler generator

In the diagram computer is used to control the process with three sensors to measure the O_2. Temperature and pressure are input to the computer and these are compared with desired temperature, pressure, O_2 and generator which are fed to the computer. Computer produces the signal for the valves of water, fuel, air to control the quantity of these. It also controls the rate of steam by throttle valve and controls the speed of turbine by speed generator and also controls the excitation of generator according to the measurement.

3. *Economic control.* The control is also applicable in non-engineering fields like economics, sociology and biology. Now we are considering a system for economic inflation which is evidenced by continuously rising prices. The model is shown in Fig. 1.14.

1.5 LAPLACE TRANSFORM

The Laplace transform is one of the mathematical tools which is used for the solution of linear ordinary differential equations. So we can say that the Laplace transform is used to transform the time domain to the s domain. The advantage of using Laplace transform for solving the linear differential equation is:

1. It converts the differential equation into an algebraic equation in s domain. The solution of algebraic equation is easy as compared to differential equation. The final solution is obtained by taking inverse Laplace transform.
2. The particular integral (P.I.) and homogeneous equation of the solution are obtained in one step.

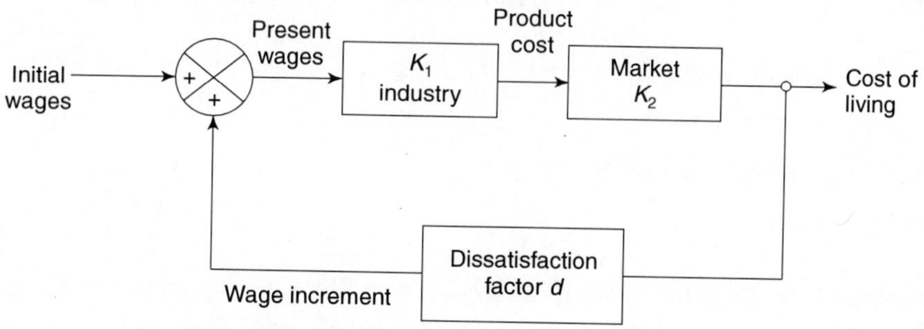

Figure 1.14 Economic inflation dynamics

Let us consider the function $f(t)$ whose Laplace transform is $F(s)$

i.e., $$L[f(t)] = F(s)\ t \geq 0$$

or $$F(s) = \int_0^\infty f(t)e^{-st}dt \quad ...(1.1)$$

The symbolic representation of Laplace transform is L.

The function $f(t)$ is obtained from $F(s)$ by a process called inverse Laplace transform. It is denoted by L^{-1}.

$$L^{-1}[L f(t)] = L^{-1}[F(s)] = f(t)$$

1.5.1 Laplace Transform of Some Signal

1. Laplace transform of e^{at}

$$L[e^{at}] = \int_0^\infty e^{at} \cdot e^{-st}dt = \int_0^\infty e^{(a-s)t}dt$$

$$L[e^{at}] = \frac{1}{s-a} \quad ...(1.2)$$

2. Laplace transform of 1

$\because$ $$L[e^{at}] = \frac{1}{s-a} \text{ as in Eq. 1.2}$$

Putting $a = 0$ $$L\,[e^0] = \frac{1}{s-0} = \frac{1}{s}$$

$$L\,(1) = \frac{1}{s} \qquad ...(1.3)$$

3. Laplace transform of sin ωt and cos ωt

We know $$L\,[e^{at}] = \frac{1}{s-a} \text{ by the Eq. 1.2}$$

Putting $a = j\omega$ $$L\,[e^{j\omega t}] = \frac{}{(s-j\omega)}$$

$$L\,[\cos \omega t + \sin \omega t] = \frac{1}{s-j\omega}$$

$$\because \quad c^{j\omega t} = \cos \omega t + j \sin \omega t = \frac{1}{s-j\omega} \times \frac{(s+j\omega)}{(s-j\omega)}$$

$$= \frac{s+j\omega}{s^2-(j\omega)^2} = \frac{s+j\omega}{s^2+\omega^2}$$

Separating into a real imaginary part

$$L\,[\cos \omega t] + [j \sin \omega t] = \frac{s}{s^2+\omega^2} + j\frac{\omega}{s^2+\omega^2}$$

So $$L\,[\cos \omega t] = \frac{s}{s^2+\omega^2} \qquad ...(1.4)$$

$$L\,[\sin \omega t] = \frac{\omega}{s^2+\omega^2} \qquad ...(1.5)$$

4. Laplace transform of e^{-at} sin ωt and e^{-at} cos ωt

We know that

$$L\,[e^{at}] = \frac{1}{s-a}$$

Putting $a = -\alpha + j\omega$ $$[e^{(-\alpha+j\omega)t}] = \frac{1}{s-(-\alpha+j\omega)}$$

$$L\,[e^{-\alpha t}.\, e^{j\omega t}] = \frac{1}{(s+\alpha)-j\omega}$$

$$\therefore \quad L\,[e^{-\alpha t}(\cos \omega t + j \sin \omega t] = \frac{1}{(s+\alpha)-j\omega} \times \frac{(s+\alpha)+j\omega}{(s+\alpha)+j\omega}$$

$$L\,[e^{-\alpha t}(\cos \omega t + j \sin \omega t] = \frac{(s+\alpha)+j\omega}{(s+\alpha)^2+\omega^2}$$

Separating into real and imaginary parts

$$L\,[e^{-\alpha t}\cos\omega t] + j\,L\,[e^{-\alpha t}\sin\omega t] = \frac{s+\alpha}{(s+\alpha)^2+\omega^2} + j\frac{\omega}{(s+\alpha)^2+\omega^2}$$

Now
$$L\,[e^{-\alpha t}\cos\omega t] = \frac{s+\alpha}{(s+\alpha)^2+\omega^2} \quad ...(1.6)$$

$$L\,[e^{-\alpha t}\sin\omega t] = \frac{\omega}{(s+\alpha)^2+\omega^2} \quad ...(1.7)$$

5. Laplace transform theorems

(a) Laplace transform of linear equation

$$L\,[a\,f_1\,(t) + b\,f_2\,(t)] = a\,F_1(s) + b\,F_2(s) \quad ...(1.8)$$

where $f_1(t)$ and $f_2(t)$ are functions of time, a and b are constants

(b) Laplace transform of differentiation.

$$L\left[\frac{df(t)}{dt}\right] = [s\,F(s) - f\,(0+)]$$

$$L\left[\frac{d^2 f(t)}{dt^2}\right] = [s^2\,F(s) - sf\,(0+) - f'\,(0+)$$

$$L\left[\frac{d^3 f(t)}{dt^3}\right] = [s^3\,F(s) - s^2 f\,(0+) - sf'\,(0+) - f''\,(0+) \quad ...(1.9)$$

where $f\,(0+)$ is value of $f\,(t)$ at $t = (0+)$

$f'\,(0+)$ is value of $\dfrac{d\,f(t)}{dt}$ at $t = (0+)$

$f''\,(0+)$ is value of $\dfrac{d^2 f(t)}{dt^2}$ at $t = (0+)$

(c) Laplace transform of intigration

$$L\left[\int f(t)dt\right] = \left[\frac{F(s)}{s} + \frac{f^{-1}(0+)}{s}\right]$$

$$L\left[\iint f(t)dt^2\right] = \left[\frac{F(s)}{s^2} + \frac{f^{-1}(0+)}{s^2} + \frac{f^{-2}(0+)}{s}\right] \quad ...(1.10)$$

$$L\left[\iiint f(t)dt^3\right] = \left[\frac{F(s)}{s^3} + \frac{f^{-1}(0+)}{s^3} + \frac{f^{-2}(0+)}{s^2} + \frac{f^{-3}(0+)}{s}\right]$$

where $f^{-1}(0+)$ is value of $\int f(t)$ at $t = (0+)$

$f^{-2}(0+)$ is value of $\iint f(t)$ at $t = (0+)$

f^{-3} (0)+) is value of $\iiint f(t)$ at = (0 +)

(d) Laplace transform of $e^{-at} f(t)$

$$L[e^{-at} f(t)] = F(s + a) \quad ...(1.11)$$

(e) Laplace transform of $t f(t)$

$$L[t f(t)] = \frac{-d}{ds} F(s) \quad ...(1.12)$$

(f) Initial value theorem

$$\lim_{t \to 0} f(t) = \lim_{s \to \infty} sL[f(t)]$$

or

$$\lim_{t \to 0} f(t) = \lim_{s \to \infty} sF(s) \quad ...(1.13)$$

(g) Final value theorem

$$\lim_{t \to \infty} f(t) = \lim_{s \to 0} s[F(s)] \quad ...(1.14)$$

1.5.2 Inverse Laplace Transform

Inverse Laplace transform is a reciprocal of Laplace transform. It converts s domain into time domain.

$$L^{-1} [F(s)] = f(t)$$

$$L^{-1} [F(s)] = \frac{1}{2\pi} \int_{-\infty}^{\infty} F(s)e^{st} ds$$

Table 1.1 Laplace transform of some functions

S. No.	$f(t)$	$F(s)$
1.	$\delta(t)$ unit impluse at $t = 0$	1
2.	$u(t)$ unit step	$1/s$
3.	$r(t)$ ramp signal	$1/s^2$
4.	t^n	$\frac{\underline{\lfloor n}}{s^n + 1}$
5.	e^{-at}	$\frac{1}{s+a}$
6.	e^{at}	$\frac{1}{s-a}$
7.	$t\,e^{-at}$	$\frac{1}{(s+a)^2}$
8.	$t^n\,e^{-at}$	$\frac{\underline{\lfloor n}}{(s+a)^{n+1}}$
9.	$\sin \omega t$	$\frac{\omega}{s^2 + \omega^2}$

Table 1.1 contd.

10.	$\cos \omega t$	$\frac{s}{s^2+\omega^2}$
11.	$e^{-at} \sin \omega t$	$\frac{\omega}{(s+a)^2+\omega^2}$
12.	$e^{-at} \cos \omega t$	$\frac{(s+a)}{(s+a)^2+\omega^2}$
13.	$\sin h\, \omega t$	$\frac{\omega}{s^2-\omega^2}$
14.	$\cos h\, \omega t$	$\frac{s}{s^2-\omega^2}$
15.	$\frac{t^n}{\lfloor n}$	$\frac{1}{s^{n+1}}$

Example 1.5.1 Find the Laplace transform of $f(t) = e^{at} \cos \omega t$.

Solution

$$L\,[e^{at} \cos \omega t] = L\left[\frac{e^{j\omega t}+e^{-j\omega t}}{2j} \cdot e^{at}\right]$$

$$= \frac{1}{2} L[e^{j\omega t} e^{at} + e^{j\omega t} e^{at}]$$

$$= \frac{1}{2}\left[\int_0^\infty e^{j\omega t} e^{at} e^{-st}\, dt + \int_0^\infty e^{-j\omega t} e^{at} e^{-st}\, dt\right]$$

$$= \frac{1}{2}\left[\int_0^\infty e^{-t(s-a-j\omega})dt + \int_0^\infty e^{-t(s-a+j\omega)}\right]$$

$$= \frac{1}{2}\left[\frac{1}{s-a-j\omega} + \frac{1}{s-a+j\omega}\right]$$

$$= \frac{s-a}{(s-a)^2+\omega^2} \quad \textbf{Ans.}$$

Example 1.5.2 Find the inverse Laplace transform of $\frac{2}{s(s+2)}$.

Solution

$$F\,(s) = \frac{2}{s(s+2)}$$

The partial fraction of $F(s)$ is

$$F\,(s) = \frac{1}{s} - \frac{1}{s+2}$$

$$f(t) = L^{-1}[F(s)] = L^{-1}\left[\frac{1}{s}\right] - L^{-1}\left[\frac{1}{s+2}\right] = 1 - e^{-2t} \quad \textbf{Ans.}$$

1.6 TRANSFER FUNCTION

"The transfer function is defined as the ratio of the Laplace transform of the output quantity to the Laplace transform of the input quantity, with all initial conditions assumed to be zero."

The transfer function of the system represents the relationship between the input and output and describes the dynamics of the system under consideration.

Then the transfer function is

$$\text{Transfer function} = \frac{\text{Output}}{\text{Input}}$$

$$G(s) = \frac{Y(s)}{R(s)} \qquad \text{...(1.15)}$$

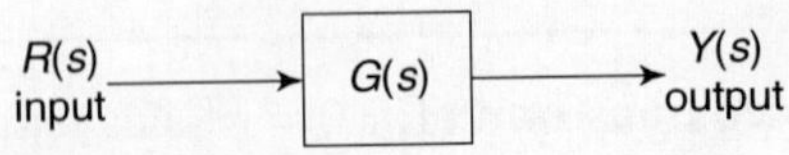

Figure 1.15 Open loop control system

The transfer function can be defined only for a linear stationary system. Stationary systems are those having constant parameter, also called time invariant system. A non-stationary system, often called a time varying system, has one or more time varying parameters, and the Laplace transformation may not be used. Thus, the transfer function does not include any information concerning the internal structure of the system and its behaviour, because it is an input and output description of the behaviour of a system.

Let us consider a linear time invariant system whose input is $r(t)$ and the output is $y(t)$. The input and output relation can be represented by the following n^{th} order differential equation.

$$a_n \frac{d^n y(t)}{dt^n} + a_{n-1}\frac{d^{n-1}y(t)}{dt^{n-1}} + ... + a_0 y(t) = b_m \frac{d^m r(t)}{dt^m} + b_{m-1}\frac{d^{m-1}r(t)}{dt^{m-1}} + ... + b_0 r(t) \qquad \text{...(1.16)}$$

Taking the Laplace transform on both side of Eq. (1.16)

$$a_n s^n Y(s) + a_{n-1}s^{n-1}Y(s) + ... + a_0 Y(s) + \text{initial condition terms of } y(t)$$

$$= b_m\, s^m\, R(s) + m_{m-1}\, s^{m-1}\, R(s) + ... + b_0\, R(s) + \text{initial condition terms of } r(t) \qquad \text{...(1.17)}$$

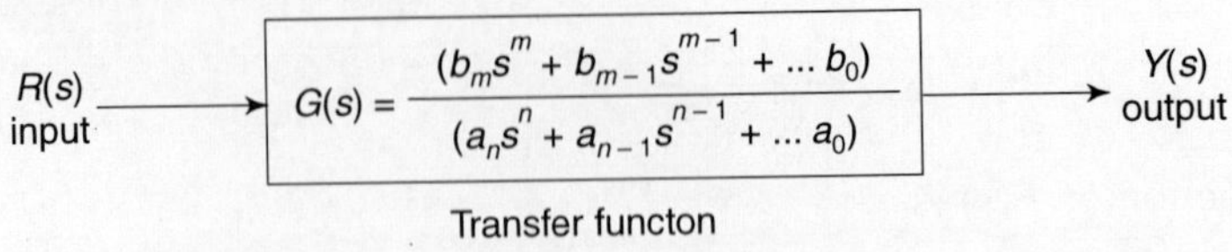

Figure 1.16 Block diagram representation of transfer function

Equation (1.17) is pure algebraic. We assume that initial conditions are zero.

$$(a_n\, s^n + a_{n-1}\, s^{n-1} + \ldots\, a_0)\; Y(s) = (b_m\, s^m + b_{m-1}\, s^{m-1} + \ldots\, b_0)\; R(s) \qquad \ldots(1.18)$$

$$\frac{(b_m\, s^m + b_{m-1}s^{m-1} + \ldots b_0)}{(b_n s^n + a_{n-1}s^{n-1} + \ldots a_0)} = \frac{Y(s)}{R(s)} = G(s) \qquad \ldots(1.19)$$

- *The denominator of the transfer function is equal to the characteristic polynomial of the differential equation. If characteristic polynomial is equal to zero then it is called characteristic equation like.*

$$a^n\, s^n + a_{n-1}\, s^{n-1} + \ldots\, a_0 = 0 \qquad \ldots(1.20)$$

Example 1.6.1 Find the transfer function for given differential equation

$$\frac{dy\,(t)}{dt} + 5\, y(t) = r(t) \qquad \ldots(1.21)$$

Solution Taking Laplace transform on both side

$$s\, Y(s) + 5\, Y(s) = R(s), \text{ assuming initial conditions are zero}$$

$$(s + 5)\; Y(s) = R(s) \qquad \ldots(1.22)$$

$$\frac{Y\,(s)}{R\,(s)} = \frac{1}{(s+5)} \qquad \ldots(1.23)$$

The transfer function is

$$G(s) = \frac{Y\,(s)}{R\,(s)} = \frac{1}{s+5} \quad \textbf{Ans.}$$

Example 1.6.2 Find the response of the system for a unit step input. The transfer function of the system is $\frac{1}{s+2}$ assuming zero initial conditions.

Solution Given that

$$G(s) = \frac{1}{s+2}$$

$$r(t) = u(t), \text{ a unit step input}$$

$$R(s) = \frac{1}{s} \text{ from Table 1.1}$$

We know that

$$G(s) = \frac{Y\,(s)}{R\,(s)}$$

$$Y(s) = R(s)\; G(s) \qquad \ldots(1.24)$$

Putting the value of $R(s)$ and $G(s)$ in Eq. (1.24)

$$Y(s) = \frac{1}{s(s+2)}$$

$$Y(s) = \frac{1}{2s} - \frac{1}{2(s+2)} \qquad \text{...(1.25)}$$

Taking the inverse Laplace transform of Eq. (1.25)

$$L^{-1}[Y(s)] = L^{-1}\left[\frac{1}{2s}\right] - L^{-1}\left[\frac{1}{2(s+2}\right]$$

$$y(t) = \frac{1}{2}L^{-1}\left[\frac{1}{s}\right] - \frac{1}{2}L^{-1}\left[\frac{1}{s+2}\right]$$

$$y(t) = \frac{1}{2}\cdot 1 - \frac{1}{2}e^{-2t} \quad \text{by Table 1.1}$$

$$y(t) = \frac{1}{2} - \frac{1}{2}e^{-2t} \qquad \text{...(1.26)}$$

1.6.1 Advantages and Disadvantages of Transfer Function

Advantages

1. It provides the gain of the system.
2. Integral and differential equations are converted to algebraic equations.
3. Differential equation of system can be obtained by replacement of variable s by $\frac{d}{dt}$ and s^2 by $\frac{d^2}{dt^2}$, etc.
4. The transfer function is dependent on the parameters of the system and independent of the input.
5. If transfer function $G(s)$ is known then any output for any given input can be known.

Disadvantages

1. Transfer function can be calculated only for linear and time invariant system.
2. Consider only when initial conditions are zero.
3. It does not give any information about physical structure of the system.

1.6.2 Poles and Zeros of a Transfer Function

The transfer function of a linear control system can be expressed in the form of a quotient of polynomials or as in Eq. 1.19.

$$G(s) = \frac{N(s)}{D(s)} = \frac{c_0 s^n + c_1 s^{n-1} + c_2 s^{n-2} + \ldots c_n}{d_0 s^m + d_1 s^{m-1} + d_2 s^{m-2} + \ldots d_m} \qquad \text{...(1.27)}$$

Now factorising the numerator and the denominator of Eq. 1.27

$$G(s) = \frac{N(s)}{D(s)} = K\frac{(s - Z_1)(s - Z_2)\ldots(s - Z_n)}{(s - P_1)(s - P_2)\ldots(s - P_m)} \quad \ldots(1.28)$$

where $K = \frac{c_0}{d_0}$ is known as the gain factor.

In Eq. (1.28), if the numerator is equal to zero then

$$(s - Z_1)(s - Z_2) \ldots (s - Z_n) = 0 \quad \ldots(1.29)$$

n roots of the equation are $Z_1, Z_2, Z_3 \ldots Z_n$

if the denominator is equal to zero then

$$(s - P_1)(s - P_2) \ldots (s - P_m) = 0 \quad \ldots(1.30)$$

m roots of the equation are $P_1, P_2, P_3 \ldots P_m$

Poles of the Transfer Function

The values of s which are substituted in the denominator of the transfer function, after substituting the value, the transfer function becomes 'infinite', these values are called *poles of the transfer function.*

Like $P_1, P_2, P_3 \ldots P_m$ are those values which make the transfer function infinite when substituted in Eq. (1.28).

Zeros of the Transfer Function

The values of s which are substituted in the numerator of the transfer function, after substituting the value, the transfer function becomes 'zero', these values are called *zeros of the transfer function.*

Like $Z_1, Z_2, Z_3, \ldots Z_n$ are those values which make the transfer function zero when substituted in Eq. (1.28).

- *When the value of poles and zeros are not repeated, such type of poles and zeros are called simple poles and zeroes. If repeated, such poles and zeros are called nultiple poles and zeros. The order of repeated poles and zeroes is eual to the number of times they are repeated.*

1.6.3 Pole-Zero Plot

If we locate all poles and zeros of the transfer function in the s plane (or complex plane) that diagram is called the *pole-zero plot.*

$$\therefore \quad s = \sigma + j\omega$$

where σ = real part and located on X axis or real axis.

$j\omega$ = imaginary part and located on Y axis or imaginary axis.

$$\text{Simple}\begin{Bmatrix}\text{Pole} = \times \text{ (Symbolic representation on } s \text{ plane} \\ \text{Zero} = \text{O (Symbolic representation on } s \text{ plane}\end{Bmatrix}$$

$$\text{Repeated}\begin{cases}\text{Pole} = \text{Repeated} \times \text{Equal to order of multiplicity} \\ \text{Zero} = \text{Repeated O equal to order of multiplicity}\end{cases}$$

Example 1.6.3 Consider the following transfer function

$$G(s) = \frac{20\,(s+2)\,(s+3)}{s\,(s+1)\,(s+5)\,(s+2-4j)\,(s+2+4j)}$$

Draw pole-zero plot.

Solution Zeros: $Z_1 = -2,\ Z_2 = -3$

Poles : $P_1 = 0,\ P_2 = -1,\ P_3 = -5,\ P_6 = -2 + 4j,\ P_7 = -2 - 4j$

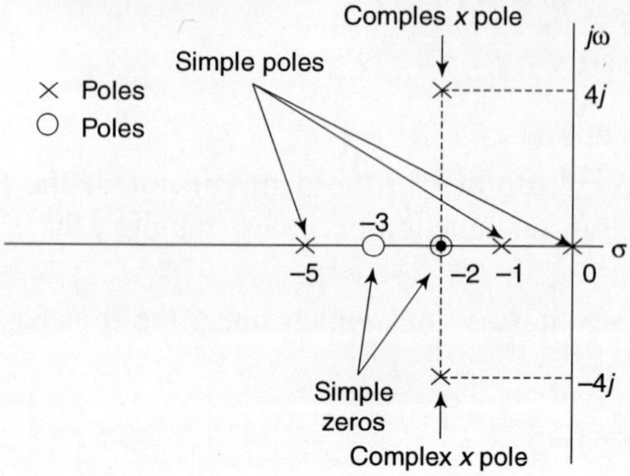

Figure Ex. 1.6.3 Pole-zero plot

• *Complex poles or complex zeros always appear in conjugate pairs.*

Example 1.6.4 Consider the following transfer function $G(s) = \dfrac{30(s+1)\,(s+2)^3}{(s+5)\,(s+3)^2}$.

Draw pole-zero plot.

Solution

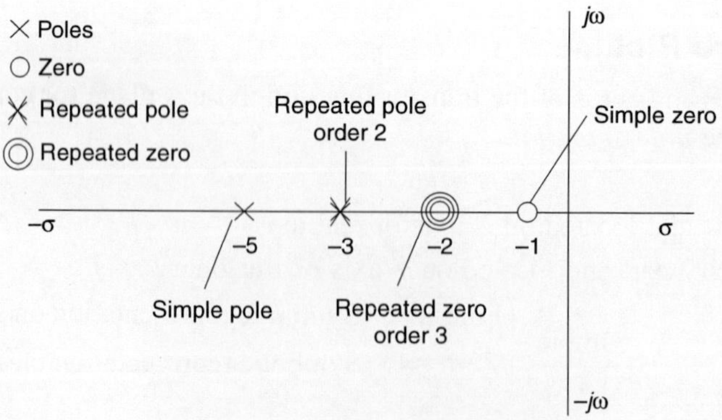

Figure Ex. 1.6.4 Pole-zero plot

Zeros: $Z_1 = -1, Z_2 = -2, Z_3 = -2, Z_4 = -2$

Poles: $P_1 = -5, P_2 = -3, P_3 = -3$

Z_2, Z_3, Z_4: Repeated zeros of order 3

P_2, P_3: Repeated poles of order 2

The above-mentioned poles and zeros are the finite one. If we consider that the s plane includes infinity. From Eq. (1.28) it is observed that:

1. If $Z > P$, i.e., number of poles is less than number of zeros, then the value of the transfer function becomes infinity for $s = \infty$. Hence, we can say that there are poles at infinity and the order of such poles is $(Z - P)$.
2. If $Z < P$, i.e., number of poles is greater than number of zeros, then the value of the transfer becomes zero for $s = \infty$. Hence, we can say that there are zeros at infinity and the order of such zeros is $(P - Z)$.

- *If poles and zeros located at zero and infinity are considered, then for a rational function the total number of zeros is found to be equal to the total number of poles.*

Example 1.6.5 Consider the following transfer function $G(s) = \dfrac{(s+2)(s+1)}{s\,(s+5)(s+3)(s+7)}$.

Draw pole-zero plot, s plane is considered up to infinity.

Solution Zeros: $Z_1 = -2, Z_2 = -1$

Poles: $P_1 = 0, P_2 = -3, P_3 = -5, P_4 = -7$

$\therefore$ number of zero = number of poles

So remaining two zeros are located at $s = \infty$

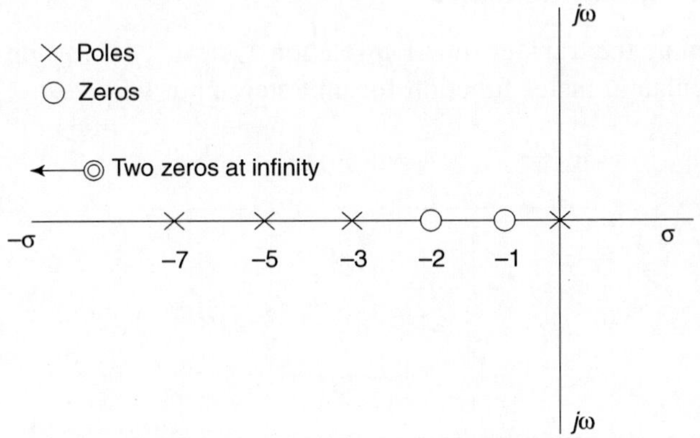

Figure Ex. 1.6.5 Pole-zero plot

1.6.4 Impulse Response

The response of the system when the input to the system is impulse, called *impulse response*.

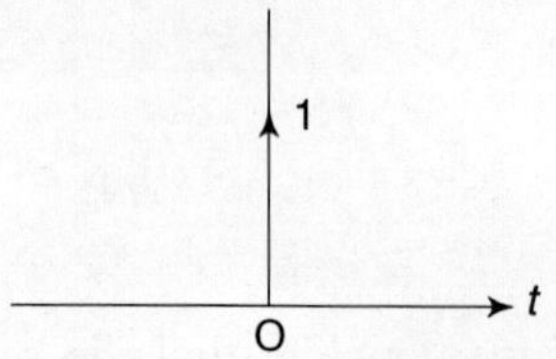

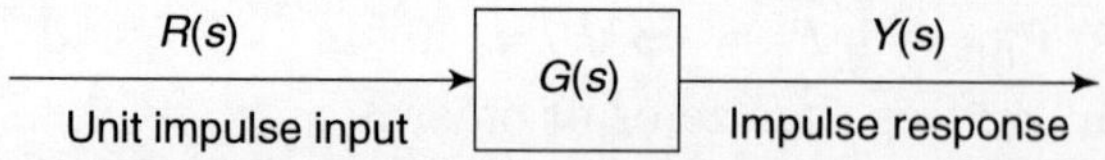

Figure 1.17

$$G(s) = \frac{Y(s)}{R(s)}$$

$$Y(s) = R(s)\ G(s)$$

∵ Laplace transform of unit impulse, $\delta(t) = 1$ by Table 1.1

$$R(s) = 1$$

$$Y(s) = G(s)$$

Taking inverse Laplace transform of eqnuation ...(1.31)

$$Y(t) = g(t) \quad ...(1.32)$$

where $g(t)$ = unit impulse response of a system.

- *Inverse Laplace transform of G(s) is called impulse response or transfer function of a system in the Laplace transform of its impulse response by Eq. (1.31).*

EXAMPLES

Transfer Function (Transform Method)

Example 1.1 Determine the transfer function relation $V_o(s)$ to $V_i(s)$ for the network shown in Fig. Ex. 1.1 (a), calculate transfer function for unit step input voltage.

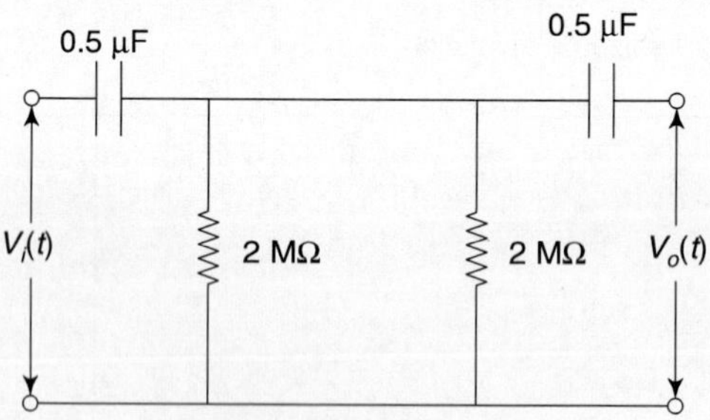

Figure Ex. 1.1 (a)

Solution Redraw the circuit as shown in Fig. 1.1. (b).

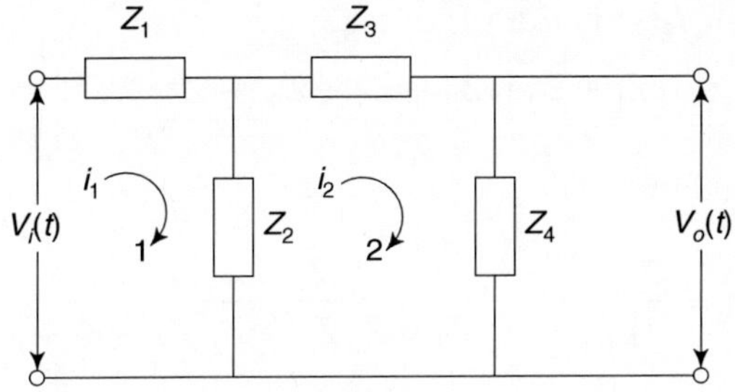

Figure Ex. 1.1 (b)

where $Z_1 = \dfrac{1}{0.5\,s}$

$Z_2 = 2$

$Z_3 = \dfrac{1}{0.5\,s}$

$Z_4 = 2$

$\because$ Applying KVL in loop 1 and loop 2, we get

$$Z_1\, I_1(s) + Z_2\,(I_1 - I_2) = V_i(s)$$

$$V_i(s) = I_1(s)\,(Z_1 + Z_2) - Z_2\, I_2(s) \qquad ...(1)$$

$$0 = Z_2\,[I_2(s) - I_1(s)] + Z_3\, I_2(s) + Z_4\, I_2(s)$$

$$0 = Z_3\, I_2(s) + Z_4\, I_2(s) + Z_2\, I_2(s) - Z_2\, I_1(s) \qquad ...(2)$$

From Eq. (2)

$$(Z_3 + Z_4 + Z_2)\, I_2(s) = Z_2\, I_1(s)$$

$$I_1(s) = \frac{(Z_2 + Z_3 + Z_4)\, I_2(s)}{Z_2} \qquad ...(3)$$

Substituting the value of $I_1(s)$ in Eq. (1)

$$V_i(s) = \frac{(Z_1 + Z_2)(Z_2 + Z_3 + Z_4)\, I_2(s)}{Z_2} - Z_2\, I_2(s)$$

$$V_i(s) = \frac{(Z_1 + Z_2)(Z_2 + z_3 + Z_4)}{Z_2} - Z_2^2\, I_2(s)$$

$$\frac{V_i(s)}{I_2(s)} = \frac{\left[(Z_1 + Z_2)(Z_2 + Z_3 + Z_4) - Z_2^2\right]}{Z_2} \qquad ...(4)$$

But we know from Fig. Ex. 1.1

$$V_o(s) = Z_4\, I_2(s)$$

$$\Rightarrow \qquad I_2(s) = \frac{V_o(s)}{Z_4} \qquad ...(5)$$

Substituting the value of $I_2(s)$ is Eq. (4), we get

$$\frac{\dfrac{V_i(s)}{V_o(s)}}{Z_4} = \frac{(Z_1 + Z_2)(Z_2 + Z_3 + Z_4) - Z_2^2}{Z_2}$$

$$\frac{V_o(s)}{V_i(s)} = \frac{Z_2 Z_4}{Z_1 Z_2 + Z_1 Z_3 + Z_1 Z_4 + Z_2 Z_3 + Z_2 Z_4}$$

$$\frac{V_o(s)}{V_i(s)} = \frac{Z_2 Z_4}{Z_1 Z_2 + Z_1 Z_3 + Z_1 Z_4 + Z_2 Z_3 + Z_2 Z_4} \qquad ...(6)$$

Substituting the value of Z_1, Z_2, Z_3 and Z_4 in equation

$$\frac{V_o(s)}{V_i(s)} = \frac{2 \times 2}{\left(\frac{1}{0.5} \times 2\right) + \left(\frac{1}{0.5s} \times \frac{1}{0.5s}\right) + \left(\frac{1}{0.5s} \times 2\right) + \left(2 \times \frac{1}{0.5s}\right) + (2 \times 2)}$$

$$\frac{V_o(s)}{V_i(s)} = \frac{4}{\frac{4}{s} + \frac{4}{s^2} + \frac{4}{s} + \frac{5}{s} + 4}$$

$$\frac{V_o(s)}{V_i(s)} = \frac{4s^2}{4s + 4 + 4s + 4s + 4s^2}$$

$$\frac{V_o(s)}{V_i(s)} = \frac{4\ ^2}{4s^2 + 12s + 4} \quad \frac{5^2}{s^2 + 3s + 1}$$

$$\frac{V_o(s)}{V_i(s)} = \frac{s^2}{s^2 + 3s + 1} \qquad ...(7)$$

For a unit step input

$$V_o(s) = \frac{s}{s^2 + 3s + 1} \quad \textbf{Ans.} \qquad \because u(s) = \frac{1}{s}$$

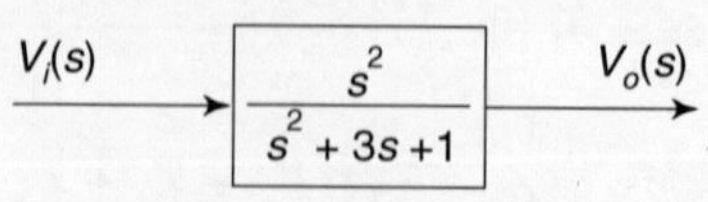

Fig. Ex. 1.1 (c)

Example 1.2 Find the transfer function of the network given in Fig. Ex. 1.2 (a).

[U.P.T.U., 2004–05, ACS]

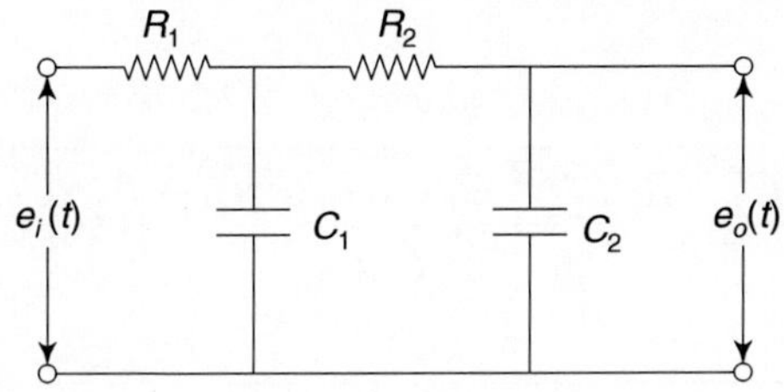

Figure Ex. 1.2 (a)

Solution Redraw Fig. Ex. 1.2 (b) as shown below.

This is similar to Example 1.1

The transfer function is

$$\frac{e_o(t)}{e_i(t)} = \frac{Z_2\, Z_4}{Z_1\, Z_2 + Z_1\, Z_3 + Z_1\, Z_4 + Z_2\, Z_4}$$

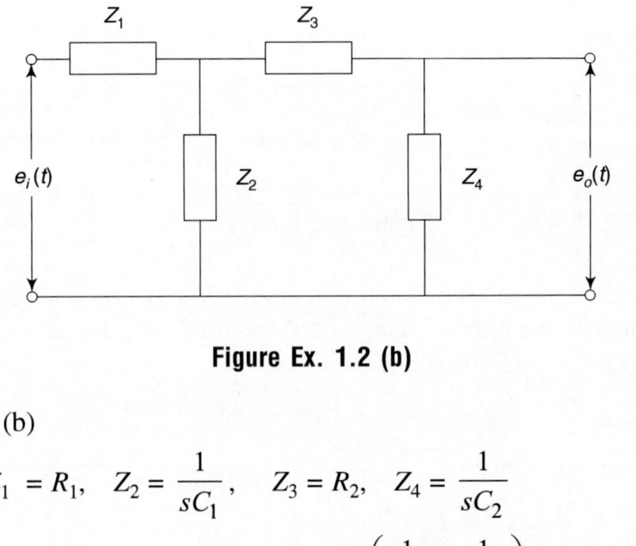

Figure Ex. 1.2 (b)

By Fig. Ex. 1.2 (b)

$$Z_1 = R_1, \quad Z_2 = \frac{1}{sC_1}, \quad Z_3 = R_2, \quad Z_4 = \frac{1}{sC_2}$$

$$\frac{e_o(s)}{e_i(s)} = \frac{\left(\frac{1}{sC_1} \times \frac{1}{sC_2}\right)}{\left(R_1 \times \frac{1}{sC_1}\right) + (R_1 \times R_2) + \left(R_1 \times \frac{1}{sC_2}\right) + \left(\frac{1}{sC_1} \times R_2\right) + \left(\frac{1}{sC_1} \times \frac{1}{sC_2}\right)}$$

$$\frac{e\ (s)}{e\ (s)} = \frac{\frac{1}{s^2 C_1\, C_2}}{\frac{R_1}{sC_1} + R_1\, R_2 + \frac{R_1}{sC_2} + \frac{R_2}{sC_1} + \frac{1}{s^2 C_1\, C_2}}$$

$$\frac{e_o(s)}{e_i(s)} = \frac{\dfrac{1}{s^2 C_1 C_2}}{\dfrac{R_1 sC_2 + R_1 R_2 s^2 C_1 C_2 + R_1 sC_1 + R_2 sC_2 + 1}{s^2 C_1 C_2}}$$

$$\frac{e_o(s)}{e_i(s)} = \frac{1}{R_1 sC_2 + R_1 R_2 s^2 C_1 C_2 + R_1 C_1 s + R_2 C_2 s + 1}$$

$$\frac{e_o(s)}{e_i(s)} = \frac{1}{R_1 R_2 C_1 C_2 s^2 + (R_1 C_1 + R_1 C_2 + R_2 C_2) s + 1}$$

Let $R_1 R_2 C_1 C_2 = A$

$R_1 C_1 + R_1 C_2 + R_2 C_2 = B$

$$\frac{e_0(s)}{e_i(s)} = \frac{1}{As^2 + Bs + 1} \quad \textbf{Ans.}$$

i.e., the network is

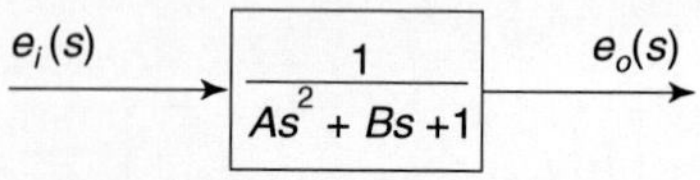

Figure Ex. 1.2 (c)

Example 1.3 Determine the transfer function relating $e_o(s)$ to $e_i(s)$ for the network shown in Fig. Ex. 1.3 (a). Calculate the output voltage $t \geq 0$ for a unit step input voltage at $t = 0$ when $C = 1\ \mu F$, $R_1 = 1\ M\Omega$, $C_2 = 0.5\ \mu F$ and $R_2 = 1\ M\Omega$.

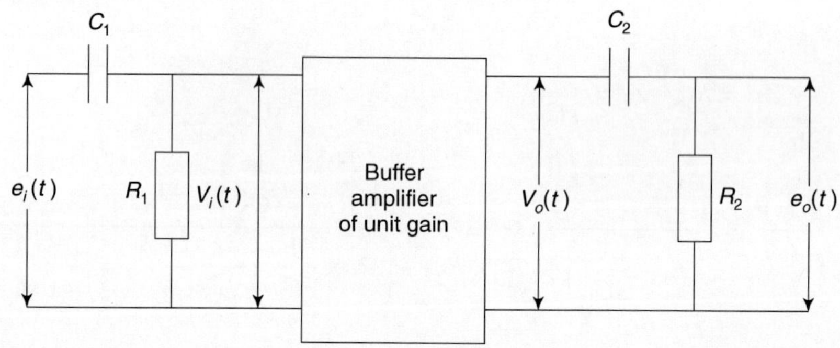

Figure Ex. 1.3 (a)

Solution Redrawing the circuit diagram as shown in Fig. Ex. 1.3 (b) and applying KVL, we get

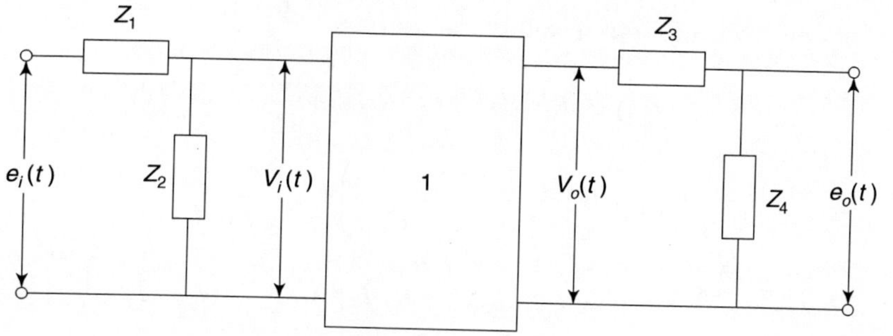

Figure Ex. 1.3 (b)

By Example 1.1, we have

$$\frac{e_o(t)}{e_i(t)} = \frac{Z_2 Z_4}{Z_1 Z_2 + Z_1 Z_3 + Z_1 Z_4 + Z_2 Z_3 + Z_2 Z_4}$$

Substituting the value of Z_1, Z_2, Z_3 and Z_4, we get

$$\frac{e_o(s)}{e_i(s)} = \frac{s^2}{s^2 + 3s + 2}$$

For the unit step input

$$e_o(s) = \frac{s^2}{s^2 + 3s + 2}$$

By the partial fraction

$$e_o(s) = \frac{s^2}{s^2 + 3s + 2} = \frac{2}{(s+2)} - \frac{1}{(s+1)}$$

$$e_o(s) = \frac{2}{(s+2)} - \frac{1}{(s+1)}$$

Taking inverse Laplace transform.

$$e_o(t) = 2e^{-3t} - e^{-t} \quad \textbf{Ans.}$$

Example 1.4 Derive the transfer function $\dfrac{e_o(s)}{e_i(s)}$ for the network shown in Fig. Ex. 1.4.

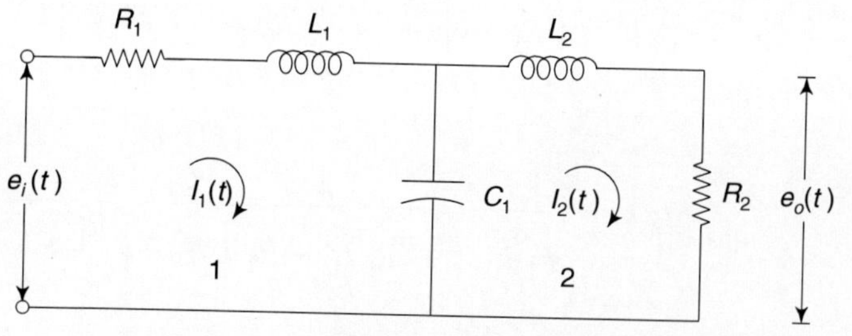

Figure Ex. 1.4

Solution Applying KVL in loops 1 and 2

$$e_i(s) = R_1 I_1(s) + L_1 s I_1(s) + \frac{1}{C_1 s}(I_1 - I_2)$$

$$e_i(s) = \left(R_1 + L_1 s + \frac{1}{C_1 s}\right) I_1 - \frac{1}{C_1 s} I_2 \qquad ...(1)$$

$$0 = L_2 s I_2(s) + R_2 I_2(s) + \frac{1}{C_1 s}(I_2 - I_1)$$

$$0 = -\frac{1}{C_1 s} I_1(s) + \left(L_2 s + R_2 + \frac{1}{C_2 (s)}\right) I_2(s) \qquad ...(2)$$

We know from Fig. Ex. 1.4

$$e_o(s) = R_2 I_2(s) \qquad ...(3)$$

Now calculating $I_2(s)$ from Eqs. (1) and (2) by Cramer rule

$$I_2(s) = \frac{\Delta_2(s)}{\Delta(s)}$$

$$\Delta_2(s) = \begin{bmatrix} R_1 + L_1 s + \frac{1}{C_1 s} & e_i(s) \\ -\frac{1}{C_1 s} & 0 \end{bmatrix}$$

$$\Delta(s) = \begin{bmatrix} R_1 + L_1 s + \frac{1}{C_1 s} & -\frac{1}{C_1 s} \\ -\frac{1}{C_1 s} & R_2 + L_2 s + \frac{1}{C_1 s} \end{bmatrix}$$

$$I_2(s) = \frac{\frac{e_i(s)}{C_1 s}}{\left(R_1 + L_1 s + \frac{1}{C_1 s}\right)\left(R_2 + L_2 s + \frac{1}{C_1 s}\right) - \frac{1}{C_1^2 s^2}}$$

$$I_2(s) = \frac{\frac{e_i(s)}{C_1 s}}{L_1 L_2 s^2 + (R_1 L_2 + R_2 L_1) s + \left(\frac{R_1}{C_1} + \frac{R_2}{C_1}\right)\frac{1}{s} + \frac{L_1}{C_1} + \frac{L_2}{C_1}}$$

Substituting the value of $I_2(s)$ in Eq. (3), we get

$$e_o(s) = \frac{\frac{R_2}{C_1 S} e_i(s)}{L_1 L_2 s^2 + (R_1 L_2 + R_2 L_1) s + \left(\frac{R_1}{C_1} + \frac{R_2}{C_1}\right)\frac{1}{s} + \frac{L_1}{C_1} + \frac{L_2}{C_1}}$$

$$\frac{e_o(s)}{e_i(s)} = \frac{\frac{R_2}{C_1 s}}{L_1 L_2 s^2 + (R_1 L_2 + R_2 L_1) s + \left(\frac{R_1 + R_2}{C_1}\right)\frac{1}{s} + \left(\frac{L_1 + L_2}{C_1}\right)}$$ **Ans.**

Example 1.5 Find the transfer function of the electrical network shown in Fig. Ex. 1.5, where $R_1 = 2\ M\Omega$ and $R_2 = 1\ M\Omega$, $C_1 = 0.2\ \mu F$ and $C_2 = 0.5\ \mu F$.

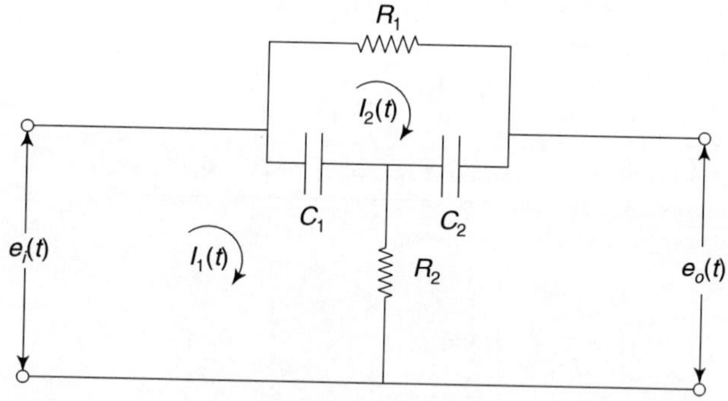

Figure Ex. 1.5

Solution Applying KVL in loops 1 and 2 and taking Laplace transform to the two loops, we get

$$e_i(s) = \frac{1}{s C_1}(I_1 - I_2) + R_2 I_1$$

$$e_i(s) = \left(\frac{1}{s C_1} + R_2\right) I_1(s) - \frac{1}{s C_1} I_2 \quad \text{...(1)}$$

In loop 2

$$0 = R_1 I_2(s) + \frac{1}{sC_1} I_2(s) + \frac{1}{sC_1}[I_2(s) - I_1(s)]$$

$$0 = R_1 I_2(s) + \frac{1}{sC_2} I_2(s) + \frac{1}{sC_1} I_2(s) - \frac{1}{sC_1} I_1(s)$$

$$0 = -\left(R_1 + \frac{1}{sC_2} + \frac{1}{sC_1}\right) I_2(s) + \frac{1}{sC_1} I_1(s) \quad \text{...(2)}$$

and from the circuit

$$e_o(s) = I_1(s) R_2 + \frac{I_2}{sC_2} \quad \text{...(3)}$$

Solving Eqs. (1), (2) and (3) by Cramer's rule, we get

$$D_1 = \begin{vmatrix} \frac{1}{sC_1} & -\left(R_1 + \frac{1}{sC_1} + \frac{1}{sC_2}\right) \\ \frac{1}{sC_1} + R_2 & -\frac{1}{sC_1} \end{vmatrix}$$

$$= \frac{1}{s^2 C_1^2} + \left(\frac{1}{sC_1} + R_2\right)\left(R_1 + \frac{1}{sC_1} + \frac{1}{sC_2}\right)$$

$$= \frac{1 + s(R_1 + R_2)C_2 + sC_1 R_2 + s^2 R_1 R_2 C_1 C_2}{s^2 C_1 C_2}$$

$$D_2 = \begin{vmatrix} 0 & -\left(\frac{1}{R_1} + \frac{1}{sC_1} + \frac{1}{sC_2}\right) \\ e_i(s) & -\frac{1}{sC_1} \end{vmatrix} = e_i(s)\left(R_1 + \frac{1}{sC_1} + \frac{1}{sC_2}\right)$$

$$= \frac{e_i(s)(sC_1 + sC_2 + s^2 R_1 C_1 C_2)}{s^2 C_1 C_2}$$

Now $$D_3 = \begin{vmatrix} \frac{1}{sC_1} & 0 \\ \frac{1}{sC_1} + R_2 & e_i(s) \end{vmatrix} = \frac{e_i(s)}{sC_1}$$

Therefore, $$I_1 = \frac{D_2}{D_1} = \frac{(sC_1 + sC_2 + s^2 C_1 C_2 R_1)}{1 + s(R_1 + R_2)C_2 + sR_2 C_1 + s^2 R_1 R_2 C_1 C_2}$$

$$I_2 = \frac{D_3}{D_1} = \frac{sC_2 e_i(s)}{1 + s(R_1 + R_2)C_2 + sR_2 C_1 + s^2 R_1 R_2 C_1 C_2}$$

Substituting the values of I_1 and I_2 in Eq. (3), we get

$$e_o(s) = \frac{e_i(s) R_2 (sC_1 + sC_2 + s^2 R_1 C_1 C_2)}{1 + s(R_1 + R_2)C_2 + sR_2 C_1 + s^2 R_1 R_2 C_1 C_2} + \frac{e_i(s)}{1 + s(R_1 + R_2)C_2 + sR_2 C_1 + s^2 R_1 R_2 C_1 C_2}$$

$$\frac{e_o(s)}{e_i(s)} = \frac{1 + s^2 R_1 R_2 C_1 C_2 + sR_2 C_1 + sR_2 C_2}{1 + s(R_1 + R_2)C_2 + sR_2 C_1 + s^e R_1 R_2 C_1 C_2}$$

Substituting the values of R_1, R_2, C_1 and C_2 we get

$$\frac{e_o(s)}{e_i(s)} = \frac{1 + s^2 \times (2 \times 1 \times 0.5 \times 0.2) + s \times (1 \times 0.2) + s(1 \times 0.5)}{1 + s(1+2) \times 0.5 + s(1 \times 0.2) + s^2 (1 \times 2 \times 0.5 \times 0.2)}$$

$$\frac{e_o(s)}{e_i(s)} = \frac{1+0.2s^2+0.2s+0.5s}{1+1.5s+0.2s+0.2s^2}$$

$$\frac{e_o(s)}{e_i(s)} = \frac{0.2s^2+0.7s+1}{0.2s^2+1.5s+1.7s+1} \quad \textbf{Ans.}$$

Example 1.6 Find the transfer function of the network shown in Fig. Ex. 1.6 (a) where $R_1 = 1\ M\Omega$, $R_2 = 2\ M\Omega$ $C_1 = 0.5\ \mu F$ and $C_2 = 0.8\ \mu F$.

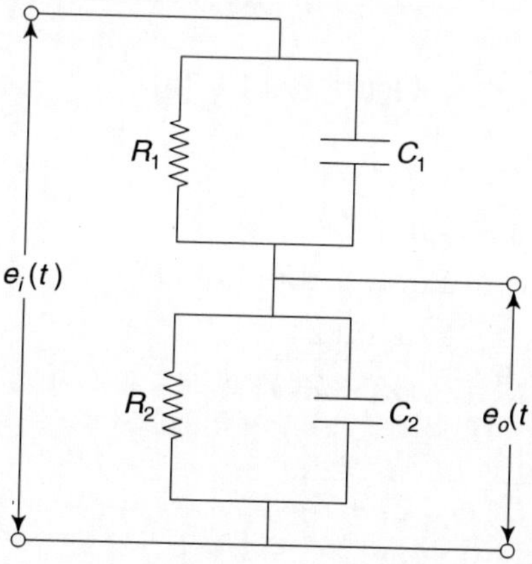

Figure Ex. 1.6 (a)

Solution Redrawing the network as shown in Fig. Ex. 1.6 (b)

where $Z_1 = R_1 \parallel C_1$

$$Z_1 = \frac{R_1 \times \dfrac{1}{sC_1}}{R_1 + \dfrac{1}{sC_1}}$$

$$Z_1 = \frac{R_1}{sR_1 C_1 + 1}$$

and $$Z_2 = R_2 \parallel C_2 = \frac{R_2}{sR_2 C_2 + 1}$$

The current flowing at the input side,

$$I(s) = \frac{e_i(s)}{Z_1 + Z_2} \qquad ...(1)$$

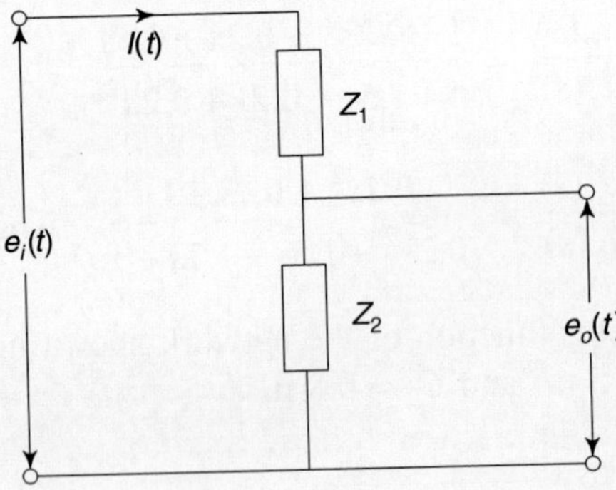

Figure Ex. 1.6 (b)

The output voltage $e_o(s)$,

$$e_o(s) = I(s)\, Z_2 \qquad \text{...(2)}$$

Substituting the value of $I(s)$ in Eq. (2) from Eq. (1)

$$e_o(s) = \frac{e_i(s)\, Z_2}{(Z_1 + Z_2)}$$

$$\frac{e_o(s)}{e_i(s)} = \frac{Z_2}{Z_1 + Z_2} \qquad \text{...(3)}$$

Substituting the values of Z_1 and Z_2

$$\frac{e_o(s)}{e_i(s)} = \frac{\dfrac{R_2}{1 + SR_2\, C_2}}{\dfrac{R_1}{1 + SC_1\, R_1} + \dfrac{R_2}{1 + SC_2\, R_2}}$$

$$\frac{e_o(s)}{e_i(s)} = \frac{R_2\,(1 + SR_1\, C_1)}{(R_1 + R_2)\left[1 + S\dfrac{R_1\, R_2\, C_1 + R_1\, R_2\, C_2}{R_1 + R_2}\right]}$$

Now substituting the values of R_1 R_2 C_1 and C_2, we get

$$\frac{e_o(s)}{e_i(s)} = \frac{2\,(1 + s \times 1 \times 0.5)}{(1+2)\left[1 + s\dfrac{(1 \times 2 \times 0.5) + (1 \times 2 \times 0.8)}{(1+2)}\right]}$$

$$= \frac{2(1 + 0.5s)}{3\left(1 + s\dfrac{2.6}{3}\right)} = \frac{s+2}{2.6s + 3}$$

$$\boxed{\frac{e_o(s)}{e_i(s)} = \frac{s+2}{2.6s+3}} \qquad \textbf{Ans.}$$

Example 1.7 Find the transfer function of the network shown in Fig. Ex. 1.7.

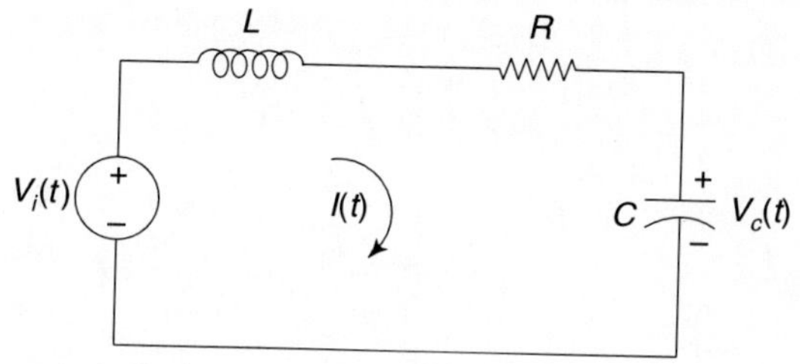

Figure Ex. 1.7

Solution Assuming zero initial conditions, yields the integral differential equation for this network as

$$L\frac{d\,I(t)}{dt} + R\,I(t) + \frac{1}{C}\int_0^t I(t)\,dt = V_i(t) \qquad ...(1)$$

Changing variables from current to charge

$$I(t) = \frac{dq(t)}{dt}$$

$$L\frac{d^2q(t)}{dt^2} + \frac{dq(t)}{dt} + \frac{1}{C}q(t) = V_i(t) \qquad ...(2)$$

We have

$$q(t) = C\,V_c(t)$$

Substituting the value of $q(t)$ in Eq. (2)

$$LC\frac{d^2V_c(t)}{dt^2} + RC\frac{dV_c(t)}{dt} + V_c(t) = V_i(t). \qquad ..(3)$$

Taking the Laplace transform of Eq. (3)

$$(LCs^2 + RCs + 1)\,V_c(s) = V_i(s)$$

Now

$$\frac{V_c(s)}{V_i(s)} = \frac{1}{LCs^2 + RCs + 1} \quad \textbf{Ans.}$$

- *The simple problem like single loop network can be solved by differential equation method as shown in Fig. Ex. 1.7 but the complex problem will be solved some other method like transform methods, mesh analysis, node analysis, etc.*

Example 1.8 Find the transfer function $G(s) = \frac{V_L(s)}{V_i(s)}$, for the circuit given in Fig. Ex. 1.8 (a). Solve the problem two ways—mesh analysis and nodal analysis. Show that the two methods yield the same result.

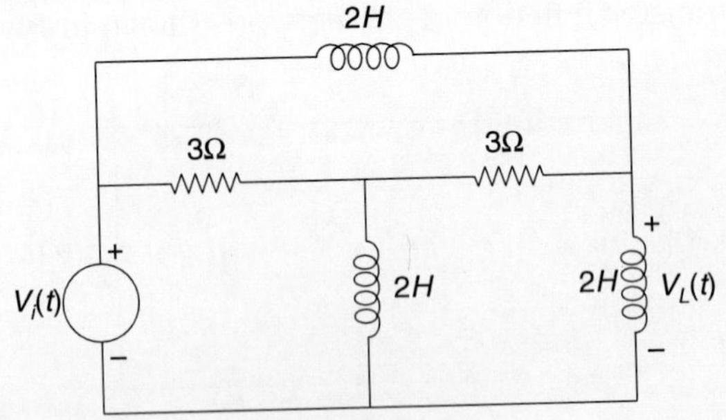

Figure Ex. 1.8 (a)

Solution Transforming the network as shown in Fig. 1.8 (b).

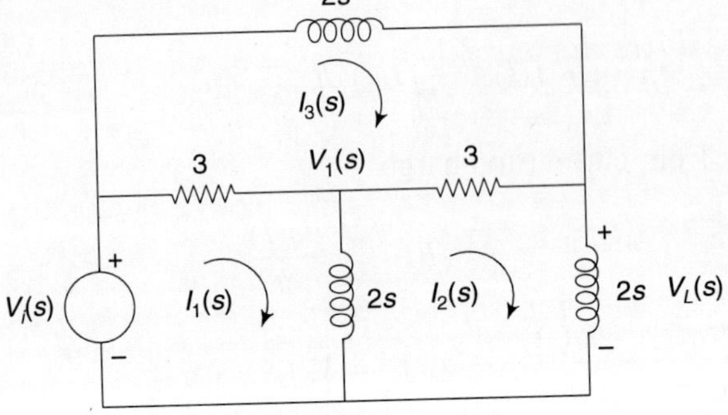

Figure Ex. 1.8 (b)

Now by the mesh analysis in loop 1

$$V_i(s) = (2s + 3)\, I_1(s) - 2s\, I_2(s) - 3\, I_3(s) \qquad \text{...(1)}$$

In loop 2

$$= 2s\, [(I_2(s) - I_1(s)] + 3\, [I_2(s) - I_3(s)] + 2s\, I_2(s) = 0$$

$$(4s + 3)\, I_2(s) - 2sI_1(s) - 3\, I_3(s) = 0$$

$$-\,2sI_1(s) + (4s + 3)\, I_2(s) - 3\, I_3(s) = 0 \qquad \text{...(2)}$$

In loop 3

$$2sI_3(s) + 3\, [I_3(s) - I_2(s)] + 3\, [I_3(s) - I_1(s)] = 0$$

$$-\,3I_1(s) - 3\, I_2(s) + (2s + 6)\, I_3(s) = 0 \qquad \text{...(3)}$$

Solving the mesh equation for $I_2(s)$ by Cramer's rule

$$I_2(s) = \frac{\begin{vmatrix} (2s+3) & V_i(s) & -3 \\ -2s & 0 & -3 \\ -3 & 0 & 2s+6 \end{vmatrix}}{\begin{vmatrix} 2s+3 & -2s & -3 \\ -2s & 4s+3 & -3 \\ -3 & -3 & 2s+6 \end{vmatrix}} = \frac{[2s\,(2s+6)+9]\, V_i(s)}{8s^2 + 60s^2 + 36s}$$

$$I_2(s) = \frac{(4s^2 + 12s + 9)\, V_i(s)}{8s^2 + 60s^2 + 36s}$$

$\because$ $$V_L(s) = 2s\, I_2(s)$$

Hence, $$V_L(s) = \frac{2s\,(4s^2 + 12s + 9)\, V_i(s)}{s\,(8s^2 + 60s + 36)}$$

$$V_L\,(s) = \frac{(8s^2 + 24s + 18)\, V_i(s)}{(8s^2 + 60\,s + 36)}$$

$$\frac{V_L(s)}{V_i(s)} = \frac{8s^2 + 24s + 18}{8s^2 + 60s + 36} \qquad \textbf{Ans.}$$

By nodal analysis

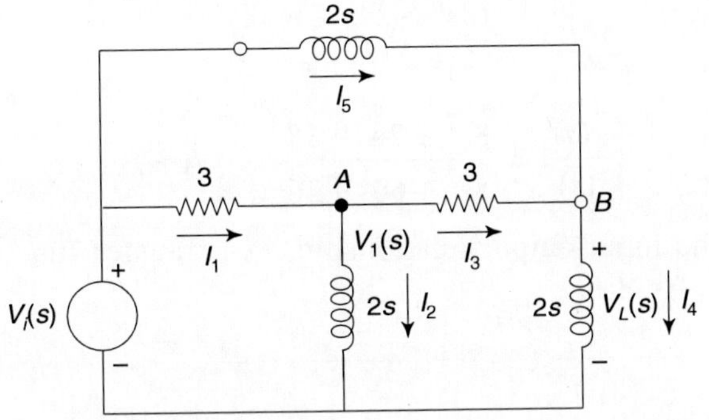

Figure Ex. 1.8 (c)

Applying KCL at node *A*

$$I_1 = I_2 + I_3$$

$$\frac{V_i(s) - V_1(s)}{3} = \frac{V_1(s)}{2s} + \frac{V_1(s) - V_2(s)}{3}$$

$$\frac{V_i(s)}{3} = \frac{2V_1(s)}{3} + \frac{V_1(s)}{2s} - \frac{V_L(s)}{3}$$

$$\frac{V_i(s)}{3} = \left(\frac{2}{3} + \frac{1}{2s}\right) V_i(s) - \frac{1}{3} V_L(s) \qquad ...(4)$$

Now applying KCL at node *B*

$$I_3 + I_5 = I_4$$

$$\frac{V_1(s) - V_L(s)}{3} + \frac{V_i(s) - V_L(s)}{2s} = \frac{V\ (s)}{}$$

$$\frac{V_1(s)}{3} - \frac{V_L(s)}{3} + \frac{V_i(s)}{2S} - \frac{2V_L(s)}{2S} = 0$$

$$-\frac{V_1(s)}{3}+\left(\frac{1}{3}+\frac{1}{s}\right)V_L(s) = \frac{V_i(s)}{2S} \quad ...(5)$$

Solving Eqs. (4) and (5) and get $V_L(s)$

$$V_L(s) = \frac{\begin{vmatrix}\left(\frac{2}{3}+\frac{1}{2s}\right) & \frac{V_i(s)}{3} \\ -\frac{1}{3} & \frac{V_i(s)}{2s}\end{vmatrix}}{\begin{vmatrix}\left(\frac{2}{3}+\frac{1}{2s}\right) & -\frac{1}{3} \\ -\frac{1}{3} & \left(\frac{1}{3}+\frac{1}{s}\right)\end{vmatrix}}$$

$$V_L(s) = \frac{(8s^2+24s+18)\,V_i(s)}{(8s^2+60s+36)}$$

$$\frac{V_L(s)}{V_i(s)} = \frac{8s^2+24s+18}{8s^2+60s+36} \quad \textbf{Ans.}$$

Example 1.9 Find the input-output relationship, i.e., transfer function of the electrical network shown in Fig. Ex. 1.9.

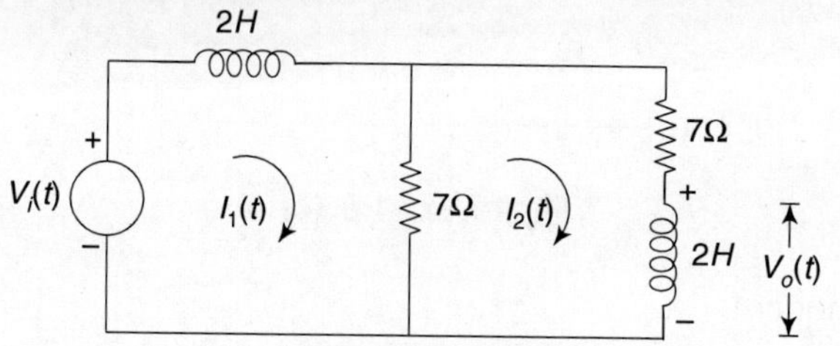

Figure Ex. 1.9

Solution The loop equation considering the Laplace transform of the network is

$$2s\,I_1(s) + 7I_2(s) - 7I_2(s) = V_i(s) \quad ...(1)$$

$$(2s+7)\,I_1(s) - 7I_2(s) = V_i(s)$$

$$7\,[I_2(s) - I_1(s)] + 7I_2(s) + 2sI_2(s) = 0 \quad ...(2)$$

$$I_1(s) = \frac{(2s+14)}{7}I_2(s) \quad ...(3)$$

Substituting the value of $I_1(s)$ in Eq. (1)

$$\frac{(2s+7)(2s+14)}{7}I_2(s) - 7I_2(s) = V_i(s)$$

$$V_i(s) = \left[\frac{(2s+7)(2s+14)}{7} - 7\right]I_2(s)$$

$$V_i(s) = \left[\frac{4s^2 28s + 14s + 98 - 49}{7}\right] I_2(s)$$

$$I_2(s) = \frac{7V_i(s)}{4s^2 + 42s + 49} \quad ...(4)$$

Substituting the value of $I_2(s)$ in

$$V_o(s) = 2s\, I_2(s)$$

$$V_o(s) = 2s\left[\frac{7V_i(s)}{4s^2 + 42s + 49}\right] \quad \textbf{Ans.}$$

$$\frac{V_o(s)}{V_i(s)} = \frac{14\, s}{4s^2 + 42\, s + 49} \quad \textbf{Ans.}$$

Example 1.10 Find the ratio of $\frac{V_o(s)}{V_i(s)}$ for the electrical network shown in Fig. Ex. 1.10

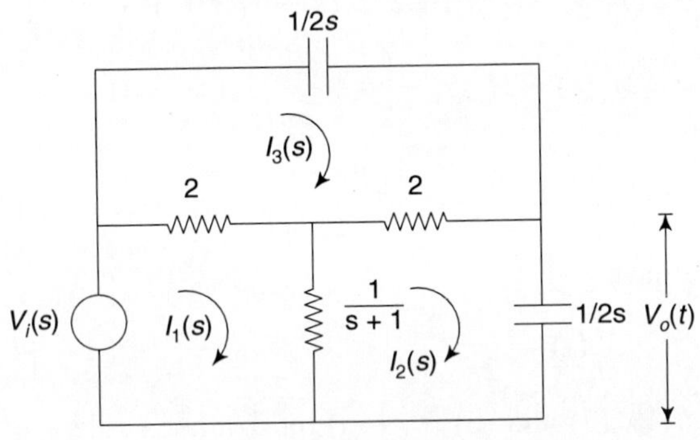

Figure Ex. 1.10

Solution Applying KVL in loop 1

$$2\,[I_1(s) - I_3(s)]\ \frac{1}{s+1}[I_1(s) - I_2(s)] = V_i(s)$$

$$\left(2 + \frac{1}{s+1}\right) I_1(s) - \frac{1}{s+1} I_2(s) - 2I_3(s) = V_i(s)$$

$$s\frac{2s+3}{s+1} I_1(s) - \frac{1}{s+1} I_2(s) - 2I_3(s) = V_i(s) \quad ...(1)$$

Applying KVL in loop 2

$$\frac{1}{S+1}[I_2(s)-I_1(s)]+2[I_2(s)-I_2(s)+\frac{1}{2s}I_2(s) = 0$$

$$-\frac{1}{s+1}I_1(s)+\left[2+\frac{1}{2s}+\frac{1}{s+1}\right]I_2(s)-2I_3(s) = 0$$

$$-\frac{1}{s+1}I_1(s)+\left[\frac{4s(s+1)+s+1+2s}{2s(s+1)}\right]I_2(s)-2I_3(s) = 0$$

$$-\frac{1}{s+1}I_1(s)=\left[\frac{4s^2+4s+s+1+2s}{2s^2+2s}\right]I_2(s)-2I_3(s) = 0$$

$$-\frac{1}{s+1}I_1(s)+\left[\frac{4s^2+7s+1}{2s^2+2s}\right]I_2(s)-2I_3(s) = 0 \quad ...(2)$$

Applying KVL in loop 3

$$\frac{1}{2s}I_3(s)+2[I_3(s)-I_2(s)]+2[I_3(s)-I_1(s)] = 0$$

$$-2I_1(s)-2I_2(s)+\left(4+\frac{1}{2s}\right)I_3(s) = 0$$

$$-2I_1(s)-2I_2(s)+\left(\frac{8s+1}{2s}\right)I_3(s) = 0 \quad ...(3)$$

From the network we have

$$V_o(s) = \frac{1}{2s}I_2(s) \quad ...(4)$$

Now solving the Eqs. (1), (2) and (3) for $I_2(s)$ by Cramer's rule

$$I_2(s) = \frac{\begin{vmatrix} \frac{2s+3}{s+1} & V_i(s) & -2 \\ -\frac{1}{s+1} & 0 & -2 \\ -2 & 0 & \frac{8s+1}{2s} \end{vmatrix}}{\begin{vmatrix} \frac{2s+3}{s+1} & -\frac{1}{s+1} & -2 \\ -\frac{1}{s+1} & \frac{4sS^2+7s+1}{2s^2+2s} & -2 \\ -2 & -2 & \frac{8s+1}{2s} \end{vmatrix}}$$

$$I_2(s) = \frac{\dfrac{8s^3+16s^2+s}{2(s+1)}V_i(s)}{\dfrac{\frac{1}{4}(16s^2 34s+3)s^2}{(s+1)}}$$

$$I_2(s) = \frac{8s^2+16s^2+s}{2(16s^4+34s^3+3s^2)}$$

$$= \frac{8s^3+16s^2+s}{32s^4+68s^3+6s^2} \qquad \text{...(5)}$$

Substituting the value of $I_2(s)$ in Eq. (4).

$$V_o(s) = \frac{1}{2s}\left[\frac{8s^3+16s^2+s}{32s^4+68x^3+6s^2}\right]V_i(s)$$

$$\frac{V_o(s)}{V_i(s)} = \frac{8s^2+16s+1}{64s^4+136s^3+12s^2} \qquad \textbf{Ans.}$$

MISCELLANEOUS PROBLEMS

Example 1.11 Derive the transfer function of the *RC* network mechanisation of the lag-lead compensator shown in Fig. Ex. 1.11 below. *(U.P.T.U., 2006–07, CS)*

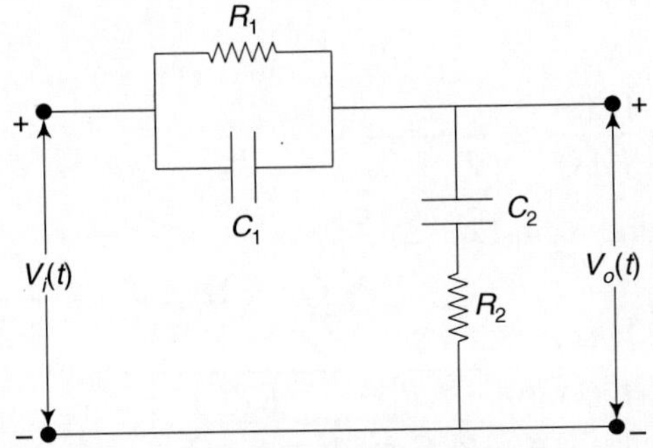

Figure Ex. 1.11 (a)

Solution The figure is redrawn as shown in Fig. Ex. 1.11. (a).

The impedance Z_1

$$Z_1 = \frac{R_1 \dfrac{1}{sC_1}}{R_1 + \dfrac{1}{sC_1}} = \frac{R_1}{R_1 C_1 s + 1}$$

$$Z_1 = \frac{R_1}{R_1 C_1 s + 1} \qquad \text{...(1)}$$

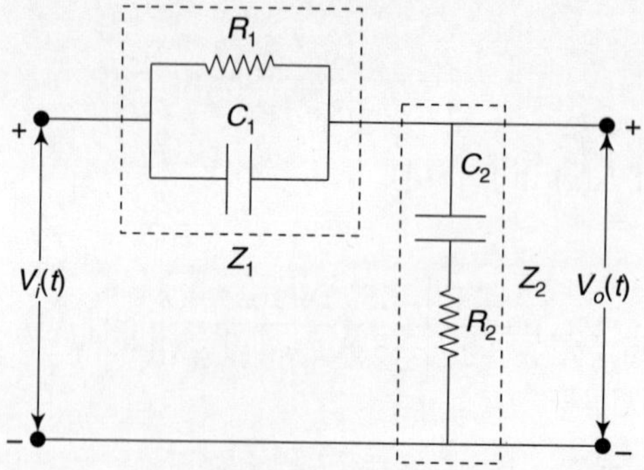

Figure Ex. 1.11 (b)

The impedance Z_2

$$Z_2 = R_2 + \frac{1}{sC_2} = \frac{R_2 C_2 s + 1}{sC_2}$$

$$Z_2 = \frac{R_2 C_2 s + 1}{sC_2} \qquad \text{...(2)}$$

The transfer function is

$$\frac{V_o(s)}{V_i(s)} = \frac{Z_2}{Z_1 + Z_2} \qquad \text{...(3)}$$

Putting the value of Z_1 and Z_2 in Eq. (3)

$$\frac{V_o(s)}{V_i(s)} = \frac{\dfrac{(R_2 C_2 s + 1)}{sC_2}}{\dfrac{R_1}{R_1 C_1 s + 1} + \dfrac{R_2 C_2 s + 1}{sC_2}}$$

$$\frac{V_o(s)}{V_i(s)} = \frac{(1 + C_2 s)(1 + R_1 C_1 s)}{(1 + R_1 C_1 s)(1 + R_2 C_2 s) + R_1 C_2 s} \qquad \textbf{Ans.}$$

Example 1.12 Write the difference equation of the following system shown in Fig. Ex. 1.12.

(U.P.T.U., 2004–05, CS)

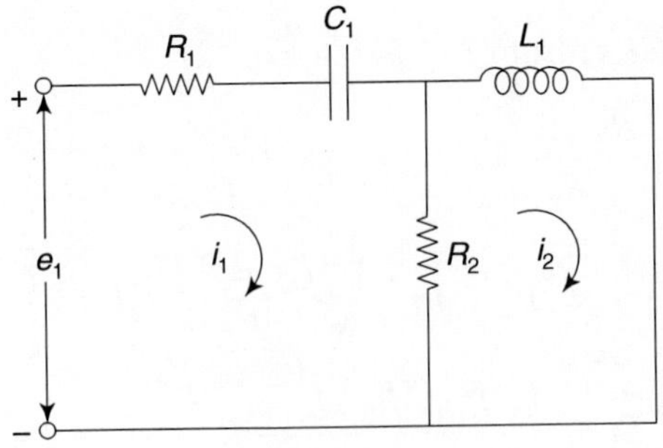

Figure Ex. 1.12

Solution Consider the current flowing in loop 1 is i_1 and in loop 2 is i_2

Applying KVL at loop 1

$$e = R_1 i_1 + \frac{1}{c}\int i_1\, dt + (i_1 - i_2) R_2 \quad \text{...(1)}$$

Taking Laplace transform of Eq. (1), we get

$$E(s) = I_1(s) + \frac{1}{C_1 s} I_1(s) + [I_1(s) - I_2(s)] R_2 \quad \text{...(2)}$$

Now applying KVL at loop 2

$$L_1 \frac{di_2}{dt} = -R_2 (i_2 - i_1)$$

$$L_1 \frac{di_2}{dt} + R_2 i_2 = i_1 R_2 \quad \text{...(3)}$$

Taking Laplace transform

$$sL_1 I_2(s) + R_2 I_2(s) = I_1(s) R_2$$

$$(sL_1 + R_2) I_2(s) = I_1(s) R_2$$

$$I_2(s) = \frac{I_1(s) R_2}{(sL_1 + R_2)} \quad \text{...(4)}$$

Putting the value of $I_2(s)$ from Eqs. (4) to (2), we get

$$E(s) = R_1 I_1(s) + \frac{1}{C_1 s} I_1(s) + \left[I_1(s) - \frac{R_2 I_1(s)}{R_2 + sL_1}\right] R_2$$

$$E(s) = I_1(s)\left[R_1 + R_2 + \frac{1}{C_1 s} - \frac{R_2^2}{R_2 + sL_1}\right]$$

$$\frac{E(s)}{I_1(s)} = R_1 + R_2 + \frac{1}{C_1 s} - \frac{R_2^2}{R_2 + sL_1} \qquad \textbf{Ans.}$$

Now putting the value of $I(s)$ from Eq. (4) in Eq. (2)

$$I_1(s) = I_2(s)\left(\frac{R_2 + sL_1}{R_2}\right)$$

$$E(s) = \frac{R_1}{R_2}(R_2 + sL_1)\, I_2(s) + \frac{1}{C_1 s}\frac{R_2 + sL_1}{R_2} I_2(s) + \left[\frac{R_2 + sL_1}{R_2} I_2(s) - I_2(s)\right] R_2$$

$$E(s) = I_2(s)\left[R_1 + \frac{sL_1 R_1}{R_2} + \frac{R_2\, sL_1}{C_1 s\, R_1} + R_2 + sL_1 - R_2\right]$$

$$E(s) = I_2(s)\left[R_1 + \frac{1}{C_1 s} + sL_1, \left(\frac{R_1}{R_2} + \frac{1}{C_1 R_2 s} + 1\right)\right]$$

$$\frac{E(s)}{I_2(s)} = R_1 + \frac{1}{C_1 s} + sL_1\left[\frac{R_1}{R_2} + \frac{1}{C_1 R_2 s} + 1\right] \qquad \textbf{Ans.}$$

Example 1.13 Develop a transfer function model for the armature controlled dc motor shown in Fig. Ex. 1.13 with the armature voltage e as input and speed θ as output.

(U.P.T.U., 2002–03, CS)

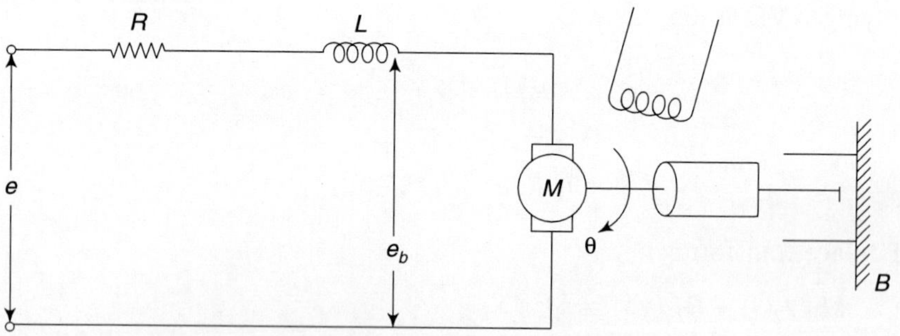

Figure Ex. 1.13 (b) Armature control dc motor

Solution Applying KVL in loop

$$e = Ri + L\frac{di}{dt} + e_b$$

Taking Laplace transform

$$E(s) = RI(s) + LsI(s) + E_b(s) \qquad \ldots(1)$$

Now from rotational mechanical system

$$T_M = J\frac{d^2\theta}{dt^2} + B\frac{d\theta}{dt}$$

Taking Laplace transform

$$T_M(s) = Js^2\,\theta(s) + Bs\,\theta(s) \qquad \text{...(2)}$$

Again the torque T_M is directly related to current i so,

$$T_M = K_1\, I$$

$$T_M(s) = K_1\, I(s) \qquad \text{...(3)}$$

and the back e.m.f. e_b directly related to rate of change of θ is $\frac{d\theta}{dt}$

$$e_b = K_2\frac{d\theta}{dt}$$

$$E_b(s) = K_2\, s\,\theta(s) \qquad \text{...(4)}$$

Putting the value of $E_b(s)$ from Eq. (4) to Eq. (1)

$$E(s) = RI(s) + LsI(s) + K_2 s\,\theta(s) \qquad \text{...(5)}$$

Now putting the value of T_M of Eq. (3) in Eq. (2), we get

$$K_1\, I(s) = Js^2\,\theta(s) + Bs\,\theta(s)$$

$$I(s) = \frac{1}{K_1}\left[Js^2\theta(s) + Bs\,\theta(s)\right] \qquad \text{...(6)}$$

Putting the value of $I(s)$ in Eq. (5), we get

$$E(s) = (R + sL)\frac{1}{K_1}[Js^2 + Bs]\,\theta(s) + K_2 s\,\theta(s)$$

$$\frac{\theta(s)}{E(s)} = \frac{1}{(R + sL)\frac{1}{K_1}[Js^2 + Bs] + K_2 s}$$

$$\frac{\theta(s)}{E(s)} = \frac{1}{(R + sL)\frac{1}{K_1}[Js^2 + Bs] + K_2 s} \qquad \textbf{Ans.}$$

MATLAB PROGRAMS

Introduction

MATLAB is a software package for high performance numerical computation and visualisation. This environment is suitable for solving scientific and engineering problems. It provides an interactive environment with hundreds of built-in functions for technical computation, graphics and animation. When it is used with routines from its companion software, the control system toolbox, MATLAB can be used to analyze and design control system problems. Basically, the MATLAB stands for Matrix Laboratory. Whatsoever work we do on the MATLAB it treats it as a matrix. Even if a single character is assigned, it will treat it as a matrix of order one by one. The software of MATLAB has different toolboxes like mechanical system, electrical system, communication system, control system, simulation, etc.

MATLAB and its toolbox are commercial software products available from the Mathworks, Inc., 3 Apple Hill Drive, and Natick MA01760-2098. Phone: (508)647-7000. Email: info@mathworks.com, http:// www.mathworks.com you must have MATLAB version 6.5 or above and the control system toolbox version 5.2 installed on your computer.

The program must be write on the M-file so the name of the file will be like filename.m, to run the M-files, first be sure the files are either added to the search path in set path:... Under the file menu or appear in the 'current directory' window, which is a part of the MATLAB window. To execute the program or to see the computer response after code the program which is written on m-file, run each program by type the M-file name on command window(>>). You may also run the files by right clicking the file name, if it appears in the current directory window, and select run.

Program 1.1

```
%Program for using different Command
'program'       % Display the string
'Hallo Dear'    % on Command window
A = -2.15       % Display scalar number
B = -4+3j       % Display complex number
C = -2-3j       % Display complex number
S = B+C         % Sum two complex number and equal to s
P = B*C         % Product of two complex number and equal to P
x = 3;          % Assign 3 to x and not display on Command window
y = 2;          % Assign 2 to y and not display on Command window
z = x+y         % Assign x+y to z and display on Command window
end
```

OUTPUT:

```
Hello Dear
A = -2.1500
B = -4.0000 + 3.0000i
C = -2.0000 - 3.0000i
S = -6
P = 17.0000 + 6.0000i
z = 5
```

- v % sign is used to show the comments on M-files, MATLAB ignores the line started with % sign and does not execute it.
- v ; (Semicolon) is used, if we don't want to display the execution of command on command window, i.e., suppress the display of command. It is used after the command

Program 1.2

```
% The way by which matrix can be inserted into the MATLAB
A = [1 2 3; 4 5 6; 7 8 9]       % Input matrix A
%OR
A = [1,2, 3; 4, 5, 6; 7, 8, 9]  % Input matrix A
%OR
A = [1 2 3                      % Input matrix A
     4 5 6
     7 8 8]
% insert a row matrix
B = [1 2 3 5 8]                 % Input row matrix
%OR
B = [1, 2, 3, 5, 8]             % Input row matrix
% Insert a column matrix
C = [1; 2; 3; 4]                % Input column matrix
%OR
C = [1 2 3 4]                   % Input column matrix
end
```

OUTPUT:

```
A =
  1 2 3
  4 5 6
  7 8 9
A =
  1 2 3
  4 5 6
  7 8 9
A =
  1 2 3
  4 5 6
  7 8 8
```

```
B =
  1  2  3  5  8
B =
  1  2  3  5  8
C =
  1
  2
  3
  4
C =
  1
  2
  3
  4
```

- *(Semicolon) is also used to separate the row and column can be separated by space or comma. Row also can be separated by carriage return.*

Program 1.3

```
% Program for representation of polynomial as row vector
% If P1 = 3s^3+5s^2+7s+8
clear all                          % Clear all the variable
clc                                % Clear the screen
P1 = [3 5 7 8]                     % Store polynomial
% OR
% if P2 = s(s+5)(s+7)
'Coefficients of polynomial P2'
P2 = poly([0 -5 -7)]               % store the polynomial
'Roots of polynomial P1'
R = roots(P1)                      % Find the roots of P1
'Multiplication of two polynomials'
c - conv(P1,P2)                    % Multiply two polynomial by convolution
```

OUTPUT

```
P1 = 3  5  7  8
Coefficients of polynomial P2
P2 = 1  12  35  0
Roots of polynomial P1
R = -1.3778
```

```
    -0.1444 + 1.38371
    -0.1444 - 1.3837i
Multiplication of two polynomials
c = 3  41  172  267  341  280  0
```

Program 1.4

```
% Program for finding the transfer function of the system
% F(s) = N(s)/D(s)
clear all                    % clear all variables
clc                          % clear the screen
numf = input ('Enter the coefficient of numerator polynomial:-');
                             % neumerator of LTI obeject
denf = input ('Enter the coefficient of denominator polynomial:-');
                             % denominator of LTI object
T = tf(numf,denf)            % Transfer function of the object
```

INPUT

```
Enter the coefficient of numerator polynomial:- [1 2 3]
Enter the coefficient of denominator polynomial:- [5 6 2 1]
```

OUTPUT

Transfer function: $\dfrac{s\text{^}2 + 2s + 3}{5s\text{^}3 + 6s\text{^}2 + 2s + 2}$

Program 1.5

```
% Program for finding the partial fraction for given function
%F(s) = 3s^3+5s^2+7s^+8/s(s+5)(s^2+10s+100)
clear all                        % Clear all the variable
clc                              % Clear the screen
numf = input('Enter the coefficient of numerator polynomial:- ');
denf = input('Enter the coefficient of denominator polynomial:-');
'T(s)'
G = tf(numf.denf)                % Create transfer function
'Partial fraction is'
[r,p,k] = residue (numf, denf)   % Find residues and assign to r,
                                 % Find roots of denominator and assign to p,
                                 % Find constant and assign to k
```

INPUT:

```
Enter the coefficient of numerator polynomial:- [1 4 4]
Enter the coefficient of denominator polynomial:- [1 5 4 0]
```

OUTPUT

```
T(s)
                          s^2 + 4s + 4
Transfer function:  ----------------------
                      s^3 + 5s ^ 2 + 4s

Partial fraction is
r =  0.3333
    -0.3333
     1.0000
p =  -4
     -1
      0
k=[]
```

l.e. T(s) = 1/s - 0.33/(s+1) + 0.33/(s+4)

v The command : input('display text') is used to take the input from the user

Program 1.6

```
% Program for finding transfer function if the function is given in the
% form of Poles, Zeros and Gain
% G = 12(s+2)(s+3)/s(s+1)(s+5)(s+7)
clear all;                    % clear all the variables
clc;                          % clear the screen
numg = input('Enter the zeros of the function:-');% input zeros like [-2 -3];
deng = input('Enter the poles of the function:-'); % input poles like
                                                              [0 -1 -5 -7];

k = input('Enter the value of k:-');% input the value of k like 12;
'Transfer function is'
G = zpk(numg,deng,k)  % create the transfer function if zeros
                        poles and k are given
```

INPUT

```
Enter the zeros of the function:- [-2 -3]
Enter the poles of the function:- [0 -1 -5 -7]
Enter the value of k:- 12
```

OUTPUT

```
Transfer function is
                        12(s + 2) (s + 3)
zero/pole/gain:  ----------------------------
                   s (s + 1) (s + 5) (s + 7)
```

EXERCISE

1.1. Explain open loop and closed loop control system with one example.

1.2. Explain the construction and working of servomechanism.

1.3. What is the use of feedback in control system to parameter variantion?

1.4. What is sensitivity? What is the effect of sensitivity in control system?

1.5. What do you understand by "disturbances"? What are various types of disturbances? Explain. Give an example of time varying control system. *(U.P.T.U., 2007–08, CS)*

1.6. Describe the operation of the biological control system consisting of a human being reaching for an object. Also draw the block diagram. *(U.P.T.U., 2006–07, CS)*

1.7. Why is closed loop control system preferred over open loop? *(U.P.T.U., 2005–06, CS)*

1.8. Explain difference between positive feedback and negative feedback. *(U.P.T.U., 2005–06, CS)*

1.9. Compare open loop and closed loop systems. Which system is advantageous? *(U.P.T.U., 2004–05, CS)*; *(U.P.T.U., 2006–07, A CS)*

1.10. Explain the following terms :

(i) node (ii) output node
(iii) input node (iv) path
(v) loop (vi) loop gain in a signal flow graph with examples

1.11. What is the need of a control system? Explain the different performance specifications, which the control systems have to meet? *(U.P.T.U., 2006–07, A CS)*

1.12. Give the general block diagram of automatic control system and explain in brief about each block with input and output signals.

1.13. Draw the diagram of a closed loop control system and indicate the following it.

(i) plant (ii) Command input
(iii) controlled output (iv) actuating signal
(v) feedback element and controlled element. *(U.P.T.U., 2003–04, A CS)*

CHAPTER

2 Reduction of Multiple Subsystems

2.1 INTRODUCTION

In this chapter we will study that how to reduce a block diagram of multiple subsystems to a single block, representing the transfer function of that block or system. In previous chapter, you have studied about individual subsystems represented by a block and about their transfer function. When the systems are interconnection of many subsystems such systems are complicated and it is difficult to find the transfer function of those systems.

Since the response of a single transfer function is easy to analyse to obtain transient response information about the entire system. Hence, it is necessary to represent multiple subsystems as a single transfer function. In this chapter multiple subsystems are represented in ways : as block diagram and as signal-flow graphs out of these block diagrams are usually used for frequency-domain analysis and design while the signal-flow graphs are used for state-space analysis.

Now we will develop techniques to reduce complicated systems to a single block with its transfer function. Block diagram algebra will be used to reduce block diagrams and Mason's rule to reduce signal-flow graphs.

2.2 BLOCK DIAGRAM ALGEBRA

In the control system engineering, we need to control specific variables. So we require, interrelationship controlled variables and the controlling variables. This interrelationship between controlled variable and controlling variables is represented by the transfer function of the system relating the input and output variables. So we can say that the transfer function is an important relation for the controlling system.

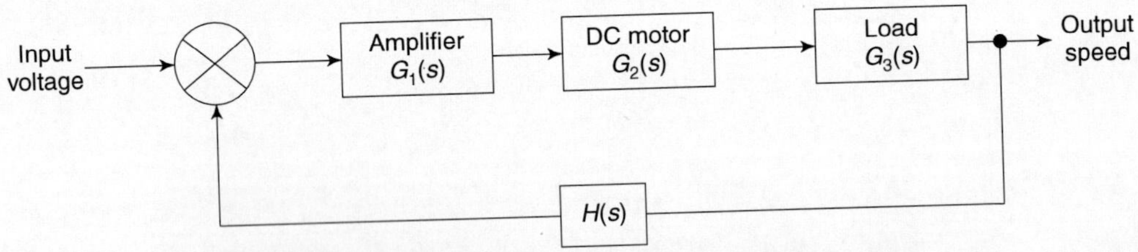

Figure 2.1 Block diagram of a dc motor control system

It is easy to derive a transfer function for a simple system but if a system is complex or large so it is not easy to derive the complete transfer function for such complex system. So we divide such complex system into subsystems. Subsystem of the system may be a single element or group of elements, the transfer function of each subsystem of a system is represented by a block. The symbol mentioned in the block represents the transfer function of the subsystem. The composition and interconnection of these blocks represent the block diagram of complete system. It relates the output and the input of a complete system. A convenient graphical representation of this behaviour (input-output) is the block diagram.

Figure 2.1 shows the block diagram of dc motor control. In diagram

$G_1(s)$ — Transfer function of amplifier circuit

$G_2(s)$ — Transfer function of dc motor circuit

$G_3(s)$ — Transfer function of load circuit

2.2.1 Basic Terms used in Block Diagram

(i) The flow of the signal is from the input to the output, i.e., unidirectional

Figure 2.2 Diagram for flow of singnal

(ii) The blocks are connected by lines with arrows indicating the flow of information in one direction, i.e., output of one block is the input to the other block as shown in Fig. 2.1.

(iii) Two or more signals entering a system are represented by a summing point. As shown in Fig. 2.3.

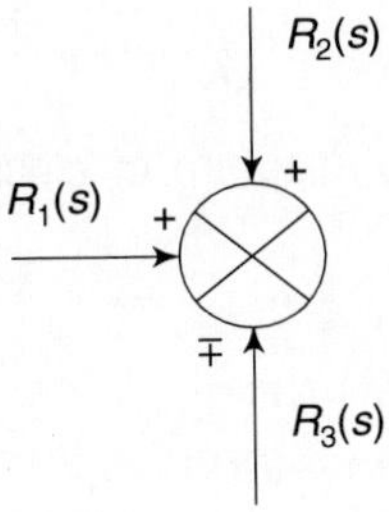

Figure 2.3 Summing point

(iv) If one input source is used to two or more systems, it is represented by a take off point as shown in Fig. 2.4. T is take-off point.

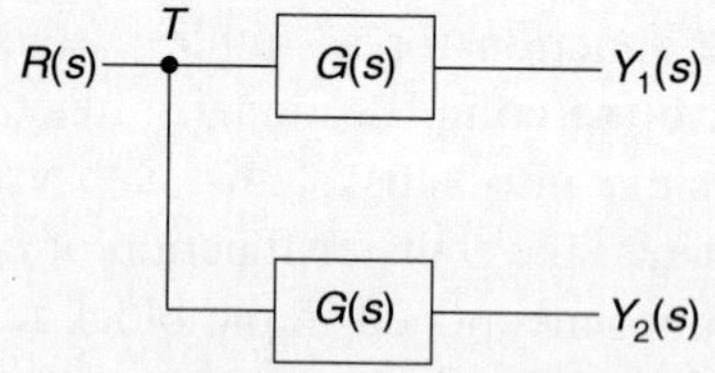

Figure 2.4 Take-off point

2.2.2 Signals Used in Closed Loop System

Block diagram of a closed loop system is shown in Fig. 2.5. With reference to this figure the signals used in block diagrams of control system are given below.

$R(s)$ = Reference input signal

$Y(s)$ = Output signal

$B(s)$ = Feedback signal [$H(s)\ Y(s)$]

$H(s)$ = Transfer function of feedback system [$B(s)/Y(s)$]

$G(s)$ = $Y(s)/E(s)$ forward path transfer function

$E(s)$ = Equating signal or error signal [$R(s) \pm B(s)$]

$T(s)$ = Transfer function of complete system

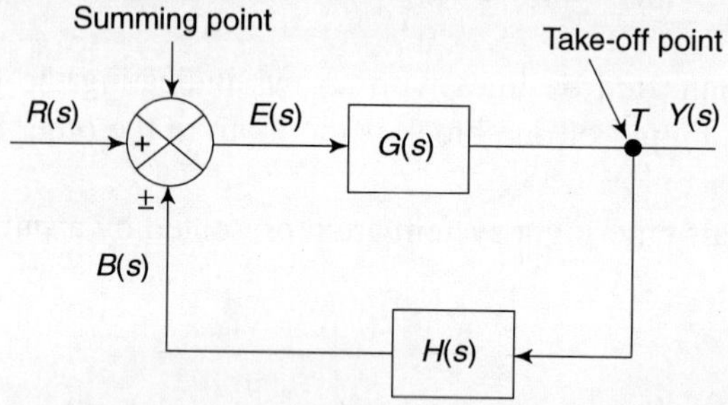

Figure 2.5 Block diagram of closed loop system

From Fig. 2.5

$$Y(s) = G(s)\ E(s) \quad ...(2.1)$$

$$E(s) = R(s) + B(s)$$

$$E(s) = R(s) \pm H(s)\ Y(s)$$

$\because$ $$B(s) = H(s)\ Y(s)$$

$$E(s) = R(s) \pm H(s)\ Y(s) \quad ...(2.2)$$

Putting the value of $E(s)$ from Eq. (2.2) to the Eq. (2.1)

$$Y(s) = G(s)\ [R(s) \pm H(s)\ Y(s)$$

$$Y(s) = G(s)\, R(s) \pm G(s)\, H(s)\, Y(s)$$

$$Y(s) \mp Y(s)\, G(s)\, H(s) = G(s)\, R(s)$$

$$[1 \mp G(s)\, H(s)]\, Y(s) = G(s)\, R(s)$$

$$\frac{Y(s)}{R(s)} = \frac{G(s)}{1 \mp G(s)\, H(s)}$$

$$\therefore \quad \frac{Y(s)}{R(s)} = T(s) \text{ transfer function of complete system}$$

$$T(s) = \frac{G(s)}{1 \mp G(s)\, H(s)} \qquad \text{...(2.3)}$$

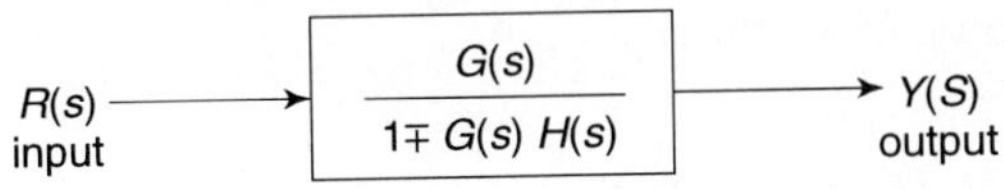

Figure 2.6

2.2.3 Rules for Reducing Block Diagram

As we discussed earlier the large system can be represented by a number of subsystems interconnected with each other and represented by blocks. The transfer function of such system can be obtained by reducing blocks. Some rules are followed for reducing the blocks.

Rule 1: Blocks connected in cascade

If the blocks are connected in cascade the equivalent diagram or equivalent transfer function can be obtained by multiplying the individual transfer function of the blocks.

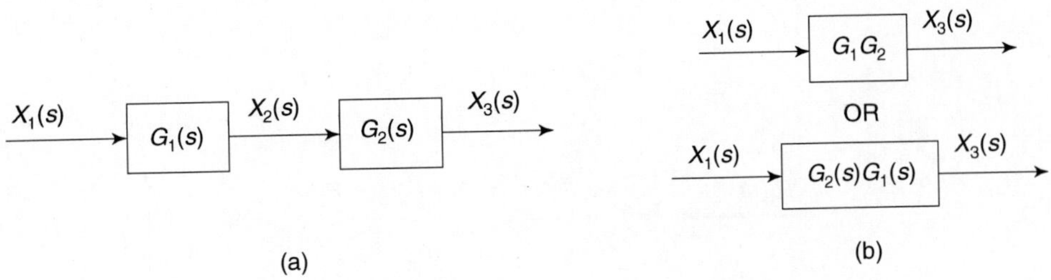

Figure 2.7 (a) Original diagram (b) Equivalent diagram

Rules 2: Block connected in parallel

If the blocks are connected in parallel the equivalent diagram or equivalent transfer function can be obtained by adding the individual transfer function of the blocks.

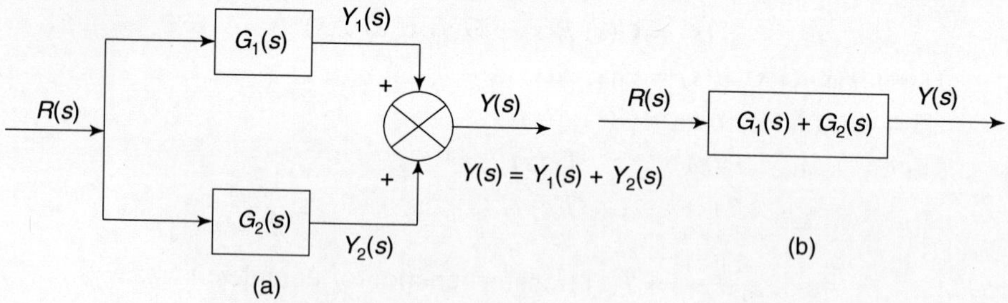

Figure 2.8 (a) Orginal diagram (b) Equivalent diagram

Rule 3: Take-off point

If input is common to two or more blocks

∵ We know that

$$Y(s) = G(s)\,R(s)$$

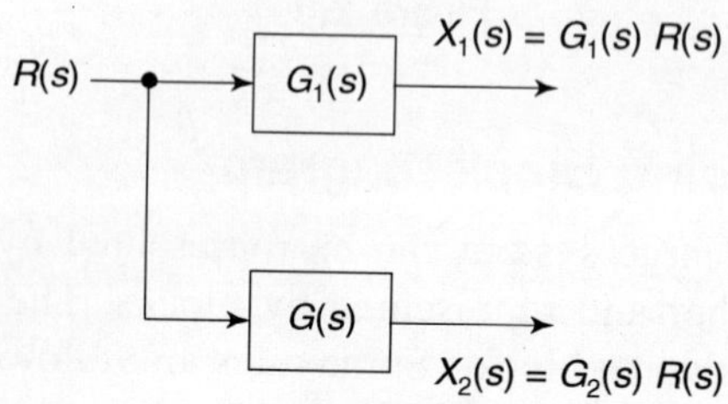

Figure 2.9 Diagram for take-off point

Rule 4: Summing point

If two or more signals are entering a system

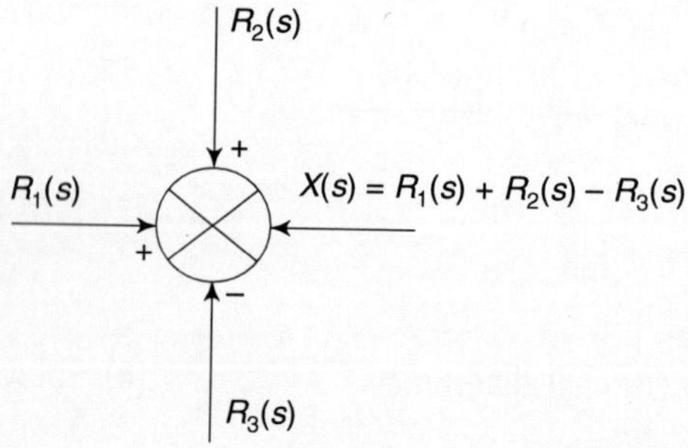

Figure 2.10 Diagram for summing point

Rule 5: Moving a summing point from a position before a block to a position after the block (behind a block)

All the input is multiplied by transfer function of the block $[G(s)]$ except the input given to the block as shown in Fig. 2.11.

$$X_3(s) = [X_1(s) \pm X_2(s)]\ G(s) \text{ from Fig. 2.11 (a)}$$
$$X_3(s) = X_1(s)\ G(s) \pm X_2(s)\ G(s) \text{ from Fig. 2.11 (b)}$$
$$X_3(s) = [X_1(s) \pm X_2(s)]\ G(s), \text{ i.e., it is equal to Fig. 2.11 (a)}$$

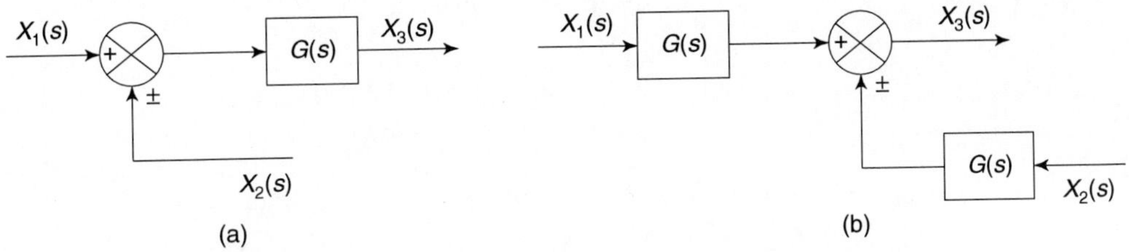

Figure 2.11 (a) Original diagram, (b) Equivalent diagram

Rule 6: Moving a summing point from a position after a block to a position before a block (ahead of a block)

All the input is multiplied by $\left[\dfrac{1}{G(s)}\right]$ except the input given to the block as shown in Fig. 2.12.

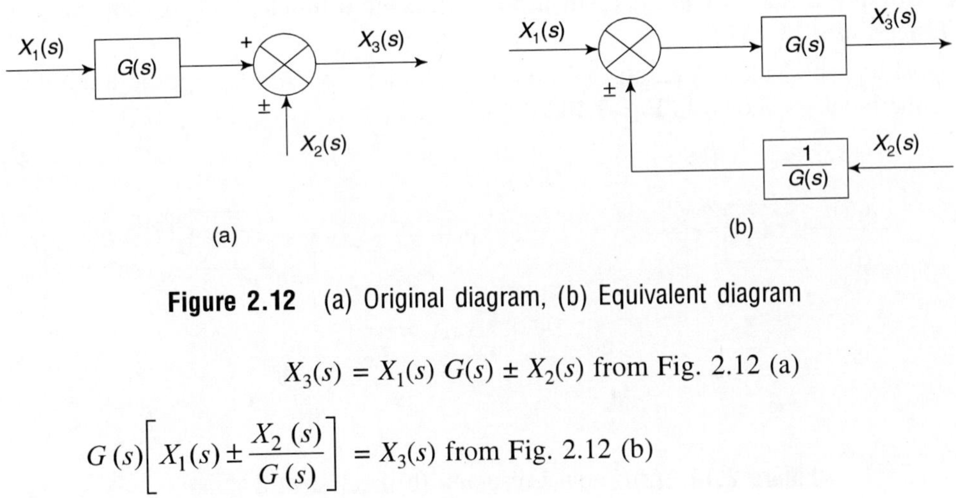

Figure 2.12 (a) Original diagram, (b) Equivalent diagram

$$X_3(s) = X_1(s)\ G(s) \pm X_2(s) \text{ from Fig. 2.12 (a)}$$

$$G(s)\left[X_1(s) \pm \frac{X_2(s)}{G(s)}\right] = X_3(s) \text{ from Fig. 2.12 (b)}$$

$$G(s)\left[\frac{X_1(s)\,G(s) \pm X_2(s)}{G(s)}\right] = X_3(s)$$

$$X_1(s)\ G(s) \pm X_2(s) = X_3(s) \quad \text{i.e., it is equal to Fig. 2.12 (a)}$$

Rule 7: Moving a take-off point from a position before a block to a position after the block (behind a block)

It is opposite to the rule 5, i.e., all the input is multiplied by $\frac{1}{G(s)}$ except the input directly given to the block as shown in Fig. 2.13.

$X_2(s) = G(s)\,X_1(s)$ from Fig. 2.13 (a)

$X_1(s) = X_1(s)$

$X_2(s) = X_1(s)\,G(s)$ from Fig. 2.13 (b)

$$X_1(s) = \frac{X_1(s)\,G(s)}{G(s)}$$

$X_1(s) = X_1(s)$

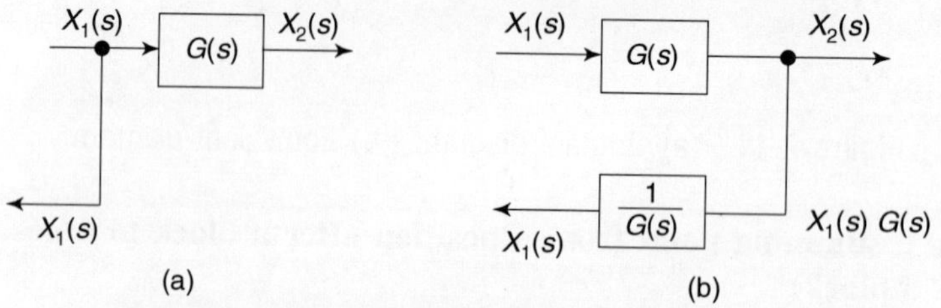

Figure 2.13 (a) Original diagram, (b) Equivalent diagram

Rule 8: Moving a take-off point from a position after a block to a position before the block (ahead of a block)

It is opposite to the Rule 6, i.e., all the input is multiplied by $G(s)$ except the input directly given to the block as shown in Fig. 2.14.

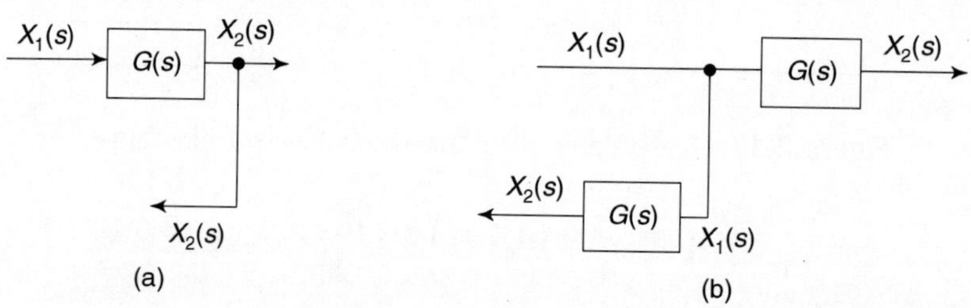

Figure 2.14 (a) Original diagram, (b) Equivalent diagram

$X_2(s) = X_1(s)$ from Fig. 2.14 (a)

$X_2(s) = X_2(s)\,G(s)$

$X_2(s) = X_1(s)\,G(s)$ from Fig. 2.14 (b)

$X_2(s) = X_1(s)\,G(s)$

Rule 9: Consecutive summing point can be interchanged

Output is not affected by interchanging the summing point as shown in Fig. 2.15.

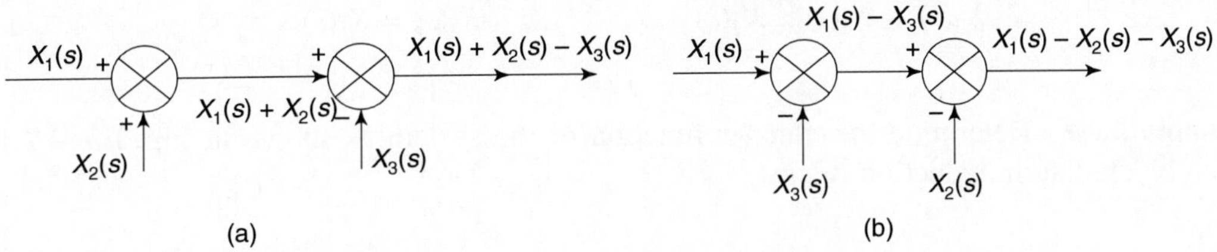

Figure 2.15 (a) Original diagram, (b) Equivalent diagram

Rule 10: Moving a take-off point from a position before a summing point to a position after the summing point (behind a summing point)

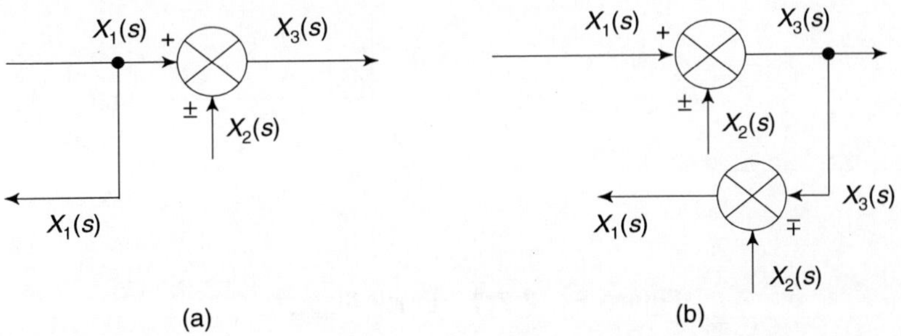

Figure 2.16 (a) Original diagram, (b) Equivalent diagram

$$X_3(s) = X_1(s) \pm X_2(s)$$
$$X_1(s) = X_1(s)$$

$$X_3(s) = X_1(s) \pm X_2(s)$$
$$X_1(s) = X_3(s) \mp X_2(s)$$
$$X_1(s) = X_1(s) \pm X_2(s) \mp X_2(s)$$
$$X_1(s) = X_1(s)$$

Rule 11: Moving a take-off point from a position after a summing point to a position before the summing point (ahead of a summing point)

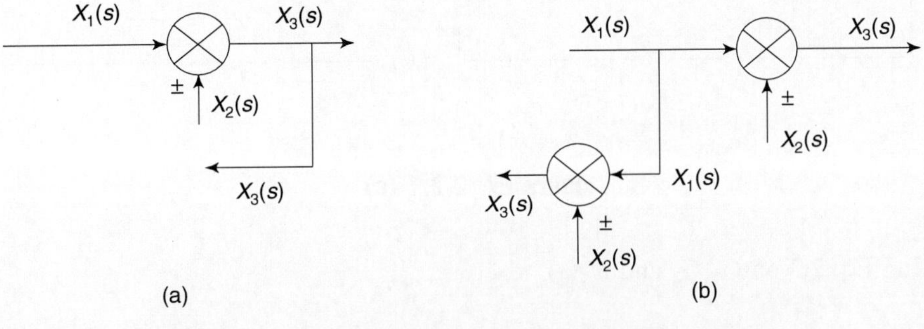

Figure 2.17 (a) Original diagram, (b) Equivalent diagram

$$X_3(s) = X_1(s) \pm X_2(s) \qquad X_3(s) = X_1(s) \pm x_2(s)$$
$$X_3(s) = X_1(s) \pm X_2(s)$$

Example 2.2.1 Determine the transfer function of the system as shown in Fig. Ex. 2.2.1 using block diagrm reduction rules.

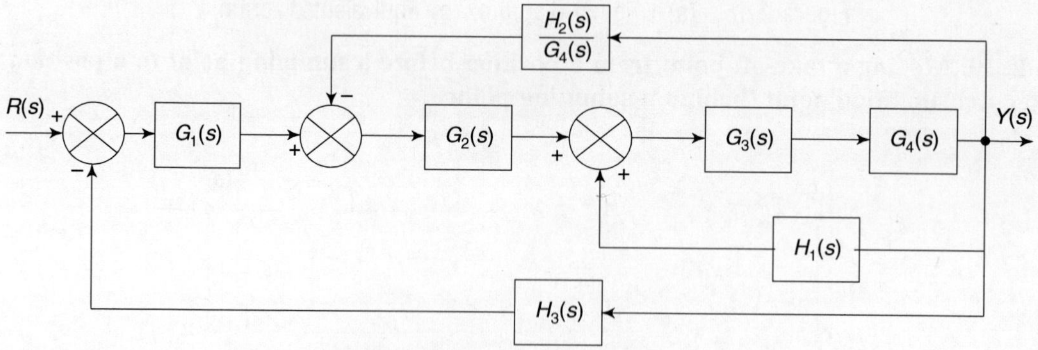

Figure Ex. 2.2.1 Block diagram

Solution Applying Rule 1 on G_3 and G_4

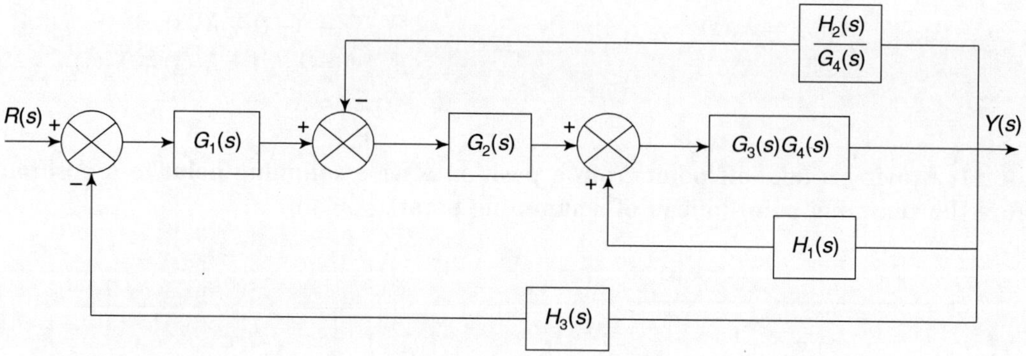

Figure Ex. 2.2.1 (a)

Applying Eq. 2.3 on G_3 G_4 and $H_1(s)$

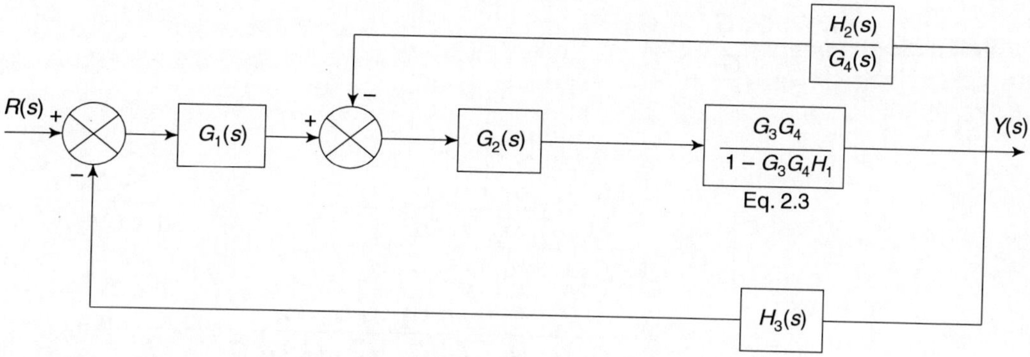

Figure Ex. 2.2.1 (b)

Applying again Rule 1 and Eq. 2.3

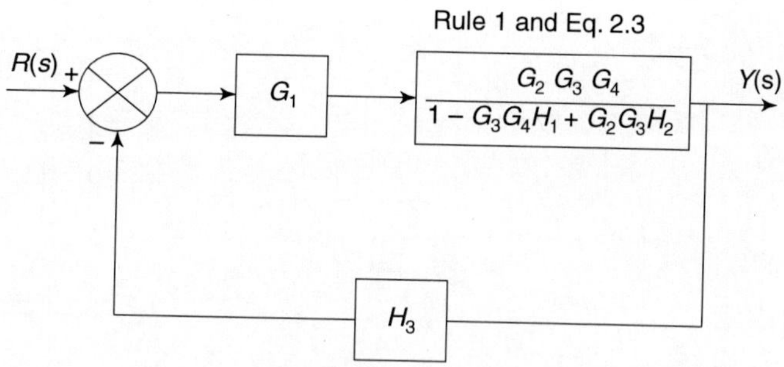

Figure Ex. 2.2.1 (c)

$$\Rightarrow \quad \frac{\dfrac{G_2\ G_3\ G_4}{1 - G_3\ G_4\ H_1}}{1 + \dfrac{G_2\ G_3\ G_4}{1 - G_3\ G_4\ H_1} \times \dfrac{H_2}{G_4}} = \frac{\dfrac{G_2\ G_3\ G_4}{1 - G_3\ G_4\ H_1}}{\dfrac{1 - G_3\ G_4\ H_1 + G_2\ G_3\ H_2}{1 - G_3\ G_4\ H_1}}$$

$$\Rightarrow \quad \frac{G_2\ G_3\ G_4}{1 - G_3\ G_4\ H_1 + G_2\ G_3\ H_2}$$

Applying Rule 1 and Eq. 2.3.

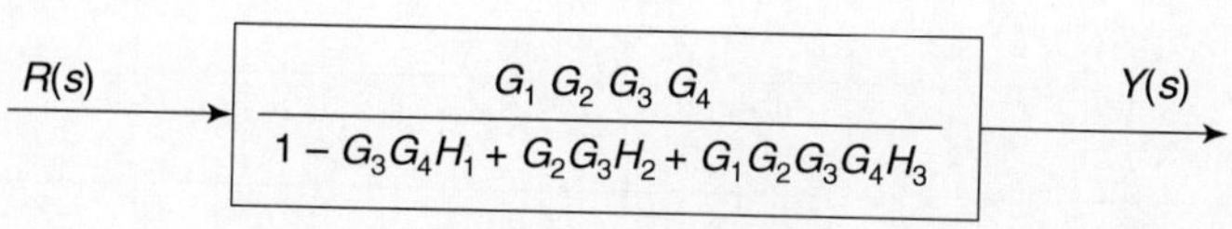

Figure Ex. 2.2.1 (d)

Rule 1: $\dfrac{G_1\,G_2\,G_3\,G_4}{1-G_3\,G_4\,H_1+G_2\,G_3\,H_2}$

by Eq. 2.3

$$\Rightarrow \qquad \frac{\dfrac{G_1\,G_2\,G_3\,G_4}{1-G_3\,G_4\,H_1+G_2\,G_3\,H_2}}{1+\dfrac{G_1\,G_2\,G_3\,G_4}{1-G_3\,G_4\,H_1+G_2\,G_3\,H_2}} \qquad \because \frac{G(s)}{1+G(s)\,H(s)}$$

$$\Rightarrow \qquad \frac{G_1\,G_2\,G_3\,G_4}{1-G_3\,G_4\,H_1+G_2\,G_3\,H_2+G_1\,G_2\,G_3\,G_4\,H_3} \quad \textbf{Ans.}$$

Example 2.2.2 Consider the system shown in Fig. Ex. 2.2.2 simplify this diagram and find the overall transfer function of the system.

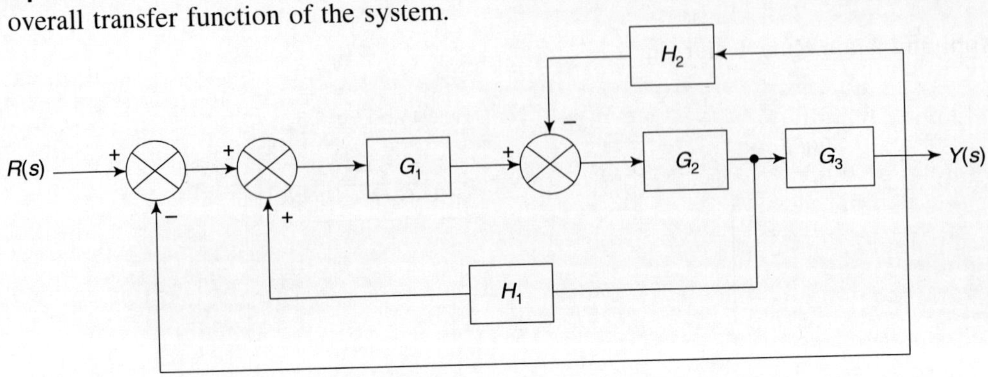

Figure Ex. 2.2.2 Block diagram

Applying Rule 6 on summing point of the negative feedback loop H_2, i.e., H_2 is multiplied by $\dfrac{1}{G_1}$.

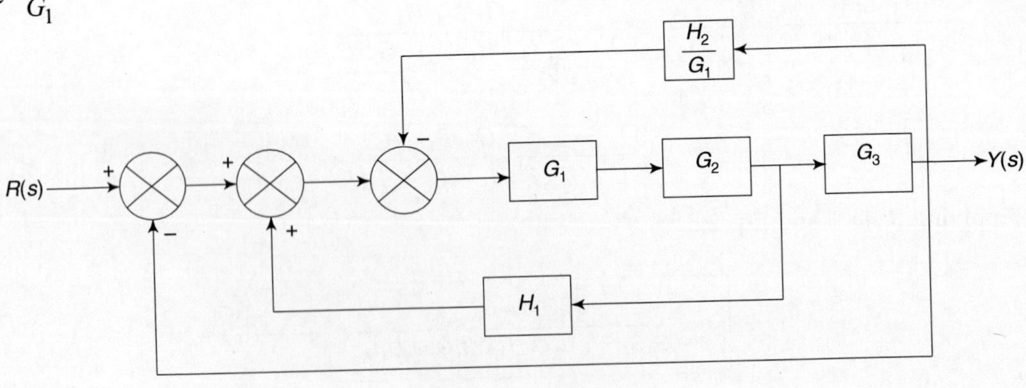

Figure Ex. 2.2.2 (a)

Applying Rule 1 on G_1 and G_2 and Rule 9 on feedback H_1 and $\dfrac{H_2}{G_1}$

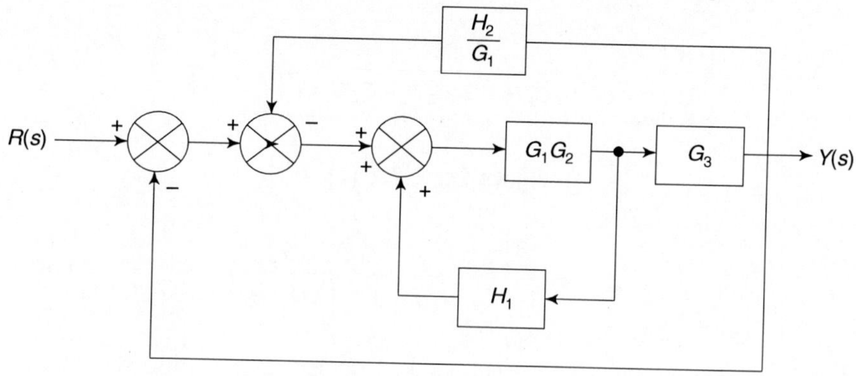

Figure Ex. 2.2.2 (b)

Applying Eq. (2.3) on closed loop having feedback H_1 and Rule 1 with G_3.

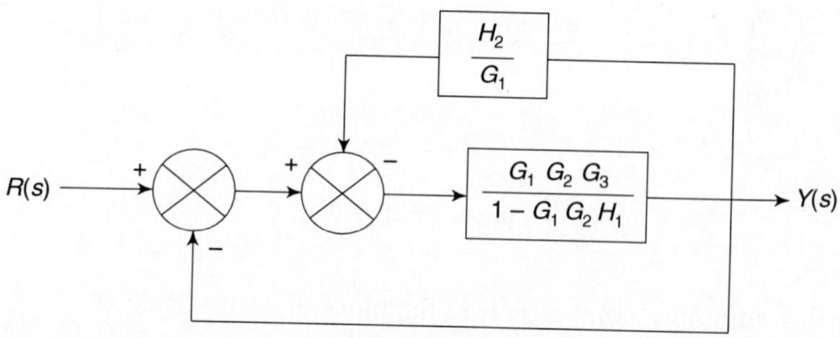

Figure Ex. 2.2.2 (c)

Applying Eq. (2.3) on closed loop system

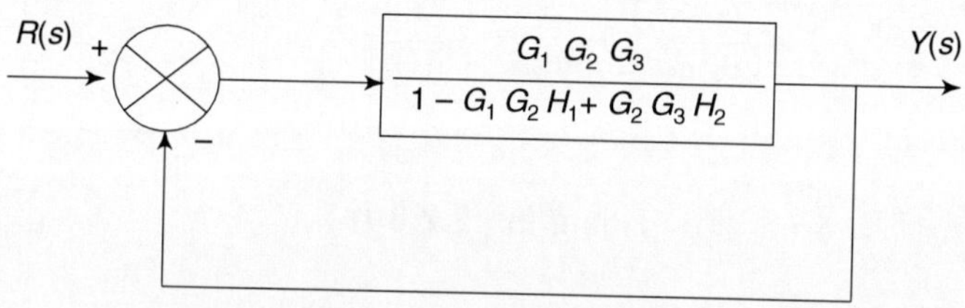

Figure Ex. 2.2.2 (d)

Applying Eq. (2.3) on closed loop system

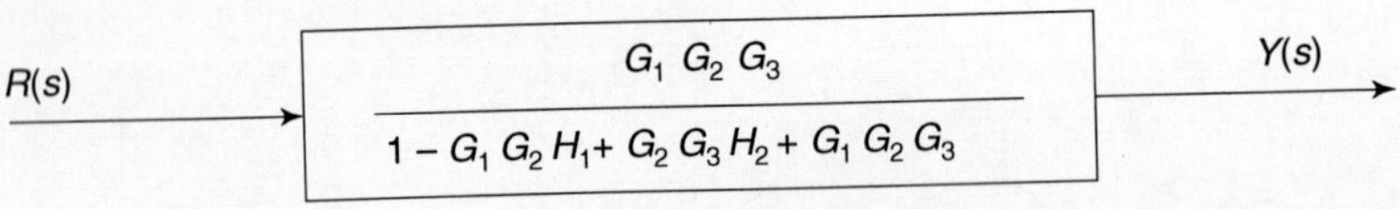

Figure Ex. 2.2.2 (e)

$$T(s) = \frac{G_1 \, G_2 \, G_3}{1 - G_1 \, G_2 \, H_1 + G_2 \, G_3 \, H_2 + G_1 \, G_2 \, G_3} \quad \textbf{Ans.}$$

Example 2.2.3 Consider the system shown in Fig. Ex. 2.2.3. Find the overall transfer function of the system using reduction rules.

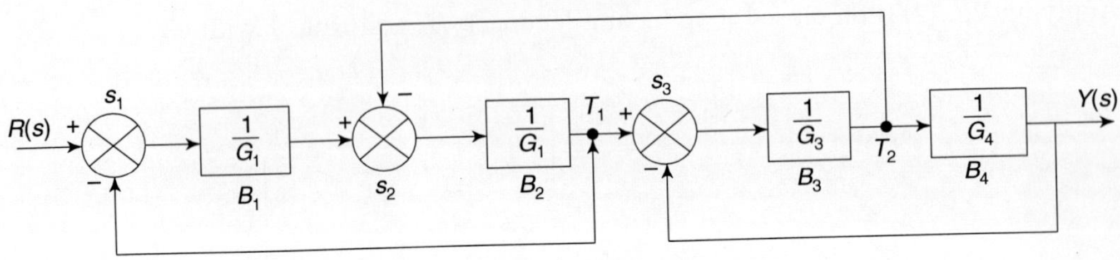

Figure Ex. 2.2.3

Solution Shifting summing point s_2 before the block B_1 using Rule 6.

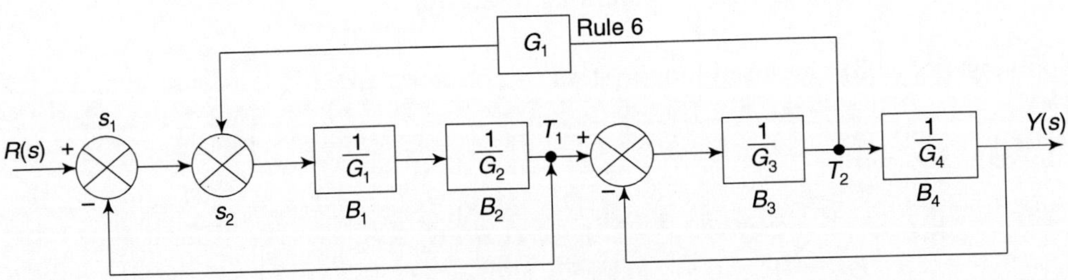

Figure Ex. 2.2.3 (a)

Shifting take-off point T_2 after the block B_4 by Rule 7 and interchanging the summing point s_1 and s_2 by Rule 9 and Rule 1 on blocks B_1 and B_2.

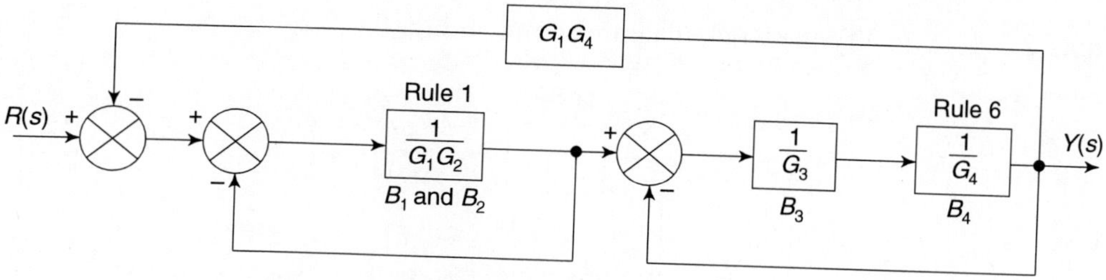

Figure Ex. 2.2.3 (b)

Applying Eq. 2.3 for closed loop system and Rule 1 on blocks B_3 and B_4.

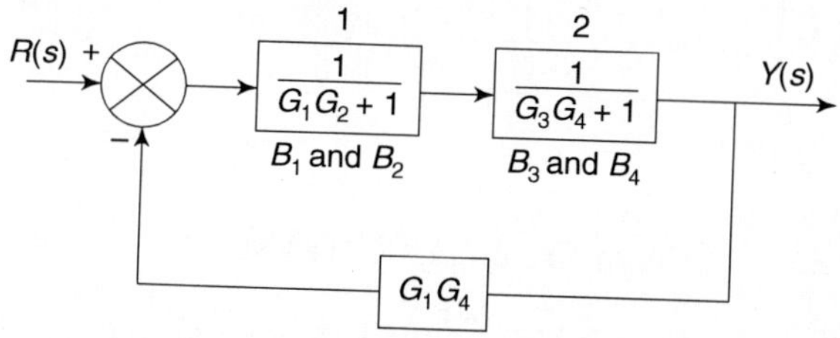

Figure Ex. 2.2.3 (c)

Applying Eq. (2.3) for closed loop system and Rule 1 on blocks 1 and 2.

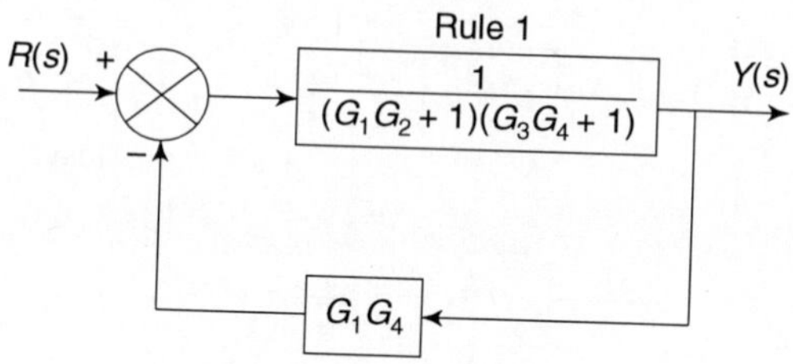

Figure Ex. 2.2.3 (d)

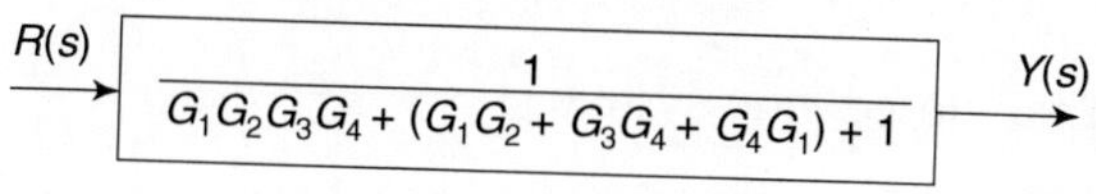

Figure Ex. 2.2.3 (e)

$$\frac{Y(s)}{R(s)} = \frac{}{G_1\ G_2\ G_3\ G_4 + (G_1\ G_2 + G_3\ G_4 + G_4\ G_1) + 1}$$ **Ans.**

Example 2.2.4 Simplify the block diagram shown in Fig. Ex. 2.2.4.

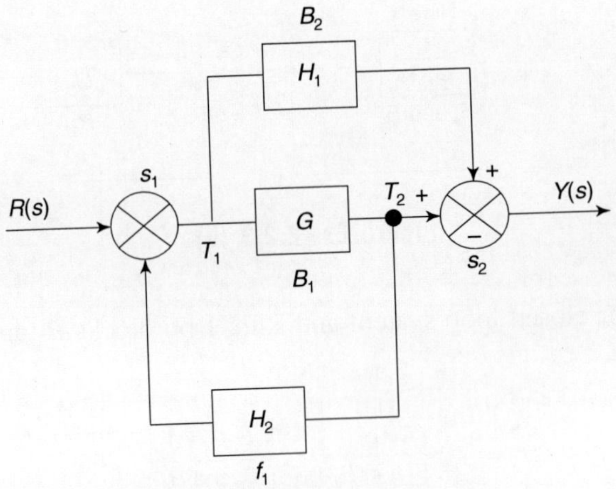

Figure Ex. 2.2.4 Block diagram

Solution Moving take-off point T_1 after the block B_1 using Rule 7.

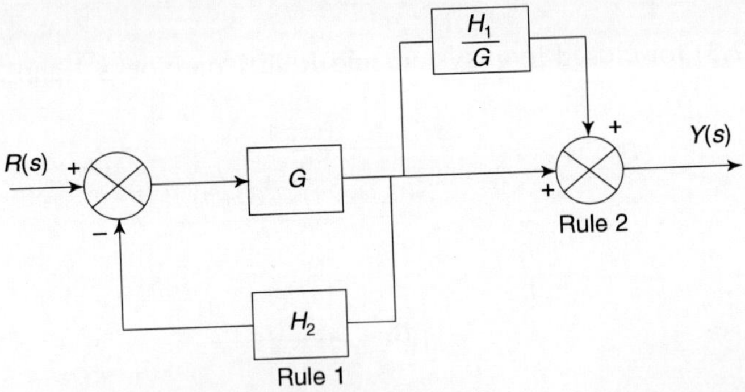

Figure Ex. 2.2.4 (a)

Applying Eq. (2.3) for closed loop system

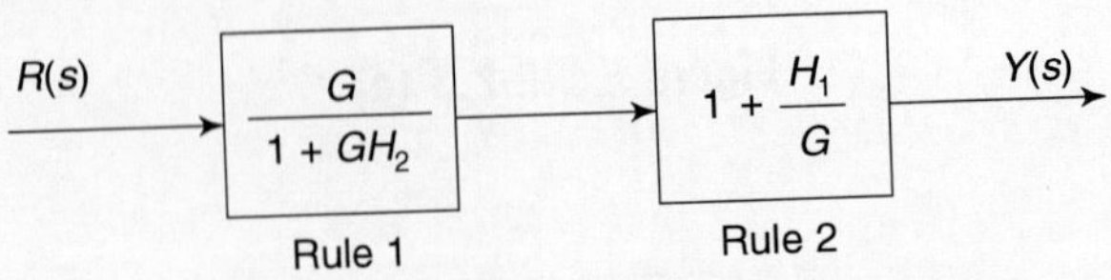

Figure Ex. 2.2.4 (b)

Applying Rule 1 and Rule 2

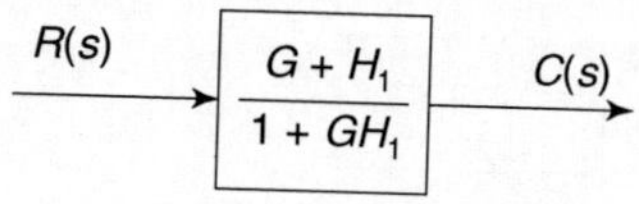

Figure Ex. 2.2.4 (c)

Example 2.2.5 Simplify the block diagram shown in Fig. Ex. 2.2.5.

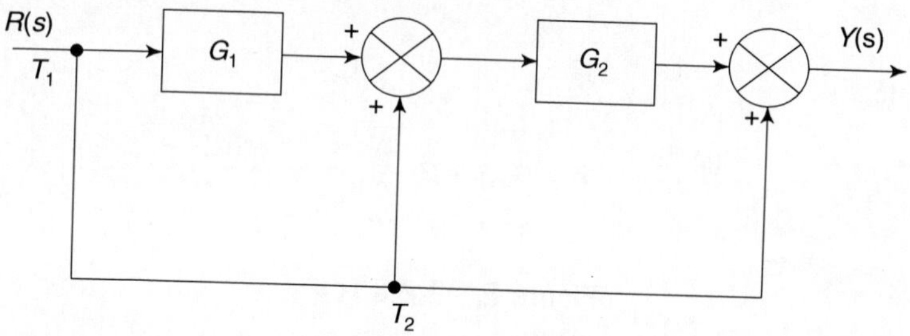

Figure Ex. 2.2.5

Solution Block diagram can be redrawn by Rule 3 at T_2

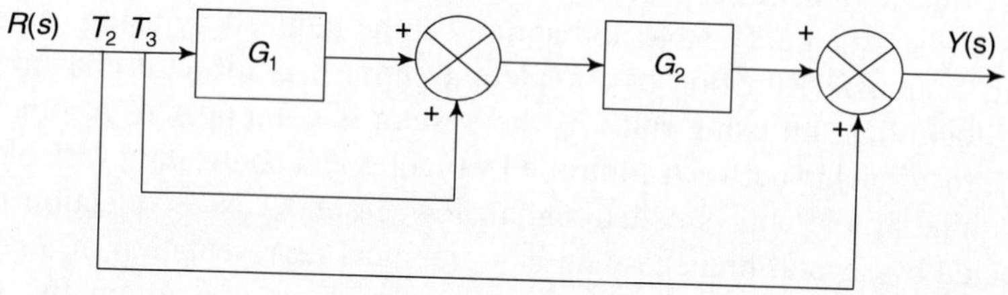

Figure Ex. 2.2.5 (a)

Applying Rule 2 on block G_1

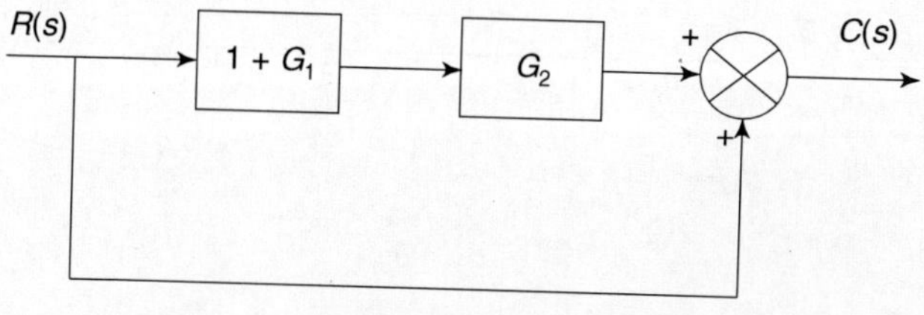

Figure Ex. 2.2.5 (b)

Applying Rule 1

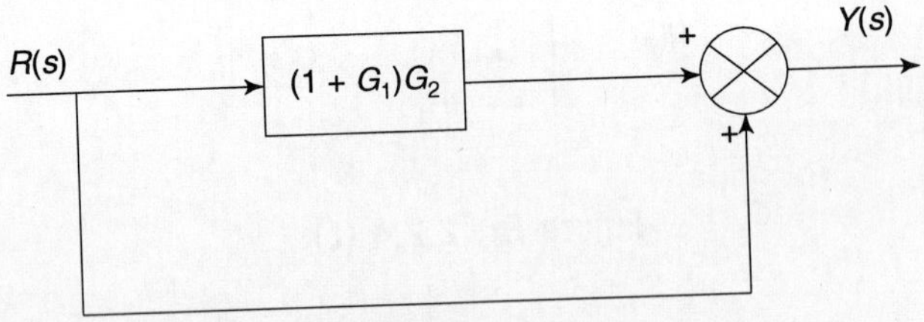

Figure Ex. 2.2.5 (c)

Applying Rule 2

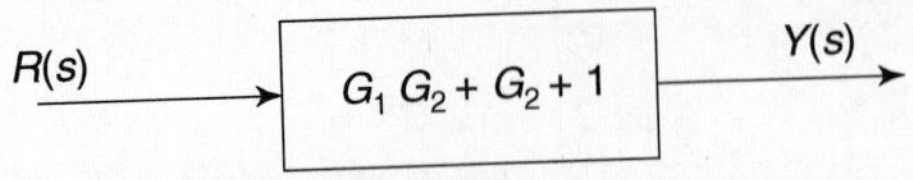

Figure Ex. 2.2.5 (d)

2.3 SIGNAL FLOW GRAPH

In previous section, we have studied block diagram algebra in which we represent the control system, in the form of blocks, i.e., any control system can be represented by the blocks those are interconnected with each other, each block of block diagram represents one or more elements of the system. This block diagram can be reduced by using some rules to get simple diagram or transfer function of complete system. It is difficult and time consuming to reduce the block diagram using rules, if the system is complicated. Another method for determining the relationship between controlled variables and controlling variables, developed by Mason, is quite simple and is called signal flow graph. It is a collection of nodes that are interconnected by several branches and is a graphical representation of a set of linear or simultaneous equations. The closed loop system has the signal flow graph given in Fig. 2.18. The terms used in signal flow graphs are:

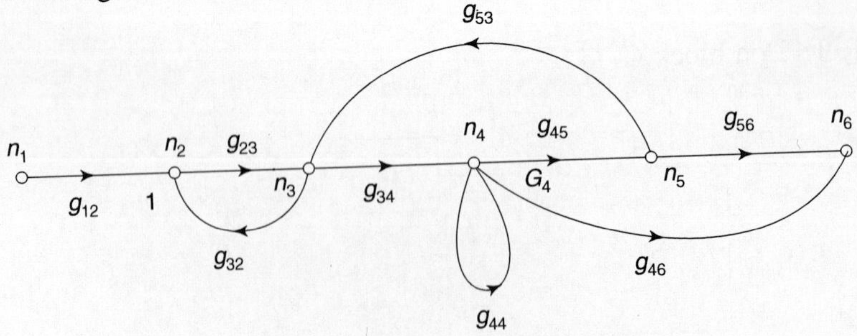

Figure 2.18 Signal flow graph

1. **Branch:** A line segment joining two nodes is called branch. Signal travels along a branch from one node to another. The direction is indicated by the branch arrow, it relates the dependency of an input and an output variable equivalent to a block of block diagram. In Fig. 2.18, n_1n_2, n_2n_3, n_3n_4, etc., represent branches.
2. **Node:** A point at which one or more branches meet is called node. According to the branches the node can be classified as follows.
   (a) *Input node.* If the node has only outgoing branches like n_1 in Fig. 2.18, i.e., the direction of arrow is away from the node.
   (b) *Output node.* If the node has only one or more incoming branches like n_6 shown in Fig. 2.18, i.e., the direction of arrow is towards the node.
   (c) *Intermediate node.* If the node has both incoming and outgoing branches, like n_2, n_3, n_4, n_5 in Fig. 2.18.
3. **Transmittance:** Transmittance is known as transfer function. It is written somewhere on the branch, normally near the arrow like g_{12}, g_{23}, etc. in Fig. 2.18.
   Transmittance is derived from trans + mittance is transfer immittance, immittance is again derived from im + mittance, i.e., **im**pedance and ad**mittance** so we can say that transmittance is nothing it is basically a function of impedance or admittance, i.e., transfer function.
4. **Path:** A path is a continuous sequence of branches that can be traversed from one node to another node such that no node is traversed more than once.
   like $n_1 - n_2 - n_3 - n_4 - n_5 - n_6$
   $n_1 - n_2 - n_3 - n_4 - n_6$ in Fig. 2.18
5. **Forward path:** It is a path from the input node to the output node, i.e., originates from input node and terminates at output node.
   like $n_1 - n_2 - n_3 - n_4 - n_5 - n_6$ in Fig. 2.18
6. **Loop:** A loop is a closed path that originates and terminates on the same node, and along the path no node meets twice like $n_2 - n_3 - n_2$ in Fig. 2.18 loop are again classified as follow:
   (a) *Self loop*: It is a closed path that originates and terminates on the same node like $n_4 - n_4$ in Fig. 2.18
   (b) *Non-touching loop*: Loops are said to be non-touching if they do not have common node, like $n_2 - n_3 - n_2$ and $n_4 - n_4$ in Fig. 2.18
7. **Forward path gain:** It is the product of transmittance or branch gain along the forward path, like the path $n_1 - n_2 - n_4 - n_6$ the gain is $g_{12}\, g_{23}\, g_{34}\, g_{46}$ as shown in Fig. 2.18.
8. **Loop gain:** It is the product of branch gains of the loop is called *loop gain* like $n_2 - n_3 - n_2$ is $g_{23}\, g_{32}$ in Fig. 2.18.

2.3.1 Construction of Signal Flow Graphs

The signal flow graph of a system can be constructed by the set of linear or simultaneous equations of the system. So, let us consider the following set of equations:

$$n_2 = g_{12}\, n_1 + g_{32}\, n_3 + g_{42}\, n_4 \qquad ...(2.4)$$

$$n_3 = g_{23}\, n_2 \qquad ...(2.5)$$

$$n_4 = g_{34}\, n_3 + g_{44}\, n_4 + g_{54}\, n_5 \qquad ...(2.6)$$

$$n_5 = g_{35}\, n_3 + g_{45}\, n_4 \qquad ...(2.7)$$

where n_1 is the input node and n_3 is the output node.

1. First draw all the node by above Eqs. (2.5 to 2.7) we have 5 nodes

$O \quad O \quad O \quad O \quad O$

$n_1 \quad n_2 \quad n_3 \quad n_4 \quad n_5$

2. Now draw the signal flow graph for Eq. (2.5) in which node n_2 is the sum of three signals like $g_{12}\, n_1$, $g_{32}\, n_3$, $g_{42}\, n_4$

   (1) = $g_{12}\, n_2$ (2) = $g_{23}\, n_2$ (3) = $g_{42}\, n_4$

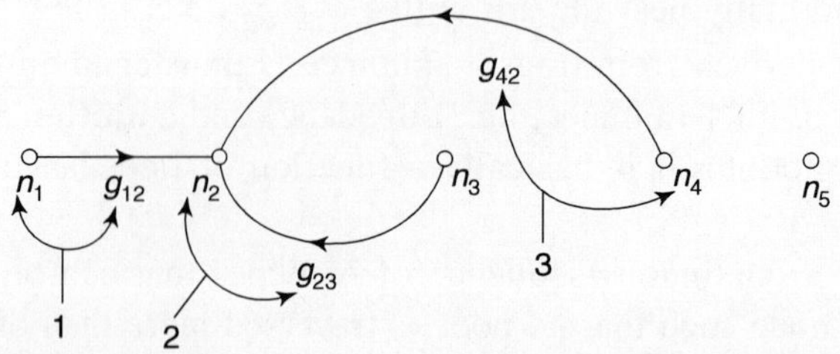

i.e., $n_2 = 1 + 2 + 3 = g_{12}\, n_1 + g_{23}\, n_2 + g_{42}\, n_4$

⇒ Gain x leaving node ⇒ $g_{12} \times n_1$

Similarly, $g_{23} \times n_2$ and $g_{42} \times n_4$

3. Now draw the signal flow graph for Eq. (2.6) in which node n_3 is equal to $g_{23}\, n_2$

   (1) = $g_{23}\, n_2$

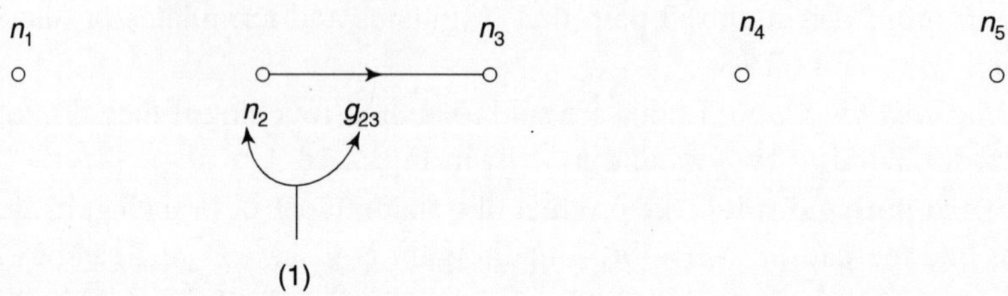

(1) = $g_{23} \times n_2$

4. Now draw the signal flow graph of Eq. (2.6) in which node n_4 is the sum of three signals.

   (1) = $g_{34}\, n_4$ (2) = $g_{44}\, n_4$ (3) = $g_{54}\, n_5$

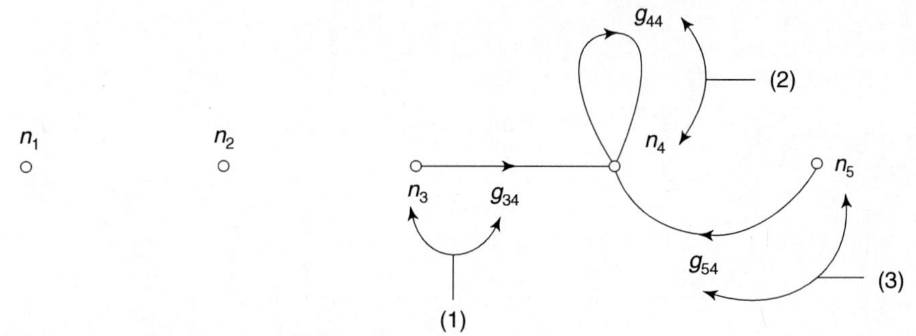

5. Now draw the signal flow graph for Eq. (2.7) in which node x_5 is the sum of two signals.

   (1) = $g_{35}\ n_3$ (2) $g_{45}\ n_4$

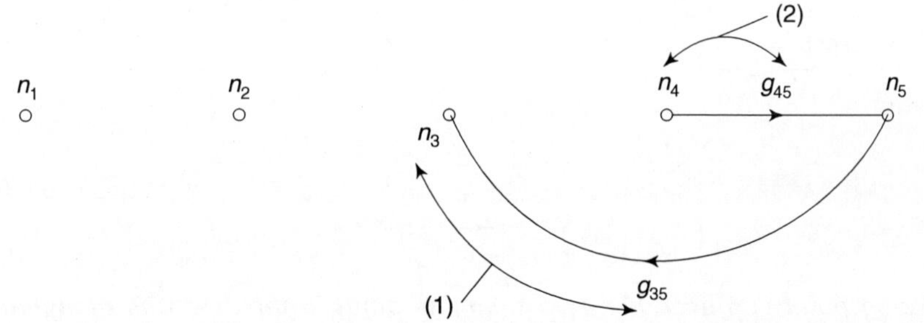

6. Now draw the complete signal flow graph by combining the above graphs.

Figure 2.19 Complete signal flow graph

2.3.2 Construction of Signal Flow Graph from Block Diagram

The signal flow graph of a system can also be constructed by the block diagram of the system. In this all the summing points and take-off points are represented by nodes.

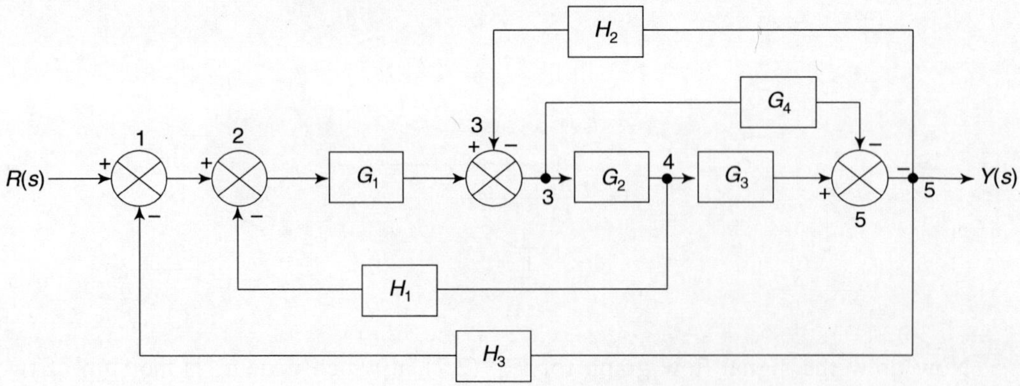

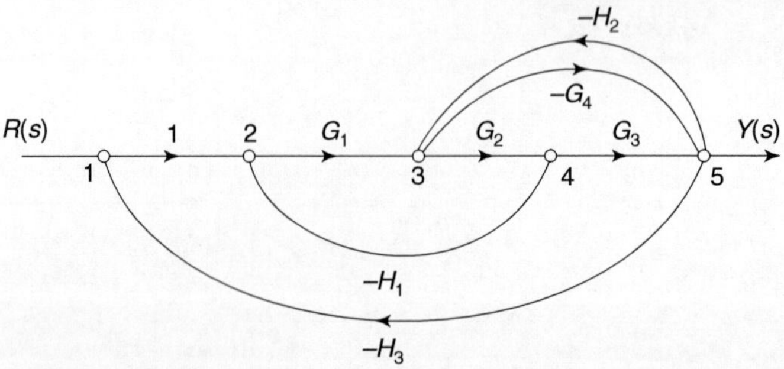

- *If a take-off point is placed before a summing point in the direction of signal flow then summing point and take-off point are represented by separate node connected by a branch having gain unity.*
- *If a take-off point is placed after a summing point in the direction of signal flow then summing point and take-off point are represented by common (single) node.*

2.3.3 Mason's Gain Formula

The relationship between input variable $R(s)$ and output variable $Y(s)$ of a signal flow graph is given by the total gain between input and output nodes and is known as the overall gain of the system. For determining the overall gain of the system, formula is derived by S.J Mason is called Mason's gain formula.

Mason's formula has several components that must be evaluated. First we must be sure that the definitions of the components are well understood. Then we must exert care in evaluating the components. We have already discussed some basic definitions applicable to signal flow graph.

The Mason's formula is

$$T_o = \frac{1}{\Delta}\left[\sum_{i=1}^{n} P_i \Delta_i\right] \qquad ...(2.8)$$

where T_o = overall transfer function

P_i = path gain of ith forward path

Δ = 1 – (sum of loop gains of all individual loops) + (sum of gain products of all possible combinations of two nontouching loops) – (sum of gain products of all possible combinations of three non-touching loops) + ...

$$\Delta = 1-\sum_j L_{j1}\sum_j L_{j2}\sum_j L_{j3}+\ldots \quad \text{where } j = 1, 2, 3 \ldots \qquad \ldots(2.9)$$

L_{jr} = gain product j^{th} possible combination of r non-touching loops.

Δ_i = it is formed by removing loop gains from Δ that is touching the ith forward path.

Example 2.3.1 Find the overall transfer function or overall gain of the signal flow graph shown in Fig. Ex. 2.3.1.

Solution

Step 1 There are two forward paths with path gain as shown in Fig. Ex. 2.3.1 (a) and (b)

$$P_1 = g_{12}\, g_{23}\, g_{34}\, g_{45}$$

$$P_2 = g_{12}\, g_{23}\, g_{35}$$

Here i = 1, 2 so $P_i = P_1\, P_2$

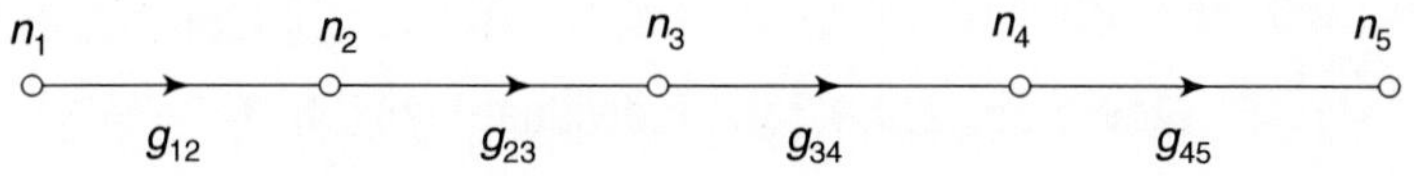

Figure Ex. 2.3.1 (a) Forward path P_1

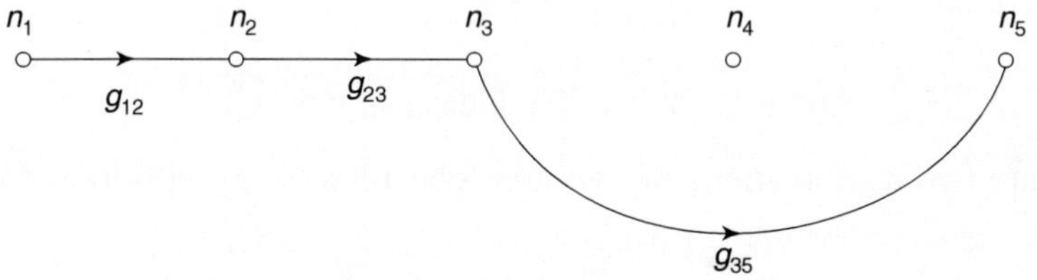

Figure Ex. 2.3.1 (b) Forward path P_2

Step 2 There are four individual loops with loop gain

$$L_{11} = g_{23}\, g_{32}$$

$$L_{21} = g_{23}\, g_{34}\, g_{42}$$

$$L_{31} = g_{44}$$

$$L_{41} = g_{45}\, g_{54}$$

Here j = 1, 2, 3, 4 and r = 1

So $L_{jr} = L_{11}, L_{21}, L_{31}, L_{41}$

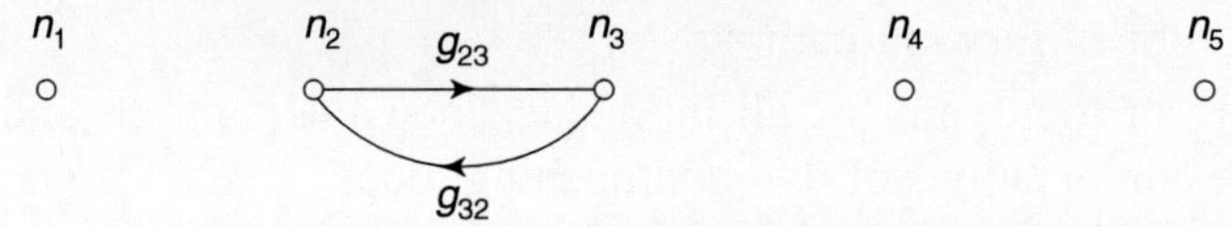

Figure Ex. 2.3.1 (c) Individual loop L_{11}

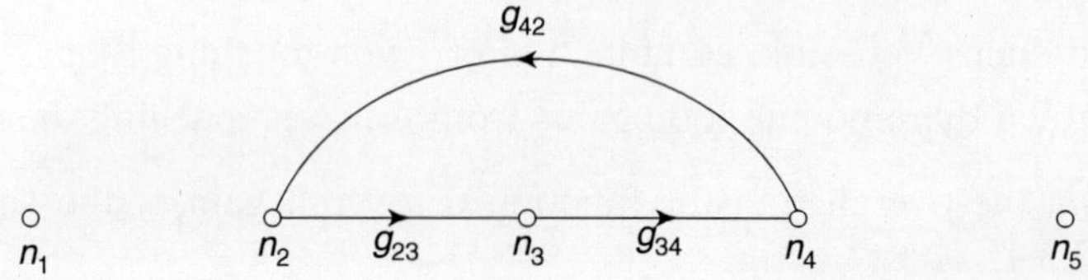

Figure Ex. 2.3.1 (d) Individual loop L_{21}

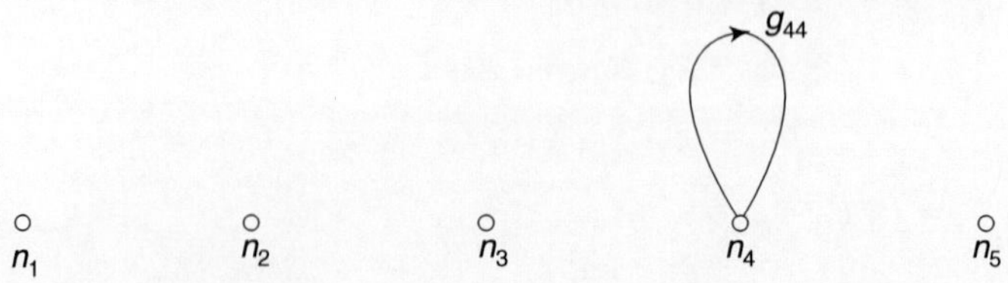

Figure Ex. 2.3.1 (e) Individual loop L_{31}

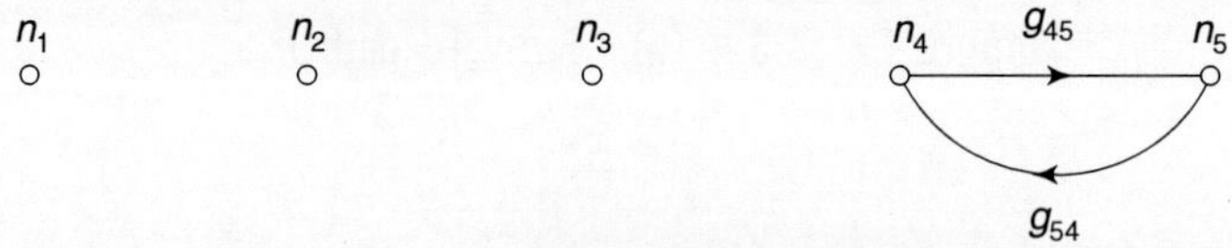

Figure Ex. 2.3.1 (f) Individual loop L_{41}

Step 3 There are two combinations of two non-tocuching loops with loop gain

$$L_{12} = g_{23}\, g_{32}\, g_{44}$$
$$L_{22} = g_{23}\, g_{32}\, g_{45}\, g_{54}$$

Here $j = 1, 2$ and $r = 2$

So $L_{jr} = L_{12}, L_{22}$

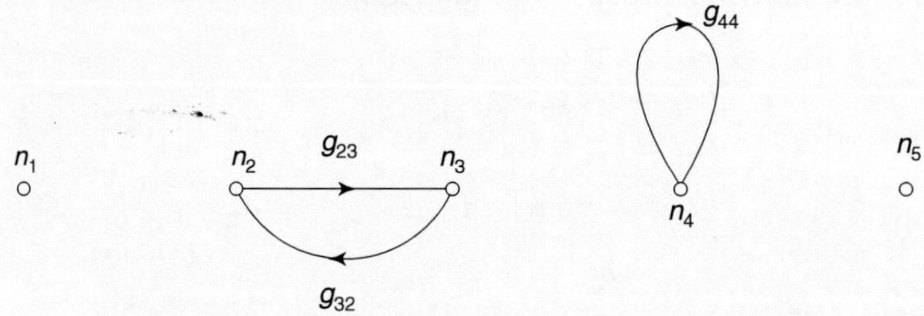

Figure Ex. 2.3.1 (g) Non-touching loop L_{12}

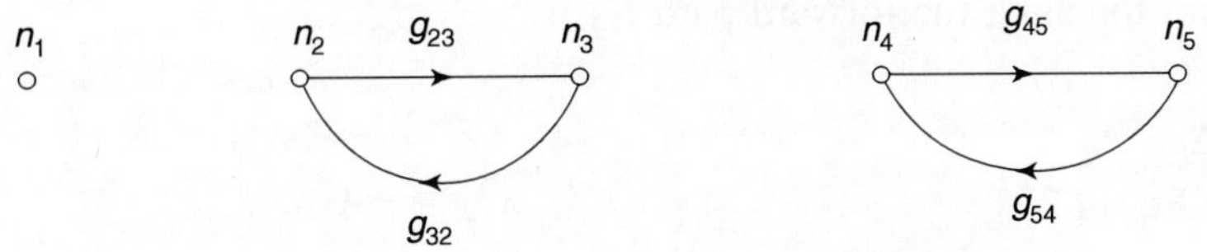

Figure Ex. 2.3.1 (h) Non-touching loop L_{22}

Step 4 There is no combination of three non-touching loops four and five non-touching loops.

i.e., $$L_{j3} = L_{j4} = L_{j5} = 0$$

Step 5 Now from the Eq. (2.9)

$$\Delta = 1 - (L_{11} + L_{12} + L_{13} + L_{14}) + (L_{12} + L_{22})$$

$$\Delta = 1 - (g_{23}\, g_{32} + g_{23}\, g_{34}\, g_{42} + g_{44} + g_{45}\, g_{54}) + g_{23}\, g_{32}\, g_{44} + g_{23}\, g_{32}\, g_{45}\, g_{54})$$

$$= 1 - g_{23}\, g_{32} - g_{23}\, g_{34}\, g_{42} - g_{44} - g_{45}\, g_{54} + g_{23}\, g_{32}\, g_{44} + g_{23}\, g_{32}\, g_{45}\, g_{54}$$

Step 6 Removing the loop gains from Δ that touching P_1 forward path like

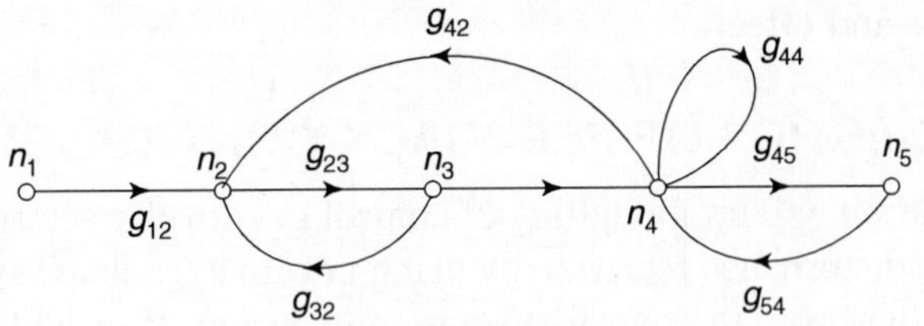

Figure Ex. 2.3.1 (i)

i.e., all the loop gain touch the forward path G_1

Now $\Delta_1 = 1$

$\because$ $i = 1$ so $\Delta_i = \Delta_1$

Removing the loop gains from Δ that touch P_2 forward path like

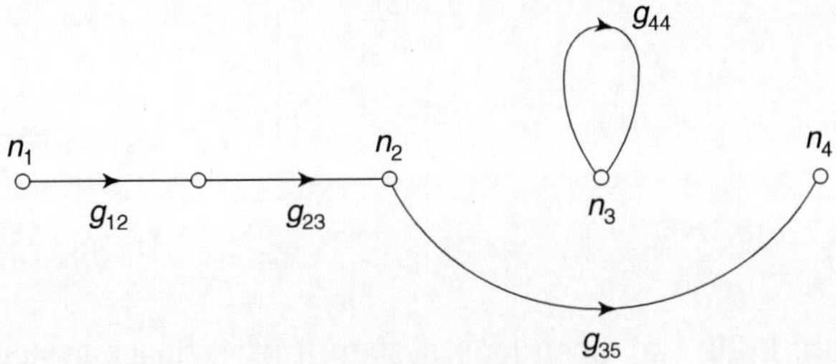

Figure Ex. 2.3.1 (j)

Only g_{44} is not touching the forward path G_2.

Now $\Delta_2 = (1 - g_{44})$

$\therefore\ i = 2$ so $\Delta_i = \Delta_2$

Now from Eq. (2.8)

$$T_0 = \frac{1}{\Delta}\sum_{i=1}^{n} P_i\Delta_i = \frac{1}{\Delta}[P_1\,\Delta_1 + P_2\,\Delta_2]$$

$$T_0 = \frac{g_{12}\,g_{23}\,g_{34}\,g_{45} + g_{12}\,g_{23}\,g_{35}(1 - g_{44})}{1 - g_{23}\,g_{32} - g_{23}\,g_{34}\,g_{42} - g_{44} - g_{45}\,g_{54} + g_{23}\,g_{32}\,g_{44} + g_{23}\,g_{32}\,g_{45}\,g_{54}}$$

T_0 is representing overall gain of the signal flow graph.

- *In Mason's gain formula, the existence of non-touching loops increases the complexity of the formula. However, many systems do not have non-touching loops. For such systems, you may find Mason's formula easier to use than block diagram algebra.*

2.3.4 Limitations of Signal Flow Graph (SFG)

1. It is applicable only on linear systems
2. The equation required for drawing a signal flow graph must be algebraic equations in the form of cause and effect.

2.4 CHARACTERISTICS OF FEEDBACK CONTROL SYSTEM

Basically in this section we will try modelling of control system characteristics, like sensitivity to model uncertainties, disturbance rejection by using negative feedback system. The feedback system plays an important role in control system engineering. It provides automatic control by using feedback system. In general, the feedback system is a system in which the output of the system is fed back to the input of the system through feedback. This feedback signal is compared with the reference input signal by the comparing circuit and error signal is generated. This error signal is used to control the process using the notion of feedback. Generally, our goal is to minimize the error signal.

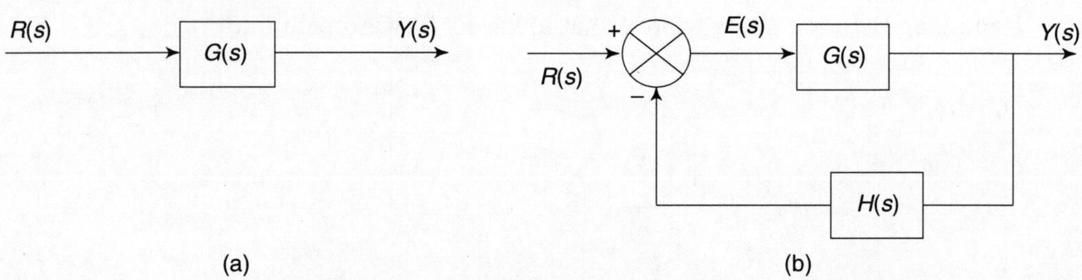

Figure 2.20 (a) Open loop system, (b) Feedback system

Basically, here we are also trying to develop the concept of the sensitivity of a system to a parameter change, i.e., if there is unwanted parameter variation and the system is more sensitive than the output of the system will be affected and the accuracy of the system will decrease so we can say that it is desirable to minimize the effect of unwanted parameter variation.

The accuracy of the system also depends upon the unwanted input signals, effect on the output signal is called disturbance.

2.4.1 Reduction of Parameter Variation using Negative Feedback

As we have seen in previous section the performance of the system depends upon the sensitivity of the system to parameter variations. So our main purpose of negative feedback system is to reduce the sensitivity of the system to parameter variations. The parameter of a system may vary by changing environment, aging, ignorance of the exact values of the process parameters other natural factors, etc. Conceptually, sensitivity is a measure of the effectiveness of feedback in reducing the influence of these variations on system performance. So our prime goal is to reduce sensitivity to changes in parameter.

Open loop system. Now we are considering the open loop system. In open loop system

$$Y(s) = G(s)\,R(s) \qquad \text{...(2.10)}$$

Due to changes in parameter the $G(s)$ changes to $[G(s) + \Delta G(s)]$ where $|G(s)| >>> |\Delta G(s)|$

So the output of open loop system changes to

$$Y(s) + \Delta Y(s) = [G(s) + \Delta G(s)]\,R(s)$$

where $\Delta Y(s)$ = small change in output due to changes in parameter

$\Delta G(s)$ = small change in $G(s)$ due to changes in parameter

$$Y(s) + \Delta Y(s) = G(s)\,R(s) + \Delta G(s)\,R(s) \qquad \text{...(2.11)}$$

So

$$\Delta Y(s) = \Delta G(s)\,R(s) \qquad \text{...(2.12)}$$

Closed loop system. Now we are considering closed loop system. In closed loop system for negative feedback

$$Y(s) = \frac{G(s)}{1 + G(s)\,H(s)}\,R(s) \qquad \text{...(2.13)}$$

Due to changes in parameter the $G(s)$ and $Y(s)$ changes to

$$Y(s) + \Delta Y(s) = \frac{G(s) + \Delta G(s)}{1 + [G(s) + \Delta G(s)]\,H(s)}\,R(s)$$

$$Y(s) + \Delta Y(s) = \frac{G(s) + \Delta G(s)}{1 + G(s)\,H(s) + \Delta G(s)\,H(s)}\,R(s) \qquad \text{...(2.14)}$$

Since $|\Delta G(s)| <<< |G(s)|$ so we have from the Eq. (2.14), the variation in the output

$$\Delta Y(s) = \frac{\Delta G(s)}{1 + G(s)\,H(s)}\,R(s) \qquad \text{...(2.15)}$$

Now from the comparison of Eqs. (2.12) and (2.13), it is seen that the output of the closed loop system is reduced by a factor of 1 + $G(s)$ $H(s)$ which is much greater than unity in practical cases.

So sensitivity of the system is used to represent the relative variation in the overall transfer function $T_o(s) = Y(s)/R(s)$ due to changes in $G(s)$ and is defined

$$\text{Sensitivity} = \frac{\%\text{ change in } T_o(s)}{\%\text{ change in } G(s)} \qquad ...(2.16)$$

i.e., the system sensitivity is defined as the ratio of the percentage change in the system transfer function [$T_o(s)$] to the percentage change of the process transfer function [$G(s)$].

So the sensitivity,

$$S = \frac{\dfrac{\Delta T_o(s)}{T_o(s)}}{\dfrac{\Delta G(s)}{G(s)}} \qquad ...(2.17)$$

For small increment changes, Eq. (2.17) becomes

$$S = \frac{\dfrac{\partial T_o}{T_o}}{\dfrac{\partial G}{G}} = \frac{\partial l_n T_o}{\partial l_n G} \qquad ...(2.18)$$

According to the Eq. (2.18), the sensitivity of the closed loop system is

$$S_G^{T_o} = \frac{\dfrac{\partial T_o}{T_o}}{\dfrac{\partial G}{G}} \Rightarrow \frac{\partial T_o}{\partial G} \times \frac{G}{T_o}$$

$$= \frac{1 + G(s)\,H(s) - G(s)\,H(s)}{[1 + G(s)\,H(s)]^2} \times \frac{G}{\dfrac{G}{[1 + G(s)\,H(s)]}}$$

$$S_G^{T_o} = \frac{1}{1 + G(s)\,H(s)} \qquad ...(2.19)$$

Hence, $S_G^{T_o}$ the sensitivity of T_o with respect to G.

Similarly, the sensitivity of the open loop system.

$$S_G^{T_o} = \frac{\partial T_o}{\partial G} \times \frac{G}{T_o} = 1 \quad \because (G = T) \qquad ...(2.20)$$

So by comparing Eqs. (2.19) and (2.20) we can say that the sensitivity of the closed loop system with respect to change in G is reduced by a factor [1 + $G(s)$ $H(s)$] as compared to that of an open loop system.

The sensitivty of T_o with respect to H,

$$S_G^{T_o} = \frac{\partial T_o}{\partial H} \times \frac{H}{T_o}$$

$$G(s) = G(s)\left[\frac{-G(s)}{[1+G(s)\,H(s)]^2}\right]\frac{H(s)}{\frac{G(s)}{1+G(s)\,H(s)}}$$

$$S_G^{T_o} = \frac{-G(s)H(s)}{1+G(s)\,H(s)} \qquad ...(2.21)$$

Equation (2.21) shows that the large value of $G(s)\,H(s)$, $S_G^{T_o}$ approaches unity. So we can say that the change in feedback (H) directly affects the output of the system. So it is important to use feedback elements which do not vary with environmental changes (like ambient temperature) or can be maintained constant.

2.4.2 Effects of Disturbance by Using Negative Feedback

Another important effect of feedback system in a control system is reduced effect of disturbance signals. Disturbance signal is an unwanted signal that enters the system and it affects the system's output signal. There are many circuits that have disturbance signals like:

1. **Electronic amplifier:** It has inherent noise generated within the transistor or integrated circuits.
2. **Radar antennas:** It is subjected to wind gusts.
3. **Non-linear system:** These systems generate unwanted signals due to non-linear elements in the systems, etc.

By the use of feedback system, the effect of distortion, noise and unwanted disturbances can be reduced. The effect of disturbance can be understood easily by the example given below.

Example 2.4.1 The closed loop control system is shown in block diagram in which speed of dc motor is controlled by tachometer as a negative feedback system.

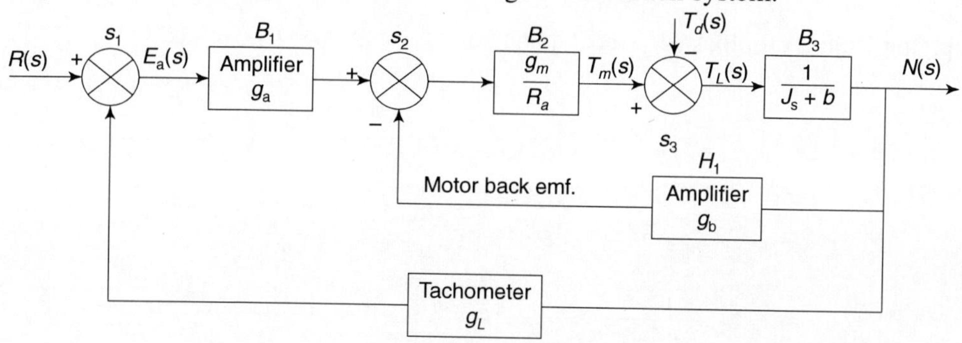

Figure 2.21 Block diagram of speed control system

where $R(s)$ = reference input

$E_a(s)$ = actuated signal or error signal

g_a = gain of amplifier or transfer function of amplifer

$\frac{g_m}{R_a}$ = gain or transfer function of motor

$\frac{1}{J_s + b}$ = gain of the load

g_b = gain of motor back emf

g_t = gain of the tachometer

$T_m(s)$ = motor torque

$T_d(s)$ = disturbance torque

$T_L(s)$ = load torque

$N(s)$ = output speed

Solution In this example we are considering the transfer function model of an armature controlled dc motor with external feedback (tachometer) and a load torque disturbance.

By the block diagram reduction process.

Apply Rule 6 on summing point S_2

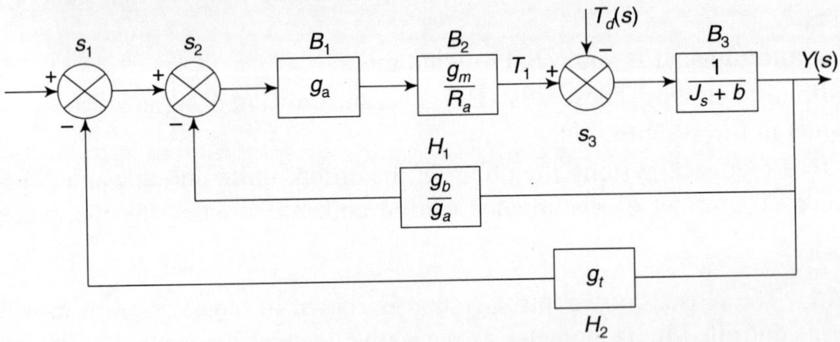

Applying Rule 2 on blocks H_1 and H_2 and Rule 1 on B_1 and B_2, we get

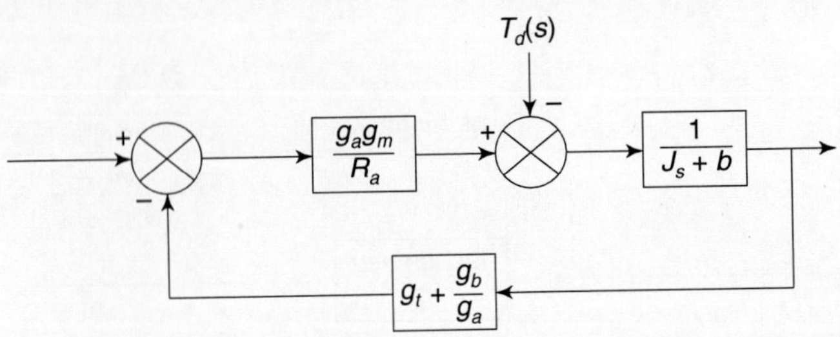

For simplicity we are assuming that

$$G_1(s) = \frac{g_a\, g_m}{R_a}$$

$$G_2(s) = \frac{1}{J_s + b}$$

$$H(s) = g_t + \frac{g_b}{g_a}$$

$$E(s) = R(s) - W(s)$$

and $R(s) = N_d(s)$, desired speed. For the calculation $R(s) = 0$ and examine

$$E_a(s) = -\,N_d(s) \quad \text{...(2.22)}$$

So the simplified diagram is shown in Fig. 2.22.

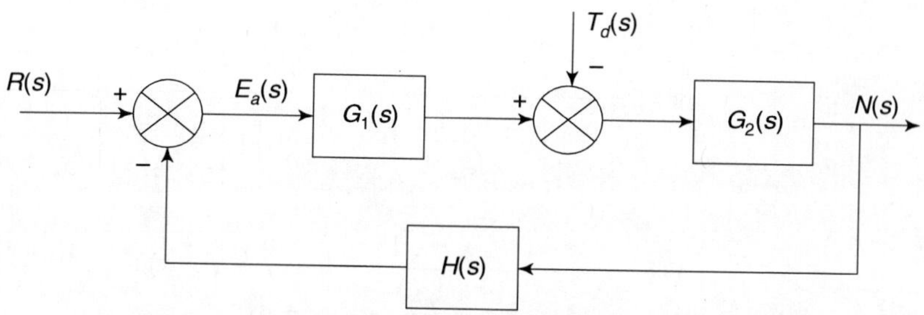

Figure 2.22 Simple block diagram of speed control system

The signal flow graph of the system is shown in Fig. 2.23.

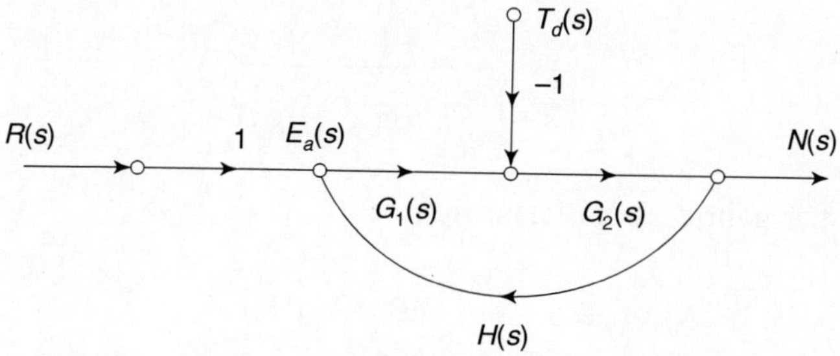

Figure 2.23 Signal flow graph of speed control system

The error signal, $E_a(s) = -\,N(s)$, of the closed loop system of Fig. Ex. 2.4.1 (b) can be obtained by Mason's gain formula.

$$E_a(s) = -\,N(s)$$

$$= \frac{G_2(s)}{1 + G_1(s)\,G_2(s)\,H(s)} T_d(s) \quad \text{...(2.23)}$$

If $G_1\ G_2\ H$ is much greater than 1 over the range of s then

$$E_a(s) = \frac{1}{G_1(s)\,H(s)} T_d(s) \qquad ...(2.24)$$

Therefore, if $G_1(s)\ H(s)$ is made sufficiently large, the effect of disturbance can be decreased by closed loop feedback.

$$G_1(s)\ H(s) = \frac{g_a\ g_m}{R_a}\left(g_t + \frac{g_b}{g_a}\right) = \frac{g_a\ g_m\ g_t}{R_a}$$

Because $g_a >> g_b$. Thus, we strive to obtain large g_n then we need to keep $R_a << 2\Omega$. The error for the system shown in Fig. Ex. 2.4.1 (a) is

$$E_a(s) = R(s) - N(s)$$

For the claculation $R(s) = 0$ is the output for the speed control system of Fig. 2.4.1(b) due to the load disturbance.

$$N(s) = \frac{\frac{1}{(J_s + b)}}{1 + \left(\frac{g_t\ g_a\ g_m}{R_a}\right)\left(\frac{1}{(J_s + b}\right) + \left(\frac{g_m\ g_b}{R_a}\right)\left[\frac{1}{(j_s + b\_}\right]} T_d\ (s)$$

$$N(s) = \frac{-1}{J_s + b + \left(\frac{g_m}{R_a}\right)(g_t\ g_a + g_b)} T_d\ (s) \qquad ...(2.25)$$

The steady state output is obtained by utilizing the final value theorem, we have

$$\lim_{t \to \infty} N(t) = \lim_{S \to 0} [s\ N(s)]$$

$$= \frac{-1}{b + \left(\frac{g_m}{R_a}\right)(g_t\ g_a + g_b)} D$$

When the amplifier gain g_a is sufficient light,

We have
$$N\ (\infty) = \frac{-R_a}{g_a\ g_m\ g_t} D = N_c(\infty)$$

The ratio of closed loop to open loop steady state speed output due to an undesirable disturbance is

$$\frac{N_c\ (\infty)}{N_o\ (\infty)} = \frac{R_a\ b + g_m\ g_b}{g_a\ g_m\ g_t} \qquad ...(2.26)$$

and usually less than 0.02.

- *The introduction of feedback to improve the control system is often necessary. It is also applicable in nature, such as biological and physiological systems, and feedback is inherent in these systems. For example, the human heart rate control system is a feedback control system.*
- *The effect of feedback on noise and disturbance depends greatly on where these extraneous signals occur in the system. No general conclusions can be reached, but in many situations, feedback can reduce the effect of noise and disturbance on system performance.*

2.4.3 Advantages and Disadvantages of Feedback System

Advantages

1. In a closed loop system the controlled variable follows the desired value accurately.
2. The effect of parameters variation (in forward path) on system performance is reduced to acceptable levels.
3. Effect on the controlled variable of external disturbances other than those associated with the feedback sensor is greatly reduced.
4. The feedback in a control system improves the speed of its response.

Disadvantages

1. Due to the feedback system, the complexity of the system is increased.
2. Due to the feedback system, there is need for much larger forward path gain.
3. There is a possibility of system instability is undesired/persistent oscillations of the output variable.

2.5 MODELING IN THE FREQUENCY DOMAIN

In previous sections, we discussed the introduction to control system included open loop and closed loop control systems and Laplace transform.

In this section our purpose is to develop mathematical models from schematics of physical systems. There are two methods : (1) transfer function in the frequency domain and (2) state equation in the time domain. First topic will be covered in this section. As we know that in every case the first step is to develop a mathematical model by applying the fundamental physical laws of science and engineering. For example, when we model.

Network	Fundamental physical laws
1. Electrical networks	Ohm's law and Kirchhoff's law
2. Mechanical systems	Newton's laws

Ohm's law and Kirchhoff's law are basic laws of electric networks, will be applied initially. According to Kirchhoff's law, we will sum voltages in a loop or sum currents at a node. Similarly, Newton's laws are basic laws of mechanical systems, will be applied initially. According to Newton's law, we will sum forces or torques. With the help of these equations we will obtain the relationship between the system's output and inputs.

Kirchhoff's vsoltage law. The sum of voltages around a closed loop or path is equal to zero.

Kirchhoff's current law. The sum of electric currents flowing away from the node and towards the node is equal to zeros.

Newton's law. The sum of forces on a body is equal to zero; the sum of movements on a body is equal to zeros.

2.6 ELECTRIC NETWORKS TRANSFER FUNCTIONS

In this section, we apply the transfer function to the mathematical equation of electric circuits which includes passive and active components like, inductor, capacitor, resistors and operational amplifier.

First, we consider an electric network that consists of three passive linear components like inductors, capacitor and resistors as shown in Fig. 2.24.

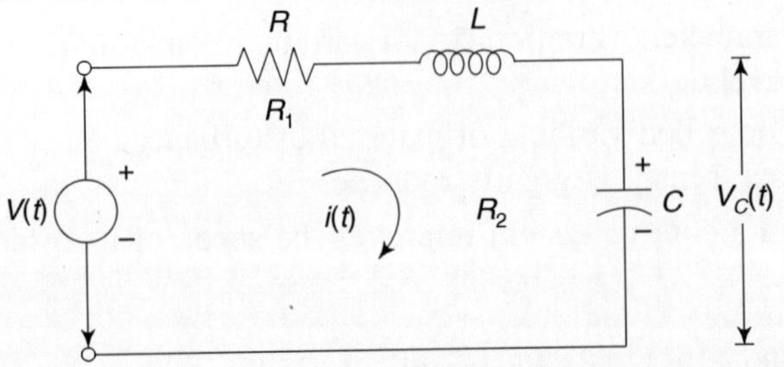

Figure 2.24 RLC network circuit

The voltage and current relationship is shown in Table 2.1. When the three components *R*, *L* and *C* are combined into electrical circuits. Based on the input and output we find the transfer function. Before finding the transfer function analysis must first decide the input and output. In this circuit several variables can be chosen as output like the resistor voltage $V_R(t)$, inductor voltage $V_L(t)$ and the capacitor voltage $V_C(t)$ or the current for these. Now we will consider capacitor voltage $V_C(t)$ as an output in Fig. 2.24.

Applying Kirchhoff's voltage law, summing the voltages around the loop, assuming zero initial conditions, yields the equation for this network as

$$L\frac{di(t)}{dt} + Ri(t) + \frac{1}{C}\int_0^t i(t)\,dt = V(t) \quad \text{...(2.27)}$$

We know that $i(t) = \dfrac{dq(t)}{dt}$

$$\text{then } L\,d^2\frac{q(t)}{dt^2} + R\frac{dq(t)}{dt} + \frac{1}{C}q(t) = V(t) \quad \text{...(2.28)}$$

From Table 2.1 we know that voltage-charge relationship for a capacitor

$$q(t) = CV_C(t) \quad \text{...(2.29)}$$

Table 2.1 Voltage current and charge relationship for R, L and C

Component	Voltage current $V-i$	Voltage charge $V-q$	Current voltage $i-V$	Impedance $Z(s) = V(s)/F(s)$	Admiltance $Y(s) = I(s)/V(s)$
Capacitor	$V_C(t) = \frac{1}{C}\int_0^t i_C(t)\,dt$ $V_C(s) = \frac{1}{Cs} I_C(s)$	$V_C(t) = \frac{1}{C} q(t)$ $V_C(s) = \frac{1}{Cs} q(s)$	$i_C(t) = C\frac{d\,V_C(t)}{dt}$ $i_C(t) = Cs\,V_C(s)$	$\frac{1}{CS}$	Cs
Inductor	$V_L(t) = L\frac{d\,i_L(t)}{dt}$ $V_L(s) = I_L(s)$	$V_L(t) = L\frac{d^2\,q(t)}{dt^2}$ $V_L(s) = Ls^2 q(s)$	$I_L(t) = \frac{1}{L}\int_0^t V_L(t)\,dt$ $I_L(s) = \frac{1}{Ls} V_L(s)$	Ls	$\frac{1}{Ls}$
Resistor	$V_R(t) = R i_R(t)$ $V_R(s) = R I_R(s)$	$V_R(t) = R\frac{d\,q(t)}{dt}$ $V_R(s) = RSq(s)$	$i_R(t) = \frac{1}{R} V_R(t)$ $I_R(s) = \frac{1}{R} V_R(s)$	R	$\frac{1}{R} = G$

Substituting Eq. (2.29) into Eq. (2.28) yields

$$Lc\frac{d^2\,V_C(t)}{dt^2} + RC\frac{d\,V_C(t)}{dt} + V_C(t) = V(t) \qquad \text{...(2.30)}$$

Taking the laplace transform of equation (2.30), rearranging terms.

$$(LC\,s^2 + RCS + 1)\;V_C(s) = V(s) \qquad \text{...(2.31)}$$

$$\frac{V_C(s)}{V(s)} = \frac{\frac{1}{LC}}{s^2 + \frac{R}{L}s + \frac{1}{LC}} \qquad \text{...(2.32)}$$

Hence, from Eq. (2.32) the block diagram of RLC network is shown in Fig. 2.25.

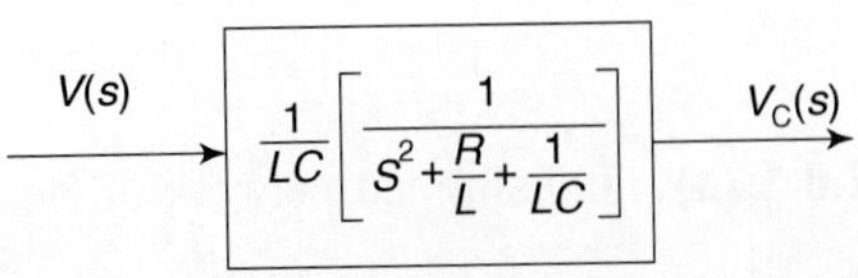

Figure 2.25 Block diagram series RLC network

The Laplace transform of Eq. (2.27), assuming initial condition

$$\left(Ls + R + \frac{1}{Cs}\right) I(s) = V(s) \quad \text{...(2.33)}$$

From Eq. (2.33) the Laplace transform network is shown in Fig. 2.26.

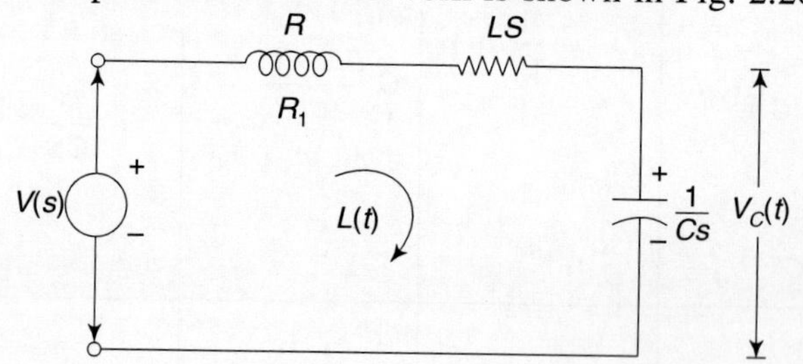

Figure 2.26 Laplace transform network or transform network

Example 2.6.1 For a given network in Fig. Ex. 2.6.1, find the transfer function $I_2(s)/V(s)$.

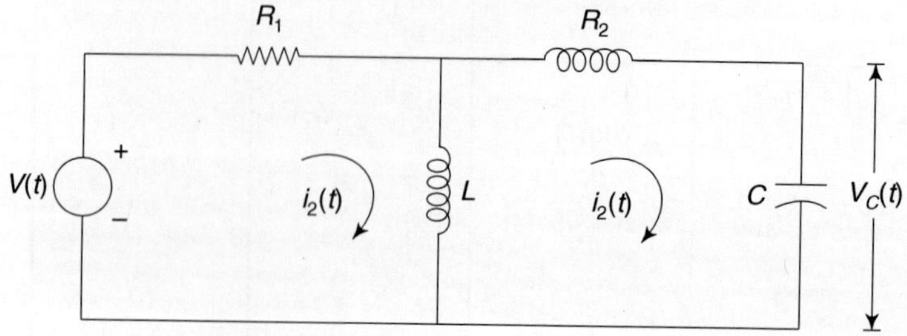

Figure Ex. 2.6.1 Electric network

Solution First, we will convert the network into Laplace transforms for different circuit variables while initial condition to be considered zero. The transformed network is shown in Fig. Ex. 2.6.1 (a).

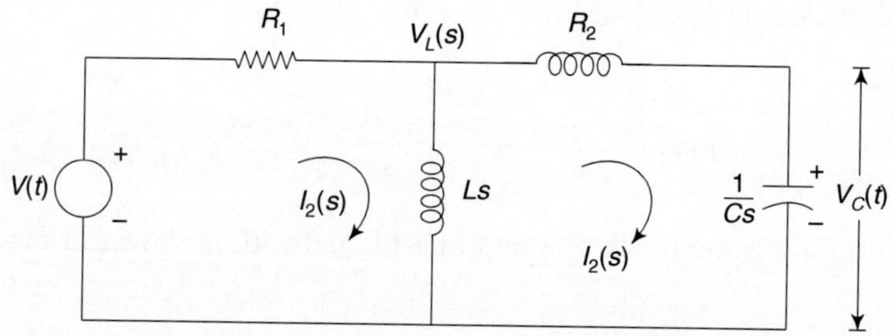

Figure Ex. 2.6.1 (a) Transformed network of Fig. Ex. 2.6.1

Applying Kirchhoff's voltage law in loop 1, we obtain

$$R_1 I_1(s) + Ls I_1(s) - Ls I_2(s) = V(s) \quad \text{...(2.34)}$$

Now applying Kirchhoffs's voltage law in loop 2, we obtain

$$Ls\, I_2(s) + R_2\, I_2(s) + \frac{1}{Cs}\, I_2(s) - Ls\, I_1(s) = 0 \qquad ...(2.35)$$

Solving Eqs. (2.34) and (2.35), we obtain

$$(R_1 + Ls)\, I_1(s) - Ls\, I_2(s) = V(s) \qquad ...(2.36)$$

$$-Ls\, I_1(s) + \left(Ls + R_2 + \frac{1}{Cs}\right) I_2(s) = 0 \qquad ...(2.37)$$

We can use Cramer's rule to solve above equations, hence

$$I_2(s) = \frac{\begin{vmatrix} R_1 + Ls & V(s) \\ -Ls & 0 \end{vmatrix}}{\Delta} = \frac{Ls\, V(s)}{\Delta} \qquad ...(2.38)$$

where $\Delta = \begin{vmatrix} (R_1 + Ls) & -Ls \\ -Ls & Ls + R_2 + \dfrac{1}{Cs} \end{vmatrix}$

Now the transfer function, $G(s)$

$$G(s) = \frac{I_2(s)}{V(s)} = \frac{Ls}{\Delta} = \frac{LC\, s^2}{(R_1 + R_2)LCs^2 + (R_1\, R_2\, C + L)\, s + R_1} \qquad ...(2.39)$$

The transfer function for a network is as follows:

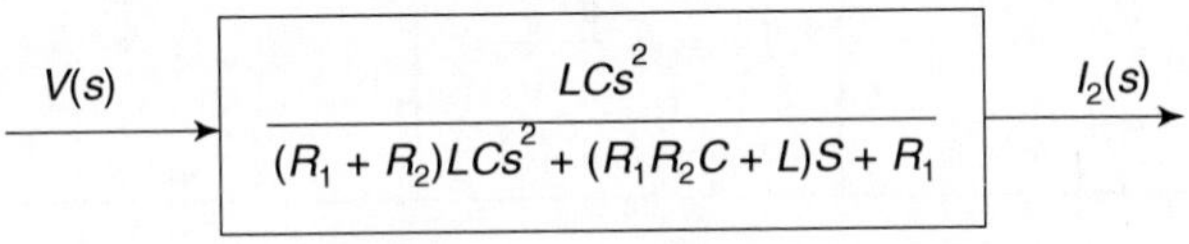

Figure Ex. 2.6.1 (b) Modelled as the transfer function of Fig. 2.6.1

2.7 MECHANICAL SYSTEM TRANSFER FUNCTIONS

In a previous section, we have studied about electrical network which is transformed into Laplace form. Now we will study same for mechanical system. We can study mechanical system as two types; (1) translational mechanical system and (2) rotational mechanical system. First topic will be covered in this section. It is to be noted that the transfer function shown in Fig. 1.16, will be mathematically indistinguishable from an electrical network. Hence, an electrical network can be inerfaced to a mechanical system by cascading their transfer functions. It is considered that one system is not loaded by the other.

Mechanical systems can be interfaced with electrical system to such an extent that there are analogies between their components and variables.

There are three passive linear components in mechanical system like electrical system. Two of them are energy storage elements; spring and mass while one of them dissipates energy, viscous damper as shown in Table 2.2.

Table 2.2

Component	Mechanical system	Electrical system
Energy storage component	1. Spring 2. Mass	Capacitor Inductor
Energy dissipate component	3. Viscous damper or friction	Resistor

The mechanical system elements are shown in Table 2.3.

Table 2.3 Force-velocity, force-displacement, and impedance translation relationship for spring, mass and viscous damper

Component	Force velocity	Force displacement	Impedance $Z_M(s) = F(s)/X(s)$	Units
Spring 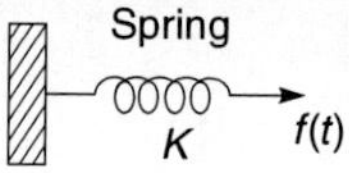	$f(s) = K\int_0^t V(t)\,dt$ $F(s) = \frac{K}{s}V(s)$	$f(t) = K\,x(t)$ $F(s) = K\,X(s)$	K	Force [$f(t)$]– *Newton* (N) displacement [$x(t)$] – *metres* (m)
Mass 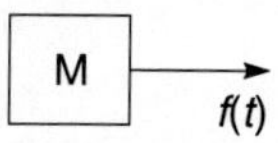	$f(t) = M\frac{dV(t)}{dt}$ $F(s) = Ms\,V(s)$	$f(t) = M\frac{d^2x(t)}{dt^2}$ $F(s) = Ms^2\,X(s)$	Ms^2	Velocity [$V(t)$] metre/second (m/s) mass (M)
Viscous 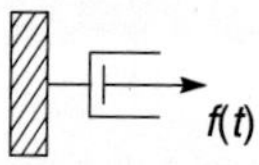	$f(t) = f_v(t)$ $F(s) = f_v\,s\,V(s)$	$f(t) = f_v\frac{dx(t)}{dt}$ $F(s) = f_v\,s\,X(s)$	$f_v\,s$	Killogram (kg)

Now we create analogies between mechanical and electrical systems by comparing Tables 2.2 and 2.3. We observe that

Force velocity column	Voltage current column
Mechanical force	Electrical voltage
Mechanical velocity	Electrical current
Force displacement column	Voltage charge column
Mechanical displacement	Electrical charge
Spring	Capacitor
Viscous damper	Resistor
Mass	Inductor

Thus, summing forces written in terms of velocity is analogous to summing voltages written in terms of current, and the resulting mechanical differential equations are analogous to mesh equations. If the forces are written in terms of displacement, the resulting mechanical equations resemble, but are not analogous to, the mesh equations.

Another comparison can be drawn between Table 2.1 and Table 2.3, in reverse order, we observe that

Force velocity column	Current voltage column
Mechanical force	Electrical current
Mechanical velocity	Electrical voltage
Spring	Inductor
Viscous damper	Resistor
Mass	Capacitor

Thus, summing forces written in terms of velocity are analogous to summing currents written in terms of voltage and the resulting mechanical differential equations are analogous to nodal equations.

First, we consider a mechanical system that consists of three passive linear components like spring, mass and viscous as shown in Fig. 2.27.

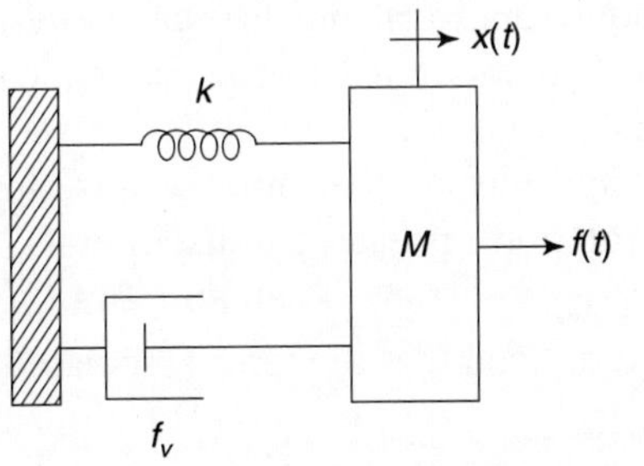

Figure 2.27 Mass, spring and damper system

The force and velcocity equation is shown in Table. 2.3. The free body diagram of Fig. 2.27 is shown in Fig. 2.28 (a) and transformed free body diagram in shown in Fig. 2.28 (b).

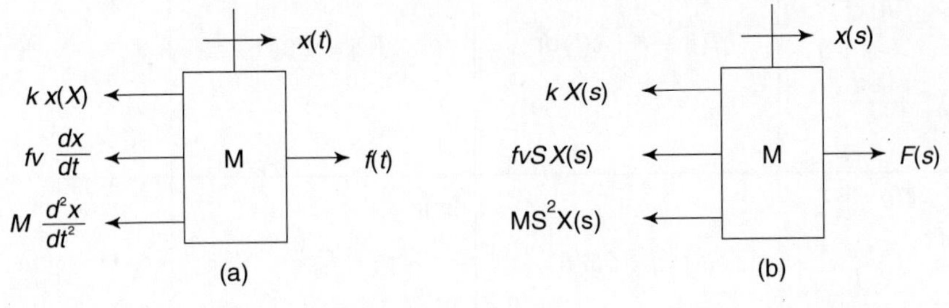

Figure 2.28 (a) Free body diagram of Fig. 2.27, (b) Transformed free body diagram

In Fig. 2.28 (a) and (b), all the forces place on the mass those, felt due to mass. We assmume mass is travelling towards the right. Thus, only the applied forces point to the right all other forces make it difficult for motion and act to oppose it. Hence, the spring, viscous damper and the force due to acceleration point to the left.

The differential equation of motion using Newton's law is

$$M\frac{d^2x(t)}{dt^2} + f_v\frac{dx(t)}{dt} + K\,x(t) = f(t) \qquad ...(2.40)$$

Taking the Laplace transform of Eq. 2.40

$$Ms^2\,X(s) + f_V\,s\,X(s) + K\,X(s) = F(s) \qquad ...(2.41)$$

$$(Ms^2 + f_v\,s + K)\,X(s) = F(s)$$

$$G(s) = \frac{X(s)}{F(s)} = \frac{1}{Ms^2 + \delta_V\,s + K} \qquad ...(2.42)$$

Equation (2.42) is shown in Fig. 2.28 (b).

2.7.1 Rotational Mechanical System

In previous sections we have covered electrical and translational mechanical systems. In this section we will study about rotational mechanical system. The rotational mechanical system is modelled in same way as the translational mechanical system, except that force is replaced by a torque and translational displacement is replaced by a angular displacement. The components for rotational systems are the same as those for translational system, except that the components undergo rotation instead of translational.

The relationship between torque and angular velocity, and torque and angular displacement is shown in Table 2.4. It is also noticed that the term associated with the mass is replaced by inertia.

Table 2.4

Component	Torque angular velocity	Torque angular displacement	Impedance $Z_m(s) = T(s)/\theta(s)$
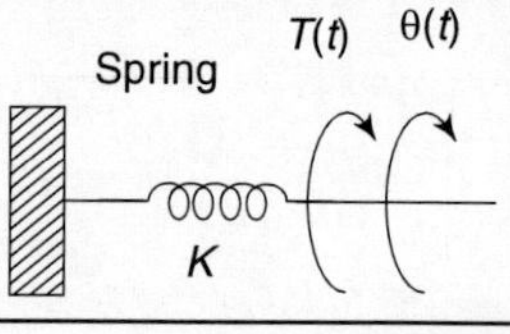	$T(t) = K\int_0^t \omega(t)\,dt$	$T(t) = K\,\theta(t)$	K
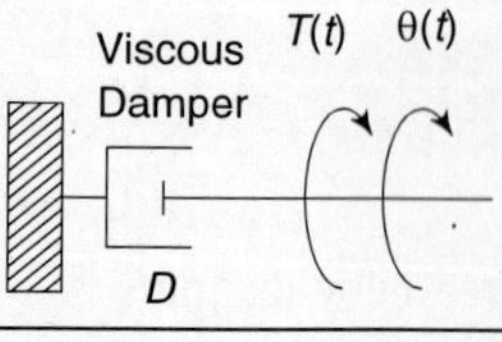	$T(s) = D\omega(t)$	$T(t) = D\frac{d\theta(t)}{dt}$	Ds
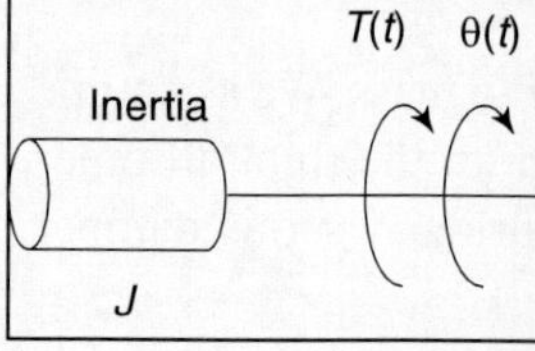	$T(t) = J\frac{d\omega(t)}{dt}$	$T(t) = J\frac{d^2\theta(t)}{dt^2}$	Js^2

The values of K, D and J are called spring constant, coefficient of viscous friction, and moment of inertia, respectively. The concept of degrees of freedom carries over to rotational system, except that we test a point of motion by rotating it while other points of motion are on hold. The number of equations of motion required to describe the system, is equals to the number of points of motion that can be rotated while all others are still.

The way to write the equations of motion for rotational system is similar to write them for translational system; the only difference is that the free body diagram consists of torques rather than forces. We obtain these torques using superposition.

1. First, we rotate a body while holding all other points still and place on it free body diagram all torques due to the body's own motion.
2. Holding the body still, we rotate adjacent points of motion one at a time and add the torques due to the adjacent motion to the free body diagram.
3. The process is repeated for each point of motion.
4. For each free body diagram, these torques are summed and set equal to zero to form the equation of motion.

Example 2.7.1 Write the laplace transform of the equations of motion for the system shown in Fig. Ex. 2.7.1.

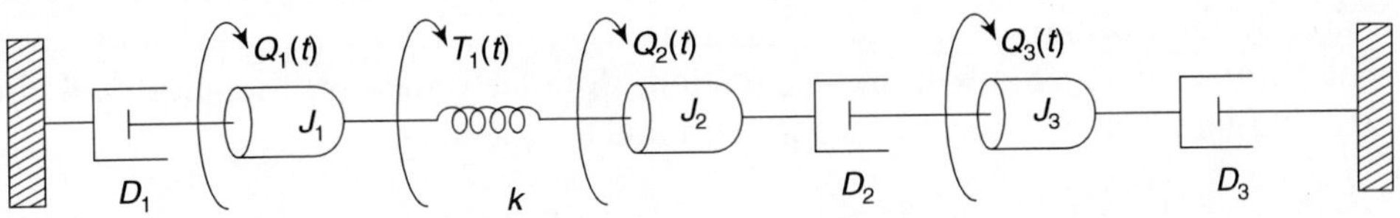

Figure Ex. 2.7.1 Rotational system

Solution According to the mesh equations of electrical system, similarly the equations will take on the following form:

[Sum of impedance at θ_1] $\theta_1(s)$ – [Sum of impedance between θ_1 and θ_2] $\theta_2(s)$

[Sum of impedances between θ_1 and θ_3] $\theta_3(s)$ = [Sum of applied torque at θ_1] ...(2.43)

Hence

$$(J_1 s^2 + D_1 s + K)\, \theta_1(s) - K\theta_2(s) - 0.\theta_3(s) = T(s)$$

$$(J_1 s^2 + D_1 s + K)\, \theta_1(s) - K\theta_2(s) = T(s) \quad ...(2.44)$$

Now

[Sum of impedance at θ_2] $\theta_2(s)$ – [Sum of impedance between θ_1 and θ_2] $\theta_1(s)$

[Sum of impendance between θ_2 and θ_3] $\theta_3(s)$ = [Sum of applied torques at θ_2] ...(2.45)

$$(J_2 s^2 + D_2 s + K)\, \theta_2(s) - K\, \theta_1(s) - D_2 s\, \theta_3(s) = 0 \quad ...(2.46)$$

Now

[Sum of impedances at θ_3] $\theta_3(s)$ – [Sum of impedances between θ_1 and θ_3] $\theta_1(s)$ –

[Sum of impedances between θ_2 and θ_3] $\theta_2(s)$ = [Sum of applied torques at θ_3) ...(2.47)

$$(J_3 s^2 + D_3 s + D_2 s)\, \theta_3(s) - 0.\theta_1(s) - D_2 s\, \theta_2(s) = 0 \quad ...(2.48)$$

$$(J_3 s_2 + D_3 s + D_2 s)\, \theta_3(s) - D_2 s\, \theta_2(s) = 0 \quad ...(2.49)$$

Equations (2.44, 2.45 and 2.49) show the equation of motion for the system shown in Fig. 2.28.

2.8 CIRCUIT ANALOGUS

In this section, we will study about the similarity of electrical, mechanical systems and systems from the various disciplines. The mechanical systems with which we worked can be represented by equivalent electrical circuits. We have already discussed about the similarity between the equations resulting from Kirchhoff's laws for electrical systems and the equations of motions of mechanical systems. We now show this analogies by providing electric circuit equivalents for mechanical systems. The variable of electric circuits behaves exactly as the analogous variable of the mechanical systems.

Analogoies can be obtained by comparing the equations, such as the equation of motions of a mechanical system, with mesh or nodal equations. The resulting electrical circuit is called a series analogy when compared with mesh equations. Similarly, the resulting electrical circuit is called a parallel analogy when compared with nodal equations.

2.8.1 Series Analogy

Consider a mechanical system (translational) system in Fig. shown 2.29, whose equation of motion is

$$(Ms^2 + f_V s + K)\, X(s) = F(s) \quad ...(2.50)$$

Kirchhoffs mesh equation for the RLC network shown in Fig. 2.30 is

$$\left(Ls + R + \frac{1}{Cs}\right) I(s) = E(s) \quad ...(2.51)$$

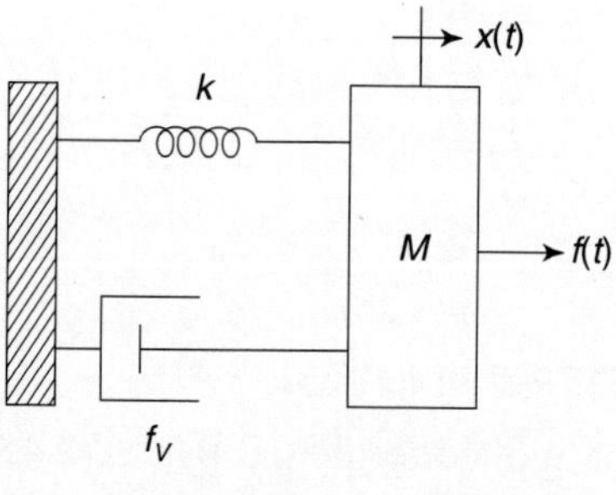

Figure 2.29 Mechanical system

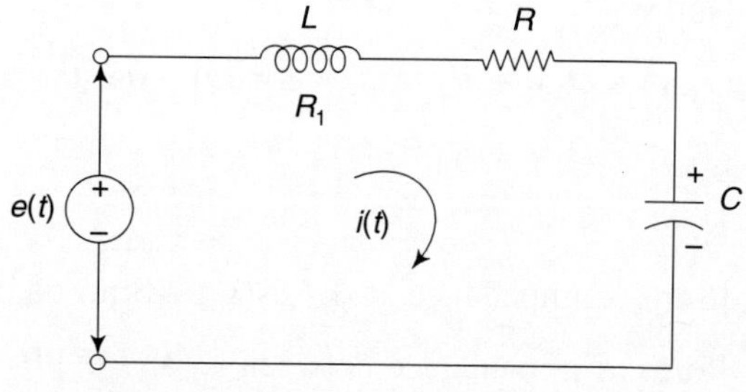

Figure 2.30 Desired electrical representation

As we know that Eq. (2.50) is not directly analogous to Eq. (2.35) because displacement and current are not analogous. Hence, we convert displacement to velocity by dividing and multiplying the left hand side of Eq. (2.50) by s.

$$\frac{Ms^2 + f_V\, s + K}{s}\, s\, X(s) = F(s)$$

$$\because \qquad sX(s) = V(s)$$

Hence

$$\frac{Ms^2 + f_V\, s + K}{s}\, V(s) = F(s)$$

$$\left(Ms + f_V + \frac{K}{s}\right) V(s) = F(s) \qquad \text{...(2.52)}$$

Now comparing Eq. (2.51) and Eq. (2.52), we observe that

Mass = $M \rightarrow$ inductor = M henries

Viscous damper = $\propto \rightarrow$ resistor = $\propto$ ohms

Spring = $K \rightarrow$ capacitor = $\frac{1}{K}$ farads

Applied force = $f(t) \rightarrow$ voltage source = $f(t)$

Velocity = $V(t) \rightarrow$ mesh current = $V(t)$

The conversions are summarized in Fig. 2.31.

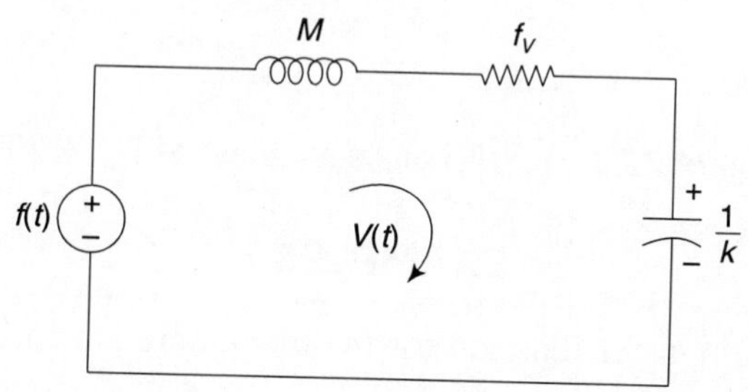

Figure 2.31 Series analogy circuits of Fig. 2.29

2.8.2 Parallel Analogy

Consider a mechanical system as shown in Fig. 2.29 and derive electrical representation as shown in Fig. 2.32 whose Kirchhoff's nodal equation for the simple parallel RLC network is

$$\left(Cs + \frac{1}{R} + \frac{1}{Ls}\right) E(s) = I(s) \qquad \text{...(2.53)}$$

Now comparing Eqs. (2.36) and (2.37). We observe that

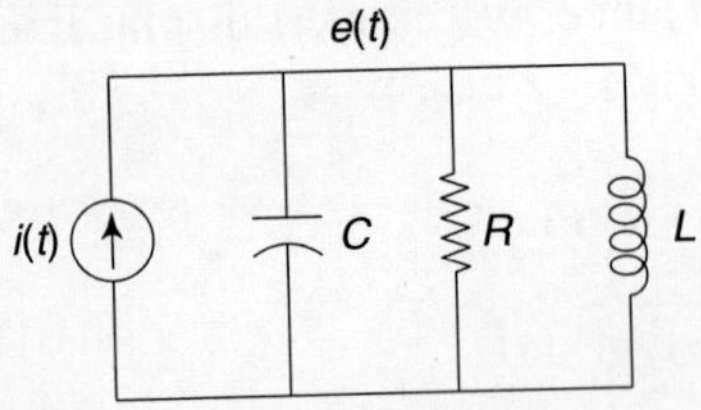

Figure 2.32 Desired electrical representation

$$\text{Mass} = M \rightarrow \text{capacitor} = M \text{ farads}$$

$$\text{Viscous damper} = \propto \rightarrow \text{resistor} = \frac{1}{f_V} \text{ ohms}$$

$$\text{Spring} = K \rightarrow \text{inductor} = \frac{1}{K} \text{ henries}$$

$$\text{Applied force} = f(t) \rightarrow \text{current source} = f(t)$$

$$\text{Velocity} = V(t) \rightarrow \text{node voltage} = V(t)$$

The conversions are summarized in Fig. 2.33

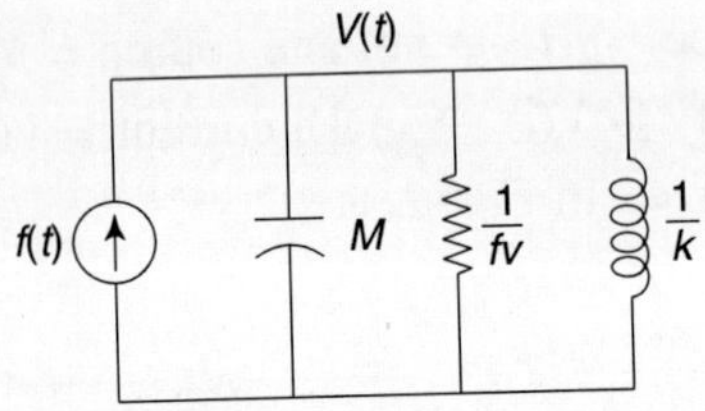

Figure 2.33 Parallel analogy circuit of Fig. 2.29

EXAMPLES

Example 2.1 Obtain the transfer function for the block diagram shown in Fig. Ex. 2.1

(*Pune University 1988*)

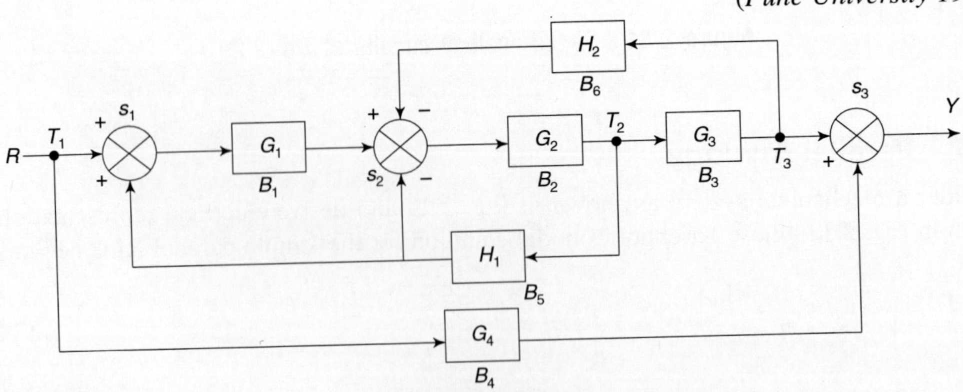

Figure Ex. 2.1

Solution Moving take-off point T_2 beyond, block B_3 (Rule 7)

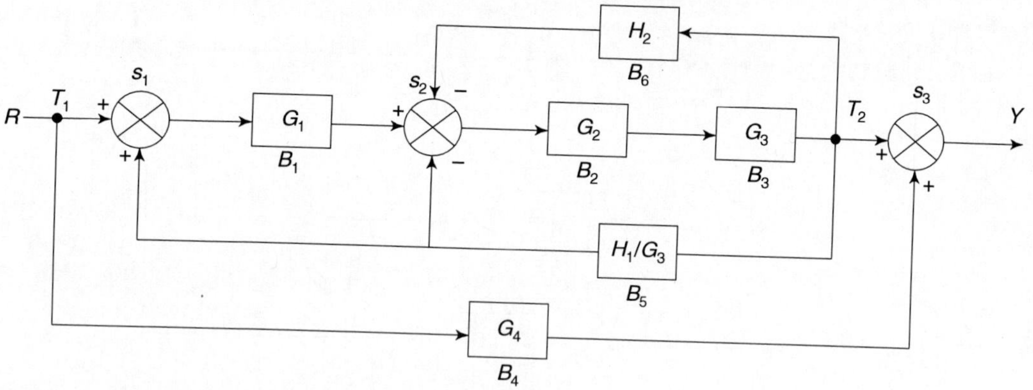

Applying Rule 1; blocks B_2 and B_3 are in Cascade.

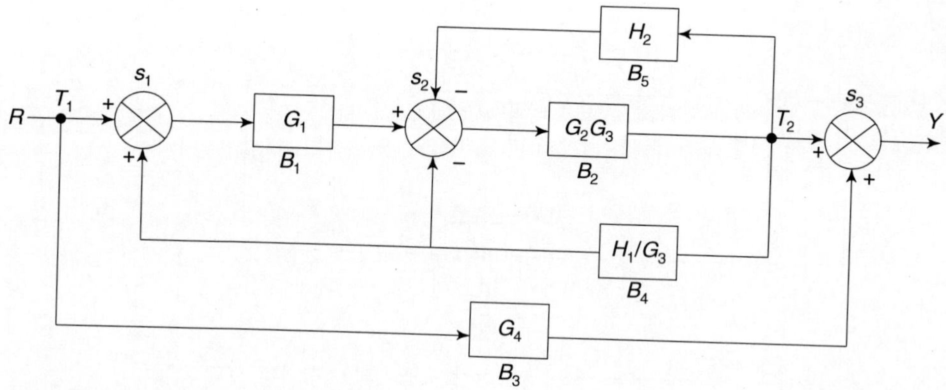

Eliminating feedback path, block B_5

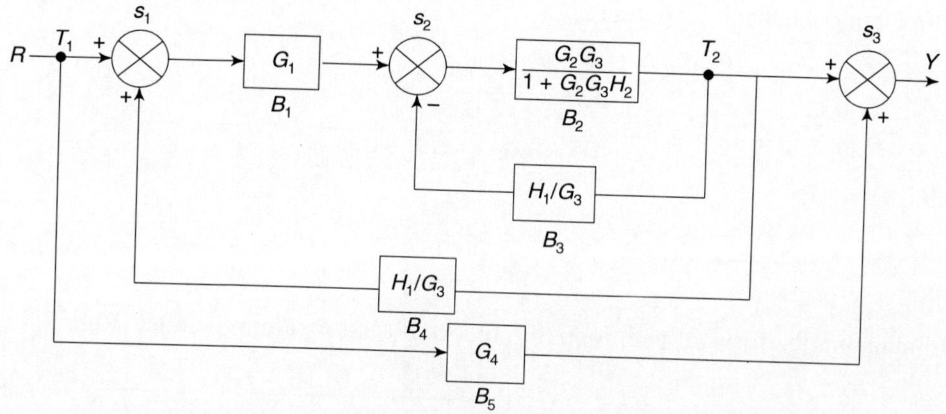

Eliminating feedback loop, block B_3

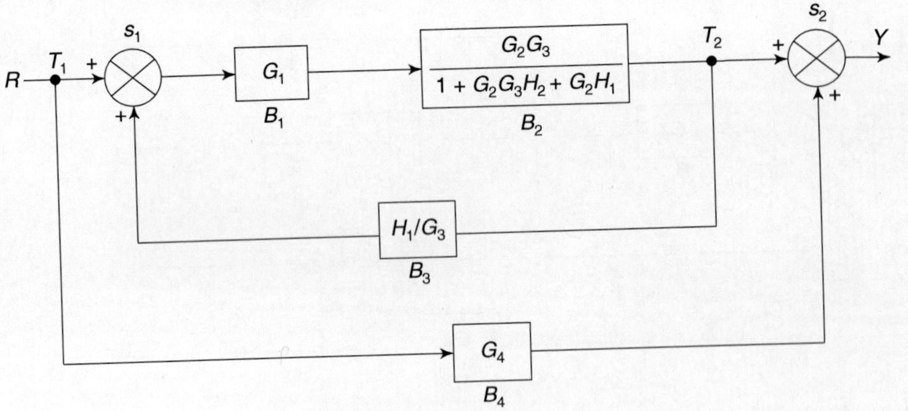

Blocks B_1 and B_2 are in cascade (Rule 1)

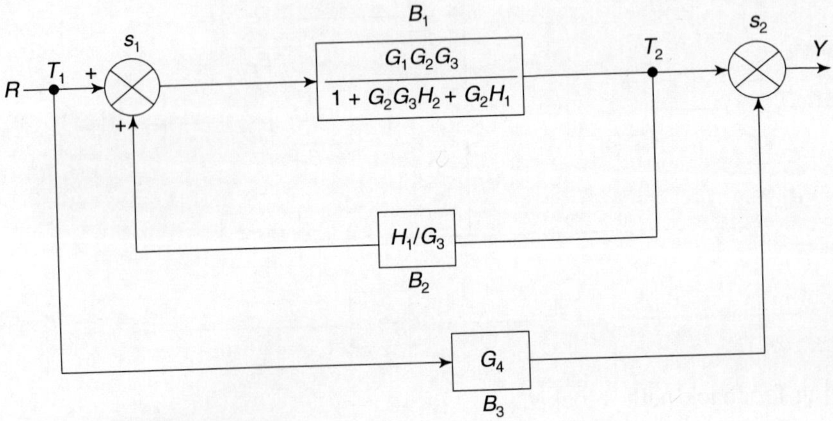

Eliminating feedback loop, block B_2

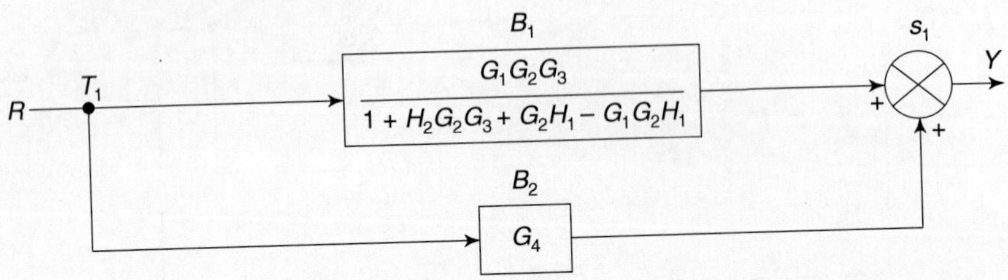

Eliminating the forward path loops B_2, blocks B_1 and B_2 are in parallel (Rule 2)

$$\frac{Y}{R} = G_4 + \frac{G_1\, G_2\, G_3}{1 + G_2\, G_3\, H_2 + G_2\, H_1 - G_1\, G_2\, H_1} \qquad \textbf{Ans.}$$

Example 2.2 Simplify the block diagram to obtain transfer function $Y(s)/R(s)$.

(*U.P.T.U., 2003–04*)

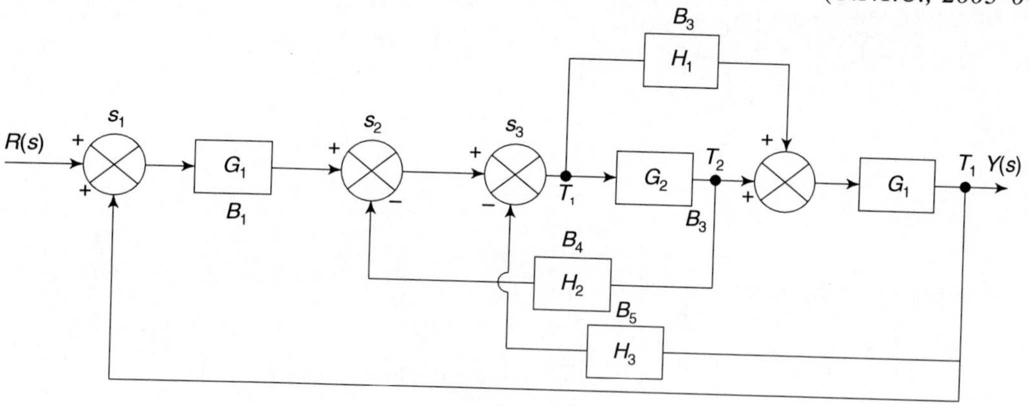

Figure Ex. 2.2

Solution Shifting take-off point T_2 before the block B_2 (Rule 8) and applying the feedback rule in blocks B_2 and B_3.

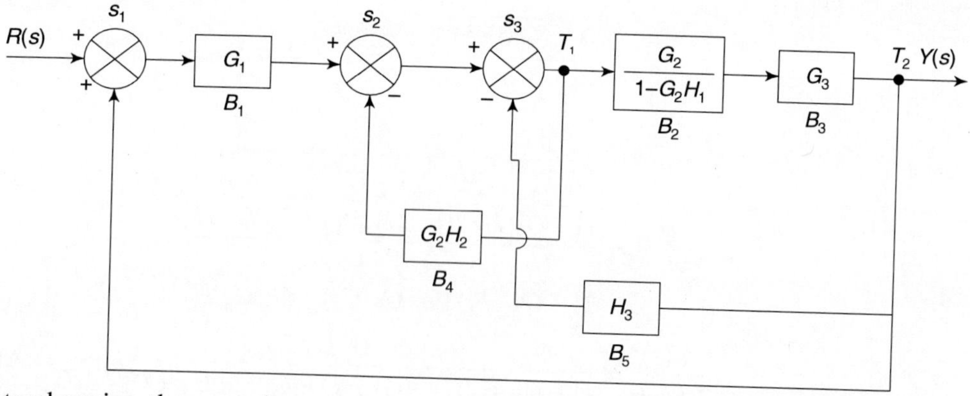

Interchanging the summing points s_2 and s_3 (Rule 9), and applying the feedback rule in block B_4.

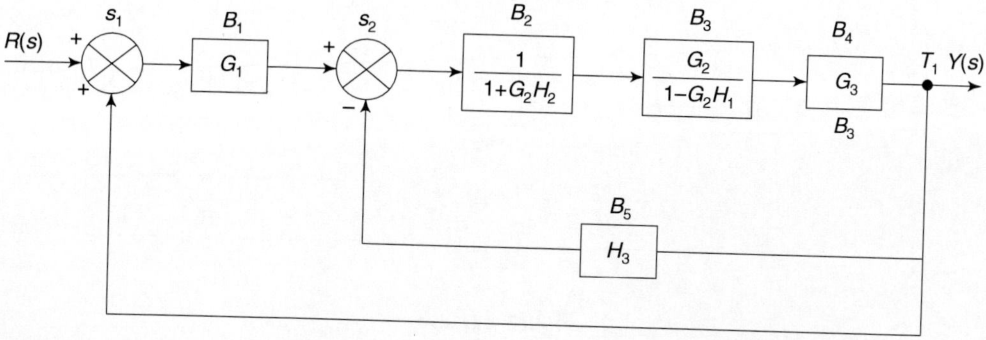

The blocks B_2, B_3 and B_4 are in cascade (Rule 1)

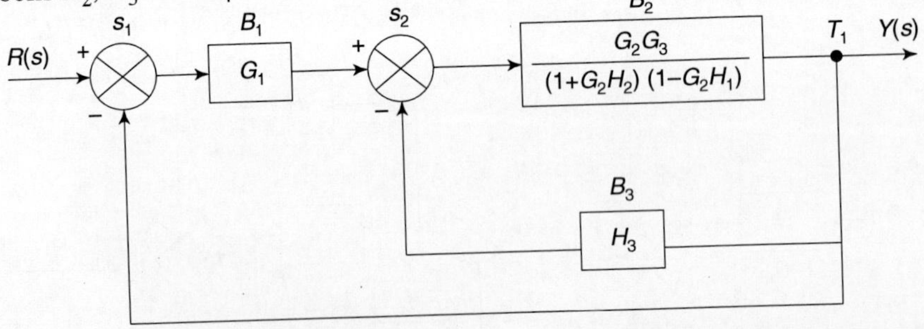

Eliminating feedback loop, block B_3

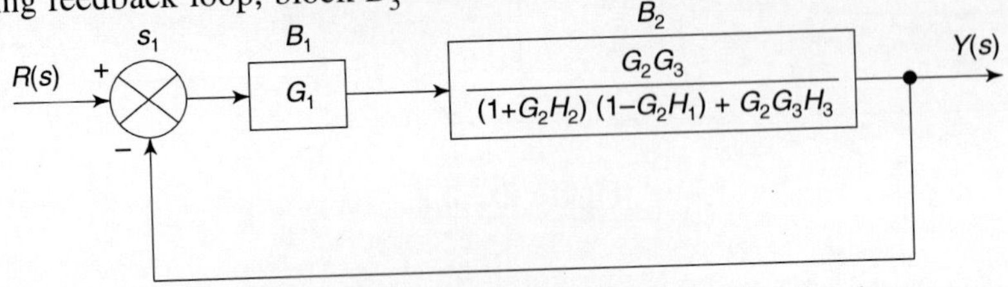

Block B_1 and B_2 are in cascade (Rule 1) and applying feedback rule.

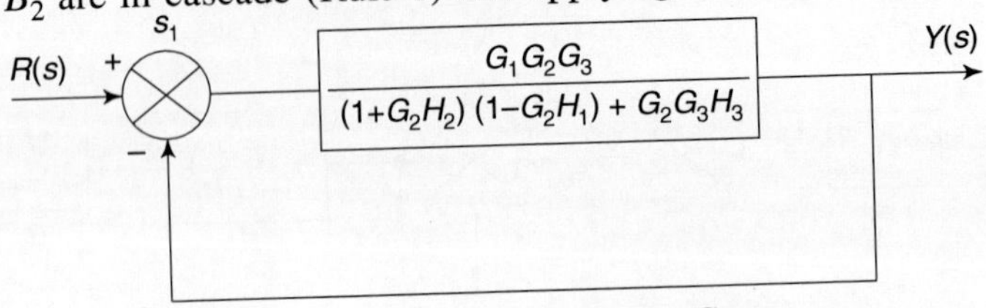

$$\frac{Y(s)}{R(s)} = \frac{\dfrac{G_1\, G_2\, G_3}{(1+G_2\, H_2)(1-G_2\, H_1)+G_2\, G_3\, H_3}}{\dfrac{G_1\, G_2\, G_3}{(1+G_2\, H_2)(1-G_2\, H_1)+G_2\, G_3\, H_3}}$$

$$\frac{Y(s)}{R(s)} = \frac{G_1\, G_2\, G_3}{(1+G_2\, H_2)(1-G_2\, H_1)+G_2\, G_3\, H_3+G_1\, G_2\, G_3} \quad \textbf{Ans.}$$

Example 2.3 Find the transfer function of the system where block diagram is given in Fig. Ex. 2.3.

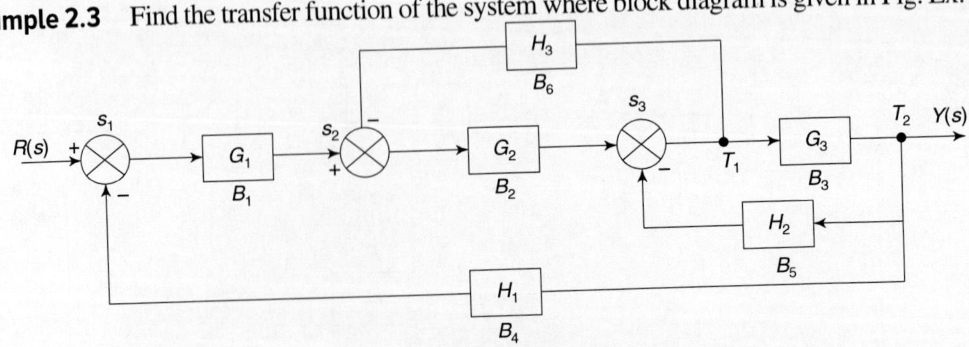

Figure Ex. 2.3

Solution Moving take-off point T_1 after the block B_3 (Rule 7)

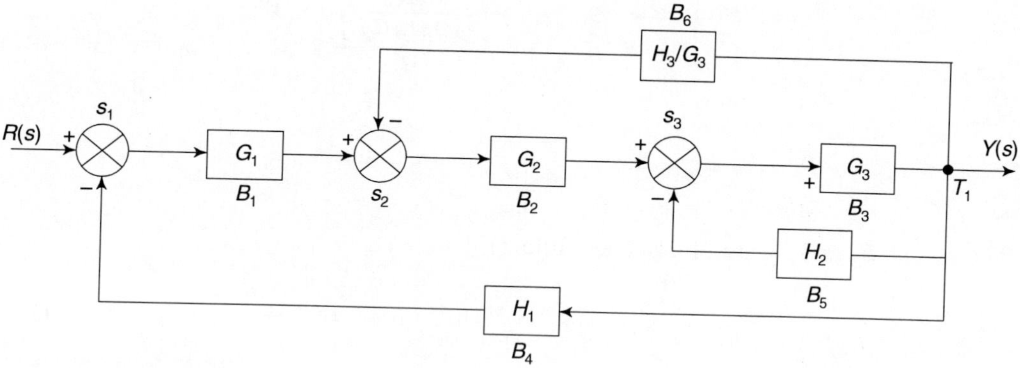

Eliminating the feedback loop, block B_5

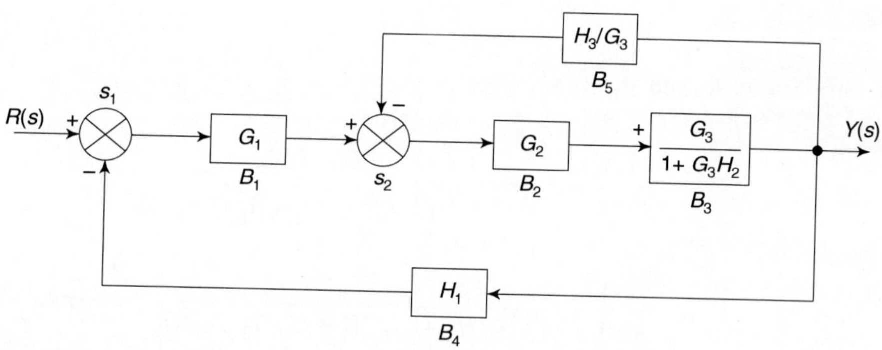

The blocks B_2 and B_3 are in Cascade (Rule 1)

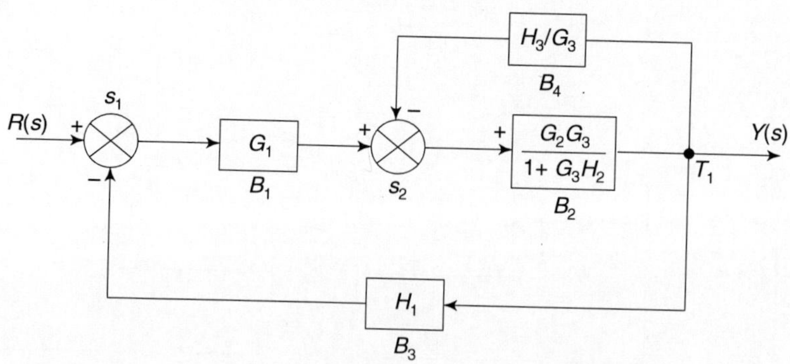

Eliminating the feedback loop, block B_4

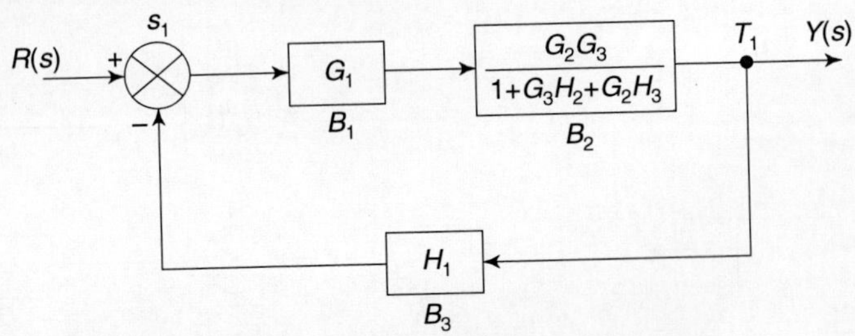

The blocks B_1 and B_2 are in cascade (Rule 1)

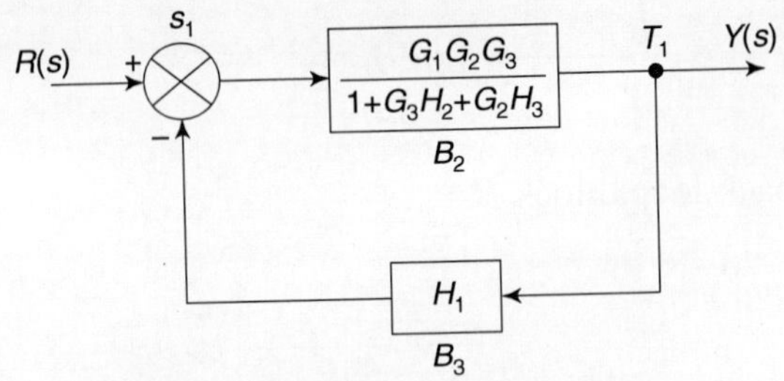

Eliminating blocks B_1 and B_2

$$\frac{Y(s)}{R(s)} = \frac{\dfrac{G_1\ G_2\ G_3}{1+G_3H_2+G_2\ H_3}}{1+\dfrac{G_1\ G_2\ G_3}{1+G_3\ H_2+G_2\ H_3}\times H_1}$$

$$\frac{Y(s)}{R(s)} = \frac{G_1\ G_2\ G_3}{1+G_3\ H_2+G_2\ H_3+G_1\ G_2\ G_3\ H_1}$$ **Ans.**

Example 2.4 Obtain overall transfer function $Y(s)/R(s)$ by block diagram reduction technique of Fig. Ex. 2.4 *(U.P.T.U., 2006–07, ACS)*

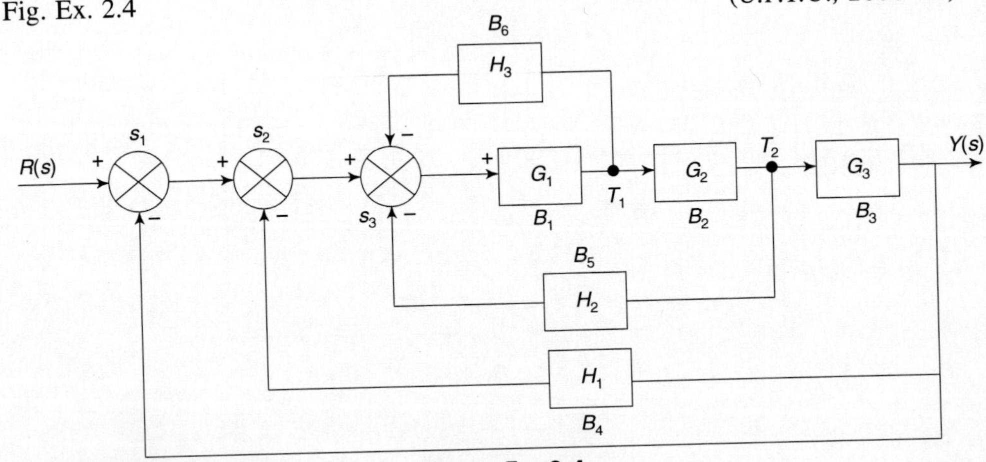

Figure Ex. 2.4

Solution Eliminating the feedback loop, block B_2 and B_1.

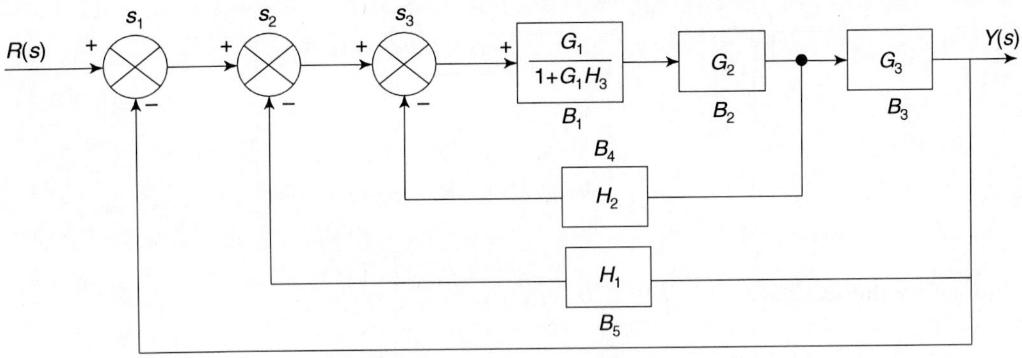

The blocks B_1 and B_2 are in cascade (Rule 1)

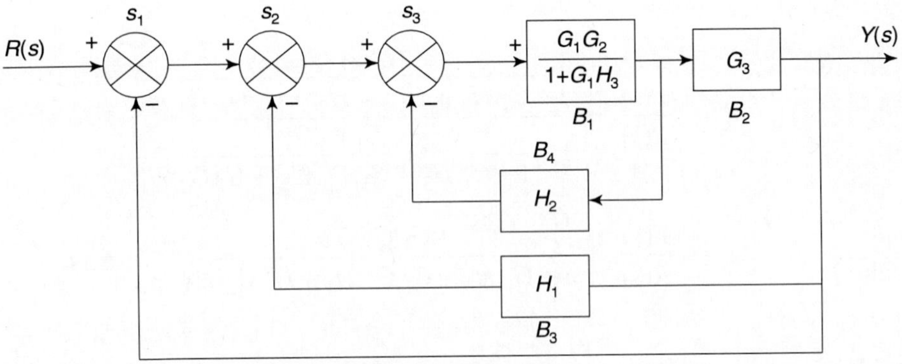

Eliminating the feedback loop, block B_4 and B_1.

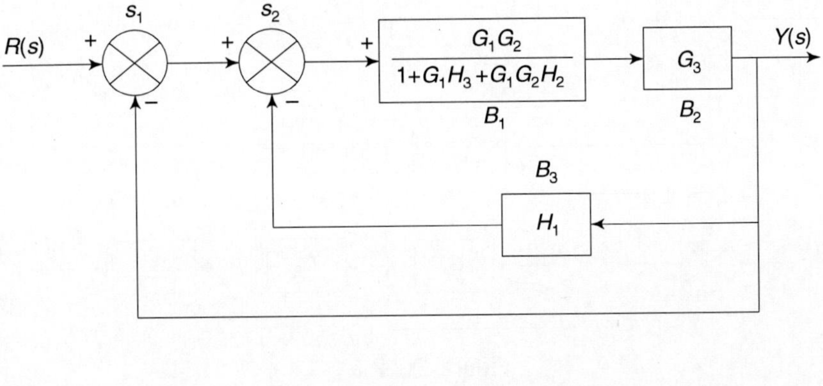

The blocks B_1 and B_2 are in cascade (Rule 1)

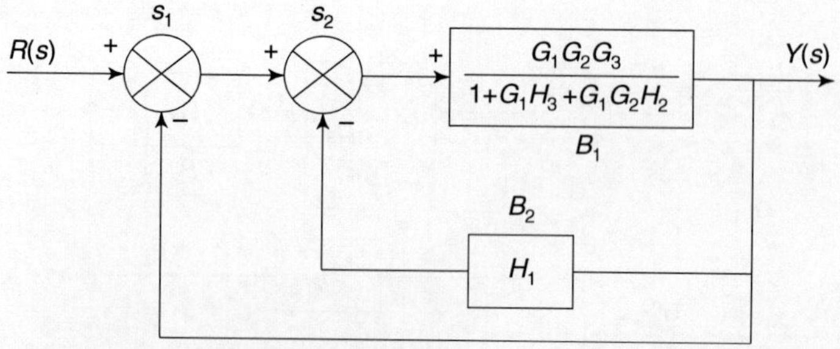

Eliminating the feedback loop, block B_1 and B_2.

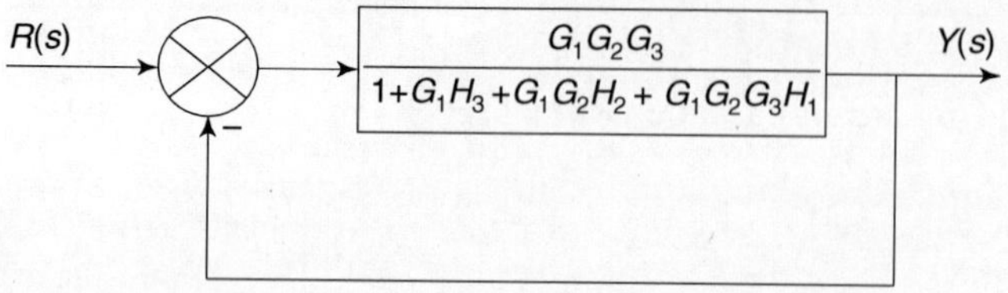

$$\frac{Y(s)}{R(s)} = \frac{\dfrac{G_1\, G_2\, G_3}{1 + G_1\, H_3 + G_1\, G_2\, H_2 + G_1\, G_2\, G_3\, H_1}}{1 + \dfrac{G_1\, G_2 G_3}{1 + G_1\, H_3 + G_1\, G_2\, H_2 + G_1\, G_2\, H_1}}$$

$$\frac{Y(s)}{R(s)} = \frac{G_1\, G_2\, G_3}{1 + G_1\, H_3 + G_1\, G_2\, H_1 + G_1\, G_2\, H_3}$$ **Ans.**

Example 2.5 Evaluate $\frac{Y_1}{R_1}$ and $\frac{Y_2}{R_2}$ for a system whose block diagram repesentation is shown in Fig. Ex. 2.5.

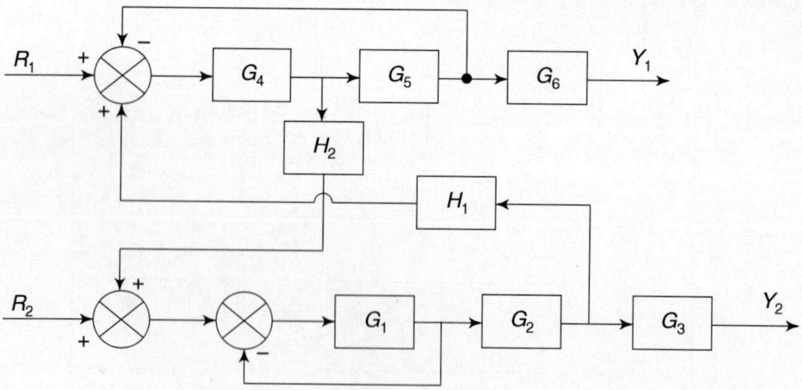

Figure Ex. 2.5

Solution Assuming $R_2 = 0$, $Y_2 = 0$ the diagram is

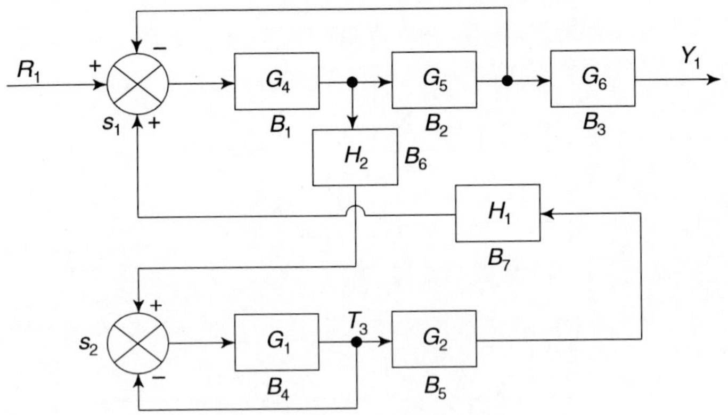

Moving take-off point T_2 before the block G_3 (Rule 8)

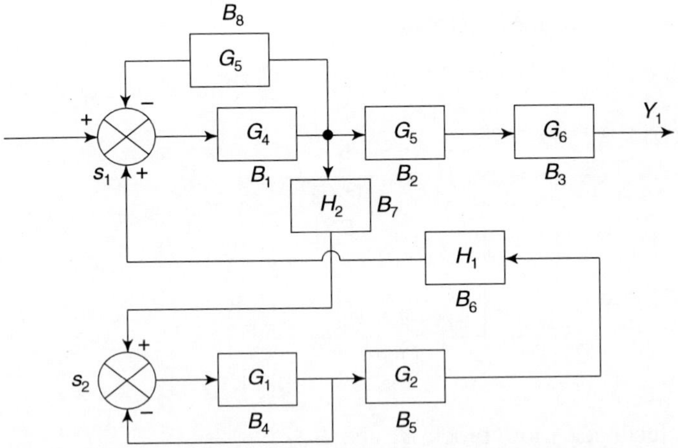

Eliminating feedback loop, blocks B_1 and B_8 and the blocks B_2 and B_3 are in cascade (Rule 1)

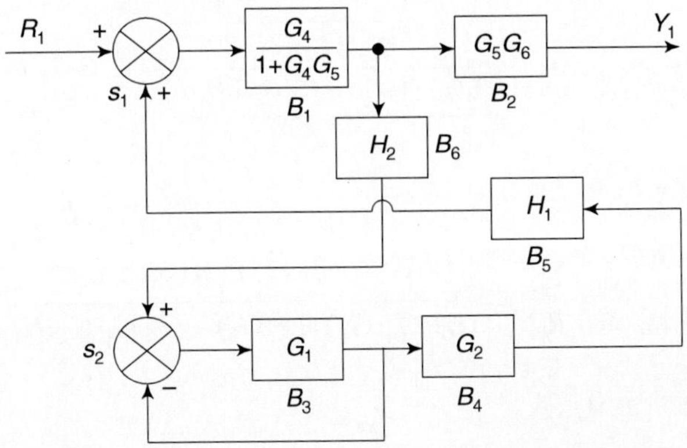

Eliminating feedback loop, block B_3 and block B_4 and B_5 are in cascade (Rule 1)

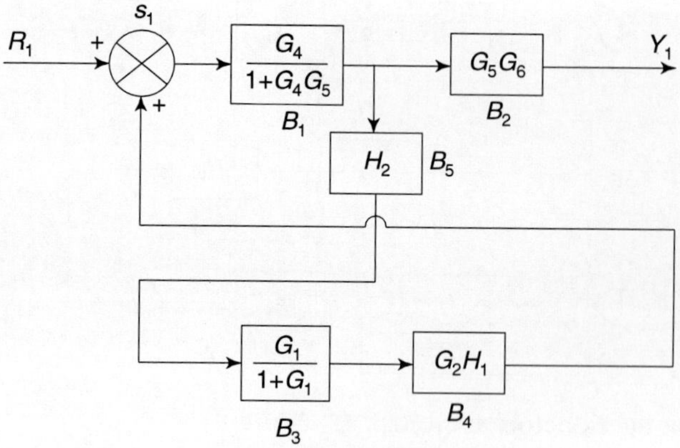

The blocks B_3, B_4 and B_5 are in cascade (Rule 1)

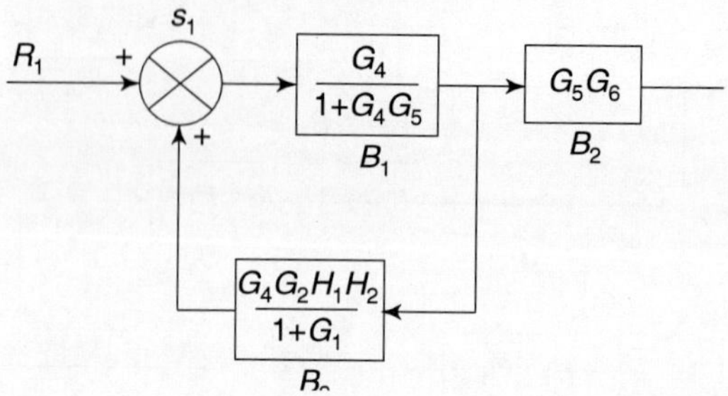

Eliminating the feedback loop, block B_1 and B_3

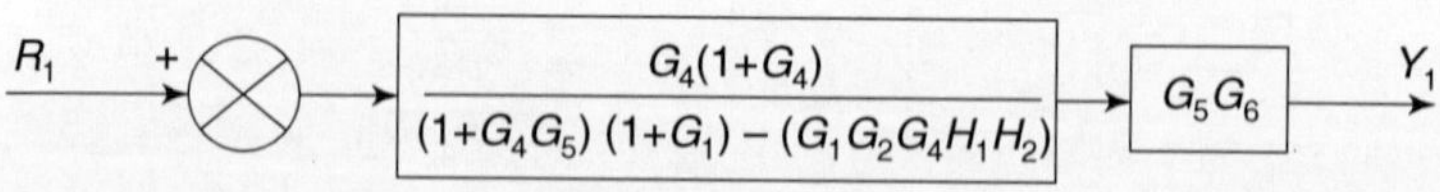

The transfer function Y_1/R_1 is

$$\frac{Y_1}{R_1} = \frac{G_4\ G_5\ G_6\ (1+G_1)}{(1+G_4\ G_5)(1+G_1) - G_4\ G_1\ G_2\ H_1\ H_2} \qquad \textbf{Ans.}$$

for Y_2/R_2 assuming $Y_1 = 0$

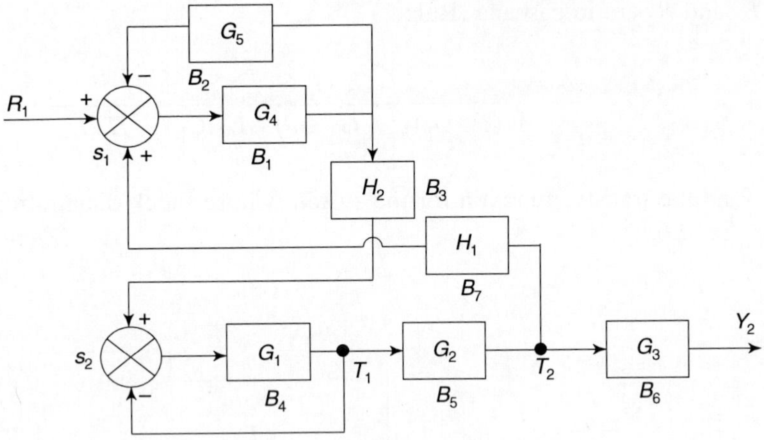

Eliminating the feedback loop, block B_1, B_2 and B_4.

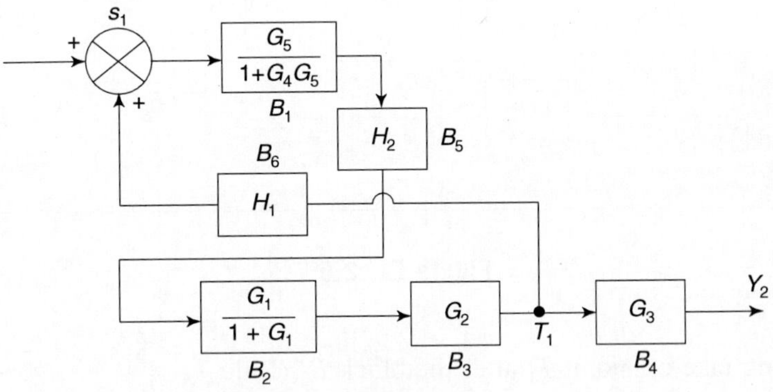

The blocks B_1, B_5, B_2, B_3 are in cascade (Rule 1)

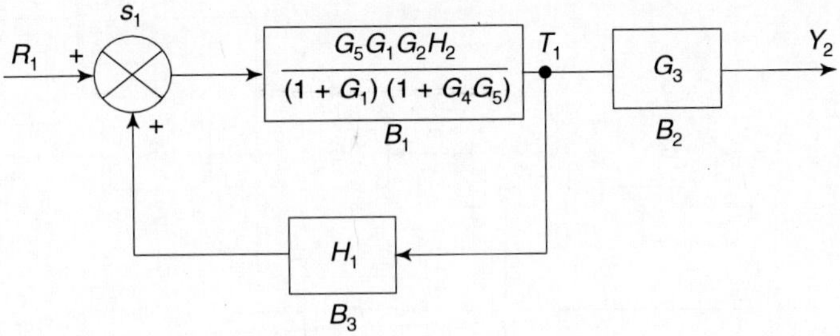

Eliminating feedback loop, block B_1 B_3

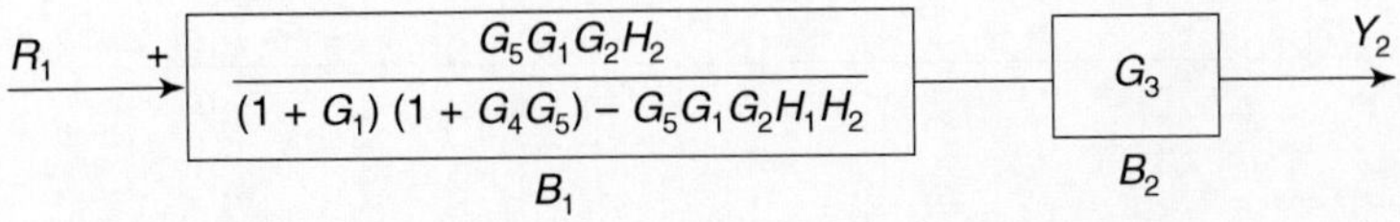

The blocks B_1 and B_2 are in cascade (Rule 1)

$$\frac{Y_2}{R_1} = \frac{G_5\,G_1\,G_2\,G_3\,H_2}{(1+G_1)(1+G_4\,G_5) - G_5\,G_1\,G_2\,H_1\,H_2} \qquad \textbf{Ans.}$$

Example 2.6 Find the transfer function for the system whose block diagam representation is shown if Fig. Ex. 2.6. *(Pune University)*

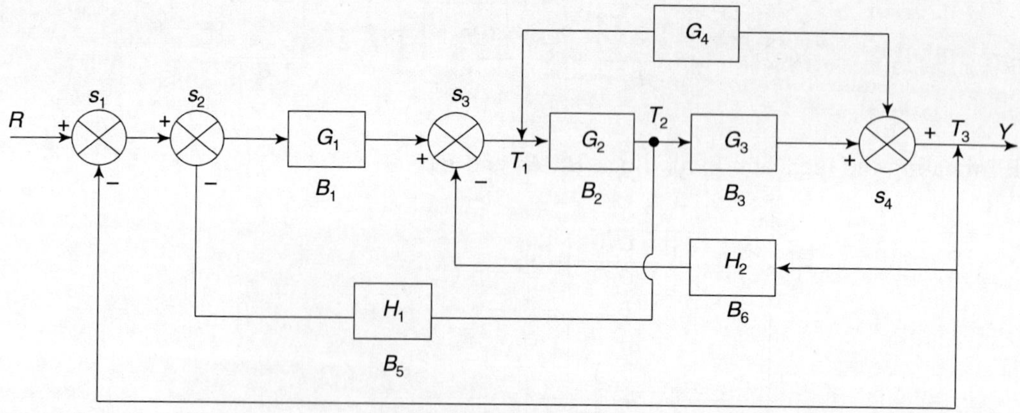

Figure Ex. 2.6

Solution Moving take-off point T_1 after the block B_2 (Rule 7)

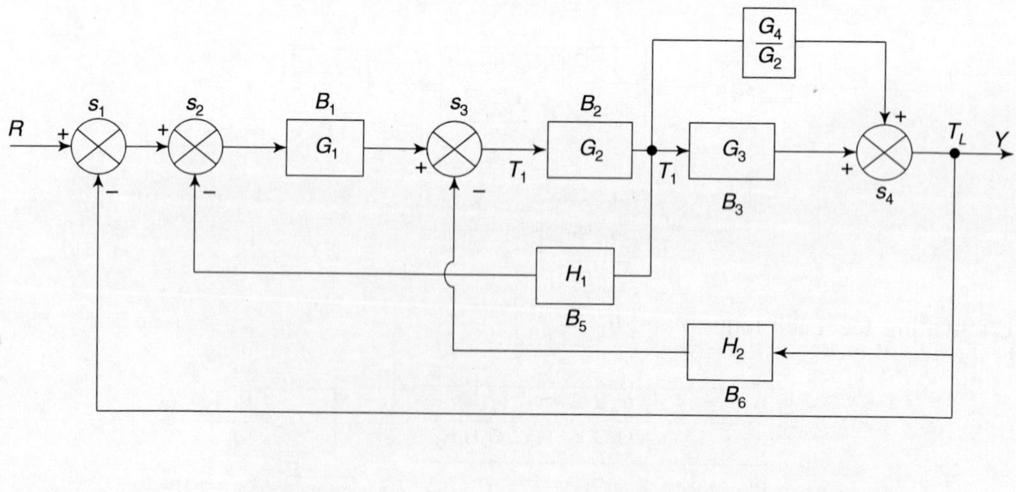

The blocks B_3 and B_4 are in parallel (Rule 2)

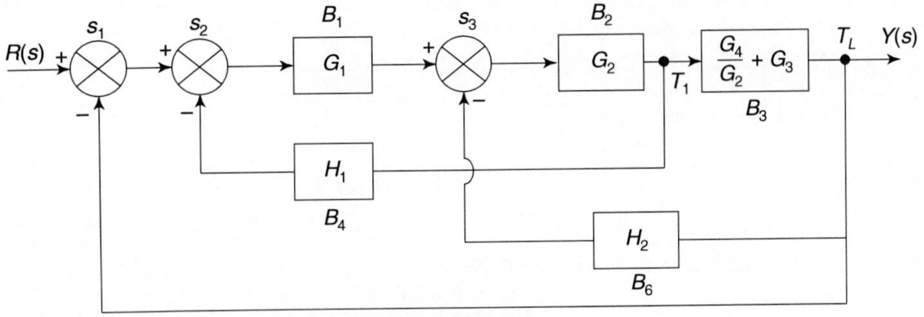

Moving take-off point T_1 after the block B_3 (Rule 7)

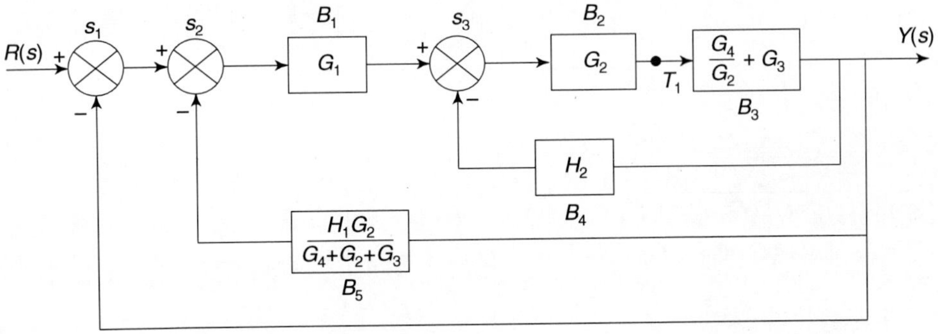

The blocks B_2 and B_3 are in cascade (Rule 1)

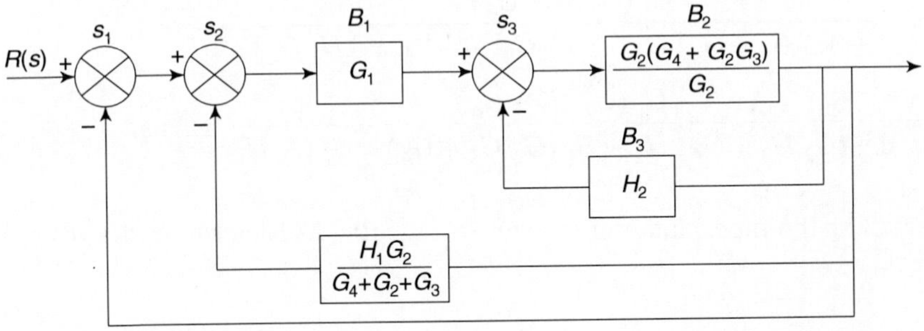

Eliminating the feedback loop, B_3

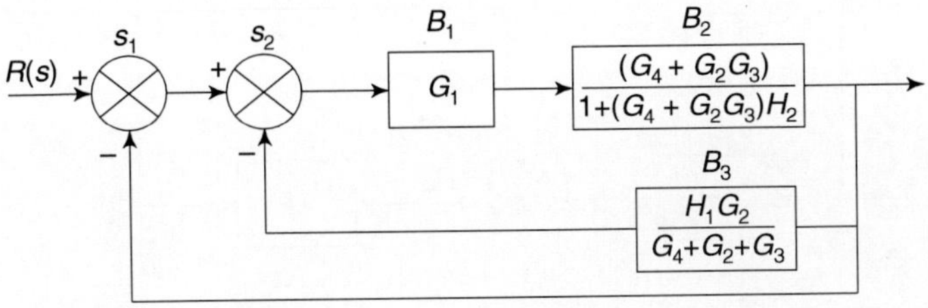

The blocks B_1 and B_2 are in cascade (Rule 1)

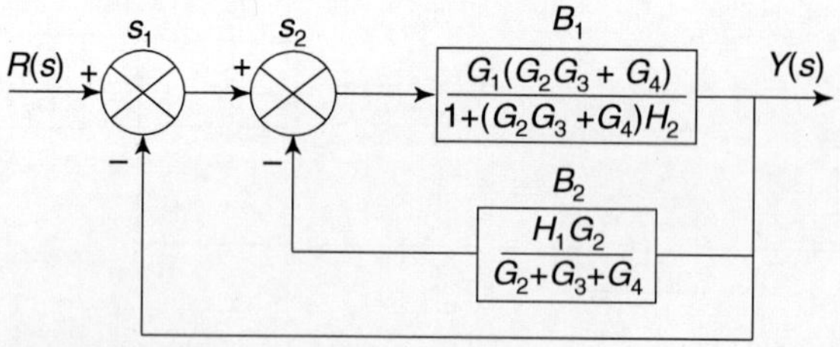

The above block diagram can be redrawn as

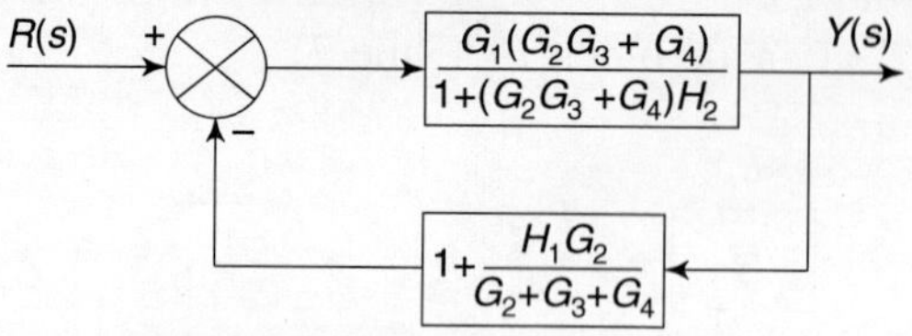

$$\frac{Y(s)}{R(s)} = \frac{\dfrac{G_1\,(G_2\,G_3+G_4)}{1+(G_2\,G_3+G_4)\,H_2}}{1+\dfrac{G_1\,(G_2\,G_3+G_4)}{1+(G_2\,G_3+G_4)\,H_2}\times\left[1+\dfrac{H_1\,G_2}{G_2\,G_3+G_4}\right]}$$

$$\frac{Y(s)}{R(s)} = \frac{G_1\,(G_2\,G_3+G_4)(G_2\,G_3+G_4)}{(G_2\,G_3+G_4)+(G_2\,G_3+G_4)^2\,H_2+G_1\,(G_2\,G_3+G_4)[(G_2\,G_3+G_4)+H_1\,G_2]}$$

$$\frac{Y(s)}{R(s)} = \frac{G_1\,(G_2G_3+G_4)}{1+(G_2\,G_3+G_4)\,H_2+G_1\,[(G_2\,G_3+G_4)+H_1\,G_2]}$$

$$\frac{Y(s)}{R(s)} = \frac{G_1\,(G_2\,G_3+G_4)}{1+(G_2\,G_3+G_4)\,H_2+G_1\,(G_2\,G_3+G_4)+G_1\,G_2\,H_1}$$ **Ans.**

Example 2.7 For the block diagram shown in Fig. Ex. 2.7 determine the overall transfer function.

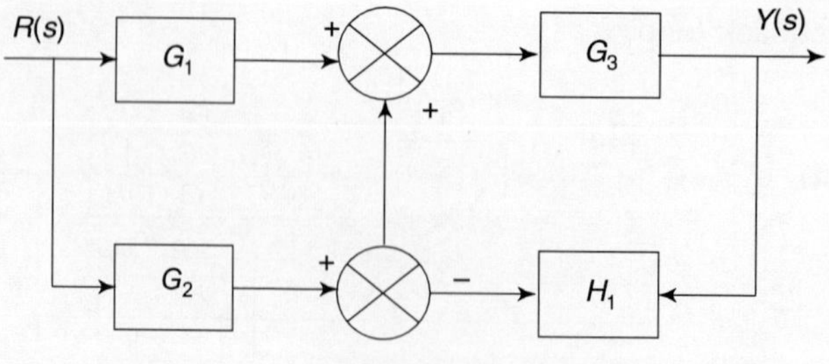

Figure Ex. 2.7

Solution The block diagram of Fig. Ex. 2.7 can be redrawn as shown in Fig. Ex. 2.7 (a).

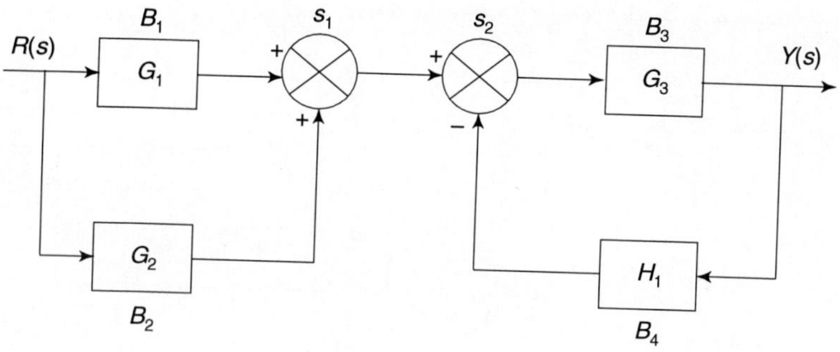

Figure Ex. 2.7 (a)

The blocks B_1 and B_2 are in parallel (Rule 2)

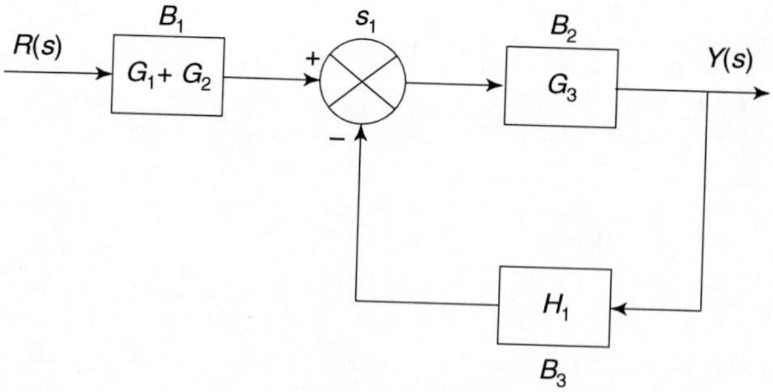

Eliminating the feedback loop, block B_3 B_2

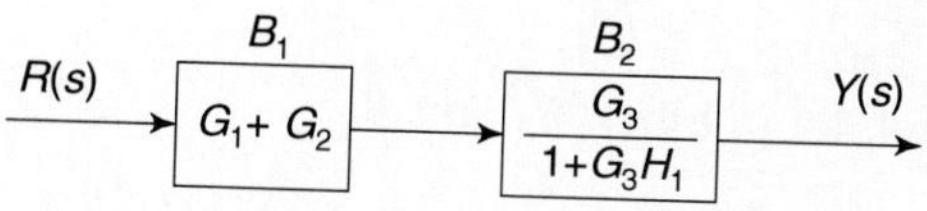

The blocks B_1 and B_2 are in cascade (Rule 1)

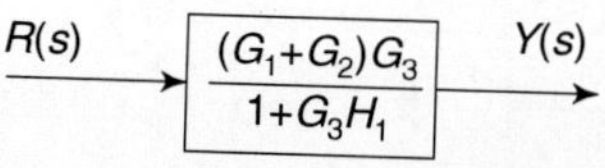

$$\frac{Y(s)}{R(s)} = \frac{(G_1 + G_2)\,G_3}{1 + G_3\,H_3} \quad \textbf{Ans.}$$

Example 2.8 Find the transfer function of the block diagram shown in Fig. Ex. 2.8.

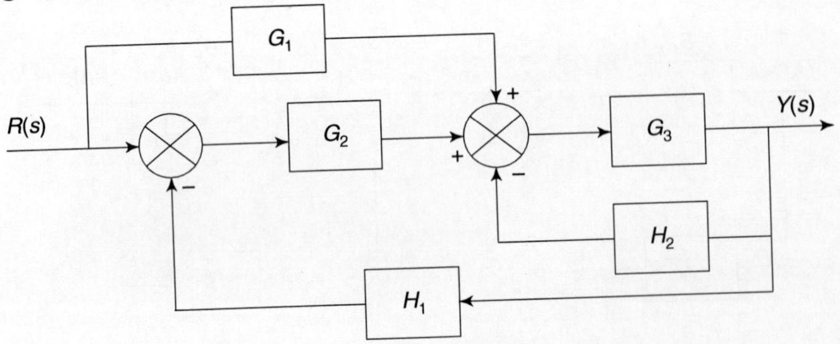

Figure Ex. 2.8

Redraw the Fig. Ex. 2.8 as shown in FIg. Ex. 2.8 (a)

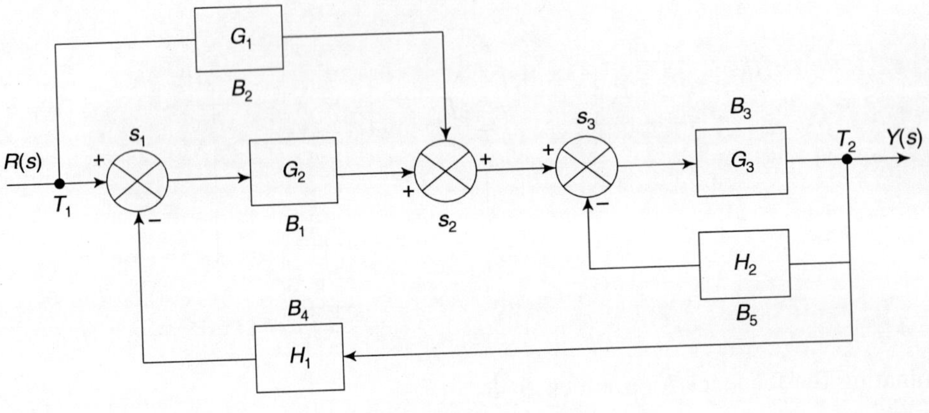

Figure Ex. 2.8 (a)

Eliminating the feedback loop, block B_3 and B_5

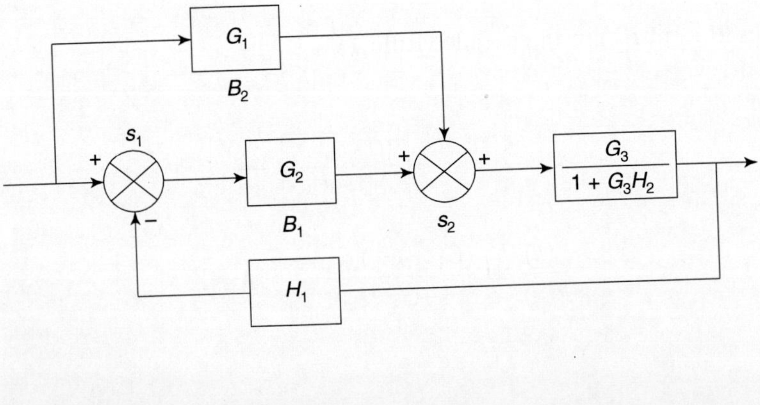

Shifting the summing point s_1 after the block B_1 (Rule 5)

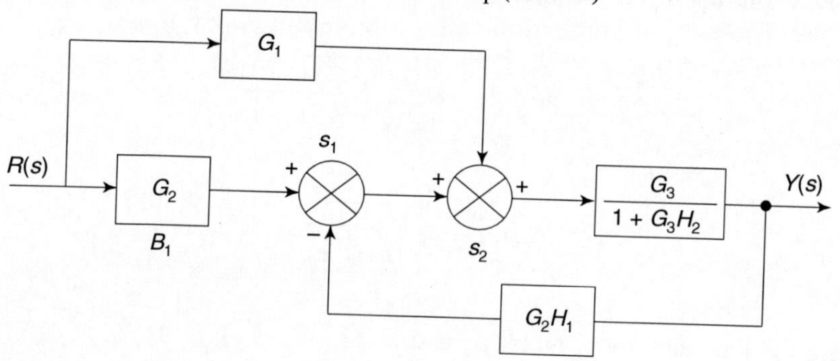

Interchanging the consecutive summing point (Rule 9)

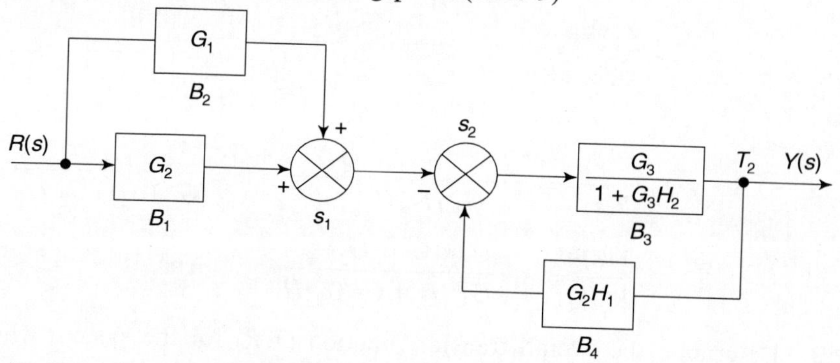

The blocks B_1 and B_2 are in parallel (Rule 2)

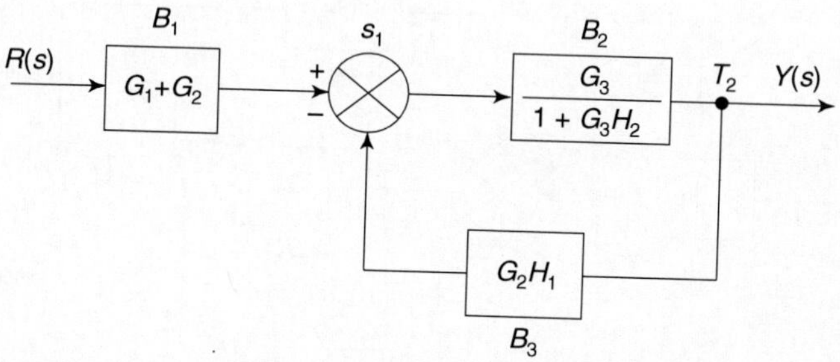

Shifting the summing point s_1 before the block B_1 (Rule 6)

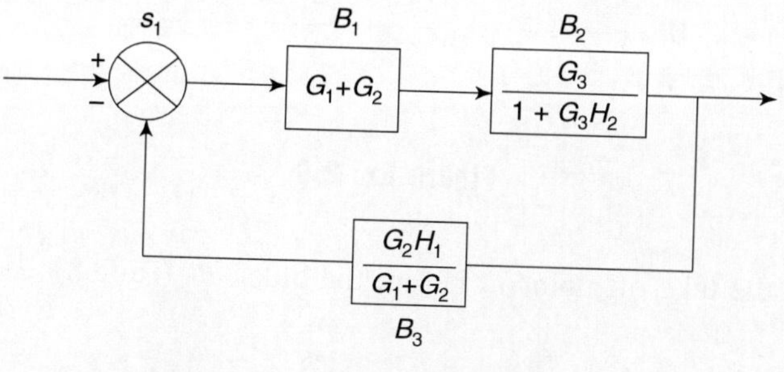

The blocks B_1 and B_2 are in cascade (Rule 1)

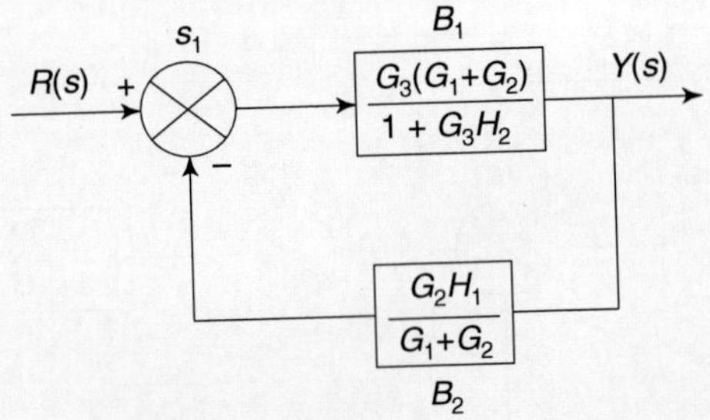

Eliminating the feedback loop, block B_1 and B_2

$$\frac{Y(s)}{R(s)} = \frac{\dfrac{G_3\,(G_1+G_2)}{1+G_3\,H_2}}{1+\dfrac{G_3\,(G_1+G_2)}{(1+G_3\,H_2)}\times\dfrac{G_2\,H_1}{(G_1+G_2)}}$$

$$\frac{Y(s)}{R(s)} = \frac{G_3\,(G_1+G_2)^2}{(1+G_3\,H_2)(G_1+G_2)+G_3\,(G_1+G_2)\,G_2\,H_1}$$

$$\frac{Y(s)}{R(s)} = \frac{G_3\,(G_1+G_2)}{1+G_3\,H_2+G_3\,G_2\,H_1} \qquad \textbf{Ans.}$$

Example 2.9 Determine the overall transfer function (Y/R) for the system whose block diagram is shown in Fig. Ex. 2.9.

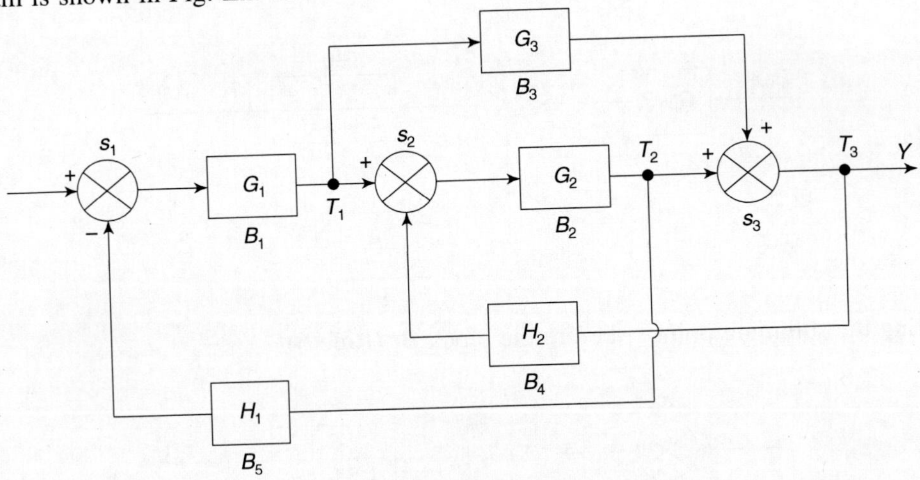

Figure Ex. 2.9

Solution Shifting the take-off point T_2 before the block B_2 (Rule 8)

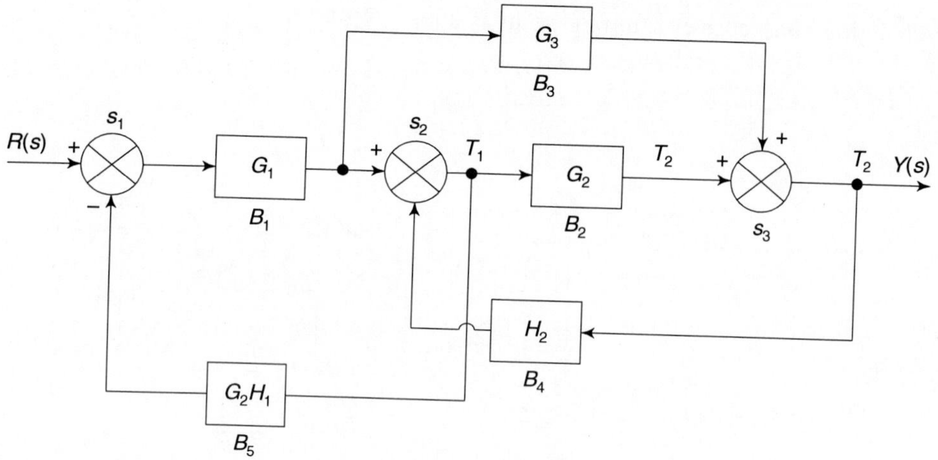

Shifting the take-off point T_1 before the summing point s_2 (Rule 11)

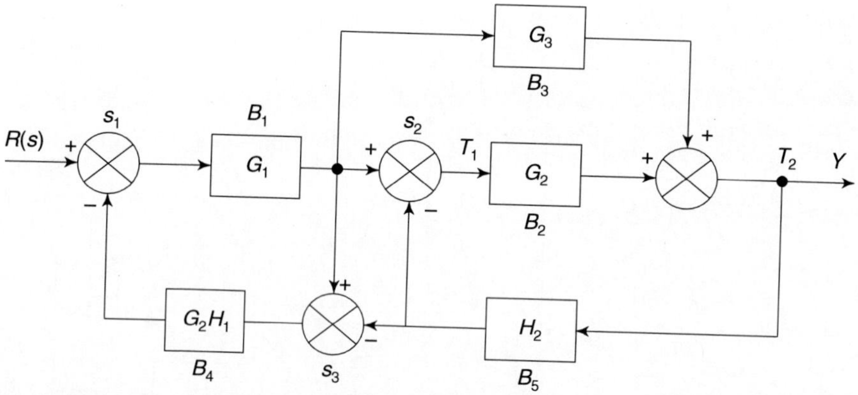

Shifting summing point s_2 after the block B_2 (Rule 5)

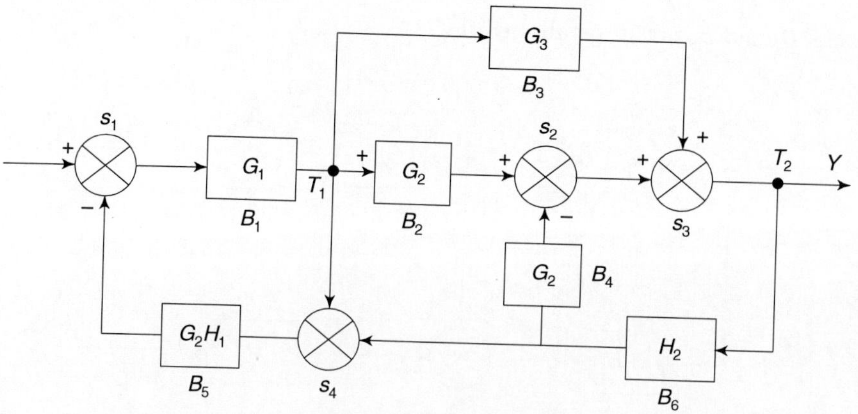

Interchanging consecutive summing point s_2 and s_3 (Rule 9)

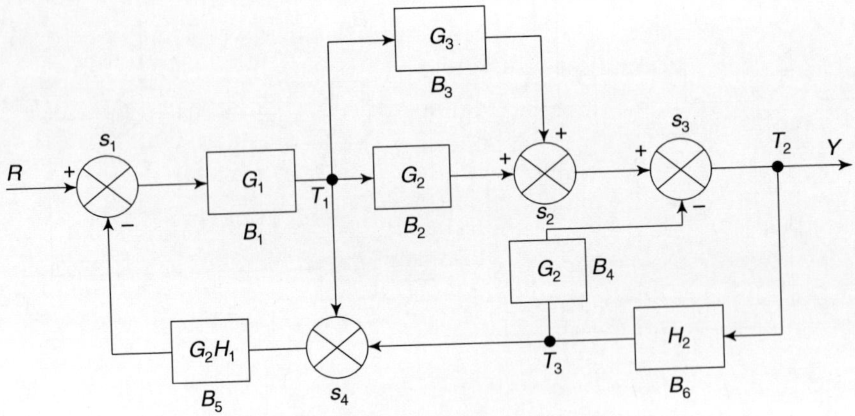

Shifting the take-off point T_3 after the block B_6 (Rule 7)

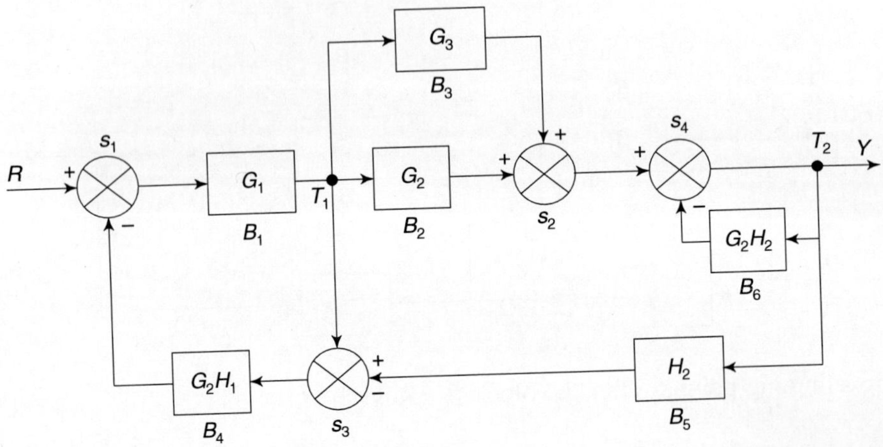

The blocks B_2 and B_3 are in parallel (Rule 2)

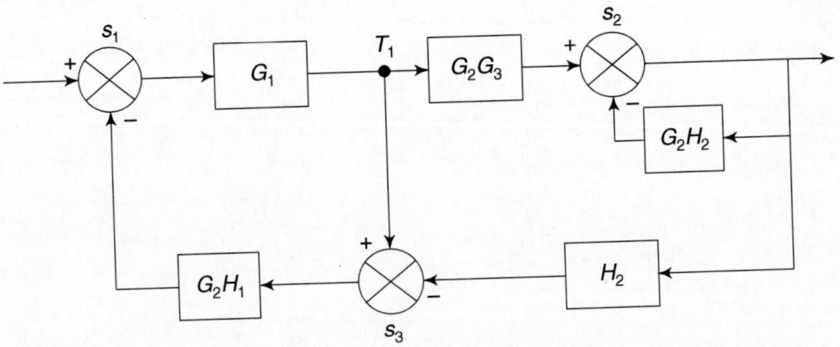

The above block diagram can be redrawn as

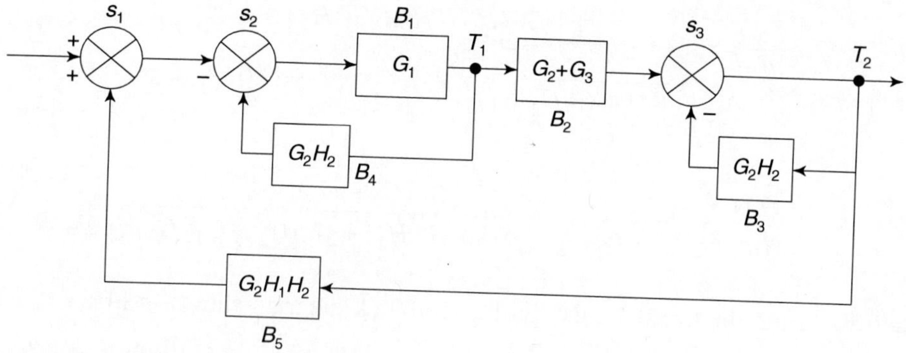

Eliminating the feedback loop, block B_3

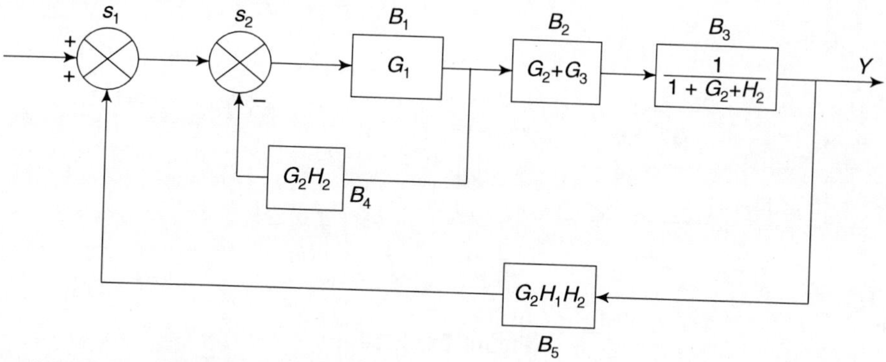

Eliminating the feedback loop, block B_4

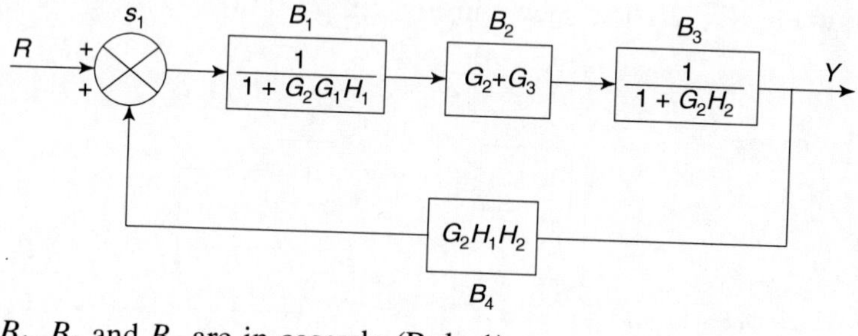

The blocks B_1, B_2 and B_3 are in cascade (Rule 1)

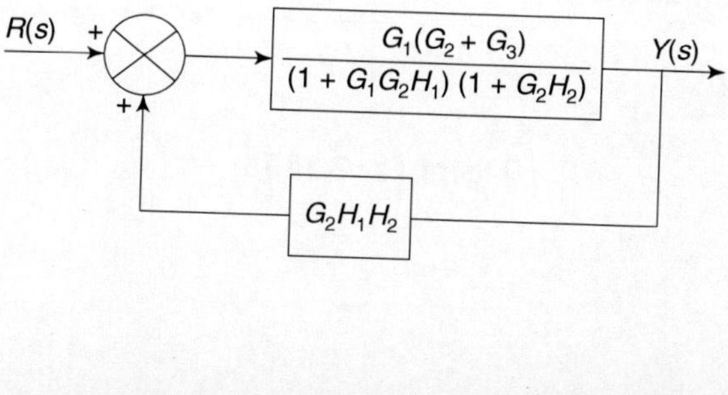

$$\frac{Y(s)}{R(s)} = \frac{\dfrac{G_1 (G_2 + G_3)}{(1 + G_1 G_2 H_1)(1 + G_2 H_2)}}{1 + \dfrac{G_1 (G_2 + G_3)}{(1 + G_1 G_2 H_1)(1 + G_2 H_2)} \times G_2 H_1 H_2}$$

$$\frac{Y(s)}{R(s)} = \frac{G_1 (G_2 + G_3)}{1(1 + G_1 G_2 H_1)(1 + G_2 H_2) + G_1 (G_2 + G_3) G_2 H_1 H_2} \quad \textbf{Ans.}$$

Example 2.10 Find the transfer function for the block diagram shown in Fig. Ex. 2.10

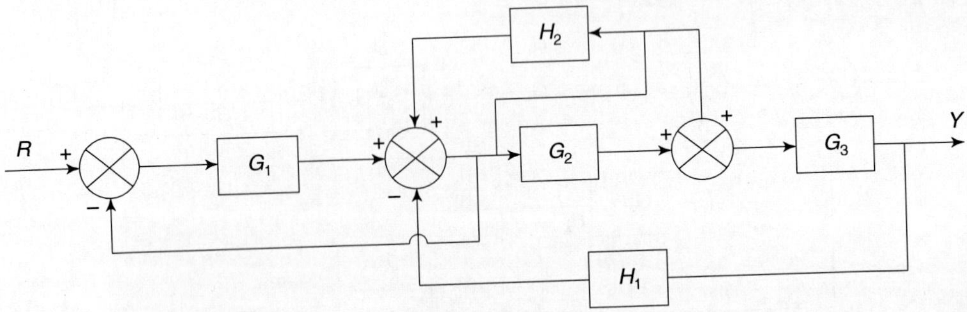

Figure Ex. 2.10

Solution Redraw Fig. Ex. 2.10 as shown in Fig. Ex. 2.10 (a)

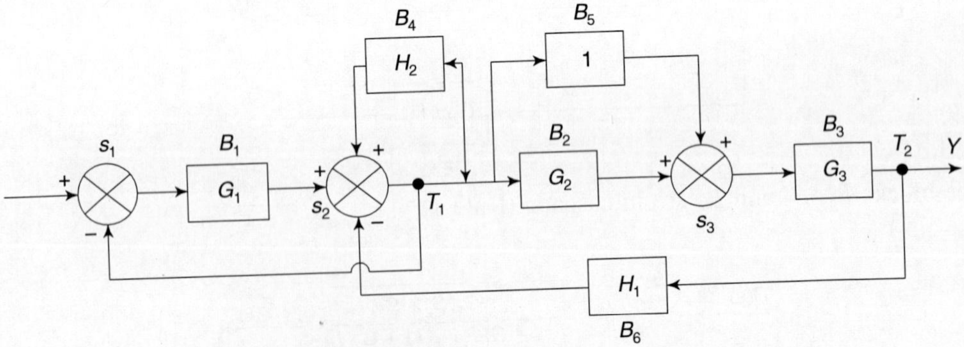

Figure Ex. 2.10 (a)

Eliminating the feedback loop, block B_4 and blocks B_5 and B_2 are in parallel (Rule 2)

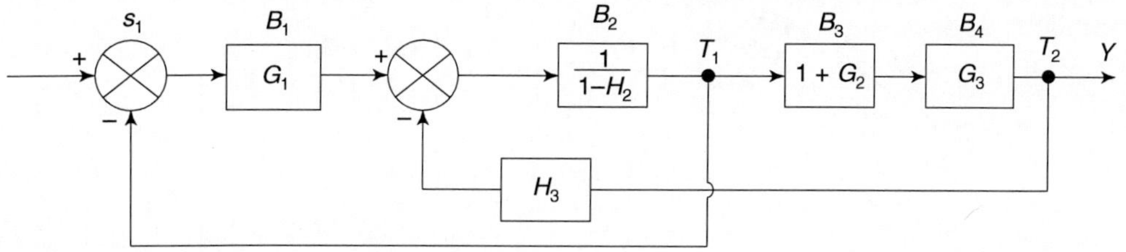

The blocks B_3 and B_4 are in cascade (Rule 1)

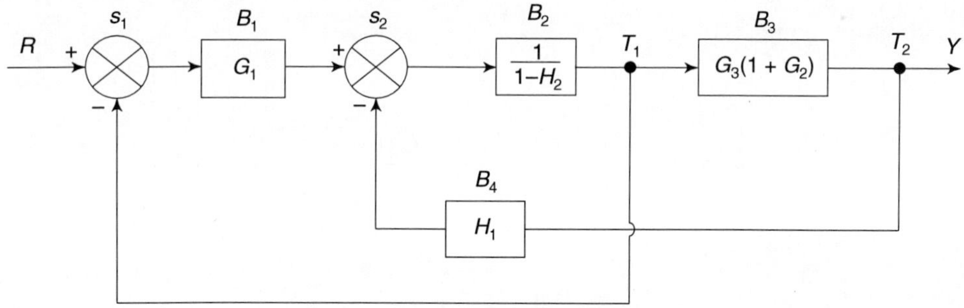

Shifting the take-off point T_1 after the block B_3 (Rule 7)

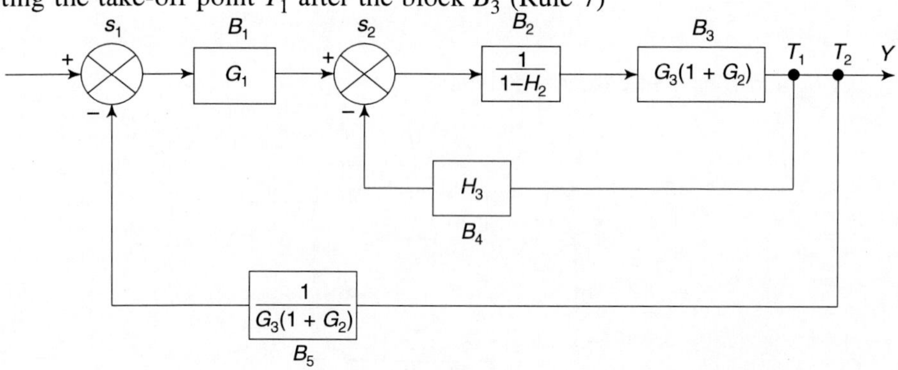

The blocks B_2 and B_3 are in cascade (Rule 1)

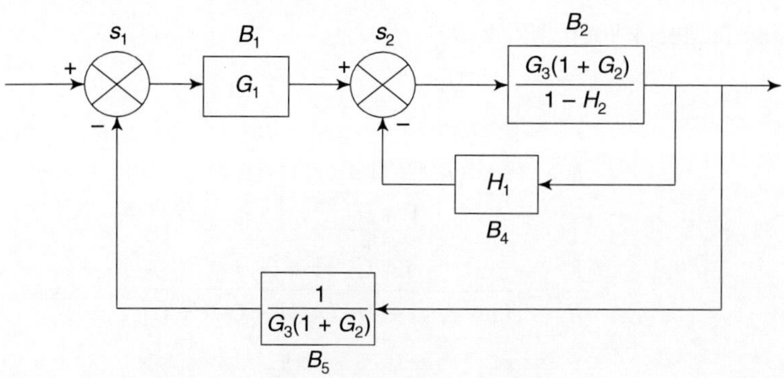

Shifting summing point s_2 before the block B_1 (Rule 6)

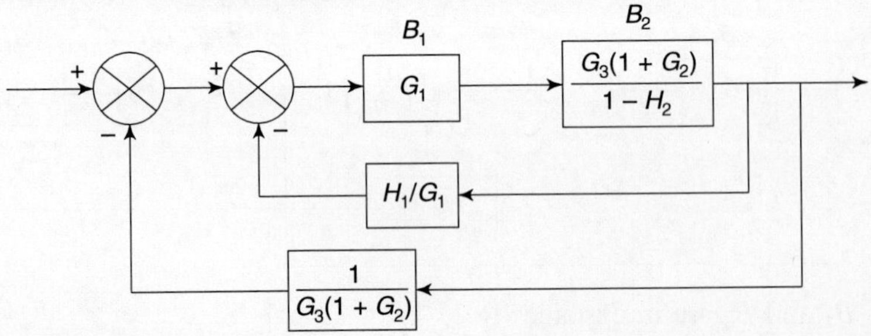

Redrawing the figure

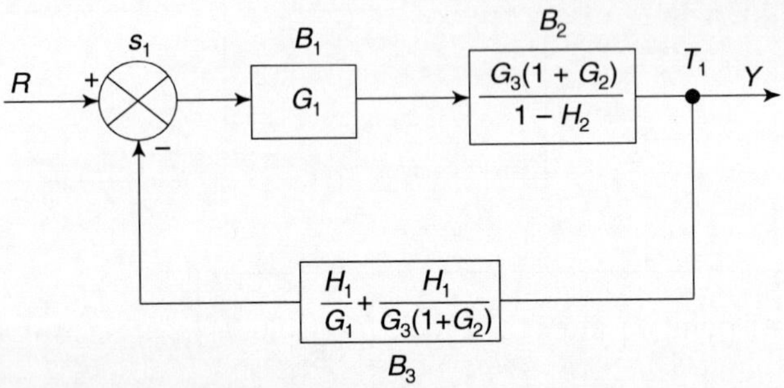

The blocks B_1 and B_2 are in cascade (Rule 1)

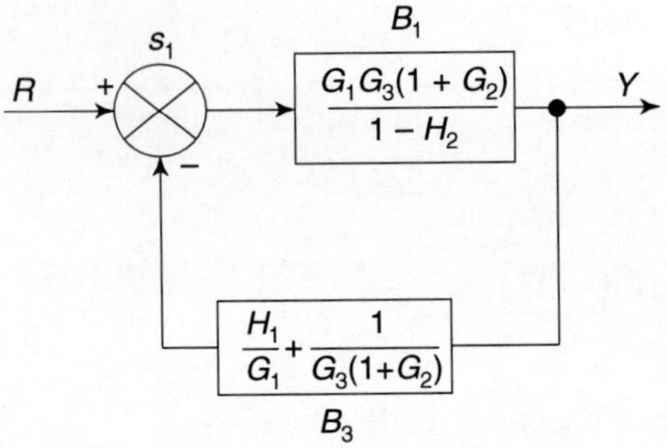

Eliminating the feedback loop, block B_2

$$\frac{Y}{R} = \frac{\dfrac{G_1\, G_3\,(1+G_2)}{1-H_2}}{1+\dfrac{G_1\, G_3\,(1+G_2)}{1-H_2}\left[\dfrac{H_1}{G_1}+\dfrac{1}{G_3\,(1+G_2)}\right]}$$

$$\frac{Y}{R} = \frac{G_1\, G_3\,(1+G_2)}{(1-H_2)+H_1\, G_3\,(1+G_2)+G_1} \qquad \textbf{Ans.}$$

Signal Flow Graph

Signal flow graph method (Mason's gain formula)

Example 2.11 Obtain the closed loop transfer function for the control system shown in Fig. Ex. 2.11.

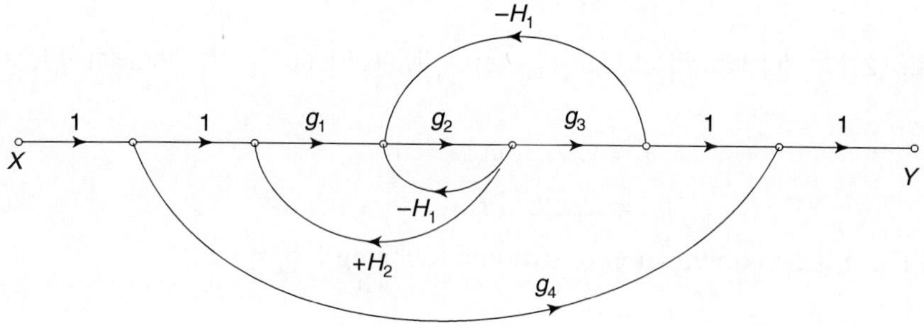

Figure Ex. 2.11

Solution **Step 1** There are two forward paths

$$P_1 = g_1\, g_2\, g_3 \text{ and } P_2 = g_4$$

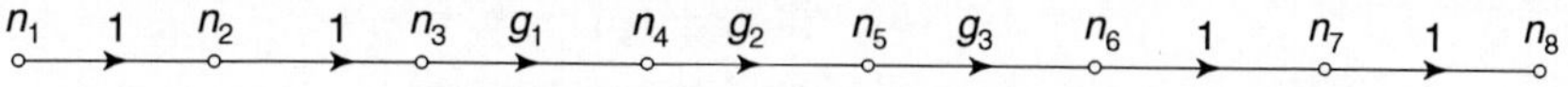

Figure Ex. 2.11 (a) Forward path P_1

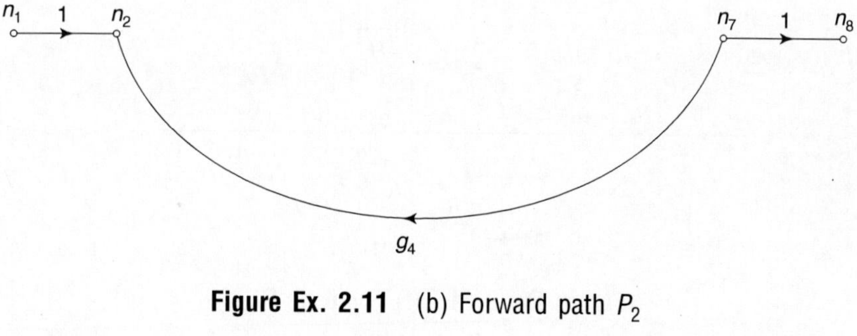

Figure Ex. 2.11 (b) Forward path P_2

Step 2 There are three individual loops with loop gain.

$$L_{11} = -\, g_2\, H_1$$
$$L_{21} = -\, g_2\, g_3\, H_1$$
$$L_{31} = -\, g_1\, g_2\, H_2$$

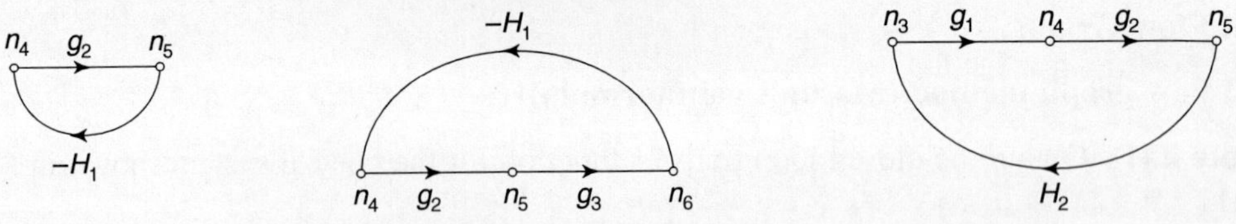

(c) (d) (e)

Figure Ex. 2.11 (c) Individual loop L_{11}, (d) Individual loop L_{21}, (e) Individual loop L_{31}

Here $$j = 1, 2, 3 \text{ and } r = 1$$

i.e., $$L_{jr} = L_{11}, L_{21}, L_{31}$$

Step 3 There is no combination of two non-touching loops, i.e.,

$$L_{j2} = 0$$

Step 4 There is no combination of three, four .... non touching loop i.e.,

$$L_{j3} = L_{j4} = L_{j5} = \ldots = 0$$

Step 5 Now by Eq. (2.9)

$$\Delta = \Sigma L_{j1} + \Sigma L_{j2} - \Sigma L_{j3} + \ldots$$

$$= 1 - (L_{11} + L_{21} + L_{31}) + 0 + 0$$

$$= 1 - (- g_2 H_1 - g_2 g_3 H_1 + g_1 g_2 H_2)$$

$$\Delta = 1 + g_2 H_1 + g_2 g_3 H_1 - g_1 g_2 H_2$$

Step 6 Removing the loop gains from Δ touching P_1 forward path like

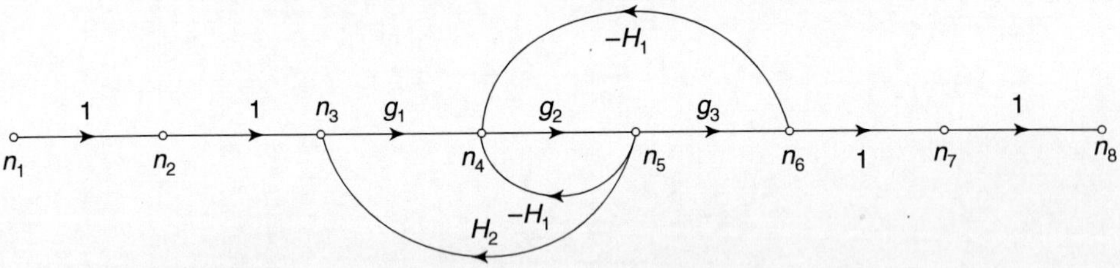

Figure Ex. 2.11 (f) For path P_1

i.e., all the loop gains touching the forward path P_1 so

$$\Delta_i = 1$$

$\because$ $$i = 1$$

$$\Delta_1 = 1$$

Removing the loop gains from Δ touching P_2 forward path like

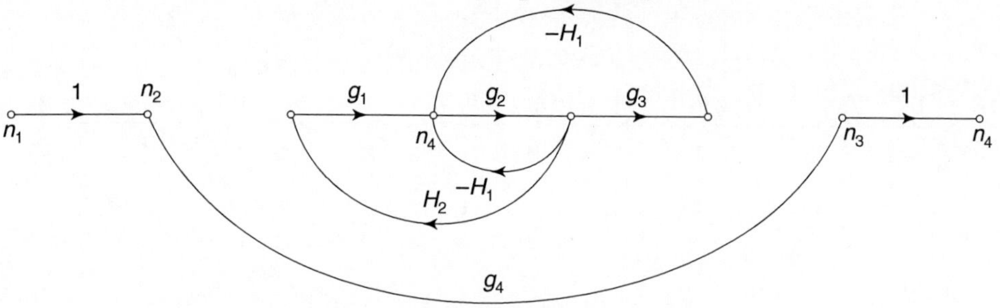

Figure Ex. 2.11 (g) For path P_2

Now $\quad \Delta_i = 1 + g_2 H_1 = g_2 g_3 H_1 - g_1 g_2 H_2$

∵ $\quad i = 2$

$$\Delta_2 = 1 + g_2 H_1 + g_2 g_3 H_1 - g_1 g_2 H_2$$

Now from Eq. (2.9)

$$T_o = \frac{1}{\Delta}\sum_i P_i \Delta_i = \frac{P_1 \Delta_1 + P_2 \Delta_2}{\Delta}$$

$$T_o = \frac{g_1 g_2 g_3 + g_4 (1 + g_2 H_1 + g_2 g_3 H_1 - g_1 g_2 H_2}{1 + g_2 H_1 + g_2 g_3 H_1 - g_1 g_2 H_2} \quad \textbf{Ans.}$$

Example 2.12 Find overall transfer function for the control system signal flow graph given in Fig. Ex. 2.12

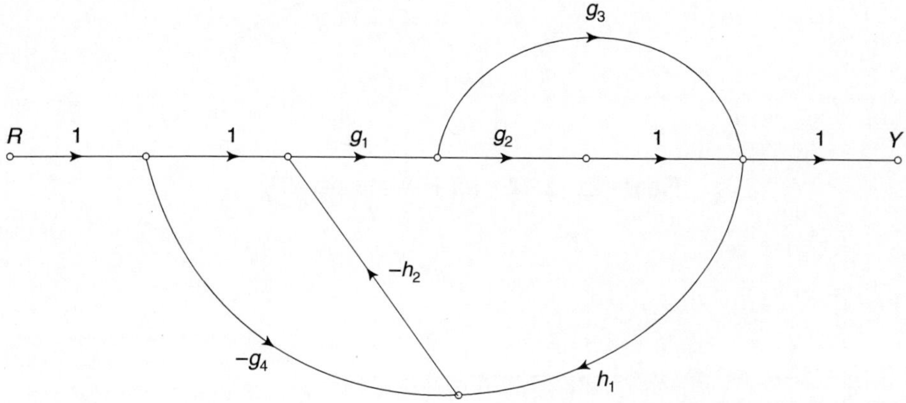

Figure Ex. 2.12

Solution Considering Fig. Ex. 2.12 as in Fig. Ex. 2.12 (a)

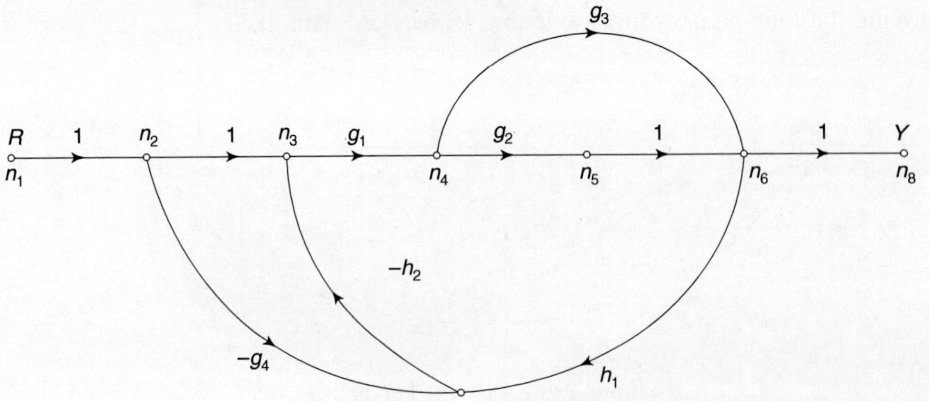

Figure Ex. 2.12 (a)

Step 1 There are four forward paths, i.e., $i = 1, 2, 3, 4$

$P_i = P_1, P_2, P_3, P_4$

$P_1 = g_1 g_2$

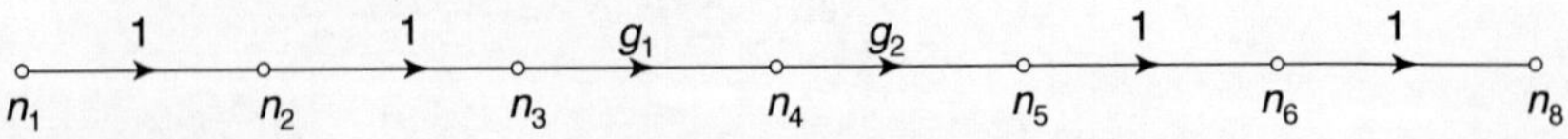

Figure Ex. 2.12 (b) Forward path P_1

$P_2 = g_1 g_3$

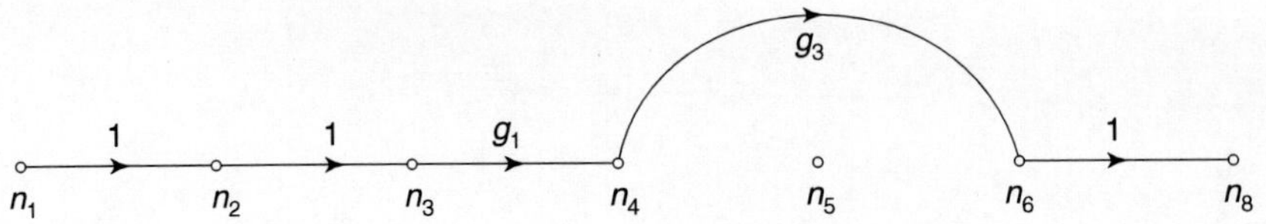

Figure Ex. 2.12 (c) Forward path P_2

$P_3 = g_4 h_2 g_1 g_2$

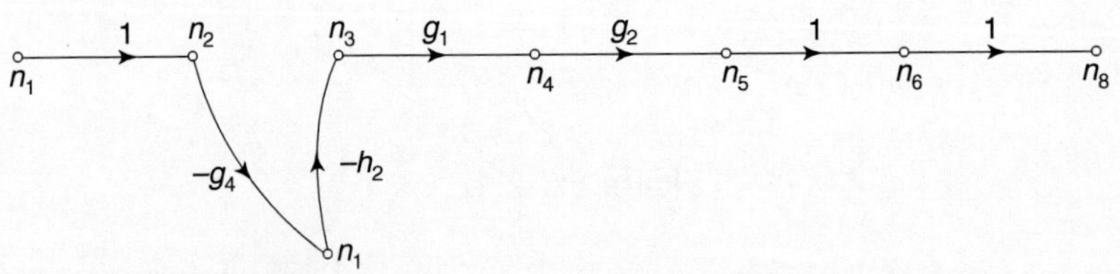

Figure Ex. 2.12 (d) Forward path P_3

$$P_4 = g_4\, h_2\, g_1\, g_3$$

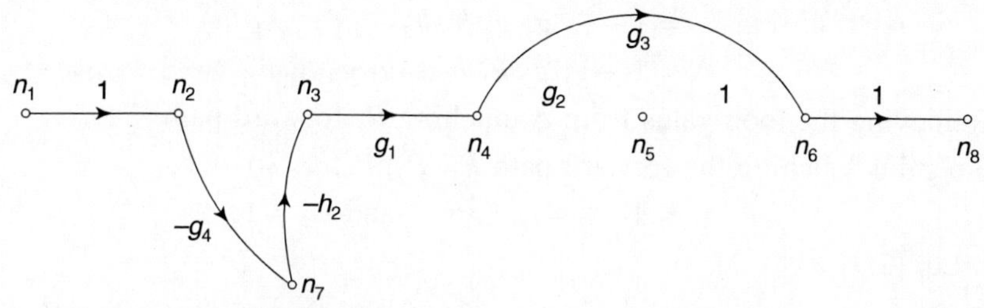

Figure Ex. 2.12 (e) Forward path P_4

Step 2 There are two individual loops, i.e.,

$$j = 1, 2 \text{ and } r = 1$$
$$L_{jr} = L_{11}, L_{21}$$
$$L_{11} = -\, g_1\, g_2\, h_1\, h_2$$

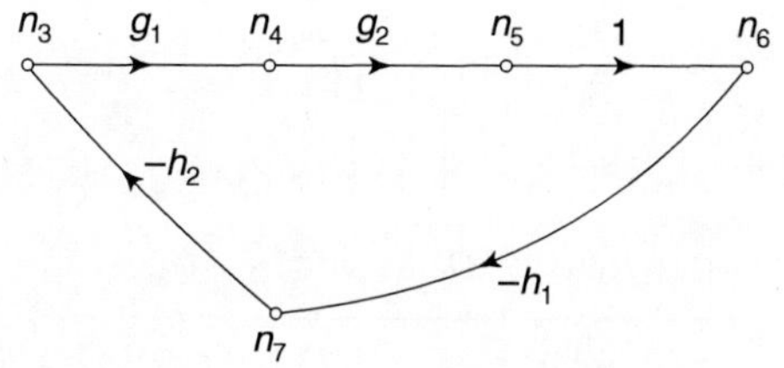

Figure Ex. 2.12 (f) Indivisual loop L_{11}

$$L_{21} = -\, g_1\, g_3\, h_1\, h_2$$

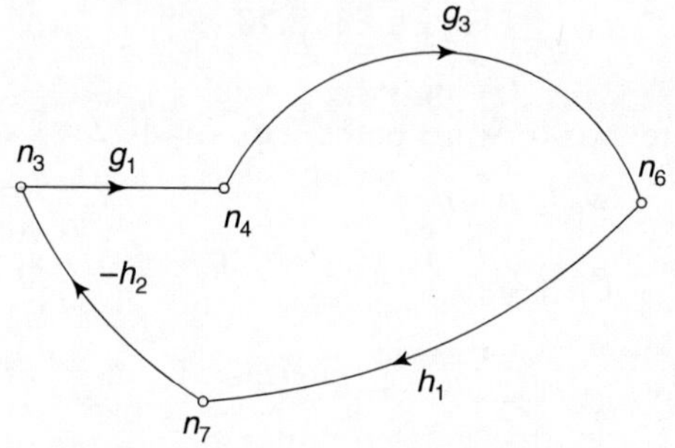

Figure Ex. 2.12 (g) Indivisual loop L_{21}

Step 3 and 4 There is no combination of two, three, four non-touching loops, i.e.,

$$L_{j2} = L_{j3} = L_{j4} = \ldots = 0$$

Step 5 Now by Eq. (2.10)

$$\Delta = 1 - \Sigma\, L_{j1} + \Sigma\, L_{j2} - \Sigma\, L_{j3} + \ldots$$

$$= 1 - (L_{11} + L_{21}) + 0 - 0 \ldots$$
$$= 1 - (- g_1\, g_2\, h_1\, h_2 - g_1\, g_3\, h_1\, h_2)$$
$$\Delta = 1 + g_1\, g_2\, h_1\, h_2 + g_1\, g_3\, h_1\, h_2$$

Step 6 Removing the loop gains from Δ touching P_1 forward path

All the loop gains touching the forward path P_1, P_2, P_3, P_4 so

$$\Delta_1 = 1,\ \Delta_2 = 1,\ \Delta_3 = 1 \text{ and } \Delta_4 = 1$$

Now from Eq. (2.9)

$$T_o = \frac{1}{\Delta}\sum_j P_i \Delta_i = \frac{g_1\, g_2 + g_1\, g_3 + g_4\, h_2\, g_1\, g_2 + g_4\, h_2\, g_1\, g_3}{1 + g_1\, g_2\, h_1\, h_2 + g_1\, g_3\, h_1\, h_2}$$

$$T_o = \frac{\mathbf{g_1\,(g_2 + g_3) + g_4\, h_3\, g_1\,(g_2 + g_3)}}{\mathbf{1 + g_1\, h_1\, h_2\,(g_2 + g_3)}} \quad \textbf{Ans.}$$

Example 2.13 From the signal flow graph given in Fig. Ex. 2.13, find the transfer function using Mason's gain formula. *(U.P.T.U., 2007–08)*

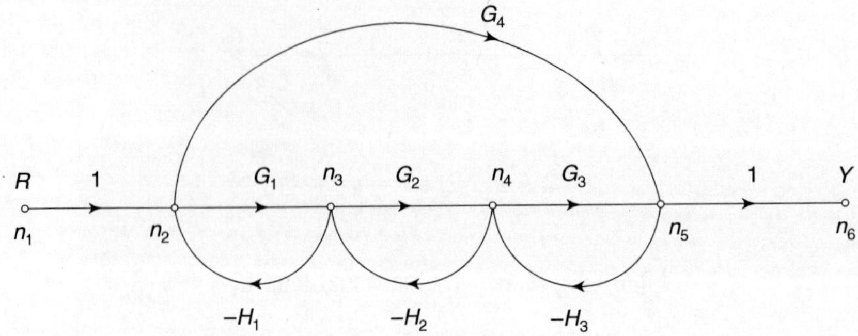

Figure Ex. 2.13

Solution **Step 1** There are two forward paths, i.e., $i = 1, 2$

$$P_i = P_1,\ P_2$$
$$P_1 = G_4$$

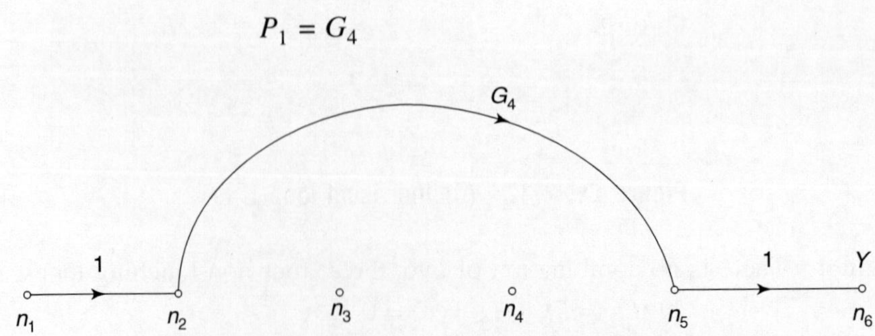

Figure Ex. 2.13 (a) Forward path P_1

$$P_2 = G_1\, G_2\, G_3$$

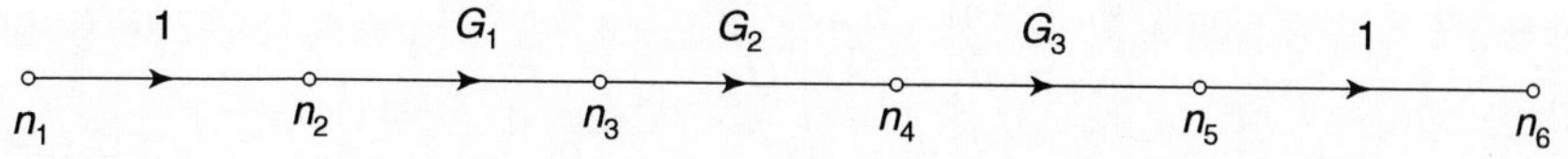

Figure Ex. 2.13 (b) Forward path P_2

Step 2 There are four individual loops

i.e., $j = 1, 2, 3, 4$, and $r = 1$

$$L_{jr} = L_{11}, L_{21}, L_{31}, L_{41}$$

$$L_{11} = -\, G_1\, H_1$$

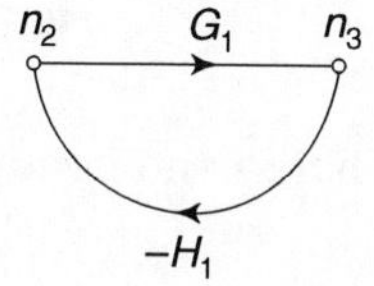

Figure Ex. 2.13 (c) Individual loop L_{11}

$$L_{21} = G_2\, H_2$$

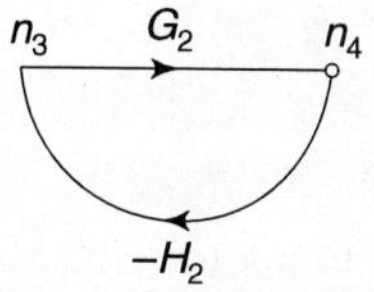

Figure Ex. 2.13 (d) Individual loop L_{21}

$$L_{31} = -\, G_3\, H_3$$

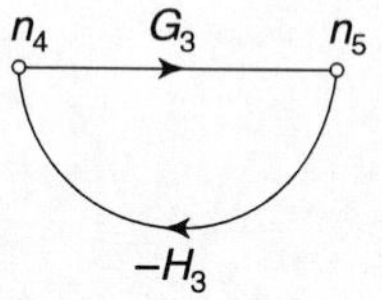

Figure Ex. 2.13 (e) Individual loop L_{31}

$$L_{41} = -\, G_4\, H_3\, H_2\, H_1$$

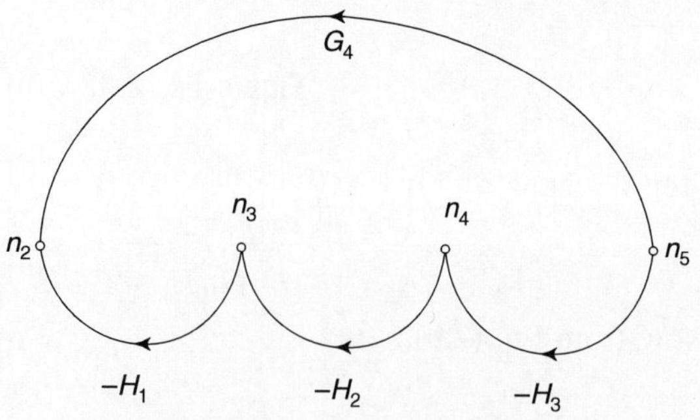

Figure Ex. 2.13 (f) Individual loop L_{41}

Step 3 There is one combination of two non-touching loops $j = 1$, $r = 2$.

$$L_{jr} = L_{12}$$
$$L_{12} = G_1 H_1 G_3 H_3$$

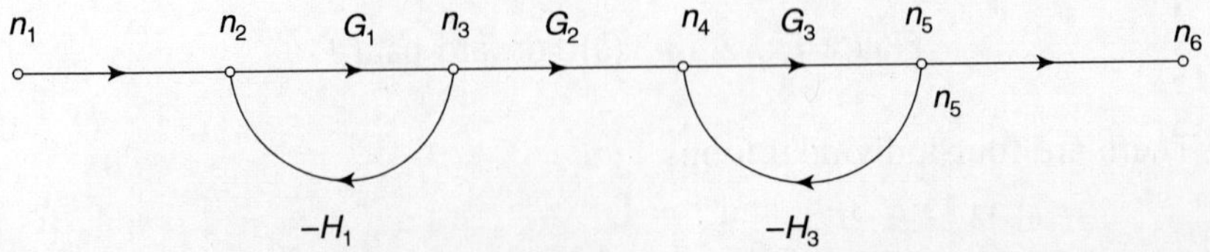

Figure Ex. 2.13 (g) Non-touching loop L_{12}

Step 4 There is no combination of three, four ... non-touching loops.

$$L_{j3} = L_{j4} = \ldots = 0$$

Step 5 Removing the loop gains from Δ touching P_1, P_2 forward path for path P_1

The loop $G_1 H_1$ and $G_3 H_3$ is touching the path P_1 so removing these from Δ

i.e.,
$$\Delta_1 = 1 + G_2 H_2$$

for path P_2

all the loop gains touching the forward path P_2 so

$$\Delta_2 = 1$$

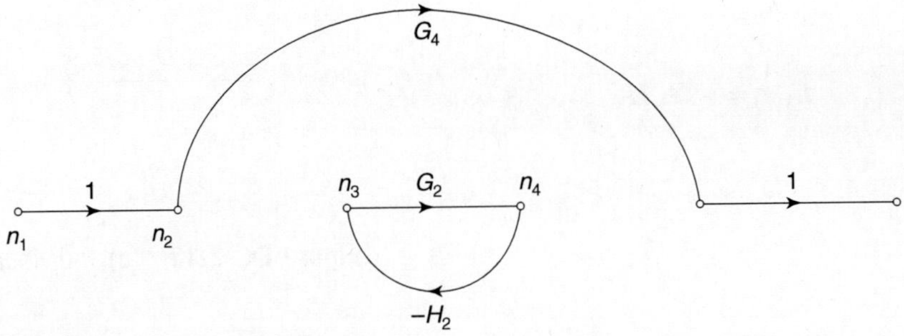

Figure Ex. 2.13 (h) For path P_1

Step 6 Now by Eq. (2.10)

$$\Delta = 1 - \Sigma L_{j1} + \Sigma L_{j2} - \Sigma L_{j3} + \ldots$$
$$\Delta = 1 - (-G_1 H_1 - G_2 H_2 - G_3 H_3 - G_4 H_3 H_2 H_1) + (G_1 H_1 G_3 H_3)$$

Now from Eq. (2.9)

$$T_o = \frac{1}{\Delta}\sum_j P_i \Delta_i$$

$$T_o = \frac{P_1 \Delta_1 + P_2 \Delta_2}{\Delta}$$

$$T_o = \frac{G_4 (1 + G_2 H_2) + G_1 G_2 G_3 (1)}{1 + G_1 H_1 + G_1 H_2 + G_3 H_3 + G_4 H_3 H_2 H_1 + G_1 H_1 G_3 H_3}$$

Ans.

Example 2.14 Find the transfer function for the control system shown in Fig. Ex. 2.14.

(U.P.T.U., 2006–07)

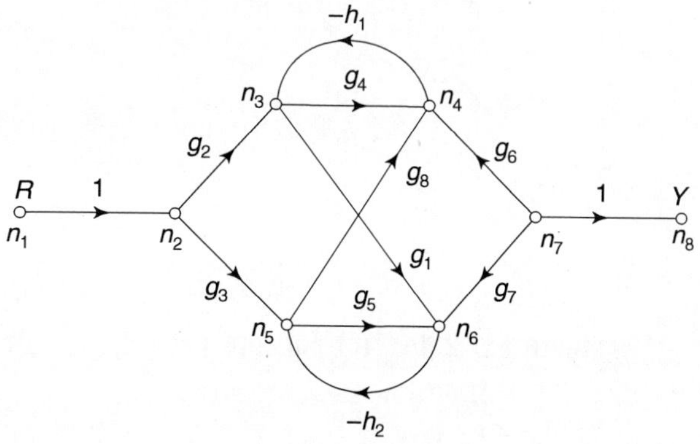

Figure Ex. 2.14

Solution **Step 1** There are six forward paths, i.e.,

$$i = 1, 2, 3, 4, 5, 6$$

$$P_j = P_1, P_2, P_3, P_4, P_5, P_6$$

So the path gains are

$$P_1 = g_2\ g_4\ g_6$$

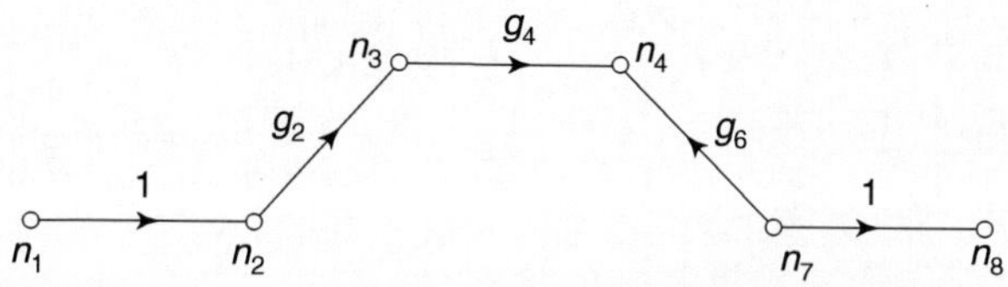

Figure Ex. 2.24 (a) Forward path P_1

$$P_2 = g_3\, g_5\, g_7$$

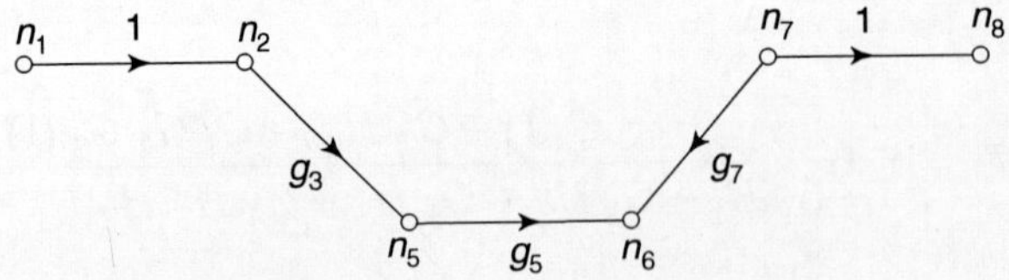

Fiure Ex. 2.14 (b) Forward path P_2

$$P_{3.} = g_2\, g_1\, g_7$$

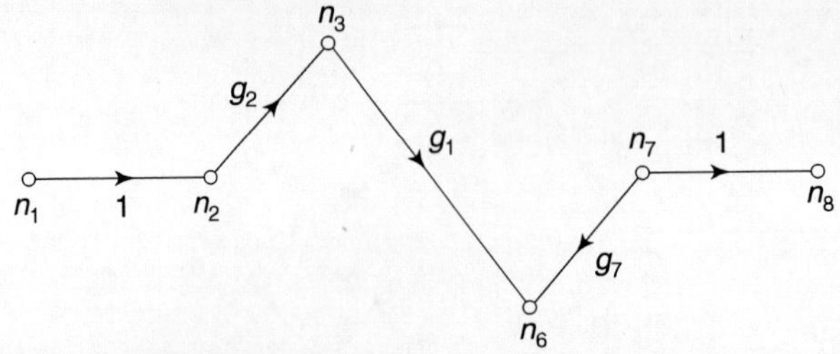

Figure Ex. 2.14 (c) Forward path P_3

$$P_4 = g_3\, g_8\, g_6$$

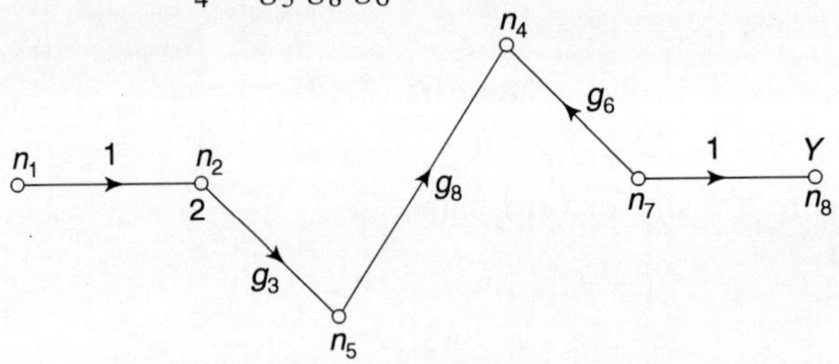

Figure Ex. 2.14 (d) Forward path P_4

$$P_5 = -\, g_2\, g_1\, h_2\, g_8\, g_6$$

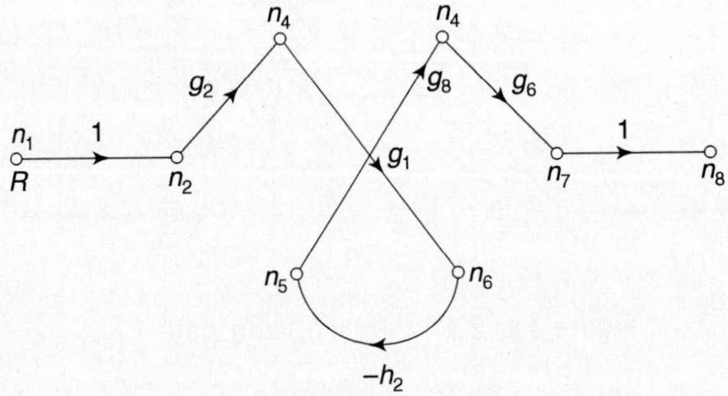

Figure Ex. 2.14 (e) Forward path P_5

$$P_6 = -\ g_3\ g_8\ h_1\ g_1\ g_7$$

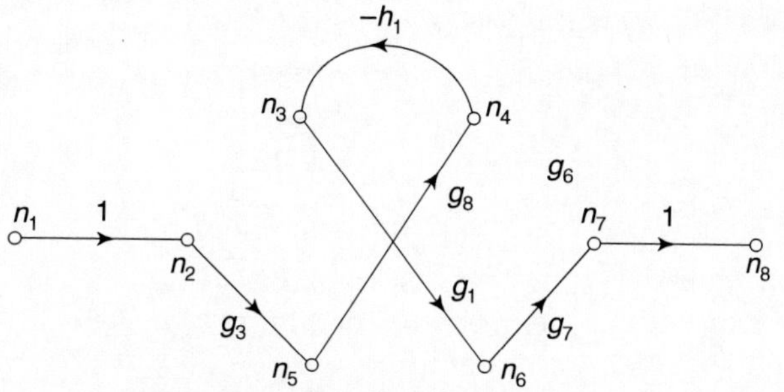

Figure Ex. 2.14 (f) Forward path P_6

Step 2 There are three individual loops, i.e.,

$$j = 1, 2, 3 \text{ and } r = 1$$

$$L_{jr} = L_{11}, L_{22}, L_{31}$$

$$L_{11} = -\ g_4\ h_1$$

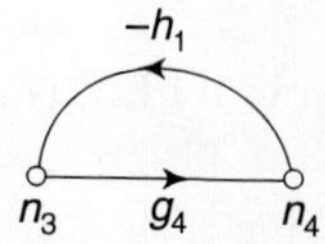

Figure Ex. 2.14 (g) Individual loop L_{11}

$$L_{21} = -\ g_5\ h_2$$

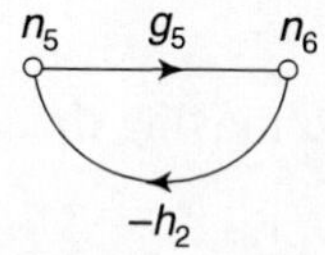

Figure Ex. 2.14 (h) Individual loop L_{21}

$$L_{31} = g_1\ h_2\ g_8\ h_1$$

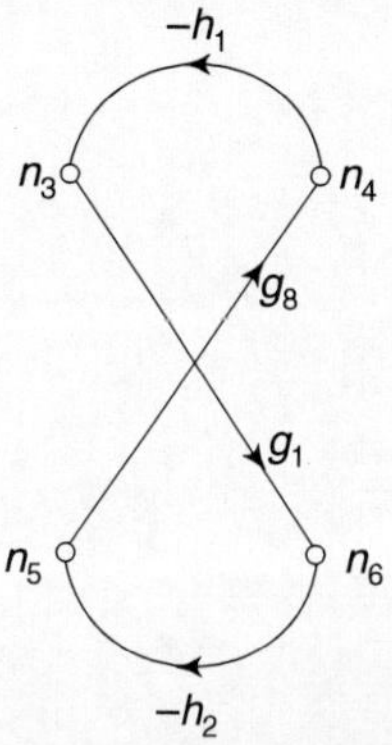

Figure Ex. 2.14 (i) Individual loop L_{31}

Step 3 There is one combination of two non-touching loop, i.e., $j = 1$, $r = 2$ so the gain product is

$$L_{jr} = L_{12}$$

$L_{12} = g_4\, h_1\, g_5\, h_2$

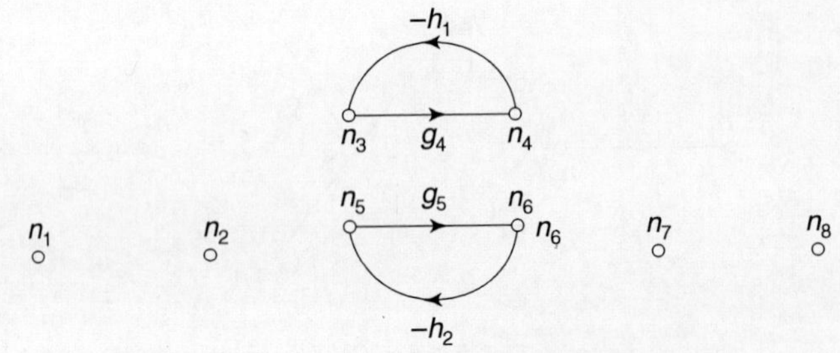

Figure Ex. 2.14 (j) Non-touching loop L_{12}

Step 4 There is no combination of three, four ... non-touching loop

$$L_{j3} = L_{j4} = 0$$

Step 5 Now by Eq. (2.10)

$$\begin{aligned}\Delta &= 1 - \Sigma L_{j1} + \Sigma L_{j2} - \Sigma L_{j2} + \ldots \\ &= 1 - (L_{11} + L_{21} + L_{31}) + (L_{12}) - (0) \\ &= 1 - (-\, g_4\, h_1 - g_5\, h_2 + g_1\, h_2\, g_8\, h_1) + (g_4\, h_1\, g_5\, h_2) \\ &= 1 + g_4\, h_1 + g_5\, h_2 - g_1\, h_2\, g_8\, h_1 + g_4\, h_1\, g_5\, h_2\end{aligned}$$

Step 6 Removing the loop gains from Δ touching P_1, P_2, P_3, P_4, P_5, P_6 forward path for path P_1

$$\Delta_1 = 1 + g_5\, h_2$$

for path P_2

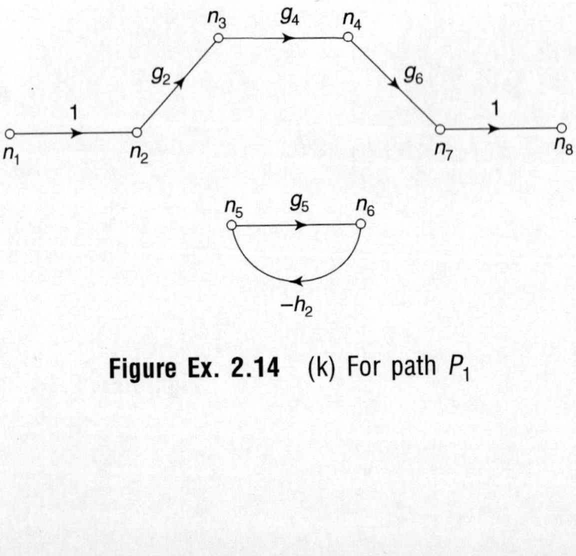

Figure Ex. 2.14 (k) For path P_1

$$\Delta_2 = 1 + g_4 h_1$$

for path P_3, P_4, P_5 and P_6

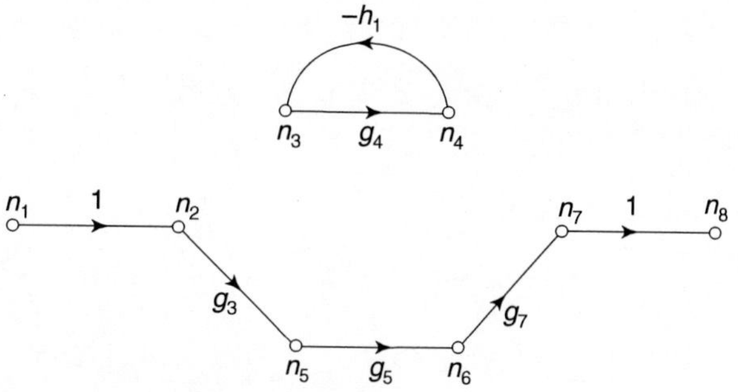

Figure Ex. 2.14 (I) For path P_2

all the loop touching the above path so $\Delta_3 = \Delta_4 = \Delta_5 = \Delta_6 = 1$

Now from Eq. (2.9)

$$T_o = \frac{1}{\Delta} \Sigma P_i \Delta_i$$

$$T_o = \frac{P_1 \Delta_1 + P_2 \Delta_2 + P_3 \Delta_3 + P_4 \Delta_4 + P_5 \Delta_5 + P_6 \Delta_6}{\Delta}$$

$$T_o = \frac{g_2 g_4 g_6 (1 + g_5 h_2) + g_3 g_5 g_7 (1 + g_4 h_2) + g_2 g_1 g_7 + g_3 g_8 g_6 - g_2 g_1 h_2 g_8 g_6 - g_3 g_8 h_1 g_1 g_7}{1 + g_4 h_1 + g_5 h_2 - g_1 h_2 g_8 h_1 + g_4 h_1 g_5 h_2}$$

Ans.

Example 2.15 Find the overall transfer function of the block diagram shown in Fig. Ex. 2.15 by signal flow graph techniques.

(U.P.T.U., C.O. 2003, ACS), (KNIT Sultanpur 99–2001)

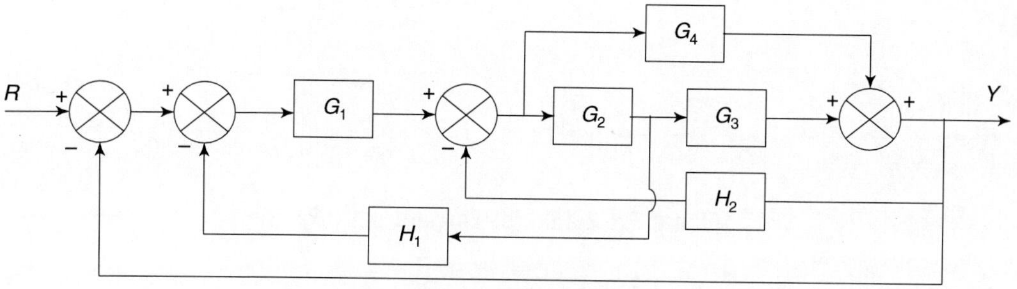

Figure Ex. 2.15

Solution The signal flow graph is shown in Fig. Ex. 2.15 (a).

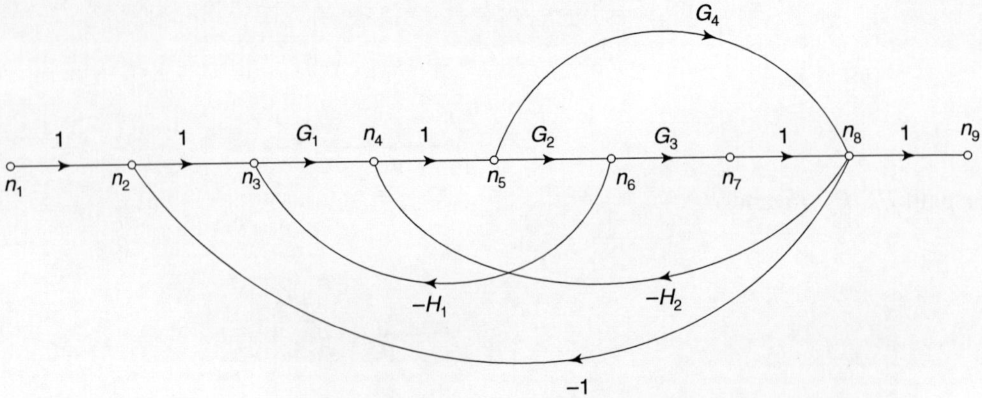

Figure Ex. 2.15 (a)

Step 1 There are two forward paths, i.e., $i = 1, 2$

$$P_i = P_1, P_2$$

so the path gains are $P_1 = G_1 \, G_2 \, G_3$

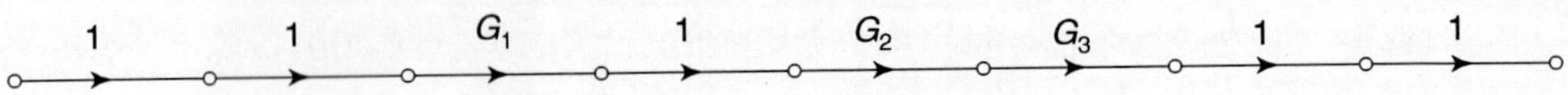

Figure Ex. 2.15 (b) Forward path P_1

$$P_2 = G_1 \, G_4$$

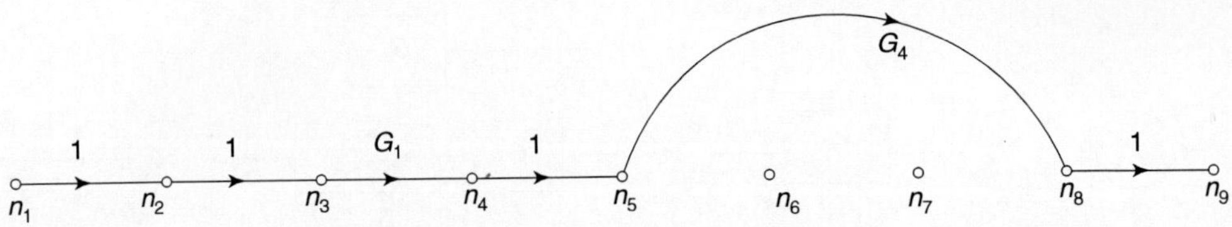

Figure Ex. 2.15 (c) Forward path P_1

Step 2 There are five individual loops, i.e.,

$$j = 1, 2, 3, 4, 5 \text{ and } r = 1$$

$$L_{jr} = L_{11}, L_{21}, L_{31}, L_{41}, L_{51}$$

$$L_{11} = - G_1 \, G_2 \, H_1$$

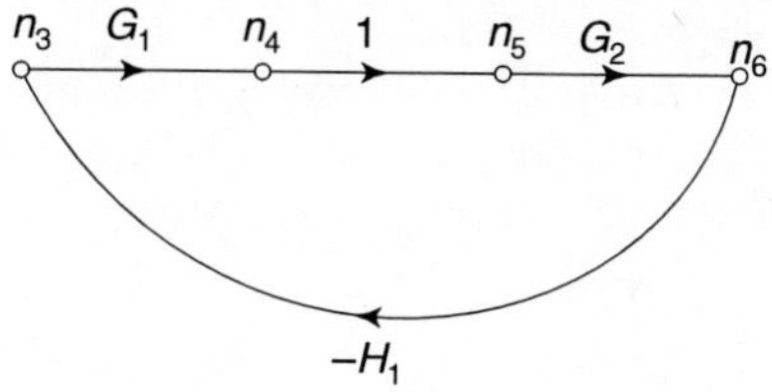

Figure Ex. 2.15 (d) Individual loop L_{11}

$$L_{21} = -\,G_1\,G_2\,G_3$$

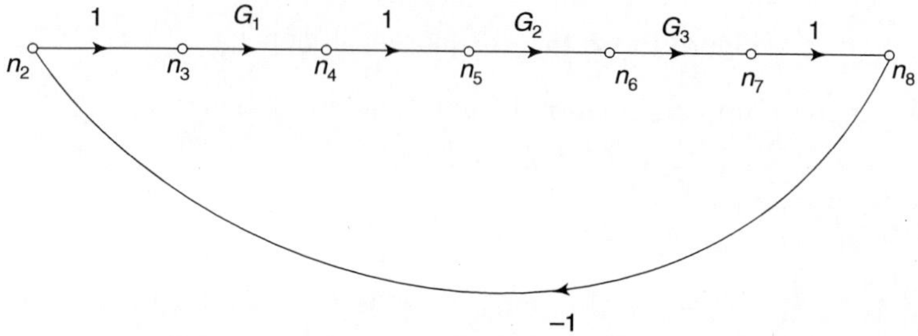

Figure Ex. 2.15 (e) Individual loop L_{21}

$$L_{31} = -\,G_2\,G_3\,H_2$$

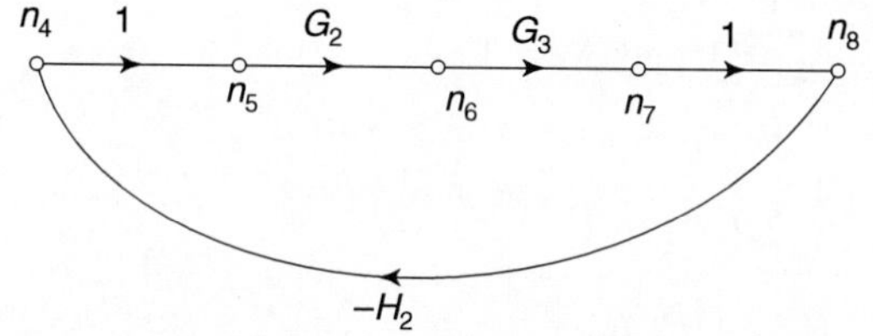

Figure Ex. 2.15 (f) Individual loop L_{31}

$$L_{41} = -\,G_4\,G_4$$

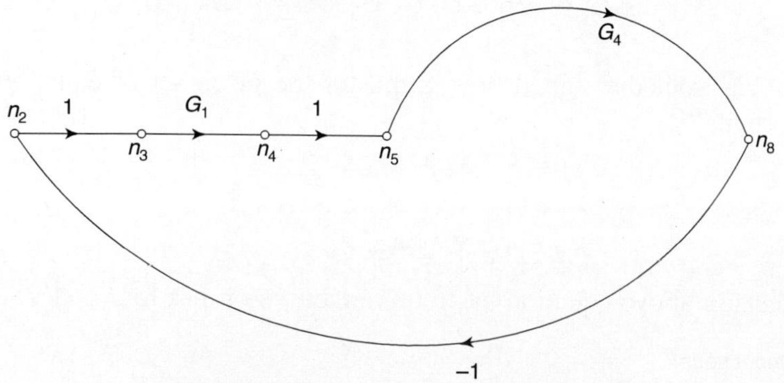

Figure Ex. 2.15 (g) Individual loop L_{41}

$$L_{51} = -G_4 H_2$$

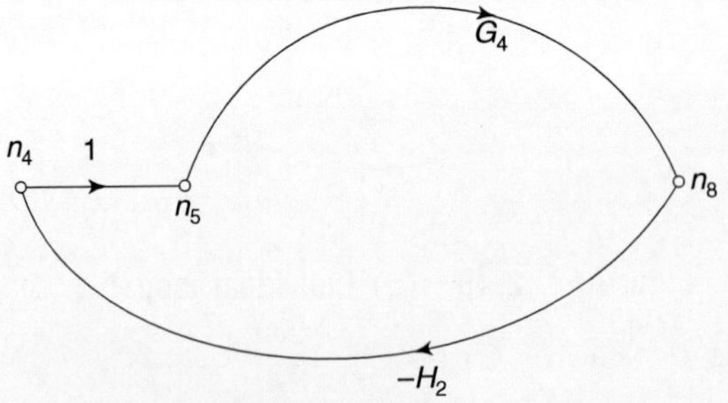

Figure Ex. 2.15 (h) Individual loop L_{51}

Step 3 and 4 There is no combination of non-touching loops, i.e.,

$$L_{j2} = L_{j3} = L_{j4} = \ldots = 0$$

Step 5 Now by Eq. (2.10)

$$\Delta = 1 - \Sigma L_{j1} + \Sigma L_{j2} - \Sigma L_{j3} + \ldots$$
$$= 1 - (L_{11} + L_{21} + L_{31} + L_{41} + L_{51}) + 0 + 0 \ldots$$
$$= 1 - (-G_1 G_2 H_1 - G_1 G_2 G_3 - G_2 G_3 H_2 - G_1 G_4 - G_4 H_2,$$
$$\Delta = 1 + G_1 G_2 H_1 + G_1 G_2 G_3 + G_2 G_3 H_2 + G_1 G_4 + G_4 H_2$$

Step 6 Removing the loop gains from Δ touching P_1, P_2 forward path

$\because$ all the loops touching path P_1 and P_2 so

$$\Delta_1 = 1 \text{ and } \Delta_2 = 1$$

Now from Eq. (2.9)

$$T_o = \frac{1}{\Delta} \Sigma P_i \Delta_i$$

$$T_o = \frac{P_1 \Delta_1 + P_2 \Delta_2}{\Delta}$$

$$T_o = \frac{G_1 G_2 G_3 + G_1 G_4}{1 + G_1 G_2 H_1 + G_1 G_2 G_3 + G_2 G_3 H_2 + G_1 G_4 + G_4 H_2}$$ **Ans.**

Example 2.16 Construct the signed flow graph for the given set of equations and define Mason's gain formula. *(U.P.T.U., 2006–07)*

$$X_2 = A_{21} X_1 + A_{23} X_3 \quad \ldots\text{(i)}$$
$$X_3 = A_{31} X_1 + A_{32} X_2 + A_{33} X_3 \quad \ldots\text{(ii)}$$
$$X_4 = A_{42} X_2 + A_{43} X_3 \quad \ldots\text{(iii)}$$

Solution 1. Because above equation has four variables so it has four nodes

Figure Ex. 2.16 (a)

2. Now draw the signal flow graph for Eq. (i) in which node X_2 is the sum of two signals like $A_{21}\ X_1 + A_{23}\ X_3$

$$(1) = A_{21}\ X_1 \text{ and } (2) = A_{23}\ X_3$$

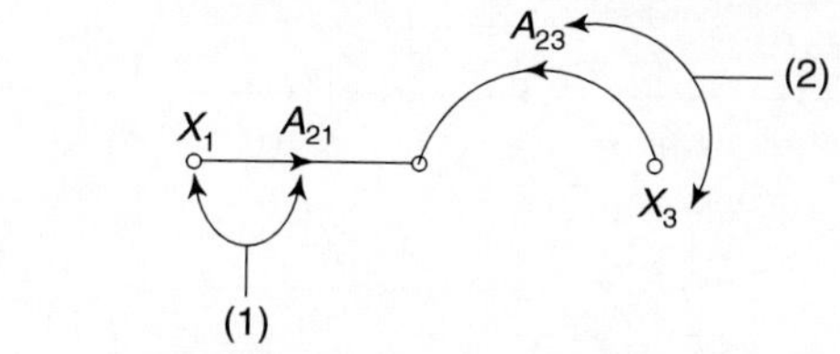

Figure Ex. 2.16 (b) SFG for Eq. (i)

3. Now draw the signal flow graph for equation (ii), in which node X_3 is the sum of three signals like $A_{31}\ X_1,\ A_{32}\ X_2,\ A_{33}\ X_3$

$$(1) = A_{31}\ X_1\ \ (2) = A_{32}\ X_2\ (3) = A_{33}\ X_3$$

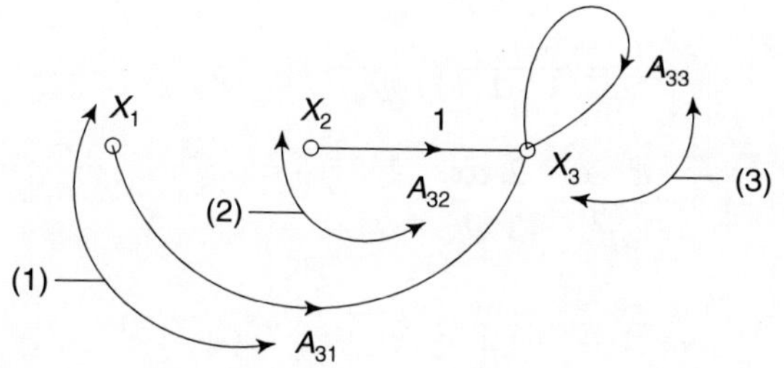

Figure Ex. 2.16 (c) SFG for Eq (ii)

4. Now draw the signal flow graph for Eq. (iii), in which node X_4 is the sum of two signals like $A_{42}\ X_2,\ A_{43}\ X_3$

$$(1) = A_{42}\ X_2\ \ (2) = A_{43}\ X_3$$

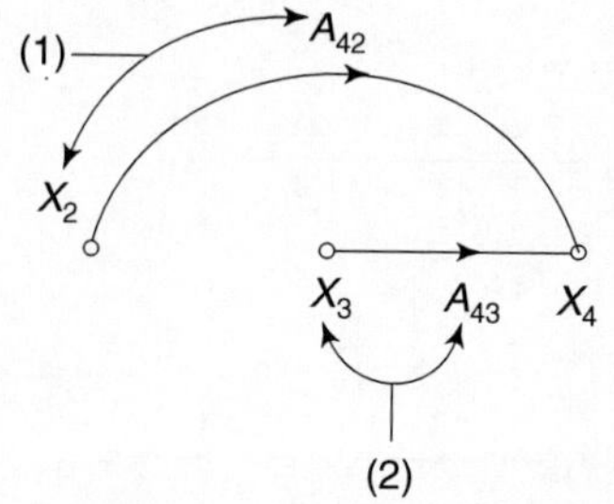

Figure Ex. 2.16 (d) SFG for Eq. (iii)

5. Now drawing the complete signal flow graph by combining graph in steps 2, 3, 4 is shown in Fig. Ex. 2.16.

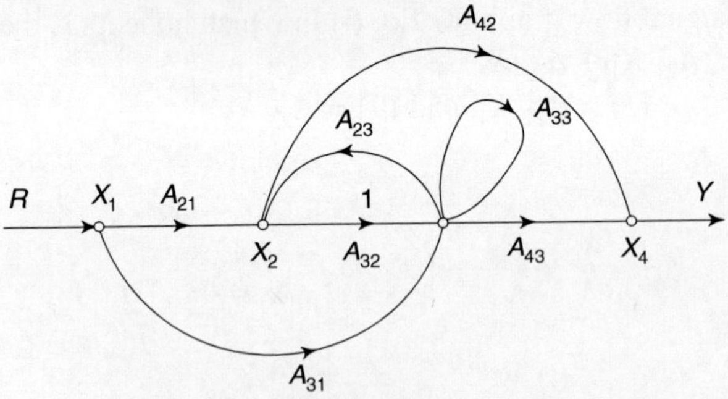

Figure Ex. 2.16 (e) Complete SFG

Example 2.17 Draw the signal flow graph from the following equation

$$X_2 = a_1 \frac{dx_1}{dt} \quad \text{...(i)}$$

$$X_3 = \frac{d^2 x_2}{dt^2} + \frac{dx_1}{dt} - x_1 \quad \text{...(ii)}$$

$$\frac{d^2 x}{dt^2} + \frac{2}{3}\frac{dx}{dt} + \frac{11}{2} x = t \quad \text{...(iii)}$$

Solution

$$x_2 = a_1 \frac{dx_1}{dt}$$

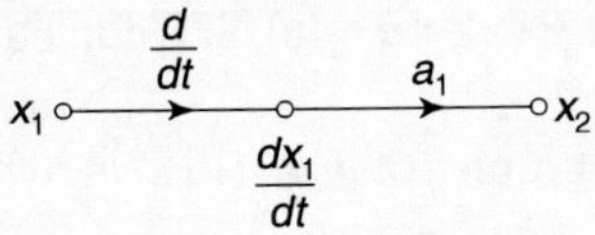

Figure Ex. 2.17 (a) SFG for Eq. (i)

$$x_3 = \frac{d^2 x_2}{dt^2} + \frac{dx_1}{dt} - x_1$$

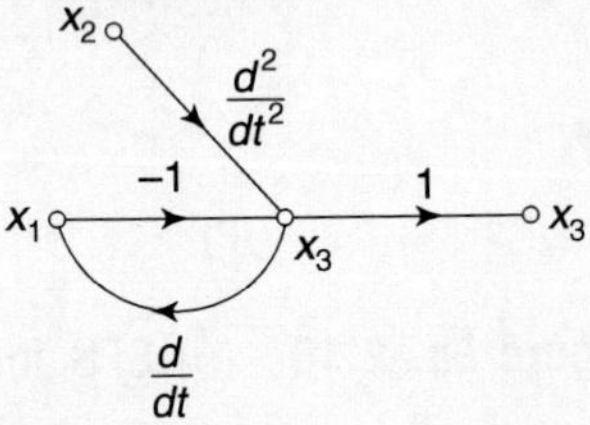

Figure Ex. 2.17 (b) SFG for Eq. (ii)

$$\frac{d^2 x}{dt^2}+\frac{2}{3}\frac{dx}{dt}+\frac{11}{2}x = t$$

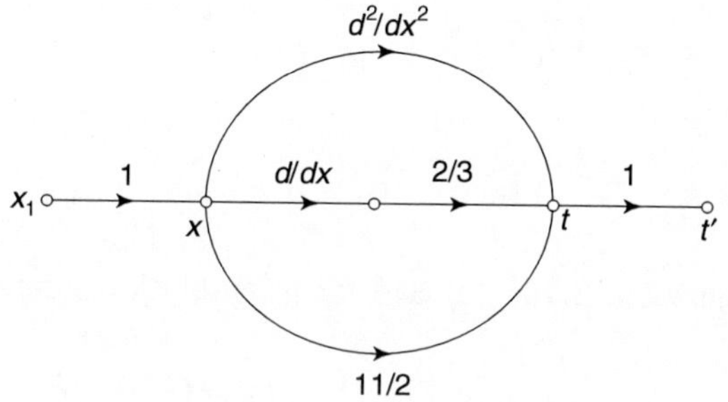

Figure Ex. 2.17 (c) SFG for Eq. (iii)

Example 2.18 Find the transfer function of the system given in Fig. Ex. 2.18 by Mason's gain formula.

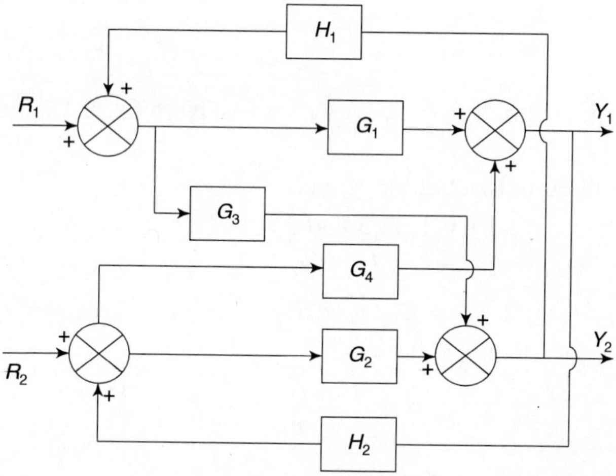

Figure Ex. 2.18

Solution The signal flow graph of the system is shown in Fig. Ex. 2.18 (a)

Assuming $\quad Y_1 = 0 = \text{and } R_2 = 0$

Step 1 There are two forward paths, i.e., $i = 1, 2$ so

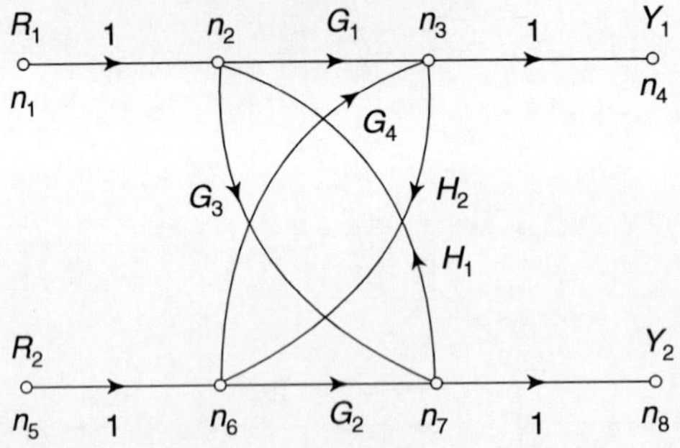

Figure Ex. 2.18 (a) SFG for given block diagram

$P_i = P_1, P_2$

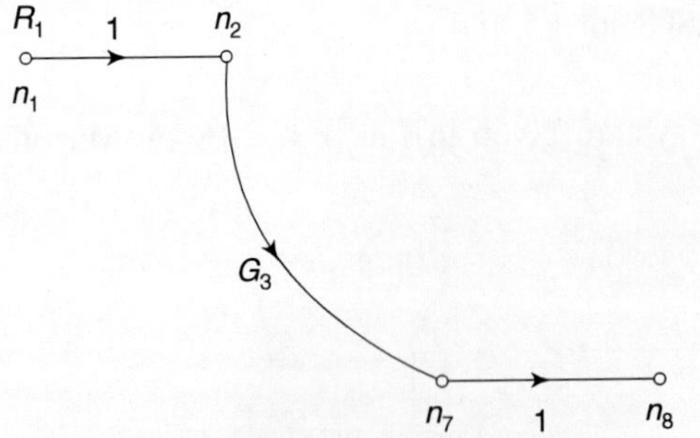

Figure Ex. 2.18 (b) Forward path P_1

$P_2 = G_1\ G_2\ H_2$

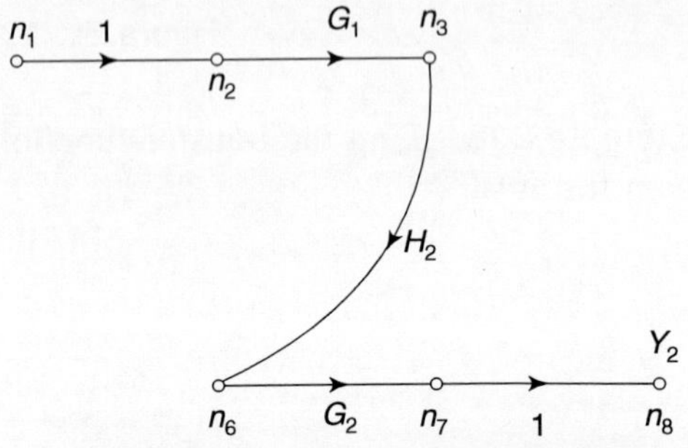

Figure Ex. 2.18 (c) Forward path P_2

Step 2 There are three individual loops, i.e.,,

$$j = 1, 2, 3 \text{ and } r = 1$$

$$L_{jr} = L_{11}, L_{21}, L_{31}$$

$L_{11} = G_3\ H_1$ $\qquad L_{21} = G_4\ H_2$ $\qquad L_{31} = G_1\ H_2\ G_2\ H_1$

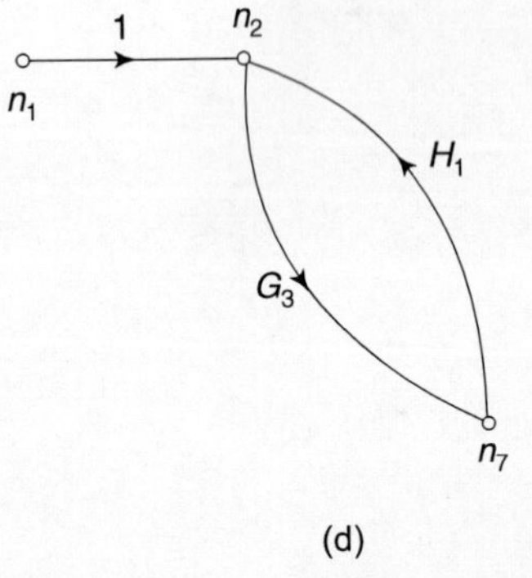

Figure Ex. 2.18 (d) Loop L_{11}

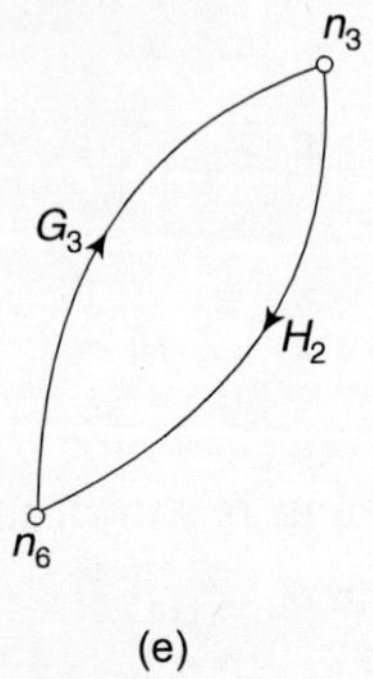

Figure Ex. 2.18 (e) Individual loop L_{31}

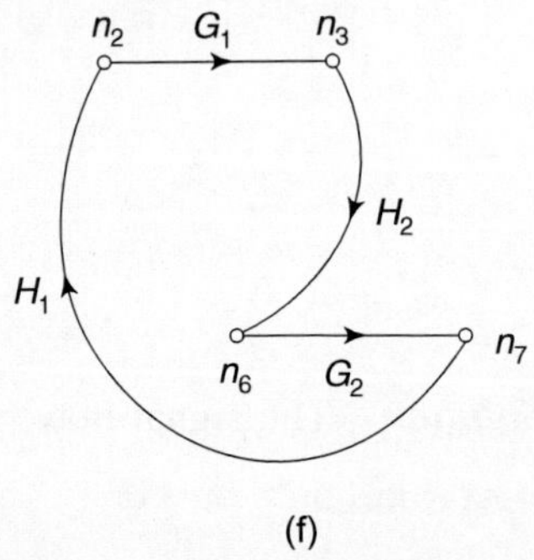

Figure Ex.2.18 (f) Individual loop L_{31}

Step 3 There is one pair of two non-touching loops, i.e.,

$$J = 1 \text{ and } r = 2$$

$$L_{jr} = L_{12}$$

$$L_{12} = G_3\, G_4\, H_1\, H_2$$

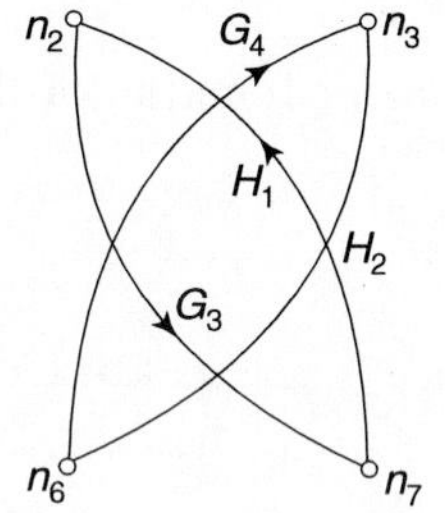

Figure Ex. 2.18 (g) Non-touching loop L_{12}

Step 4 There is no combination of three, four ... non-touching loops is $L_{j3} = L_{j4} = L_{j5} = ... = 0$

Step 5 Now by Eq. (2.10)

$$\Delta = 1 - \Sigma\, L_{j1} + \Sigma\, L_{j2} - \Sigma\, L_{j3} + ...$$

$$\Delta = 1 - (L_{11} + L_{21} + L_{31}) + (L_{12}) + 0$$

$$\Delta = 1 - (G_3\, H_1 + G_4\, H_2 + G_1\, H_2\, G_2\, H_1) + (G_3\, G_4\, H_1\, H_2)$$

$$\Delta = 1 - G_3\, H_1 - G_4\, H_2 - G_1\, H_2\, G_2\, H_1 + G_3\, G_4\, H_1\, H_2$$

Step 6 Removing the loop gains from Δ touching P_1, P_2 forward path

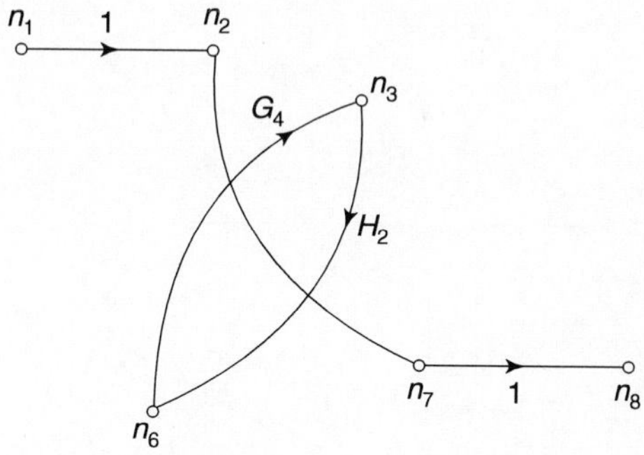

Figure Ex. 2.18 (h) For path P_1

For path P_2 $\quad \Delta_2 = 1$

Now by Eq. (2.9)

$$T_o = \frac{1}{\Delta} \Sigma P_i\, \Delta_i$$

$$T_o = \frac{P_1 \Delta_1 + P_2 \Delta_2}{\Delta}$$

$$T_o = \frac{G_3 (1 - G_4 H_2) + G_1 G_2 H_2}{1 - G_3 H_1 - G_4 H_2 - G_1 H_2 G_2 H_1 + G_3 G_4 H_1 H_2} \quad \textbf{Ans.}$$

Exmple 2.19 Find the closed loop transfer function of the control system by Mason's gain formula as shown in Fig. Ex. 2.19.

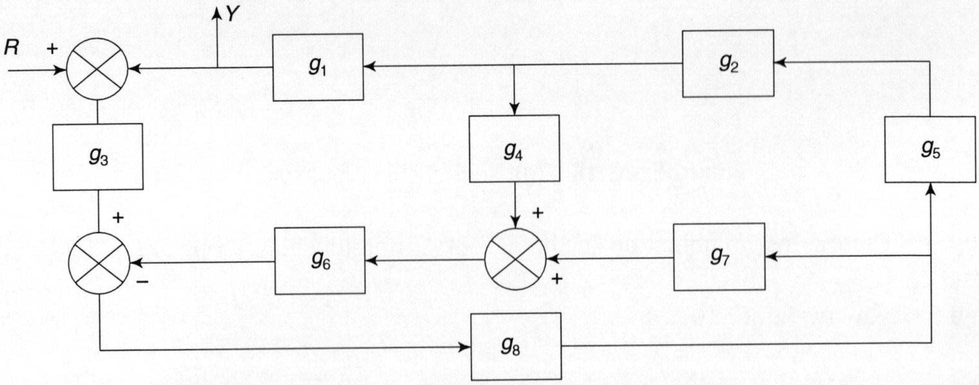

Figure Ex. 2.19

Solution The signal flow graph for the system is shown in Fig. Ex. 2.19 (a)

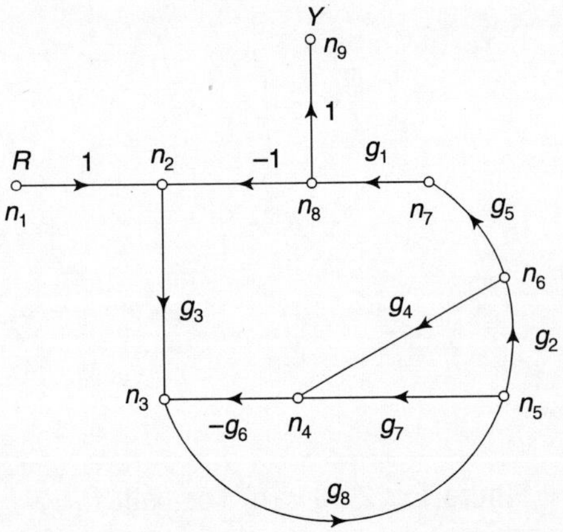

Figure Ex. 2.19 (a) SFG for given block diagram

Step 1 There is one forward path, i.e., $i = 1$, $P_i = P_1$ so the path gain is

$$P_1 = g_3\, g_8\, g_2\, g_5\, g_1$$

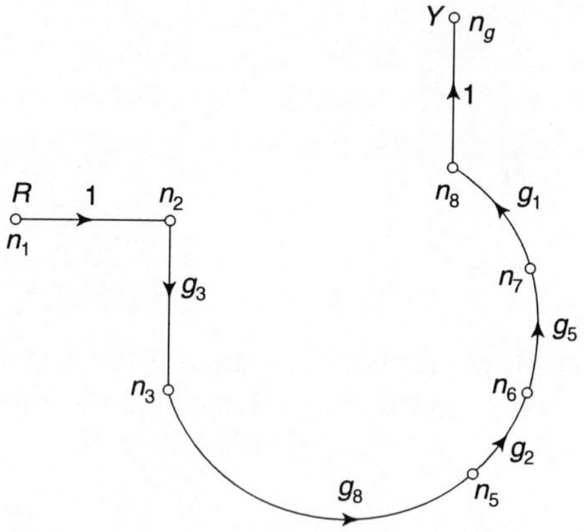

Figure Ex. 2.19 (b) Forward path P_1

Step 2 There are individual loops, i.e., $j = 1, 2, 3$ and $r = 1$, $L_{jr} = L_{11}, L_{21}, L_{31}$ so the loop gain is

$L_{11} = g_3\, g_8\, g_2\, g_5\, g_1$ $\quad$ $L_{21} = g_6\, g_8\, g_7$ $\quad$ $L_{31} = -\, g_6\, g_8\, g_2\, g_5\, g_4$

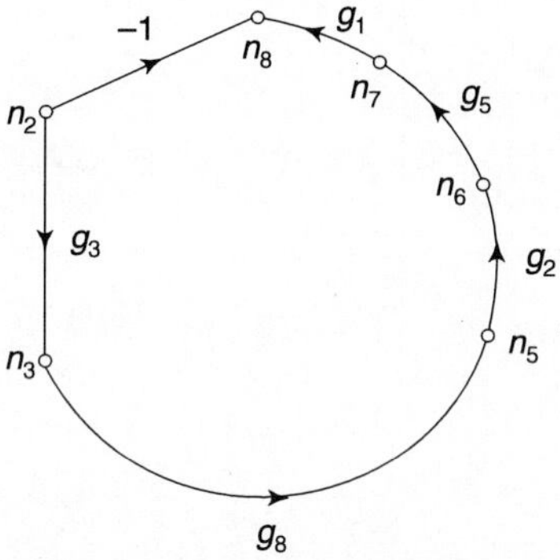

Figure Ex. 2.19 (c) Individual Loop L_{11}

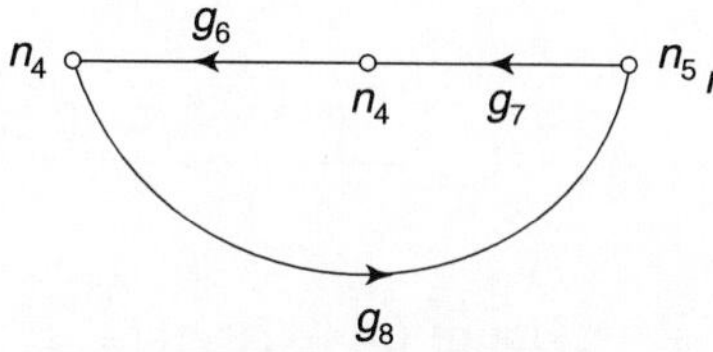

Figure Ex. 2.19 (d) Individual loop L_{21}

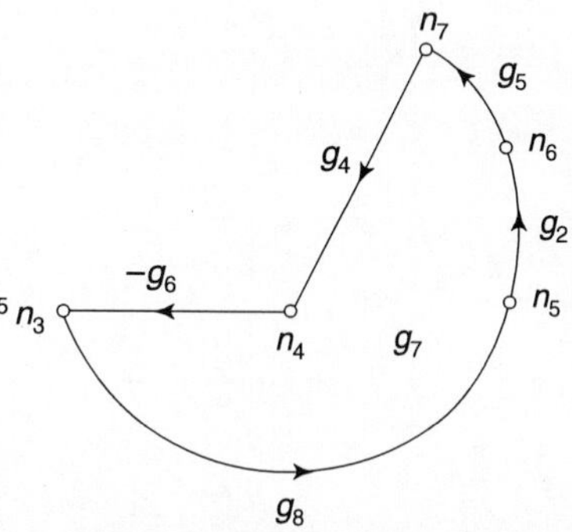

Figure Ex. 2.19 (e) Individual loop L_{31}

Step 3 and 4 There is no combination of two, three, four ... non touching loops so

$$L_{j2} = L_{j3} = L_{j4} = L_{j5} = \ldots = 0$$

Step 5 Now by Eq. (2.10)

$$\Delta = 1 - \Sigma\, L_{j1} + \Sigma\, L_{j2} - \Sigma\, L_{j3} + \ldots$$
$$\Delta = 1 - (L_{11} + L_{21} + L_{31}) + 0 + 0$$
$$\Delta = 1 - (-\, g_3\, g_8\, g_2\, g_5\, g_1 - g_6\, g_8\, g_7 - g_6\, g_8\, g_2\, g_5\, g_4)$$
$$\Delta = 1 + g_3\, g_8\, g_2\, g_5\, g_1 + g_6\, g_8\, g_7 + g_6\, g_8\, g_2\, g_5\, g_4$$

Step 6 Removing the gain from Δ touching path P_1. Because all the loops are touching path P_1, so $\Delta_1 = 1$

Now by Eq. (2.9)

$$T_o = \frac{1}{\Delta} \sum P_i \Delta_i$$

$$T_o = \frac{P_1 \Delta_1}{\Delta}$$

$$T_o = \frac{g_3 g_8 g_2 g_5 g_1}{1 + g_3 g_8 g_2 g_5 g_1 + g_6 g_8 g_7 + g_6 g_8 g_2 g_5 g_4} \quad \textbf{Ans.}$$

Example 2.20 Obtain signal flow graph representation for a system whose block diagram is given below and using Mason's gain formula determine the ratio *Y*/*R*.

(R.M.L., Univ. Faizabad)

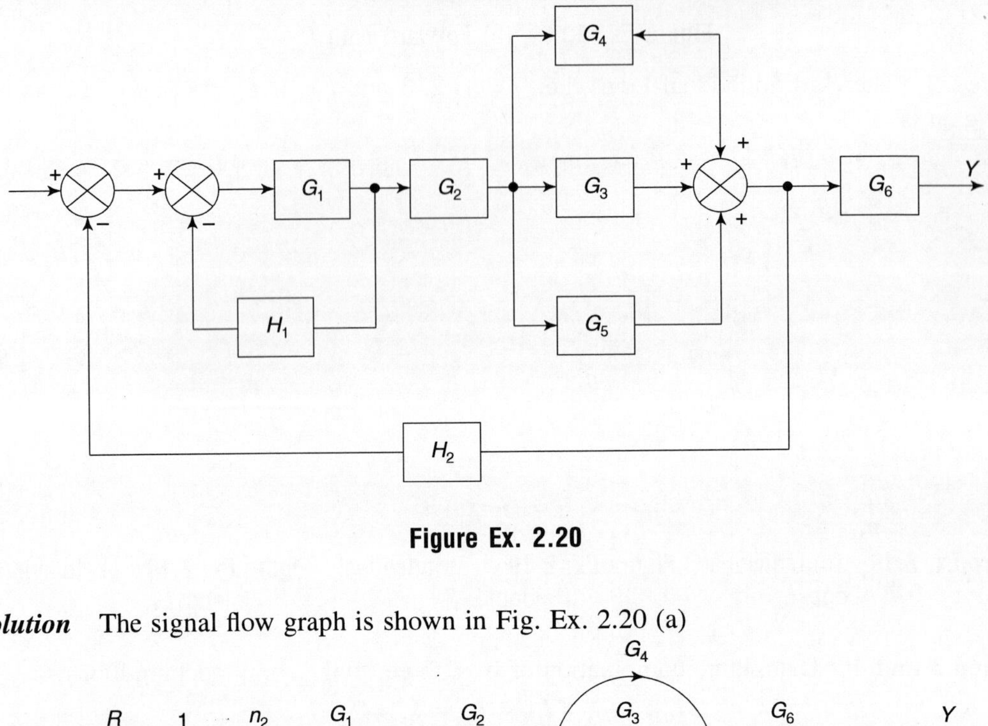

Figure Ex. 2.20

Solution The signal flow graph is shown in Fig. Ex. 2.20 (a)

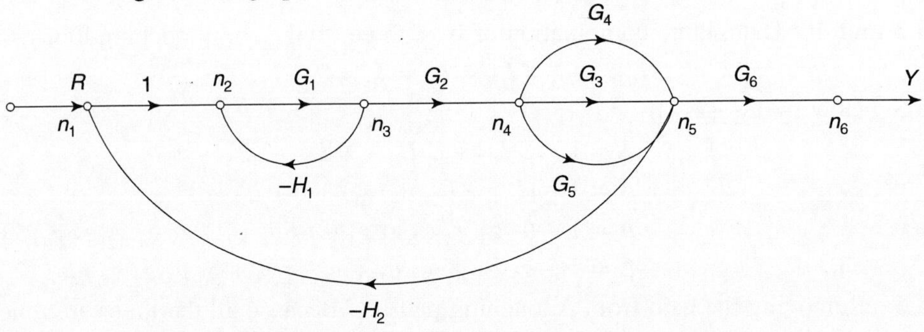

Figure Ex. 2.20 (a) SFG for given block diagram

Step 1 There are three forward paths, i.e., $i = 1, 2, 3$

$$P_i = P_1, P_2, P_3 \text{ so the path gain is}$$

$$P_1 = G_1\, G_2\, G_3\, G_6$$

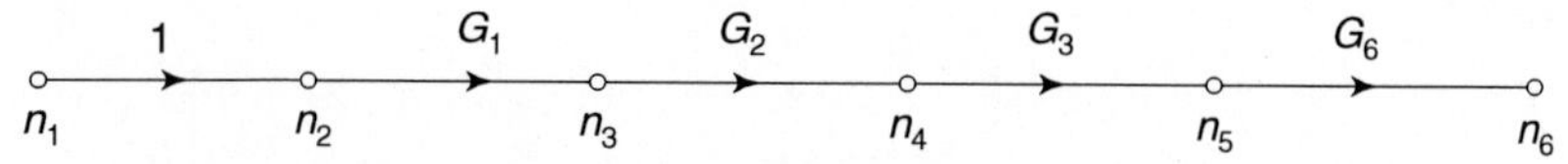

Figure Ex. 2.20 (b) Forward path P_1

$$P_2 = G_1\, G_2\, G_4\, G_6$$

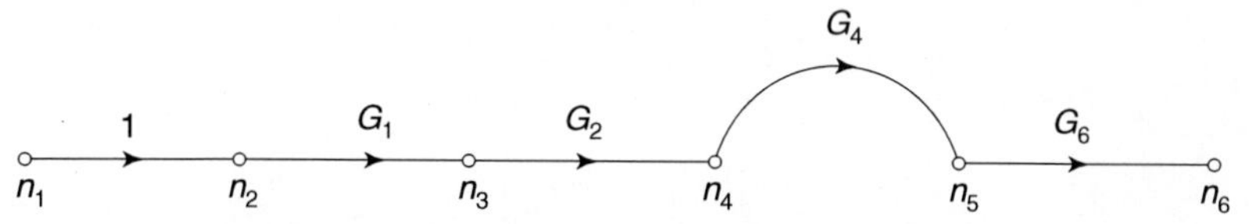

Figure Ex. 2.20 (c) Forward path P_2

$$P_3 = G_1\, G_2\, G_5\, G_6$$

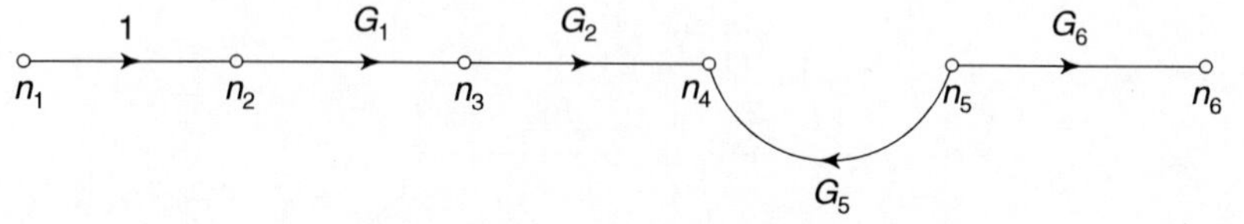

Figure Ex. 2.20 (d) Forward path P_3

Step 2 There are four individual loops, i.e., $j = 1, 2, 3$ and $r = 1$

$$L_{jr} = L_{11}, L_{21}, L_{31}, L_{41}$$

$$L_{11} = -\, G_1\, H_1$$

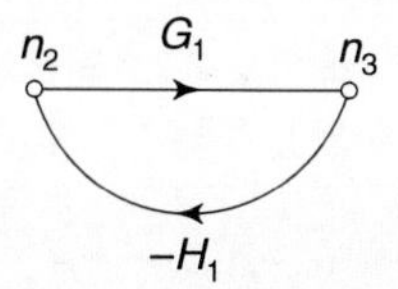

Figure Ex. 2.20 (e) Individual loop L_{11}

$$L_{21} = -G_1 G_2 G_3 H_2$$

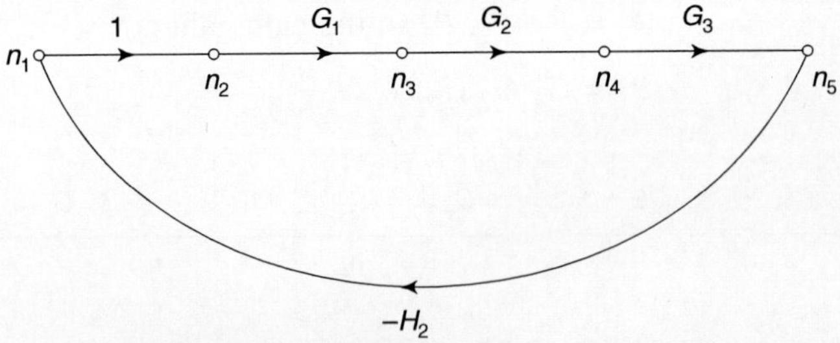

Figure Ex. 2.20 (f) Individual loop L_{21}

$$L_{31} = -G_1 G_2 G_4 H_2$$

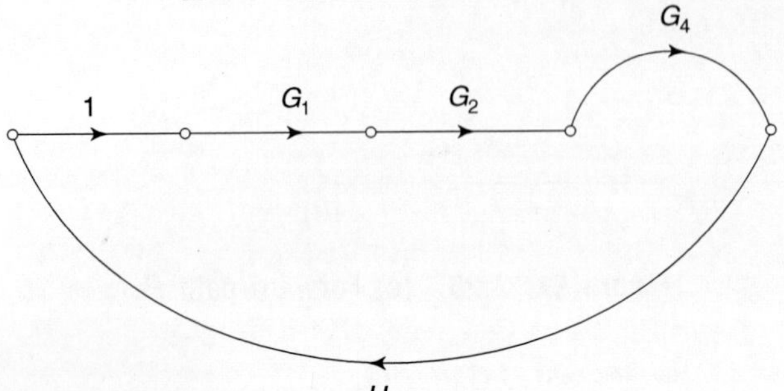

Figure Ex. 2.20 (g) Individual loop L_{31}

$$L_{41} = -G_1 G_2 G_5 H_2$$

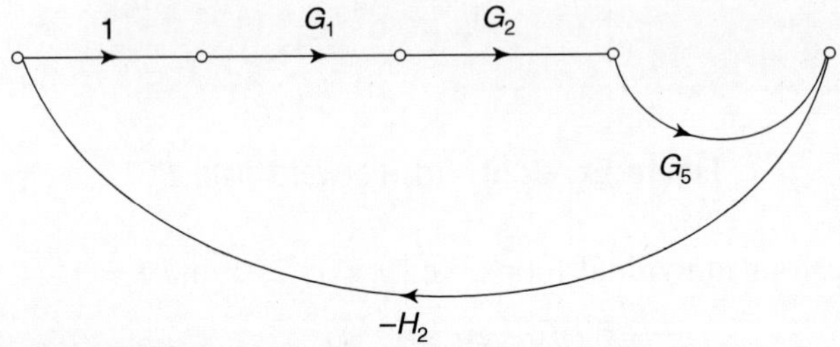

Figure Ex. 2.20 (h) Individual loop L_{41}

Step 3 and 4 There is no combination of two, three, four ... non-touching loops so

$$L_{j2} = L_{j3} = L_{j4} = \dots = 0$$

Step 5 Now by Eq. (2.10)

$$\Delta = 1 - \Sigma L_{j1} + \Sigma L_{j2} - \Sigma L_{j3} + \dots$$

$$\Delta = 1 - (L_{11} + L_{21} + L_{31} + L_{41}) + 0 + 0$$

$$\Delta = 1 - (-G_1 H_1 - G_1 G_2 G_3 H_2 - G_1 G_2 G_4 H_2 - G_1 G_2 G_5 H_2)$$

$$\Delta = 1 + G_1 H_1 + G_1 G_2 G_3 H_2 + G_1 G_2 G_4 H_2 + G_1 G_2 G_5 H_2$$

Step 6 Removing the loop gains from Δ that are touching forward path P_1, P_2, P_3

Because all the loops touching the forward path gain P_1, P_2, P_3, so

$$\Delta_1 = \Delta_2 = \Delta_3 = 1$$

Now by Eq. (2.9)

$$T_o = \frac{1}{\Delta}\sum P_i \, \Delta_i$$

$$T_o = \frac{1}{\Delta}\left[P_1 \, \Delta_1 + P_2 \, \Delta_2 + P_3 \, \Delta_3\right]$$

$$T_o = \frac{G_1 \, G_2 \, G_3 \, G_6 + G_1 \, G_2 \, G_4 \, G_6 + G_1 \, G_2 \, G_5 \, G_6}{1 + G_1 \, H_1 + G_1 \, G_2 \, G_3 \, H_2 + G_1 \, G_2 \, G_4 \, H_2 + G_1 \, G_2 \, G_5 \, H_2}$$ **Ans.**

Example 2.21 Obtain the transfer function for $\frac{Y(s)}{R(s)}$ given block diagram as shown in Fig. Ex. 2.21 by Mason's gain formula. *(R.M.L. Univ., Faizabad 2001, Control System)*

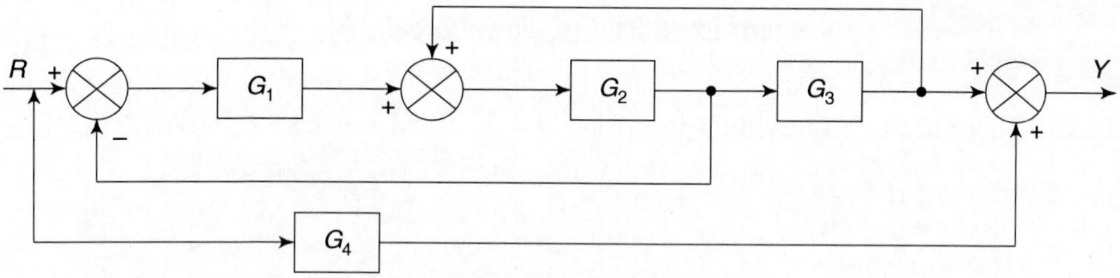

Figure Ex. 2.21

Solution The signal flow graph is shown in Fig. Ex. 2.21 (a).

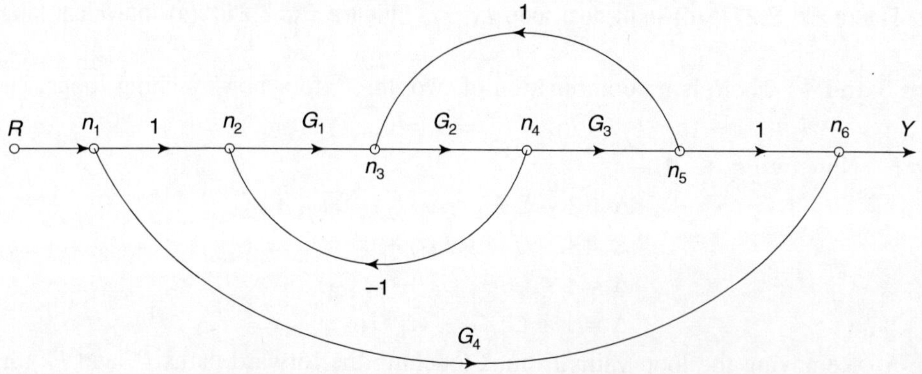

Figure Ex. 2.21 (a) SFG for given block diagram

Step 1 There are two forward paths, i.e., $i = 1, 2$

So
$$P_i = P_1, P_2$$
$$P_1 = G_1\, G_2\, G_3$$

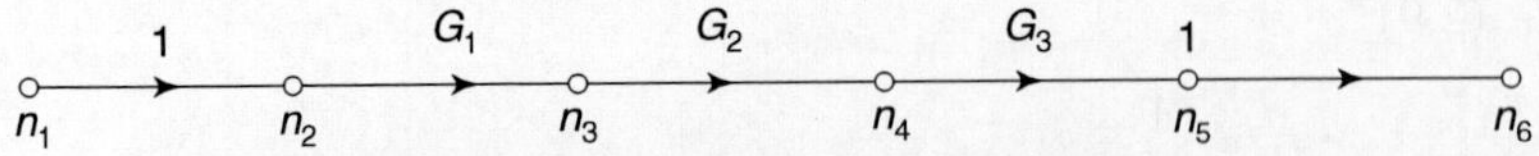

Figure Ex. 2.21 (b) Forward path P_1

$$P_2 = G_4$$

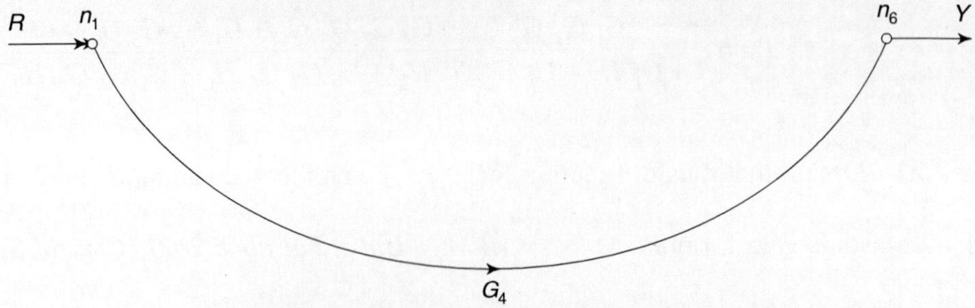

Figure Ex. 2.21 (c) Forward path P_2

Step 2 There are two individual loops i.e., $j = 1, 2$ and $r = 1$ so $L_{j1} = L_{11}, L_{21}$

$$L_{11} = -\,G_1\, G_2 \qquad\qquad L_{21} = G_2\, G_3$$

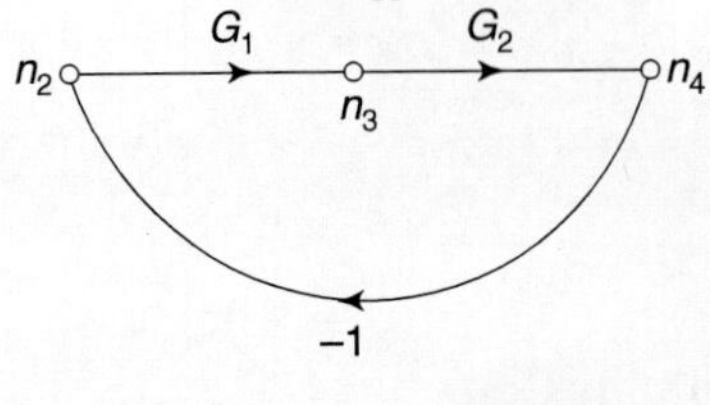

Figure Ex. 2.21 (d) Individual loop L_{11}

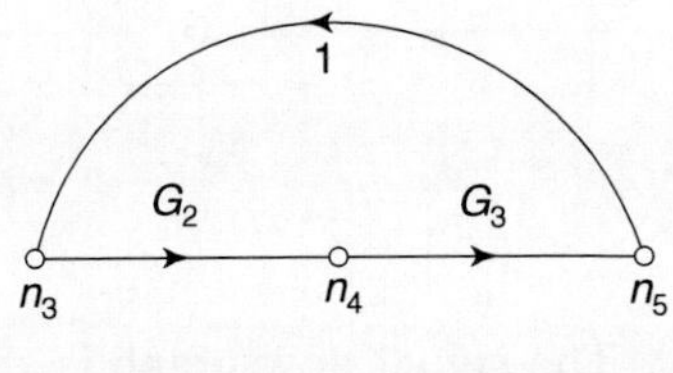

Figure Ex. 2.21 (e) Individual loop L_{21}

Steps 3 and 4 There is no combination of two, three, four non-touching loops, i.e.,
$$L_{j2} = L_{j3} = L_{j4} = \ldots = 0$$

Step 5 Now by Eq. (2.10)
$$\Delta = 1 - \Sigma\, L_{j1} + \Sigma\, L_{j2} - \Sigma\, L_{j3} + \ldots$$
$$\Delta = 1 - (L_{11} + L_{21}) + 0$$
$$\Delta = 1 - (-\,G_1\, G_2 + G_2\, G_3)$$
$$\Delta = 1 + G_1\, G_2 - G_2\, G_3$$

Step 6 Removing the loop gains from Δ touching the forward paths P_1 and P_2 for path P_1 all the loops are touching

for path P_1

$$\Delta_1 = 1$$

for path 2

loops $G_1\ G_2$ and $G_2\ G_3$ are not touching path P_2 so

$$\Delta_2 = 1 + G_1\ G_2 - G_2\ G_3$$

Now by Eq. (2.9)

$$T_o = \frac{1}{\Delta}\Sigma P_i\ \Delta_i = \frac{1}{\Delta}[P_1\ \Delta_1 + P_2\ \Delta_2]$$

$$T_o = \frac{G_1\ G_2\ G_3 + G_4\ (1 + G_1\ G_2 - G_2\ G_3)}{1 + G_1\ G_2 - G_2\ G_3} \quad \textbf{Ans.}$$

MISCELLANEOUS PROBLEMS

Example 2.22 For the system represented by the given equations find the transfer function x_5/x_1 with the help of signal flow graph technique.

$$x_2 = a_{12}\ x_1 + a_{32}\ x_3 + a_{42}\ x_4 + a_{52}\ x_5 \ \ldots \qquad \text{...(i)}$$

$$x_3 = a_{23}\ x_2 \qquad \text{...(ii)}$$

$$x_4 = a_{34}\ x_3 + a_{44}\ x_4 \qquad \text{...(iii)}$$

$$x_5 = a_{35}\ x_3 + a_{45}\ x_4 \qquad \text{...(iv)}$$

where $x_1 \rightarrow$ input variable

$x_5 \rightarrow$ output variable *(R.M.L. Univ., Faizabad, 2002)*

Solution 1. Because there are five variables so five nodes will exist in SFG

x_1	x_2	x_3	x_4	x_5
0	0	0	0	0

2. Now drawing the SFG for Eq. (i) which is the sum of four signals like $a_{12}\ x_1$, $a_{32}\ x_3$, $a_{42}\ x_4$, $a_{52}\ x_5$

   (1) = $a_{12}\ x_1$; (2) = $a_{32}\ x_3$; (3) = $a_{42}\ x_4$; (4) = $a_{52}\ x_5$

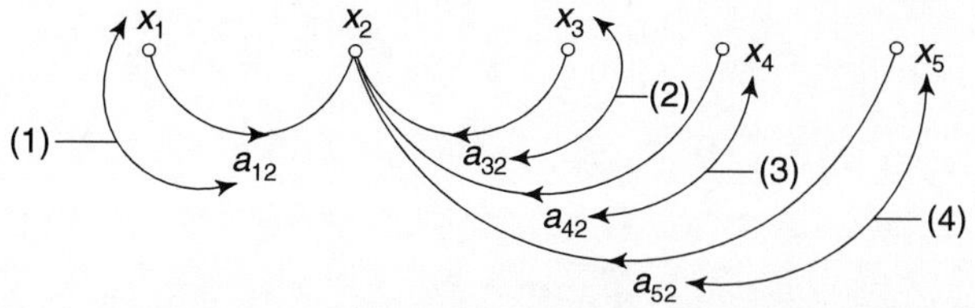

Figure Ex. 2.22 (a) SFG for Eq. (i)

3. Now drawing the SFG for Eq. (ii)

   $(1) = a_{23}\, x_2$

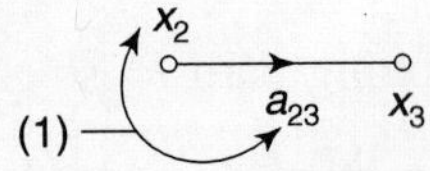

Figure Ex. 2.22 (b) SFG for Eq. (ii)

4. Now drawing the SFG for equation (iii) which is the sum of two signals like $a_{34}\, x_3$, $a_{44}\, x_4$.

   $(1) = a_{34}\, x_3$; $(2) = a_{44}\, x_4$

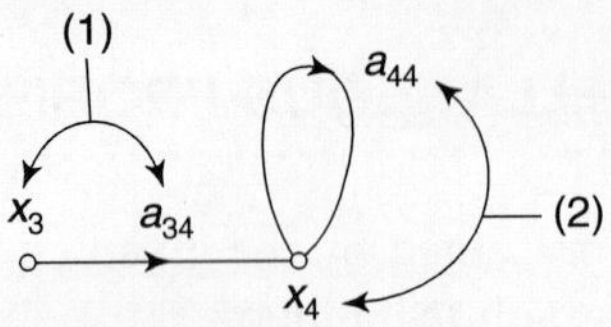

Figure Ex. 2.22 (c) SFG for Eq. (iii)

5. Now drawing the SFG for Eq. (iv) which is the sum of two signals like $a_{35}\, x_3$, $a_{45}\, x_4$

   $(1) = a_{35}\, x_3$; $(2) = a_{45}\, x_4$

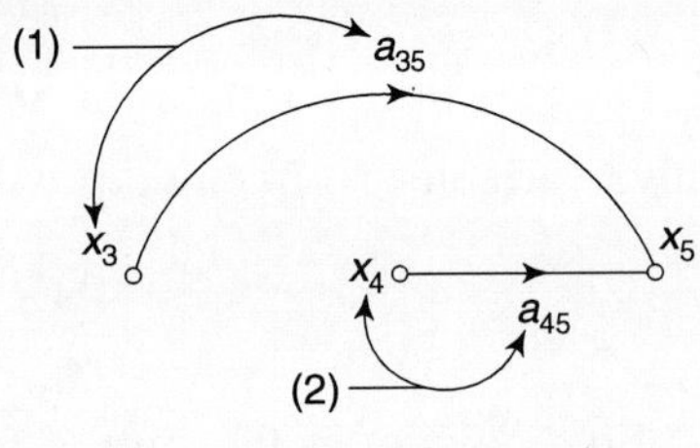

Figure Ex. 2.22 (d) SFG for Eq. (iv)

6. Now combining the SFG in steps 2 to 5 as shown in Fig. Ex. 2.22.

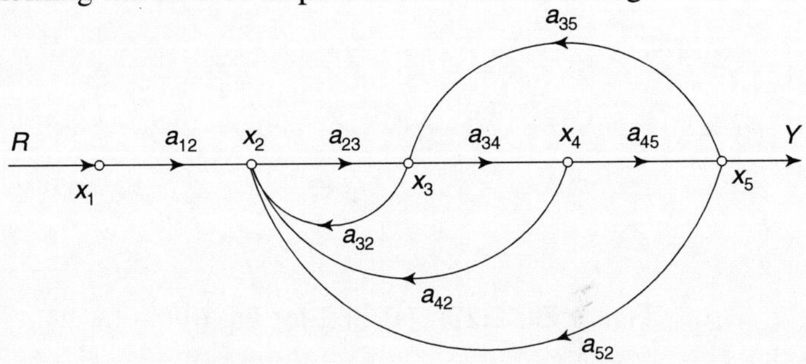

Figure Ex. 2.22 (e) Complete SFG

Step 1 There are two forward paths, i.e., $i = 1, 2$

$$P_i = P_1, P_2$$
$$P_1 = a_{12}\, a_{23}\, a_{34}\, a_{45}$$

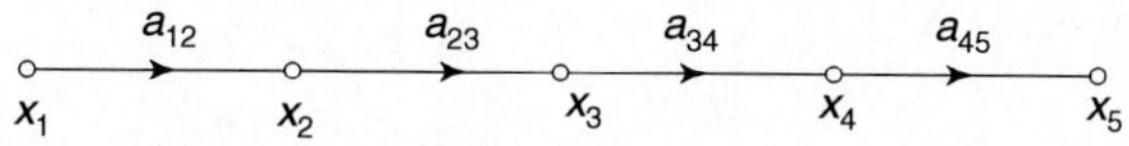

Figure Ex. 2.22 (f) Forward path P_1

$$P_2 = a_{12}\, a_{23}\, a_{35}$$

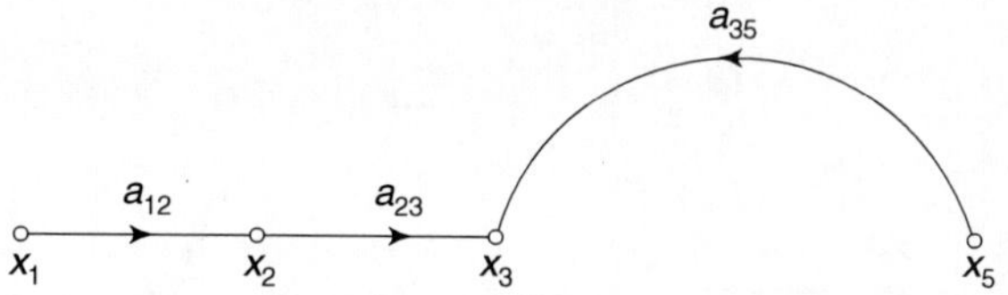

Figure Ex. 2.22 (g) Forward path P_2

Step 2 There are five individual loops, i.e.,

$j = 1, 2, 3, 4, 5$ and $r = 1$ so the loop gain

$$L_{jr} = L_{11}, L_{21}, L_{31}, L_{41}, L_{51}$$
$$L_{11} = a_{23}\, a_{32}$$

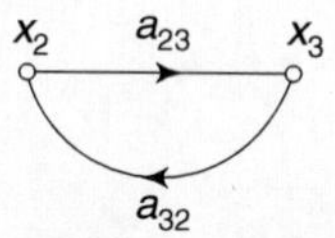

Figure Ex. 2.22 (h) Individual loop L_{11}

$$L_{21} = a_{23}\, a_{34}\, a_{32}$$

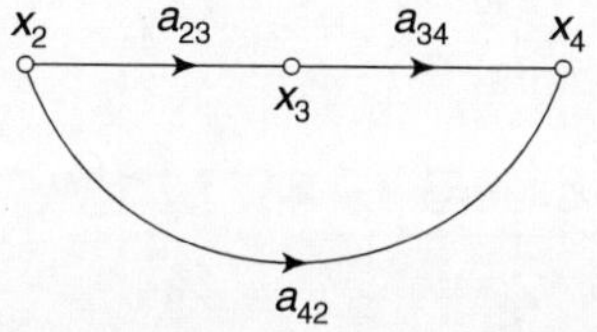

Figure Ex. 2.22 (i) Individual loop L_{21}

$$L_{31} = a_{23}\, a_{34}\, a_{45}\, a_{52}$$

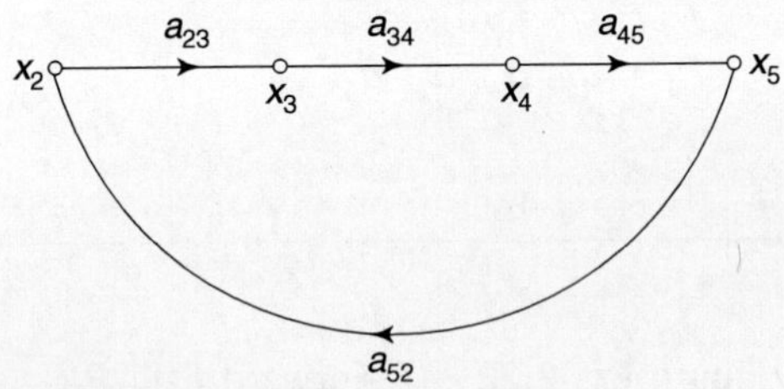

Figure Ex. 2.22 (j) Individual loop L_{31}

$$L_{41} = a_{23}\, a_{35}\, a_{52}$$

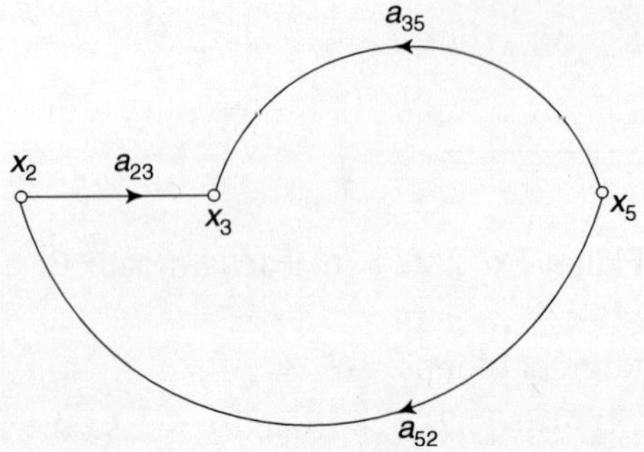

Figure Ex. 2.22 (k) Individual loop L_{41}

$$L_{51} = a_{44}$$

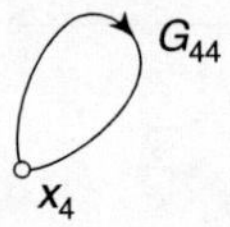

Figure Ex. 2.22 (l) Individual loop L_{51}

Step 3 There are two combinations of two non-touching loops, i.e., $j = 1, 2$ and $r = 2$

$$L_{j2} = L_{12},\, L_{22}$$
$$L_{12} = a_{23}\, a_{32}\, a_{44}$$

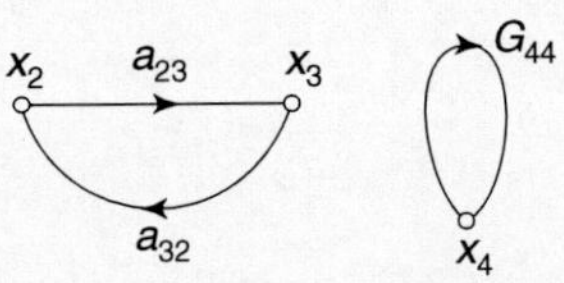

Figure Ex. 2.22 (m) Non-touching loop L_{12}

$$P_{22} = a_{23}\, a_{35}\, a_{52}\, a_{44}$$

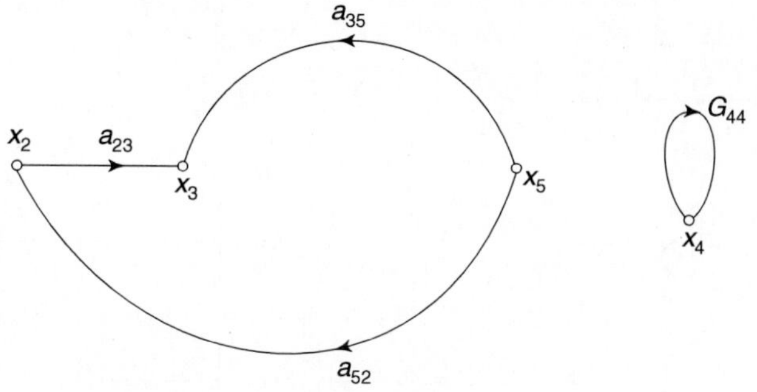

Figure Ex. 2.22 (n) Non-touching loop L_{22}

Step 4 There is no combination of three, four ... non-touching loops so

$$L_{j3} = L_{j4} = \dots = 0$$

Step 5 Now by Eq. (2.10)

$$\Delta = 1 - \Sigma L_{j1} + \Sigma L_{j2} - \Sigma L_{j3} + \dots$$

$$\Delta = 1 - (L_{11} + L_{21} + L_{31} + L_{41} + L_{51}) + (L_{12} + L_{22}) + 0$$

$$\Delta = 1 - (a_{23}\, a_{32} + a_{23}\, a_{34}\, a_{42} + a_{23}\, a_{34}\, a_{45}\, a_{52} + a_{23}\, a_{35}\, a_{52} + a_{44}) + (a_{23}\, a_{32}\, a_{44} + a_{23}\, a_{35}\, a_{52}\, a_{44})$$

$$\Delta = 1 - a_{23}\, a_{32} - a_{23}\, a_{34}\, a_{42} - a_{23}\, a_{34}\, a_{45}\, a_{52} - a_{23}\, a_{35}\, a_{52} - a_{44} + a_{23}\, a_{32}\, a_{44} + a_{23}\, a_{35}\, a_{52}\, a_{44}$$

Step 6 Removing the loop gains of Δ touching the path P_1 and P_2

for path P_1

$$\Delta_1 = 1$$

$\because$ touching all the loops

for path P_2

$$\Delta_2 = 1 - a_{44}$$

$\because$ a_{44} loop is not touching the path P_2

Now

$$\frac{x_5}{x_1} = \frac{1}{\Delta}\Sigma P_i \Delta_i$$

$$\frac{x_5}{x_1} = \frac{1}{\Delta}[P_1\, \Delta_1 + P_2\, \Delta_2]$$

$$\frac{x_5}{x_1} = \frac{a_{12}a_{23}a_{34}a_{45} + a_{12}a_{23}a_{35}\,(1 - a_{44})}{1 - a_{23}a_{32} - a_{23}a_{34}a_{42} - a_{23}a_{34}a_{45}a_{52} - a_{23}a_{35}a_{52} - a_{44} + a_{23}a_{32}a_{44} + a_{23}a_{35}a_{52}a_{44}}$$ **Ans.**

Example 2.23 Obtain the transfer function C/R of the block diagram shown in Fig. Ex. 2.23.

(U.P.T.U., 2006–07, 2002–03, CS)

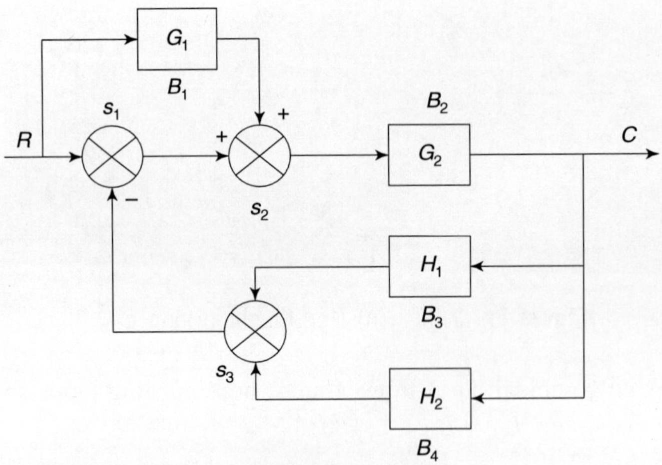

Figure Ex. 2.23

Solution Eliminating blocks B_3 and B_4

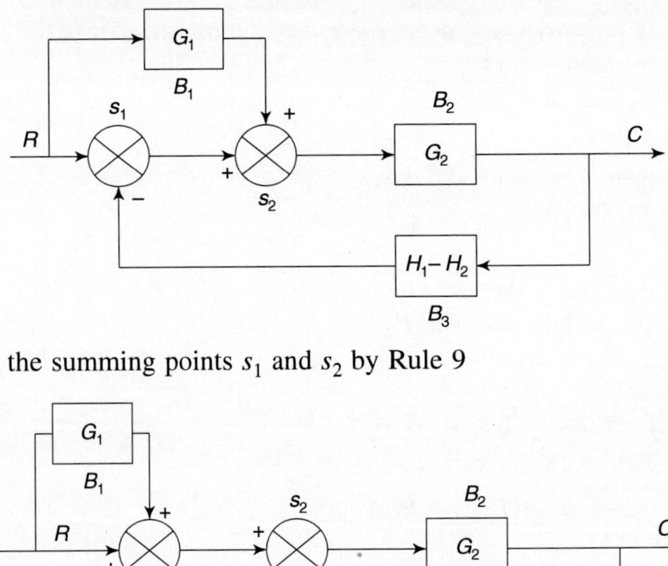

Interchanging the summing points s_1 and s_2 by Rule 9

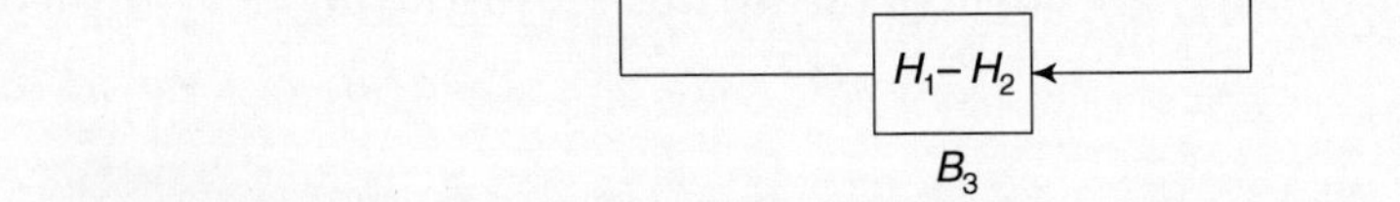

Eliminating block B_1

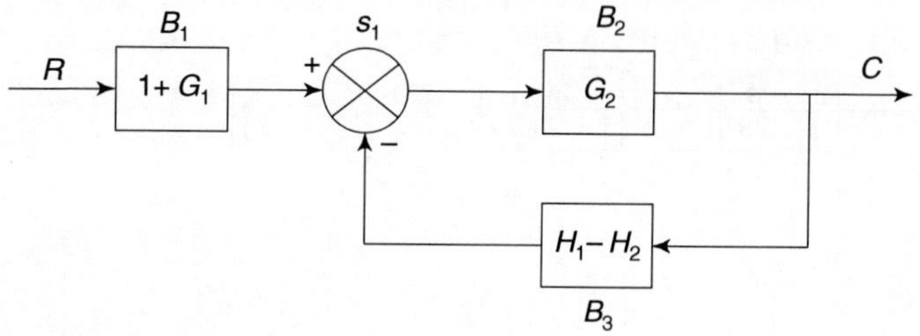

Eliminating blocks B_2 and B_1, we get

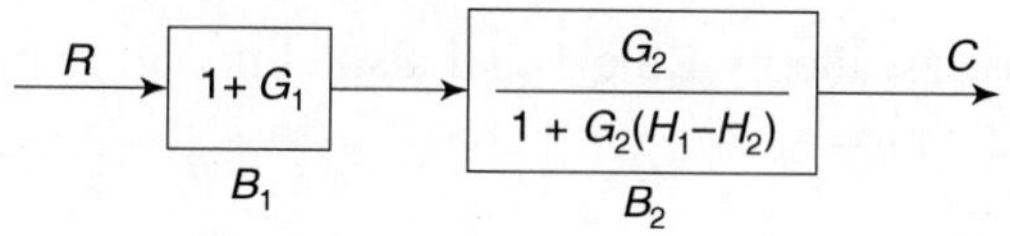

Blocks B_1 and B_2 are in cascade (Rule 1)

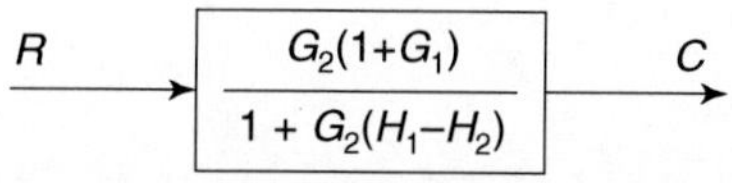

$$\frac{C(s)}{R(s)} = \frac{G_2\,(1+G_1)}{1+G_2\,(H_1-H_2)} \qquad \textbf{Ans.}$$

Example 2.24 For a given block diagram shown in Fig. Ex. 2.24 take $K_f = 1$, $T_F = 0$, $K_F = 1$, $T_E = 1$, $V_T = 1$ and $T_T = 1$

(a) Find $\dfrac{E_{fd}(s)}{V_{ref}(s)}$ at $V_T = 0$

(b) Find $\dfrac{E_{fd}(s)}{V_T(s)}$ at $V_{ref.} = 0$ *(U.P.T.U., 2005–06, CS)*

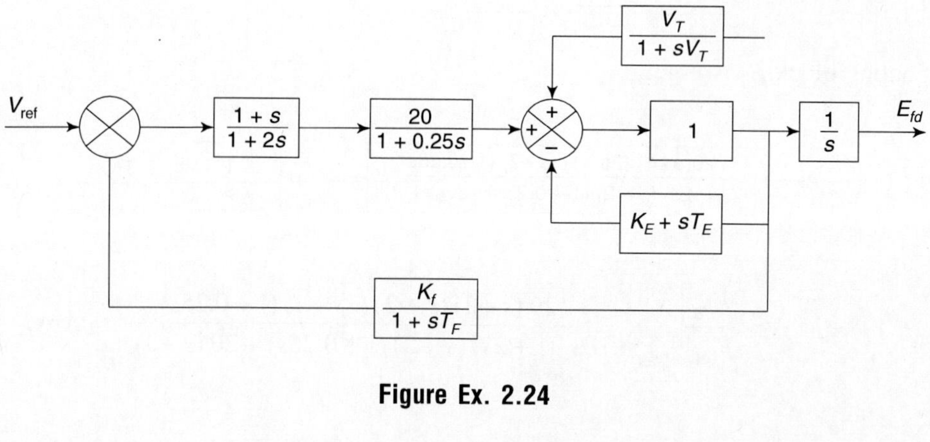

Figure Ex. 2.24

Solution Putting the value of K_E, T_f, T_E, K_f

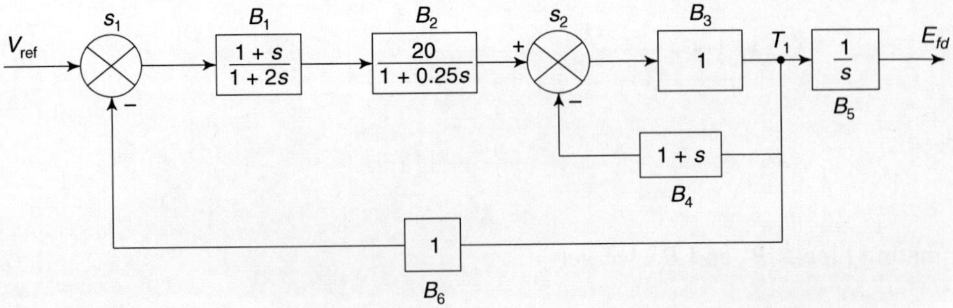

Blocks B_1 and B_2 are in cascade by Rule 1 and also eliminating block B_3 and B_4

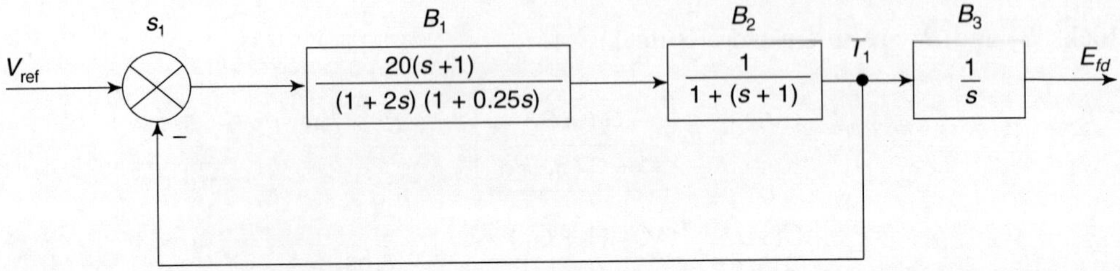

Blocks B_1 and B_2 are in cascade by Rule 1

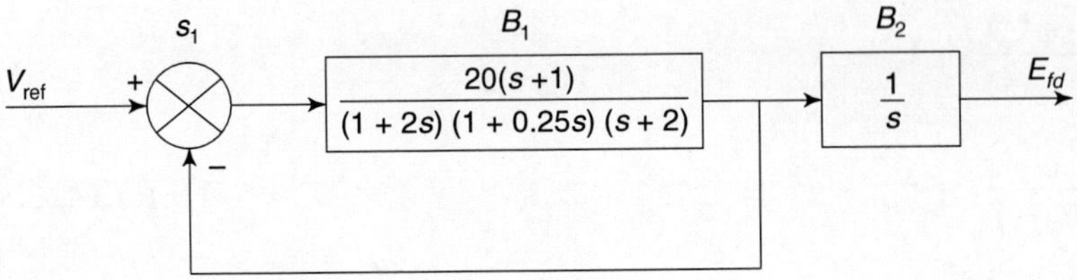

Eliminating block B_1, we get

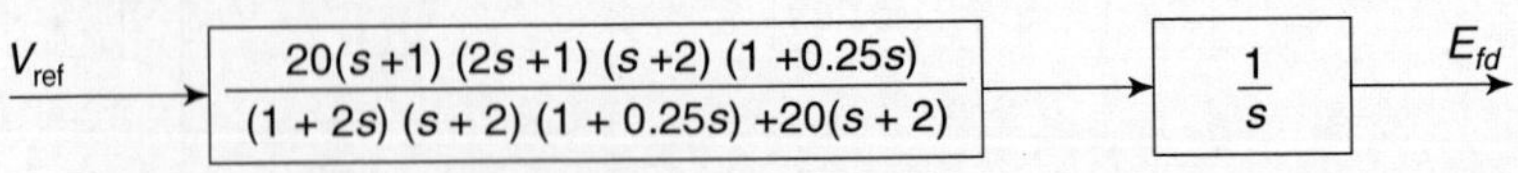

$$\left.\frac{E_{fd}}{V_{\text{ref.}}}\right|_{V_T=0} = \frac{20\,(s+1)\,(s+2)\,(2s+1)\,(1+0.25s)}{s\,[(1+2s)\,(s+2)\,(1+0.25s)+20\,(s+1)}\qquad \textbf{Ans.}$$

(b) At $V_{ref.} = 0$ given block diagram reduces to as shown below.

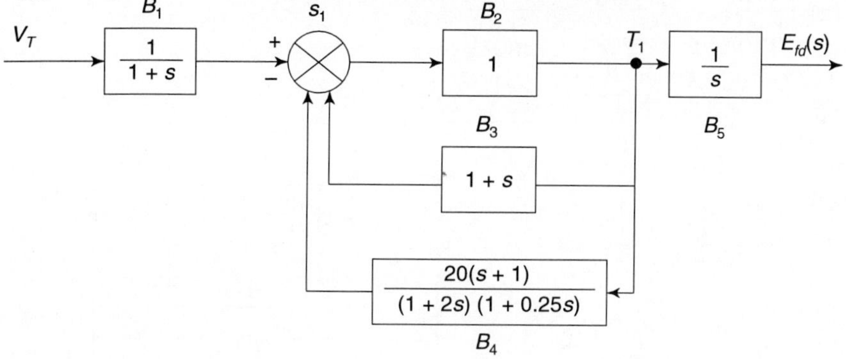

Eliminating blocks B_2 and B_3

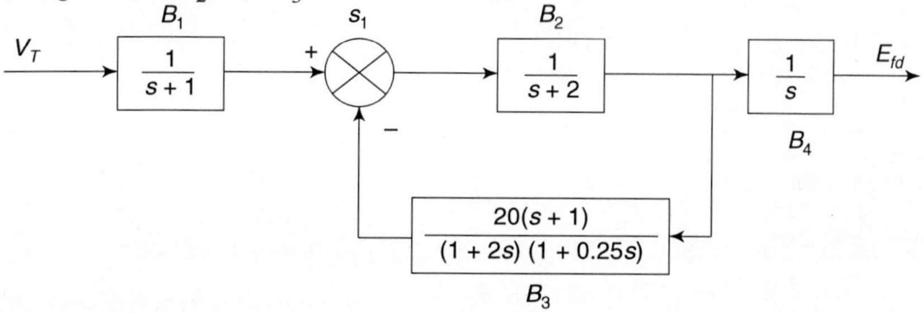

Eliminating blocks B_2 and B_3

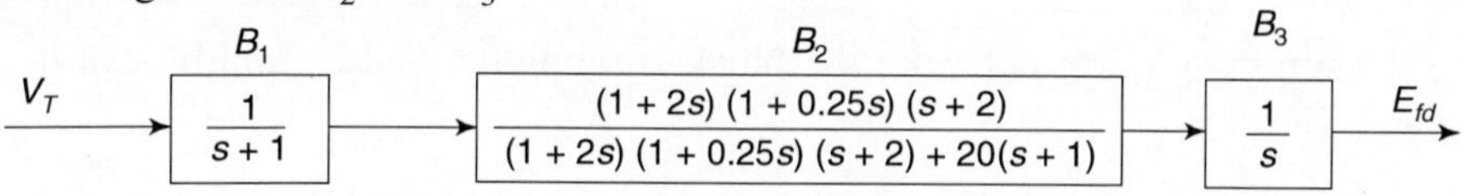

Blocks B_1, B_2 and B_2 are in cascade applying Rule 1

$$\frac{E_{fd}}{V_T} = \frac{(1+2s)(1+0.25s)(s+2)}{s(s+1)[(1+2s)(s+0.25s)(s+2)+20(s+1)]} \quad \textbf{Ans.}$$

Example 2.25 The block diagram of a feedback control system is shown in Fig. Ex. 2.25. Find the following transfer function

(a) $\left.\dfrac{Y(s)}{R(s)}\right|_{N=0}$ (b) $\left.\dfrac{Y(s)}{E(s)}\right|_{N=0}$ (c) $\left.\dfrac{Y(s)}{N(s)}\right|_{R=0}$

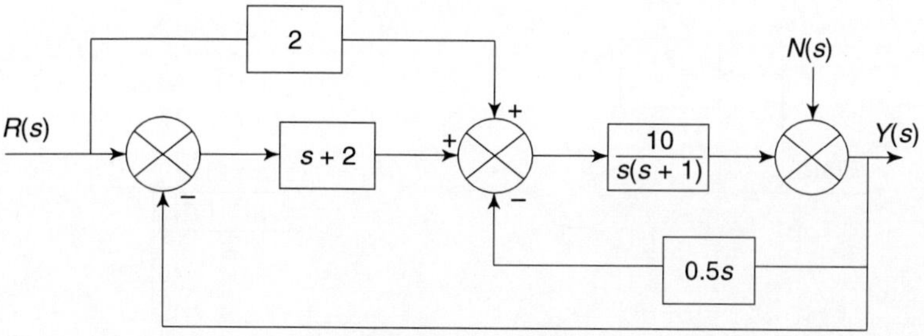

Figure Ex. 2.25

Solution (a) $\left.\dfrac{Y(s)}{R(s)}\right|_{N=0}$

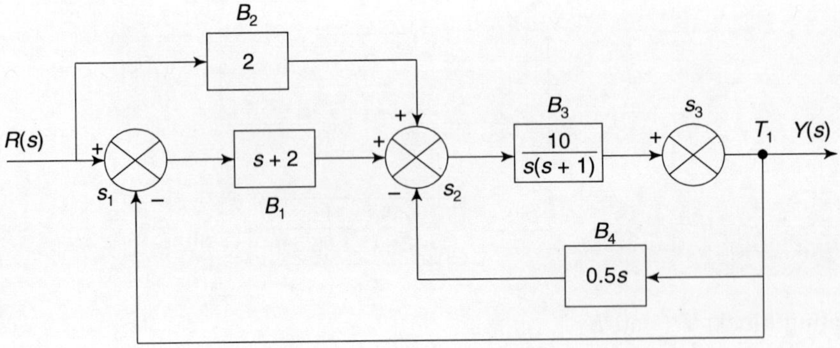

Eliminating blocks B_3 and B_4, we get

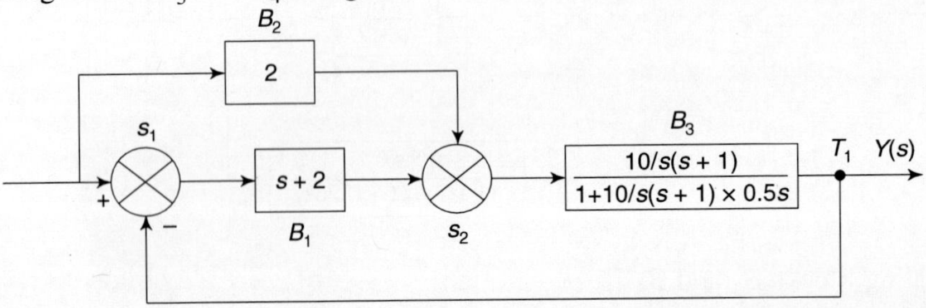

Moving the summing point s_2 before the block B_1 applying Rule 6 and blocks B_1 and B_3 are in cascade Rule 1

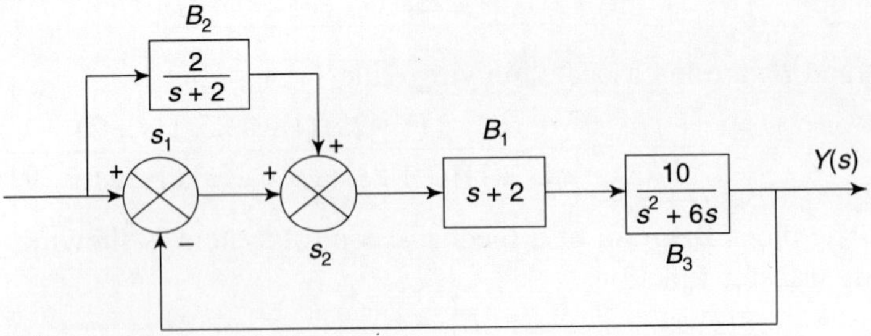

Interchaning the consecutive summing points s_1 and s_2, Rule 9

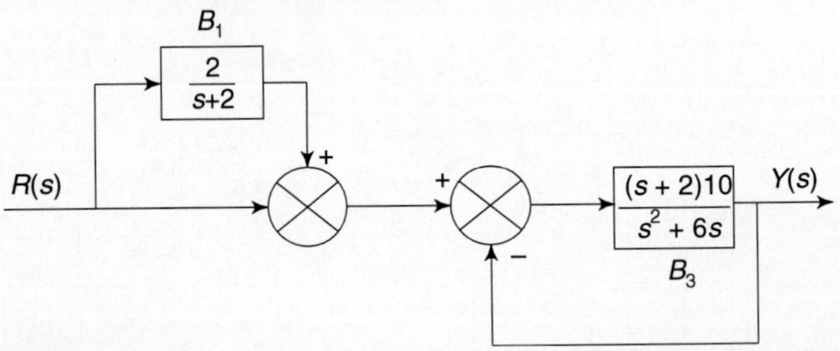

Elminating blocks B_1 and B_2

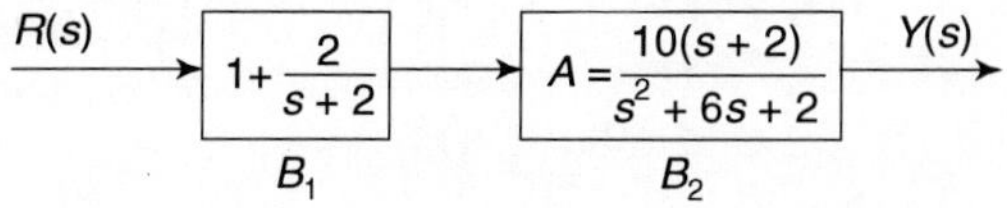

$$A = \frac{\dfrac{(s+2)\,10}{s^2+6s}}{1+\dfrac{(s+2)\,10}{s^2+6s}\times 1} = \frac{(s+2)\,10}{s^2+6s+10\,(s+2)}$$

$$= \frac{10\,(s+2)}{s^2+16s+2}$$

Blocks B_1 and B_2 are in cascade (Rule 1)

$$\frac{Y(s)}{R(s)} = \left(\frac{s+4)}{s+2}\right)\left[\frac{10\,(s+2)}{s^2+16s+2}\right]$$

$$\frac{Y(s)}{R(s)} = \frac{10\,(s+4)}{s^2+16s+2} \qquad \textbf{Ans.}$$

(b) $\left.\dfrac{Y(s)}{E(s)}\right|_{N=0}$

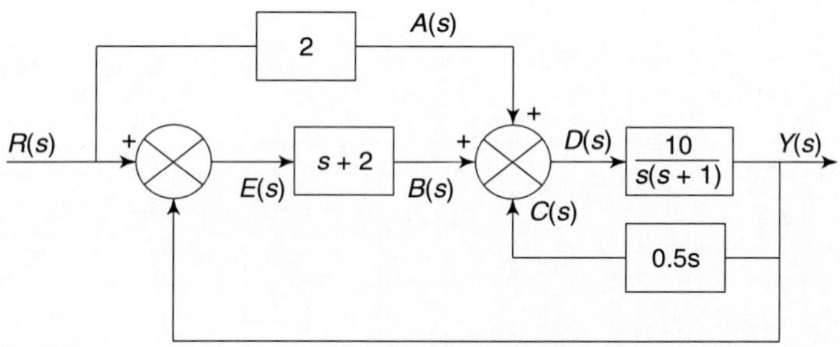

By the above figure

$$D(s) = A(s) + B(s) - C(s)$$
$$= 2\,.\,R(s) + (s+2)\,E(s) - 0.5s\,Y(s)$$
$$= 2\,[E(s) + Y(s)] + (s+2)\,E(s) - 0.5s\,Y(s)$$
$$= [2 + s + 2)\,E(s) + (2 - 0.5s)\,Y(s)$$
$$D(s) = (s+4)\,E(s) + [2 - 0.5s]\,Y(s)$$

Now
$$Y(s) = \frac{10}{s\,(s+1)}\,D(S)$$

$$Y(s) = \frac{10}{s\,(s+1)}[(s+4)\,E(s) + (2-0.5s)\,Y(s)]$$

$$Y(s) = \left[1 - \frac{10}{s(s+1)}(2 - 0.5s)\right] = \frac{10}{s(s+1)}[(s+4)\,E(s)]$$

$$\frac{Y(s)}{E(s)} = \frac{\dfrac{10(s+4)}{s(s+1)}}{1 - \dfrac{(20-5s)}{s(s+1)}} = \frac{10s+40}{s^2+s-20+55}$$

$$\frac{Y(s)}{E(s)} = \frac{10s+40}{s^2+6s-20} \quad \textbf{Ans.}$$

(c) $\left.\dfrac{Y(s)}{N(s)}\right|_{R=0}$

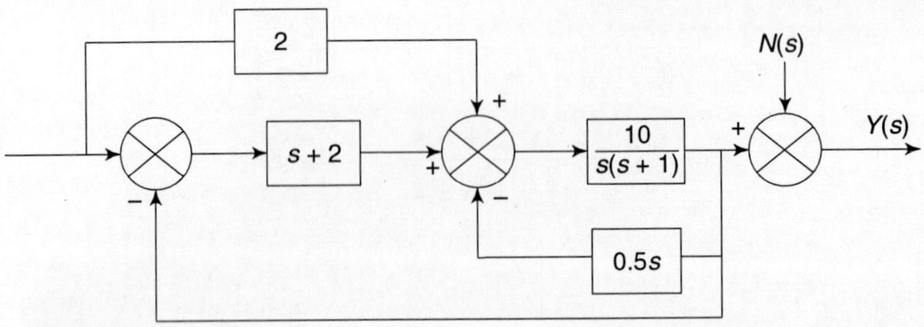

When $R(s) = 0$

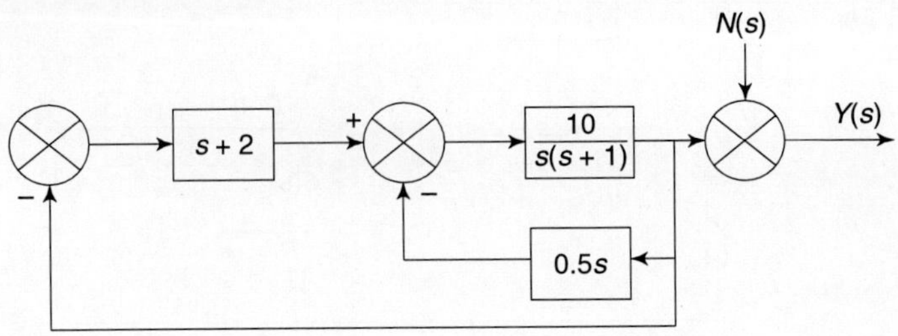

The above diagram can be redrawn as

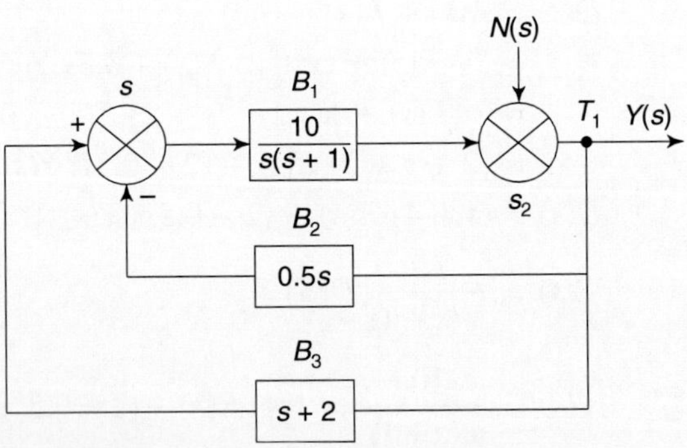

Eliminating blocks B_2 and B_3, we get

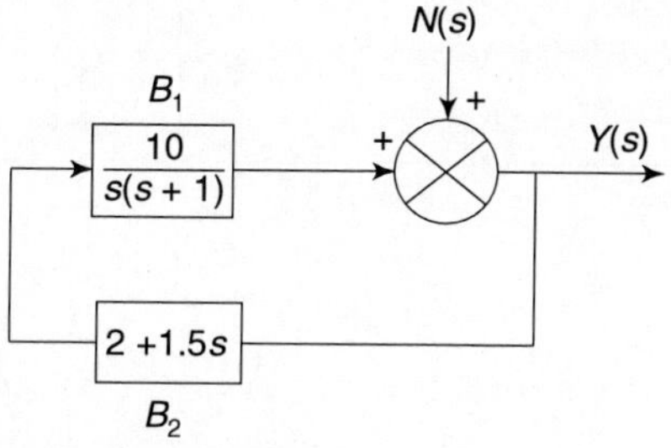

Blocks B_1 and B_2 are in cascade (Rule 1)

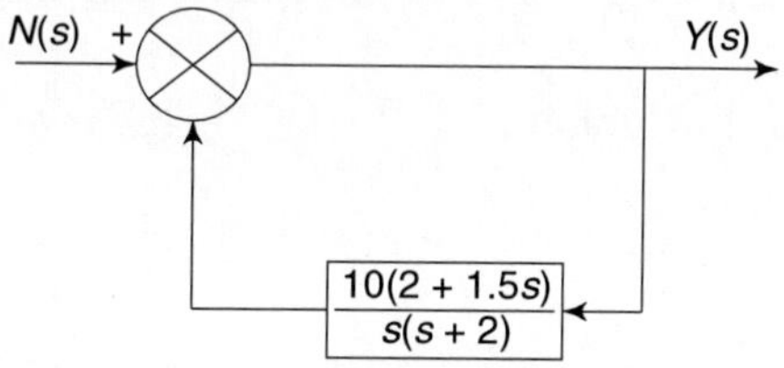

$$\frac{Y(s)}{N(s)} = \frac{1}{1+\dfrac{10\,(2+1.5s)}{s\,(s+1)}}$$

$$\frac{Y(s)}{N(s)} = \frac{s\,(s+1)}{s\,(s+1)+10\,(2+1.55)}$$

$$\frac{Y(s)}{N(s)} = \frac{s\,(s+1)}{s^2+16s+20} \qquad \textbf{Ans.}$$

Example 2.26 Draw the signal flow graph of the following system. Also find overall gain using Mason's gain formula with $T_o(s) = 0$. *(U.P.T.U., 2005–05, CS)*

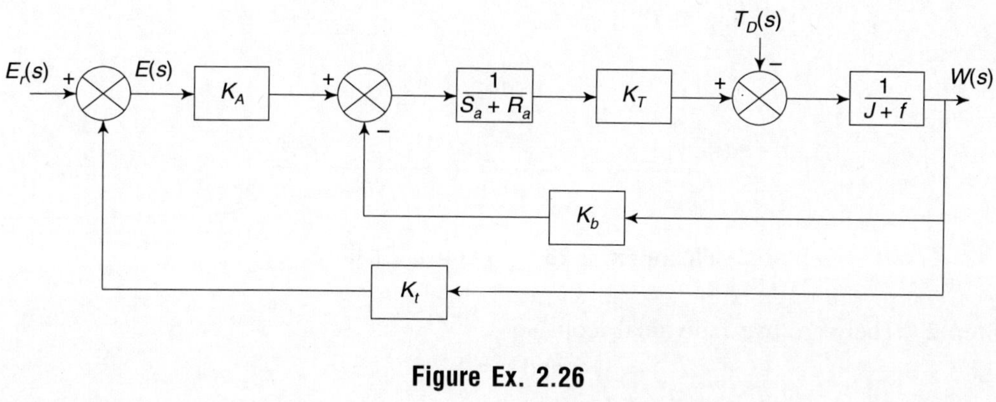

Figure Ex. 2.26

Solution

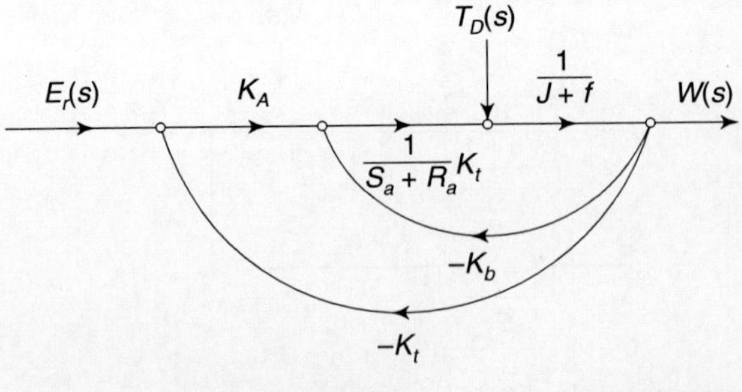

Figure Ex. 2.26 (a) SFG for given block diagram

When $T_D(s) = 0$ the signal flow graph is

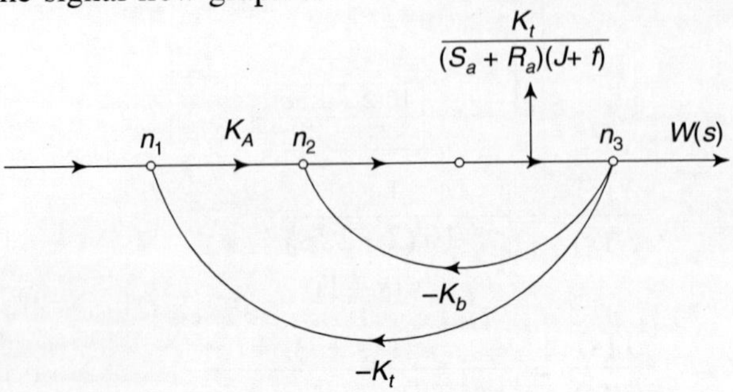

Figure Ex. 2.26 (b)

Step 1 There is one forward path, i.e., $i = 1$

$$P_i = P_1$$

$$P_1 = \frac{K_A\, K_T}{(S_a + R_a)(J + f)}$$

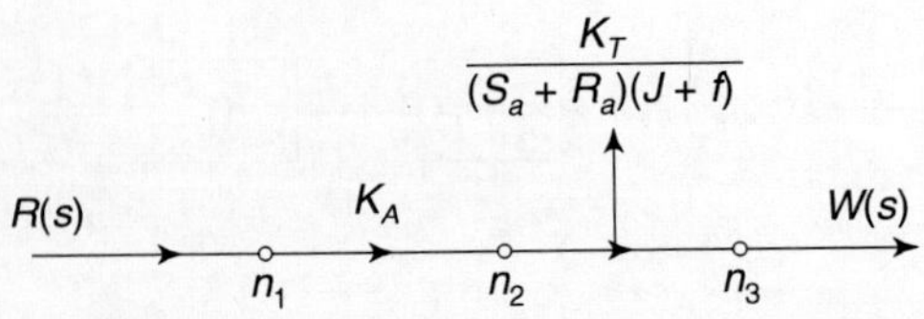

Figure Ex. 2.26 (c) Forward path P_1

Step 2 There are two individual loops so

$$j = 1, 2 \text{ and } r = 1$$

$$L_{jr} = L_{11}, L_{21}$$

$$L_{11} = \frac{-K_b\,K_T}{(S_a + R_a)(J + f)} \qquad\qquad L_{21} = \frac{-K_t\,K_A\,K_T}{(S_a + R_a)(J + f)}$$

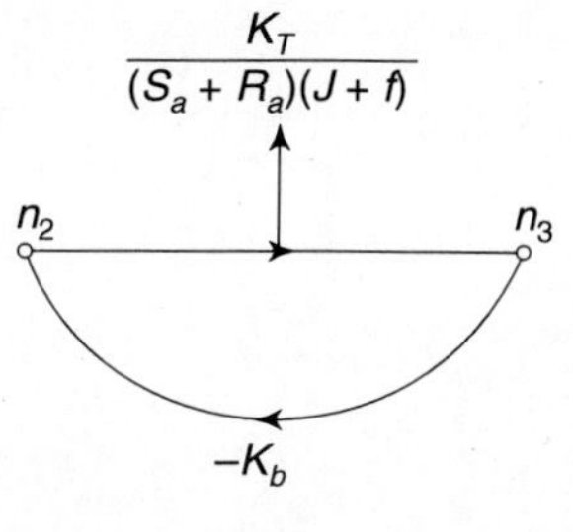

Figure Ex. 2.26 (d) Individual loop L_{11}

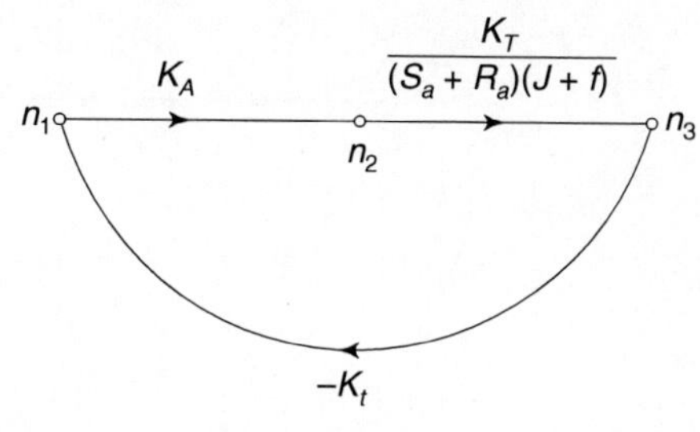

Figure Ex. 2.26 (e) Individual loop L_{21}

Step 3 and 4 There is no combination of two, three, four, non-touching loops

Step 5 Now by Eq. (2.9)

$$\begin{aligned}\Delta &= 1 - \Sigma L_{j1} + \Sigma L_{j2} - \Sigma L_{j3} + \ldots \\ &= 1 - (L_{11} + L_{21}) + 0 \\ &= 1 - \left[\frac{-K_b\,K_T}{(S_a + R_a)(J + f)} - \frac{K_A K_T K_t}{(S_a + R_a)(J + g)}\right] \\ &= = 1 + \frac{1}{(S_a + R_a)(J + f)}[K_b K_T + K_A K_T K_t]\end{aligned}$$

Step 6 Removing the loop gains of Δ touching the forward path P_1

$$\Delta_1 = 1$$

Now by Eq. (2.9)

$$T_o = \frac{W(s)}{E_r(s)} = \frac{P_1\,\Delta_1}{\Delta}$$

$$\frac{W(s)}{E_r(s)} = \frac{\dfrac{K_A\,K_T}{(S_a + R_a)(J + f)} \times 1}{1 + \dfrac{1}{(S_a + R_a)(J + f)}[K_b\,K_T + K_A\,K_T\,K_t]}$$

$$\frac{W(s)}{E\;(s)} = \frac{K_A\,K_T}{(S_a + R_a)(J + f) + K_b\,K_T + K_A\,K_T\,K_t} \qquad \textbf{Ans.}$$

Example 2.27 Draw the signal flow graph for one blcok diagram of system shown in Fig. Ex. 2.27 and find $\left.\dfrac{Y(s)}{R(s)}\right|_{N(s)=0}$ using Mason's gain formula. *(U.P.T.U., 2003–04, CS)*

Solution

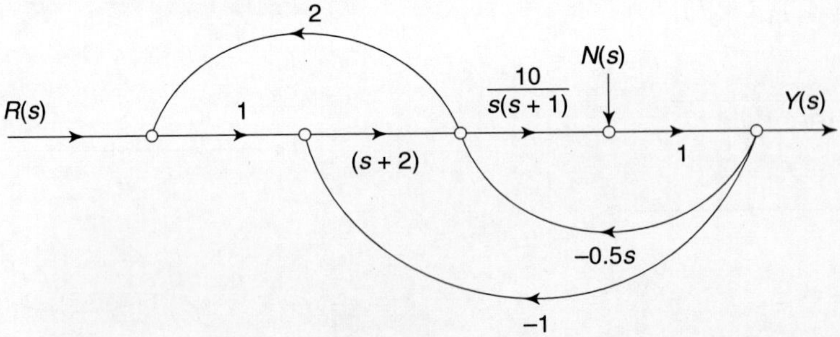

Figure Ex. 2.27

When $N(s) = 0$

The graph is redrawn

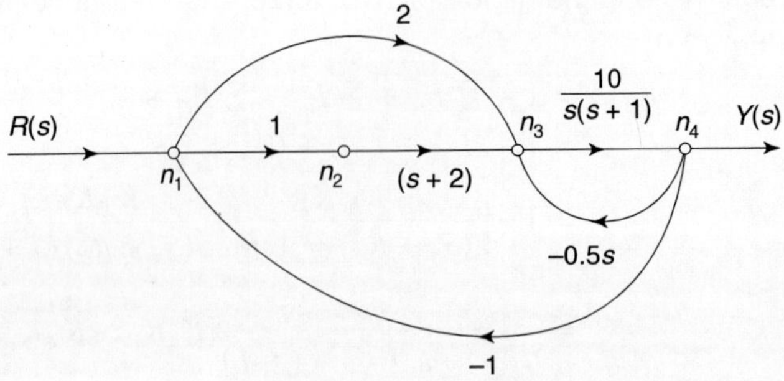

Figure Ex. 2.27 (a)

Step 1 There are two forward paths, i.e., $i = 1, 2$

$$P_i = P_1, P_2$$

$$P_2 = \frac{(s+2)\,10}{s\,(s+1)}$$

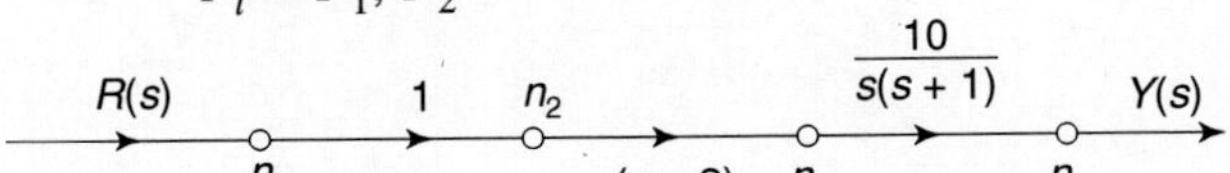

Figure Ex. 2.27 (b) Forward path P_1

$$P_2 = 2\cdot\left(\frac{10}{s\,(s+1)}\right)$$

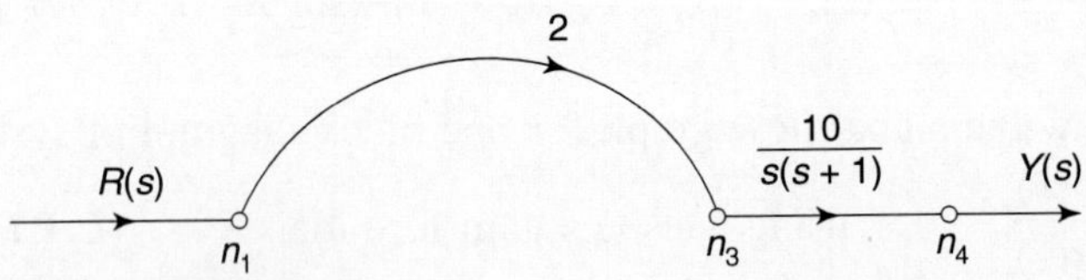

Figure Ex. 2.27 (c) Forward path P_2

Step 2 There are two individual loops, i.e.,

$$j = 1, 2 \text{ and } r = 1$$

$$L_{jr} = L_{11}, L_{21}$$

$$L_{11} = \frac{10}{s(s+1)}(-0.5s) = \frac{-5s}{s(s+1)}$$

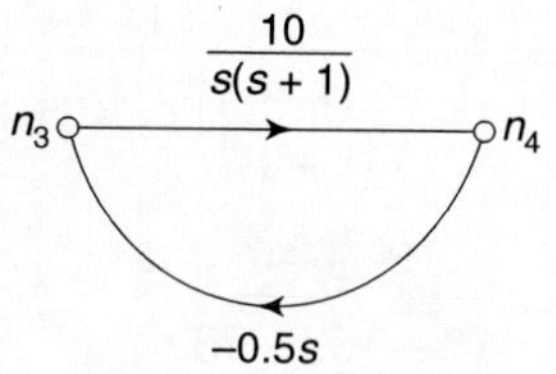

Figure Ex. 2.27 (d) Individual loop L_{11}

$$L_{21} = (s+2)\left(\frac{10}{s(s+1)}\right)(-1)$$

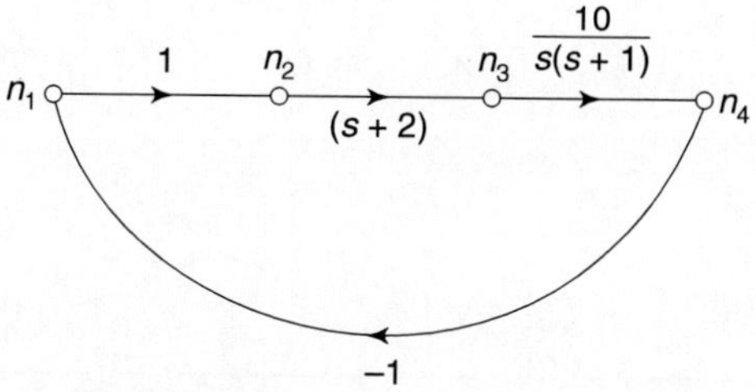

Figure Ex. 2.27 (e) Individual loop L_{21}

Step 3 and 4 There is no combination of two, three, four ... non-touching loops.

Step 5 Now by Eq. (2.10)

$$\Delta = 1 - \Sigma L_{j1} + \Sigma L_{j2} - \Sigma L_{j3} + \ldots$$

$$\Delta = 1 - (L_{11} + L_{21}) + 0$$

$$\Delta = 1 - \left[-\frac{5s}{s(s+1)} - \frac{10s+20}{s(s+1)}\right]$$

$$\Delta = 1 + \frac{1}{s(s+1)}[15s+20]$$

Step 6 Removing the loop gain from Δ following the forward paths P_1 and P_2 for path P_1 and P_2

$$\Delta_1 = 1 \text{ and } \Delta_2 = 1$$

Now by Eq. (2.9)

$$T_o = \frac{1}{\Delta}\sum P_i\,\Delta_i$$

$$T_o = \frac{P_1\,\Delta_1 + P_2\,\Delta_2}{\Delta}$$

$$\frac{Y(s)}{R(s)} = \frac{\dfrac{10\,(s+2)}{s\,(s+1)} + \dfrac{20}{s\,(s+1)}}{1 + \dfrac{1}{s\,(s+1)}[15s+20]}$$

$$\frac{Y(s)}{R(s)} = \frac{10s+40}{s^2+16s+20} \quad \textbf{Ans.}$$

Example 2.28 For the system represented by the following equations, draw the signal flow graph. *(U.P.T.U., 2002–03, CS)*

$$X = X_1 + \alpha_3 U$$
$$\dot{X}_1 = -\beta_1 X_1 + X_2 + \alpha_2 U$$
$$\dot{X}_2 = -\beta_2 X_1 + \alpha_1 U$$

Solution

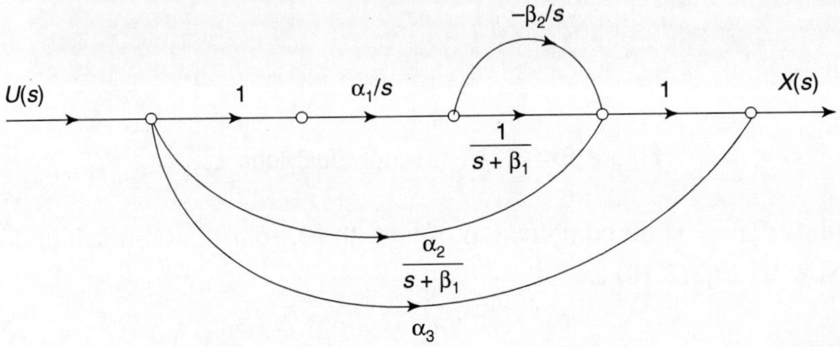

Figure Ex. 2.28 SFG for given equation

Given $\quad X = X_1 + \alpha_3 U$

Taking Laplace transform

$$X(s) = X_1(s) + \alpha_3\,U(s) \quad \text{...(1)}$$
$$X_1 = -\beta_1 X_1 + X_2 + \alpha_2 U$$

Taking Laplace transform

$$sX_1(s) = -\beta_1 X_1(s) + X_2(s) + \alpha_2\,U(s) \quad \text{...(2)}$$
$$(s+\beta_1)\,X_1(s) = X_2(s) + \alpha_2\,U(s)$$

$$X_1(s) = \frac{1}{s+\beta_1} X_2(s) + \frac{\alpha_2}{s+\beta_1} U(s) \qquad ...(3)$$

$$\dot{X}_2 = \beta_2 X_1 + \alpha_1 U$$

Taking Laplace transform

$$sX_2(s) = -\beta_2 X_1(s) + \alpha_1 U(s)$$

$$X_2(s) = \frac{\beta_2}{s} X_1(s) + \frac{\alpha_1}{s} U(s) \qquad ...(4)$$

The signal flow graph for Eqs. (1), (2) and (4) is shown in Fig. Ex. 2.28.

EXERCISE

2.1. Using block diagram reduction techniques, find the closed loop transfer function of the system whose block diagram is given in Fig. E. 2.1 (a) and (b)

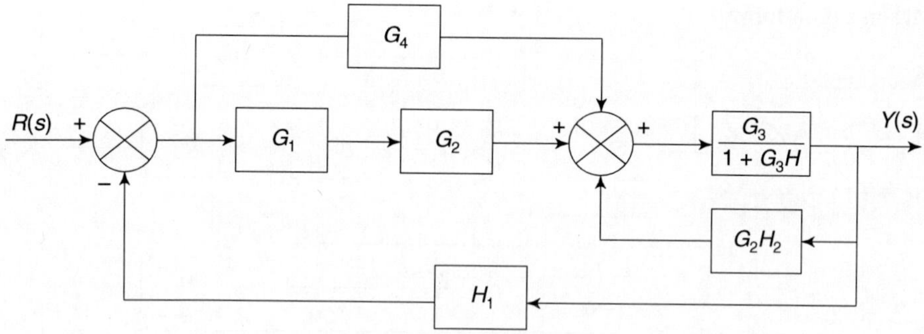

Figure E. 2.1 (a)

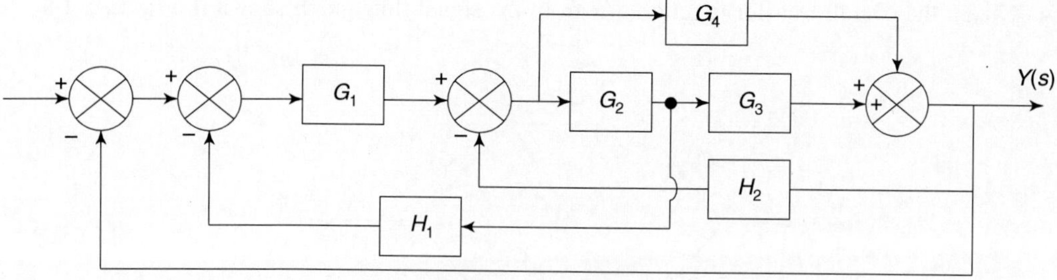

Figure E. 2.1 (b)

2.2. For the control system shown in Fig. E. 2.2, find the ratio *Y*/*R* using Mason's gain formula.

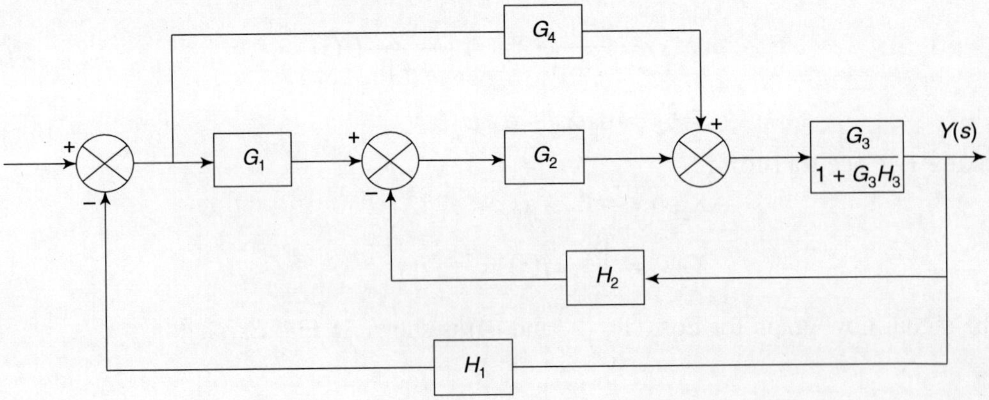

Figure E. 2.2

2.3. For the control system shown in Fig. E. 2.3, find the transfer function of the system using Mason's gain formula.

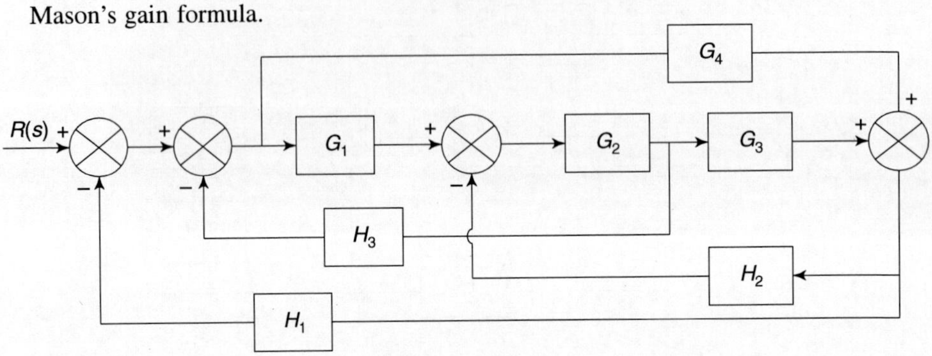

Figure E. 2.3

2.4. Obtain the overall trnasfer function *Y*/*R* from the signal flow graph shown in Fig. E. 2.4.

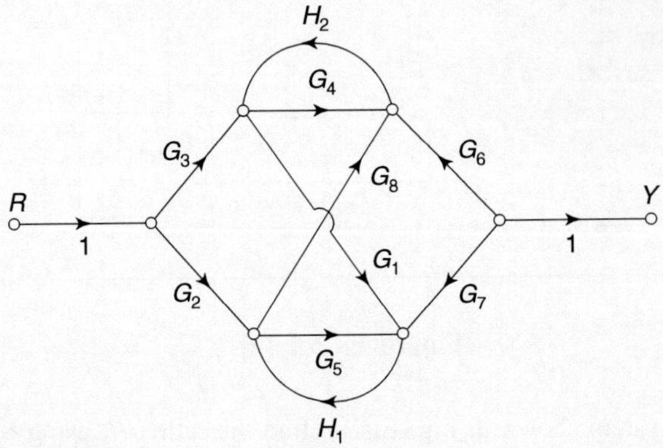

Figure E. 2.4

2.5. The system represented by the following equations, find the transfer function $Y(s)/U(s)$ by the signal flow graph.

$$x = x_1 + \beta_2 u$$
$$\dot{x}_1 = -a_1 x_1 + x_2 + \beta_3 u$$
$$\dot{x}_2 = -a_2 x_1 + \beta_1 u$$

2.6. Find Y/R for the following system using Mason's gain formula shown in Fig. E. 2.6.

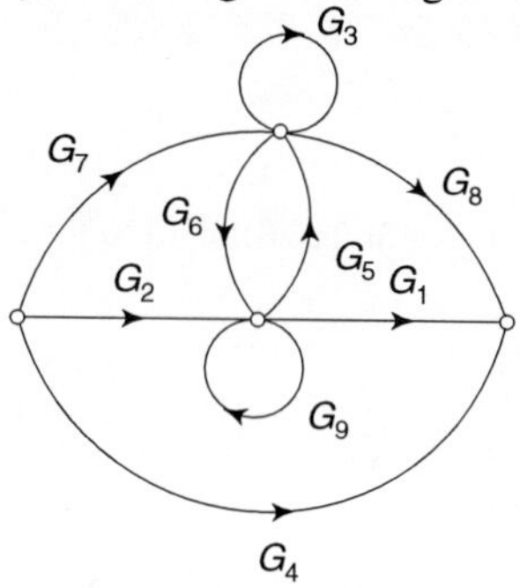

Figure E. 2.6

2.7. Find the transfer function of the system shown in Fig. E. 2.7 using block diagram reduction technique.

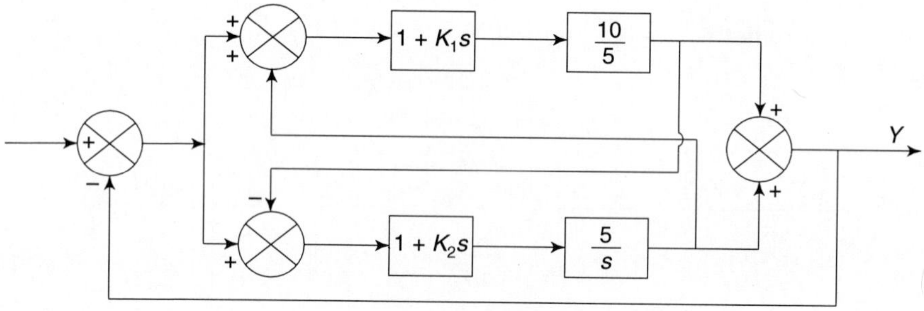

Figure E. 2.7

2.8. Why do transfer functions for mechanical networks look identical to transfer functions for electrical network?

2.9. What do we call the mechanical equations written in order to evaluate the transfer function?

2.10. Write short notes on the following:

(a) Modeling in the frequency domain

(b) Series analog circuits

(c) Parallel analog circuits

2.11. Draw a series and parallel analog for the mechanical system of Fig. E. 2.11.

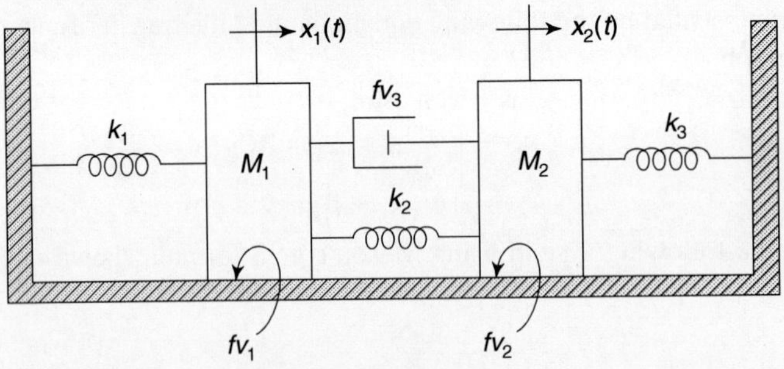

Figure E. 2.11 Translational mechanical system

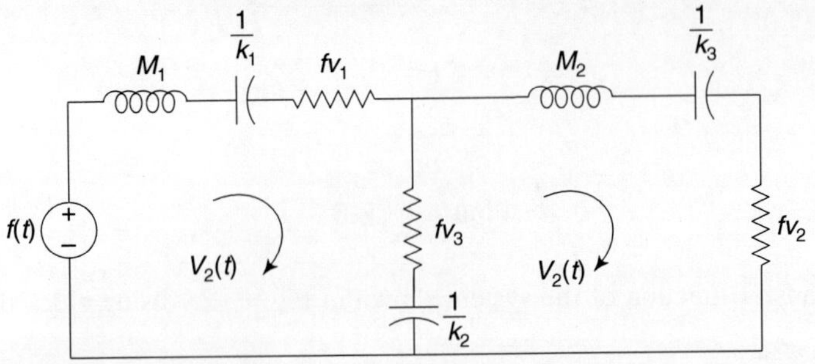

Figure E. 2.11 (a) Series analogy of Fig. E. 2.11

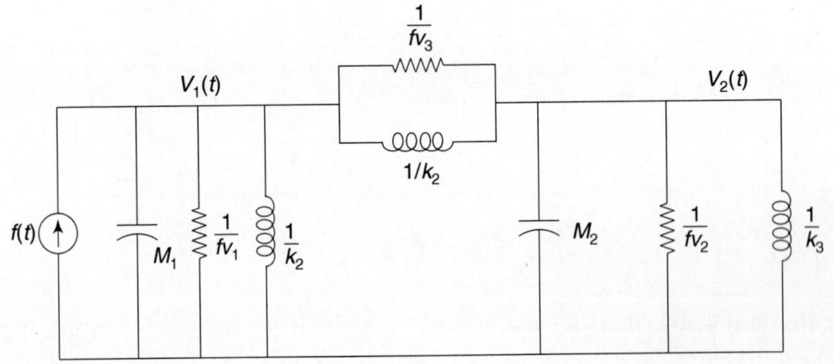

Figure E. 2.11 (b) Series analogy of Fig. E. 2.11

ANSWERS

2.1 (a) $$\frac{G_1\, G_2\, G_3 + G_3\, G_4}{1 + G_3\, H_3 + G_2\, G_3\, H_2 + G_3\, G_4\, H_1 + G_1\, G_2\, G_3\, H_1}$$

(b) $$\frac{G_1\,(G_2\, G_3 + G_4)}{1 + (G_2\, G_3 + Gu)\,(G_1 + H_2) + G_1\, H_1\, G_2}$$

2.2 $$\frac{G_3\,(G_1 G_2 + G_4)}{1 + G_3\,(G_2 H_2 + H_3 + G_4 H_1 + G_1 G_2 H_1)}$$

2.3 $$\frac{G_1\, G_2\, G_3 + G_4}{1 + G_1\, G_2\, G_3\, H_1 + G_1\, G_2\, H_3 + G_2\, G_3\, H_2 + G_4\, H_1 - G_2\, G_4\, H_2\, H_3}$$

2.4 $[G_3\, G_4\, g_6\,(1 + G_5\, H_1) + G_2\, G_5\, G_7\,(1 + G_4\, H_2) + G_2\, G_8\, G_6 + G_3\, G_1\, G_7 +$

$$\frac{[-\,G_2\, G_8\, H_2\, G_1\, G_7 - G_3\, G_1\, H_1\, G_8\, G_6]}{1 + G_4\, H_2 + G_5\, H_1 - H_2\, G_1\, H_1\, G_8 + G_4\, H_2\, G_5\, H_1}$$

2.5 $$\frac{\beta_2\,(s^2 + a_1 s + a_2) + \beta_3 s + \beta_1}{s^2 + a_1 s + a_2}$$

2.6 $$\frac{G_2 G_1\,(1 - G_3)\, G_4\,(1 - G_3 - G_9 - G_5 G_6 + G_3 G_9) + G_7 G_8\,(1 - G_9) + G_2 G_5 G_8 + G_7 G_6 G_1}{1 - G_9 - G_3 - G_5\, G_6 + G_3\, G_9}$$

2.7 $$\frac{5s\,(K_2 + 2K_1) + 15}{s\,(10\,K_1 + 5\,K_2 + 1) + 15}$$

2.11. $$\left[M_1 s + (f_{V_1} + f_{V_3}) + \frac{(K_1 + K_2)}{s}\right] V_1(s) - \left(f_{V_3} + \frac{K_2}{s}\right) V_2(s) = F(s)$$

$$-\left(f_{V_3} + \frac{K_2}{s}\right) V_1(s) + \left[M_2 s + (f_{V_2} + f_{V_3}) + \frac{(K_2 + K_3)}{s}\right] V_2(s) = 0$$

CHAPTER

3 Time Response Analysis

3.1 INTRODUCTION

The time response analysis is a analysis of the control system in which we analyze the response of the system with respect to time. We provide some test signals to the system and study the response of the system. The response of the system is generally divided in two parts: one is transient (dynamic) and other one is steady state response. The feedback control system has the ability to adjust the transient and study state performance. Basically, to analyze and design a control system, we must define and measure its performance. The parameters may be adjusted to provide the desired response based on the desired performance. The performance of the control system is usually specified in terms of both the transient response and the steady state response.

The time response is the response that disappears with time. The steady state response is that which exists a long time following any input signal.

There are certain steps to analyze the time response of the system,

(i) First obtain mathematical model of the system for particular input signal.

(ii) Time response expression can be obtained through the inverse Laplace transform.

It is nacessary to test the stability of the system before proceeding the time response analysis of a control system. If the system is unstable then there is no need to proceed with its transient response.

3.2 TIME RESPONSE

The time response of a control system is generally divided into two parts: The transient response and the steady state response.

Let $y\,(t)$ represent the time response of a continuous data system, then

$$y(t) = y_t(t) + y_{ss}(t) \qquad \text{...(3.1)}$$

where $y_t(t)$ = Transient response

$y_{ss}(t)$ = Steady state response

3.2.1 Transient Response

Transient response is defined as the part of the time response that goes to zero as time becomes very large, i.e., transient response appears only for a small period of time, thus

$$\lim_{t\to\infty} y_t(t) = 0 \quad ...(3.2)$$

In other words we can say that the transient response of a control system is the response that starts with the initial state of time and oscillatory or overdamped in nature, i.e., reveals the nature of response for a specified input test signal. The standard test signal we will discuss in the next section.

3.2.2 Steady State Response

The steady state response is simply the part of the total response that remains after the transient has disappeared, i.e., the steady state response appears a large period of time.

In other words we can say that the steady state response of control system is the response that starts with the final state of the time, of transient response. It reveals the accuracy of control system.

The time response of a control system for a specified input test signal in shown in Fig. (3.1)

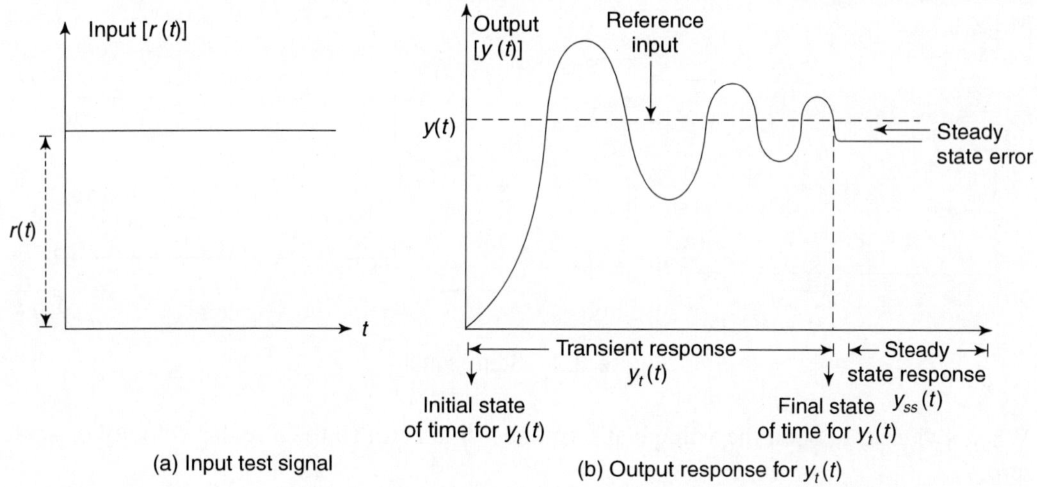

Figure 3.1 Time response

3.3 STANDARD TEST SIGNALS

Basically, time is used as an independent variable in most control systems, so generally it evaluates the state and output response with respect to time or simply the time response. The design specifications for control systems normally include several time response indices for a specified input command as well as a desired steady state accuracy. In the analysis of control

system, a reference input signal is applied to a system and the performance of the system is evaluated by studying the system response in the time domain.
The input to electrical networks and communication systems are exactly known ahead of time but to the many practical control systems are not exactly known. In many systems, the actual inputs of a control system may vary in random manner with respect to time, for example, in radar tracking system for aircraft missile, the speed and position of the aircraft to be tracked may vary in random manner, so that they cannot be determined before.

The problem discussed above regarding varying input it creates a problem for the designer that it is difficult to design system which will perform satisfactory process for all possible forms of input signals. So that for the purpose of analysis and design, it is necessary to assume some standard test inputs so that the performance of a system can be evaluated.

3.3.1 Step Signal

The step signal represents an instantaneous change in the reference inputs. For example, if the input is an angular velocity of a shaft. The step input represents the sudden rotation of shaft.

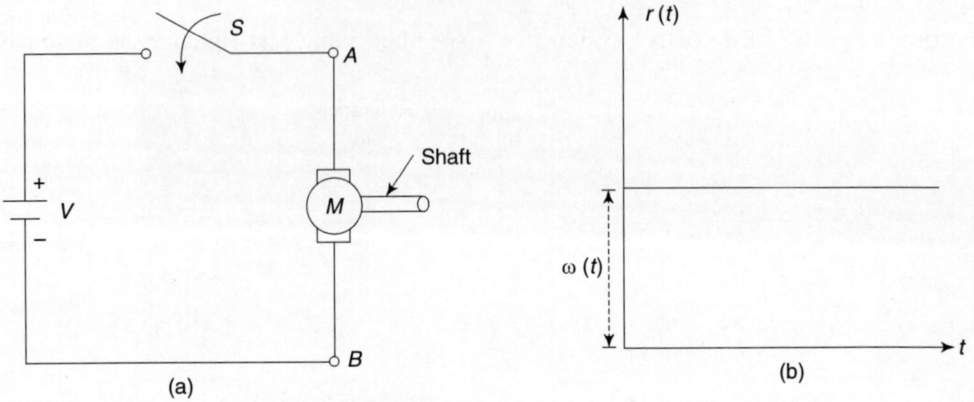

Figure 3.2 Step signal

When switch σ is open the voltage at terminal AB is zero, so the angular velocity of shaft is zero.

$$\omega(t) = 0 \qquad -\infty < t < 0 \qquad ...(3.3)$$

when switch is closed at $t = 0$

$$\omega(t) = k \qquad 0 < t < \infty \qquad ...(3.4)$$

where k = constant speed of shaft

By Eqs. (3.3) and (3.4)

$$\omega(t) = r(t) = \begin{cases} 0 & -\infty < t < 0 \\ k & 0 < t < \infty \end{cases}$$

The unit step signal is represented by $u(t)$
when $k = 1$ in Eq. (3.4) then

$$u(t) = \begin{cases} 0 & -\infty < t < 0 \\ k & 0 < t < \infty \end{cases} \quad \text{...(3.5)}$$

Laplace transformation of unit step signal is

$$L\{[u(t)\} = 1/s \quad \text{By the Table 1.1}$$

$$R(s) = 1/s$$

3.3.2 Ramp Signal

The rmp signal changes linearly with time mathematically

$$r(t) = \begin{cases} 0; & 0 < t \\ kt; & t > 0 \end{cases}$$

where k = slope of line

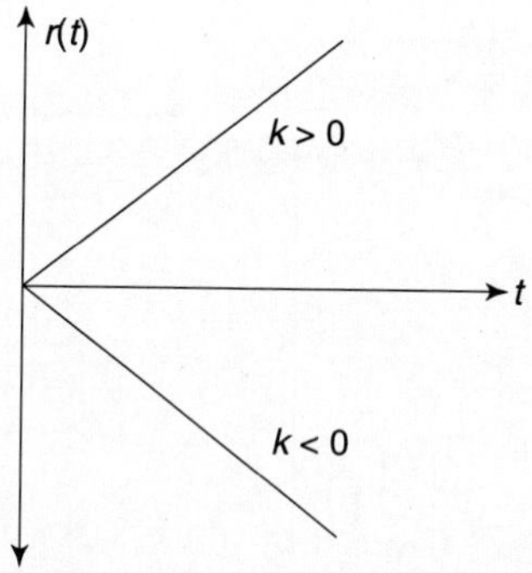

Figure 3.3 Ramp signal

Laplace of transform of ramp signal is

$$r(t) = \begin{cases} 0; & 0 < t \\ kt; & t > 0 \end{cases} \quad \text{...(3.6)}$$

$$R(s) = L[r(t)] = \frac{k}{s^2}$$

for unit ramp $k = 1$

$$\boxed{R(s) = \frac{1}{s^2}}$$

3.3.3 Impulse Signal

Unit impulse signal is defined as the signal which has zero value everywhere except $t = 0$, where its magnitude is infinite. It is symbolically represented by $\delta(t)$

$$\delta(t) = \begin{cases} 0; & t \neq 0 \\ \infty; & t = 0 \end{cases} \qquad \text{...(3.7)}$$

$$\int_{-\infty}^{\infty} \delta(t)\, dt = 1$$

$$\delta(t) = \dot{u}(t)$$

The Laplace transform of unit impulse is

$$L[\delta(t)] = L\frac{d}{dt}[u(t)]$$

$$= sL\ [\text{unit step function}]$$

$$= s \cdot \frac{1}{s} = 1$$

$$\boxed{R(s) = 1} \qquad \text{...(3.8)}$$

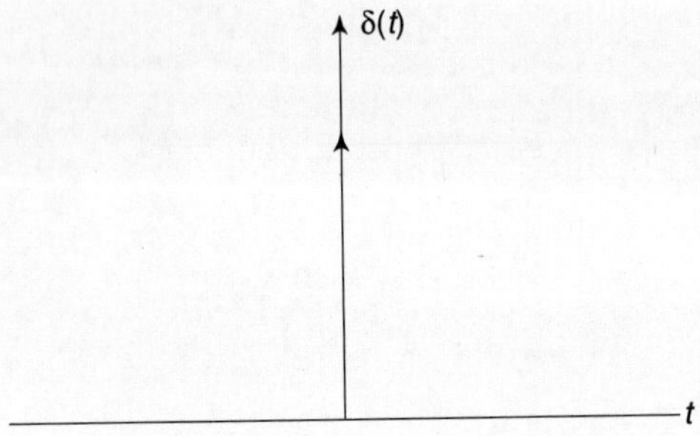

Figure 3.4 Imuulse signal

- *An Impulse function is a derivative of a step function.*

3.3.4 Parabolic Signal

The mathematical form of parabolic signal is

$$r(t) = \begin{cases} A\dfrac{t^2}{2}; & t > 0 \\ 0; & t < 0 \end{cases} \qquad \text{...(3.9)}$$

By Eq. (3.9) we can say that the value of $r(t)$ is zero for $t < 0$ and is quadratic of time for $t > 0$.

The Laplace transform of parabolic signal is

$$L\,[r(t)] = L\left[\frac{At^2}{2}\right]$$

$$\boxed{R(s) = \frac{A}{s^3}}$$

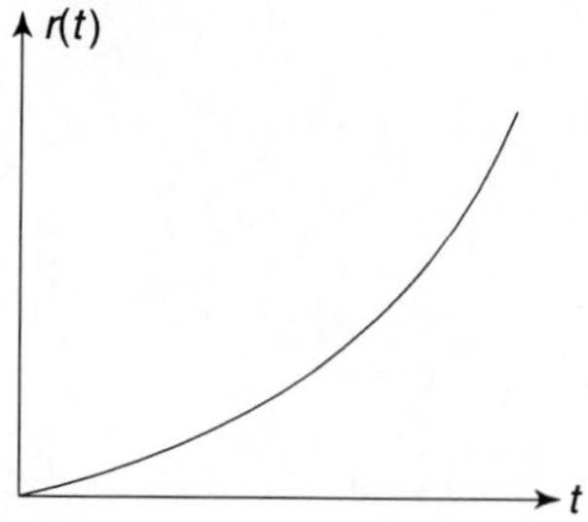

Figure 3.5 Parabolic signal

3.4 TIME RESPONSE OF THE FIRST ORDER CONTROL SYSTEM

If the highest power of s in the denominator of its transfer function is equal to one, such type of systems are called the first order control systems.

Let us consider the first order system of Fig. 3.6 with unity feedback.

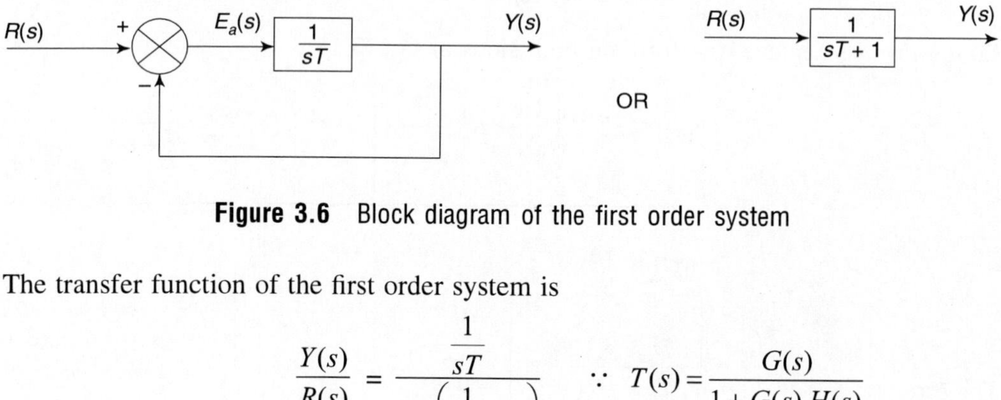

Figure 3.6 Block diagram of the first order system

The transfer function of the first order system is

$$\frac{Y(s)}{R(s)} = \frac{\frac{1}{sT}}{1+\left(\frac{1}{sT}\times 1\right)} \qquad \because\ T(s) = \frac{G(s)}{1+G(s)\,H(s)}$$

$\because$ $\quad H(s) = 1$ i.e., unity feedback

$$G(s) = \frac{1}{sT}$$

$$\frac{Y(s)}{R(s)} = \frac{1}{sT+1} \qquad \text{...(3.10)}$$

3.4.1 Time Response Subjected to Unit Step Input

By Eq. (3.10) the output of the system is

$$Y(s) = R(s)\frac{1}{sT+1} \quad ...(3.11)$$

If the input is unit step, then

$$r(t) = 1$$

Laplace transform of $r(t)$ is

$$L\,[r(t)] = L\,\{1\}$$

$$R(s) = \frac{1}{s}$$

Substituting the value of $R(s)$ in Eq. (3.11)

$$Y(s) = \frac{1}{s}\cdot\frac{1}{sT+1} \quad ...(3.12)$$

Solving Eq. (3.11) by partial fractions

$$Y(s) = \frac{1}{s} - \frac{T}{sT+1}$$

$$Y(s) = \frac{1}{s} - \frac{1}{s+\frac{1}{T}} \quad ...(3.13)$$

Taking inverse Laplace transform on both sides of Eq. (3.13).

$$L^{-1}\,[Y(s)] = L^{-1}\left[\frac{1}{s}\right] - L\left[\frac{1}{s+\frac{1}{T}}\right] \qquad \because L\,[e^{-at}] = \frac{1}{s+a}$$

$$y(t) = 1 - e^{-\frac{t}{T}}$$

$$\boxed{y(t) = 1 - e^{-\frac{t}{T}}} \quad ...(3.14)$$

The error is

$$e_a(t) = r(t) - y(t)$$

$$e_a(t) = 1 - \left(1 - e^{-\frac{t}{T}}\right)$$

$$e_a(t) = e^{-\frac{t}{T}} \quad ...(3.15)$$

The steady state error is

$$e_{ss}(t) = \lim_{t\to\infty} e^{-\frac{t}{T}}$$

$$e_{ss}(t) = 0 \quad ...(3.16)$$

where $e_{ss}(t)$ = Steady state error

T = Time constant

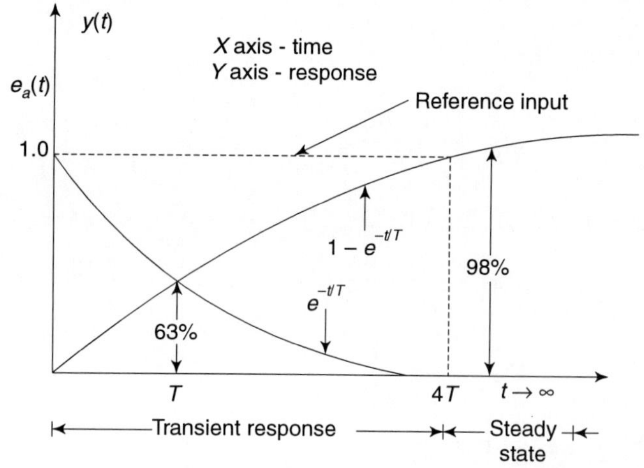

Figure 3.7 Unit step response of first order system

Considering two control systems having time constants T_1 and T_2 respectively if $T_1 > T_2$ that means the control system one is slower than the second one. We can say that the second will reach earlier at steady state, i.e., system two is faster than the system one as shown in Fig. 3.8.

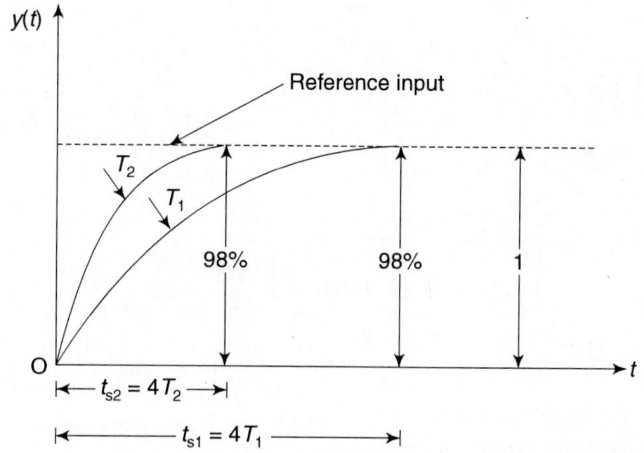

Figure 3.8 Comparison of time response of two control systems having constants T_1 and T_2, $T_2 < T_1$

3.4.2 Time Response Subjected to Unit Ramp Input

The output of the system is

$$Y(s) = R(s)\frac{1}{sT+1} \quad \text{...(3.17)}$$

If the input is unit ramp then

$$r(t) = t$$

$$R(s) = \frac{1}{s^2}$$

Substitute the value of $R(s)$ in Eq. (3.17)

$$Y(s) = \frac{1}{s^2}\cdot\frac{1}{sT+1}$$

$$Y(s) = \frac{1}{s^2(sT+1)} \quad \text{...(3.18)}$$

Solving Eq. (3.18) by partial fraction

$$Y(s) = \frac{1-sT}{s^2}+\frac{T^2}{sT+1}$$

$$Y(s) = \frac{1}{s^2}-\frac{sT}{s^2}+\frac{T^2}{T\left(s+\frac{1}{T}\right)}$$

$$Y(s) = \frac{1}{s^2}-T\frac{1}{s}+T\frac{1}{s+\frac{1}{T}} \quad \text{...(3.19)}$$

Taking the inverse Laplace transfer of Eq. (3.19)

$$L^{-1}[Y(s)] = L^{-1}\left[\frac{1}{s^2}\right]-TL^{-1}\left[\frac{1}{s}\right]+TL^{-1}\left[\frac{1}{s+\frac{1}{T}}\right]$$

$$\boxed{y(t) = t-T+Te^{-\frac{t}{T}}} \quad \text{...(3.20)}$$

The error is

$$e_a(t) = r(t)-y(t)$$

$$= t-\left(t-T+Te^{-\frac{t}{T}}\right)$$

$$e_a(t) = T-Te^{-\frac{t}{T}}$$

$$\boxed{e_a(t) = T\left(1-e^{-\frac{t}{T}}\right)} \quad \text{...(3.21)}$$

The steady state error is

$$e_{ss}(t) = \lim_{t \to \infty} \left(t - Te^{-\frac{t}{T}} \right)$$

$$e_{ss}(t) = T \qquad \text{...(3.22)}$$

The time response in relation to Eq. (3.22)

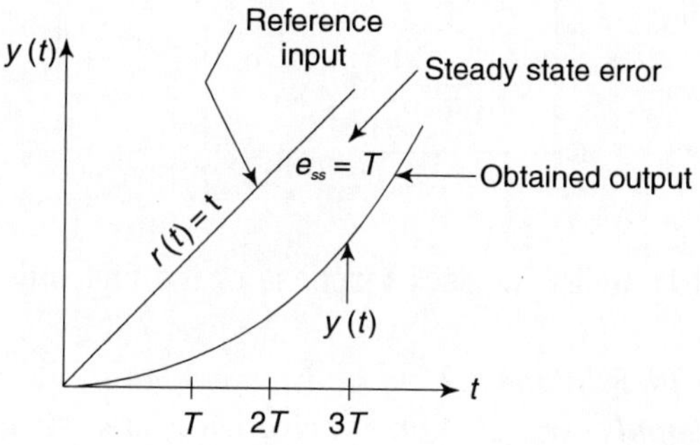

Figure 3.9 Unit ramp response of the first order system

3.4.3 Time Response Subjected to Unit Impulse Input

The output of the system is

$$Y(s) = R(s)\frac{1}{sT + 1} \qquad \text{...(3.23)}$$

If the input is unit impulse

$$R(s) = 1$$

Substituting $R(s)$ in Eq. (3.23)

$$Y(s) = 1 \cdot \frac{1}{sT + 1} \qquad \text{...(3.24)}$$

Taking inverse Laplace transform on both sides of Eq. (3.24)

$$L^{-1}[Y(s)] = L^{-1}\left[\frac{1}{sT+1}\right]$$

$$y(t) = \frac{1}{T}L^{-1}\left[\frac{1}{s+\frac{1}{T}}\right]$$

$$\boxed{y(t) = \frac{1}{T}e^{-\frac{t}{T}}} \qquad \text{...(3.25)}$$

The time response is

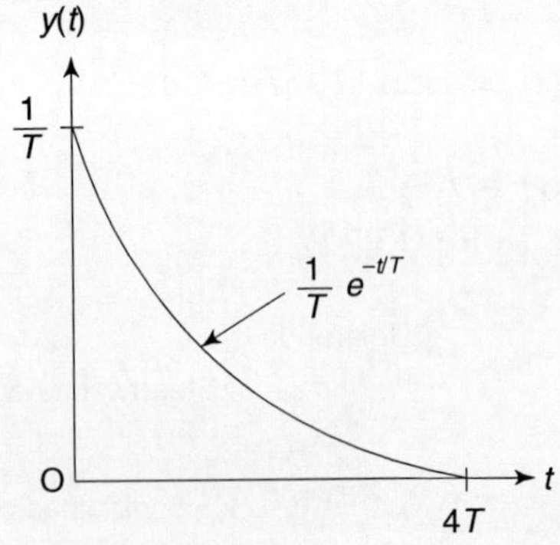

Figure 3.10 Unit imupulses response of the first order system

- *By examining Eqs. 3.20, 3.14 and 3.25, it is observed that step signal is the first derivative of ramp signal and impulse signal is first derivative of step signal and second derivative of ramp signal, so we can say that the output time response also follows the same sequence as that of input.*

3.5 TIME RESPONSE OF THE SECOND ORDER CONTROL SYSTEM

If the highest power of s in the denominator of its transfer function is equal to two, such type of systems are called the second order control systems.

The block diagram of the second order control system is shown in Fig. 3.11.

$$\because \qquad G(s) = \frac{\omega_n^2}{s\,(s+2\,\xi\omega_n)}$$

$$H(s) = 1 \text{ (Unity feedback)}$$

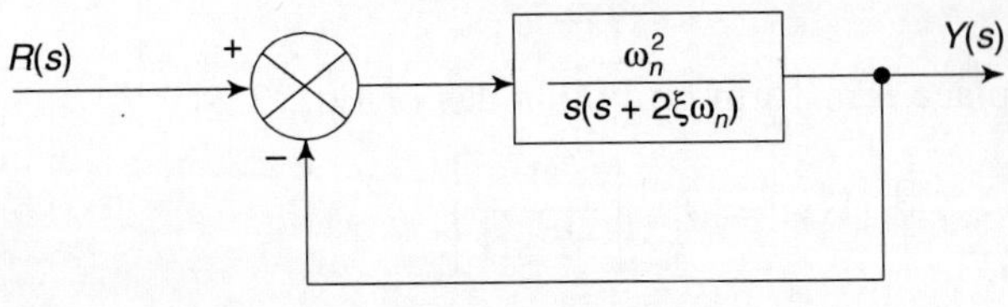

Figure 3.11 Second order control system

The transfer function of the second order control system

$$T_o(s) = \frac{Y(s)}{R(s)} = \frac{G(s)}{1+G(s)\,H(s)}$$

$$\frac{Y(s)}{R(s)} = \frac{\dfrac{\omega_n^2}{s(s+2\xi\omega_n)}}{1+\left(\dfrac{\omega_n^2}{s(s+2\xi\omega_n}\right)\times 1}$$

$$\frac{Y(s)}{R(s)} = \frac{\dfrac{\omega_n^2}{s(s+2\xi\omega_n)}}{\dfrac{s^2+2\xi\omega_n s+\omega_n^2}{s(s+2\xi\omega_n)}}$$

$$\frac{Y(s)}{R(s)} = \frac{\omega_n^2}{s^2+2\xi\omega_n s+\omega_n^2} \quad \text{...(3.26)}$$

3.5.1 Time Response Subjected to Unit Step Input

By Eq. (3.26) the output of the system is

$$Y(s) = R(s)\frac{\omega_n^2}{s^2+2\xi\omega_n s+\omega_n^2} \quad \text{...(3.27)}$$

If the input is a unit step then

$$r(t) = 1$$

Laplace transform of $r(t)$ is

$$L\,[r(t)] = L\,\{1\}$$

$$R(s) = \frac{1}{s}$$

Substituting the value of $R(s)$ in Eq. (3.27)

$$Y(s) = \frac{1}{s}\cdot\frac{\omega_n^2}{s^2+2\xi\omega_n s+\omega_n^2} \quad \text{...(3.28)}$$

$$Y(s) = \frac{1}{s}\cdot\frac{\omega_n^2}{s^2+2\xi\omega_n s+\omega_n^2+\xi^2\omega_n^2-\xi^2\omega_n^2}$$

$$Y(s) = \frac{1}{s}\cdot\frac{\omega_n^2}{s^2+2\xi\,\omega_n s+\xi^2\omega_n^2+\omega_n^2-\xi^2\omega_n^2}$$

$$Y(s) = \frac{1}{s}\cdot\frac{\omega_n^2}{(s+\xi\,\omega_n)+\omega_n^2(1-\xi^2)} \quad \text{...(3.29)}$$

Solving the Eq. (3.29) by partial fracrtion

$$Y(s) = \frac{1}{s} - \frac{s + 2\xi\omega_n}{[(s+\xi\omega_n)^2 + \omega_n^2(1-\xi^2)]} \qquad ...(3.30)$$

Let $\omega_d = \omega_n\sqrt{1-\xi^2}$ then

$$Y(s) = \frac{1}{s} - \frac{s + 2\xi\omega_n}{(s+\xi\omega_n)^2 + \omega_d^2} \qquad ...(3.31)$$

Equation (3.31) can be rewritten as follows :

$$Y(s) = \frac{1}{S} - \left[\frac{s + \xi\omega_n + \xi\omega_n}{(s+\xi\omega_n)^2 + \omega_d^2}\right]$$

$$Y(s) = \frac{1}{s} - \left[\frac{s + \xi\omega_n}{(s+\xi\omega_n)^2 + \omega_d^2} + \frac{\xi\omega_n}{(s+\xi\omega_n)^2 + \omega_d^2}\cdot\frac{\omega_d}{\omega_d}\right]$$

$$Y(s) = \frac{1}{s} - \frac{s + \xi\omega_n}{(s+\xi\omega_n)^2 + \omega_d^2} - \frac{\xi\omega_n}{\omega_d}\cdot\frac{\omega_d}{(s+\xi\omega_n)^2 + \omega_d^2} \qquad ...(3.32)$$

Taking inverse Laplace transform of Eq. (3.32)

$$L^{-1}[Y(s) = L^{-1}\left[\frac{1}{s}\right] - L^{-1}\left[\frac{s + \xi\omega_n}{(s+\xi\omega_n)^2 + \omega_d^2} - \frac{\xi\omega_n}{\omega_d}\cdot\frac{\omega_d}{(s+\xi\omega_n)^2 + \omega_d^2}\right]$$

$$y(t) = 1 - e^{-\xi\omega_n t}\cos\omega_d t - \frac{\xi\omega_n}{\omega_d}e^{-\xi\omega_{nt}}\sin\omega_d t \qquad ...(3.33)$$

$\because$ $\omega_d = \omega_n\sqrt{1-\xi^2}$ so the Eq. (3.33) is

$$c(t) = 1 - \frac{e^{-\xi\omega_n t}}{\sqrt{1-\xi^2}}\left[\sqrt{1-\xi^2}\cos\omega_d t + \xi\sin\omega_d t\right] \qquad ...(3.34)$$

Let $\sin\theta = \sqrt{1-\xi^2}$ $\therefore \cos\theta = \xi$

$\therefore$

$$c(t) = 1 - \frac{e^{-\xi\omega_n t}}{\sqrt{1-\xi^2}}(\sin\theta\cos\omega_d t + \cos\theta\sin\omega_d t)$$

$$= 1 - \frac{e^{-\xi\omega_{nt}}}{\sqrt{1-\xi^2}}[\sin(\omega_{dt} + \theta)] \qquad ...(3.35)$$

$$\tan\theta = \frac{\sqrt{1-\xi^2}}{\xi}$$

$$\theta = \tan^{-1}\frac{\sqrt{1-\xi^2}}{\xi}$$

Equation (3.35) can be rewritten as

$$c(t) = c(t) = 1 - \frac{e^{-\xi\omega_n t}}{\sqrt{1-\xi^2}} \sin\left[\omega_n\sqrt{1-\xi^2}\, t + \tan^{-1}\frac{\sqrt{1-\xi^2}}{\xi}\right] \quad ...(3.36)$$

The error is $e_a(t) = r(t) - y(t)$

$$e_a(t) = 1 - \left[1 - \frac{e^{-\xi\omega_n t}}{\sqrt{1-\xi^2}} \sin[\omega_n\sqrt{1-\xi^2}\, t + \tan^{-1}\frac{\sqrt{1-\xi^2}}{\xi}\right]$$

$$\boxed{e_a(t) = \frac{e^{-\xi\omega_n t}}{\sqrt{1-\xi^2}} \sin\left[\omega_n\sqrt{1-\xi^2}\, t + \tan^{-1}\frac{\sqrt{1-\xi^2}}{\xi}\right]} \quad ...(3.37)$$

The steady state error is

$$\boxed{e_{ss}(t) = \lim_{t\to\infty} \frac{e^{-\xi\omega_n t}}{\sqrt{1-\xi^2}} \sin\left[\omega_n\sqrt{1-\xi^2}\, t + \tan^{-1}\frac{\sqrt{1-\xi^2}}{\xi}\right]} \quad ...(3.38)$$

where ξ = damping ratio

ω_n = natural frequency of oscillations

$\omega_d = \omega_n\sqrt{1-\xi^2}$ is damped frequency of oscillations

$\xi\omega_n$ = damping factor or actual damping or damping coefficient.

The time response of a second order control system given by Eq. (3.36) is based on its damping ratio (ξ).

Case 1 Underdamped ($\xi < 1$) Response

From Eq. (3.36), it is clear that the time constant is $\frac{1}{\xi\,\omega_n}$ and response having damped oscillation with undershoot and overshoot, such response is called underdamped response.

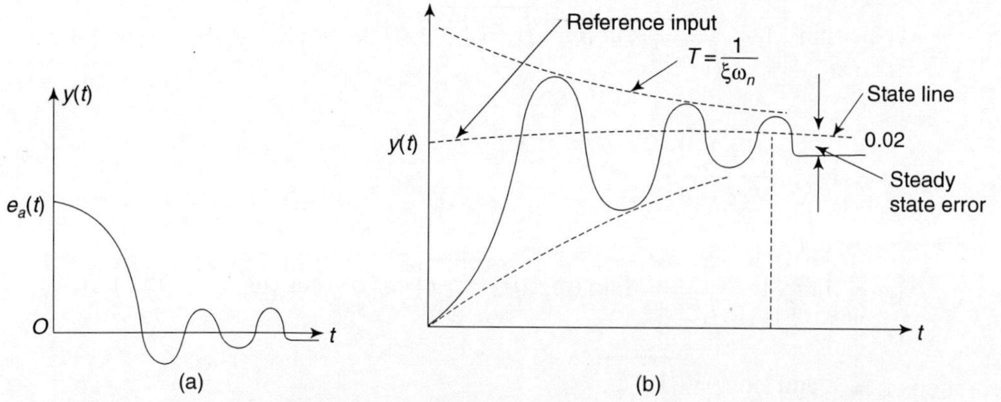

Figure 3.12 (a) Error of a second order system for $\xi < 1$, (b) Time response

Case 2 Undamped Oscillation ($\xi = 0$)

Substituting $\xi = 0$ in Eq. (3.36)

$$y(t) = 1 - \frac{e^{-0\omega_n t}}{\sqrt{1-0^2}} \sin\left[\omega_n \sqrt{1-0^2}\, t \tan^{-1} \frac{\sqrt{1-0^2}}{0}\right]$$

$$y(t) = 1 - \sin(\omega_n t + \tan_{\infty}^{-1})$$

$$y(t) = 1 - \sin\left(\omega_{nt} + \frac{\pi}{2}\right)$$

$$y(t) = (1 - \cos\omega_n t) \qquad ...(3.39)$$

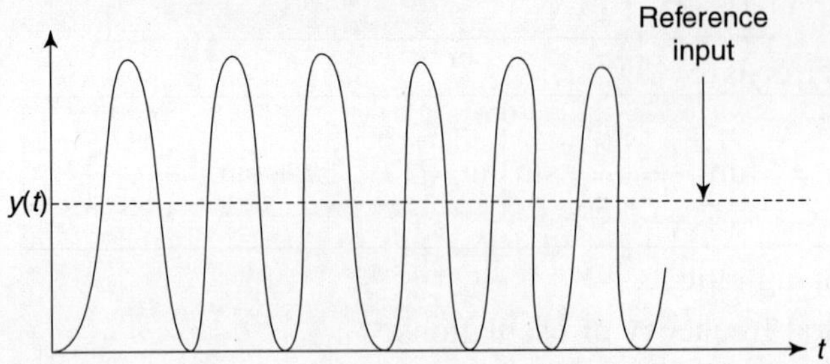

Figure 3.13 Time response of second order system for $\xi = 0$

Case 3 Critically Damped ($\xi = 1$)

Taking the limits of Eq. (3.36) as the value of $\xi \to 1$.

$$y(t) = \lim_{\xi \to 1}\left[1 - \frac{e^{-\xi\omega_n t}}{\sqrt{1-\xi^2}} \sin\left(\omega_n \sqrt{1-\xi^2}\, t + \tan^{-1}\frac{\sqrt{1-\xi^2}}{\xi}\right)\right.$$

$$= \lim_{\xi \to 1}\left[1 - \frac{e^{-\xi\,\omega_n t}}{\sqrt{1-\xi^2}} \sin\left(\omega_n \sqrt{1-\xi^2}\, t + \phi\right)\right.$$

$$\because \quad \phi = \tan^{-1}\frac{\sqrt{1-\xi^2}}{\xi}$$

$$y(t) = \lim_{\xi \to 1}\left\{1 - \frac{e^{-\xi\,\omega_n t}}{\sqrt{1-\xi^2}}\left(\sin\left(\omega_n \sqrt{1-\xi^2}\, t\right)\cos\phi + \cos\left(\omega_n \sqrt{1-\xi^2}\, t\right)\sin\phi\right\}\right.$$

$$\therefore \quad \cos\phi = \xi \text{ and } \sin\phi = \sqrt{1-\xi^2}$$

Substituting sin ϕ and cos ϕ in the above equation

$$y(t) = \lim_{\xi \to 1} \left\{ 1 - \frac{e^{-\xi \omega_n t}}{\sqrt{1-\xi^2}} \left(\sin\left(\omega_n \sqrt{1-\xi^2}\, t\right) \xi + \cos\left(\omega_n \sqrt{1-\xi^2}\, t\right) \sqrt{1-\xi^2} \right) \right\} \quad ...(3.40)$$

$$y(t) = \lim_{\xi \to 1} \left[1 - \frac{e^{-\xi \omega_n t}}{\sqrt{1-\xi^2}} \left(\omega_n \sqrt{1-\xi^2}\, t + 1 \cdot \sqrt{1-\xi^2} \right) \right]$$

$$y(t) = \lim_{\xi \to 1} \left[1 - \frac{e^{-\xi \omega_n t}}{\sqrt{1-\xi^2}} \sqrt{1-\xi^2}\, (\omega_n t + 1) \right]$$

$$\boxed{y(t) = 1 - e^{-\xi \omega_n t} (\omega_n t + 1)} \quad ...(3.41)$$

The time response related to Eq. (3.41) is

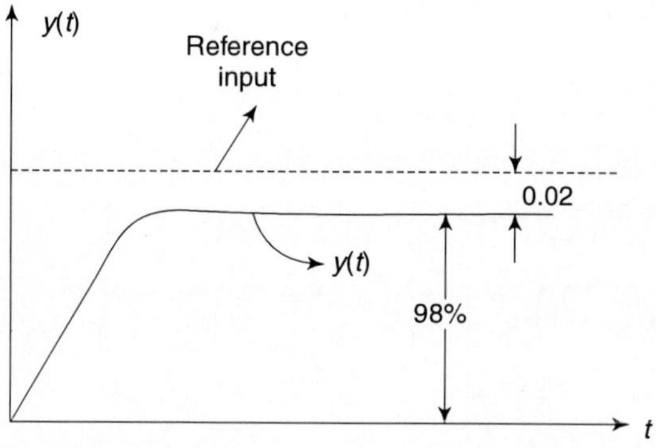

Figure 3.14 Time response of second order system for $\xi = 1$

Case 4 Overdamped ($\xi > 1$) response

For $\xi > 1$ Eq. (3.42) can be written as

$$y(s) = \frac{1}{s} \cdot \frac{\omega_n^2}{(s + \xi\omega_n)^2 - \omega_n^2 (\xi^2 - 1)}$$

$$y(s) = \frac{1}{s} \cdot \frac{\omega_n^2}{\underbrace{\left[s + \left(\xi + \sqrt{\xi^2 - 1}\right)\omega_n\right]}_{x_1} \underbrace{\left[s + \left(\xi - \sqrt{\xi^2 - 1}\right)\omega_n\right]}_{x_2}} \quad ...(3.42)$$

Solving Eq. (3.42) by partial fraction

$$= \frac{\omega_n^2}{s\left[s + \left(\xi + \sqrt{\xi^2 - 1}\right)\omega_n\right]\left[s + \left(\xi - \sqrt{\xi^2 - 1}\right)\omega_n\right]}$$

$$= \frac{A}{s} + \underbrace{\frac{B}{s + \left(\xi + \sqrt{\xi^2 - 1}\right)\omega_n}}_{x_1} + \underbrace{\frac{C}{s + \left(\xi - \sqrt{\xi - 1}\ \right)\omega_n}}_{x_2} \qquad \text{...(3.43)}$$

Solving the above Eq.

$$A = 1$$

$$B = \frac{1}{\underbrace{2\sqrt{\xi^2 - 1}\left[\xi + \sqrt{\xi^2 - 1}\right]}_{y_1}}$$

$$C = \frac{1}{\underbrace{2\sqrt{\xi^2 - 1}\left[\xi - \sqrt{\xi^2 - 1}\right]}_{y_2}}$$

Putting the values of A, B and C in Eq. (3.44)

$$y(s) = \frac{1}{s} + \frac{1}{x_1\, y_1} - \frac{1}{x_2\, y_2} \qquad \text{...(3.44)}$$

where $\quad x_1 = s + \left(\xi + \sqrt{\xi^2 - 1}\right)\omega_n$

$$x_2 = s + \left(\xi - \sqrt{\xi^2 - 1}\right)\omega_n$$

$$y_1 = 2\sqrt{\xi^2 - 1}\left[\xi + \sqrt{\xi^2 - 1}\right]$$

$$y_2 = 2\sqrt{\xi^2 - 1}\left[\xi - \sqrt{\xi^2 - 1}\right]$$

Taking inverse Laplace transform of Eq. (3.44)

$$y(t) = 1 + \frac{e^{-\left(\xi + \sqrt{\xi^2 - 1}\right)\omega_n t}}{2\sqrt{\xi^2 - 1}\left[\xi + \sqrt{\xi^2 - 1}\right]} - \frac{e^{-\left(\xi - \sqrt{\xi^2 - 1}\right)\omega_n t}}{2\sqrt{\xi^2 - 1}\left[\xi - \sqrt{\xi^2 - 1}\right]} \qquad \text{...(3.45)}$$

From Eq. (3.45) we get two time costants

$$T_1 = \frac{1}{\left(\xi + \sqrt{\xi^2 - 1}\right)\omega_n},$$

$$T_2 = \frac{1}{\left(\xi - \sqrt{\xi^2 - 1}\right)\omega_n}$$

From Eq. (3.45) when $\xi > 1$ in two time constants $T_1 < T_2$

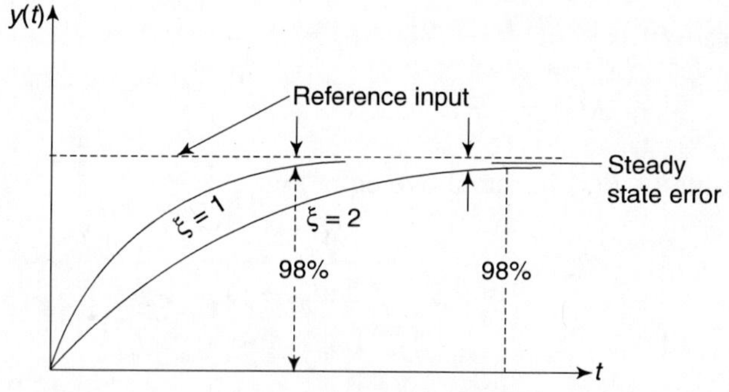

Figure 3.15 Time response of the second order system for $\xi > 1$

3.5.2 Location of Poles for a Second Order System

The general equation for the transfer function of a second order system is

$$\frac{Y(s)}{R(s)} = \frac{\omega_n^2}{s^2 + 2\,\xi\,\omega_n s + \omega_n^2} \qquad ...(3.46)$$

The characteristic equation for the poles is

$$s^2 + 2\xi\,\omega_n s + \omega_n^2 = 0 \qquad ...(3.47)$$

The roots of Eq. (3.47) are

$$P_1, P_2 = -\xi\,\omega_n \pm j\,\omega_n\sqrt{1-\xi^2}$$

P_1, P_2 are the poles of transfer function as shown in Eq. (3.47). The study of roots P_1, P_2 gives a prediction about the nature of time response. The real part of the roots represents the damping (ξ) and the imaginary part represents damped frequency (ω_d) of oscillations.

Equation (3.47) is called characteristic equation.

Let us consider the ω_n is constant and ξ is varying from 0 to ∞.

As ξ increases the poles move away from the imaginary axis along a circular path of radius ω_n meeting at the point $\sigma = -\,\omega_n$ and then separating and travelling along the real axis, one towards zero and the other towards infinity.

For $0 < \xi < 1$, the poles are complex conjugate pairs making an angle of $\theta = \cos^{-1}\xi$ with the negative real axis.

3.5.3 Time Response Subjected to Unit Ramp Input

By Eq. (3.46) the output of the system is

$$Y(s) = R(s)\frac{\omega_n^2}{s^2 + 2\xi\,\omega_n s + \omega_n^2}$$

If the input is a unit ramp then

$$r(t) = t$$

Laplace transform g $r(t)$ is

$$R(s) = \frac{1}{s^2}$$

Substituting the value of $R(s)$ in the above equation.

$$Y(s) = \frac{1}{s^2} \cdot \frac{\omega_n^2}{s^2 + 2\xi\,\omega_n s + \omega_n^2}$$

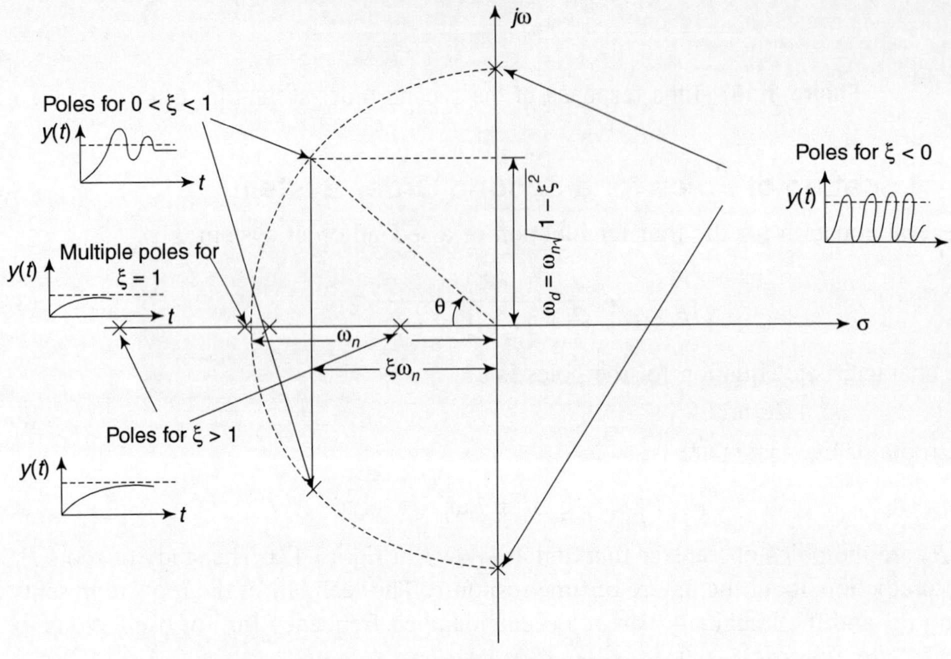

Figure 3.16 Pole's location for a second order control system and corresponding time response

Rewriting the term $s^2 + 2\xi\,\omega_n\,s + \omega_n^2$ as in Eq. (3.29)

$$Y(s) = \frac{1}{s^2} \cdot \frac{\omega_n^2}{(s + \xi\,\omega_n)^2 + \omega_n^2\left(1 - \xi^2\right)} \qquad \text{...(3.48)}$$

Solving Eq. (3.48) by partial fraction

$$\frac{\omega_n^2}{s^2\left[(s + \xi\,\omega_n)^2 + \omega_n^2\left(1 - \xi^2\right)\right]} = \frac{A}{s} + \frac{B}{s^2} + \frac{C_s + D}{(s + \xi\,\omega_n)^2 + \left(1 - \xi^2\right)\omega_n^2} \qquad \text{...(3.49)}$$

$$\omega_n^2 = As(s^2 + 2\xi\omega_n s + \omega_n^2) + B\,[(s^2 + \xi\,\omega_n)^2 + \omega_n\,(1 - \xi^2)] + (Cs + D)\,s^2 \qquad \text{...(3.49a)}$$

Equating the coefficient of s^3

$$A + C = 0 \quad \text{...(3.50)}$$

Equating the coefficient of s^2

$$2\xi\omega_n A + B + D = 0 \quad \text{...(3.51)}$$

Equating the coefficient of s

$$A\omega_n^2 + 2\xi\omega_n B = 0 \quad \text{...(3.52)}$$

Equating the constant term

$$B\omega_n^2 = \omega_n^2 \quad \text{...(3.53)}$$

From Eqs. (3.50) to (3.53)

$$A = \frac{-2\xi}{\omega_n}$$

$$C = \frac{2\xi}{\omega_n}$$

$$B = 1$$

$$D = -1 + 4\xi^2$$

Putting the value of A, B, C and D in Eq. (3.49)

$$Y(s) = \frac{-2\xi}{\omega_n s} + \frac{1}{s^2} + \frac{\frac{2\xi}{\omega_n} s + (4\xi^2 - 1)}{s^2 + 2\xi\omega_n s + \omega_n^2} \quad \text{...(3.54)}$$

$$Y(s) = \frac{1}{s^2} - \frac{2\xi}{\omega_n s} + \frac{2\xi}{\omega_n} \cdot \frac{(s + \xi\omega_n - \xi\omega_n)}{(s + \xi\omega_n)^2 + \omega_n^2(1 - \xi^2)} + \frac{4\xi^2 - 1}{(s + \xi\omega_n)^2 + \omega_n^2(1 - \xi^2)}$$

$$Y(s) = \frac{1}{s^2} - \frac{2\xi}{\omega_n s} + \frac{2\xi}{\omega_n} \cdot \frac{(s + \xi\omega_n)}{(s + \xi\omega_n)^2 + \omega_n^2(1 - \xi^2)} - \frac{2\xi}{\omega_n}$$

$$\times \frac{\xi\omega_n}{(s + \xi\omega_n)^2 + \omega_n(1 - \xi^2)} + \frac{4\xi^2 - 1}{(s + \xi\omega_n)^2 + \omega_n^2(1 - \xi^2)}$$

$$Y(s) = \frac{1}{s^2} - \frac{2\xi}{\omega_n s} + \frac{2\xi}{\omega_n} \cdot \frac{s + \xi\omega_n}{(s + \xi\omega_n)^2 + \omega_n\left(1 - \xi^2\right)} + \frac{2\xi^2 - 1}{(s + \xi\omega_n)^2 + \omega_n^2(1 - \xi^2)}$$

Put $\quad \omega_d^2 = \omega_n^2(1 - \xi^2)$

$$Y(s) = \frac{1}{s^2} - \frac{2\xi}{\omega_n s} + \frac{2\xi}{\omega_n} \cdot \frac{s + \xi\omega_n}{(s + \xi\omega_n)^2 + \omega_d^2} + \frac{2\xi^2}{(s + \xi\omega_n)^2} \cdot \frac{\omega_d}{\omega_d} \quad \text{...(3.55)}$$

Taking inverse Laplace transform of Eq. (3.55)

$$y(t) = t - \frac{2\xi}{\omega_n} + \frac{2\xi}{\omega_n} e^{-\xi\omega_n t} \cos \omega_d t + \frac{2\xi^2 - 1}{\omega_d} e^{-\xi\omega_n t} \sin \omega_d t \quad ...(3.56)$$

$$= t - \frac{2\xi}{\omega_n} + \frac{2\xi\sqrt{1-\xi^2}}{\omega_n\sqrt{1-\xi^2}} e^{-\xi\omega_n t} \cos \omega_d t + \frac{2\xi^2 - 1}{\omega_n\sqrt{1-\xi^2}} e^{-\xi\omega_n t} \sin \omega_d t$$

$$= t - \frac{2\xi}{\omega_n} + \frac{e^{-\xi\omega_n t}}{\omega_n\sqrt{1-\xi^2}}\left[2\xi\sqrt{1-\xi^2} \cos \omega_d t + \left(2\xi^2 - 1\right) \sin \omega_d t\right]$$

Let $2\xi\sqrt{1-\xi^{\omega}} = \sin\theta$

$$2\xi^2 - 1 = \cos\theta$$

$$y(t) = t - \frac{2\xi}{\omega_n} + \frac{e^{-\xi\omega_n t}}{\omega_n\sqrt{1-\xi^2}}[\sin\theta \cos \omega_d t + \cos\theta \sin \omega_d t]$$

$$y(t) = t - \frac{2\xi}{\omega_n} + \frac{e^{-\xi\omega_n t}}{\omega_n\sqrt{1+\xi^2}} \sin(\omega_d t + \theta)$$

$$y(t) = t - \frac{2\xi}{\omega_n} + \frac{e^{-\xi\omega_n t}}{\omega_n\sqrt{1-\xi^2}} \sin\left[\left(\omega_n\sqrt{1-\xi^2}\right)t + \theta\right] \quad ...(3.57)$$

where $\theta = \tan^{-1}\dfrac{2\xi\sqrt{1-\xi^2}}{2\xi^2 - 1}$

The error signal is

$$e_a(t) = r(t) - y(t)$$

$$e_a(t) = t - \left\{t - \frac{2\xi}{\omega_n} + \frac{e^{-\xi\omega_n t}}{\omega_n\sqrt{1-\xi^2}} \sin\left[\left(\omega_n\sqrt{1-\xi^2}\right)t + \theta\right]\right.$$

$$e_a(t) = \frac{2\xi}{\omega_n} - \frac{e^{-\xi\omega_n t}}{\omega_n\sqrt{1-\xi^2}} \sin\left[\left(\omega_n\sqrt{1-\xi^2}\right)t + \theta\right] \quad ...(3.58)$$

The steady state error is

$$e_{ss}(t) = \lim_{t\to\infty}\left[\frac{2\xi}{\omega_n} - \frac{e^{-\xi\omega_n t}}{\omega_n\sqrt{1-\xi^2}} \sin\left[\left(\omega_n\sqrt{1-\xi^2}\right)t + \theta\right]\right] \quad ...(3.59)$$

When $t \to \infty$. The second term in Eq. (3.59) approaches zero, therefore

$$e_{ss}(t) = \frac{2\xi}{\omega_n} \quad ...(3.60)$$

The time response of the second order system when input is unit ramp as shown in Fig. 3.17.

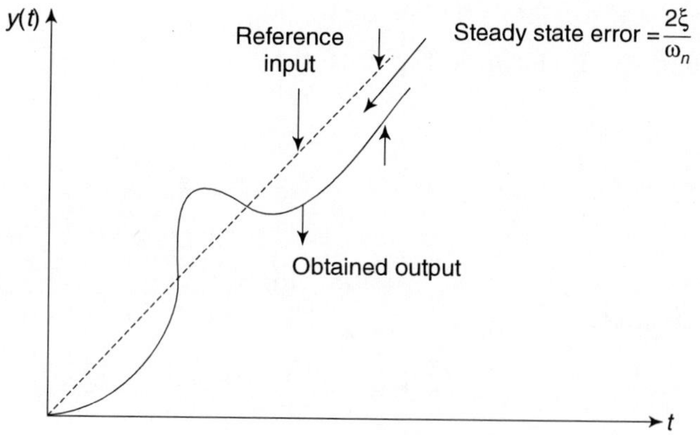

Figure 3.17 Time response of second order system for unit ramp input

3.5.4 Time Response Subjected to Unit Impulse Input

By Eq. (3.26) the output of the second order system is

$$Y(s) = R(s)\frac{\omega_n^2}{s^2 + 2\xi\omega_n s + \omega_n^2} \quad \text{...(3.61)}$$

If the input is a unit impulse then

$$R(s) = 1$$

Substituting the value of $R(s)$ in Eq. (3.61)

$$Y(s) = \frac{\omega_n^2}{s^2 + 2\xi\omega_n s + \omega_n^2}$$

Rewriting $s^2 = 2\xi\omega_n s + \omega_n^2$ as in Eq. (3.29)

$$Y(s) = \frac{\omega_n^2}{(s^2 + \xi\omega_n)^2 + \omega_n^2(1-\xi^2)} \quad \text{...(3.62)}$$

$$Y(s) = \frac{\omega_n^2}{\omega_n\sqrt{1-\xi^2}} \cdot \frac{\omega_n\sqrt{1-\xi^2}}{(s + \xi\omega_n)^2 + \omega_n^2(1-\xi^2)} \quad \text{...(3.62a)}$$

Taking inverse transform on both sides of Eq. (3.62)

$$y(t) = \frac{\omega_n}{\sqrt{1-\xi^2}} e^{-\xi\omega_n t} \sin\left[\left(\omega_n\sqrt{1-\xi^2}\right)t\right] \quad \text{...(3.63)}$$

Case 1 When $\xi < 1$

$$y(t) = \frac{\omega_n}{\sqrt{1-\xi^2}} e^{-\xi\omega_n t} \sin\left[\left(\omega_n \sqrt{1-\xi^2}\right)\right] \quad ...(3.64)$$

Time response for $\xi < 1$ is shown in Fig. 3.18 (b)

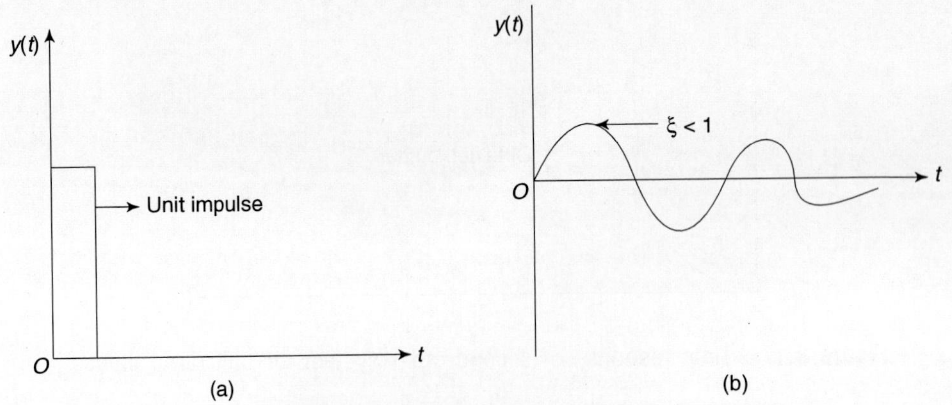

Figure 3.18 Time response for $\xi < 1$

Case 2 When $\xi = 1$

$$y(t) = \lim_{\xi \to 1} \frac{\omega_n}{\sqrt{1-\xi^2}} e^{-\xi\omega_n t} \sin\left[\left(\omega_n \sqrt{1-\xi^2}\right) t\right]$$

$$y(t) = \frac{\omega_n}{\sqrt{1-\xi^2}} e^{-\xi\omega_n t} \left(\omega_n \sqrt{1-\xi^2}\right) t$$

$$y(t) = t\,\omega_n^2\, e^{-\omega_n t} \quad ...(3.65)$$

The response for $\xi = 1$ is shown in Fig. 3.19

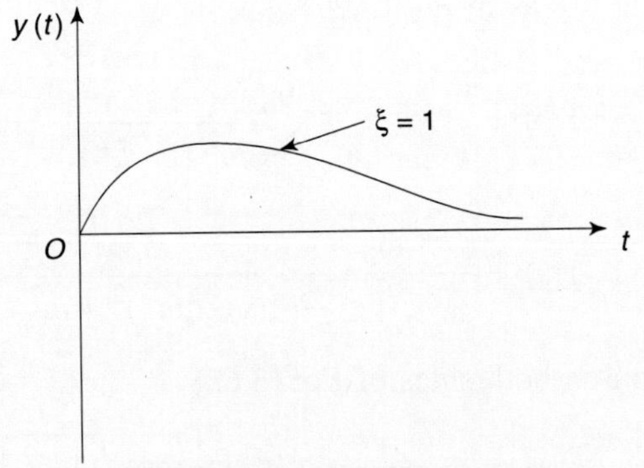

Figure 3.19 Time response for $\xi = 1$

Case 3 When $\xi > 1$

By Eq. (3.62)

$$y(s) = \frac{\omega_n^2}{(s+\xi\,\omega_n)^2 - \omega_n^2\,(\xi^2 - 1)} \qquad ...(3.66)$$

Solving Eq. (3.66) by partial fraction and putting $\omega_d^2 = \omega_n^2\sqrt{\xi^2 - 1}$

$$\frac{\omega_n^2}{(s+\xi\,\omega_n)^2 - \omega_d^2} = \frac{\omega_n^2}{(s+\xi\omega_n + \omega_d)\,(s+\xi\omega_n - \omega_d)}$$

$$= \frac{A}{(s+\xi\,\omega_n + \omega_d)} + \frac{B}{(s+\xi\,\omega_n - \omega_d)} \qquad ...(3.67)$$

Solving Eq. (3.67)

$$A = \frac{-\,\omega_n}{2\sqrt{\xi^2 - 1}}$$

and

$$B = \frac{\omega_n}{2\sqrt{\xi^2 - 1}}$$

Substituting the values of A and B in Eq. (3.67)

$$Y(s) = \frac{-\,\omega_n}{2\sqrt{\xi^2 - 1}}\cdot\frac{1}{(s+\xi\,\omega_n + \omega_d)} + \frac{\omega_n}{2\sqrt{\xi^2 - 1}}\cdot\frac{1}{s+\xi\omega_n - \omega_d} \qquad ...(3.68)$$

Taking inverse Laplace transform

$$y(t) = \frac{\omega_n}{2\sqrt{\xi^2 - 1}}\left[-\,e^{(\xi\omega_n + \omega_d)\,t} + e^{-(\xi\omega_n - \omega_d)\,t}\right]$$

$$y(t) = \frac{\omega_n}{2\sqrt{\xi^2 - 1}}\left[-\,e^{-\left(\xi\omega_n + \omega_n\sqrt{\xi^2 - 1}\right)} + e^{-\left(\xi\omega_n - \omega_n\sqrt{\xi^2 - 1}\right)t}\right]$$

$$y(t) = \frac{\omega_n}{2\sqrt{\xi^2 - 1}}\left[e^{\left(\xi - \sqrt{\xi^2 - 1}\right)\omega_n t} + e^{-\left(\xi + \sqrt{\xi^2 - 1}\right)\omega_n t}\right]$$

The time response of the second order system for unit impulse input and for $\xi > 1$

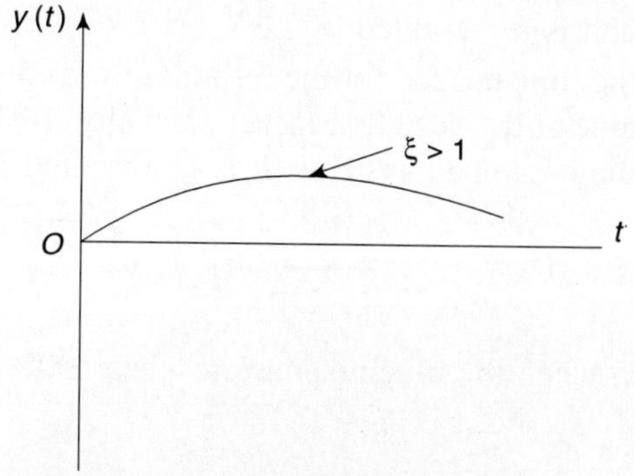

Figure 3.20 Time response for $\xi < 1$

3.5.5 Time Response Specifications

In the previous section we have studied the first and second order stystems. Generally, the second order system is used, it has two poles and no zero the response of such system we have seen earlier. High order control systems usually have a pair of complex conjugate poles with damping less than one. These complex conjugate poles dominate over other poles. Therefore, the time response of the second and higher order systems is generally damped oscillatory when input is unit step.

The specifications pertaining to time response during transient part are shown in Fig. 3.21.

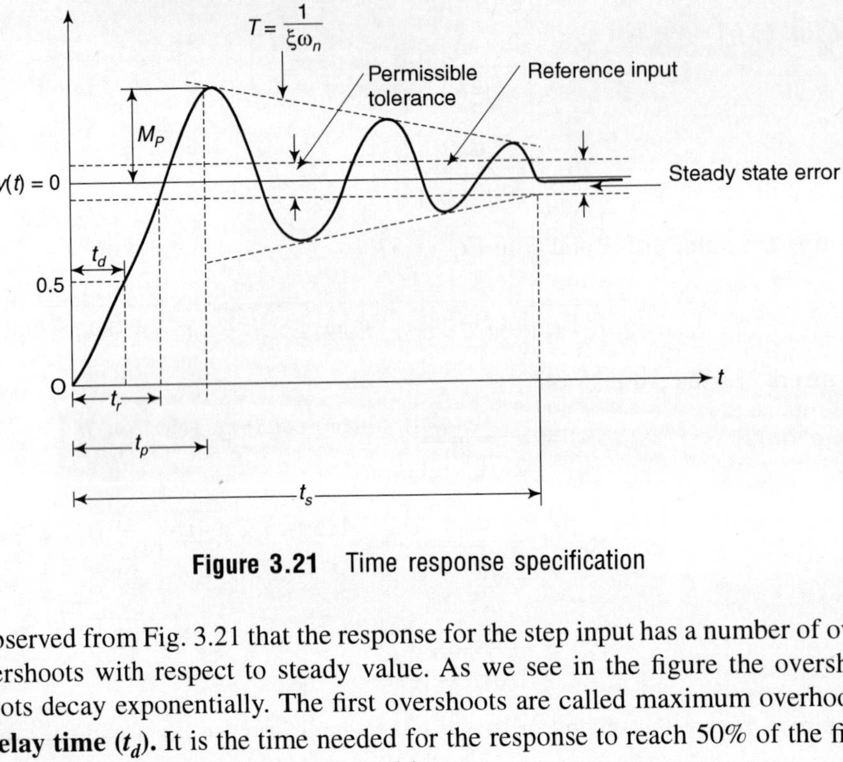

Figure 3.21 Time response specification

It is observed from Fig. 3.21 that the response for the step input has a number of overshoots and undershoots with respect to steady value. As we see in the figure the overshoots and undershoots decay exponentially. The first overshoots are called maximum overhoots.

1. **Delay time (t_d).** It is the time needed for the response to reach 50% of the final value in the first attempt and is represented by t_d.
2. **Rise time (t_r).** It is the time needed for the response to reach from 10% to 90% or 0 to 100% of the final value at the very first instant. Usually, 10% to 90% for overdamped and 0 to 100% for underdamped systems. It is represented by t_r, by Eq. (3.26)

$$y(t) = 1 - \frac{e^{-\xi\omega_n t}}{\sqrt{1-\xi^2}} \sin\left[\omega_n \sqrt{1-\xi^2}\, t + \theta\right] \qquad ...(3.69)$$

for underdamped system when the time response reaches 100% of the desired value, i.e.,

$$y(t) = 1$$

$$t = t_r$$

Substituting $y(t)$ and t in Eq. (3.69)

$$1 = 1 - \frac{e^{-\xi\omega_n t_r}}{\sqrt{1-\xi^2}} \sin\left[\omega_n \sqrt{1-\xi^2}\, t_r + \theta\right]$$

$$= \frac{e^{-\xi\omega_n t_r}}{\sqrt{1-\xi^2}} \sin\left[\omega_n \sqrt{1-\xi^2}\, t_r + \theta\right]$$

$\because \quad \dfrac{e^{-\xi\omega_n t_r}}{\sqrt{1-\xi^2}}$ is finite.

So $\quad \sin\left[\left(\omega_n \sqrt{1-\xi^2}\right) t_r + \theta\right] = 0$

$$\omega_n \sqrt{1-\xi^2}\, t_r + \theta = \pi$$

$$\boxed{t_r = \frac{\pi - \theta}{\omega_n \sqrt{1-\xi^2}}} \qquad \text{...(3.70)}$$

where $\quad \theta = \tan^{-1}\left(\dfrac{1-\xi^2}{\xi}\right)$

3. **Maximum overshoot (M_p).** The maximum positive deviation of the output with respect to the steady output is known as maximum overshoot or peak overshoot.

$\because$ The desired output is unity

$\therefore$

$$M_p = y(t)_{\max} - 1 \qquad \text{...(3.71)}$$

$$\%M_p = \frac{y(t)_{\max} - 1}{1} \times 100$$

4. **Peak time (t_p).** It is the time needed to reach the maximum overshoot is called peak time and is represented by t_p.

It is observed in figure that M_p occurs when the slope of the time response curve after the initiation of the input signal is zero. Then

$$\frac{dy(t)}{dt} = 0 \qquad \text{...(3.72)}$$

By Eq. (3.21)

$$y(t) = 1 - \frac{e^{-\xi\omega_n t}}{\sqrt{1-\xi^2}} \cdot \sin \cdot \left[\left(\omega_n \sqrt{1-\xi^2}\right) t + \theta\right]$$

$$\frac{dy(t)}{dt} = \frac{e^{-\xi\omega_n t}}{\sqrt{1-\xi^2}} \cdot \omega_n \sqrt{1-\xi^2} \cdot \cos\left[\left(\omega_n \sqrt{1-\xi^2}\right) t + \theta\right]$$

$$= \frac{(-\xi\omega_n)\, e^{-\xi\omega_n t}}{\sqrt{1-\xi^2}} \sin\left[\left(\omega_n \sqrt{1-\xi^2}\right)t + \theta\right]$$

By Eq. (3.72)

$$= \frac{e^{-\xi\omega_n t}}{\sqrt{1-\xi^2}} \left\{-\omega_n \sqrt{1-\xi^2} \cdot \cos\left[\omega_n \sqrt{1-\xi^2}\; t + \theta\right] + \xi\omega_n \sin\left[\left(\omega_n \sqrt{1-\xi^2}\; t + \theta\right)\right] = 0\right. \quad \text{...(3.73)}$$

In Eq. (3.73)

$$\frac{e^{-\xi\omega_n t}}{\sqrt{1-\xi^2}} \text{ is finite}$$

$$\therefore \quad \omega_n \sqrt{1-\xi^2} \cdot \cos\left[\left(\omega_n \sqrt{1-\xi^2}\right)t + \theta\right] = \xi\omega_n \sin\left[\left(\omega_n \sqrt{1-\xi^2}\right)t + \theta\right]$$

$$\tan\left[\left(\omega_n \sqrt{1-\xi^2}\right)t + \theta\right] = \frac{\sqrt{1-\xi^2}}{\xi} \quad \text{...(3.74)}$$

$$\because \quad \theta = \tan^{-1}\frac{\sqrt{1-\xi^2}}{\xi}, \text{ substituting in Eq. (3.74)}$$

$$\tan\left[\left(\omega_n \sqrt{1-\xi^2}\right)t + \tan^{-1}\frac{\sqrt{1-\xi^2}}{\xi}\right] = \frac{\sqrt{1-\xi^2}}{\xi} \quad \text{...(3.75)}$$

The general solution of Eq. (3.75) is

$$\left(\omega_n \sqrt{1-\xi^2}\right)t = n\pi$$

where $n = 0, 1, 2, 3 \ldots$

for maximum overshoot $n = 1$

$$\boxed{t_p = \frac{\pi}{\omega_n \sqrt{1-\xi^2}} = \frac{\pi}{\omega_d}} \quad \text{...(3.76)}$$

$y(t)_{max}$ can be obtained by putting $t = t_p$ in Eq. (3.76)

$$y(t)_{max} = 1 - \frac{e^{-\xi\omega_n t_p}}{\sqrt{1-\xi^2}} \sin\left[\left(\omega_n \sqrt{1-\xi^2}\right)t_p + \theta\right]$$

$$y(t)_{max} = 1 - \frac{e^{-\xi\omega_n \frac{\pi}{\omega_n\sqrt{1-\xi^2}}}}{\sqrt{1-\xi^2}} \sin\left[\left(\omega_n \sqrt{1-\xi^2}\right)\frac{\pi}{\omega_n \sqrt{1-\xi^2}} + \theta\right]$$

$$\because \qquad t_p = \frac{\pi}{\omega_n \sqrt{1-\xi^2}}$$

$$y(t)_{max} = 1 - \frac{e^{-\frac{\xi\pi}{\sqrt{1-\xi^2}}}}{\sqrt{1-\xi^2}} \cdot \sin(\pi + \theta) = 1 - \frac{e^{-\frac{\xi\pi}{\sqrt{1-\xi^2}}}}{\sqrt{1-\xi^2}} (-\sin\theta)$$

$$\because \qquad \theta = \tan^{-1} \frac{\sqrt{1-\xi^2}}{\xi} \qquad \text{Hence, } \sin\theta = \sqrt{1-\xi^2}$$

So

$$y(t)_{max} = 1 + \frac{e^{-\frac{\xi\pi}{1-\xi^2}}}{\sqrt{1-\xi^2}} \cdot \sqrt{1-\xi^2}$$

$$y(t)_{max} = 1 + e^{-\frac{\xi\pi}{\sqrt{1-\xi^2}}} \qquad \text{...(3.77)}$$

$$\because \qquad M_p = y(t)_{max} - 1$$

$$\therefore \qquad M_p = \left[1 + e^{-\frac{\xi\pi}{\sqrt{1-\xi^2}}}\right] - 1$$

$$\boxed{M_p = e^{-\frac{\xi\pi}{\sqrt{1-\xi^2}}}} \qquad \text{...(3.78)}$$

5. **Settling time (t_s).** It is the time needed to settle down aforesaid oscillation within 2% of final value of the output. It is known as settling time and is represented by t_s. The settling time for the second order control system is four times, the time constant (T) of the system on 2% basis

$$t_s = 4T \qquad \text{...(3.79)}$$

$$\boxed{t_s = 4\frac{1}{\xi\omega_n}}$$

$$\because \qquad T = \frac{1}{\xi\omega_n}$$

On 5% basis

$$t_s = 3T$$

$$\boxed{t_s = 3\frac{1}{\xi\omega_n}} \qquad \text{...(3.80)}$$

- *Step response of the second order control system for underdamped response is characterized by the performance indices like the speed of the system moves to follow the input and the type of oscillation and how much time it takes to practically reach the final value.*

3.6 SYSTEM RESPONSE OF HIGHER ORDER SYSTEM

In the previous sections, we have studied the systems with one or two poles only. It should be noted that the formulae describing percent overshoot, settling time and peak time were derived only for a system with two complex poles and no zeros, i.e., for the second order system. If a system has more than two poles or has zeros, the above derived formulae cannot be used to calculate the performance specification.

Under certain conditions, it is approximated that a system with more than two poles or with zeros can be considered a second order system that has only two complex dominant poles, then the above derived formulae can be applied to these higher order systems by using the location of the dominant poles for the performance specification.

3.6.1 System Response with Additional Poles

The second order system has been explained in the earlier section. Now the results of the second order system could be extended to higher order systems in which the real part of the poles is much less, whereas it has a dominant pair of complex poles (i.e., much higher than the order poles). Now the time response of higher order systems can be approximated by that obtained by neglecting all the system poles other than the complex poles.

Let us consider the third order system with transfer function

$$\frac{Y(s)}{R(s)} = T_o(s) = \frac{\omega_n^2}{(1+sT)(s^2+2\xi\,\omega_n s+\omega_n^2)}$$

$$\frac{Y(s)}{R(s)} = \frac{\omega_n^2}{T\left(s+\dfrac{1}{T}\right)(s^2+2\xi\,\omega_n s+\omega_n^2)} \qquad ...(3.81)$$

$$\because \qquad P = \frac{1}{T}$$

Substituting the value of P in Eq. (3.81)

$$\frac{Y(s)}{R(s)} = \frac{\omega_n^2\,P}{(s+P)(s^2+2\xi\,\omega_n s+\omega_n^2)} \qquad ...(3.82)$$

The unit step response is

$$Y(s) = \frac{1}{s}\,\frac{\omega_n^2\,P}{s\,(s+P)(s^2+2\xi\,\omega_n s+\omega_n^2)}$$

$$Y(s) = \frac{\omega_n^2\,P}{(s+P)(s^2+2\xi\,\omega_n s+\omega_n^2)} \qquad ...(3.83)$$

$$\frac{\omega_n^2 P}{s(s+P)(s^2+2\xi\,\omega_n s+\omega_n^2)} = \frac{A}{s}+\frac{B}{(s+P)}+\frac{Ds+E}{s^2+2\xi\,\omega_n s+\omega_n^2}$$

Solving by partial fraction of Eq. (3.83) and taking inverse Laplace transform

$$y_p(t) = 1 - C_1 e^{-Pt} + C_2 e^{-\xi\omega_n t}\left[\omega_n \sqrt{(1-\xi^2)}\, t - \phi\right] \quad ...(3.84)$$

where C_1 and C_2 are constants as

$$C_1 = \frac{\omega_n^2}{\omega_n^2 - 2\xi\,\omega_n P + P^2}$$

$$C_2 = \frac{1}{\sqrt{1-\xi^2\left(1-\frac{2\xi\omega_n}{P}+\frac{\omega_n^2}{P^2}\right)}}$$

$$\phi = \tan^{-1}\frac{\sqrt{1-\xi^2}}{-\xi} + \tan^{-1}\frac{\omega_n\sqrt{1-\xi^2}}{P-\xi\omega_n}$$

Now we study Eq. 3.82 in which the pole $s = -P$ will depend upon the magnitude of P. If $P \ll \xi\omega_n$, the effect of the term e^{-Pt} in Eq. (3.84) will last much longer than that of the term whose magnitude is governed by $e^{-\xi\gamma_n t}$. Further from Eq. 3.84.

$$\lim_{P\to 0} C_1 = \lim_{P\to 0}\frac{\omega_n^2}{\omega_n^2 - 2\xi\,\omega_n P + P^2}$$

$$= \lim_{P\to 0}\frac{\omega_n^2}{\omega_n^2} = 1$$

and

$$\lim_{P\to 0} C_2 = \lim_{P\to 0}\frac{1}{\sqrt{1-\xi^2\left(1-\frac{2\xi\omega_n}{P}+\frac{\omega_n^2}{P}\right)}}$$

$$\lim_{P\to 0} C_2 = 0$$

Therefore, as the pole at $s = -P$ moves towards the origin (i.e., closer to the origin), the response $y_p(t)$ in Eq. (3.84) is dominated by this pole and the effect of the two complex poles at $-\xi\omega_n \pm j\omega_n\sqrt{1-\xi^2}$ diminishes.

In the second case if $P \gg \xi\omega_n$, the effect of the term e^{-Pt} in Eq. (3.84) will be over much more faster than the term whose magnitude is governed by $e^{-\xi\omega_n t}$. Now from Eq. (3.84) we observe that

$$\lim_{P \to 0} C_1 = \lim_{P \to \infty} \frac{\omega_n^2}{\omega_n^2 - 2\xi\, \omega_n\, P + P^2}$$

$$C_1 = 0$$

$$\lim_{P \to \infty} = \lim_{P \to \infty} \frac{1}{\sqrt{1-\xi^2 \left(1 - \frac{2\xi\omega_n}{P} + \frac{\omega_n^2}{P}\right)}}$$

$$C_2 = \frac{1}{\sqrt{1-\xi^2}}$$

Therefore, at the pole at $s = -P$ moves far away from the origin, the response $y_p(t)$ in Eq. (3.84) is dominated by the complex poles at $-\xi\omega_n \pm j\omega_d$

$\because$ $$\omega_d = \omega_n \sqrt{1-\xi^2}$$

Figure 3.22 shows unit step response of a third order system of Eq. 3.82 with $\xi = 0.3$ it is observed that for $\frac{P}{\xi\omega_n} = 6$, response of the third order system is, for all practical purposes the response $y(t)$ of the system with the transfer function.

The relative location of dominant poles of the third order system is shown in Fig. 3.22.

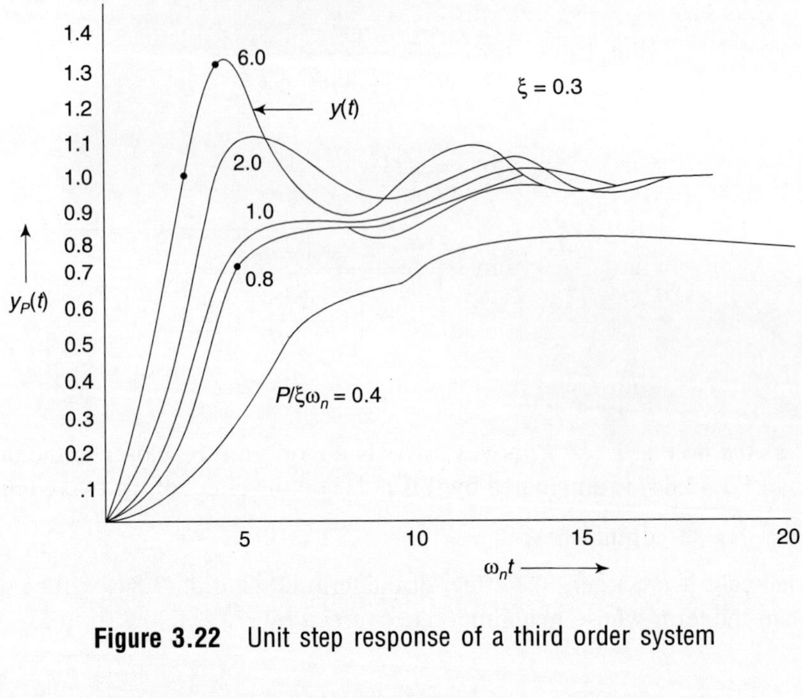

Figure 3.22 Unit step response of a third order system

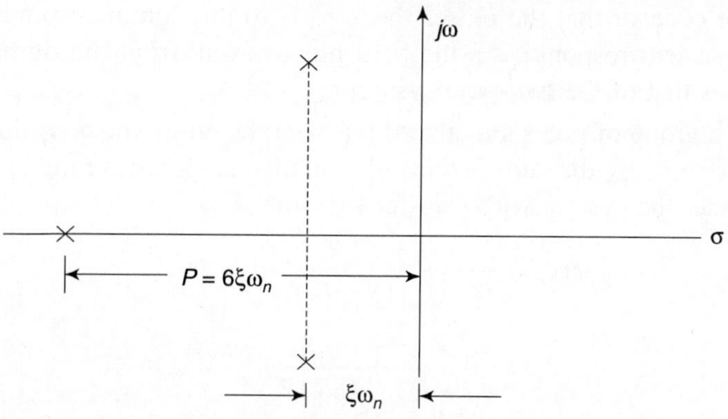

Figure 3.23 Location of dominant poles of the third order system

3.6.2 System Response with Zeros

In the previous section we have studied the systems having additionl poles and we have seen that the formulae describing percent overshoot, settling time, and peak time were derived only for a system with two complex poles and no zeros. Now we will see the effect of additional zeros to the second order system. As we know that the zeros of a response affect the residue or amplitude, of a response component but it does not affect the nature of the response.

We will analyze the effect of a additional zero when added first in the left half-plane and then in the right half-plane. Let us consider that the complex poles of two-pole system are at $-1 \pm j\,2.8$ and we add zeros consecutively at -2, -4, -10 and the results, normalized to the steady state value, are shown in Fig. 3.24.

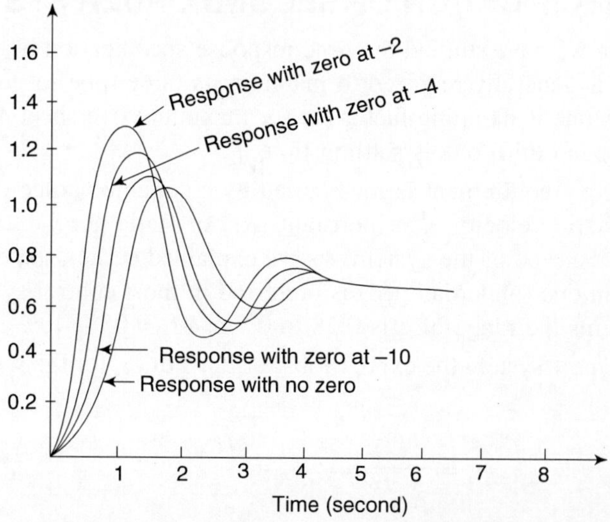

Figure 3.24 Effect of addding a zero to a two-poles system

In Fig. 3.24, we observe that the closer the zero is to the dominant poles, the greater is its effect on the transient response. As the zero moves away from the dominant poles, the response approaches that of the two-pole system.

Let us consider a group of poles and a zero far from the poles, the amplitude of each pole will be affected the same by the zero. Hence, the relative amplitudes remain appreciably the same. Let us consider the system with transfer function

$$T(s) = \frac{(s+Z)}{(s+P_1)(S+P_2)}$$

$$= \frac{A}{(S+P_1)} + \frac{B}{(S+P_2)}$$

$$= \frac{(-P_1+Z)/(-P_1+P_2)}{(S+P_1)} + \frac{(-C+Z)/(-P_2+P_1)}{(S+P_2)}$$

If the zero is far from the poles, then Z is larger compared to P_1 and P_2

and $$T(s) = Z\left[\frac{1/(-P_1+P_2)}{S+P_1} + \frac{1/(-P_2+P_1)}{S+P_2}\right]$$

$$T(s) = \frac{Z}{(S+P_1)(S+P_2)}\text{t}$$

Hence, the zero looks like a simple gain factor and does not change the relative amplitudes of the components of the response.

3.7 DESIGN SPECIFICATION OF SECOND ORDER SYSTEMS

In the previous section we have studied the time response specification by which we can say that a control system is generally required to maintain its three-time response specifications within desirable limit, one is damping factor (ξ) (or maximum oversheet M_p), second one is steady state error (e_{ss}) and third one is settling time, t_s.

The steady state error requirement is maintained by a suitable choice of static positional error constant (K_p), static velocity error constant (K_v) or static acceleration error constant (K_a) depending upon the type of the system. As we explained earlier, the damping factor (ξ) is sufficiently less than one (underdamped) is preferred in most control systems. Practically, for most control systems the range of ξ is 0.28 to 0.7 or M_p = 05 – 40% is considered.

Now considering type 1 system the expressions for e_{ss}, ξ ω_n, t_s for the second order system are

$$\xi = \frac{1}{2\sqrt{K_v\,\tau}};\ \omega_n = \frac{\sqrt{K_v}}{\tau} \qquad ...(3.85)$$

$$t_s = 4T = \frac{4}{\xi\omega_n} \qquad ...(3.86)$$

$$e_{ss} = \frac{2\xi}{\omega_n} = \frac{1}{K_v} \qquad \text{...(3.87)}$$

By the above three equations we can conclude that only two specifications out of three (ξ, e_{ss} and t_s) can be maintained by the suitable choice of K_v, i.e., static velocity error constant. The time constant τ is generally fixed. Thus, K_v or the open loop gain $G(s)$ is the only adjustable parameter of the system. So the second order system can then maintain only one of the specifications which is steady state error. In most of the systems when $G(s)$ is adjusted to maintain the steady state error (e_{ss}) upto specified accuracy then ξ reduces below the specified.

So we can say that even to maintain two independent specifications the second order system requires to be modified. So this modification can be done by some modification schemes called additional control as discussed below.

3.7.1 Proportional Control

The schematic diagram for proportional control is shown in Fig. 3.25.

In Fig. 3.25, actuating signal, $E_a(s)$, is proportional to the error signal, $E(s)$, so the system is called proportional control system.

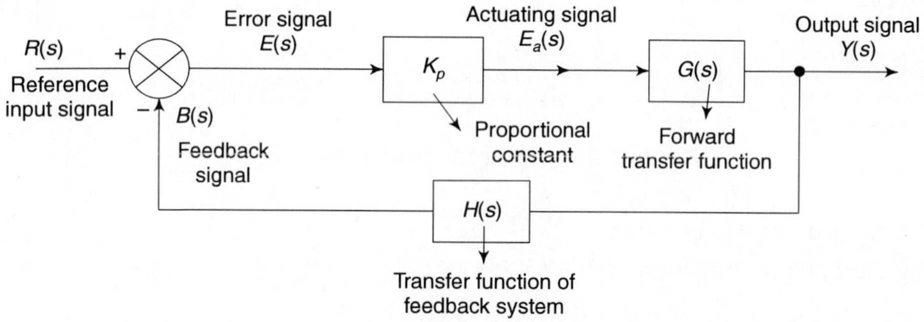

Figure 3.25 Block diagram of proportional control

We know that, if the control system is underdamped then the response of the system will be fast and it will give exponentially decaying oscillation in the time response during the transient period as discussed in section 3.5. If the control system is overdamped then the response of the system can be made faster by increasing forward path gain of the system. By increasing the forward path gain will reduce the steady state error, but at the same time maximum overshoot is increased. So far satisfactory performance, the maximum overshoot (M_p) can be reduced without sacrificing the steady state accuracy by modifying the actuating signal.

3.7.2 Derivative Control (PD Control)

This type of modification is used for modifying the actuating signal, $E_a(s)$. In this control the actuating signal consists of proportional error signal added with derivative of the error signal as shown in Fig. 3.26 so the actuating signal for derivative control is given by the expression

$$e_a(t) = e(t) + K_d \frac{de(t)}{dt} \quad ...(3.88)$$

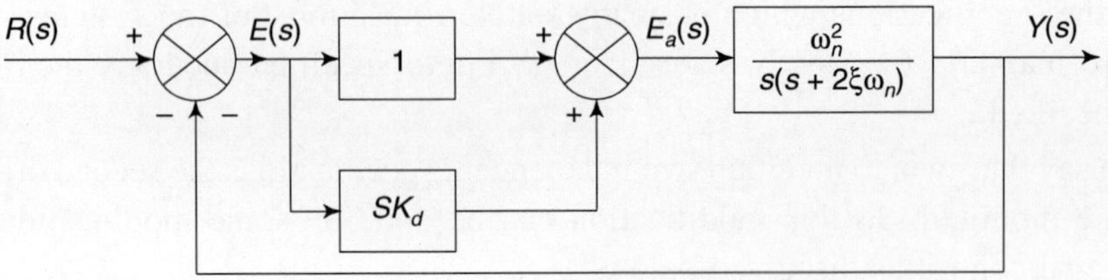

Figure 3.26 Proportional derivative control

Redrawing the block diagram 3.26

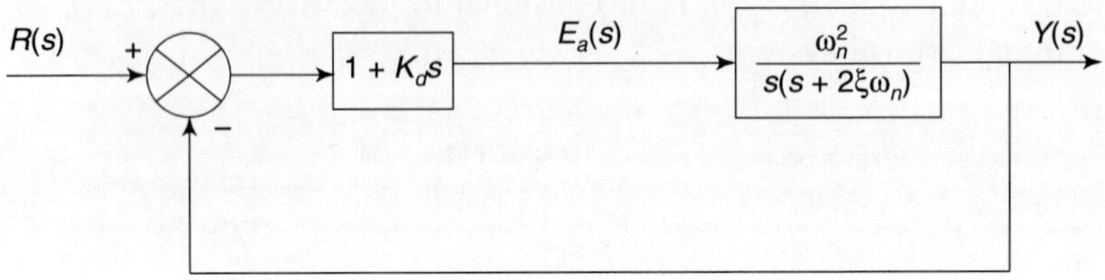

Figure 3.27 Reduced form of Fig. 3.26

where K_d = derivative constant

Taking the Laplace transform of Eq. (3.85)

$$K_a(s) = E(s) + s\, K_d\, E(s) \quad ...(3.89)$$

The overall transfer function of the system in Fig. 3.27 is

$$\frac{Y(s)}{R(s)} = \frac{(1 + K_d\, s) \cdot \dfrac{\omega_n^2}{s\,(s + 2\xi\, \omega_n)}}{1 + \dfrac{\omega_n^2}{s\,(s + 2\xi\, \omega_n)}(1 + K_d\, s)} \quad ...(3.90)$$

$$= \frac{\dfrac{(1 + K_d\, s)\, \omega_n^2}{s\,(s + 2\xi\, \omega_n)}}{\dfrac{s\,(s + 2\xi\, \omega_n) + \omega_n^2\,(1 + K_d\, s)}{s\,(s + 2\xi\, \omega_n)}}$$

$$\frac{Y(s)}{R(s)} = \frac{(1+K_d\, s)\,\omega_n^2}{s^2 + 2\xi\,\omega_n\, s + \omega_n^2 + K_d\, s\,\omega_n^2}$$

$$\frac{Y(s)}{R(s)} = \frac{(1+K_d\, s)\,\omega_n^2}{s^2 + (2\xi\,\omega_n + K_d\,\omega_n^2)\, s + \omega_n^2} \quad ...(3.91)$$

The characteristic equation is

$$s^2 + (2\xi\,\omega_n + K_d\,\omega_n^2)\, s + \omega_n^2 = 0 \quad ...(3.92)$$

Comparing Eq. (3.92) with the standard characteristic equation of the second order system.

$$2\xi\omega_n = 2\xi\omega_n + K_d\,\omega_n^2$$

$$\xi' = \frac{2\xi\omega_n + K_d\,\omega_n^2}{2\,\omega_n}$$

$$\xi' = \xi + \frac{K_d\,\omega_n}{2} \quad ...(3.93)$$

By Eq. 3.7.9 we can say that the effective damping (ξ') is increased, using derivative control. So the maximum overshoot (M_p) is also reduced.

The overall transfer function of the second order system is

$$\frac{Y(s)}{R(s)} = \frac{\omega_n^2}{s^2 + 2\xi\,\omega_n s + \omega_n^2} \quad ...(3.94)$$

In Fig. 3.27, the forward path transfer function is

$$G(s) = (1 + s K_d)\cdot\frac{\omega_n^2}{s\,(s + 2\xi\,\omega_n)} \quad ...(3.95)$$

and the feedback path transfer function

$$H(s) = 1 \quad ...(3.96)$$

The transfer function relating to $E(s)$ and $R(s)$ is given by

$$E(s) = R(s) - Y(s)$$

$$E(s) = R(s)\left[1 - \frac{Y(s)}{R(s)}\right]$$

$$\frac{E(s)}{R(s)} = \left[1 - \frac{G(s)}{1+G(s)\,H(s)}\right]$$

$\because$ $$\frac{Y(s)}{R(s)} = \frac{G(s)}{1+G(s)\,H(s)}$$

$$H(s) = 1 \text{ by Eq. (3.96)}$$

Then $$\frac{E(s)}{R(s)} = 1 - \frac{G(s)}{1+G(s)}$$

$$\frac{E(s)}{R(s)} = \frac{1+G(s)-G(s)}{1+G(s)}$$

$$\frac{E(s)}{R(s)} = \frac{1}{1+G(s)} \qquad \text{...(3.97)}$$

Substituting the value of $G(s)$ in Eq. (3.97) from Eq. (3.95)

$$\frac{E(s)}{R(s)} = \frac{1}{1+\dfrac{(1+s\,K_d)\,\omega_n^2}{s\,(s+2\xi\,\omega_n)}}$$

$$\frac{E(s)}{R(s)} = \frac{s\,(s+2\xi\,\omega_n)}{s^2+(2\xi\,\omega_n+\omega_n^2\,K_d)\,s+\omega_n^2} \qquad \text{...(3.98)}$$

For a unit ramp input

$$R(s) = \frac{1}{s^2}$$

Substituting the value of $R(s)$ in Eq. (3.98)

$$\therefore \qquad E(s) = \frac{1}{s^2}\cdot\frac{s\,(s+2\xi\,\omega_n)}{s^2+(2\xi\,\omega_n+K_d\,\omega_n^2)\,s+\omega_n^2} \qquad \text{...(3.99)}$$

The steady state error for unit input is

$$e_{ss} = \lim_{s\to 0} s\,E(s) = \lim_{s\to 0} s\cdot\frac{1}{2}\cdot\frac{s\,(s+2\xi\,\omega_n)}{s^2+(2\xi\,\omega_n+K_d\,\omega_n^2)\,s+\omega_n^2}$$

$$e_{ss} = \frac{2\xi\,\omega_n}{\omega_n^2}$$

$$\boxed{e_{ss} = \frac{2\xi}{\omega_n}} \qquad \text{...(3.100)}$$

Comparing Eqs. (3.100) and Eq. (3.60) it is concluded that the steady state error is not affected by derivative control.

3.7.3 Integral Control (PI Control)

This is another type of modified scheme, which is used for modifying the actuating signal, $E_a(s)$. In this control the actuating signal consists of proportional error signal added with integral of the error signal as shown in Fig. 3.28, the actuating signal for integral control is given by the expression,

$$e_a(t) = e(t) + K_i \int e(t)\,dt \qquad \text{...(3.101)}$$

where K_i = integral constant

Taking the Laplace transform of Eq. (3.101)

$$E_a(s) = E(s) + K_i \frac{E(s)}{s} \qquad ...(3.102)$$

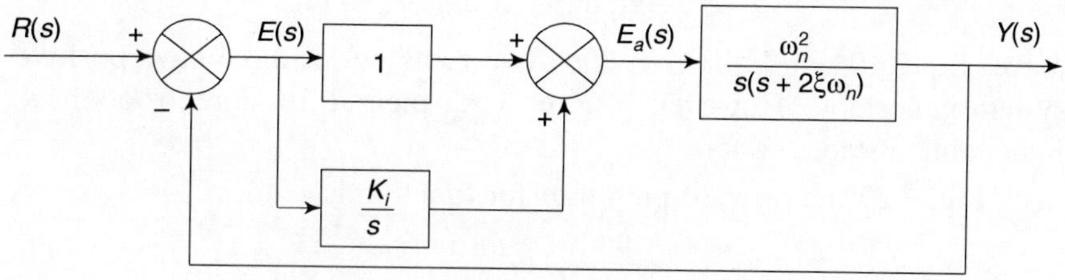

Figure 3.28 Block diagram of proportional integral control

Redrawing the block diagram

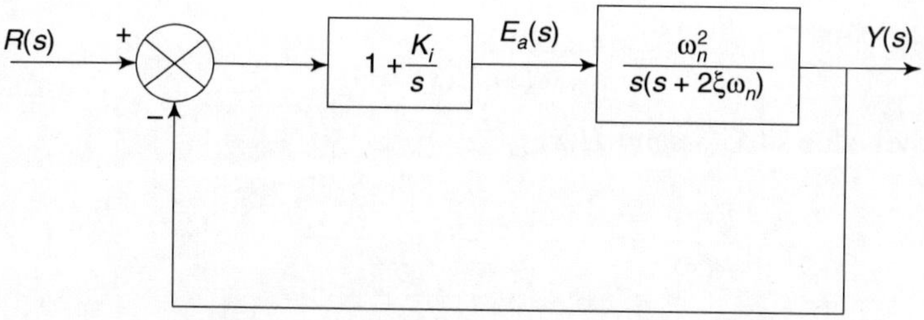

Figure 3.29 Reduced form of Fig. 3.28

The overall transfer function of the system shown in Fig. 3.28 is

$$\frac{Y(s)}{R(s)} = \frac{\left(1+\frac{K_i}{s}\right)\left[\frac{\omega_n^2}{s\,(s+2\xi\,\omega_n)}\right]}{1+\left(1+\frac{K_i}{s}\right)\left[\frac{\omega_n^2}{s\,(s+2\xi\,\omega_n}\right]\times 1}$$

$$\frac{Y(s)}{R(s)} = \frac{\left(1+\frac{K_i}{s}\right)\left[\frac{\omega_n^2}{s\,(s+2\xi\,\omega_n)}\right]}{\frac{s\,(s+2\xi\,\omega_n)+\left(1+\frac{K_i}{s}\right)\omega_n^2}{s\,(s+2\xi\,\omega_n)}}$$

$$\frac{Y(s)}{R(s)} = \frac{\left(1+\frac{K_i}{s}\right)\omega_n^2}{s\,(s+2\xi\,\omega_n)+\left(1+\frac{K_i}{s}\right)\omega_n^2}$$

$$\frac{Y(s)}{R(s)} = \frac{(s+K_i)\,\omega_n^2}{s^3 + 2\xi\,\omega_n\, s^2 + \omega_n^2\, s + K_i\,\omega_n^2} \quad ...(3.103)$$

The characteristic equation is

$$s^3 + 2\xi\,\omega_n\, s^2 + \omega_n^2\, s + K_i\,\omega_n^2 \quad ...(3.104)$$

In the above Eq. 3.104, if $2\xi\omega_n < K_i$, then two roots out of three have positive real part and such system is unstable. However, if $2\xi\omega_n > K_i$, then all the three roots have negative real parts indiciating a stable system.

In the block Fig. 3.29 the forward path transfer function is

$$G(s) = \frac{(s+K_i)\,\omega_n^2}{s^2\,(s+2\xi\,\omega_n)} \quad ...(3.105)$$

and $$H(s) = 1 \quad ...(3.106)$$

The transfer function relating $E(s)$ and $R(s)$ is

$$\frac{E(s)}{R(s)} = \frac{1}{1+G(s)\,H(s)} \quad ...(3.107)$$

Substituting the value of $G(s)$ and $H(s)$

$$\frac{E(s)}{R(s)} = \frac{1}{1+\dfrac{(s+K_i)\,\omega_n^2}{s^2\,(s+2\xi\,\omega_n)}\cdot 1}$$

$$E(s) = R(s)\cdot\frac{s^2\,(s+2\xi\,\omega_n)}{s^3 + 2\xi\,\omega_n\, s^2 + \omega_n^2\, s + K_i\,\omega_n^2} \quad ...(3.108)$$

If the input is unit ramp then

$$R(s) = \frac{1}{s^2}$$

The steady state error is

$$e_{ss} = \lim_{s\to 0} s\cdot\frac{1}{s^2}\,\frac{s^2\,(s+2\xi\,\omega_n)}{s^3 + 2\xi\,\omega_n\, s^2 + \omega_n\, s + K_i\,\omega_n^2}$$

$$\boxed{e_{ss} = 0} \quad ...(3.109)$$

By comparing Eq. (3.109) and Eq. (3.60) it is concluded that the steady state error is reduced.

If the input is unit parabolic then

$$R(s) = \frac{1}{s^3}$$

Then the steady state error is

$$e_{ss} = \lim_{s \to 0} s\,E(s)$$

$$= \lim_{s \to 0} s \cdot \frac{1}{s^3} \cdot \frac{s^2\,(s + 2\xi\,\omega_n)}{s^3 + 2\xi\,\omega_n\,s^2 + \omega_n s + K_i\,\omega_n^2}$$

$$e_{ss} = \frac{2\xi\,\omega_n}{K_i\,\omega_n^2}$$

$$\boxed{e_{ss} = \frac{2\xi}{K_i\,\omega_n}} \qquad \text{...(3.110)}$$

3.7.4 Proportional Plus Derivative Plus Integral Control (PID Control)

This is a modified scheme, which is used for modifying the actuating signal, $E_a(s)$. In this control the actuating signal consists of proportional error signal added with integral and derivative of the error signal as shown in Fig. 3.29. This type of scheme combines the advantages of proportional, derivative and integral control actions.

The actuating signal for integral and derivative control is given by the expression

$$e_a(t) = e(t) + K_d \frac{de(t)}{dt} + K_i \int e(t)\,dt \qquad \text{...(3.111)}$$

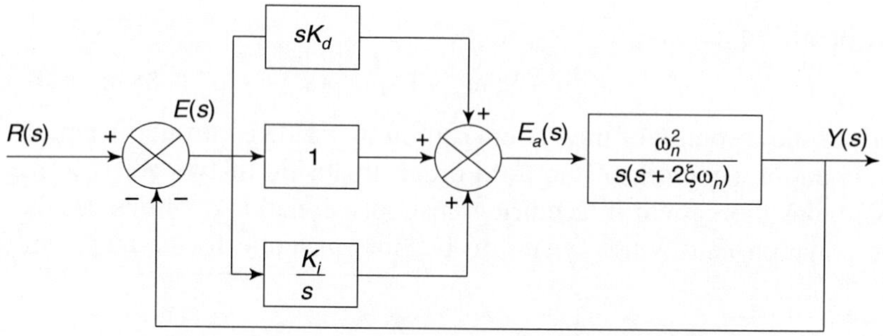

Figure 3.30 Block diagram of PID control

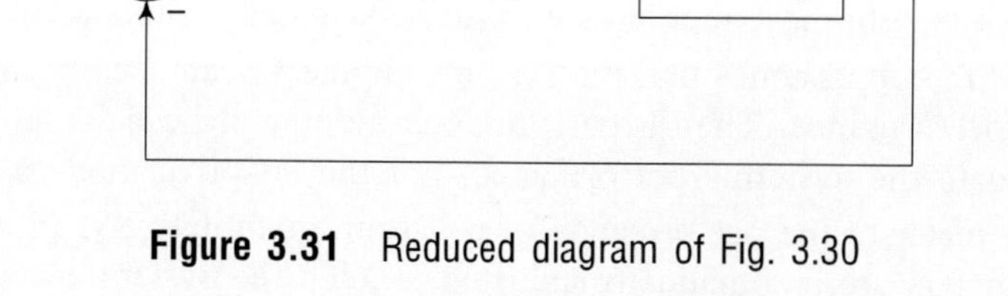

Figure 3.31 Reduced diagram of Fig. 3.30

From the block diagram

$$\frac{Y(s)}{R(s)} = \frac{\left(1 + sK_d + \frac{K_i}{s}\right) \times \left(\frac{\omega_n^2}{s\,(s + 2\xi\,\omega_n}\right)}{1 + \left(1 + s\,K_d + \frac{K_i}{s}\right)\left(\frac{\omega_n^2}{s\,(s + 2\xi\,\omega_n)}\right)}$$

$$\frac{Y(s)}{R(s)} = \frac{\left(1 + s\,K_d + \frac{K_i}{s}\right)\frac{\omega_n^2}{s\,(s + 2\xi\,\omega_n)}}{\frac{s\,(s + 2\xi\,\omega_n) + \left(1 + s\,K_d + \frac{K_i}{s}\right)\omega_n^2}{s\,(s + 2\xi\,\omega_n)}}$$

$$\frac{Y(s)}{R(s)} = \frac{(s + s^2\,K_d + K_i)\,\omega_n^2}{s^3 + 2\xi\,\omega_n\,s^2 + (s + s^2\,K_d + K_i)\,\omega_n^2}$$

$$\frac{Y(s)}{R(s)} = \frac{(s^2\,K_d + s + K_i)\,\omega_n^2}{s^3 + 2\xi\,\omega_n\,s^2 + (s^2\,K_d + s + K_i)\,\omega_n^2}$$

$$\frac{Y(s)}{R(s)} = \frac{(s^2\,K_d + s + K_i)\,\omega_n^2}{s^3 + (2\xi\,\omega_n + K_d\,\omega_n^2)\,s^2 + s\,\omega_n^2 + K_i\,\omega_n^2} \qquad \text{...(3.112)}$$

The characteristic equation

$$s^3 + (2\xi\,\omega_n + K_d\,\omega_n^2)\,s^2 + s\,\omega_n^2 + K_i\,\omega_n^2 = 0$$

Effect on dynamic response of the system caused by a PID controller is not obvious as the characteristic equation becomes of the third order. Inclusion of two zeros in the numerator ($K_d s^2 + s + K_i$) helps the system in acquiring desirably dynamic qualities but the third order characteristic equation causes the system to become unstable for certain combinations of gains (K_d, K_i, K_p).

3.8 PERFORMANCE INDICES

As we discussed earlier, we need to meet some specifications like maximum overshoot (M_p), rise time (t_r) settling time (t_s), peak time (t_p) and damping ratio (ξ) for designing a control system. These terms have been introduced in respect of dynamic response for step input and the steady state error (e_{ss}). These measures define the overall performance of the system. For designing a control system these measures have to be satisfied simultaneously.

Modern control theory assumes that the systems engineers can specify quantitatively the required system performance. Then a performance index can be calculated or measured and used to evaluate the system's performance. For the adaptive modern control system it is necessary to meausre the performance of a system quantitatively. In adaptive control sytems, the parameters are automatically adjusted to keep the system at an optimum level of performance under varying inputs and varying conditions of operation. That means there is no need to change the system's parameters manually according to change in inputs and conditions of operation. Adaptive control systems require a performance index which is a function of the variable system parameters. The desirable feature of the performance index is its maximum or minimum value corresponding to the optimum set of parameter values, its selectivity (i.e., its power to clearly distinguish between an optimum and non-optimum system) and its sensitivity to parameter variations and its simplicity to analytical computations or its on-line analog or digital determination.

A suitable performance index is the integral of the square of the error, ISE which is defined as

$$ISE = \int_0^T e^2(t)\,dt \quad \text{...(3.113)}$$

where ISE = Integral of the square of the error
$e(t)$ = Error
T = Finite time

In Eq. (3.113), the upper limit (T) is a finite time, it is arbitrarily chosen so that the integral approaches a steady state value. It is usually convenient to chose $T = t_s$, settling time.

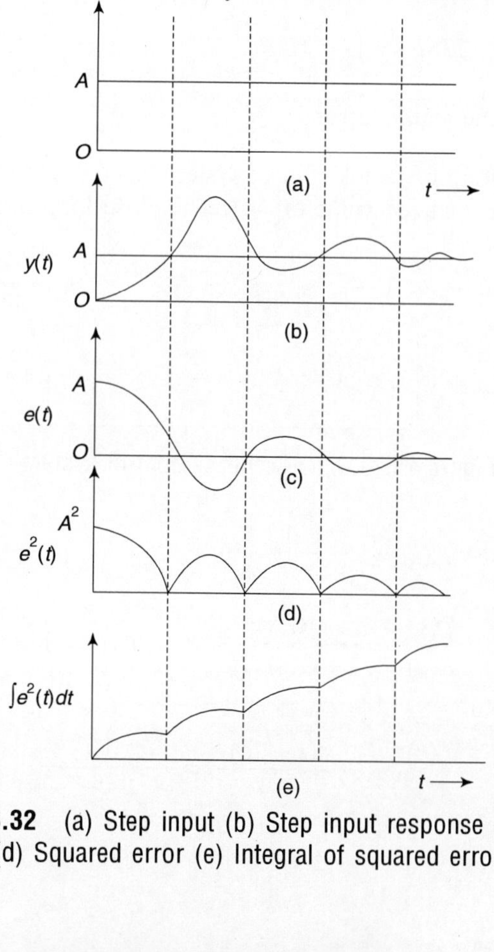

Figure 3.32 (a) Step input (b) Step input response (c) Error (d) Squared error (e) Integral of squared error

The step response for a particular feedback control system is shown in Fig. 3.32 (b), and the error in Fig. 3.32 (c). The error squared is shown in Fig. 3.32 (d), and the integral of the error squared in Fig. 3.32 (e). The performance index of Eq. (3.113) is easily adapted for practical mesurements because a squaring circuit is easily obtained. The squared error is mathematically convenient for analytical and computational purposes.

Another easily instrumented performance index is the integral of the absolute magnitude of the error, *IAE*, which is

$$IAE = \int_0^T |e(t)|\, dt \qquad ...(3.114)$$

This (*IAE*) index is particularly useful for computer simulation studies.

To reduce the contribution of the large initial error and to penalize the small errors occurring later in the response, the following index has been proposed.

$$ITAE = \int_0^T t|e(t)|\, dt \qquad ...(3.115)$$

where *ITAE* = Integral time – Absolute error

$$ITSE = \int_0^T t\, e^2(t)\, dt \qquad ...(3.116)$$

where *ITSE* = integral time square error

Example 3.8.1 A signle loop feedback control system is shown in Fig. Ex. 3.8.1, where the natural frequency is the normalized value $\omega_n = 1$. The close loop transfer function is then

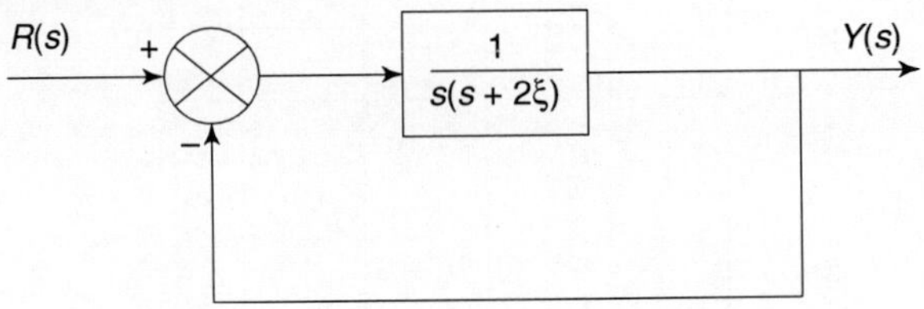

Figure Ex. 3.8.1 Feedback control system

$$T_o(s) = \frac{1}{s^2 + 2\xi\, s + 1} \qquad ...(3.117)$$

$$\frac{Y(s)}{R(s)} = \frac{G(s)}{1 + G(s)\, H(s)} \qquad \because H(s) = 1$$

$$\frac{Y(s)}{R(s)} = \frac{\dfrac{1}{s\,(s+2\xi)}}{1 + \dfrac{1}{s\,(s+2\xi)}} = \frac{1}{s^2 + 2\xi s + 1}$$

$$\boxed{T_o(s) = \frac{1}{s^2 + 2\xi s + 1}}$$

Three performance indices – *ISE*, *ITSE*, and *ITAE*—calculated for various values of the damping ratio and for a step input are shown in Fig. Ex. 3.8.1 (a). These curves show the selectivity of the *ITAE*. Index in comparison with the *ISE* index. By the curve we can say that the *ITAE* performance index is a better choice.

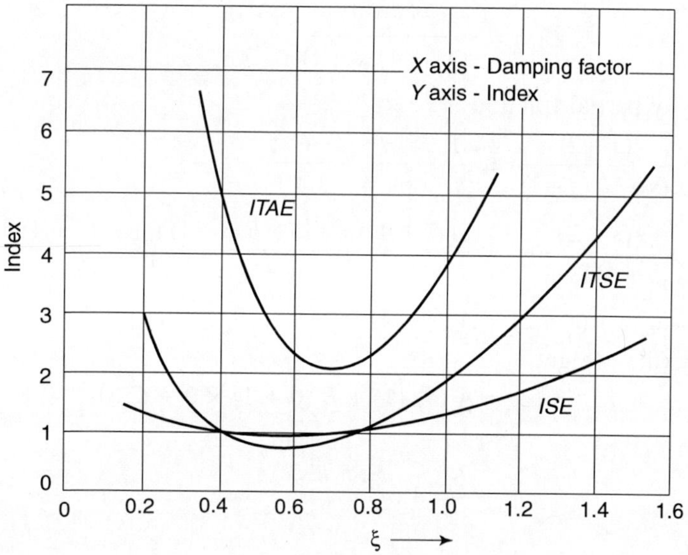

Figure Ex. 3.8.1 (a) Performance indices for second order system

- *A performance index is a quantitative measure of the performance of a system and is chosen so that emphasis is given to the important system specification.*

EXAMPLES

Time Response Analysis

Example 3.1 For the closed loop transfer function $\frac{C(s)}{R(s)} = \frac{3(s+2)}{(s+4)(s+1)}$ of a control system.

Find the system response to a unit step input. *(U.P.T.U., 2006–07, CS)*

Solution

$$\frac{C(s)}{R(s)} = \frac{3(s+2)}{(s+4)(s+1)} \quad ...(i)$$

For a unit step input

$$R(t) = u(t)$$

$$L\{R(t)\} = L\{u(t)]$$

$$R(s) = \frac{1}{s}$$

Substituting the value of $R(s)$ in Eq. (i)

$$C(s) = \frac{1}{s} \cdot \frac{3(s+2)}{(s+4)(s+1)}$$

$$C(s) = \frac{3(s+2)}{s(s+4)(s+1)} \qquad \text{...(ii)}$$

Solving Eq. (ii) by partial fraction

$$\frac{3(s+2)}{s(s+4)(s+1)} = \frac{A}{s} + \frac{B}{s+4} + \frac{C}{s+1}$$

$$\frac{3(s+2)}{s(s+4)(s+1)} = \frac{A(s+4)(s+1) + B(s+1)s + C(s+4)s}{s(s+4)(s+1)}$$

$$3(s+2) = A(s+4)(s+1) + B(s+1)s + C(s+4)s \qquad \text{...(iii)}$$

Put $s = 0$ in Eq. (iii), we get

$$3(2) = A(4)(1) + B(0+1) \times 0 + C(0+4)0$$

$$6 = 4A$$

$$A = \frac{6}{4} = 1.5$$

$$A = 1.5$$

Now putting $s = -4$ in Eq. (iii), we get

$$3(-4+2) = A(0) + B(-4+1)(-4) + C(0)$$

$$-6 = 12B$$

$$B = -\frac{6}{12} = -0.5$$

$$B = -0.5$$

Now putting $s = -1$ in Eq. (iii), we get

$$3(-1+2) = A(0) + B(0) + C(-1+4)(-1)$$

$$3 = -3C$$

$$C = -1$$

Now substituting the values of A, B, C in Eq. (ii), we get

$$C(s) = \frac{1.5}{s} - \frac{0.5}{s+4} - \frac{1}{s+1}$$

Now taking the inverse Laplace transform

$$L^{-1}\{C(s)\} = L^{-1}\left\{\frac{1.5}{s}\right\} - L^{-1}\left\{\frac{0.5}{s+4}\right\} - L^{-1}\left\{\frac{1}{s+1}\right\}$$

$$L^{-1}\{C(s)\} = 1.5\,L^{-1}\left\{\frac{1}{s}\right\} - 0.5\,L^{-1}\left\{\frac{1}{s+4}\right\} - L^{-1}\left\{\frac{1}{s+1}\right\}$$

$$c(t) = 1.5 - 0.5\,e^{-4t} - e^{-t} \quad \textbf{Ans.}$$

Example 3.2 A control system is shown in Fig. Ex. 3.2. Determine the transfer function and derive an expression relating the output time if the input is a step having a magnitude of 2 units.

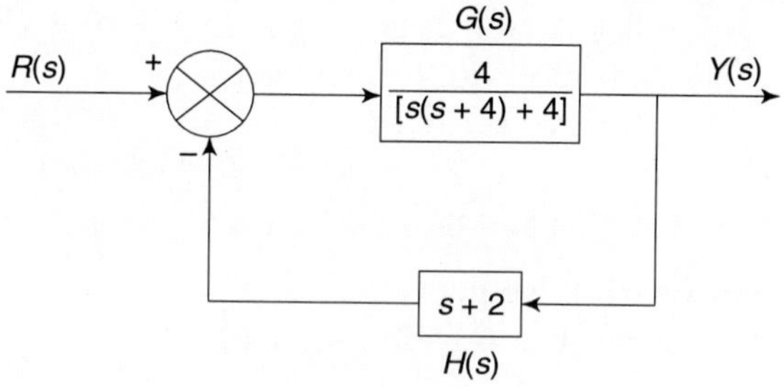

Figure Ex. 3.2

Solution

$$\frac{Y(s)}{R(s)} = \frac{G(s)}{1 + G(s)\,H(s)}$$

$$\frac{Y(s)}{R(s)} = \frac{\dfrac{4}{[s\,(s+4)+4]}}{1 + \dfrac{4}{[s\,(s+4)+4]}(s+2)}$$

$$= \frac{\dfrac{4}{s^2+4s+4}}{\dfrac{(s^2+4s+4)+4\,(s+2)}{s^2+4s+4}}$$

$$= \frac{4}{s^2+4s+4+4s+8} = \frac{4}{s^2+8s+12}$$

$$\frac{Y(s)}{R(s)} = \frac{4}{(s+2)\,(s+6)} \quad \textbf{Ans.}$$

$$Y(s) = R(s)\cdot\frac{4}{(s+2)\,(s+6)} \qquad \text{...(i)}$$

As $r(t) = 2$

$$R(s) = \frac{2}{s}$$

Substituting the value of $R(s)$ in Eq. (i)

$$Y(s) = \frac{2}{s}\cdot\frac{4}{(s+2)\,(s+6)}$$

$$Y(s) = \frac{8}{s(s+2)(s+6)} \qquad \text{...(ii)}$$

Solving Eq. (ii) by partial fraction

$$\frac{8}{s(s+2)(s+6)} = \frac{A}{s} + \frac{B}{s+2} + \frac{C}{s+6} \qquad \text{...(iii)}$$

$$8 = A(s+2)(s+6) + Bs(s+6) + Cs(s+2) \qquad \text{...(iv)}$$

Now putting $s = 0$ in Eq. (iv), we get

$$8 = A(2)(6) + 0$$

$$A = \frac{8}{12} = \frac{2}{3}$$

Now putting $s + 2 = 0$ in Eq. (iv), we get

$$8 = 0 + B(-2)(-2+6) + 0$$

$$B = -\frac{8}{8} = -1$$

and now putting $s + 6 = 0$ in Eq. (iv), we get

$$8 = 0 + 0 + C(-6)(-6+2)$$

$$8 = 24\,C$$

$$C = \frac{8}{24} = \frac{1}{3}$$

Now putting the value of *A*, *B*, *C* in Eq. (iii), we get

$$Y(t) = \frac{2}{3}\cdot\frac{1}{s} - \frac{1}{s+2} + \frac{1}{3}\cdot\frac{1}{s+6}$$

Taking inverse Laplace transform

$$L^{-1}\{Y(s)\} = \frac{2}{3}L^{-1}\left\{\frac{1}{s}\right\} - L^{-1}\left\{\frac{1}{s+2}\right\} + \frac{1}{3}L^{-1}\left\{\frac{1}{s+C}\right\}$$

$$y(s) = \frac{2}{3} - e^{-2t} + \frac{1}{3}e^{-6t} \quad \textbf{Ans.}$$

Example 3.3 A unit feedback control system has open loop transfer function as $G(s) = \dfrac{K(s+2)}{s^3+\beta s^2+4s+1}$. Determine the values of *K* and β, so that the closed loop unit step response has $\xi = 0.2$ and $\omega_n = 3$ rad./sec. *(U.P.T.U., 2007–08, CS)*

Solution Given that

$$G(S) = \frac{K(s \quad 2)}{s^3+\beta s^2+4s+1}$$

$$\xi = 0.2 \text{ and } \omega_n = 3 \text{ rad./sec.}$$

The characteristic equation is

$$1 + G(s)H(s) = 0$$

$\because$ System has unity feedback so $H(s) = 1$

$$1 + \frac{K(S+2)}{S^3 + \beta S^2 + 4S + 1} = 0$$

$$s^3 + \beta s^2 + 4s + 1 + K(s + 2) = 0$$

$$s^3 + \beta s^2 + (K + 4)s + 2K + 1 = 0 \quad \text{...(A)}$$

Let $$(s + a)(s^2 + 6s + C) \quad \text{...(i)}$$

$$s^3 + (a + b)s^2 + (ab + c)s + ac = 0 \quad \text{...(B)}$$

The characteristic equation for the second order

$$s^2 + 2\xi\omega_n + \omega_n^2 = 0 \quad \text{...(ii)}$$

Comparing Eq. (ii) with Eq. (i), we get

$$\omega_n^2 = C$$

$$3^2 = C$$

$$C = 9$$

$$2\xi\omega_n s = bs$$

$$2\xi\omega_n = b$$

$$2 \times (0.2)(3) = b \Rightarrow b = 1.2$$

Now from Eq. A and B, we get

$$a + b = \beta$$

$$K + 4 = ab + C \text{ and } 2K + 1 = ac$$

$$a = \frac{2K+1}{c} \Rightarrow \frac{2K+1}{9}$$

$\because$ $$C = 9$$

Now $$ab + c = K + 4$$

$$a(1.2) + 9 = K + 4$$

$$\frac{2K+1}{9} \cdot (1.2) + 9 = K + 4$$

$$6.6K = 45$$

$$6.6.\ K = \frac{45}{6.6} = 6.81$$

$\because$ $$a = \frac{2k+1}{9} = \frac{(2 \times 6.81) + 1}{9} = 1.51$$

$$a = 1.51$$

$$\beta = a + b = 1.51 + 1.2$$

$$\beta = 2.71 \quad \textbf{Ans.}$$

Example 3.4 Find the time response, initial value and final values of the following

$$F(s) = \frac{s(s+10)}{(s+2)(s+4)(s+6)}.$$

Solution The function is given

$$F(s) = \frac{s\,(s+10)}{(s+2)\,(s+4)\,(s+6)}$$

Solving the function with partial fraction

$$\frac{s\,(s+10)}{(s+2)\,(s+4)\,(s+6)} = \frac{A}{s+2} + \frac{B}{s+4} + \frac{C}{s+6} \quad \text{...(i)}$$

$$\frac{S\,(S+10)}{(S+2)\,(S+4)\,(S+6)} = \frac{A\,(s+4)\,(s+6) + B\,(s+2)\,(s+6) + C\,(s+2)\,(s+4)}{(s+2)\,(s+4)\,(s+6)}$$

$$s\,(s + 10) = A\,(s + 4)\,(s + 6) + B\,(s + 2)\,(s + 6) + C\,(s + 2)\,(s + 4) \quad \text{...(ii)}$$

Now putting $s + 2 = 0$ in Eq. (ii), we get

$$-2\,(-2 + 10) = A\,(-2 + 4)\,(-2 + 6) + 0 + 0$$

$$-16 = 8\,A$$

$$A = -2$$

Now putting $s + 4 = 0$ in Eq. (ii), we get

$$-4\,(-4 + 10) = 0 + B\,(-4 + 2)\,(-4 + 6) + 0$$

$$-24 = -4\,B$$

$$B = 6$$

Now putting $s + 6 = 0$ in Eq. (iii), we get

$$-6\,(-6 + 10) = 0 + 0 + C\,(-6 + 2)\,(-6 + 4)$$

$$-24 = 8\,C$$

$$C = -3$$

Now putting the value of A, B, C in Eq. (ii), we get

$$F(s) = -\frac{2}{s+2} + \frac{6}{s+4} - \frac{3}{s+6} \quad \text{...(iii)}$$

Taking inverse Laplace transform of Eq. (iii)

$$L^{-1}\,\{F(s)\} = -2\,L^{-1}\left\{\frac{1}{s+2}\right\} + 6\,L^{-1}\left\{\frac{1}{s+4}\right\} - 3\,L^{-1}\left\{\frac{1}{s+6}\right\}$$

$$F(t) = -2\,e^{-2t} + 6\,e^{-4t} - 3\,e^{-6t} \quad \textbf{Ans.}$$

$$\text{Initial value} = \lim_{t\to 0} f(t) = -2 + 6 - 3 = 1 \quad \textbf{Ans.}$$

$$\text{Final value} = \lim_{s\to 0} sF(s) = \lim_{s\to 0} \frac{s^2\,(s+10)}{(s+2)\,(s+4)\,(s+6)}$$

$$= 0 \quad \textbf{Ans.}$$

Example 3.5 Calculate the performance indices for a unit step input, second order system given below $\frac{1000}{s^2 + 3H0.5s + 1000}$. (*U.P.T.U., 2004–05, CS*)

Solution Given that

$$\frac{Y(s)}{R(s)} = \frac{1000}{s^2 + 3.5s + 1000} \quad \text{...(i)}$$

Comparing Eq. (i) with second order system equation below

$$\frac{Y(s)}{R(s)} = \frac{\omega_n^2}{s^2 + 2\xi\,\omega_n s + \omega_n^2}$$

We get

$$\omega_n^2 = 1000$$

$$\omega_n = 31.6$$

and

$$2\,\xi\omega_n = 3.5$$

$$2\xi = \frac{3.5}{\omega_n} = \frac{3.5}{31.6} = 0.11$$

$$2\xi = 0.11$$

$$\xi = \frac{0.11}{2} = 0.06$$

Integral square error

$$ISE = \frac{1}{\omega_n}\left[\xi + \frac{1}{4\xi}\right] = \frac{1}{31.6}\left[0.06 + \frac{1}{4 \times 0.06}\right]$$

$$ISE = \frac{4.23}{31.6} = 0.134 \quad \textbf{Ans.}$$

Integral of time multiplied square error

$$ITSE = \frac{1}{\omega_n^2}\left[\xi^2 + \frac{1}{8\xi^2}\right] = \frac{1}{(31.6)^2}\left[(0.06)^2 + \frac{1}{8\,(0.06)^2}\right]$$

$$ITSE = 0.0348 \quad \textbf{Ans.}$$

Example 3.6 Certain measurements were conducted on a servomechanism which show the system response as $c(t) = 1 + 0.2e^{-60t} - 1.2\,e^{-10t}$

When subjected to a unit step input

(i) Find the expression for closed loop transfer function

(ii) Obtain the undamped natural frequency and damping ratio. (*U.P.T.U, 2002–03, CS*)

Solution Given that

$$c(t) = 1 + 0.2\,e^{-160t} - 1.2\,e^{-10t}$$

Taking the Laplace transform of Eq. (i)

$$L\,\{c(t)\} = L\,\{1\} + 0.2\,L\,\{e^{-60t}\} - 1.2\,L\,\{e^{-10t}\}$$

$$C(s) = \frac{1}{s} + 0.2 \cdot \frac{1}{s+60} - 1.2 \cdot \frac{1}{s+10}$$

$$C(s) = \frac{(s+60)(s+10) + 0.2\,(s+10)\,s - 1.2\,s\,(s+60)}{s\,(s+60)\,(s+10)}$$

$$C(s) = \frac{s^2 + 70s + 600 + 0.2 5^2 + 2s - 1.2 5^2 - 7.2 5}{s\,(s+60)\,(s+10)}$$

$$C(s) = \frac{64.8\,s + 600}{s\,(s^2 + 70\,s + 600)}$$

If $R(s) = \frac{1}{s}$, i.e., unit step input, then

$$C(s) = \frac{64.8\,s + 600}{(s^2 + 70\,s + 600)} R(s)$$

$$\frac{C(s)}{R(s)} = \frac{64.8\,s + 600}{s^2 + 70\,s + 600}$$

So, (i) The expression for closed loop transfer function

$$\frac{C(s)}{R(s)} = \frac{64.8\,s + 600}{s^2 + 70\,s + 600}$$ **Ans.**

(ii) The undamped natural frequency is given by

$$\omega_n = \sqrt{600} = 24.5$$

and $2\xi\omega_n = 70$

$$2\xi = \frac{70}{24.5} = 2.9$$

$$\xi = \frac{2.9}{2} = 1.43$$

$\xi = 1.43$ **Ans.**

Example 3.7 A unity feedback system is characterised by an open loop transfer function, $G(s) = \frac{K}{s\,(s+100)}$. Determine the gain k so that the system has a damping ratio of 0.5. For this value of K, determine settling time, peak overshoot and time to peak overshoot for a unit step input. *(U.P.T.U., 2002–03, CS, U.P.T.U., 2006–07, ACS)*

Solution Given $G(s) = \frac{K}{s\,(s+10)}$

and $H(s) = 1$

The characteristic equation is

$$1 + G(s)\,H(s) = 0$$

$$1+\frac{K}{s\,(s+10)}\times 1 = 0$$

$$s\,(s + 10) + K = 0$$

$$s^2 + 10\,s + K = 0 \qquad \text{...(ii)}$$

Comparing Eq. (ii) with equation of the second order system equation

$$s^2 + 2\xi\omega_n\,s + \omega_n^2$$

we get

$$2\xi\omega_n = 10$$

$$\omega_n = \frac{10}{2\xi} = \frac{10}{2\times 0.5} = \frac{10}{1} = 10$$

$$\omega_n^2 = K$$

$$K = (10)^2 = 100$$

Settling time

$$t_s = \frac{4}{\xi\omega_n}$$

$$t_s = \frac{4}{0.5\times 10} = 0.8\ \text{sec.} \quad \textbf{Ans.}$$

Maximum overshoot, M_P

$$M_P = e^{-\frac{-\pi\xi}{\sqrt{1-\xi^2}}} = e^{-\frac{0.5\,\pi}{\sqrt{1-0.50^2}}} = e^{-\frac{1.57}{0.87}}$$

$$M_P = e^{-1.8} = 16.5\% \quad \textbf{Ans.}$$

Peak time, t_p

$$t_P = \frac{\pi}{\omega_n\sqrt{1-\xi^2}} = \frac{3.14}{10\sqrt{1-0.5^2}}$$

$$t_P = 0.326\ \text{sec.} \quad \textbf{Ans.}$$

Example 3.8 The open loop transfer function of a unity feedback system is given by $G(s) = \dfrac{K}{s\,(1+Ts)}$ where T and K are constants having positive values. By what factor the amplifier gain be reduced so that (a) the peak overshoot of unit step response of the system is reduced from 75% to 25% (b) the damping ratio increases from 0.1 to 0.6.

Solution (a)

$$G(s) = \frac{K}{s\,(1+Ts)}$$

Let ξ_1 be damping ratio at $M_P = 75\%$

ξ_2 be damping ratio at $M_P = 25\%$

We have $$M_P = e^{-\frac{\pi\xi}{\sqrt{1-\xi^2}}}$$

when $$M_P = 75\% = e^{-\frac{\pi\xi_1}{1-\xi_1^2}}$$

$$\xi_1 = 0.091$$

when $$M_P = 25\% = e^{-\frac{-\pi\xi_2}{1-\xi_2^2}}$$

$$\xi_2 = 0.4037$$

The transfer function of the system is

$$\frac{C(s)}{R(s)} = \frac{G(s)}{1+G(s)\,H(s)}$$

$$\frac{C(s)}{R(s)} = \frac{K}{Ts^2+s+K} = \frac{\frac{K}{T}}{s^2+\frac{1}{T}s+\frac{K}{T}} \qquad \text{...(i)}$$

Comparing Eq. (i) with standard equation of the second order system

$$\omega_n^2 = \frac{K}{T} \Rightarrow \omega_n = \sqrt{\frac{K}{T}}$$

and $$2\xi\omega_n = \frac{1}{T}$$

Let the value of $K = K_1$ when $\xi = \xi_1$

and $K = K_2$ when $\xi = \xi_2$

$$2\xi\omega_n^2 = \frac{1}{T}$$

$$\xi = \frac{1}{2\,T\omega_n} = \frac{1}{2\sqrt{KT}}$$

$$\frac{\xi_1}{\xi_2} = \sqrt{\frac{K_2\,T}{K_1\,T}} = \sqrt{\frac{K_2}{K_1}}$$

$$\frac{K_2}{K_1} = \frac{0.091}{0.4037} = 0.0508$$

$$K_2 = 0.0508\,K_1$$

The amplifier gain has to be reduced by a factor $\frac{1}{0.0508}$ = 20 Ans.

(b) Let $\xi = 0.1$, where gain is K_1 and $\xi = 0.6$ when gain is K_2

$$\frac{0.1}{0.6} = \sqrt{\frac{K_2}{K_1}}$$

$$K_2 = \frac{1}{36} K_1$$

∴ The amplifier gain should be reduced by a factor 36. **Ans.**

Example 3.9 A unity feedback control system has its open loop transfer function given by $G(s) = \frac{4s+1}{4s^2}$. Determine an expresson for the time response when the system is subjected to

(a) Unit impulse input function
(b) Unit step input function

Solution The overall transfer function of the system is

$$\frac{C(s)}{R(s)} = \frac{G(s)}{1+G(s)\,H(s)} = \frac{\frac{4s+1}{4s^2}}{1+\frac{(4s+1)}{4s^2}\times 1}$$

$$\frac{C(s)}{R(s)} = \frac{4s+1}{4s^2+4s+1}$$

$$\frac{C(s)}{R(s)} = \frac{(s+).25)}{(s^2+s+0.25)} = \frac{(s+0.25)}{(s+0.5)^2}$$

(a) For unit impulse response $R(s) = 1$

$$C(s) = R(s)\cdot\frac{s+0.25}{(s+0.5)^2}$$

$$C(s) = 1\cdot\frac{s+0.25}{(s+0.5)^2}$$

By the partial fraction expression

$$\frac{s+0.25}{(s+0.5)^2} = \frac{A}{(s+0.5)} + \frac{B}{(s+0.5)^2}$$

$$\frac{s+0.25}{(s+0.5)^2} = \frac{A\,(s+0.5)+B}{(s+0.5)^2}$$

$$(s + 0.25) = A\,(s + 0.5) + B$$

Now putting $s + 0.5 = 0$

$$-0.5 + 0.25 = 0 + B$$

$$B = -0.25$$

Now putting

$$A = 1$$

Now substituting the values of A and B

$$C(s) = \frac{1}{s+0.5} - \frac{0.25}{(s+0.5)^2}$$

Taking the inverse Laplace transform on both sides

$$L^{-1}\{C(s)\} = L^{-1}\left\{\frac{1}{s+0.5}\right\} - 0.25\,L^{-1}\left\{\frac{1}{(s+0.5)^2}\right\}$$

$$C(t) = e^{-0.5t} - 0.25\,t\,e^{-0.5t}$$

$$C(t) = e^{-0.5t}\,(1 - 0.25\,t) \quad \textbf{Ans.}$$

(b) For unit step input

$$R(s) = \frac{1}{s}$$

$$C(s) = \frac{1}{s}\cdot\frac{s+0.25}{(s+0.5)^2}$$

By the partial fraction

$$\frac{1}{s}\cdot\frac{s+0.25}{(s+0.5)^2} = \frac{A}{s} + \frac{B}{(s+0.5)} + \frac{C}{(s+0.5)^2}$$

$$= \frac{A\,(s+0.5)^2 + B\,(s+0.5)\,s + C(s)}{s\,(s+0.5)^2}$$

$$s + 0.25 = A\,(s + 0.5)^2 + Bs\,(s + 0.5) + C(s) \quad \text{...(i)}$$

Solving Eq. (i) we get

$$A = 1$$

$$B = 2$$

and

$$C = 1$$

Now

$$C(s) = \frac{1}{s} - \frac{2}{s+2} + \frac{1}{s+2}$$

Taking inverse Laplace transform on both side

$$C(t) = (1 - 2\,e^{-2t} + e^{-2t}) \quad \textbf{Ans.}$$

Example 3.10 A load is controlled by an electric motor. The position of the input shaft is adjusted to θ_r. The motor torque constant is 360 N_m per radian of error between the input and output position. The moment of inertia and the coefficient of viscous friction are 10 kg m^2 and 60 N_m rad/sec. respectively. The system is initially at rest. Determine an expression relating time and angular position of the output following a sudden shift of 50 in the input shaft.

Solution The block diagram for the system is shown in Fig. Ex. 3.10.

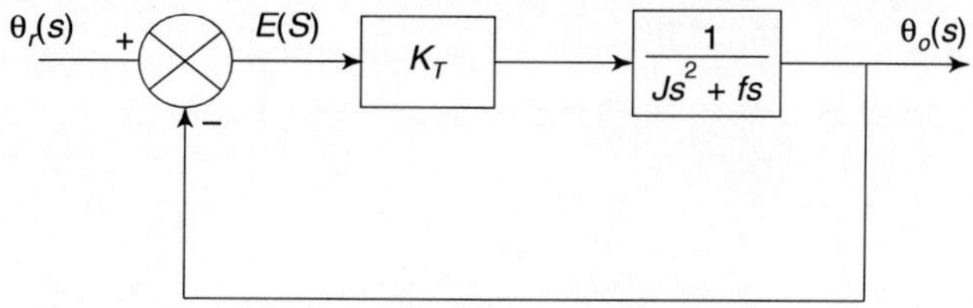

Figure Ex. 3.10

The overall transfer function is

$$\frac{\theta_o(s)}{\theta_r(s)} = K_T \frac{\dfrac{1}{Js^2 + fs}}{1 + K_T \cdot \dfrac{1}{Js^2 + fs} \times 1}$$

$$\frac{\theta_o(s)}{\theta_r(s)} = \frac{K_T}{Js^2 + fs + K_T} \quad \text{...(i)}$$

We have

$$K_T = 360 \quad \text{and } J = 10 \text{ kg-m}^2$$
$$f = 60 \text{ Nm/(rad/sec.)}$$

Substituting the value in Eq. (i)

$$\frac{\theta_o(s)}{\theta_r(s)} = \frac{360}{10s^2 + 60s + 360}$$

$$\frac{\theta_o(s)}{\theta_r(s)} = \frac{36}{s^2 + 6s + 36}$$

$$\theta_o(s) = \theta_r(s) \cdot \frac{36}{s^2 + 6s + 36}$$

Since

$$\theta_r = 50 \times \frac{\pi}{180} = \frac{5\pi}{18} \text{ rad}$$

$$\theta_r(s) = \frac{5\pi}{18} \cdot \frac{1}{s}$$

$\therefore$

$$\theta_o(s) = \frac{5\pi}{18} \cdot \frac{1}{s} \cdot \frac{36}{s^2 + 6s + 36} \quad \text{...(ii)}$$

By Eq. (ii)

$$\omega_n^2 = 36$$

$$\Rightarrow \quad \omega_n = 6 \text{ rad/sec.}$$

$$2\,\xi\omega_n = 6$$

$$\therefore \quad \xi = \frac{6}{2\times 6} = 0.5$$

The time response expression is given by

$$\theta_o(s) = \frac{5\pi}{18}\left\{1 - \frac{e^{-\xi\omega_n t}}{\sqrt{1-\xi^2}} \sin\left[\left(\omega_n \sqrt{1-\xi^2}\; t + \tan^{-1}\left(\frac{\sqrt{1-\xi^2}}{\xi}\right)\right)\right]\right\}$$

$$\theta_o(t) = \frac{5\pi}{18}\left\{\frac{e^{-0.5\times 6t}}{\sqrt{1-(0.5)^2}} \sin\left[\left(6\sqrt{1-0.5^2}\; t + \tan^{-1}\frac{\sqrt{1-0.5^2}}{0.5}\right]\right\}$$

$$\theta_o(t) = \theta_o(t) = \frac{5\pi}{18}\left\{1 - \frac{e^{-3t}}{0.86} \sin\left[(5.19\,t + \tan^{-1} 1.732)\right]\right\} \quad \textbf{Ans.}$$

Example 3.11 A two phase a.c. servomotor having a torque constant of 0.045 N-m/V control a position load through a gear ratio of 10 : 1. The effective moment of inertia and coefficient of viscous friction referred to load side are 0.25 kg m^2 and 1.0 N-m / (rad/sec). The servomotor produces an error signal of 0.1 V per degree error in misalignment. Develop the block diagram representation of the control system and therefore obtain the transfer function.

Calculate the natural frequency of oscillation and damaping ratio.

Solution The error detector gain is

$$K_e = 0.1 \text{ V/degree} = 0.1 \times \frac{180}{\pi} = 5.73 \text{ V/rad}$$

Referred to motor side

$$J = 0.25\left[\frac{N_1}{N_2}\right]^2 = 0.25 \times \frac{1}{10^2}$$

$$J = 0.0025 \text{ Kg m}^2$$

$$f = 1 \times \left(\frac{N_1}{N_2}\right)^2 = 1 \times \frac{1}{10^2}$$

$$f = 0.001 \text{ N-m/(rad/sec.)}$$

The block diagram of the system is shown in Fig. Ex. 3.11.

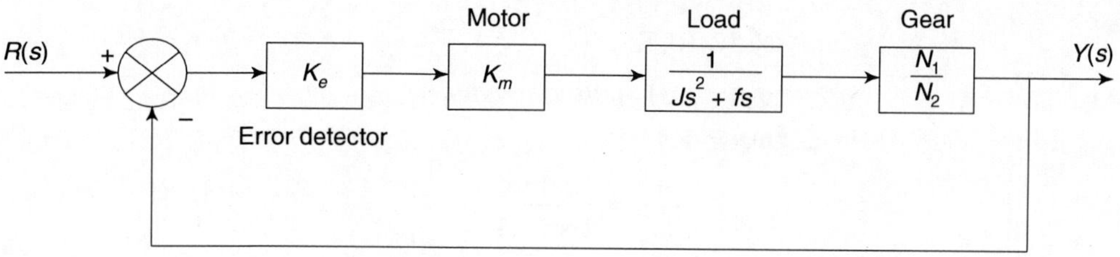

Figure Ex. 3.11

The simplified block diagram is

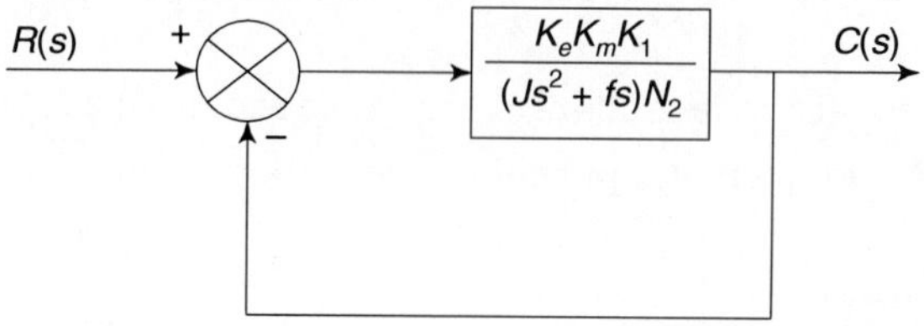

Figure Ex. 3.11 (a)

The transfer function of the system is

$$\frac{Y(s)}{R(s)} = \frac{G(s)}{1 + G(s)\,H(s)} = \frac{K_e\,K_m\,N_1\,(Js^2 + fs)\,N_2}{1 + \dfrac{K_e\,K_m\,N_1}{(Js^2 + fs)\,N_2} \cdot 1}$$

$$\frac{Y(s)}{R(s)} = \frac{K_e\,K_m\,N_1}{(Js^2 + fs)\,N_2 + K_e\,K_m\,N_1}$$

$$\frac{Y(s)}{R(s)} = \frac{(5.73)\,(0.045)\,(1)}{\dfrac{(0.25\,s^2 + 1.0\,s)\,10}{100} + (0.573)\,(0.045)\,(1)}$$

$$\frac{Y(s)}{R(s)} = \frac{0.258 \times 10}{(0.25\,s^2 + 1.05) + 0.258 \times 100}$$

$$\frac{Y(s)}{R(s)} = \frac{2.58}{0.25\,s^2 + 1.0\,s + 2.58} = \frac{10.32}{s^2 + 4s + 10.32}$$

$$\frac{Y(s)}{R(s)} = \frac{2.58}{0.25\,s^2 + 1.0\,s + 2.58} = \frac{10.32}{s^2 + 4\,s + 10.32}$$

The characteristic equation is

$$s^2 + 4s + 10.32 = 0$$

(a)
$$\omega_n^2 = 10.32$$
$$\Rightarrow \qquad \omega_n = 3.21 \text{ rad/sec.} \quad \textbf{Ans.}$$
$$2\,\xi\omega_n = 4$$
$$\xi = \frac{4}{2\omega_n} = \frac{4}{2 \times 3.21}$$
$$\xi = 0.623 \quad \textbf{Ans.}$$

MATLAB PROGRAMS

Program 3.1

```
% Program to calculate characteristics of second order system like damping
% ratio, natural frequency, percent overshoot, settling time and peak
% time
% F(s) = 1/(s-(2+5j))(s-(2-5j))
Clear all;
clc;
p1 = [1 2+5j];                      % Define polynomial containing Pole 1
p2 = [1 2-5j];                      % Define polynomial containing Pole 2
deng = conv(p1,p2);                 % Multiplying to polynomial
'Natural frequency'                 % display string
omegan = sqrt(deng(3)/deng(1))      % omegan = sqrt(2^2+5^2) Or sqrt(c/a)
'Damping ratio'                     % Display string
zeta = (deng(2)/deng(1)/(2*omegan)  % Zeta = (b/a)/2*omegan
'Settling time'                     % Display string
Ts = 4/(zeta*omegan)                % Settling time = 4/zeta*omegan
'Peak time'                         % Display string
Tp = pi/(omegan*sqrt(1-zeta^2))     % peak time
'Percent overshoot'                 % Display string
Mp = 100*exp(-zeta*pi/sqrt(1-zeta^2)) % Percent overshoot
```

Output

```
zeta = 0.3714
Settlling time
Ts = 2.0000
Peak time
Tp = 0.628
Percent overshoot
Mp = 28.4610
```

Program 3.2

```
% Program for step response of the system
% T(s)  = 24/s^2+4s+20
% numt  = input ('Enter the numerator polynomials:- ');
% dent  = input ('Enter the denominator polynomials:- ');
numt  = input ('Enter the coefficient of numerator polynomial:- ');
                       % Define numerator
                       % of T(s) like 24
dent  = input ('Enter the coefficient of denominator polynomial:- ');
                       % Define denominator of % T(s) like [1 4 20]
To = tf (numt,dent)    % Create Transfer Function
step(To)               % Step response
```

Input

```
Enter the coefficient of numerator polynomial:- 24
Enter the coefficient of denominator polynomial:- [1 4 20]
```

Output

```
                            24
Transfer function:  ─────────────────
                     s^2 + 4s + 20
```

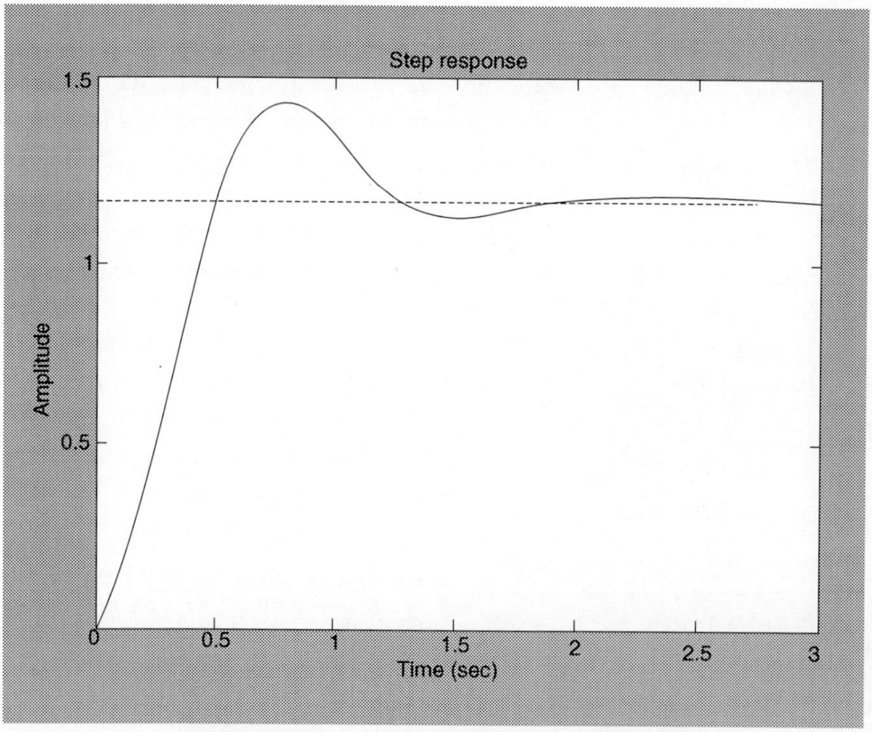

Figure P. 3.2 Step response of funcion $T_0(s)$

Program 3.3

```
% Program for step response of the following system
% (i) T1(s) = 20/(s+8)(s^2+4s+24)
% (ii) T2(s) = 25/(s+5)(s^2+4s+15)
% (i) for T1(s)
clear all
clc
numt1 = input ('Enter the coefficient of numerator polynomial of T1(5):- ')
                                % Define numerator of T1(s) like 20
P11 = input ('Enter the coefficient of first polynomial of
      denominator of T1(s);- ')
         % Define coefficient of first polynomial in denominator of T1(s)
P12 = input ('Enter the coefficient of second polynomial of
      denominator of T1(s):- ');
        % Define coefficient of second polynomial in denominator of T1(s)
dent1 = conv(P11,P12);   % Multuplying P11 & P12
'T1(s)'                  % Display string
T1 = tf(numt1,dent1)     % Transfer function
[y1 t1] = step(T1);      % Step response
% (ii) for T2(s)
numt2 = input ('Enter the coefficient of numerator polynomial of T2(s):- ')
                                % Define numerator of T2(s) like 25
P21 = input ('Enter the coefficient of first polynomial of
      denominator of T2(s):- ')
         % Define coefficient of first polynomial in denominator of T2(s)
P22 = input ('Enter the coefficient of second polynomial of
      denominator of T2(s):- ')
        % Define coefficient of second polynomial in denominator of T2(s)
dent2 = conv(P21,P22):      % Multiplying P11 & P12
'T2(s)'                     % Display function
T2 = tf(numt2,dent2)        % Transfer function
[y2 t2] = step(T2):         % Step response
clf                         % clear graph
plot(t1, y1,t2,y2)          % plot response
xlabel{'Time(second)')      % Add time axis label
Ylabel('Normalized response') % Add response axis label
Title ('Time Response')     % add title
```

Input

```
Enter the coefficient of numerator polynomial of T1(s):- 20
Enter the coefficient of first polynomial of denominator of T1(s):- [1 8]
Enter the coefficient second polynomial of denominator of T1(s):- [1 4, 24]
Enter the coefficient of numerator polynomial of T2(s):- 25
Enter the coefficient of first polynomial of denominator of T2(s):- [1 5]
Enter the coefficient of second polynomial of denominator of T2(s):- [1 4 15]
```

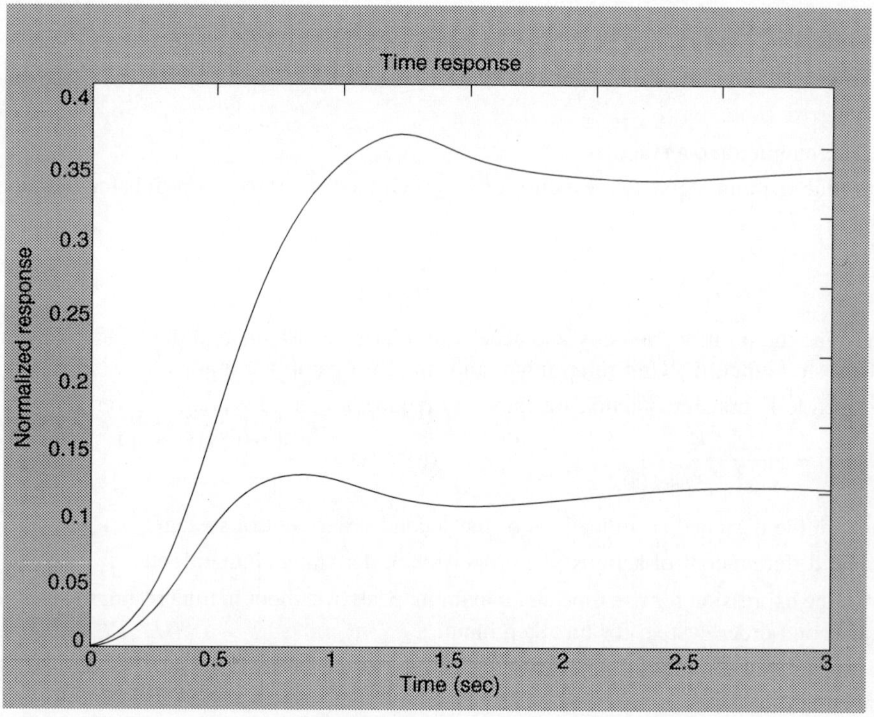

Figure P. 3.3 Time response of function $T1(s)$ and $T2(s)$

Output

```
                                    20
Transfer function:T1(s)  ----------------------------
                          s^3 + 12s^2 + 56s + 192

                                   25
Transfer function:T2(s)  --------------------------
                          s^3 + 9s^2 + 35s + 75
```

EXCERCISE

3.1 What is the need of standard test signal in control engineering? Give Graphical representation. *(U.P.T.U., 2006–07, CS)*

3.2 What are the performance specifications for the first order system?

3.3 What does the performance specification for a first order system tell us?

3.4 What are the advantages of using time domain techniques for the solution of the response?

3.5 A unity feedback control system has forward path transfer function given by $G(s) = \dfrac{s+2}{s\,(s+1)}$

Determine

(a) The expression of unit step response

(b) The rise time

(c) The maximum overshoot

3.6 Determine the unit step time response of a system having closed loop transfer function given below :

$$\frac{Y(s)}{R(s)} = T_o(s) = \frac{1}{(s+1)(s^2+1)}$$

Determine the position, velocity and acceleration error constants and the steady state error for (i) Unit step input, (ii) Unit ramp input, and (iii) Unit parabolic input.

The open loop transfer function for the unity feedback control system is given below :

(a) $G(s) = \dfrac{K}{s\,(s^2+3s+100)}$ (b) $G(s) = \dfrac{K(1+0.5s)(1+2s)}{s^2\,(s^2+4s+5)}$

3.7 What are the designed specifications of the second order control system?

3.8 Explain different control actions like proportional, derivative, integral, etc.

3.9 Derive the expression for rise time and maximum peak overshoot in time response specifications for a second order system for unit step input. *(U.P.T.U., 2006–07, CS)*

3.10 Discuss proportional plus derivative (PD) controller. Also describe the derivative error compensation. *(U.P.T.U., 2006–07, CS)*

3.11 With respect to transient response of a second order system for a step input, define the following:

(i) Time delay (ii) Rise time (iii) Settling time

(iv) Relative peak overshoot (v) Peak time

(U.P.T.U., 2005–06, CS, U.P.T.U., 2006–07, ACS, U.P.T.U., 2004–05, ACS)

Define

(i) Position error constant (ii) Velocity error constant

(ii) Acceleration error constant *(U.P.T.U., 2004–05, CS)*

3.12 Drive the expression for performance of time response for a second order system with step input and discuss. *(U.P.T.U., 2004–05, CS)*

3.13 What are the standard test signals for time domain analysis? Plot them as a function of time and give their Laplace transforms. *(U.P.T.U., 2002–03, CS, 2004–05, CS, 2005–06, CS)*

3.14 For a standard second order system, find the expression of maximum overshoot in the unit step response. *(U.P.T.U., 2003–04, CS)*

3.15 What are time domain performance criteria? *(U.P.T.U., 2003–04, CS)*

3.16 What is the effect of adding a pole to a system? *(U.P.T.U., 2004–05, ACS)*

ANSWERS

3.6 $C(t) = 1 - \frac{e^{-2t}}{2} - \frac{1}{\sqrt{2}} \sin (t + 45°)$

3.7 (a) (i) $K_P = \infty$, $e_{ss} = 0$

(ii) $K_v = \frac{K}{100}$; $e_{ss} = \frac{100}{K}$

(iii) $K_a = 0$, $e_{ss} = \infty$

(b) (i) $K_{\ell} = 0$, $e_{ss} = 0$

(ii) $K_v = \infty$; $e_{ss} = 0$

(iii) $K_a = K/s$; $e_{ss} = 5/K$

CHAPTER 4

Steady State Errors

4.1 INTRODUCTION

The steady state errors play an important role for a system's performance. The steady state error can be defined, if the actual output of a control system deviates from the desired output or reference input, during the steady state. So the difference between actual output and reference input is called steady state error. Basically, for a control system, it is desirable that the output follows its input. So, we can state that the steady state error is a measure of accuracy of control system. These errors happen due to the non-linear system, type of inputs and type of system.

4.2 STEADY STATE ERROR AND TEST INPUTS

As we discussed earlier that the steady state error for prescribed test input as $t \to \infty$ can be defined as the difference between actual output and reference input. Test inputs used for steady state error analysis and design are given in Table 4.1.

Tabel 4.1 Test signals for evaluating steady state errors of a position control system

Signal	Name	Physical interpretation	Time function	Laplace transfer
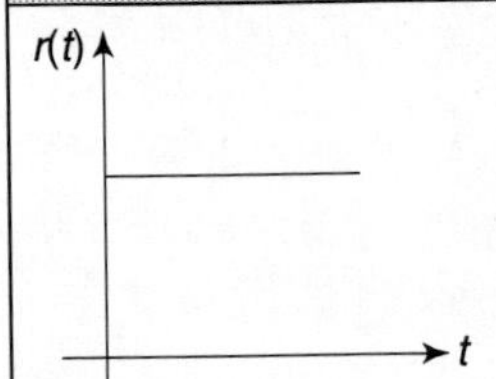	Step	Constant position	1	$\frac{1}{s}$
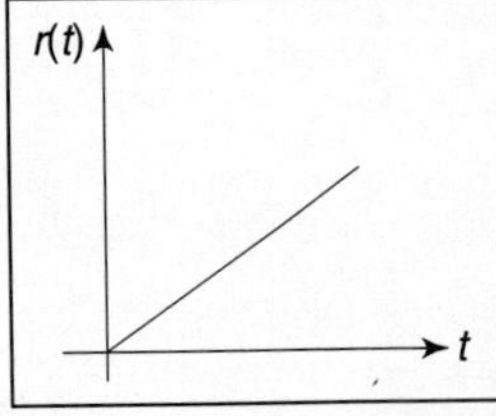	Ramp	Constant velocity	t	$\frac{1}{s^2}$

Contd.

Table 4.1 *Contd.*

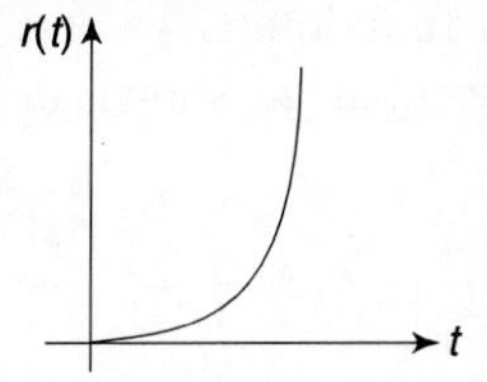	Parabola	Constant acceleration	$\frac{1}{2}t^2$	$\frac{1}{s^3}$

4.3 EVALUATING STEADY STATE ERRORS

Now we will examine the concept of steady state error. A step input and two possible outputs are shown in Fig. 4.1 (a), and a ramp input and three possible outputs are shown in Fig. 4.1 (b). In Fig. 4.1 output 1 has zero steady state error, and output 2 has a finite steady state error, i.e., e_2 (∞). For the ramp input, the output 3, shown in Fig. 4.1 (b), results due to the output's slope which is different from that of the input. While the transients have died down, and t approaches infinity then the steady state error is infinite as measured vertically between the input and output 3.

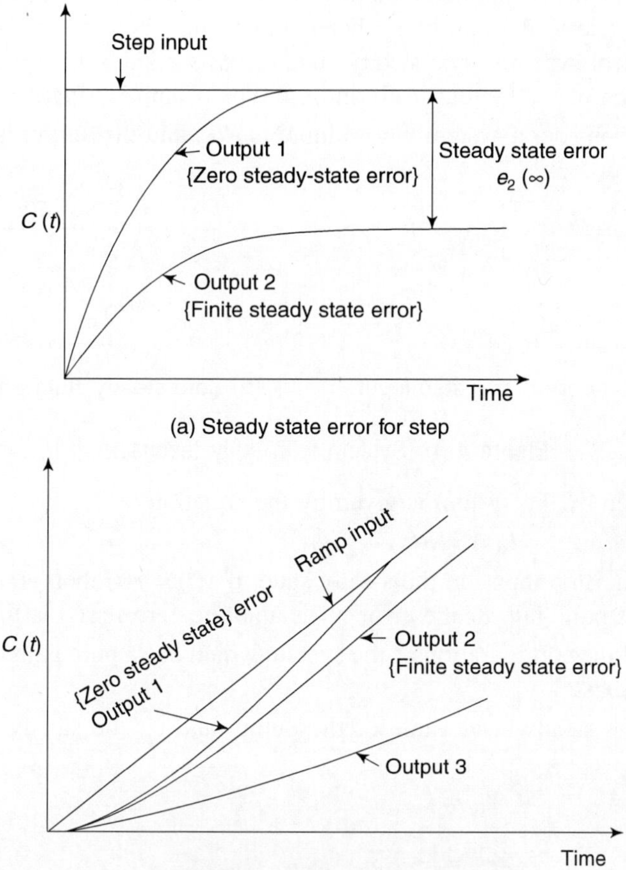

(a) Steady state error for step

(b) Steady state error for ramp input

Figure 4.1 Steady state error

From Fig. 4.1, it is clear that the error is the difference between the input and the output of the system, so we assume a closed loop system which has the transfer function, $G(s)$, and form the error, $E(s)$, by taking the difference between the input and the output, as shown in Fig. 4.2.

(a) General system (b) Unity feedback system

Figure 4.2 Closed loop control system error

4.3.1 Sources of Steady State Error

Steady state error in control system arises from non-linear sources, like backlash in gear and/or a motor that will not move unless the input voltage exceeds a threshold. The detailed study of steady state error due to the non-linear behaviour of the sources, is beyond the scope of a text on linear control systems. The steady state errors we study here are errors that arise from the configuration of the system itself and the type of applied input.

For example, we consider a system whose input is $R(s)$ and the output is $Y(s)$ as shown in Fig. 4.3. (a).

(a) Finite steady state error for a step input (b) Zero steady state error for a step input

Figure 4.3 System with unity feedback

The error produced by the system is given by the equation

$$E(s) = R(s) - Y(s) \quad ...(4.1)$$

Let us consider the step input, in the steady state, if $y(t) = r(t)$ then $e(t)$ will be zero. But when the system has pure gain K, the error $e(t)$ cannot be zero if $y(t)$ is finite and non-zero.

Thus, by virtue of the configuration of the system which has a pure gain of k in the forward path, an error must exist.

Let us call Yss the steady state value of the output and e_{ss} the steady state value of the error, then

$$Y_{ss} = k\, e_{ss} \quad ...(4.2)$$

or

$$e_{ss} = \frac{1}{k} Y_{ss} \quad ...(4.3)$$

From Eq. 4.3, it is clear that the larger the value of k, the smaller the value of Y_{ss} would have to be yield a similar value of y_{ss}.

- *From the above discussion, the conclusion can be made that with a pure gain in the forward path, there will always be a steady state error (e_{ss}) for a step input. This error diminishes as the value of k increases.*

In Fig. 4.4 (b), the forward path gain is replaced by an integrator $\left(\frac{k}{s}\right)$, the error produced by the system for step input is zero. Because

$$e(t) = r(t) - c(t) \quad \text{...(4.4)}$$

From Eq. 4.4 $c(t)$ increases, $e(t)$ will decrease. This decrease will continue until there is zero error, but there will still be a value of $c(t)$ since an integrator can have a constant output without any input.

Example: A motor can be represented simply as an integrator. A voltage applied to the motor will cause rotation. When the input voltage is removed, the motor will stop and remain at its present output position. Since it does not return to its initial position, we have are angular displacement output without an input to the motor.

4.4 STEADY STATE ERROR AND ERROR CONSTANT FOR UNITY FEEDBACK SYSTEM

The steady state errors play an important role for system's performance. The steady state error can be defined, if the actual output of a control system during steady state deviates from the desired output or reference input. So the difference between actual output and reference input is called steady state error, basically for the control system it is desirable that the output is following of its input. So we can say that the steady state error is a measure of accuracy of control system. These errors happen due to the non-linear system, type of inputs and type of system. As we discussed earlier that the steady state performance of a stable control system is generally assessed by its magnitude of steady state error to step, ramp and parabolic inputs.

Let us consider unity feedback system as shown in Fig. 4.4.

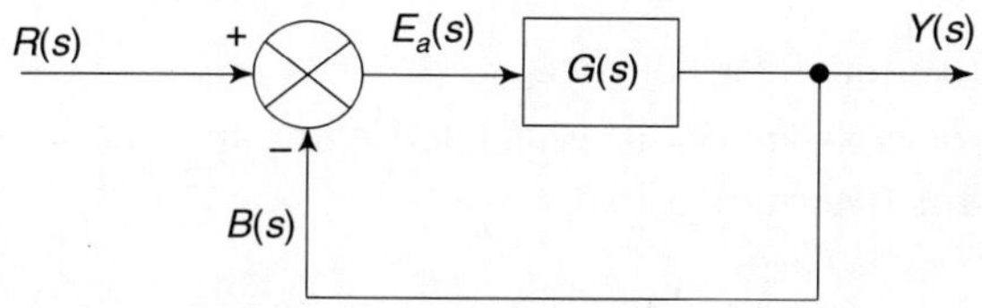

Figure 4.4 Unity feddback system

where $R(s)$ = Reference input

$Y(s)$ = Output

$B(s)$ = Feedback signal

$E_a(s)$ = Error signal or actuated signal

From Fig. 4.4, we see that

$$\frac{Y(s)}{R(s)} = \frac{G(s)}{1+G(s)\,H(s)}$$

$\because$ $H(s) = 1$

$$\therefore \quad \frac{Y(s)}{R(s)} = \frac{G(s)}{1+G(s)} \quad \text{...(4.5)}$$

By the final value theorem, the steady state error is

$$e_{ss} = \lim_{t \to \infty} e_a(t) = \lim_{s \to 0} sE(s)$$

$$e_{ss} = \lim_{s \to 0} \frac{s\,R(s)}{1+G(s)} \quad \text{...(4.6)}$$

By Eq. 4.5, we see that the steady state error depends upon the $R(s)$ and the $G(s)$. So the steady state error for different inputs are:

1. Unit step input: For step input

$$r(t) = u(t)$$

$$R(s) = \frac{1}{s} \text{ by Table 1.1}$$

Substituting the value of $R(s)$ in Eq. (4.6)

$$e_{ss} = \lim_{s \to 0} \frac{1}{s} \cdot \frac{s}{1+G(s)} \cdot = \lim_{s \to 0} \frac{1}{1+G(s)}$$

$$e_{ss} = \frac{1}{1+G(0)} = \frac{1}{1+K_P} \quad \text{...(4.7)}$$

where $G(s)$ = forward transfer function

$G(o)$ = forward transfer function when

$s = 0 = K_P$

K_P = Static position error constant

i.e., static positional error constant K_P is associated with unit step input applied to a closed loop control system and is determined like

$$e_{ss} = \lim_{s \to 0} = sR(s) \frac{1}{1+G(s)\,H(s)}$$

$$\because \quad \text{input } R(s) = \frac{1}{s}$$

$$e_{ss} = \lim_{s \to 0} s \cdot \frac{1}{s} \frac{1}{1+G(s)\,H(s)}$$

$$e_{ss} = \frac{1}{1+\lim\limits_{s \to 0} G(s)\,H(s)}$$

Let $$K_P = \lim_{s \to 0} G(s)\,H(s)$$

2. Unit ramp input: For ramp input

$$r(t) = t$$

$$R(s) = \frac{1}{s^2} \quad \text{by Table 1.1}$$

Substituting the value of $R(s)$ in Eq. (4.6)

$$e_{ss} = \lim_{s \to 0} \frac{s}{s^2\,[1 + G(s)]}$$

$$e_{ss} = \lim_{s \to 0} \frac{1}{s + s\,G(s)} = \lim_{s \to 0} \frac{1}{s\,G(s)}$$

$$e_{ss} = \frac{1}{K_V} \quad \text{...(4.8)}$$

where K_v = static velocity error constant, i.e., static velocity error constant K_v is associated with unit ramp input applied to a closed loop control system and is determined like

$$e_{ss} = \lim_{s \to 0} s \frac{R(s)}{1 + G(s)\,H(s)}$$

$\because$ $$\text{input } R(s) = \frac{1}{s^2}$$

$$e_{ss} = \lim_{s \to 0} s \cdot \frac{1}{s^2} \frac{1}{1 + G(s)\,H(s)}$$

$$e_{ss} = \lim_{s \to 0} \frac{1}{s + s\,G(s)\,H(s)} = \lim_{s \to 0} \frac{1}{s\,G(s)\,H(s)}$$

Let $$K_v = \lim_{s \to 0} s\,G(s)\,H(s)$$

3. Unit parabolic input: For unit parabolic input

$$r(t) = \frac{t^2}{2}$$

$$R(s) = \frac{1}{s^3}$$

Substituting the value of $R(s)$ in Eq. (4.6)

$$e_{ss} = \lim_{s \to 0} s \cdot \frac{1}{s^3} \cdot \frac{1}{1 + G(s)}$$

$$e_{ss} = \lim_{s \to 0} \frac{1}{s^2 + s^2\,G(s)} = \lim_{s \to 0} \frac{1}{s^2\,G(s)}$$

$$e_{ss} = \frac{1}{K_a} \qquad ...(4.9)$$

where K_a = static acceleration error constant, i.e., static acceleration constant (K_a) is associated with unit parabolic input applied to a closed loop control system and is determined like

$$e_{ss} = \lim_{s \to 0} s\,R(s) \frac{1}{1+G(s)\,H(s)}$$

$\because$ $$\text{input } R(s) = \frac{1}{s^3}$$

$$e_{ss} = \lim_{s \to 0} s \cdot \frac{1}{s^3} \frac{1}{1+G(s)\,H(s)}$$

$$e_{ss} = \lim_{s \to 0} \cdot \frac{1}{s^2 + s^2\, G(s)\,H(s)}$$

$$e_{ss} = \lim_{s \to 0} \frac{1}{s^2\, G(s)\,H(s)}$$

Let $$K_a = \lim_{s \to 0} s^2\, G(s)\,H(s)$$

4.4.1 Steady State Error in Terms of *G*(*s*)

From Fig. 4.3 (a), the error $E(s)$ between the input $R(s)$ and the output $Y(s)$, we write

$$E(s) = R(s) - Y(s) \qquad ...(4.10)$$

or $$Y(s) = R(s)\ G(s) \qquad ...(4.11)$$

Substituting Eq. (4.11) into Eq. (4.5), we get

$$E(s) = R(s) - R(s)\ T(s)$$

$$E(s) = R(s)\ [1 - T(s)] \qquad ...(4.12)$$

Although Eq. (4.12) allows us to solve for $e(t)$ at any time, t, we are interested in the final value of the error, $e(\infty)$. Applying the final value theorem, which allows us to use the final value of $e(t)$ without taking the inverse Laplace transform of $E(s)$, and then limit t approaches infinity, we obtain

$$e(\infty) = \lim_{t \to \infty} e(t) = \lim_{s \to 0} s\,E(s) \qquad ...(4.13)$$

$$e(\infty) = \lim_{s \to 0} S\,R(s)\,[1 - G(s)] \qquad ...(4.14)$$

Example 4.4.1 Find the steady state error for the system of Fig. 4.3 (a) if $G(s) = \dfrac{10}{(s^2 + 9s + 20)}$ and the input is a unit step.

Solution Given that

$$R(s) = \frac{1}{s}$$

$$G(s) = \frac{10}{s^2 + 9s + 20}$$

From Eq. (4.12), we get

$$E(s) = R(s)\,[1 - T(s)]$$

$$= \frac{1}{s}\left[1 - \frac{10}{s^2 + 9s + 20}\right]$$

$$= \frac{1}{s}\left[\frac{s^2 + 9s + 10 - 10}{s^2 + 9s + 20}\right]$$

$$= \frac{1}{s}\left[\frac{s^2 + 9s + 10}{s^2 + 9s + 20}\right]$$

$$E(s) = \frac{s^2 + 9s + 10}{s\,(s^2 + 9s + 20)}$$

From Eq. (4.13), we know that

$$e_{ss} = \lim_{s \to 0} s\,E(s)$$

$$= \lim_{s \to 0} s\left[\frac{s^2 + 9s + 10}{s\,(s^2 + 9s + 20}\right]$$

$$= \lim_{s \to 0}\left[\frac{s^2 + 9s + 10}{s^2 + 9s + 20}\right]$$

$$e_{ss} = \frac{1}{2} \quad \textbf{Ans.}$$

Example 4.4.2 Find the steady state error for inputs of $5u(t)$, $5tu(t)$ and $5t^2u(t)$ to the system shown in Fig. Ex. 4.4.2. The function $u(t)$ is the unit step.

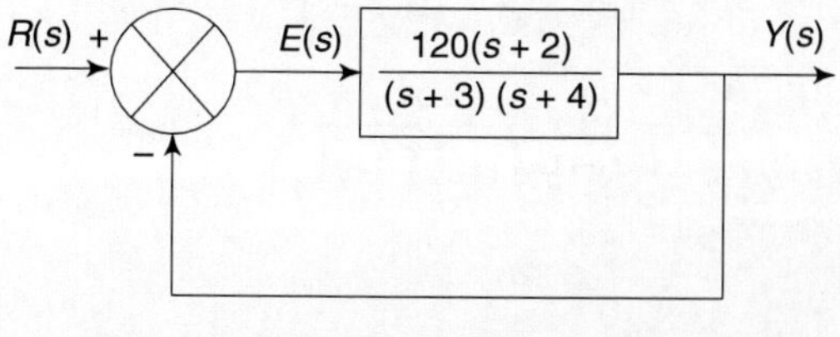

Figure Ex. 4.4.2 Feedback control system

Solution Given that

$$R(s) = \frac{5}{s}, \frac{5}{s^2}, \frac{10}{s^3}$$

$$G(s) = \frac{120\,(s+2)}{(s+3)\,(s+4)}$$

(i) When input in 5 $u(t)$, i.e., unit step input

The Laplace transform of 5 $u(t) = \frac{5}{s}$

From Eq. 4.6 we know that the steady state error will be five times as given by Eq.

$$e_{ss} = \lim_{s \to 0} \frac{SR(s)}{1+G(s)}$$

$$= \lim_{s \to 0} \frac{5}{s} \cdot \frac{s}{1+G(s)} \qquad \because R(s) = \frac{5}{s}$$

$$= \lim_{s \to 0} \frac{5}{1 + \dfrac{120\,(s+2)}{(s+3)\,(s+4)}}$$

$$= \frac{5}{1+20} = \frac{5}{21}$$

$$e_{ss} = \frac{5}{21} \quad \textbf{Ans.}$$

(ii) When input is 5 $tu(t)$ i.e., ramp input

The Laplace transform of 5 $tu(t) = \frac{5}{s^2}$

From Eq. (4.6), we know that

$$e_{ss} = \lim_{s \to 0} \frac{S\,R(s)}{1+G(s)}$$

$$= \lim_{s \to 0} \frac{5}{s^2} \cdot \frac{s}{1+G(s)} \qquad \because R(s) = \frac{5}{s^2}$$

$$= \lim_{s \to 0} \frac{5}{s\,[1+G(s)]}$$

$$= \frac{5}{0} = \infty$$

$$e_{ss} = \infty \quad \textbf{Ans.}$$

(iii) When input is $5t^2u(t)$, i.e., parabolic input

The Laplace transform of $5t^2u(t) = \dfrac{10}{s^3}$

From Eq. (4.6), we know that

$$e_{ss} = \lim_{s\to 0} \frac{S\,R(s)}{1+G(s)}$$

$$= \lim_{s\to 0} \frac{10}{s^3}\cdot\frac{s}{1+G(s)}$$

$$= \lim_{s\to 0} \frac{10}{s^2\,[1+G(s)]}$$

$$= \frac{10}{0} = \infty$$

$e_{ss} = \infty$ **Ans.**

4.5 STATIC ERROR CONSTANTS AND SYSTEM TYPE

In this section, we focus on unity negative feedback systems and define parameters that we can use as steady state error performance specifications, the performance parameters for the transient response are defined like damping ratio, natural frequency, settling time, percent overshoot, and so on as performance specification. These steady state error performance specifications are called static error constants. Let us see how they are defined, how to calculate them, and in the next section, how to use them for diagrams.

4.5.1 Static Error Constants

In section 4.4, we derived the following relationship for steady state error for a step input, $u(t)$

$$e_{ss} = \frac{1}{1+\lim\limits_{s\to 0} G(s)} \qquad ...(4.15)$$

For a ramp input, $tu(t)$

$$e_{ss} = \frac{1}{\lim\limits_{s\to 0} s\,G(s)} \qquad ...(4.16)$$

For a parabolic input, $\dfrac{1}{2}t^2\,u(t)$

$$e_{ss} = \lim_{s\to 0} \frac{1}{s^2\,G(s)} \qquad ...(4.17)$$

The three terms in the denominator of Eqs. (4.15), (4.16) and (4.17) that are taken to the limit determine the steady state error. We call these limits static error constants. Individually, their names are:

Position constant, K_P,

where $$K_P = \lim_{s \to 0} G(s) \qquad \text{...(4.18)}$$

Velocity constant, K_v,

where $$K_v = \lim_{s \to 0} S\,G(s) \qquad \text{...(4.19)}$$

Acceleration constant, K_a,

where $$K_a = \lim_{s \to)} S^2\,G(s) \qquad \text{...(4.20)}$$

From Eqs. (4.15) to (4.17), the values of the steady state error decrease as the static error constant increases because the static error constant appears in the denominator of the steady state error.

Example 4.5.1 Calculate the static error constants and find the expected error for the system given in Fig. Ex. 4.5.1 with the inputs are standard step, ramp and parabolic.

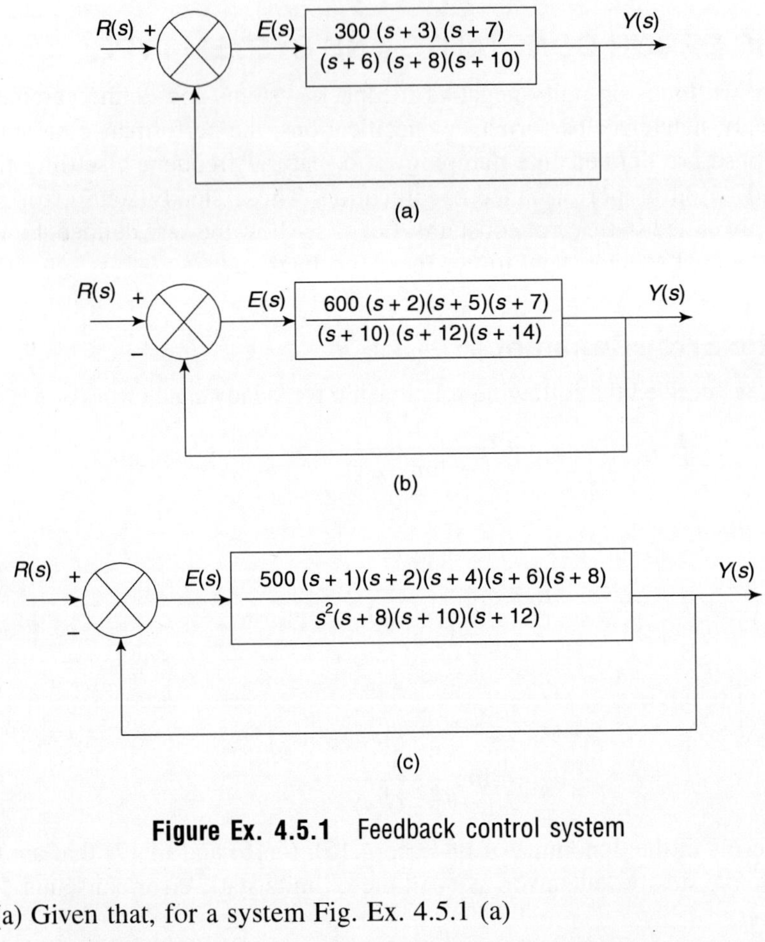

Figure Ex. 4.5.1 Feedback control system

Solution (a) Given that, for a system Fig. Ex. 4.5.1 (a)

$$G(s) = \frac{300\,(s+3)\,(s+7)}{(s+6)\,(s+8)\,(s+10)}$$

From Eq. 4.18, we know that

$$\text{Position constant, } K_p = \lim_{s\to 0} G(s) = \lim_{s\to 0} \frac{300\,(s+3)\,(s+7)}{(s+6)\,(s+8)\,(s+10)}$$

$$K_p = \frac{300\,(3\times 7)}{6\times 8\times 10} = 13.125$$

From Eq. 4.19, we know that

$$\text{Velocity constant, } K_v = \lim_{s\to 0} S\,G(s)$$

$$= \lim_{s\to 0} s\,\frac{300\,(s+3)\,(s+7)}{(s+6)\,(s+8)\,(s+10)}$$

$$\boxed{K_v = 0}$$

From Eq. 4.20, we know that

$$\text{Acceleration constant, } K_a = \lim_{s\to 0} s^2\,G(s)$$

$$= \lim_{s\to 0} s^2\,\frac{300\,(s+3)\,(s+7)}{(s+6)\,(s+8)\,(s+10)}$$

$$\boxed{k_a = 0}$$

Thus, for a step input,

$$e_{ss} = \frac{1}{1+K_p} \quad \text{(From Eq. 4.15)}$$

$$= \frac{1}{1+13.125}$$

$e_{ss} = 0.071$ **Ans.**

For a ramp input,

$$e_{ss} = \frac{1}{K_v} \quad \text{(From Eq. 4.16)}$$

$$e_{ss} = \frac{1}{0}$$

$e_{ss} = \infty$ **Ans.**

For a parabolic input,

$$e_{ss} = \frac{1}{K_a} \quad \text{(From Eq. 4.17)}$$

$e_{ss} = \infty$ **Ans.**

(b) For a system Fig. Ex. 4.5.1 (b), given that

$$G(s) = \frac{600\,(s+2)\,(s+5)\,(s+7)}{s\,(s+10)\,(s+12)\,(s+14)}$$

From Eq. 4.18, we know that

$$\text{Position constant, } K_p = \lim_{s \to 0} G(s)$$

$$= \lim_{s \to 0} \frac{600\,(s+2)\,(s+5)\,(s+7)}{s\,(s+10)\,(s+12)\,(s+14)}$$

$$= \frac{600\,(2 \times 5 \times 7)}{0 \times 10 \times 12 \times 14}$$

$$\boxed{K_p = \infty}$$

Now,

$$\text{velocity constant, } K_v = \lim_{s \to 0} S\,G(s)$$

$$= \lim_{s \to 0} s \cdot \frac{600\,(s+2)\,(s+5)\,(s+7)}{s\,(s+10)\,(s+12)\,(s+14)}$$

$$= \frac{600\,(2 \times 5 \times 7)}{10 \times 12 \times 14}$$

$$K_v = \frac{42000}{1680} = 25$$

$$\boxed{K_v = 25}$$

$$\text{and} \quad \text{acceleration constant, } K_a = \lim_{s \to 0} s^2 G(s)$$

$$K_a = \lim_{s \to 0} s^2 \cdot \frac{600\,(s+2)\,(s+5)\,(s+7)}{s\,(s+10)\,(s+12)\,(s+14)}$$

$$\boxed{K_a = 0}$$

Thus, for a step input,

$$e_{ss} = \frac{1}{1 + K_p} = 0$$

$e_{ss} = 0$ **Ans.**

For a ramp input,

$$e_{ss} = \frac{1}{K_v} = \frac{1}{25}$$

$$e_{ss} = 0.04$$

For a parabolic input,

$$e_{ss} = \frac{1}{K_a} + \frac{1}{0}$$

$$e_{ss} = \infty \quad \textbf{Ans.}$$

(c) For a system Fig. Ex. 4.5.1 (c), given that

$$G(s) = \frac{500(s+1)(s+2)(s+4)(s+6)(s+8)}{s^2(s+8)(s+10)(s+12)}$$

From Eq. 4.18, we know that

$$\text{Position constant, } K_P = \lim_{s \to 0} G(s)$$

$$K_P = \lim_{s \to 0} \frac{500(s+1)(s+2)(s+4)(s+6)(s+8)}{s^2(s+8)(s+10)(s+12)}$$

$$\boxed{K_P = \infty}$$

Now, velocity constant, $K_v = \lim_{s \to 0} s\,G(s)$

$$K_v = \lim_{s \to 0} s \cdot \frac{500(s+1)(s+2)(s+4)(s+6)(s+8)}{s^2(s+8)(s+10)(s+12)}$$

$$\boxed{K_v = \infty}$$

and acceleration constant, $K_a = \lim_{s \to 0} s^2\,G(s)$

$$K_a = \lim_{s \to 0} s^2 \cdot \frac{500(s+1)(s+2)(s+4)(s+6)(s+8)}{s^2(s+8)(s+10)(s+12)}$$

$$= \frac{500(2 \times 1 \times 4 \times 6 \times 8)}{(8 \times 10 \times 12)}$$

$$= \frac{192000}{960}$$

$$\boxed{K_a = 200}$$

Thus, for a step input,

$$e_{ss} = \frac{1}{1+K_p} = 0$$

$$e_{ss} = 0 \quad \textbf{Ans.}$$

For a ramp input,

$$e_{ss} = \frac{1}{K_v} = \frac{1}{\infty}$$

$$e_{ss} = 0 \quad \textbf{Ans.}$$

For a parabolic input,

$$e_{ss} = \frac{1}{K_a} = \frac{1}{200}$$

$$e_{ss} = 0.005 \quad \textbf{Ans.}$$

4.5.2 System Type

Now we focus on the negative feedback system. The values of the static error constants depend upon the form of $G(s)$, especially the number of pure integrations in the forward path. Since steady state errors depend upon the number of integrations in the forward path, we give a name to the system attribute. Let us consider a negative feedback system as in Fig. 4.5, we define system type to be the value of n in the denominator or, equivalently, the number of pure integration in the forward path. Therefore, a system with $n = 0$ is a type 0 system.

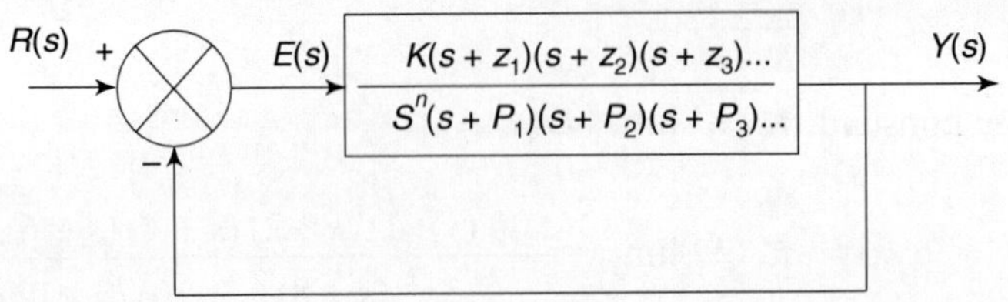

Figure 4.5 Negative feedback control system

The concept of steady state error, static error constant and system type is given in Table 4.2. The table shows the static error constants and the steady state errors and functions of input waveform and system type.

Table 4.2 Relation between input system type, static error constant and steady state error

Inputs	e_{ss}	Type 0		Type 1		Type 2	
		Static error constant	Error	Static error constant	Error	Static error constant	Error
Step $u(t)$	$\frac{1}{1+K_p}$	K_p = constant	$\frac{1}{1+K_p}$	$K_p = \infty$	0	$K_p = \infty$	0
Ramp $tu(t)$	$\frac{1}{K_v}$	$K_v = 0$	∞	K_v = constant	$\frac{1}{K_v}$	$K_v = \infty$	0
Parabolic $\frac{1}{2}t^2 u(t)$	$\frac{1}{K_a}$	$K_a = 0$	∞	$K_a = 0$	∞	K_a = constant	$\frac{1}{K_a}$

4.6 STEADY STATE ERROR SPECIFICATIONS

Static error constant explained in earlier section can be used to specify the steady state error characteristics of control system. Static error constants (K_p, K_v, K_a) can be used as specifications for a control system's steady state error in the same way as the damping ratio, ξ, settling time, T_s, peak time, T_p, and percent overshoot, % OS are used as specifications for a control system's transient response. We will observe that a wealth of information is contained within the specification of a static error constant.

For example, if a control system has the specification, $K_v = 500$, we can give several conclusions:

(i) The system is stable.

(ii) The velocity constant is finite so the system is type 1 because only type 1 system has K_v's that are finite, whereas the $K_v = 0$ for type 0 and $K_v = \infty$ for type 2 systems.

(iii) The input signal is ramp signal. Since K_v is specified as a finite constant, and the steady state error for a ramp input is inversely proportional to K_v, we know that the test input is a ramp.

(iv) The steady state error between the input ramp and the output ramp is $1/K_v$ per unit of input slope.

Example 4.6.1 Given the control system in Fig. Ex. 4.6.1, find the value of k so that there is 20% error in the steady state.

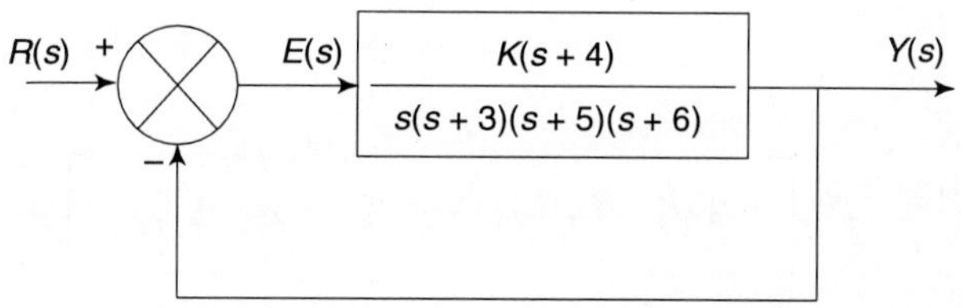

Figure Ex. 4.6.1 Feedback control system

Solution Given that

$$G(s) = \frac{K\,(s+4)}{s\,(s+3)\,(s+5)\,(s+6)}$$

Since the system is type 1, the error stated in the problem must apply to a ramp input : only a ramp yields a finite error in a type 1 system. Thus,

$$e_{ss} = \frac{1}{K_v} = 0.05$$

Therefore,

$$K_v = 20 = \lim_{s\to 0} S\,G(s)$$

$$= \lim_{s\to 0} s \cdot \frac{K(s+4)}{s\,(s+3)\,(s+5)\,(s+6)}$$

$$20 = \frac{K \times 4}{3 \times 5 \times 6}$$

$$20 = \frac{4K}{90}$$

$$4K = 1800$$

$$K = \frac{1800}{4} = 450$$

$$K = 450 \quad \textbf{Ans.}$$

4.7 STEADY STATE ERROR FOR DISTURBANCES

In control systems, we observe that sometimes disturbances or unwanted inputs enter a system. Hence, feedback control systems are used to compensate disturbances or unwanted inputs. The advantage of using feedback is that regardless of these advantages, the system can be designed to follow the input with small or zero error. Figure 4.6 shows a feedback control system with a disturbance, $D(s)$, injected between the controller and the plant. Now we derive the expression for steady state error with the disturbance included.

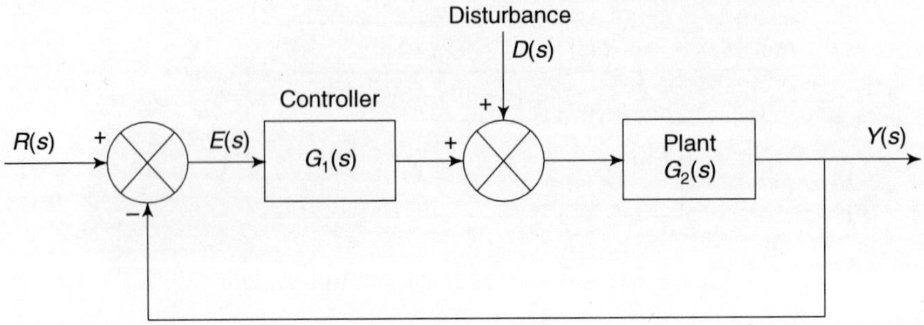

Figure 4.6 Feedback control system with disturbance

The output of the system is given by

$$Y(s) = E(s)\, G_1(s)\, G_2(s) + D(s)\, G_2(s) \qquad \text{...(4.21)}$$

But

$$R(s) - Y(s) = E(s)$$

$$Y(s) = R(s) - E(s) \qquad \text{...(4.22)}$$

Substituting Eq. 4.22 into Eq. 4.21, we get

$$R(s) - E(s) = E(s)\, G_1(s)\, G_2(s) + D(s)\, G_2(s)$$

$$E(s) + E(s)\, G_1(s)\, G_2(s) = R(s) - D(s)\, G_2(s)$$

$$E(s)\, [1 + G_1(s)\, G_2(s)] = R(s) - D(s)\, G_2(s)$$

$$E(s) = \frac{1}{1 + G_1(s)\, G_2(s)} R(s) - \frac{G_2(s)}{1 + G_1(s)\, G_2(s)} D(s) \qquad \text{...(4.23)}$$

From Eq. (4.23), we can think of $\frac{1}{[1+G_1(s)\,G_2(s)]}$ as a transfer function relating $E(s)$ to $R(s)$ and $\frac{-G_2(s)}{[1+G_1(s)\,G_2(s)]}$ as a transfer function relating to $E(s)$ to $D(s)$.

To find the steady state value of the error, we apply the final value theorem's to Eq. 4.23, we obtain.

$$e_{ss} = \lim_{s\to 0} s\,E(s) = \lim_{s\to 0} \frac{s}{1+G_1(s)\,G_2(s)} R(s) - \lim_{s\to 0} \frac{s\,G_2(s)}{1+G_1(s)\,G_2(s)} D(s)$$

$$e_{ss} = e_R + e_D \qquad \text{...(4.24)}$$

where $e_R = \lim_{s\to 0} \frac{s}{1+G_1(s)\,G_2(s)} R(s)$

and $$e_D = \lim_{s\to 0} \frac{S\,G_2(s)}{1+G_1(s)\,G_2(s)} D(s)$$

In Eq. (4.24), the first term, e_R, is the steady state error due to $R(s)$, which we have already obtained. The second term, e_D, is the steady state error due to the disturbance. Let us explore the conditions on $e_D(\infty)$ that must exist to reduce the error due to the disturbance.

Let us make some assumptions about $D(s)$, the controller, and the plant. First we assume a step disturbance, $D(s) = \frac{1}{s}$. Substituting this value into the second term of Eq. 4.24, e_D, the steady state error component due to a step disturbance is found to be

$$e_D = \frac{1}{\lim_{s\to 0} \frac{1}{G_2(s)} + \lim_{s\to 0} G_1(s)} \qquad \text{...(4.25)}$$

From Eq. 4.25, it is to be noted that the steady state error produced by a step disturbance can be reduced by increasing the dc gain of $G_1(s)$ or decreasing the dc gain of $G_2(s)$.

This concept is shown in Fig. 4.7, where the system of Fig. 4.6 has been rearranged so that the disturbance, $D(s)$, is depicted as the input and the error, $E(s)$, as the output, with $R(s)$ set equal to zero.

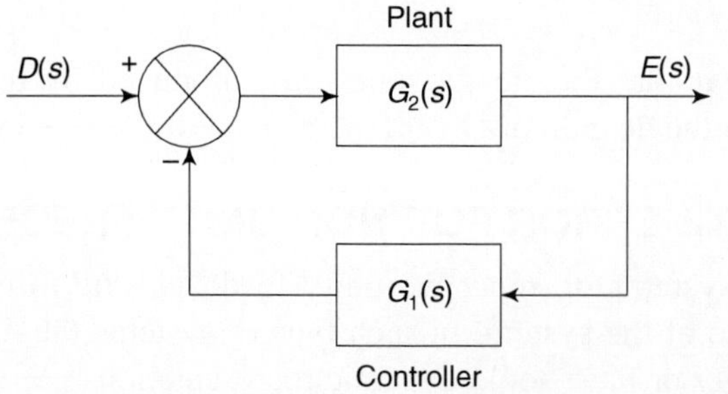

Figure 4.7 System that shows disturbance an input and error as output

If we want to minimize the steady state error of $E(s)$, shown as the output in Fig. 4.7, we must either increase the dc gain of $G_1(s)$ so that a lower value of $E(s)$ will be feedback to match the steady state value of $D(s)$, or decrease the dc value of $G_2(s)$, which then yields a smaller value of e_{ss} as predicted by the feedback formula.

Example 4.7.1 Find the steady state error component due to a step disturbance for the system of Fig. Ex. 4.7.1.

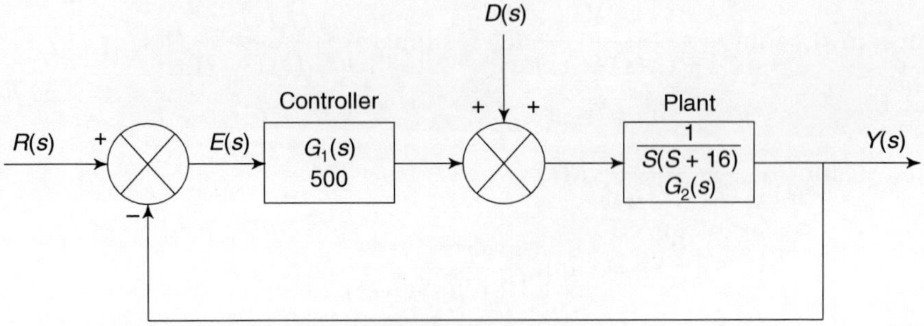

Figure Ex. 4.7.1

Solution Given that

$$G_1(s) = 500 \quad \text{and } G_2(s) = \frac{1}{s\,(s+16)}$$

From Eq. (4.25), we know that

$$e_D = \frac{1}{\lim\limits_{s\to 0} \dfrac{1}{G_2(s)} + \lim\limits_{s\to 0} G_1(s)}$$

$$e_D = \frac{1}{0+500} = -\frac{1}{500}$$

$$e_D = -\frac{1}{500} = -\,0.002$$

$$e_D = -\,0.002 \quad \textbf{Ans.}$$

From result, it is clear that the steady state error produced by the step disturbance is inversely proportional to the dc gain of $G_1(s)$. The dc gain of $G_2(s)$ is infinite in this example.

4.8 STEADY STATE ERROR FOR NON-UNITY FEEDBACK SYSTEMS

Generally, the control systems do not have unity feedback due to compensation used to improve the performance of the system. In such type of systems the feedback path can be a pure gain other than unity or have some dynamic representation.

Generally, feedback system is shown in Fig. 4.8 (a) in which controller, $G_1(s)$, plant $G_2(s)$, and feedback, $H_1(s)$ are used.

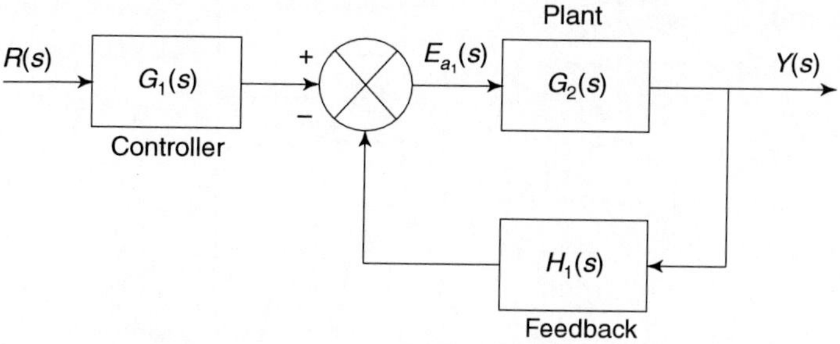

Figure 4.8 (a) Non-unity feedback system

Now simplifying the non-unity feedback system by block diagram reduction technique as explained in previous chapter.

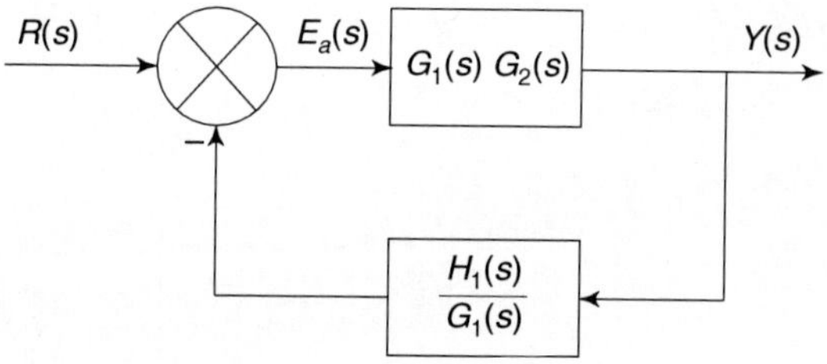

i.e., shifting the summing point before the block

Let $G(s) = G_1(s)\ G_2(s)$

and $H(s) = H_1(s)/G_1(s)$

Hence, the general non-unity feedback system is shown in Fig. 4.8 (b).

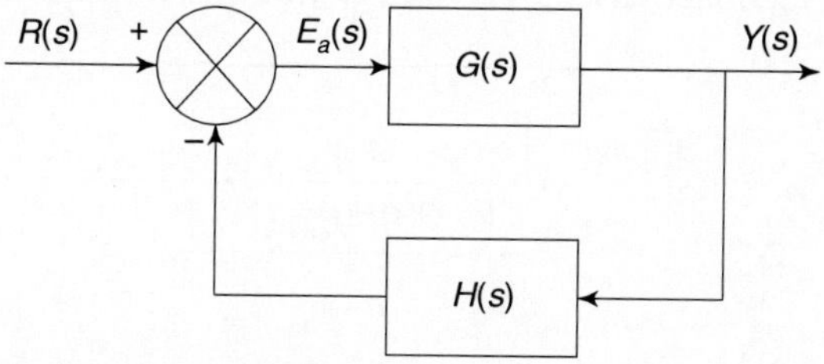

Figure 4.8 (b) Simple non-unity feedback system

Notice that unlike a unity feedback system, where $H(s) = 1$, the error is not the difference between the input and the output. So we call the signal at the output of the summing junction the actuating signal, $E_a(s)$.

If $R(t)$ and $C(t)$ have the same units, we can find the steady state error, $e_{ss} = R(s) - Y(s)$.

Now unity feedback system is formed from non-unity feedback system, shown in Fig. 4.8 (b) by adding and subtracting unity feedback path, as shown in Fig. 4.8 (c).

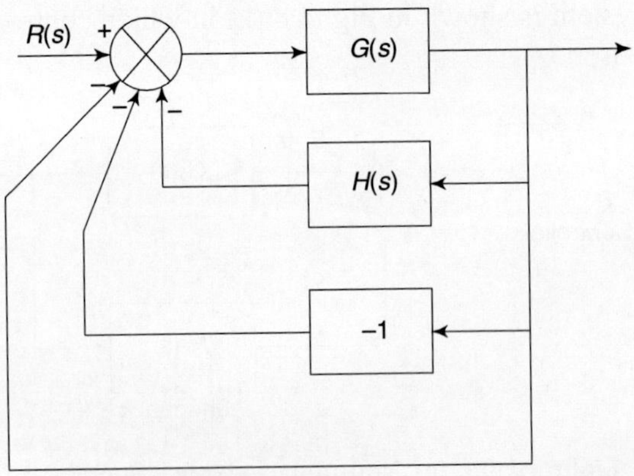

Figure 4.8 (c) Feedback system with adding and subtracting unity feedback

By simplifying Fig. 4.8 (c), we obtained the simplified system as shown in Fig. 4.8 (d).

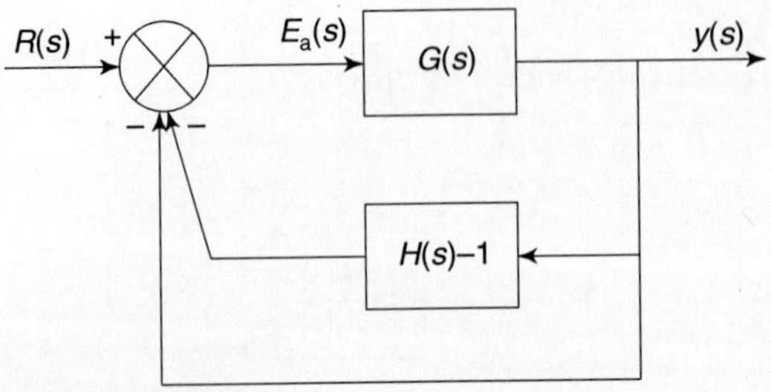

Figure 4.8 (d) Simplification of Fig. 4.8 (c)

Finally, combine the feedback system consisting of $G(s)$ and $[H(s) - 1]$, leaving an equivalent forward path and a unity feedback, as shown in Fig. 4.8 (e). Notice that the final figure shows $E(s) = R(s) - Y(s)$ explicitly.

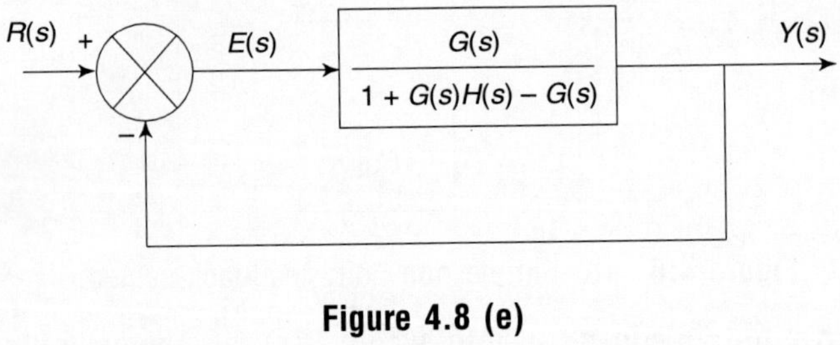

Figure 4.8 (e)

$$G_{NFS}(s) = \frac{G(s)}{1 + G(s)\,H(s) - G(s)} \qquad ...(4.26)$$

where $G(s) = G_1(s)\, G_2(s)$

and $H(s) = H_1(s)/G_1(s)$

$G_{NFS}(s)$ = Transfer function of unity feedback system when converted from non-unity feedback system.

Example 4.8.1 Evaluate the steady state error for the system given in Fig. Ex. 4.8.1 for a unity step input. Assume that input and output units are the same. Also find the system type, approximate error constant associated with the system type.

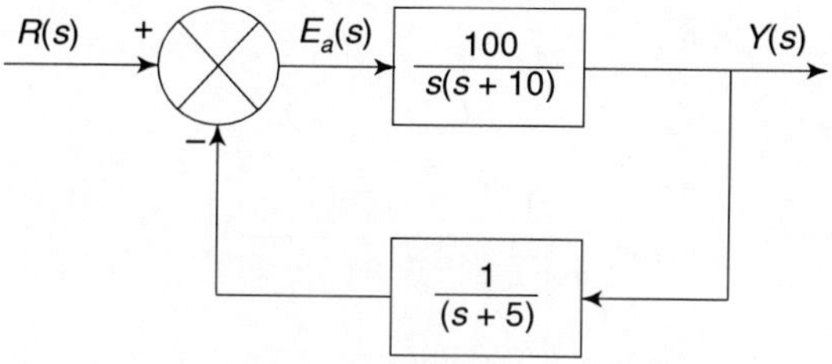

Figure Ex. 4.8.1 (a) Non-unity feedback control system

Solution Given that

$$G(s) = \frac{100}{s\,(s+10)} \quad \text{and } H(s) = \frac{1}{(s+5)}$$

From $G(s)$, it is observed that the system is type 1.

Since the system shown in Fig. Ex. 4.8.1 has a non-unity feedback element, and the plant's actuating signal is not the difference between the input and the output.

The first step is solving the problem to convert the system of Fig. Ex. 4.8.1 into an equivalent unity feedback system.

From Eq. (4.26), we know that

$$G_{NFS}(s) = \frac{G(s)}{1 + G(s)\,H(s) - G(s)}$$

$$= \frac{\dfrac{100}{s\,(s+10)}}{1 + \left[\dfrac{100}{s\,(s+10)} \times \dfrac{1}{(s+5)}\right] - \dfrac{100}{s\,(s+10)}}$$

$$= \frac{100\,(s+5)}{s\,(s+5)\,(s+10) + 100 - 100\,(s+5)}$$

$$G_{NFS}(s) = \frac{100\,(s+5)}{s^3 + 15s^2 - 50s - 400} \qquad \text{...(4.27)}$$

Thus, the system is type 0, since there is no pure integration in Eq. 4.27. The appropriate static error constant is then K_p, whose value,

$$K_p = \lim_{s\to 0} G_{NFS}(s) = \frac{100\times 5}{-400} = -\frac{5}{4}$$

$$K_p = -\frac{5}{4}$$

The steady state error,

$$e_{ss} = \frac{1}{1+K_p} = \frac{1}{1+(-5/4)}$$

$$e_{ss} = -4 \quad \textbf{Ans.}$$

The negative value for steady state error implies that the output step is larger than the input step.

Now the steady state error for systems with non-unity feedback, let us look at the general system of Fig. Ex. 4.8.1, which has both a disturbance and non-unity feedback.

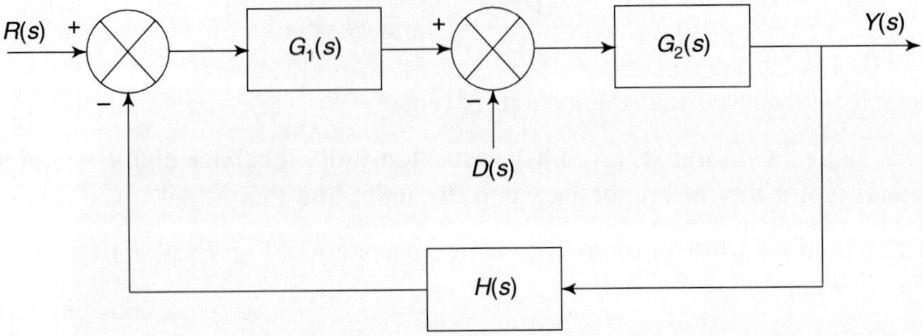

Figure Ex. 4.8.1 (b) Non-unity feedback system with disturbance

Now we will derive a general equaion for the steady state error.

The steady state error for this system

$$e_{ss} = R(s) - C(s) \quad \text{(From Eq. 4.22)}$$

$$= \lim_{s\to 0} s\,E(s) \quad \text{(From Eq. 4.24)}$$

$$e_{ss} = \lim_{s\to 0} s\left\{\left[1-\frac{G_1(s)\,G_2(s)}{1+G_1(s)\,G_2(s)\,H(s)}\right]R(s)-\left[\frac{G_2(s)}{1+G_1(s)\,G_2(s)\,H(s)}\right]\right\}D(s)$$

$$= \left\{1-\frac{\lim_{s\to 0}[G_1(s)\,G_2(s)]}{\lim_{s\to 0}[1+G_1(s)\,G_2(s)\,H(s)]}\right] - \left[-\frac{\lim_{s\to 0} G_2(s)}{\lim_{s\to 0}[1+G_2(s)\,G_2(s)\,H(s)]}\right] \quad ...(4.28)$$

From zero error

$$\frac{\lim_{s\to 0}[G_1(s)\,G_2(s)]}{\lim_{s\to 0}[1+G_1(s)G_2(s)\,H(s)]} = 1 \quad \text{and} \quad \frac{\lim_{s\to 0} G_2(s)}{\lim_{s\to 0}[1+G_1(s)\,G_2(s)H(s)]} = 0 \qquad ...(4.29)$$

Eq. (4.29) can always be satisfied if

(i) The system is stable

(ii) $G_1(s)$ is a type 1 system

(iii) $G_2(s)$ is a type 0 syatem

(iv) $H(s)$ is a type 0 system with a dc gain of unity

4.9 SENSITIVITY

Sensitivity is the degree to which changes in system parameters affect system transfer function, and hence performance is called sensitivity. As we know that the parameter changes due to heat or other causes should not appreciably affect a system's performance. A system with zero sensitivity is an ideal system. It means, greater is the sensitivity, less desirable is the effect of parameter change.

Let us have an example, assume the function

$$F = \frac{K}{(K+a)} \quad \text{if } K = 50 \text{ and } a = 200$$

then
$$F = \frac{50}{50+200} = 0.2$$

If parameter doubles to 400

then
$$F = \frac{50}{50+400} = 0.11$$

We see that a fractional change in parameter, a

$$= \frac{(400-200)}{200} = 1 \quad \text{i.e., 100\% change}$$

and change in the function, F

$$= \frac{(0.11-0.20)}{0.20} = -\,0.45$$

$$= -\,45\%$$

From the above discussion it has been observed that the function (F) has reduced sensitivity to change in parameter, a.

Based upon the above discussion, the sensitivity can be defined as: It is the ratio of the fractional change in the function to the fractional change in the parameter as the fractional change of the parameter approaches zero, hence

$$S_{F,P} = \lim_{\Delta P \to 0} \frac{\text{Fractional change in the function, } F}{\text{Fractional change in the parameter, } P}$$

$$S_{F,P} = \lim_{\Delta \to 0} \frac{\Delta F/F}{\Delta P/P}$$

$$= \lim_{\Delta \to 0} \frac{P\,\Delta F}{F\,\Delta P}$$

$$\boxed{S_{F,P} = \lim_{\Delta P \to 0} \frac{P\,\Delta F}{F\,\Delta P}} \qquad \text{...(4.30)}$$

where $S_{F,P}$ = Function to parameter sensitivity

δF = Fractional change in function

δP = Fractional change in parameter

Example 4.9.1 Calculate the sensitivity of the closed loop transfer function to changes in the parameter a. How would you reduce the sensitivity? The system is given in Fig. Ex. 4.9.1.

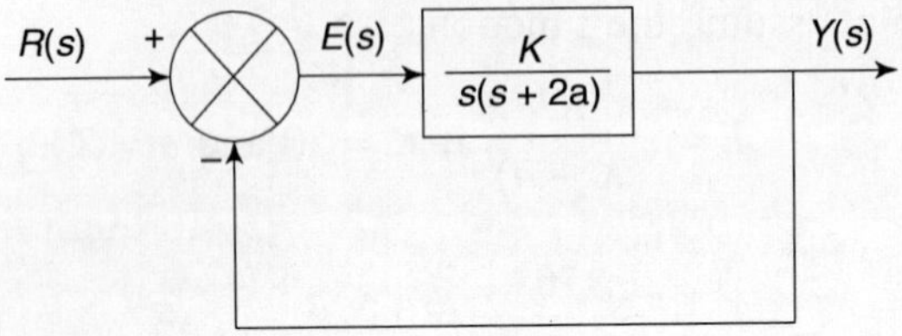

Figure Ex. 4.9.1 Feedback control system

Solution Given that

$$G(s) = \frac{K}{s(s+2a)} \text{ and } H(s) = 1$$

We know that the transfer function of the system is

$$T(s) = \frac{G(s)}{1+G(s)\,H(s)}$$

$$T(s) = \frac{\dfrac{K}{s(s+2a)}}{1+\left[\dfrac{K}{s(s+2a)}\times 1\right]} = \frac{\dfrac{K}{s(s+2a)}}{\dfrac{s^2+2as+K}{s(s+2a)}}$$

$$T(s) = \frac{K}{s(s+2a)+K}$$

$$T(s) = \frac{K}{s^2 + 2a\,s + K}$$

From Eq. (4.30), the sensitivity is given by

$$S_{T,a} = \frac{a\,\delta T}{T\,\delta a} = \left[\frac{\frac{a}{K}}{(s^2 + 2a\,s + K)}\right]\left[\frac{-2\,K\,s}{(s^2 + 2a\,s + K)^2}\right]$$

$$S_{T,\,a} = \frac{-2a\,s}{s^2 + 2a\,s + K} \qquad \text{...(4.31)}$$

Equation (4.30) is a function of the value of s. For any value of s, however, an increase in k reduces the sensitivity of the closed loop transfer function ($T(s)$) to changes in the parameter a.

Example 4.9.2 For the system of Fig. Ex. 4.9.1, find the sensitivity of the steady state error to changes in parameter k and parameter a with ramp inputs.

Solution For the system Fig. Ex. 4.9.1, we know that

$$G(s) = \frac{K}{s\,(s + 2a)}$$

From Eq. 4.19, we know that the velocity constant,

$$K_v = \lim_{s \to 0} s\,G(s)$$

$$= \lim_{s \to 0} s \cdot \frac{K}{s\,(s + 2a)}$$

$$\boxed{K_v = \frac{K}{2a}}$$

The steady state error for the system is given by

$$e_{ss} = \frac{1}{K_v} = \frac{1}{K/2a}$$

$$e_{ss} = \frac{2a}{K}$$

The sensitivity of e_{ss} to changes in parameter K is

$$S_{e,\,K} = \frac{K}{e}\frac{\delta e}{\delta K} = \frac{K}{2a/K} \cdot \left[\frac{-2a}{K^2}\right]$$

$$S_{e,\,K} = -2 \quad \textbf{Ans.} \qquad \text{...(4.32)}$$

The sensitivity of e_{ss} to changes in parameter a is

$$S_{e,\,a} = \frac{a}{e}\frac{\delta e}{\delta a} = \frac{a}{\frac{2a}{K}}\left[\frac{2}{K}\right] = 1 \qquad \text{...(4.33)}$$

Thus, changes in either parameter a or parameter K are directly reflected in e_{ss}, and there is no reduction or increase in sensitivity. The negative sign in Eq. (4.32) indicates a decrease in e_{ss} for an increase in K.

EXAMPLES

Example 4.1 Calculate K_p, K_v, K_a for $\dfrac{s^2 (s+5)}{s(s^3 + 21s^2 + 147s + 343)}$. Write clearly what are the inputs used for calculating each of these error constants. Calculate steady state error for each of these inputs. *(U.P.T.U., 2005–06, CSI)*

Solution Given that

$$G(s) = \frac{s^2 (s+5)}{s(s^3 + 21s^2 + 147s + 343)}$$

(i) Position error constant (K_p)
Input used : Step input

$$K_P = \lim_{s \to 0} G(s)$$

$$K_P = \lim_{s \to 0} \frac{s^2 (s+5)}{s(s^3 + 21s^2 + 147s + 343)}$$

$$\boxed{K_P = 0}$$

$$e_{ss} = \frac{1}{1+K_p} = \frac{1}{1+0} = 1$$

(ii) Velocity error constant (K_v)
Input used : ramp

$$K_v = \lim_{s \to 0} s\,G(s)$$

$$K_v = \lim_{s \to 0} s \cdot \frac{s^2 (s+5)}{s(s^3 + 21s^2 + 147s + 343)}$$

$$\boxed{K_v = 0}$$

$$\therefore \quad e_{ss} = \frac{1}{K_v} = \frac{1}{0} = \infty$$

(iii) Acceleration error constant (K_a) input used: Parabolic input

$$K_a = \lim_{s \to 0} s^2\,G(s)$$

$$K_a = \lim_{s\to 0} s^2 \cdot \frac{s^2 (s+s)}{s(s^3+21s^2+147s+343)}$$

$$K_a = 0$$

$$\therefore \quad e_{ss} = \frac{1}{K_a} = \frac{1}{0} = \infty$$

Example 4.2 For a forward path, trnsfer function is given by $G(s) = \dfrac{20(s+2)}{s(s+3)(s+4)}$.

(i) Find the static error constants

(ii) Find the steady state error for unity feedback when $Y(t) = 3\ U(t) + 5\ tU(t)$.

(U.P.T.U., 2002–03, CS)

Solution Given

$$G(s) = \frac{20(s+2)}{s(s+3)(s+4)}, H(s) = 1$$

Now position error constant (K_P)

$$K_P = \lim_{s\to 0} G(s)\,H(s)$$

$$K_P = \lim_{s\to 0} \frac{20(s+2)}{s(s+3)(s+4)} = \infty$$

Velocity error constant (K_v)

$$K_v = \lim_{s\to 0} s\,G(s)\,H(s)$$

$$= \lim_{s\to 0} \frac{20\,s(s+2)}{s(s+3)(s+4)} = \frac{40}{12}$$

$$K_v = 3.33$$

Acceleration error constant (K_a)

$$K_a = s^2\,G(s)\,H(s)$$

$$= \lim_{s\to 0} s^2 \frac{20(s+2)}{s(s+3)(s+4)}$$

$$K_a = 0$$

Given input $\quad Y(t) = 3\ U(t) + 5t\ U(t) \quad$ ...(i)

Taking Laplace transform of Eq. (i)

$$L\{Y(t)\} = 3\,L\{U(t)\} + 5\,L\{t\,U(t)\}$$

$$Y(s) = \frac{3}{s} + \frac{5}{s^2}s \quad \text{...(ii)}$$

Now

$$\frac{E(s)}{Y(s)} = \frac{1}{1+G(s)\,H(s)}$$

$$\frac{E(s)}{Y(s)} = \frac{1}{1+\dfrac{20\,(s+2)}{s\,(s+3)\,(s+4)}\times 1}$$

$$\frac{E(s)}{Y(s)} = \frac{s\,(s+3)\,(s+4)}{s\,(s+3)\,(s+4)+20\,(s+2)}$$

$$E(s) = \frac{s\,(s+3)\,(s+4)}{s\,(s+3)\,(s+4)+20\,(s+2)}\cdot Y(s) \qquad \text{...(iii)}$$

Substituting the value of $Y(s)$ in Eq. (iii)

We get
$$E(s) = \frac{s\,(s+3)\,(s+4)}{s\,(s+3)\,(s+4)+20\,(s+2)}\times\left(\frac{3}{s}+\frac{5}{s^2}\right)$$

$$E(s) = \frac{\left(\dfrac{3}{s}+\dfrac{5}{s^2}\right)(s^2+7s^2+12\,s)}{s^3+7s^2+12s+20s+40}$$

$$E(s) = \frac{3s^2+21s+36}{s^3+7s^2+32s+40}+\frac{5s^2+35s+60}{s\,(s^3+7s^2+32s+40)}$$

So steady state error,
$$e_{ss} = \lim_{s\to 0} sE(s)$$

$$e_{ss} = \lim_{s\to 0} s\left[\frac{3s^2+21s+36}{s^2+7s^2+32s+40}\right]+\lim_{s\to 0} s\left[\frac{5s^2+35s+60}{s\,(s^3+7s^2+32s)+40}\right]$$

$$e_{ss} = 0+\frac{60}{40}=1.5$$

$$\boxed{e_{ss} = 1.5}\ \textbf{Ans.}$$

Example 4.3 A unity feedback system is given by $G(s) = \dfrac{8}{s^2\,(s^2+4s+8)\,(s^2+3s+12)}$.

Determine the error coefficients with the input by (i) $r(t) = 5$ (ii) $r(t) = 2t$.

(U.P.T.U., 2006–07, ACS)

Solution
$$G(s) = \frac{8}{s^2\,(s^2+4s+8)\,(s^2+3s+12)}$$

For input
$$r(t) = 5$$

$$R(s) = \frac{5}{s}$$

The characteristics equation is

$$1+G(s)\,H(s) = 1+\frac{8}{s^2\,(s^2+4s+8)\,(s^2+3s+12)}\times 1$$

$$= s^2 (s^2 + 4s + 8)(s^2 + 3s + 12) + 8$$

$$K_P = \lim_{s \to 0} G(s) H(s) = \lim_{s \to 0} \frac{8}{s^2 (s^2 + 4s + 8)(s^2 + 3s + 12)}$$

$$\boxed{K_P = \infty}$$

$$e_{ss} = \lim_{s \to 0} \frac{5}{1 + K_P} = \frac{5}{1 + \infty} = 0$$

$$\boxed{e_{ss} = 0}$$

For input $r(t) = 2t$

$$R(s) = \frac{2}{s^2}$$

So called static velocity constant

$$K_v = \lim_{s \to 0} s \cdot G(s) H(s)$$

$$K_v = \lim_{s \to 0} s \cdot \frac{8}{s^2 (s^2 + 4s + 8)(s^2 + 3s + 12)}$$

$$K_v = \infty$$

Static error $e_{ss} = \lim_{s \to 0} \frac{1}{K_v} = \frac{1}{\infty}$

$$\boxed{e_{ss} = 0} \quad \textbf{Ans.}$$

Example 4.4 For the stable system shown in Fig. Ex. 4.4, find:

(i) Open loop transfer function

(ii) Position, velocity and acceleration error constant. *(U.P.T.U., 2003–04, ACSI)*

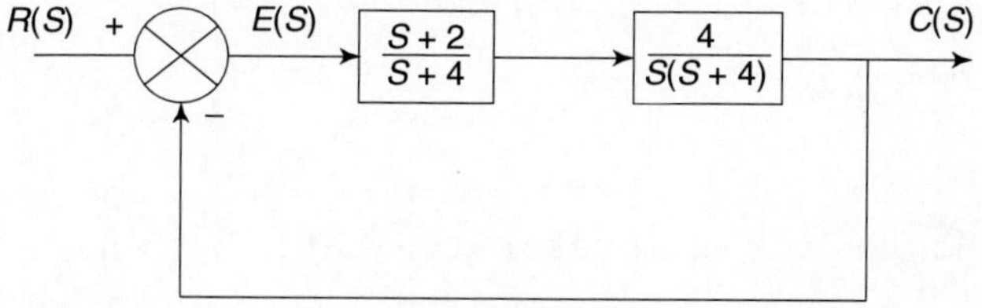

Figure Ex. 4.4

Solution (i) Open loop transfer function

$$G(s) H(s) = \frac{s+2}{s+4} \cdot \frac{4}{s(s+1)} \cdot 1$$

$$G(s)\,H(s) = \frac{4\,(s+2)}{s\,(s+1)\,(s+4)}$$ **Ans.**

(ii) Position error constant, K_P

$$K_P = \lim_{s \to 0} G(s)\,H(s) = \lim_{s \to 0} \frac{4\,(s+2)}{s\,(s+1)\,(s+4)}$$

$$K_P = \infty$$ **Ans.**

Velocity error constant, K_v

$$K_v = \lim_{s \to 0} s \cdot G(s)\,H(s) = \lim_{s \to 0} s \cdot \frac{4\,(s+2)}{s\,(s+1)\,(s+4)}$$

$$K_v = \frac{8}{(1)\,(4)} = 2$$

$$K_v = 2$$ **Ans.**

Acceleration error constant, K_a

$$K_a = \lim_{s \to 0} s^2\,G(s)\,H(s) = \lim_{s \to 0} s^2 \cdot \frac{4\,(s+2)}{s\,(s+1)\,(s+4)}$$

$$K_a = 0$$ **Ans.**

MATLAB PROGRAMS

Program 4.1

```
% Program for evaluating the static error
% constants and find the expected error for
% the standard step, ramp and parabolic
% input. while G(s) is given and system is
% unity feedback
clear all
clc
con = input('Enter the numerator constant K :-');
                                  % Enter the numerator constant k like 100
numpoly = input('Enter the zeros of G(s)  :-');
                                  % Enter the zeros of G(s) like [- 2 - 5 - 6]
denpoly = input('Enter the poles of G(s)  :-');
                                  % Enter the poles of G(s) like [0 - 6 - 8]
numg = com.*poly(numpoly);  % Evaluate numerator of G(s)
deng = Poly(denpoly);       % Evaluate denominator of G(s)
G = tf(numg,deng)           % Form G(s)
```

```
'Step input'                  % Display lables
Kp = dcgain(G)                % Evaluate Kp for s = 0
ess = 1/(1 + Kp)              % Evaluate steady state error
                              % for step input
'ramp input'                  % Display labels
numsg = conv([1  0],numg);    % Define numerator of SG(s)
densg = Poly(denpoly);        % Define denominator of SG(s)
SG = tf(numsg,densg);         % Evaluate SG(s)
SG = minreal(SG);             % Cancel common s in nume
                              % rater and denominator
Kv = dcgain(SG);              % Evaluate Kv = sG(s) for s = 0
ess = 1/Kv                    % Evaluate steady state error for
                              % ramp input
'Parabolic input'             % Display labels
nums2g = conv ([1 0 0],numg); % Define numerator of s2G(s)
dens2g = Poly(denpoly);       % Define denominator of s2G(s)
s2G = tf(nums2g,dens2g);      % Evaluate s2G(s)
s2G = minreal(s2G);           % Cancel common s in numerator
                              % and denominator
Ka = dcgain(s2G);             % Evaluate Ka = s2G(s) for s = 0
ess = 1/Ka                    % Evaluate steady state error for
                              % parabolic input
'Check stability'             % Display labels
T = feedback (G,1)            % Evaluate transfer function of
                                unit feedback system T(s).
Poles = Pole(T)               % Display close loop
```

Program 4.2

```
% Program for evaluating the gain K, required to
% meet steady state error specification for type 0 to type 2
% system while the system is unity feedback
clear all
clc
numpoly = input('Enter the zeros of G(s): -');
            % Enter the zeros of G(s) like [- 2 - 5 - 6]
denpoly = input('Enter the poles of G(s): -'); % Enter the
         poles of G(s) like [0 - 4 - 6]
numGdivideByK = Poly(numpoly); % Evaluate numerator of G(s)/K
```

```
denGdivideByK = Poly(denpoly);% Evaluate denominator of G(s)/K
GdivideByK = tf(numGdivideByK, denGdivideByK); % Form G(s)/K
Type = input('Enter the type of system :- '); % Enter feedback % system type
if type == 0
'Step input; only a step yields a finite error in a type 0 system'
                                          % Display labels
G DivideByK = minreal(GdivideByK);% Cancel common s in numerator
                                          % for and denominator
KpdivideByK = dcgain(GdivideByK) % Evaluate Kp = G(s) for s = 0
ess = input('Enter the value of steady state error :- ')
        % Enter % steady state error for step input in %
K = (1 - ess)/(ess.*KpdividebyK)  % Evaluate value of K
end
if type == 1
'ramp input; only a ramp yield a finite error in type 1 system'
numsg = conv[(1 0)], numGdivideByK); % Define numerator of S G(s)
densg = denGdivideByK;                % Define denominator of s G(s)
SG = tf(numsg, densg);                % Evaluate s G(s)
SG = minreal (SG);                    % Cancel common s in numerator
                                      % and denominator
KvdivideByK = dcgain (SG)             % Evaluate Kv/K = s G(s) for s = 0
ess = input('Enter the value of steady state error in %:- ')
                                      % Enter the value of steady state
                                      % error for ramp input in percentage
K = 1/(ess.*KvdivideByK)              % Evaluate value of K
end
if Type == 2
'Parabolic input; only a parabolic yield a finite error in a type 2 system'
nums2g = conv([1 0 0],numGdivideByK);% Define numerator of s^2G(s)
dens2g = denGdivideByK; % Define denominator of s^2G(s)
S2G = tf(nums2g,dens2g);% Define denominiator of s^2G(s)
S2g = minreal(S2G);     % Cancel common 'S' in numerator
                        % and denominator
KaDivideByK = dcgain(S2G); % Evaluate Ka = S^2G(s) for s = 0
ess = 1/(ess.*KadivideByK); % Evaluate value of K
end
'Check stability'       % Define labels
```

```
T = feedback(GdivideByK, 1);   % Evaluate the transfer
                               % function of unity feedback
                               % system
Poles = Pole(T)                % Display close loop poles
```

EXERCISE

4.1 Explain the test inputs used to evaluate steady state error.

4.2 Determine the position, velocity and acceleration error constants and the steady state error for:
  (i) Unit step input
  (ii) Unit ramp input
  (iii) Unit parabolic input

4.3 What effect upon the steady state error? While increasing system gain.

4.4 What effect does feedback have upon disturbances?

4.5 Define ramp error constant for a control system. Derive the values of ramp error constants and steady state error due to ramp input for type 0, 1, 2 and 3 control system.

4.6 How are non-unity feedback systems analyzed and designed steady state error?

4.7 Define, in words, sensitivity and descibe the goal of feedback control system engineering as it applies to sensitivity.

4.8 Find the sensitivity of the steady state error to change in K for the system of Fig. E 4.8.

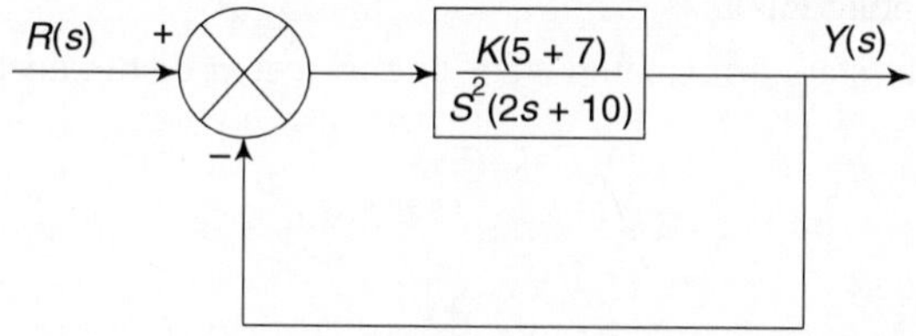

Figure E 4.8

4.9 What is steady state error? Discuss the position of velocity and acceleration error constants.

(U.P.T.U., 2006–07)

4.10 Find the sensitivity of the steady state error to changes in parameter K and parameter a for the system shown in Fig. E 4.10 with a step input.

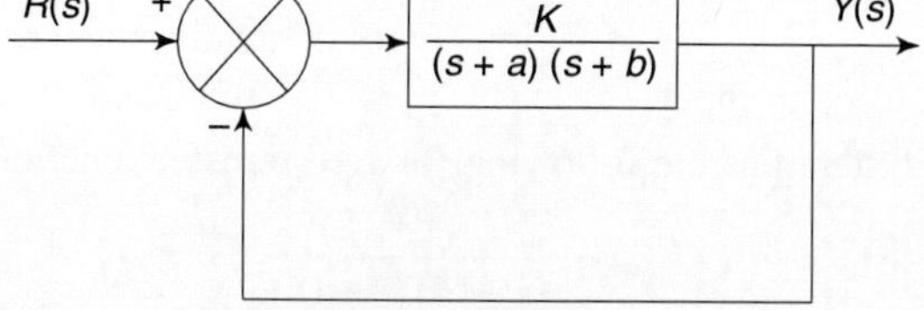

Figure E 4.10

4.11 Define

(i) Position error constant

(ii) Velocity error constant

(iii) Acceleration error constant

Explain how are these error constants useful in finding steady state errors with an example. *(U.P.T.U., 2004–05)*

4.12 Find the steady state error, $e_{ss} = R(s) - Y(s)$, for a unit step input given the non-unity feedback system of Fig. E 4.12. Repeat for a unit ramp input. Assume input and output units are the same.

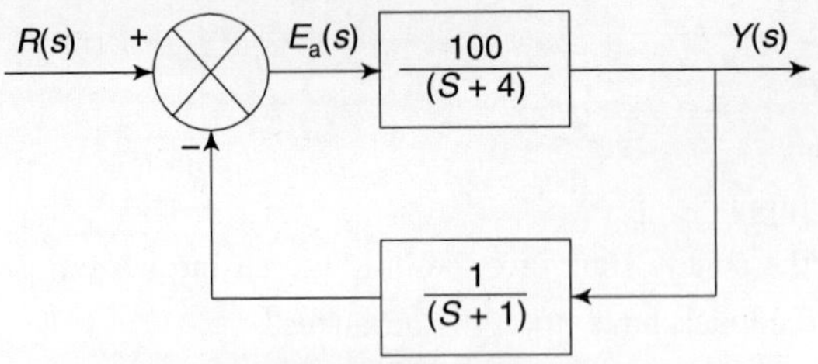

Figure E 4.12

4.13 For a step input disturbance at the input to the plant, describe the effect of controller and plant gain upon minimizing the effect of the disturbance.

4.14 What is the exact relationship between the static error constants and the steady state errors for ramp and parabolic inputs?

4.15 Evaluate the steady state error component due to a step disturbance for the system of Fig. E 4.15.

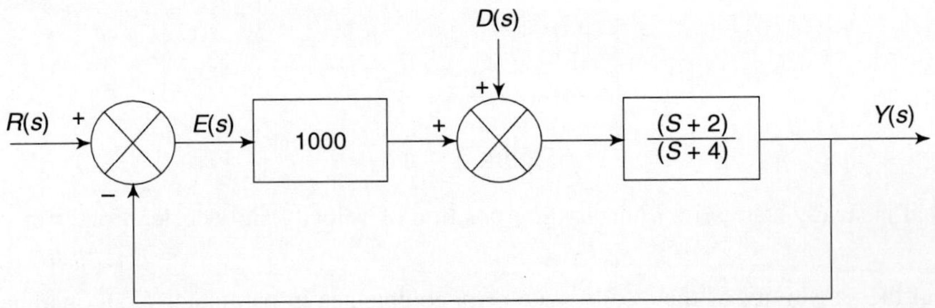

Figure E 4.15

4.16 Explain the physical significance of various types of steady state errors. Also find the various steady state errors to ramp input. *(U.P.T.U., 2005–06)*

4.17 A unity feedback system has the following forward transfer function:

$$G(s) = \frac{K(s+12)}{(s+14)(s+18)}$$

Find the value of K to yield a 10% error in the steady state.

4.18 Name two sources of steady state error.

4.19 A unity feedback system has the following forward transfer function :

$$G(s) = \frac{1000\,(s+8)}{(s+7)\,(s+9)}$$

(a) Evaluate system type, K_p, K_v and K_a.

(b) Use your answer to step (a) to find the steady state errors for the standard step, ramp and parabolic inputs.

4.20 A unity feedback system has the following forward transfer function :

$$G(s) = \frac{10\,(S+20)\,(S+30)}{S(S+25)\,(S+35)}$$

(a) Find the steady state error for the following inputs: 15 $u(t)$, 15 $tu(t)$ and 15 $t^2u(t)$

(b) Repeat for

$$G(s) = \frac{10\,(S+20)\,(S+30)}{S^2\,(S+25)\,(S+35)\,(S+50)}$$

CHAPTER 5

Control System Components

5.1 INTRODUCTION

The block diagram of a closed loop control system is given in Fig. 5.1. This diagram contains four basic parts: Error detector, the feedback element, controller, process or controlled system.

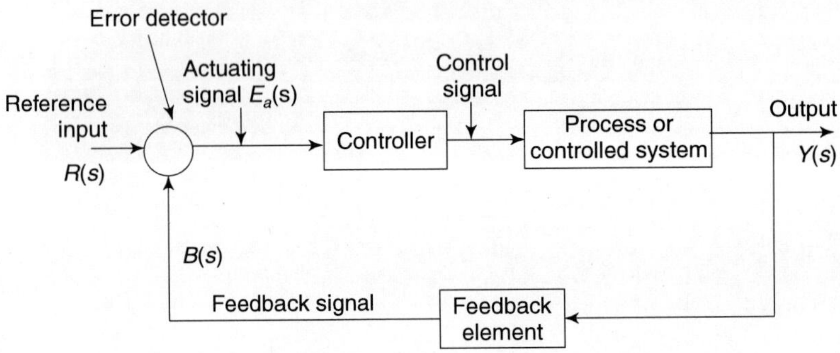

Figure 5.1 General block diagram of closed loop system

1. **Error detector.** Error detector is a device which is used to compare the feedback signal [$B(s)$] obtained from the output [$Y(s)$] with the reference input [$R(s)$] signal and gives deviation and is called actuating signal, $E_a(s)$.

$$E_a(s) = R(s) - B(s) \quad ...(5.1)$$

2. **Controller.** The controller is a device which is used to produce the control signal as the input to the process or controlled system. So in other words we can say that controller is used for signal conditioning of actuating signal because actuating signal is low power level so it is needed to mainpulate the signal by the controller or control elements. The manipulation may be like amplification, generation of a suitable function of the actuating signal and a power stage.

3. Feedback elements. The feedback signal is a device which is used to convert the output variable, $Y(s)$, into the some other variable, $B(s)$, which is compared with reference input.

4. Process or controlled system. It is a device which is used to produce the desired output, the input to the system is control signal.

5.2 TYPES OF CONTROL SYSTEM COMPONENTS

There are three types of controller elements.

1. Actuators. An actuator is a device that provides mechanical motion to the process. The mechanical motion may be translator or rotary but rotary motion is convenient and preferred. For example, the dc motor shown in Fig. 5.2 is a power actuator that delivers energy to a load. So we can say that the actuators are characterised by power output and speed-torque relationship to match the load. The actuator may be electrical, hydraulic or pneumatic. An actuator performs a number of tasks to manipulate the process.

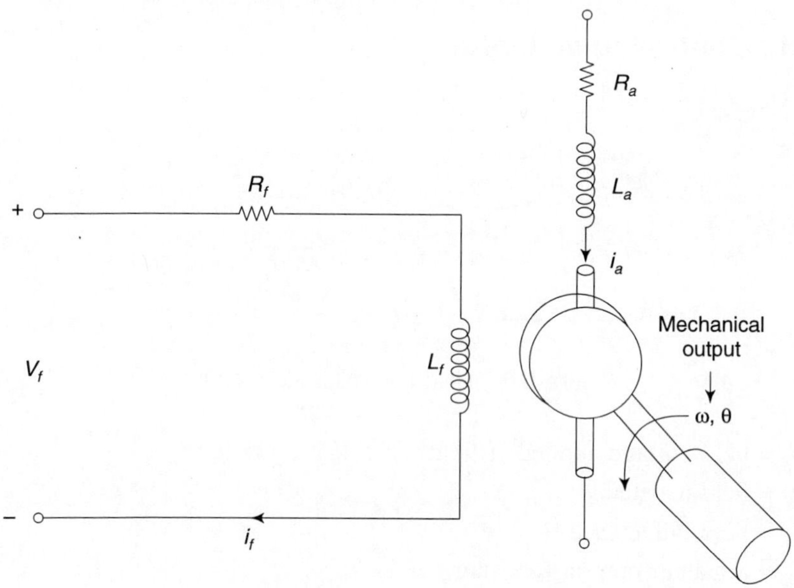

Figure 5.2 Diagram of dc motor

Due to the flexibility in transmitting electrical power and desirable speed-torque characteristics which are linear, electric actuators (motors) are widely used in control systems. But for low speed and high torque applications hydraulic actuators are used.

2. Sensors. Since the dynamic and static characteristics of the sensor or measuring element affect the indication of the actual value of the output variable, the sensor plays an important role in determining the overall performance of the control system. So we can say that the sensor or measuring element is a device that converts the output variable into another suitable variable like transducer such as displacement, pressure, or voltage that can be used to compare the output to the reference input signal. This element is in the feedback path of the closed loop system. The output signal of the sensor is invariably in electrical form; analog or digital.

3. Differencing and amplification. The error signal or differencing is produced by the comparator or error elector which is generally an electronic device. This is done by the operational amplifier (OPAMs). This error signal is further amplified by the use of power transistors, SCR's. In amplification, the suitable level in magnitude and power are most conveniently carried out. These electronic devices are out of scope of this book.

5.3 SERVOMOTORS

Servomotors are used to control the angular shift in the shaft with the change in input field voltage. Servomotors are basically of two types one is field controlled dc motor and second is armature controlled dc motor.

5.3.1 Field Controlled dc Motor

The circuit diagram of a field controlled dc motor is shown in Fig. 5.3.

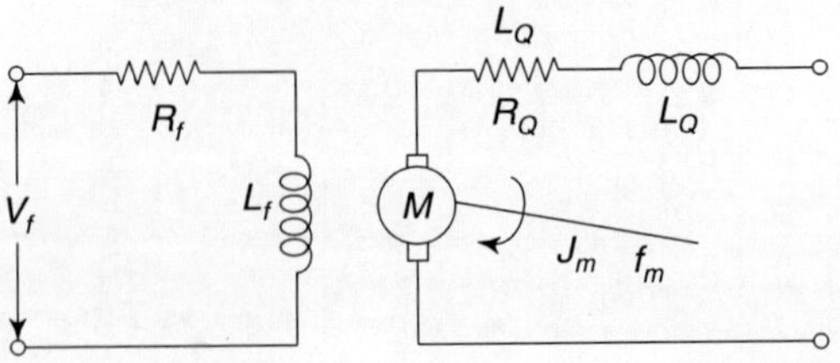

Figure 5.3 Field controlled dc motor

where V_f = field voltage applied at field winding

R_f = field resistance

L_f = field inductance

i_a = armature current (constant)

J_m = moment of inertia

f_m = coefficient of viscous friction

Since the armature current is kept constant, the relation between field current i_f and developed motor torque is given by

$$T_M \propto i_f$$

$$T_M = K_f\, i_f \qquad \text{...(5.2)}$$

where K_f = motor torque constant (Nm/A)

The dynamic equations are as follows

$$V_f = R_f\, i_f + L_f \frac{di_f}{dt} \qquad \text{...(5.3)}$$

Taking Laplace transform of Eq. (5.3)

$$V_f(s) = R_f\, I_f(s) + s\, L_f\, I_f(s) \qquad ...(5.4)$$

Taking Laplace transform of Eq. (5.2), we get

$$T_M(s) = K_f\, I_f(s) \qquad ...(5.5)$$

Now

$$T_M = J_M \frac{d^2\,\theta_m}{dt^2} + f_m \frac{d\theta_m}{dt} \qquad ...(5.6)$$

$$T_M = J_m \frac{d^2\omega_m}{dt} = f_m\, \omega_m \qquad ...(5.7)$$

where θ_m = angular shift

ω_m = angular velocity

Taking Laplace transform of Eqs. (5.6) and (5.7), we get

$$T_M(s) = s^2\, J_m\, \theta_m(s) + s\, f_m\, \theta_m(s) \qquad ...(5.8)$$

$$T_M(s) = s\, J_m\, \omega_m(s) + f_m\, \omega_m(s) \qquad ...(5.9)$$

From equation 5.3, 5.4, 5.5, 5.8 and 5.9 the block diagram for Fig. 5.3 is shown in Fig. 5.4. in which $\theta_m(s)$ is the output for the $V_f(s)$ as input.

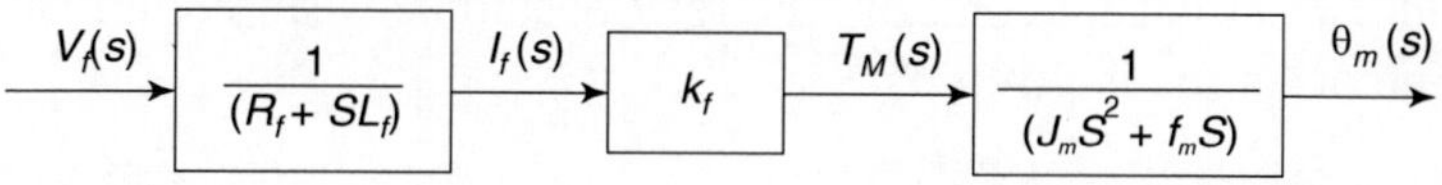

Figure 5.4 Block diagram of dc motor (field controlled)

Now

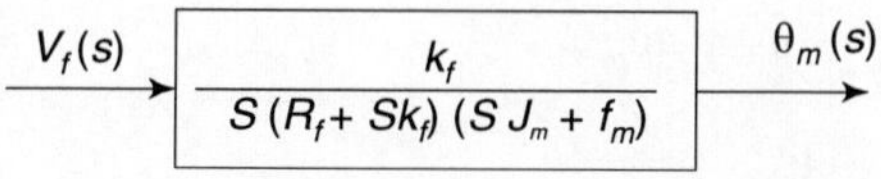

Figure 5.5 Reduced block diagram of Fig. 5.4

Hence, the transfer function is given by

$$\boxed{\frac{\theta_m(s)}{V_f(s)} = \frac{K_f}{S(R_f + SL_f)(S\,J_m + f_m)}} \qquad ...(5.10)$$

Similarly, when the $\omega_m(s)$ is taken as output then the transfer function is given by

$$\boxed{\frac{\omega_m(s)}{V_f(s)} = \frac{K_f}{(R_f + SL_f)(S\,J_m + f_m)}} \qquad ...(511)$$

5.3.2 Armature Controlled dc Motor

In this section the transfer function of an armature controlled dc motor relating to angular shift in the shaft and the input armature voltage is derived.

The circuit diagram of an armature controlled D.C. motor is shown in Fig. 5.6.

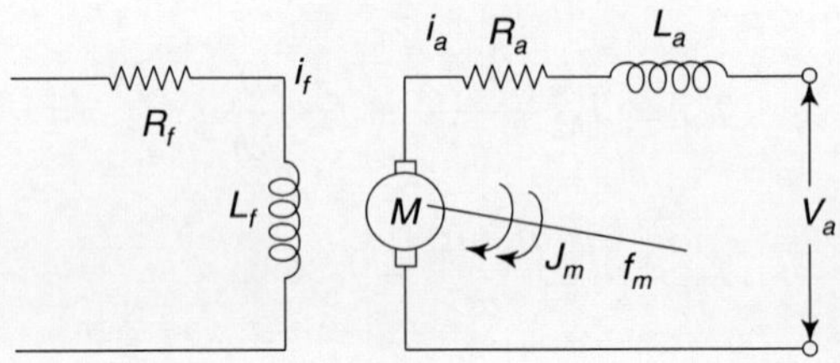

Figure 5.6 Armature controlled dc motor

where V_a = armature voltage applied at armature winding

R_a = armature resistance

L_a = armature inductance

i_f = field current (constant)

J_m, f_m = moment of inertia and coefficient of viscous friction

Since the field current is kept constant, the relationship between the armature current (i_a) and developed motor torque is given by

$$T_M \propto i_a$$

$$T_M = K_a\, i_a \quad \text{...(5.12)}$$

where K_a = motor torque constant

The applied voltage (V_a) is opposed by the back emf (e_b) developed in the armature. The relationship between (e_b) and the motor speed (ω_m) is given by

$$e_b \propto \omega_m$$

$\because$ $$\omega_m = \frac{d\theta_m}{dt}$$

Hence $$e_b = K_b \frac{d\theta_m}{dt} \quad \text{...(5.13)}$$

where K_b = Back emf constant

Related other dynamic equations are as follows

$$V_a - e_b = R_a\, i_a + L_a \frac{di_a}{dt} \quad \text{...(5.14)}$$

$$T_M = J_m \frac{d^2\,\theta_m}{dt^2} + f_m \frac{d\theta_m}{dt} \quad \text{...(5.15)}$$

$$T_M = J_m \frac{d\,\omega_m}{dt} + f_m\, \omega_m \quad \text{...(5.16)}$$

Taking Laplace transfer of Eqs. (5.12) to (5.16), we get

$$\left.\begin{aligned} T_M(s) &= K_a\, I_a(s) \\ E_b(s) &= K_b\, \omega_m(s) \end{aligned}\right\} \qquad \begin{aligned} &...(5.17) \\ &...(5.18) \end{aligned}$$

$$= s\, K_b\, \theta_m(s)$$

$$V_a(s) - E_b(s) = R_a\, I_a(s) + sL_a\, I_a(s) \qquad ...(5.19)$$

$$T_M(s) = s^2\, J_m\, \theta_m(s) + s\, f_m\, \theta_m(s) \qquad ...(5.20)$$

$$T_M(s) = s\, J_m\, \omega_m(s) + f_m\, \omega_m(s) \qquad ...(5.21)$$

From Eqs. (5.17) to (5.21), the block diagram, the block diagram for Fig. 5.6 is shown in Fig. 5.7 in which $\theta_m(s)$ in the output for $V_a(s)$ as input.

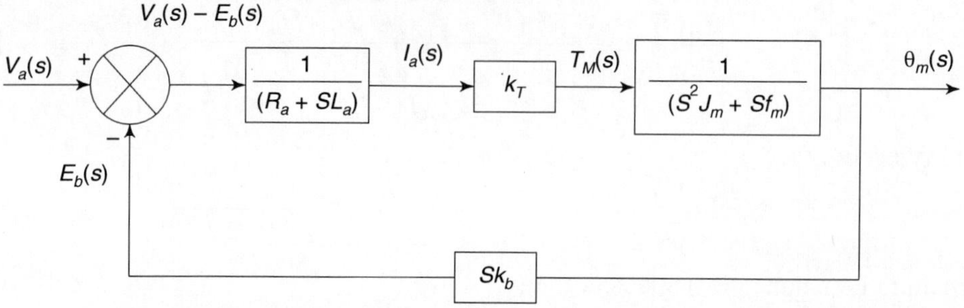

Figure 5.7 Block diagram of dc motor (armature controlled)

Now, the reduced block diagram for Fig. 5.7 is shown in Fig. 5.8.

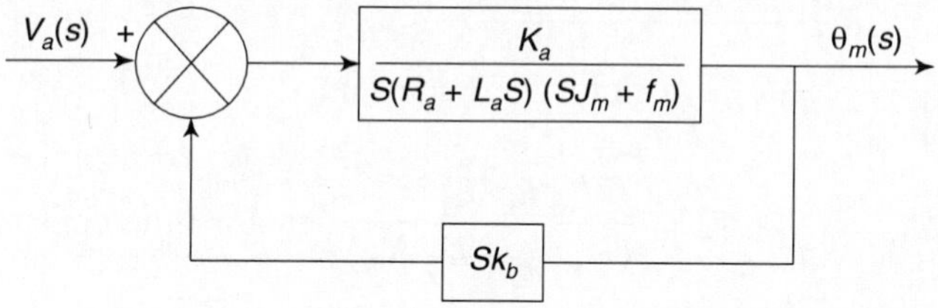

Figure 5.8 Reduced block diagram of Fig. 5.7

Similarly, when $\omega_m(s)$ is taken as output then the reduced block diagram is given in Fig. 5.9.

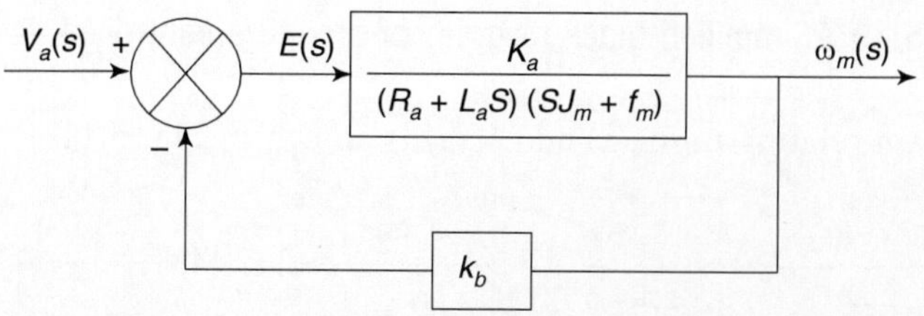

Figure 5.9 Reduced block diagram for $\omega_m(s)$ is an output

Hence, the forward path transfer function for Fig. 5.8,

$$G(s) = \frac{K_a}{s\,(R_a + sL_a)\,(s\,J_m + f_m)}$$

and the feedback path transfer function

$$H(s) = sK_b$$

Hence, the overall transfer function relating the output $\theta_m(s)$ and the input $V_a(s)$ is given by

$$\frac{\theta_m(s)}{V_a(s)} = \frac{G(s)}{1 + G(s)\,H(s)}$$

$$\frac{\theta_m(s)}{V_a(s)} = \frac{\dfrac{K_a}{s\,(R_a + sL_a)\,(s\,J_m + f_m)}}{1 + \dfrac{K_a}{s\,(R_a + sL_a)\,(s\,J_m + f_m)} \cdot sK_b} \quad \text{...(5.22)}$$

Simplifying Eq. (5.22)

$$\frac{\theta_m(s)}{V_a(s)} = \frac{K_a}{s\,(R_a + sL_a)\,(s\,J_m + f_m) + (s\,K_a\,K_b)} \quad \text{...(5.23)}$$

If armature inductance is neglected, then

$$\frac{\theta_m(s)}{V_a(s)} = \frac{K_a}{s\,R_a\,(s\,J_m + f_m) + s\,K_a\,K_b} \quad \text{...(5.24)}$$

$$\boxed{\frac{\theta_m(s)}{V_a(s)} = \frac{K_m}{s\,(1 + s\,T_m)}} \quad \text{...(5.25)}$$

where $K_m = \dfrac{K}{(R_a\,f_m + K_a\,K_b)}$, motor gain constant

and $T_m = \dfrac{R_a\,J_m}{(R_a\,f_m + K_a\,K_b)}$, motor time constant

The block diagram representing Eq. (5.25) is shown in Fig. 5.10.

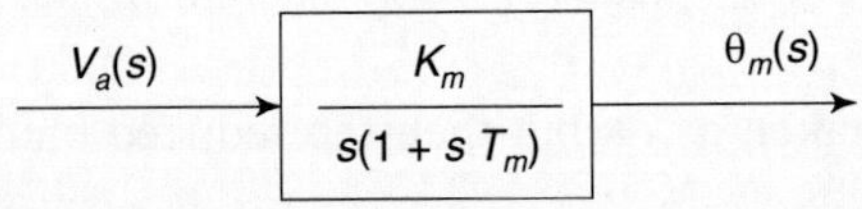

Figure 5.10 Simplified block diagram of armature controlled dc motor

The block diagram relating to $\omega_m(s)$ and $V_a(s)$ is shown in Fig. 5.11.

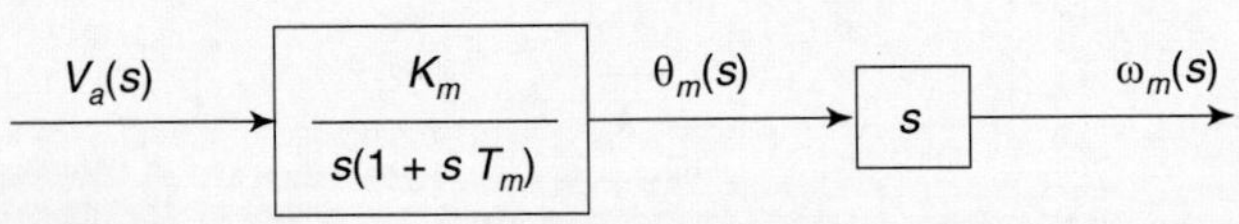

Figure 5.11 Block diagram of armature controled dc motor for $\omega_m(s)$ as output

The simplified block diagram of Fig. 5.11 is shown in Fig. 5.12.

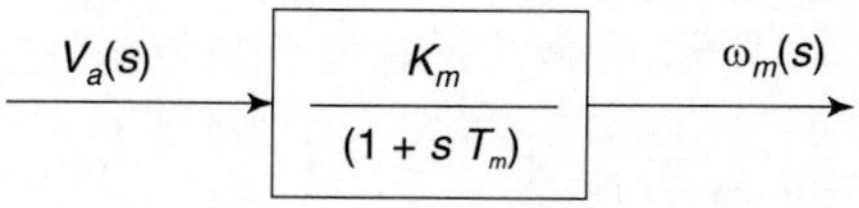

Figure 5.12 Simplified block diagram of Fig. 5.11

The transfer function relating to $\omega_m(s)$ and $V_a(s)$ is

$$\frac{\omega_m(s)}{V_a(s)} = \frac{K_m}{(1+sT_m)} \quad ...(5.26)$$

5.4 ac SERVOMOTOR

dc and ac motors are two type of electric actuators. When we use these motors for low power ratings then these are known as servomotors. dcmotors are costlier than ac motors due to the additional cost of commutation gear. The dc servomotors are out of scope of this book, so we will study only ac servomotor.

Construction. The ac servomotor is a two-phase induction motor. It has special designs feature in which, two stator windings are used because it has two pole constructions. The two stator windings have equal voltages in rms magnitude and 90° phase difference as shown in Fig. 5.13.

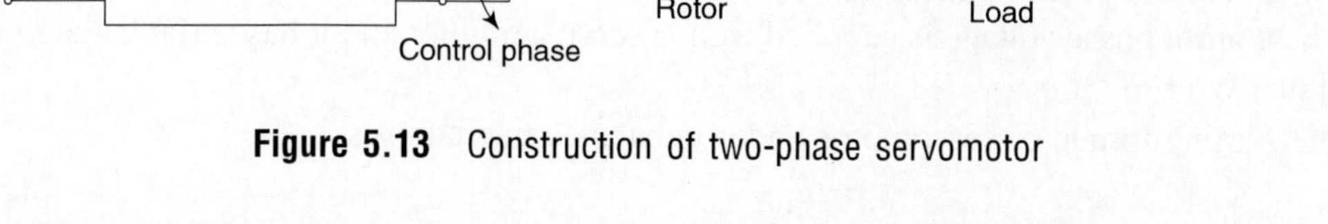

Figure 5.13 Construction of two-phase servomotor

In two-stator winding one winding is called control winding and second one is reference winding, electrically 90° apart from each other. Servo amplifier is used to amplify the actuating signal coming from error detector and passes it to the control phase. The rotor of the servomotor is built with high resistance so that its low X/R ratio and the torque-speed characteristic as shown by the curve (b) in Fig. 5.14 is nearly linear.

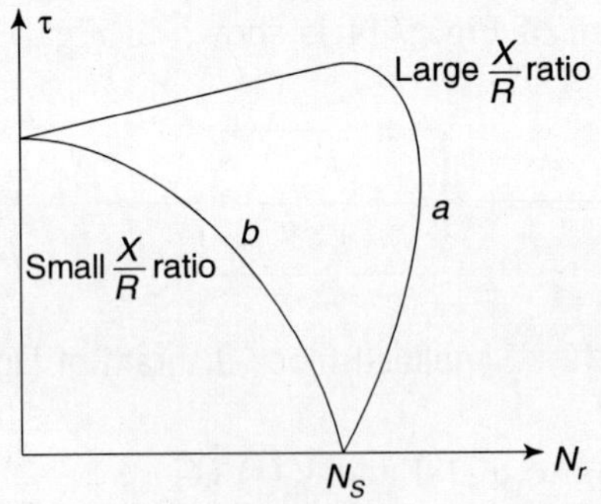

Figure 5.14 Torque-speed characteristics of induction motor

where N_r = rotor speed

N_s = synchronous speed

T = torque

$\frac{X}{R}$ ratio = rotor reactance to rotor resistance ratio

In Fig. 5.13, the τ_g is called generated torque in rotor, the inertia load is used in which J represents the inertia of load and to represent the viscous friction, V_r represents the voltage applied to reference phase. θ represents the speed of the motor.

Working. When the two voltages of equal magnitude and 90° phase difference are applied to the two stator windings resulting in a magnetic field which is constant in magnitude and relating at synchronous speed. The direction of rotation depends upon the reference voltage (V_r) which is applied on reference phase and control voltage, $e(t)$, which is applied on control phase. This rotating field cuts the conductor and emf is produced due to the rate of change of flux due to this, current is produced in the short circuited rotor. The rotating magnetic field interacts with these currents producing a torque on the rotor in the direction of field rotation.

In servo applications, the voltages excited applied to the two-stator windings are seldom balanced. One phase known as reference phase is by a constant voltage (V_r) and the other phase known as control phase is energized by a voltage which is 90° out of phase with respect to V_r. The control phase voltage is supplied from a servo amplifier and it has variable magnitude and polarity (± 90°).

The starting torque of servomotor under unbalance operation is

$$\tau_s \propto E_{\text{rms}} \qquad ...(5.27)$$

where τ_s = starting torque

E_{rms} = rms value of control voltage [$e(t)$]

The torque speed curve for varying rms control voltage is shown in Fig. 5.15 (a)

As we see the characteristics of a servomotor in Fig. 5.15 (a) that all the curves have negative slope. The curves are non-linear but in the low speed region, the curves are nearly linear and equidistant, i.e., the torque varies linearly with speed as well as control voltage. Since servomotor operates at high speeds, these curves can be linearized about the operating point.

Generated torque by the motor is a function of speed N and rms control voltage, E_{rms}.

i.e., $$\tau_g = f(N, E_{rms}) \quad ...(5.28)$$

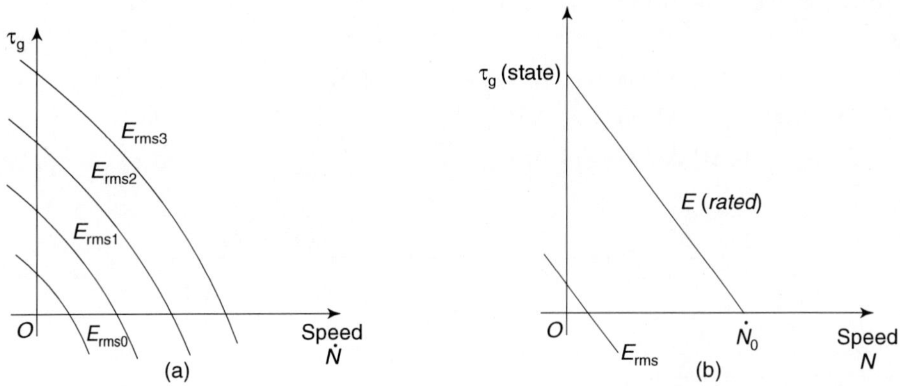

Figure 5.15 Servomotor characterisitics

Expanding Eq. (5.28) by Taylor's series about the normal operating point (τ_{go}, E_{rms}, N_o) and neglecting the second and higher order terms.

We get $$\tau_g = \tau_{go} + \left.\frac{\partial \tau_g}{\partial E_{rms}}\right|_{\substack{E_{rms} = E_{o\,rms} \\ \dot{N} = \dot{N}_o}} (E_{rms} - E_{o\,rms}) + \left.\frac{\partial \tau_g}{\partial \dot{N}}\right|_{\substack{E_{rms} = E_{o\,rms} \\ \dot{N} = \dot{N}_o}} (\dot{N} - \dot{N}_o) \quad ...(5.29)$$

$$\tau_g - \tau_{go} = C_1 (E_{rms} - E_{o\,rms}) - f_o (\dot{N} - \dot{N}_o)$$

$$\Delta\tau_g = C_1 \,\Delta E_{rms} - f \,\Delta \,\dot{N} \quad ...(5.30)$$

where $$\left.\begin{aligned} C &= \left.\frac{\partial \tau_g}{\partial E_{rms}}\right|_{\substack{E_{rms} = E_{o\,rms} \\ \dot{N} = \dot{N}_o}} \\ f &= \left.\frac{\partial \tau_g}{\partial \dot{N}}\right|_{\substack{E_{rms} = E_{o\,rms} \\ \dot{N} = \dot{N}_o}} \end{aligned}\right\} \text{constants}$$

If the load consists of inertia (J) and viscous friction (f_o) as shown in Fig. 5.13, the torque equation is

$$\Delta\tau_g = J\Delta\ddot{N} + f_o \,\Delta\dot{N} \quad ...(5.31)$$

Substituting Eq. (5.31) in Eq. (5.30).

$$J\Delta\ddot{N} + f_o\Delta\dot{N} = C_1\Delta \,E_{rms} - f_o \,\Delta\dot{N} \quad ...(5.32)$$

Equation 5.32 is valid for varying ΔE is slow as compared to sin $\omega_c t$, i.e., in frequency domain $\omega << \omega_c$.

where ω = frequency of the signal ΔE

and ω_c = carrier frequency

Control system using such type of device is called carrier control systems.

Taking the Laplace transform of Eq. (5.32)

We have

$$(Js^2 + f_o s)\,\Delta N(s) = c_1\,\Delta E(s) - f_o\,s\,\Delta N(s)$$

$$(Js^2 + f_o s)\,\Delta N(s) + f_o\,s\,\Delta N(s) = c_1\,\Delta E(s)$$

$$[Js^2 + (f_o + f)s]\,\Delta N(s) = c_1\,\Delta E(s)$$

$$\frac{\Delta N(s)}{\Delta E(s)} = \frac{c_1}{Js^2 + (f_o + f)s} = \frac{\dfrac{c_1}{(f_o + f)}}{s\left[\dfrac{J}{(f_o + f)}s + 1\right]}$$

$$\frac{\Delta N(s)}{\Delta E(s)} = \frac{C_m}{s\,(\tau_m\,s + 1)} \qquad \text{...(5.33)}$$

where $C_m = \dfrac{c_o}{(f_0 + f)}$ = motor gain constant

$\tau_m = \dfrac{J}{(f_o + f)}$ = motor time constant

When we study Eq. (5.33) in which $\Delta N(s)$ is representing the incremental Laplace transform of servo motor output, i.e., speed N and $\Delta E(s)$ is representing the incremental Laplace transform of servomotor input, i.e., control voltage, E. So we can say that Eq. 5.32 represents the incremental motor transfer function, $G_M(s)$. So

$$G_M(s) = \frac{C_m}{s\,(\tau_m\,s + 1)} \qquad \text{...(5.34)}$$

When ac servomotor is used in position constant, the operating point is $E_{o\,\text{rms}} = 0$ and $\dot{N}_o = 0$

So that $\Delta N = N$ and $\Delta E_{\text{rms}} = E_{\text{rms}}$

So $$G(s) = \frac{N(s)}{E_{\text{rms}}(s)} = \frac{C_m}{s\,(\tau_m\,S + 1)}$$

The motor constant C and f can be easily estimated by two tests: no load and block rotor tests. The no load speed ($\tau_g = 0$) and block rotor torque $(\dot{N} = \theta)$ at voltage applied to control phase are shown in Fig. 5.15. (b).

In Fig. 5.15 (b) the line joining $\dot{N}_o$ and τ_g (stall) points represents the torque speed characteristic of a servomotor.

So $$C = \frac{\text{Block rotor torque (rated voltage)}}{\text{Rated control phase voltage}} \quad \text{...(5.35a)}$$

$$f = \frac{\text{Block rotor torque (rated voltage)}}{\text{No load speed (rated voltage)}} \quad \text{...(5.35b)}$$

But when we observe Fig. 5.15 (a) that the slope of the characteristics reduces as control phase voltage decreases. In motor used in practice, the torque speed slope in low speed region is nearly one-half the slope at rated voltage. So far servomotors used in position control systems.

$$f = \frac{1}{2}\frac{\text{Block rotor torque (rated voltage)}}{\text{No load speed (rated voltage)}} \quad \text{...(5.36)}$$

- *The negative slope of the torque-speed characteristic of the servomotor contributes the viscous friction term (f) to the load friction (f_o) if the slope is positive, then the effective viscous friction ($f_o + f$) may become negative. The damping of such control system may be negative and unstable. That's why the conventional induction motors are not used as servo motor because the slope of the characteristic of conventional induction motor is positive.*

5.5 TACHOGENERATOR

A tachogenerator is used to convert a rotational speed into a proportional voltage. In tachogenerator rotational speed is used as an input and porportional voltage is as output. As the output is produced by the tachogenerator it is two types are is dc generator and second one is ac generator.

5.5.1 dc Tachogenerator

dc tachogenerator is shown in Fig. 5.16. It is used to produce dc voltage as an output when rotational speed is provided as an input. It uses a permanent magnet for producing magnetic field.

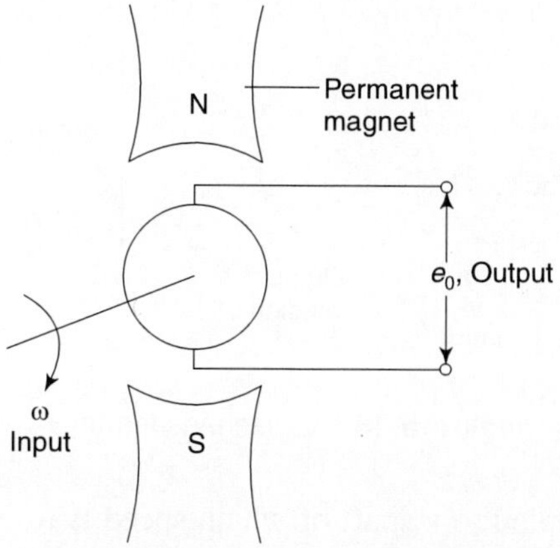

Figure 5.16 dc tachogenerator

The tachogenerator is coupled to a shaft of which speed is to be measured. Due to use of permanent magnet, the magnetic flux is constant. Hence, the induced voltage across the armature is proportional to the speed of the shaft. The polarity of induced voltage is depends upon the direction of rotation so,

$$e \propto \omega \quad ...(5.37)$$

$$e = K_t\, \omega \quad ...(5.38)$$

where K_t = tachogenerator constant, $V/(\text{rad/sec})$

The Laplace transform of Eq. (5.38) is given by

$$E(s) = K_t\, \omega(s) \quad ...(5.39)$$

The block diagram representation of Eq. (5.39) is shown in Fig. 5.17.

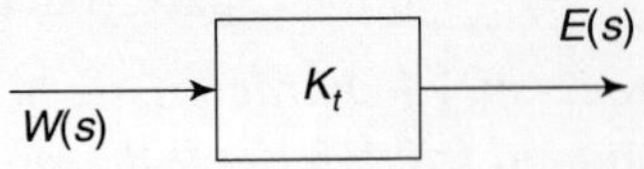

Figure 5.17 Block diagram of dc tachogenerator

The transfer function is given by

$$\boxed{\frac{E(s)}{\omega(s)} = K_t} \quad ...(5.40)$$

5.5.2 ac Tachogenerator

ac tachogenerator is shown in Fig. 5.18. It is used to produce ac voltage as an output when rotational speed is provided as an input. It works on the principle of induction generator.

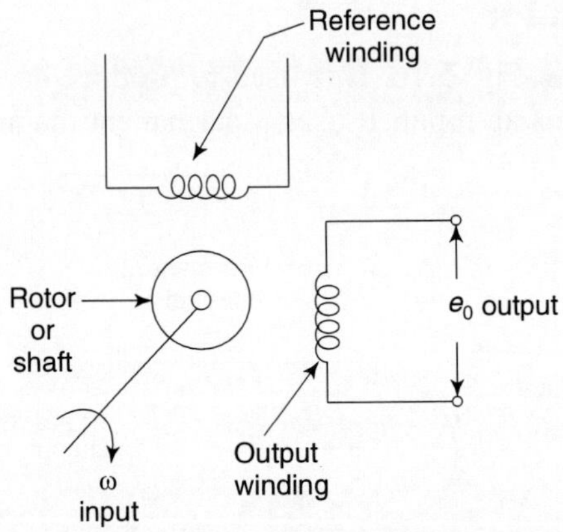

Figure 5.18 ac tachogenerator

The tachogenerator is coupled to a shaft of which speed is to be measured. The reference winding is used to supply a reference voltage and the induced voltage is developed across the output winding.

Hence, the induced voltage across the output winding is proportional to the speed of the shaft. The amplitude and the phase of the output voltage depends upon the direction of rotation so,

$$e \propto \omega$$

or $$e = K_t\,\omega \quad ...(5.41)$$

where K_f = tachogenerator constant V/(rad./sec)

The Laplace transform of the Eq. (5.41) is

$$E(s) = K_t\,\omega(s) \quad ...(5.42)$$

Hence, the block diagram related to Eq. (5.41) is shown in Fig. 5.19.

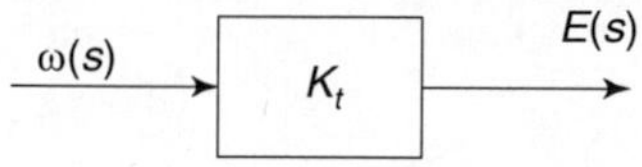

Figure 5.19 Blcok diagram of ac tachogenerator

The transfer function is given by

$$\frac{E(s)}{\omega(s)} = K_t \quad ...(5.43)$$

- *A dc tachogenerator can also be used for ac system and ac tachogenerator can also be used on dc systems by using a modulator and phase sensitive rectifier respectively.*

5.6 SYNCHROS

A device called Selsyn was developed about in the year 1925. This comprised of a system whereby a generator and a motor so connected by wire that angular rotation or position in generator is transmitted to motor. The generator and motor are also called, a transmitter and receiver. About in 1942 or 1943 the term, synchro, became the general term, replacing the word, Selsyn.

So synchro is commercially known as a selsyn, synchrotic, Autosyn, and Telegon. In other words we can say that "a synchro is an electromagnetic transducer commonly used to convert angular position of a shaft into an electric signal".

The synchros employed for transmitting angular position and rotation over a distance so far. For example, remote signalling systems, such as signalling from switchboard to generator room. Automatic or remote positioning control, remote indication of position etc.

Construction. As we discussed earlier, a synchro has two parts: one is synchro transmitter and second one is synchro receiver. The basic part of synchro is synchro transmitter. The construction of synchro transmitter is similar to 3 – φ alternator. The power synchros are always of three-phase construction, whereas the indicating synchros are constructed 1 – φ. The stator of 3φ alternator is of laminated silicon steel and is slotted to accommodate a balanced 3 – φ winding which is usually of concentric coil type, i.e., three identical coils are placed in the stator with their axis 120° apart, and is star connected. The rotor is of dumbbell construction and is wound with a concentric coil. The construction of a synchro transmitter is shown in Fig. 5.20 (a) and (b).

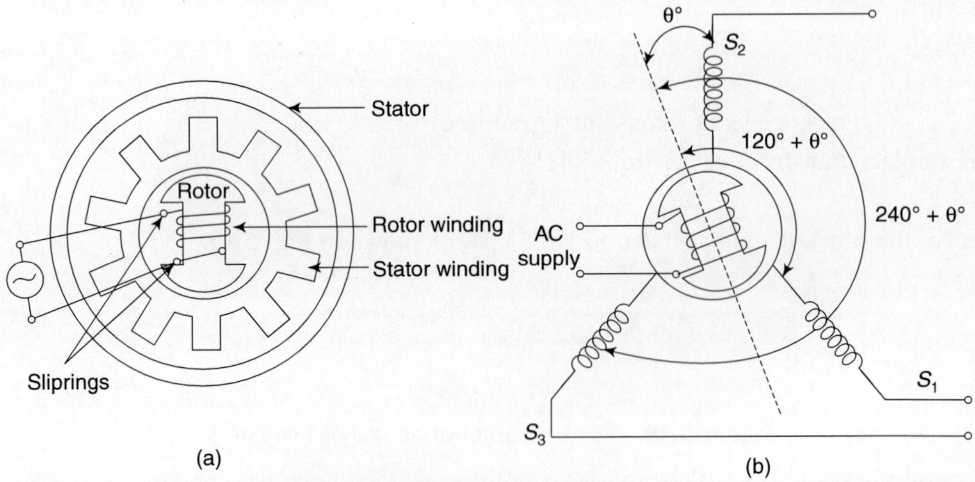

Figure 5.20 Construction of synchro transmitter

Working. The synchro is a type of rotary electrical transformer that is used for measuring the angle of a rotating machine. The primary winding of the transformer, fixed to the rotor is excited by a sinusoidal electric current (ac).

Let an ac voltage

when $$V_{rt}(t) = V_r \sin \omega_{ct}\, t$$

$$V_{rt} = \text{rotor voltage of synchro transmitter}$$

The voltage $V_r(t)$ causes a flow of magnetizing current in the rotor coil which produces a sinusoidal time varying flux $\left(\frac{d\phi}{dt}\right)$ directed along its axis and distributed nearly sinusoidal in the airgap along the stator periphery, i.e., by electromagnetic induction causes currents to flow in three-star connected secondary windings fixed at 120° to each other on the stator. The flux linking any stator coil is proportional to the cosine of the angle between the rotor and stator coil axes and so is the voltage induced in each stator coil. The stator coil voltages are of course, in time phase with each other. Thus, we can say that the synchro transmitter acts like a 1 – ϕ transformer in which the rotor coil is the primary and the stator coils form the three secondaries.

Let V_{s1}, V_{s2} and V_{s3} respectively be the voltages induced in stator coil s_1, s_2, s_3 with respect to neutral.

For the rotor position of the synchro transmitter

$$V_{s1} = CV_r \sin \omega_c t \cos (\theta + 120°) \quad ...(5.44)$$

$$V_{s2} = CV_r \sin \omega_c t \cos (\theta) \quad ...(5.45)$$

$$V_{s3} = CV_r \sin \omega_c t \cos (\theta + 240°) \quad ...(5.46)$$

The terminal voltages of the stator are

$$V_{s1}, V_{s2} = V_{s1} - V_{s2}$$
$$= CV_r \sin \omega_c t \cos (\theta + 120°) - (V_r \sin \omega_c t \cos \theta)$$
$$= CV_r \sin \omega_c t [\cos (\theta + 120°) - \cos \theta]$$
$$V_{s1}, V_{s2} = \sqrt{3}\, CV_r \sin \omega_c t \sin (\theta + 240°) \quad ...(5.47)$$

and

$$V_{s2}, V_{s3} = V_{s2} - V_{s3}$$
$$= CV_r \sin \omega_c t [\cos \theta - \cos (\theta + 240°)$$
$$V_{s2}, V_{s3} = \sqrt{3}\, CV_r \sin \omega_c t \sin (\theta + 120°) \quad ...(5.48)$$

Similarly,

$$V_{s3}\, V_{s1} = V_{s3} - V_{s1}$$
$$= CV_r \sin \omega_c\, t [\cos (\theta + \cos (\theta + 240°) - \cos \theta]$$
$$V_{s3}\, V_{s1} = \sqrt{3}\, CV_r \sin \theta \sin \omega_c t \quad ...(5.49)$$

When $\theta = 0$

By Eq. (5.44) to (5.46)

$$\left.\begin{aligned} V_{s1} &= CV_r \sin \omega_c t \cos 120 &&= -\frac{1}{2} CV_r \sin \omega_c t \\ V_{s2} &= CV_r \sin \omega_c t &&= CV_r \sin \omega_c t \\ V_{s3} &= CV_r \sin \omega_c t \cos 240° &&= -\frac{1}{2} CV_r \sin \omega_c t \end{aligned}\right\} \quad ...(5.50)$$

By Eq. (5.50), the maximum voltage is induced in the stator coil s_2, and the terminal voltage $V_{s3}V_{s1}$, = 0. This position of the rotor is defined as the electrical zero of the transmitter and is used as reference for specifying the angular position of the rotor.

So we can say that the input to the synchro transmitter is angular position of shaft and the output is three voltages $V_{s1}\ V_{s2}$, $V_{s2}\ V_{s3}$ and $V_{s3}\ V_{s1}$ given by Eqs (5.47) to (5.49), the magnitude of these voltages are functions of the shaft position, θ.

The output of the synchro transmitter is given input to the stator of control transformer. The construction of control transformer is similar to the synchro transmitter, only difference is in rotor. The rotor of control transformer is made cylindrical in shape so that the air gap is uniform. The pair of synchro transmitter and control transformer as shown in Fig. 5.21 is known as error detector. Circulating currents of the same phase but different magnitudes flow through the two sets of stator coils. Due to this there is the establishment of an identical flux pattern in the air gap of the control transformer as the leakage reactance and the voltage drop in resistance of the two sets of stator coils are usually small. The voltage induced in control transformer rotor is proportional to the cosine of the angle between the two rotor.

Let angle between two rotors be ϕ

then $$V_{rc} = CV_r \cos \phi \sin \omega_c t \quad ...(5.51)$$

when $$\phi = 90°$$

i.e., two rotors are at right angle, the voltage induced in control transformer rotor is zero

$$V_{rc}\,(t) = 0 \quad ...(5.52)$$

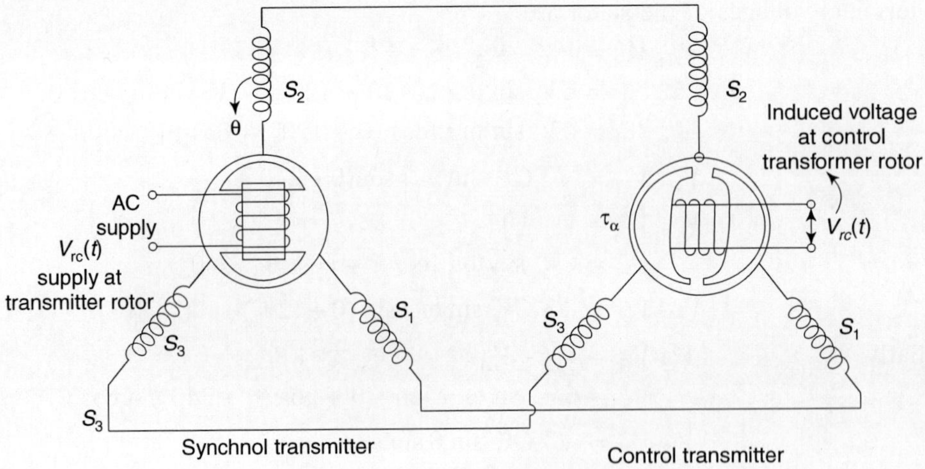

Figure 5.21 Synchro error detector

The position is known as electrical zero position for control transformer.

Let the rotor of the transmitter rotate by angle θ in clockwise and the rotor of control transformer rotate in same direction by an angle α so the resultant angle is

$$\phi = (90 - \theta + \alpha)$$

Now substituting the value of ϕ in Eq. (5.51)

$$V_{rc}(t) = CV_r \cos [90 - \theta + \alpha) \sin \omega_n t$$

$$V_{rc}(t) = CV_r \sin (\theta - \alpha) \sin \omega_t t \qquad ...(5.53)$$

if (θ – α) is very less than sin (θ – α) will be less, so

$$V_{rc}(t) = CV_r (\theta - \alpha) \sin \omega_n t \qquad ...(5.54)$$

So we can say that the synchro transmitter control transformer pair acts as an error detector giving a voltage signal at the rotor terminals of the control transformer proportional to the angular difference between the synchro transmitter and control transformer shaft position θ and α.

- *The rotor terminals of control transformer are usually connected to an amplifier, so the rotor of control transformer is made cylindrical to produce uniform air grap. Due to this the change in rotor output impedance with rotation of the shaft is minimum. The stator winding of the control transformer has higher impedance per phase by this feature the several control transformers to be fed from a signle transformer.*

5.7 STEPPER MOTOR

A stepper motor is a brushless synchronous motor that can divide a full rotation into a large number of steps. The motor position can be controlled precisely. The step size depends upon electrical pulse received from its control unit. Generally used step sizes are 7.5°, 15° or larger. Stepping motors can be used in simple open loop control systems. These are generally adequate for systems that operate at low acceleration with static loads, but closed loop control may be essential for high accelerations, particularly if they involve variable loads.

Construction. A four-phase, 2-pole shown in Fig. 5.22. The rotor used is permanent magnet type, according to construction there are three types of stepper motor:

1. Permanent magnet
2. Variable reductance stepper
3. Hybrid synchronous stepper

It has effectively multiple "toothed" electromagnets arranged around a central gear-shaped piece of iron.

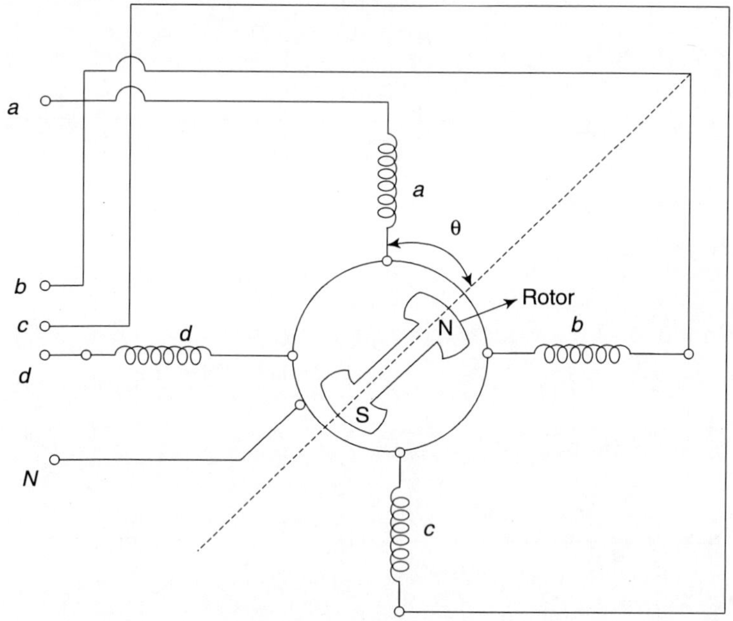

Figure 5.22 4-phase 2-pole stepper motor

Operation. To make the motor shaft turn, first one electromagnet is given power, which makes the gear's teeth magnetically attracted to the electromagnet's teeth. When the gear's teeth are thus aligned to the first electromagnet, they are slightly offset from the next electromagnet. So when the next electromagnet is turned on and the first is turned off, the gears rotate slightly to align with the next one and from there the process is repeated. Each of those slight rotations is called a "step" In the way, the motor can be turned by a precise angle. Permanent magnet stepper motor: The rotor aligns with the axis of the stator field with torque being

$$\tau \ \alpha \ \sin \theta \qquad ...(5.55)$$

where θ = angle of displacement between the rotor axis and stator field.

When phases a and b are excited, the rotor thus assumes positions 9 = 0°, 45°, 90° as phase a, $a + b$, b... are excited. It is easily observed that the stable position of the rotor corresponds to that angle at which the torque is zero and is positive for smaller angles and negative for large angles. Thus, with phase *a* the stable or locked position is $\theta = 0°$ but not $\theta = 180°$ and the torque has a maximum value at $\theta = 90°$. It is easily concluded that each excitation pattern of phases corresponds to a unique position of the rotor. Patterns for phase winding excitations can be easily visualised for steps of 22.5°, 11.25°. The advantage of a permanent magnet stepper motor is that when excited it seeks a preferred position in certain applications.

Torque angle characteristics of stepper motor is shown in Fig. 5.23 (a) and (b).

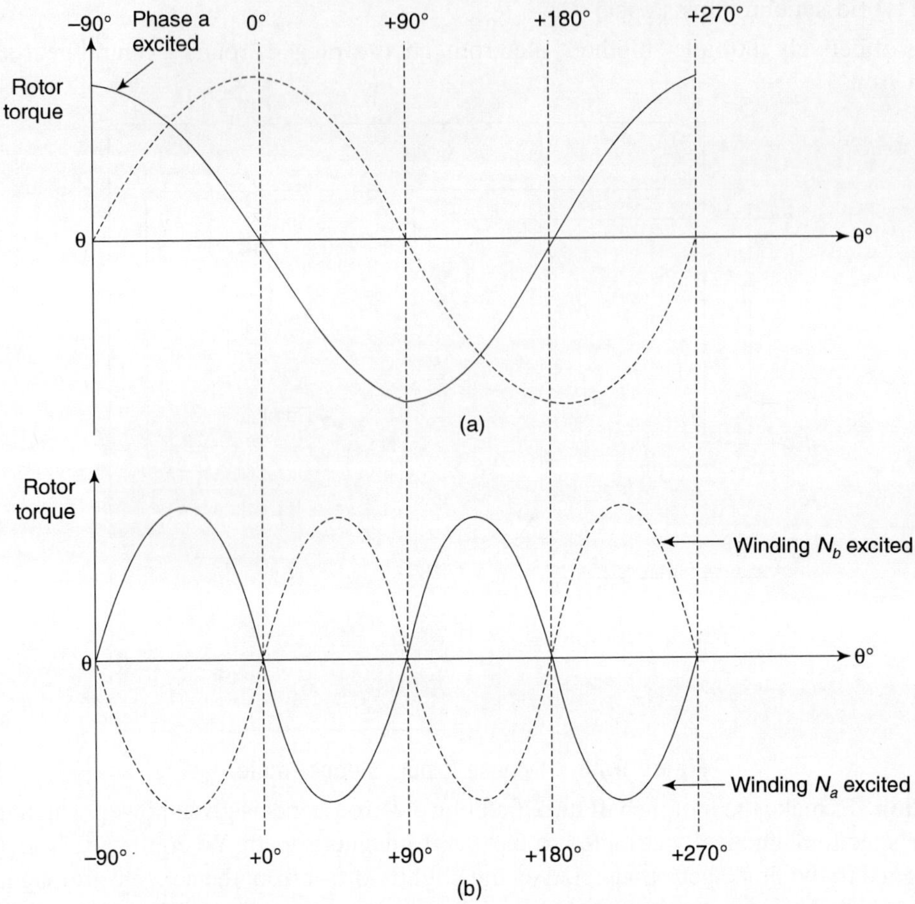

Figure 5.23 (a) Torque angle characteristic of PM motor (b) torque angle characteristic of variable reluctance motor

5.7.1 Variable Reluctance Stepper Motor

This motor is constructionally differs from permanent magnet stepper motor. It consists of single or multiple stacks of stators and rotors. These multiple stacks of stator mounted on common frame and stacks of rotor is mounted on same shaft as shown in Fig. 5.24.

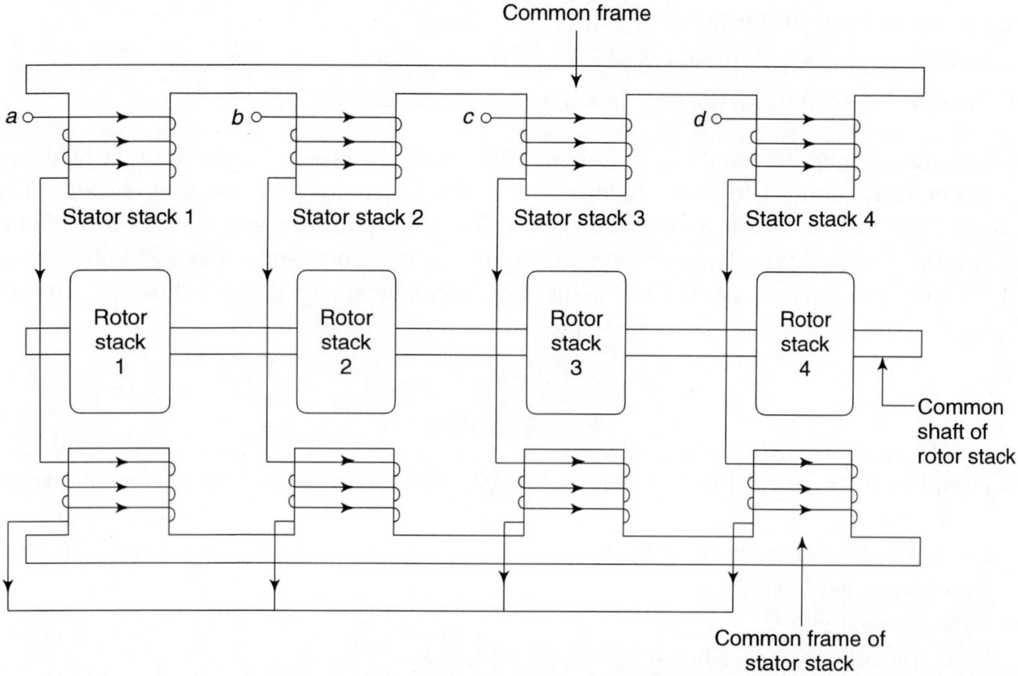

Figure 5.24 4-stack variable reluctance motor

5.7.2 Hybrid Stepper Motor

Hybrid stepper motor with constructional features of toothed and stacked rotor adopted from the variable reluctance motor. The stator has only one set of winding excited poles which interact with two rotor stacks. The hybrid stepper motor is shown in Fig. 5.25. The permanent magnet is placed axially along the rotor in the form of an angular cylinder over the motor shaft.

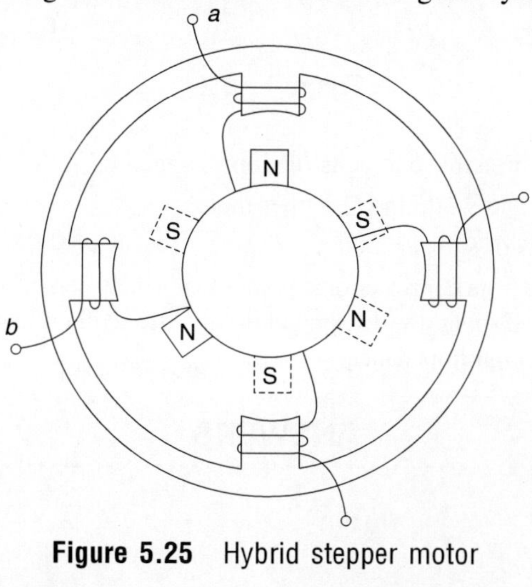

Figure 5.25 Hybrid stepper motor

Advantages of Stepper Motor

1. It is comptitable with digital system
2. There is no need of sensors for position and speed sensing.

Applications. Computer controlled stepper motors are one of the most versatile forms of positioning systems. They are digitally controlled as part of an open loop system. The industrial applications are in high speed pick and place equipment and multiaxis machine, CNC machine, and also used in precision positioning equipment such as linear actuator, linear stages, rotation stages, etc. Other uses are in packaging machinery, paper fed motors, in type writers and printers, pens in *XY* plotters, etc.

EXERCISE

5.1 Explain the construction and working of AC servomotor. Also discuss its torque speed characteristics. *(U.P.T.U., 2006–07, CS)*

5.2 A servo system is represented by the signal flow graph shown in Fig. E.5.2. The variable *T* is the torque and *E* is the error.

Determine

(a) The overall transmission if $K_1 = 1$, $K_2 = 5$ and $K_3 = 5$

(b) The sensitivity of the system to change in K_1 for $\omega = 0$

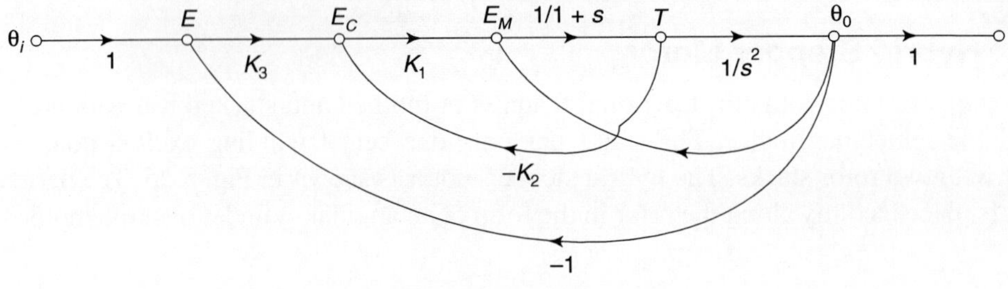

Figure E. 5.2

5.3 Explain the working principle and construction of synchros.

5.4 Explain construction and working of stepper motor.

5.5 Write the applications of stepper motor.

5.6 What is the difference from constructional point of view between the roots of synchro transmitter and synchro receiver. Explain the reason behind it. Also explain that why synchro transmitters are designed to have high impedance. *(U.P.T.U., 2007–08, CS)*

ANSWERS

5.2 (a) $\dfrac{5}{s^2(s+6)+10}$ (b) 0.5

CHAPTER 6

Stability of Control System

6.1 INTRODUCTION

In the previous chapter we have seen that there are three basic requirements for designing a control system: one is transient response, second one is steady state error and third one is stability of the system. Thus, far we have covered transient response and steady state error in chapter 3. Now we are discussing next requirement, stability.

So we can say that among the many forms of performance specifications used in design, the most important requirement is that the system is stable. An unstable system is generally considered to be useless.

6.2 CONCEPT OF STABILITY

Stability is an important characteristic of the transient performance of the system. The system is stable if there is small change in system parameters or in input initial conditions there is no large change in system output. So generally the systems is designed to be stable.

The stability can be defined that a stable system is one which gives bounded (limited) system response for a bounded input (BIBO). The stable and unstable system is shown in Fig. 6.1 (a) and (b) respectively for a bounded input.

The total response of the system is the sum of the forced and natural response i.e.,

$$Y(t) = Y_{\text{forced}}(t) + y_{\text{natural}}(t) \qquad \text{...(6.1)}$$

By Eq. (6.1), we can define the stability of the system according to the natural response.

- A linear time invariant (LTI) system is stable if the natural response approaches zero for sufficiently large time as time approaches infinity as shown in Fig. 6.1(a).
- A linear time invariant (LTI) system is unstable if the natural response approaches infinity for sufficiently large time as time approaches infinitiy as shown in Fig. 6.1 (b).

- A linear time invariant (LTI) system is marginally stable if the natural response neither approaches zero nor infinity but remains constant or oscillates as time approaches infinity.

Now we can define the stability of the system according to the total response.

- If the output of the system is bounded for the bounded input, such system is called a stable system.
- If the output of the system is unbounded for the bounded input, such system is called a unstable system.

For analysis and design purpose, we can classify stability as absolute stability and relative stability.

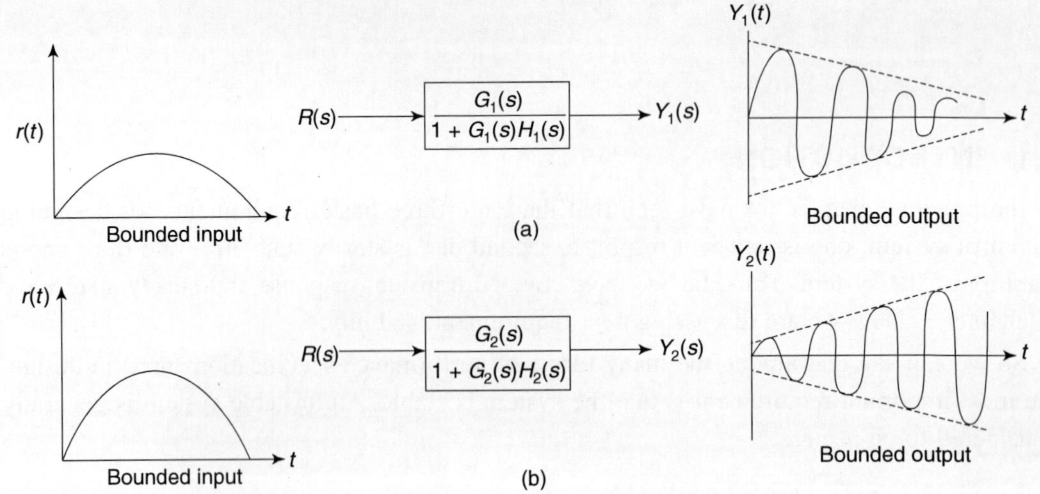

Figure 6.1 (a) Stable system, (b) unstable system

1. **Absolute stability.** Absolute stability only tells about the condition of the system whether it is stable or unstable, i.e., it will provide only Yes or No about the system stability.

2. **Relative stability.** Once the system is found to be stable, it is of interest to determine how stable it is, and this degree of stability is a measure of stability.

The measure of stability are maximum overshoot, damping ratio described earlier and gain margin, phase margin to be described later.

Let us consider $r(t)$ is input, $y(t)$ is the output of the system and $g(t)$ is the impulse response of the LTI system. When the initial condition of the system is zero, the system is said to be bounded input and bounded output system.

The transfer function of the system is

$$\frac{Y(s)}{R(s)} = G(s) \quad \text{...(6.2)}$$

$$Y(s) = R(s)\,G(s)$$

Taking inverse Laplace transform

$$y(t) = L^{-1}\,[R(s)\;G(s)]$$

The convolution integral

$$y(t) = \int_0^{\infty} r\,(t-\tau)\,g\,(\tau)\,d\tau \qquad ...(6.3)$$

where $g(\tau)$ = impulse response

Taking absolute value of both sides of the equation, we get

$$|y(t)\}| = \int_0^{\infty} r\,(t-\tau)\,g\,(\tau)\,d\tau\Big| \qquad ...(6.4)$$

Since the absolute value of integral $y(t)$ is not greater than the integral of the absolute value of the integrand $[r(t-\tau)\;g\;(\tau)]$

So
$$|Y(t)| = \int_0^{\infty} |r\,(t-\tau)|\,|g\,(\tau)|\,d\tau \qquad ...(6.5)$$

If input $r(t)$ is bounded then

$$|r(t)| \le N_1 \qquad ...(6.6)$$

where N_1 = finite positive number

$$|Y(t)| = N_1 \int_0^{\infty} |g(\tau)|\,d\tau \qquad ...(6.7)$$

If output $y(t)$ is bounded then

$$|Y(t)| \le N_2 < \infty \qquad ...(6.8)$$

where N_2 = finite positive number

So the following condition must hold

$$|Y(t)| \le N_1 \int_0^{\infty} |g(\tau)|\,d\tau \le N_2 \qquad ...(6.9)$$

The condition given in Eq. 6.9 implies that the area under the $|g(t)|$ versus τ curve must be finite.

The nature of $g(t)$ depends on the poles of the transfer function $G(s)$ which are the roots of the characteristic equation. The roots may be real, complex or multiple roots, (repeated roots), if the transfer function of the system is

$$\frac{Y(s)}{R(s)} = K\,\frac{(s-Z_1)(s-Z_2)(s-Z_3)\ldots(s-Z_n)}{(s-P_1)(s-P_2)(s-P_3)\ldots(s-P_m)} \qquad ...(6.10)$$

$Z_1, Z_2, Z_3 \ldots Z_n$ are zeros of the system and $P_1, P_2, P_3 \ldots P_m$ are poles of the system so $g(t)$ depends on poles $(P_1, P_2, P_3 \ldots P_n)$ of the transfer function.

(a) If all the roots of the characteristic equations have negative real parts then $g(t)$ is bounded so $\int_0^{\infty} |g(\tau)|\,d\tau$ is finite therefore the system is stable.

(b) If any root of the characteristics equation has positive real part then $g(t)$ is unbounded so $\int_0^{\infty} |g(\tau)|\, d\tau$ is infinite therefore the system is unstable.

(c) If the characteristic equation has repeated roots on the imaginary axis ($j\omega$-axis) the $g(t)$ is unbounded so $\int_0^{\infty} |g(\tau)|\, d\tau$ is infinite therefore the system is unstable.

(d) If one or more non-repeated roots of the characteristic equation are on the imaginary axis, then $g(t)$ is bounded but $\int_0^{\infty} |g(\tau)|\, d\tau$ is infinite therefore the system is unstable.

The stability of the linear time invariant (LTI) SISO system can be determined by checking the location of the roots of the characteristic equation of the system. For all practical purposes, there is no need to compute the complete system response to determine stability. The regions of stability and instability in the s plane are illustrated in Fig. 6.2.

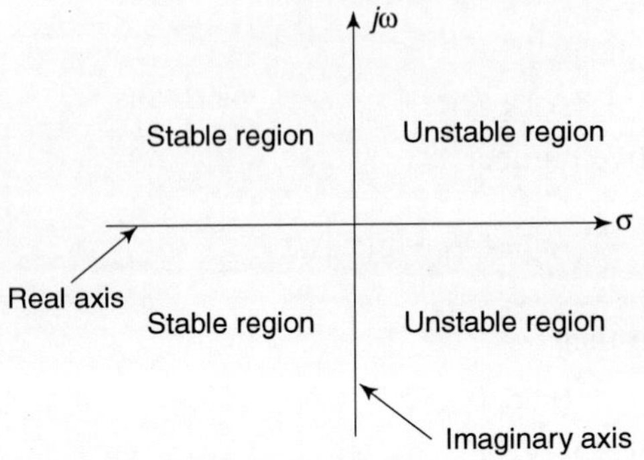

Figure 6.2 Stable and unstable regions in *s*-plane

Example 6.2.1 Explain whether the following systems are stable or unstable.

(a) $T(s) = \dfrac{20}{(s+1)(s+3)(s+5)}$

(b) $T(s) = \dfrac{10(s+1)}{(s-2)(s^2+2s+2)}$

(c) $T(s) = \dfrac{20(s-1)}{(s+2)(s^2+4)}$

(d) $T(s) = \dfrac{10}{(s^2+2)^2(s+10)}$

(e) $T(s) = \dfrac{10}{s^5+30s^2+s^2+10s}$

Solution

(a) All the roots are negative and real so the system in stable.
(b) One pole at s = 2 is positive real so the system is unstable.
(c) One root is in complete conjugate pair, i.e., $s = \pm 2j$ so the system is marginally stable.
(d) There is multiple or repeated pole at $s = \pm 2j$ on imaginary axis so the system is unstable.
(e) Stable if the pole at $s = 0$ is placed intentionally.

6.3 NECESSARY CONDITION FOR STABILITY

The stability of the system can be obtained by inspecting the characteristic equation in polynomial form. In this section we will explain the necessary but not sufficient condition for stability of a linear system.

1. All the coefficients of its characteristic equation $q(s) = 0$, should be real and have the same sign. Let the characteristic equation be

$$q(s) = a_n s^n + a_{n-1} s^{n-1} + a_{n-2} s^{n-2} + \ldots a_0 = 0 \qquad \ldots(6.11)$$

By Eq. (6.3) all the coefficients (a_j) must be positive

where $i = 1, 2, 3, \ldots n$ so we can say that all the coefficients of characteristic equation must be real and positive.

2. If some coefficients are zero or negative it can be concluded that system is not stable. The condition 1 is necessary and also sufficient condition for up to second order system. But condition 2 is not sufficient for the stability of the system if the system is of third or higher order because in the case of third and higher order it is not necessary that the positiveness of the coefficients of the characteristic equation ensure the negativeness of real roots but does not ensure the negativeness of the real parts of the complex roots. So the second condition cannot be a sufficient condition for stability of third and higher order systems.

Let us consider the system

$$s^3 + s^2 + 2s + 8 = 0 \qquad \ldots(6.12)$$

factor Eq. (6.12) is

$$(s+2)\left[s - \frac{1}{2} + j\frac{\sqrt{15}}{2}\right]\left[s - \frac{1}{2} + j\frac{\sqrt{15}}{2}\right] = 0 \qquad \ldots(6.13)$$

In Eq. (6.13) the real part of the complex roots is positive indicating instability of the system.

6.4 ROUTH–HURWITZ CRITERION

In the previous section we have seen the necessary condition for the stable system but these conditions are not sufficient for third and higher order systems. So we need some other conditions for the higher order systems for the stability of the system. So the Routh–Hurwitz criterion is a necessary and sufficient criterion for the stability of the linear systems. The method was actually developed in terms of determinants, but we shall utilize the more convenient array formulation. This array is based upon the characteristic equation. Let us consider the characteristic equation for the polynomial is

$$a_n s^n + a_{n-1} s^{n-1} + a_{n-2} s^{n-2} + \ldots\ldots a_0 \qquad \text{...(6.14)}$$

The above equation consists of the following alternate coefficients if n is even/odd

$$\begin{array}{llll} a_n & a_{n-2} & a_{n-4} & \text{(even/odd coefficients)} \\ a_{n-1} & a_{n-3} & a_{n-5} & \text{(odd/even coefficients)} \end{array}$$

The array will be formed which is known as Routh array.

$$\begin{array}{lllll} s_n & a_n & a_{n-2} & a_{n-4} & \ldots\ldots\ldots \\ s^{n-1} & a_{n-1} & a_{n-3} & a_{n-5} & \ldots\ldots\ldots \\ s^{n-2} & b_1 & b_3 & b_5 & \ldots\ldots\ldots \\ s^{n-3} & c_1 & c_3 & c_5 & \ldots\ldots\ldots \\ | & & & & \\ s^1 & d_1 & d_3 & d_5 & \ldots\ldots\ldots \\ s^0 & e_1 & e_3 & e_5 & \ldots\ldots\ldots \end{array} \qquad \text{...(6.15)}$$

In the above array the coefficients b_1, b_3, b_5 and c_1, c_3, c_5, etc. are related to the coefficient

$$b_1 = \frac{a_{n-1}\, a_{n-2} - a_n\, a_{n-3}}{a_{n-1}}$$

$$b_3 = \frac{a_{n-1}\, a_{n-4} - a_n\, a_{n-5}}{a_{n-1}},\ b_5 = 0$$

Similarly,

$$c_1 = \frac{b_1\, a_{n-3} - a_{n-1}\, b_3}{b_1}$$

and

$$c_3 = \frac{b_1\, a_{n-5} - a_{n-1}\, b_5}{b_1},\ c_5 = 0 \text{ etc.}$$

Now by examining Eq. (6.15) we conclude that:

1. All the terms of first column of Routh array must be positive for a stable system.
2. When we examine the first column of the array if there is change of sign, then the number of changes of sign indicates the number of positive real part roots of the polynomial so the system will be unstable.
3. While applying Routh–Hurwitz criterion it is imperative that no power of s in the characteristic equation is absent. Any absence of such powers indicate at least one positive real parts root so the system will be unstable.

4. If the characteristic equation contains either only odd or even powers of s, this indicates that the roots have no real parts and possess only imaginary parts and therefore the system has sustained oscillations in its output response.

Example 6.4.1 Consider the fourth order system with the characteristic equation

$$s^4 + 10s^3 + 16s^2 + 14s + 8 = 0$$

Determine the stability of the system.

Solution The Routh array is

s^4	1	16	8
s^3	(10)	14	0
s^2	b_1	b_3	b_5
s^1	c_1	c_3	c_5
s^0	d_1	d_3	d_5

So

$$b_1 = \frac{(10\times16)-(1\times14)}{10} = \frac{160-14}{10} = 14.6$$

$$b_3 = \frac{(10\times8)-(1\times0)}{10} = 8,\ b_5 = 0$$

Similarly,

$$C_1 = \frac{(b_1\times14)-(10\times b_3)}{b_1} = \frac{(14.6\times14)-(10\times8)}{14.6}$$

$$C_1 = \frac{204.4-80}{14.6} = \frac{124.4}{14.6} = 8.5$$

$$C_3 = \frac{(b_1\times0)-(10\times b_5)}{b_1} = 0,\ C_5 = 0$$

Similarly,

$$d_1 = \frac{c_1\,b_3 - b_1\,c_3}{c_1} = \frac{(8.5\times8)-(14.6\times0)}{8.5} = 8$$

$$d_3 = \frac{c_1\,b_5 - b_1\,c_5}{c_1} = 0$$

$$d_5 = 0$$

Now the array is

s^4	1	16	8
s^3	10	14	0
s^2	14.6	8	0
s^1	8.5	0	0
s^0	8	0	0

All elements of first column of the array are positive so the system is stable.

Example 6.4.2 Determine the stability of the system whose overall transfer function is given below.

$$T_o(s) = \frac{5s+2}{2s^5+2s^4+2s^3+4s^2+5s+10}$$

Solution The characteristic equation is $2s^5 + 2s^4 + 2s^3 + 4s^2 + 5s^1 + 10 = 0$

The Routh array is

$$\begin{array}{llll} s^5 & 2 & 2 & 5 \\ s^4 & \textcircled{2} & 4 & 10 \\ s^3 & b_1 & b_3 & b_5 \\ s^2 & c_1 & c_3 & c_5 \\ s^1 & d_1 & d_3 & d_5 \\ s^0 & e_1 & e_3 & e_5 \end{array}$$

Now $$b_1 = \frac{(2\times 2)-(2\times 4)}{2} = \frac{4-8}{2} = -2$$

$$b_3 = \frac{(2\times 5)-(2\times 10)}{2} = \frac{10-20}{2} = -5$$

$$b_5 = 0$$

Similarly, $$c_1 = \frac{(b_1\times 4)-(2\times b_3)}{b_1} = \frac{-8+10}{-2} = -1$$

$$c_3 = \frac{(b_1\times 10)-(2\times b_5)}{b_1} = \frac{-20-0}{-2} = 10$$

$$c_5 = 0$$

Similarly, $$d_1 = \frac{(c_1\, b_3 - b_1\, c_3)}{c_1} = \frac{(-1\times -5)-(-2\times 10)}{-1}$$

$$d_1 = \frac{5+20}{-1} = -25$$

$$d_3 = \frac{c_1\, b_5 - b_1\, c_5}{c_1} = \frac{(-1\times 0)-(-5\times 0)}{-1} = 0$$

$$d_5 = 0$$

and $$e_1 = \frac{d_1\, c_3 - c_1\, d_3}{d_1} = \frac{(-25\times 10)-(-1\times 0)}{-25} = 10$$

$$e_3 = \frac{d_1\, c_5 - c_1\, d_5}{d_1} = \frac{(-25\times 0)-(-1\times 0)}{-25} = 0$$

So now the array is

s^5	2	2	5
s^4	2	4	10

Sign change

s^3	– 2	– 5	0
s^2	– 1	10	0
s^1	– 25	0	0

Sign change

s^0	10	0	0

Now by examining the first column of the array in this sign changes 2 times so there is two positive real roots so the system is unstable.

6.5 LIMITATIONS OF ROUTH–HURWITZ CRITERION

When we apply Routh–Hurwitz criterion certain difficulties arise casuing the breakdown or failure of the Routh's test. The difficulties occurred are generally of the following type :

- All the elelmets in one of the rows of Routh array are zero.
  This situation is overcome by replacing the row of zeros in the Routh array by a row of coefficients of the polynomial generated by taking the first derivative of the auxiliary equation.

$$\frac{dA(s)}{ds} = \text{derivative of auxiliary Eq. } A(s)$$

Example 6.5.1 Determine the stability of the system having following characteristic equation.

$$s^6 + s^5 + 5\,s^4 + 3\,s^3 + 2\,s^2 - 4s - 8 = 0$$

Solution The Routh array is

s^6	1	5	2	– 8	
s^5	1	3	– 4	0	
s^4	2	6	– 8	0	Auxiliary equation
s^3	0	0	0	0	All elements are zero
s^2					
s^1					
s^0					

The auxiliary equation is

i.e., $$A(s) = 2\,s^4 + 6\,s^2 - 8$$

$$\frac{dA(s)}{ds} = 8\,s^3 + 12\,s - 0$$

The coefficients of the fourth row are 8, 12 and 0 and calculate the elements of other row as previous method. So the modified array is

$$\begin{array}{lcccc}
s^6 & 1 & 5 & 2 & -8 \\
s^5 & 1 & 3 & -4 & 0 \\
s^4 & 2 & 6 & -8 & 0 \\
s^3 & \textcircled{8} & 12 & 0 & 0 \\
s^2 & 3 & -8 & 0 & 0 \\
s^1 & 33.3 & 0 & 0 & 0 \\
s^0 & -8 & 0 & 0 & 0
\end{array}$$

By examining the first colum in which the sign changes one time so there is one root with positive real part so the system is unstable.

2. When the first term is any row of the Routh array is zero while the rest of the row has at least one non zero term. This difficulty can be overcome by replacing this zero by small positive number ε and examine the first column of Routh array by taking lim ε → 0.

Example 6.5.2 Determine the stability of a closed loop control system whose characteristic Eq. is $s^5 + s^4 + 2s^3 + 2s^2 + 11s + 10 = 0$

Solution The Routh array is formed

$$\begin{array}{lccc}
s^5 & 1 & 2 & 11 \\
s^4 & 1 & 2 & 10 \\
s^3 & b_1 & b_3 & b_5 \\
s^2 & c_1 & c_3 & c_5 \\
s^1 & d_1 & d_3 & d_5 \\
s^0 & e_1 & e_3 & e_5
\end{array}$$

Now $$b_1 = \frac{(1\times 2)-(1\times 2)}{1} = 0$$

$$b_3 = \frac{(1\times 11)-(1\times 10)}{1} = 1$$

$$b_5 = 0$$

In above array the first element of third row is zero so replacing it with ε and continue the same method.

s^5	1	2	11
s^4	1	2	10
s^3	ε	1	0
s^2	c_1	c_3	c_5
s^1	d_1	d_3	d_5
s^0	e_1	e_3	e_5

in the above array

$$c_1 = \frac{(\varepsilon \times 2) - (1 \times 1)}{\varepsilon} = \lim_{\varepsilon \to 0} \left(\frac{2\varepsilon_1 - 1}{\varepsilon} \right) = -\infty$$

$$c_3 = \frac{(10\,\varepsilon - 0) - (1 \times 1)}{\varepsilon} = 10$$

$$c_5 = 0$$

Now
$$d_1 = \frac{c_1 \times 1 - \varepsilon\, c_3}{c_1} = \lim_{\varepsilon \to 0} \left(1 - \frac{10\,\varepsilon^2}{2\,\varepsilon - 1} \right) = 1$$

$$d_3 = 0$$
$$d_5 = 0$$

Similarly,
$$e_1 = \frac{d_1\, c_3 - c_1\, d_3}{d_1} = \frac{(1 \times 10) - \left(\frac{2\varepsilon - 1}{\varepsilon} \right) \times 0}{1} = 10$$

Now the array is

s^5	1	2	11
s^4	1	2	10
s^3	ε	1	0
s^2	$-\infty$	10	0
s^1	1	0	0
s^0	10		

By examining the first column in which the sign changes two times so there are two positive real roots so the system is unstable.

3. It cannot be applied to discrete time system.

- *Routh–Hurwitz criterion tells about the absolute stability that means it tells only whether the system is stable or unstable. It does not give any idea about relative stability, i.e., degree of stability.*

EXAMPLES

Routh–Hurwitz Criterion

Example 6.1 Determine the stability of the system whose characteristic equation is

$$s^4 + 2s^3 + 3s^2 + 4s + 5 = 0$$

Solution The Routh's tabulation is

s^4	1	3	5
s^4	(2)	4	5
s^2	(a_1)	a_2	a_3
s^1	(b_1)	b_2	b_3
s^0	c_1	c_2	c_3

Now
$$a_1 = \frac{(2\times 3)-(4\times 1)}{2} = \frac{6-4}{2} = 1$$

$$a_2 = \frac{(2\times 5)-0}{2} = \frac{10}{2} = 5$$

$$a_3 = 0$$

Similarly,
$$b_1 = \frac{(a_1\times 4)-(a_2\times 2)}{a_1} = \frac{(1\times 4)-(5\times 2)}{1}$$

$$b_1 = \frac{4-10}{1} = -6$$

$$b_2 = \frac{(a_1\times 0)-(a_3\times 2)}{\varepsilon_1} = 0$$

$$b_3 = 0$$

Similarly,
$$c_1 = \frac{(b_1\times a_2)-(b_2\times a_1)}{b_1} = \frac{(-6\times 5)-0}{-6} = 5$$

$$c_2 = 0$$

$$c_3 = 0$$

Now the Routh's tabulation is

	s^4	1	3	5
	s^3	2	4	0
	s^2	1	5	0
Sign change	s^1	– 6	0	0
	s^0	5	0	0

As the number of changes of sign in the first column of the tabulation is two, therefore two roots of the given equation have positive real parts or located in the right half of s-plane. Hence, the system is unstable.

Example 6.2 Determine the stability of the system whose characteristic equation is

$$s^4 + 8s^3 + 18\,s^2 + 16s + 5 = 0$$

Solution Routh's tabulation is

s^4	1	18	5
s^3	8	16	0
s^2	a_1	a_2	a_3
s^1	b_1	b_2	b_3
s^0	c_1	c_2	c_3

Now
$$a_1 = \frac{(8\times 18)-(16\times 1)}{8} = \frac{144-16}{8} = 16$$

$$a_2 = \frac{(8\times 5)-(1\times 0)}{8} = \frac{40}{8} = 5$$

$$a_3 = 0$$

Similarly,
$$b_1 = \frac{(a_1\times 16)-(a_2\times 8)}{a_1} = \frac{(16\times 16)-(5\times 8)}{16} = \frac{256-40}{16}$$

$$b_1 = \frac{27}{2}$$

$$b_2 = \frac{(a_1\times 0)-(a_2\times 8)}{16} = 0$$

Similarly,
$$c_1 = \frac{(b_1\times a_2)-(b_2\times a_1)}{b_1} = \frac{\left(\frac{27}{2}\times 5\right)-(0\times 16)}{\frac{27}{2}}$$

$$c_1 = 5$$

$$c_2 = 0$$

Now the Routh's tabulation is

s^4	1	18	5
s^3	8	16	0
s^2	16	5	0
s^1	27/2	0	0
s^0	5	0	0

As all the elements of the first column are positive, so all the roots of equation are located in the left half plane, hence the system is stable.

Example 6.3 Determine the stability of a closed loop control system whose characteristic equation is $2s^5 + s^4 + 6s^3 + 3s^2 + s + 1 = 0$.

Solution The Routh's tabulation is

s^5	2	6	1
s^4	(1)	3	1
s^3	a_1	a_2	a_3
s^2	b_1	b_2	b_3

Now
$$a_1 = \frac{(1\times 6)-(3\times 2)}{1} = \frac{6-6}{1} = 0$$

$$a_2 = \frac{(1\times 1)-(1\times 2)}{1} = -1$$

$$a_3 = 0$$

While forming the Routh table as above the third element in the first column is zero and thus the Routh's criterion fails at this stage. The difficulty is solved if zero in the third row of first column is replaced by a symbol ε and table is formed as follows:

s^5	2	6	1
s^4	1	3	1
s^3	ε	-1	0
s^2	b_1	b_2	b_3
s^1	c_1	c_2	c_3
s^0	d_1	d_2	d_3

Now
$$b_1 = \frac{3\varepsilon+1}{\varepsilon} = \lim_{\varepsilon\to 0}\frac{3\varepsilon+1}{\varepsilon} = \infty$$

$$b_2 = \frac{\varepsilon-0}{\varepsilon} = 1$$

$$c_1 = \frac{(b_1\times -1)-(b_2\times\varepsilon)}{b_1} = \frac{\left(\frac{3\varepsilon+1}{\varepsilon}\times -1\right)-(1\times\varepsilon)}{\frac{3\varepsilon+1}{\varepsilon}}$$

$$c_1 = \frac{-3\varepsilon-1-\varepsilon^2}{3\varepsilon+1}$$

As $\lim \varepsilon \to 0$ $\quad c_1 = \lim_{\varepsilon \to 0} \dfrac{-3\varepsilon - 1 - \varepsilon^2}{3\varepsilon + 1} = -1$

$$c_2 = 0$$

and $\quad c_3 = 0$

Similarly, $\quad d_1 = \dfrac{(c_1 \times b_2) - (b_1 \times c_2)}{c_1} = \dfrac{\left(\dfrac{-3\varepsilon - 1 - \varepsilon^2}{3\varepsilon + 1}\right) \times 1 - \left(\dfrac{3\varepsilon + 1}{\varepsilon}\right) \times 0}{\dfrac{-3\varepsilon - 1 - \varepsilon^2}{3\varepsilon + 1}}$

$$d_1 = 1$$

Now the Routh's tabulation is

s^5	2	6	1
s^4	1	3	1
s^3	ε	-1	0
s^2	∞	1	0
s^1	-1	0	0
s^0			

Sign changes (between s^2 and s^1, and between s^1 and s^0)

There are two sign changes in the first column elements, Hence, the system is unstable.

Example 6.4 The open loop transfer function of a unity feedback control system is given by

$$G(s) = \frac{K}{(s+2)(s+4)(s^2+6s+25)}$$

By applying the Routh criterion, discuss the stability of the closed loop system as a function of K which will cause sustained oscillation in the system. Find corresponding oscillations frequencies. *(U.P.T.U., 2006–07, CS)*

Solution $\quad G(s) = \dfrac{K}{(s+2)(s+4)(s^2+6s+25)}$

The characteristic equation is given by

$$1 + G(s)\,H(s) = 0$$

$$1 + \frac{K}{(s+2)(s+4)(s^2+6s+25)} = 0$$

$$(s+2)(s+4)(s^2+6s+25) + K = 0$$

$$s^4 + 12s^3 + 69s^2 + 198s + 200 + K = 0$$

The Routh's tabulation is

$$\begin{array}{llll} s^4 & 1 & 69 & 200+K \\ s^3 & \textcircled{12} & 198 & 0 \\ s^2 & a_1 & a_2 & a_3 \\ s^1 & b_1 & b_2 & b_3 \\ s^0 & c_1 & c_2 & c_3 \end{array}$$

Now $$a_1 = \frac{(12 \times 69) - (198 \times 1)}{12} = \frac{828 - 198}{12} = 52.5$$

$$a_2 = \frac{12\,(200 + K) - 0}{12} = 200 + K$$

$$a_3 = 0$$

Similarly, $$b_1 = \frac{(a_1 \times 198) - (a_2 \times 12)}{a_1}$$

$$= \frac{(52.5 \times 198) - (200 + K)\,12}{52.5}$$

$$b_1 = 198 - \frac{12\,(200 + K)}{52.5}$$

$$b_2 = 0$$

Similarly, $$c_1 = \frac{(b_1 \times a_2) - (a_1 \times b_2)}{b_1} = \frac{\left[198 - \frac{12\,(200+K)}{52.5}\right] \times (220 + K)}{198 - \frac{12\,(200+K)}{52.5}}$$

$$c_1 = 200 + K$$

Now the Routh's tabulation is

$$\begin{array}{llll} s^4 & 1 & 69 & 200+K \\ s^3 & 12 & 198 & 0 \\ s^2 & 52.5 & 200+K & 0 \\ s^1 & 198 - \frac{12\,(200+K)}{52.5} & & \\ s^0 & 200+K & & \end{array}$$

The system will be stable with

$$200 + K > 0$$

$$K > -200$$

$$198 - \frac{12\,(200 + K)}{52.5} > 0$$

$$\frac{12\,(200 + K)}{52.5} < 198$$

$$200 + K < \frac{198 \times 52.5}{12}$$

$$200 + K < 866.25$$

$$K < 866.25 - 200$$

$$K < 666.25$$

Oscillation will occur when

$$K = 666.25$$

Auxiliary equation

$$52.5\,s^2 + (200 + K) = 0$$

$$52.5\,s^2 + (200 + 666.25) = 0$$

$$52.5\,s^2 = -\,866.25$$

$$s^2 = -\,16.5$$

$$s = \pm\, j\,4.06$$

$$j\omega = \pm\, j\,4.06$$

$\therefore$ Frequency of sustained oscillation is 4.06 rad/sec. **Ans.**

Example 6.5 For the following characteristic equation of systems find their stability using Routh's criterion.

(i) $13s^7 + 5s^6 + 8s^5 + 4s^3 + 20s^2 + 15s + 1 = f(s)$

(ii) $3s^4 + 10s^3 + 5s^2 + 5s + 2 = f(s)$

(iii) $3s^4 + 10s^3 + 20s + 10\,(K + 1) = f(s)$ *(U.P.T.U., 2005–06, CS)*

Solution

(i) By the inspection of polynomial characteristics it is clear that system is unstable due to the coefficient of element s^4 is zero.

(ii) $f(s) = 3s^4 + 10s^3 + 5s^2 + 5s + 2$

The Routh's tabulation is

s^4	3	5	2
s^3	(10)	5	0
s^2	a_1	a_2	a_3
s^1	b_1	b_2	b_3
s^0	c_1	c_2	c_3

Now $$a_1 = \frac{(10 \times 5) - (5 \times 3)}{10} = \frac{50 - 15}{10} = 3.5$$

$$a_2 = \frac{(10 \times 2) - (0 \times 3)}{10} = 2$$

$$a_3 = 0$$

Similarly, $$b_1 = \frac{(a_1 \times 5) - (a_2 \times 10)}{a_1} = \frac{(3.5 \times 5) - (2 \times 10)}{3.5} = 0.71$$

$$b_2 = 0$$

Similarly, $$c_1 = \frac{(b_1 \times c_2) - (b_2 \times a_1)}{b_1} = \frac{(-0.71 \times 2) - (0 \times 3.5)}{-0.71} = 2$$

$$c_2 = 0$$

and $$c_3 = 0$$

Now the Routh's tabulation is

	s^4	3	5	2
	s^3	10	5	0
	s^2	3.5	2	0
Sign changes	s^1	− 0.71	0	0
	s^0	2	0	0

By examining the first column the sign changes twice so the two roots lie in right half s plan so the system is unstable.

(iii) $f(s) = 3s^4 + 10s^3 + 20s + 10\,(1 + K)$

In this characteristic equation coefficient of s^2 is zero so the system is unstable.

Example 6.6 Determine the stability of a system having following characteristic equation

$$s^5 + 4s^4 + 8s^3 + 8s^2 + 7s + 4 = 0$$

Solution The Routh's tabulation is

s^5	1	8	7
s^4	4	8	4
s^3	a_1	a_2	a_3
s^2	b_1	b_2	b_3
s^1	c_1	c_2	c_3
s^0	d_1	d_2	d_3

Now $$a_1 = \frac{(4 \times 8) - (8 \times 1)}{4} = \frac{32 - 8}{4} = 6$$

$$a_2 = \frac{(4\times 7)-(4\times 1)}{4} = \frac{28-4}{4} = 6$$

$$a_3 = 0$$

Similarly, $$b_1 = \frac{(a_1\times 8)-(a_2\times 4)}{a_1} = \frac{(6\times 8)-(6\times 4)}{6} = \frac{48-24}{6}$$

$$b_1 = 4$$

$$b_2 = \frac{(a_1\times 4)-(a_3\times 4)}{a_1} = \frac{(6\times 4)-(0\times 4)}{6} = 4$$

$$b_3 = 0$$

Similarly, $$c_1 = \frac{(b_1\times a_2)-(a_1\times b_2)}{b_1} = \frac{(4\times 6)-(6\times 4)}{4}$$

$$c_1 = 0$$

$$c_2 = \frac{(b_1\times a_3)-(b_3\times a_1)}{b_1} = \frac{(4\times 0)-(0\times 6)}{4}$$

$$c_2 = 0$$

and $$c_3 = 0$$

The Routh's tabulation is

s^5	1	8	7
s^4	4	8	4
s^3	6	6	0
s^2	4	4	0
s^1	0	0	0
s^0	K		

It is observed that all the elements in the fifth row are zero, and the application of Routh criterion fails. This situation occurs when the array has two consecutive rows having the same ratio of corresponding elements.

This difficulty is overcome by forming an auxiliary equation using elements of the last auxiliary equatio is taken w.r t. s and the coefficients of the differentiated equation are taken as the elements of the following row.

The auxiliary equation is

$$A(s) = 4s^2 + 4 = 0$$

$$\frac{dA(s)}{ds} = 8s$$

So now the Rout tabulation is

$$\begin{array}{llll} s^5 & 1 & 8 & 7 \\ s^4 & 4 & 8 & 4 \\ s^3 & 6 & 6 & 0 \\ s^2 & 4 & 4 & 0 \\ s^1 & 8 & 0 & 0 \\ s^0 & 4 & & \end{array}$$

Since there are no sign changes in the first column of the entire Routh's tabulation and the auxiliary equation is:

$$4s^2 + 4 = 0 \text{ gives two roots at } s = \pm j$$

Thus, the characteristic equation has two roots on the $j\omega$-axis, Hence, the system is marginally stable.

Example 6.7 The characteristic equation of a control system is given as

$$s^3 + 4Ks^2 + (K + 5)\, s + 10 = 0.$$

Determine the range of values of K for stability. *(U.P.T.U., 2003–04, CS)*

Solution The Routh's tabulation is

$$\begin{array}{lll} s^3 & 1 & K + 5 \\ s^2 & \boxed{4K} & 10 \\ s^1 & a_1 & a_2 \\ s^0 & b_1 & b_2 \end{array}$$

$$a_1 = \frac{4K\,(K+5) - 10}{4K}$$

$$a_2 = 0$$

Similarly,

$$b_1 = \frac{(a_1 \times 10) - (a_2 \times 4K)}{a_1} = \frac{\left[\frac{[4K\,(K+5) - 10}{4K} \times 10\right] - 0}{\frac{4K\,(K+5) - 10}{4K}}$$

$$b_1 = 10$$

$$b_2 = 0$$

Now the Routh's tabulation is

$$\begin{array}{lcc} s^3 & 1 & K + 5 \\ s^2 & 4\,K & 10 \\ s^1 & \dfrac{4\,K\,(K+5) - 10}{4K} & 0 \\ s^0 & 10 & 0 \end{array}$$

For the stable system the coefficient of the first column should be all positive or all negative.

So, $$4K > 0$$

$$\Rightarrow \quad K > 0$$

and $$\frac{4K(K+5)-10}{4K} > 0$$

$$4K(K + 5) - 10 > 0$$

$$4K(K + 5) - 10 > 0$$

$$4K^2 + 20K - 10 > 0$$

$$K^2 + 5K - 2.5 > 0$$

$$K = \frac{-b \pm \sqrt{b^2 - 4ac}}{2a}$$

$$K = \frac{-5 \pm \sqrt{25+10}}{2} = \frac{-5 \pm \sqrt{35}}{2}$$

$$K = 0.45$$

and $$K = -5.45$$

If $K = -5.45$, then term $4K$ becomes (–) ve, so it will become unstable.

If $0 < K < 0.45$, then term $\frac{4K(K+1)-10}{4K}$ will become (–ve), so the system will become unstable.

If $K > 0.45$, then term $4K$ and $\frac{4K(K+5)-10}{4K}$ will be (+ve), so the system will be stable so far stability the range of K is

$$K > 0.45 \quad \textbf{Ans.}$$

Example 6.8 For a unity feedback system with a forward path transfer function

$$G(s) = \frac{2K}{s^3 + 4s^2 + 5s + 2}$$

Find the range of K to make the system stable. *(U.P.T.U., 2002–03, CS)*

Solution The characteristic equation is

$$1 + G(s)H(s) = 0$$

$$1 + \frac{2K}{s^3 + 4s^2 + 5s + 2} = 0$$

$$s^3 + 4s^2 + 5s + 2 + 2K = 0$$

The Routh's tabulation is

s^3	1	5
s^2	4	$2 + 2K$
s^1	a_1	a_2
s^0	b_1	b_2

Now

$$a_1 = \frac{(4 \times 5) - (2 + 2K) \times 1}{4}$$

$$a_1 = \frac{20 - (2 + 2K)}{4}$$

$$a_2 = 0$$

$$b_1 = \frac{\left[\dfrac{20 - (2 + 2K)}{4} \times (2 + 2K)\right] - 0}{\dfrac{20 - (20 + 2K)}{4}}$$

$$b_1 = 2 + 2K$$

$$b_2 = 0$$

Now the tabulation is

s^3	1	5
s^2	4	$2 + 2K$
s^1	$\dfrac{20 - (2 + 2K)}{4}$	0
s^0	$2 + 2K$	0

For stable system

$$2 + 2K > 0$$

$$2K > -2$$

$$K > -1$$

and

$$\frac{20 - (2 + 2K)}{4} = 0$$

$$2 + 2K < 20$$

$$2K < 20 - 2$$

$$2K < 18$$

$$K < 9$$

The range of value of K for stability is

$$-1 < K < 9 \quad \textbf{Ans.}$$

Example 6.9 Using Routh's stability criterion ascertain stability for the system whose characteristic equation is $s^6 + 2s^5 + 8s^4 + 12s^3 + 20s^2 + 16s + 16 = 0$.

Solution The Routh's array is

s^6	1	8	20	16
s^5	2	12	16	0
s^4	a_1	a_2	a_3	a_4
s^3	b_1	b_2	b_3	b_4
s^2	c_1	c_2	c_3	c_4

Now $$a_1 = \frac{(2\times 8)-(12\times 1)}{2} = \frac{16-12}{2} = 2$$

$$a_2 = \frac{(2\times 20)-(16\times 1)}{2} = \frac{40-16}{2} = 12$$

$$a_3 = \frac{(2\times 16)-(0\times 1)}{2} = 16$$

Similarly, $$b_1 = \frac{(a_1\times 12)-(a_2\times 2)}{a_1} = \frac{(2\times 12)-(12\times 2)}{2}$$

$$b_1 = \frac{24-24}{2} = 0$$

$$b_2 = \frac{(a_1\times 16)-(a_3\times 2)}{a_1} = \frac{(2\times 16)-(16\times 2)}{2} = 0$$

Now all the elements of fourth row are zero so the auxiliary equation is

$$A(s) = 2s^4 + 12s^2 + 16$$

$$\frac{dA(s)}{d(s)} = 8s^3 + 24s$$

So, $$b_1 = 8$$

and $$b_2 = 24$$

Now $$c_1 = \frac{(b_1\times a_2)-(b_2\times a_1)}{b_1} = \frac{(8\times 12)-(24\times 2)}{8}$$

$$c_1 = \frac{96-48}{8} = 6$$

$$c_2 = \frac{(b_1\times a_3)-(b_3\times a_1)}{b_1} = \frac{(8\times 16)-(0)}{8}$$

$$c_2 = 16$$

$$c_3 = 0$$

Similarly, $$d_1 = \frac{(c_1\times b_2)-(c_2\times b_1)}{c_1} = \frac{(6\times 24)-(16\times 8)}{6}$$

$$d_1 = \frac{144 - 128}{6} = \frac{16}{6} = \frac{8}{3}$$

Similarly, $$e_1 = \frac{(d_1 \times c_2) - (d_2 \times c_1)}{d_1} = \frac{\left(\frac{8}{3} \times 16\right) - (0 \times 6)}{\frac{8}{3}}$$

$$e_1 = 16$$

There is no sign change in the first column so the roots do not lies in right half s-plane.

The auxiliary equation is

$$2s^4 + 12s^2 + 16 = 0$$

$$s^4 + 6s^2 + 8 = 0$$

$$(s^2 + 4)(s^2 + 2) = 0$$

$$s = \pm 2j$$

and $$s = \pm\sqrt{2}j$$

It is seen that there is roots of equation lie on $j\omega$ axis. Therefore the system is marginally stable.

Example 6.10 Determine the values of K and b, so that the system whose open loop transfer function is

$$G(s) = \frac{K(s+1)}{s^3 + bs^2 + 3s + 1}$$

Assume unity feedback oscillates at a frequency of oscillation of 2 rad/sec.

Solution The characteristic equation is

$$1 + G(s)H(s) = 0$$

$$1 + \frac{K(s+1)}{s^3 + bs^2 + 3s + 1} \times 1 = 0$$

$$s^3 + bs^2 + 3s + 1 + K(s + 1) = 0$$

$$s^3 + bs^2 + (K + 3)s + (K + 1) = 0$$

The Routh's array is

s^3	1	$K + 3$
s^2	b	$K + 1$
s^1	$(K+3) - \frac{(K+1)}{b}$	0
s^0	$K + 1$	

The system will have sustained oscillation if row s^1 is zero.

i.e., $$\frac{b(K+3)-(K+1)}{b} = 0$$

$$b = \frac{K+1}{K+3} \qquad ...(i)$$

The auxiliary equation of row s^2 is

$$bs^2 + K + 1 = 0$$

$$s^2 = \frac{-(K+1)}{b}$$

$$(j\omega)^2 = \frac{-(K+1)}{b}$$

∴ $$\omega = 2 \text{ rad/sec.}$$

$$(j\omega)^2 = \frac{-(K+1)}{b}$$

$$-4 = \frac{-(K+1)}{b}$$

⇒ $$b = \frac{K+1}{4}$$

Putting $$b = \frac{K+1}{b}$$

in Eq. (i), we get

$$\frac{K+1}{4} = \frac{K+1}{K+3}$$

$$K + 3 = 4$$

$$K = 1 \quad \textbf{Ans.}$$

$$b = \frac{K+1}{4} = \frac{1+1}{4} = \frac{2}{4} = 0.5$$

$$b = 0.5 \quad \textbf{Ans.}$$

Example 6.11 The characteristic equation of a system in differential form is

$$\ddot{X} - (K+2)\dot{X} + (2K+5)X = 0$$

(i) Find the value of K for which the system is
 (1) stable (2) partially stable (3) unstable

(ii) For the stable case for what values of K system is
 (1) Underdamped (2) Overdamped (*U.P.T.U., 2005–06, ACS*)

Solution The characteristic equation is

$$\ddot{X} - (K+2)\dot{X} + (2K+5)x = 0$$

Taking Laplace transform

$$L\left\{\frac{d^2X}{ds}\right\} - (K+2)\,L\left\{\frac{dX}{ds}\right\} + (2K+5)\,L\,\{X\} = 0$$

$$S^2\,X(s) - (K+2)\,s + (2K+5)\,X(s) = 0$$

$$s^2 - (K+2)\,s + (2K+5) = 0$$

Now the Routh's array is

s^2	1	$2K + 5$
s^1	$-(K+2)$	0
s^0	a_1	a_2

Now
$$a_1 = \frac{-(K+2)(2K+5) - 0}{-(K+2)}$$

$$a_1 = 2\,K + 5$$

The Routh's tabulation is

s^2	1	$2\,K + 5$
s^1	$-(K+2)$	0
s^0	a_1	0

(A) 1. For stable system

$$-(K+2) > 0$$

$$K < -2$$

and
$$2\,K + 5 > 0$$

$$2\,K > -5$$

$$-5/2 < K < -2 \quad \textbf{Ans.}$$

2. For partial stable system

$$K = -5/2 \quad \text{or } K = -2 \quad \textbf{Ans.}$$

3. For unstable system

$$-5/2 > K > -2 \quad \textbf{Ans.}$$

(B) The roots of the characteristic equation are expressed as

$$s_1,\, s_2 = \frac{1}{2}\left\{(K+2) \pm \sqrt{(K+2)^2 - 4\,(2K+5)}\right.$$

For critically damped case

$$(K+2)^2 - 4\,(2K+5) = 0$$

$$K = 6.47, \quad -2.47$$

As per stability condition, $K = 6.47$ is unstable. So for critically damped case

$$K = -2.47$$

So,

1. Underdampled case

$$-2 > K > -2.47 \quad \textbf{Ans.}$$

2. Overdamped case

$$-2.47 > K > -2.5 \quad \textbf{Ans.}$$

Example 6.12 Use Routh's criterion to investigate the stability of the open loop system given by transfer function with unity feedback.

$$G(s) = \frac{K}{s(s+2)(s^2+3)} \qquad (U.P.T.U., 2004–05, ACS)$$

Solution The characteristic equation is

$$1 + G(s)\,H(s) = 0$$

$$1 + \frac{K}{s\,(s+2)\,(s^2+3)} \times 1 = 0$$

$$s\,(s+2)\,(s^2+3) + K = 0$$

$$s^4 + 2s^3 + 3s^2 + 6s + K = 0$$

Now the Routh's array is

s^4	1	3	K
s^3	2	6	0
s^2	a_1	a_2	a_3

So,

$$a_1 = \frac{(2\times 3)-(6\times 1)}{2} = \frac{6-6}{2} = 0$$

Because the first element in the Routh array is zero. Let us assume a small positive quantity, $\varepsilon \to 0$

s^4	1	3	K
s^3	2	6	0
s^2	ε	$\frac{2K}{2}$	0
s^1	$\frac{6\varepsilon - 2K}{\varepsilon}$	0	
s^0	K		

The

$$b_1 = \lim_{\varepsilon \to 0} \frac{6\varepsilon - 2K}{\varepsilon} = -2K$$

For any value of K whether it is positive or negative the system is unstable.

Example 6.13 Open loop transfer function of certain unity feedback system is

$$G(s) = \frac{K(s+1)}{s(s-1)(s+6)}$$

Determine:

(i) The range of value of K for which the system is stable.

(ii) The value of K that will result in the system being marginally stable.

(iii) The location of roots of the characteristic equation for the value of K found in terms (ii). *(U.P.T.U., 2003–04, ACS)*

Solution The characteristic Eq.

$$1 + G(s)\,H(s) = 0$$

$$1 + \frac{K(s+1)}{s(s-1)(s+6)} = 0$$

$$s(s-1)(s+6) + K(s+1) = 0$$

$$s^3 + 5s^2 - 6s + Ks + 1 = 0$$

$$s^3 + 5s^2 + (K-6)\,s + 1 = 0$$

The Routh's array is

s^4	1	$K-6$	K
s^3	5	1	0
s^2	$\dfrac{5(K-6)-1}{5}$	0	0
s^1	$\dfrac{\left[\dfrac{5(K-6)-1}{5}\right]\times 1}{\dfrac{5(K-6)-1}{5}} = 1$		

$$\frac{5(K-6)-1}{5} > 0$$

$$\Rightarrow \qquad \frac{5K-30-1}{5} > 0$$

$$5K > 31$$

$$K > 31/5$$

(i) For $31/5 < K < \infty$, the system is stable. **Ans.**

(ii) $K = 31/5$, the system is marginally stable. **Ans.**

(iii) $5\,s^2 + 1 = 0$

$$s^2 = -1/5$$

$$s = \pm j\,\frac{1}{\sqrt{5}}$$

The location of roots in the imaginary axis are $+\dfrac{j}{\sqrt{5}}$ and $-\dfrac{j}{\sqrt{5}}$ **Ans.**

Example 6.14 Find the range of K for stability of the closed loop system shown in Fig. Ex. 6.14 by Routh-Hurwitz criterion. *(U.P.T.U., 2004 (C.O), ACS)*

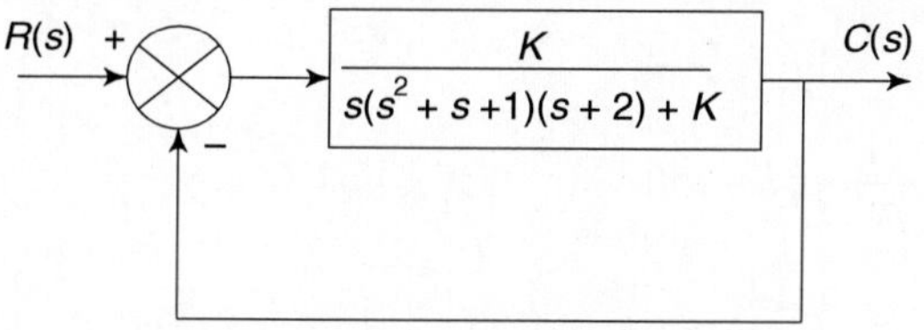

Figure Ex. 6.14

Solution

$$\frac{C(s)}{R(s)} = \frac{G(s)}{1+G(s)\,H(s)} = \frac{\dfrac{K}{s\,(s^2+s+1)\,(s+2)+K}}{1+\dfrac{K}{s\,(s^2+s+\;1)\,(s+2)+K}\times 1}$$

$$\frac{C(s)}{R(s)} = \frac{K}{s\,(s^2+s+1)\,(s+2)+K+K}$$

$$\frac{C(s)}{R(s)} = \frac{K}{s\,(s^3+s^2+s)\,(s+2)+2K}$$

$$\frac{C(s)}{R(s)} = \frac{K}{s^4+3s^3+3s^2+2s+2K}$$

So, the characteristic equation of the system is $s^4 + 3s^3 + 3s^2 + 2s + 2K = 0$

The Routh's array is

s^4	1	3	$2K$
s^3	3	2	0
s^2	7/3	$2K$	0
s^1	$\dfrac{14/3-6K}{7/3}$	0	
s^0	$2K$		

$\therefore$

$$a_1 = \frac{(3\times 3)-(2\times 1)}{3} = \frac{9-2}{3} = \frac{7}{3}$$

$$a_2 = \frac{(3\times 2K)-0}{3} = 2K$$

Similarly, $$b_1 = \frac{(7/3 \times 2) - (2K \times 3)}{7/3} = \frac{14/3 - 6K}{7/3}$$

$$b_2 = 0$$

Similarly, $$c_1 = 2K$$

For stable system

$$K > 0$$

and $$\frac{14}{3} - 6K > 0$$

$$\frac{14}{3} > 6K$$

$$K < \frac{14}{18}$$

So the range of K for stability is

$$0 < K < \frac{14}{18} \quad \textbf{Ans.}$$

Example 6.15 Find the number of poles in LHP and RHP and $j\omega$ axis for the system shown in Fig. Ex. 6.15 by Routh-Hurwitz criterion and comment on it stability.

(U.P.T.U., 2002–03, ACS)

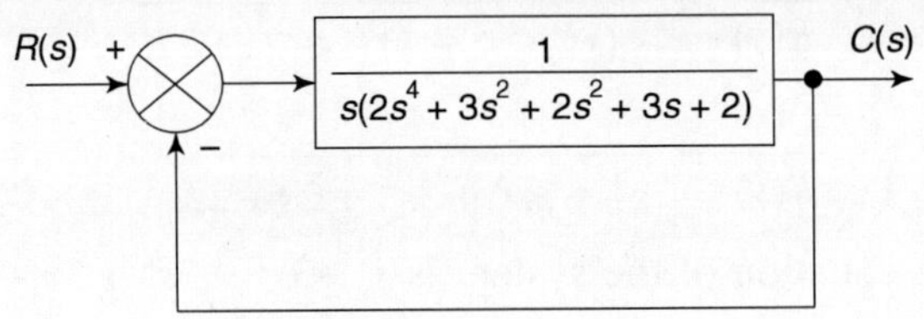

Figure Ex. 6.15

Solution $$G(s) = \frac{1}{S\,(2S^4 + 3S^3 + 2S^2 + 3S + 2)}$$

The characteristic equation is

$$1 + G(s)\,H(s) = 0$$

$$1 + \frac{1}{s\,(2s^4 + 3s^3 + 2s^2 + 3s + 2)} \times 1 = 0$$

$$2s^5 + 3s^4 + 2s^3 + 3s^2 + 2s + 1 = 0$$

Now the Routh's array is

$$
\begin{array}{lcccc}
s^5 & 2 & 2 & 2 & 0 \\
s^4 & \textcircled{3} & 3 & 1 & 0 \\
s^3 & \varepsilon & 4/3 & 0 & 0 \\
s^2 & \dfrac{3\varepsilon - 4}{\varepsilon} & 1 & 0 & 0 \\
s^1 & 4/3 - \dfrac{\varepsilon^2}{3\varepsilon - 4} & 0 & 0 & 0 \\
s^0 & 1 & 0 & 0 & 0
\end{array}
$$

$\therefore$
$$a_1 = \frac{(3 \times 2) - (3 \times 2)}{3} = \frac{6-6}{3} = 0 \text{ replace by } \varepsilon$$

$$a_2 = \frac{(3 \times 2) - (1 \times 2)}{3} = \frac{4}{3}$$

Similarly,
$$b_1 = \frac{(a_1 \times 3) - 4}{a_1} = \frac{3\varepsilon - 4}{\varepsilon}$$

$$b_2 = \frac{\varepsilon - 0}{\varepsilon} = 1$$

Similarly,
$$c_1 = \frac{(b_1 \times a_2) - (b_2 \times a_1)}{b_1} = \frac{\left(\dfrac{3\varepsilon - 4}{\varepsilon} \times \dfrac{4}{3}\right) - (1 \times \varepsilon)}{\dfrac{3\varepsilon - 4}{\varepsilon}}$$

$$c_1 = \frac{4}{3} - \frac{\varepsilon^2}{3\varepsilon - 4}$$

and
$$d_1 = \frac{(c_1 \times b_2) - (b_1 \times c_2)}{c_1} = \frac{(c_1 \times b_2) - 0}{c_1}$$

$$d_1 = b_2$$
$$d_1 = 1$$

Now
$$\lim_{\varepsilon \to 0} \left(\frac{3\varepsilon - 4}{\varepsilon}\right) = -4 \text{ (–ve)}$$

$$\lim_{\varepsilon \to 0} \left(\frac{4}{3} - \frac{\varepsilon^2}{3\varepsilon - 4}\right) = \frac{4}{3}$$

Number of poles = 5

Right hand poles = Number of sign changes = 2

Left hand poles = 5 – 2 = 3

$j\omega$ axis pole = 0

The system is unstable.

MATLAB PROGRAMS

Program 6.1

```
% Program to determine the poles of a transfer function in order to
% determine stability
% G(s) = 5/s(3s^4+6s^3+2s^2+12s+10)
clear all
clsc
numg = input('Enter the coefficient of polynomial of numerator:- ');
                                % input coefficient of
                                % polynomial of numerator like 5
deng = input('Enter the coefficient of polynomial of denominator:- ');
                                % input coefficient of
                                % denominator like [3 6 2 12 10 0]
G = tf(numg,deng)               % Create transfer function
'Closed-Loop'                   % Display String
T = feedback(G,1)               % Calculate close loop transfer function
poles = pole(T)                 % Find poles of T(s)
```

Input

```
Enter the coefficient of polynomial of numerator:- 5
Enter the coefficient of polynomial of denominator:- [3 6 12 10 0]
```

Output

```
                                   5
Transfer function:  ---------------------------------
                    3s^5 + 6s^4 + 2s^3 + 12s^2 + 10s

Closed-Loop

                                     5
Transfer function:  -------------------------------------
                    3s^5 + 6s^4 + 2s^3 + 12s^2 + 10s + 5

poles = -2.2625
         0.5654 + 1.2341i
        0.5654  - 1.2341i
        -0.4341 + 0.4597i
        -0.4341 - 0.4597i
```

Program 6.2

```
% Program to find the range of gain for stability
% G(s) =k/s(s+5)(s+7)
Clear all;
clc;
k = 1:1:1000;                 % Define range of k in step 1
for n = 1:length(k);          % set the length not loop
dent1 = input('Enter the roots of denominator:- ')
   dent2 = poly(dent1); % Define the denominator s(s+5)(s+7)
   dent = [dent2 k(n)];  % Define the denominator for nth order
   poles = roots(dent);
      r = real(poles)     % form a vector containing the real
                           parts of the poles for K(n)
        if max(r)>=0
        poles
        k==k(n)
        break
          end
          end
```

Input

```
Enter the roots of denominator:- [0 -5 -7]
```

Output

```
poles = -6.9897
        -5.0200
          0.0049 + 0.1687i
          0.0049 - 0.1687i
          K = 1
```

EXERCISE

6.1 Explain the concept of stability and necessary conditions for the stability.

6.2 A feedback system has an open loop transfer function of $G(s)\,H(s) = \dfrac{K(1-s)}{s(s^2+5s+9)}$.

Determine by the use of Routh criterion, the maximum value of K for the system to be stable.

6.3 Define the stability of the system in details.

6.4 What information about stability is provided by the Routh-Hurwitz criterion.

6.5 What are the limitations of Routh-Hurwitz criterion and how can we overcome by these limitations?

6.6 The characteristic equation of the system is given $s^4 + 10s^3 + 30s^2 + 100s + 25 = 0$. Determine the stability of the system.

6.7 Determine whether the system is stable or not whose characteristic equation is given

(a) $s^5 + 13s^4 + 54s^3 + 82\,s^2 + 170\,s + 330 = 0$

(b) $s^4 + 5^3 + 5s^2 + 4s + 4 = 0$

6.8 The open loop transfer function of a unity feedback control system is given by

$$G(s) = \frac{K}{(s+2)(s+4)(s^2+6s+2s)}.$$

Discuss the stability of the system by Routh criterion as a function of K. Determine K due to which sustained oscillation will occur. What are the corresponding oscillation frequencies.

6.9 Explain the relationship between pole positions and stability.

6.10 Using Routh's criterion, determine the number of roots in the right half s-plane for the characteristic equation $2s^4 + s^3 + 3s^2 + 5s + 10 = 0$.

6.11 Determine the range of K for which the system given by the following characteristic equations $s^3 + 10s^2 + (21 + K)\,s + 13\,K = 0$.

6.12 Determine the number of roots with positive real parts, zero real parts, negative real parts for the system characterized by the equation $s^5 + 2s^4 + 24\,s^3 + 48s^2 - 25s - 50 = 0$.

6.13 With the help of an example explain how Routh-Hurwitz criterion can be used in designing a linear control system. What are the situations under which Routh-Hurwitz criterion cannot be used to determine stability. *(U.P.T.U., 2007–08, CS)*

6.14 Explain Routh-Hurwitz criterion of stability. Also explain special cases in Routh criterion. *(C.P.T.U., 2004–05, CS)*

ANSWERS

6.6 Stable

6.7 (a) Unstable, (b) Marginally stable

6.8 $K = 666.25$; $\omega = 4.06$ rad/sec.

6.11 $0 < K < 70$

6.12 One, two and two

CHAPTER

7

Root Locus Technique

7.1 INTRODUCTION

In the previous chapter, we have seen the importance of the poles and zeros of the closed loop transfer function of a linear control system on the transient performance of the system. We have seen that the location of poles in s-plane plays an important role in the stability of the system. By the Routh-Hurwitz criterion we have got idea about the absolute stability of the system on the basis of the location of poles in s-plane but this criterion did not give any idea about relative stability, i.e., degree of stability.

So now we are proceeding in forward direction to knowing the stability. The root locus method was introduced by W.R. Evans in 1948, 1950 and has been developed and utilized extensively in control engineering. The root locus technique is a graphical representation for sketching the locus of the closed loop roots in the s-plane as a system parameter is varied. Basically, the root locus method provides a measure of the sensitivity of the roots of the system when the parameter is varied. The root locus technique may be used to great advantage in conjunction with the Routh-Hurwitz criterion.

This method provides graphical information and therefore an approximate sketch can be used to obtain qualitative information concerning the stability and performance of the system. The root locus of the characteristic equation of a multi-loop system may be investigated as readily as for a single loop system. If the locations of root are not satisfactory, the necessary parameter adjustments often can be done from the root locus. So we can say that the root locus is a powerful tool for designing and analyzing feedback control systems. Besides transient response, the root locus also gives a graphic representation of a system's stability. We can clearly see ranges of stability, ranges of instability and the conditions that cause a system to break in to oscillation.

7.2 THE ROOT LOCUS CONCEPTS

For understanding the concept of root locus, consider the simple second order system as shown in Fig. 7.1. The open loop transfer function is

$$G(s) = \frac{K}{s\,(s+a)} \qquad \text{...(7.1)}$$

by Eq. (7.1) the transfer function has two poles at $s = 0$ and $s = -a$

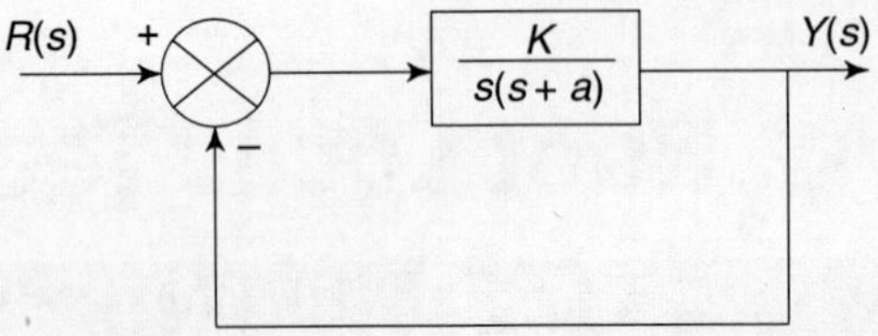

Figure 7.1 Second order system

The closed loop transfer function of this system is

$$\frac{Y(s)}{R(s)} = \frac{G(s)}{1+G(s)\,H(s)}$$

$$= \frac{\frac{K}{s}(s+a)}{1+\frac{K}{s\,(s+a)}\times 1} = \frac{\frac{K}{s\,(s+a)}}{\frac{s\,(s+a)+K}{s\,(s+a)}}$$

$$\frac{Y(s)}{R(s)} = \frac{K}{s^2+as+K} \qquad \text{...(7.2)}$$

From Eq. (7.2) the characteristic equation is

$$s^2 + as + K = 0 \qquad \text{...(7.3)}$$

where K and a are constant

By Eq. (7.3) if a and K have the positive value then the system is always stable. The dynamic behaviour is controlled by the roots of Eq. (7.3). The roots of the Eq. (7.3) are

$$s = -\frac{a}{2} \pm \sqrt{\left(\frac{a}{2}\right)^2 - K}$$

i.e.,

$$s_1 = -\frac{a}{2} + \sqrt{\left(\frac{a}{2}\right)^2 - K} \qquad \text{...(7.4)}$$

$$s_2 = -\frac{a}{2} - \sqrt{\left(\frac{a}{2}\right)^2 - K}$$

By Eq. (7.4), we can say that when K or a varies, the roots of the characteristic equation changes. Now let us consider that gain K is variable while the parameter a is held fixed. When K is varied from zero to infinity, the roots s_1, s_2 desirable loci in s-plane. The root location for various ranges of values of K are :

1. $0 \le K \le a^2/4$, the roots are real and distinct.

when $K = 0$ the roots are $s_1 = 0$ and $s_2 = -a$, i.e., it is equivalent to the open loop pole of the system.

2. When $K = a^2/4$, the roots are real and equal in value i.e., $s_1 = s_2 = -\dfrac{a}{2}$

3. When $a^2/4 \leq K \leq \infty$, the roots are complex conjugate with real part is $-\dfrac{a}{2}$, i.e., unvarying real part.

The root locus for varying K is plotted in Fig. 7.2.

- The plot has two branches starting at the two open loop poles ($s = 0$ and $s = -a$) for $K = 0$.
- When K is increased from 0 to $a^2/4$ the roots move towards point P in opposite direction. Both the roots locus on the negative real axis which represents that the system is overdampled.
- When $K = a^2/4$. The system is critically damped, and the poles meet at point P.
- K is further increased from $a^2/4$ towards the roots break away from the real axis, become complex conjugate, the roots move along the line $\sigma = -a/2$ because the real part of both the roots is fixed at $-a/2$ so the system becomes underdamped.
- The root locus in Fig. 7.2. is plotted by the direct solution of the characteristic equation because the system is of second order. If the system is of third or higher order, the direct solution is tedious so Evans developed a graphical method by which the root locus of third and higher order systems can be constructed easily.

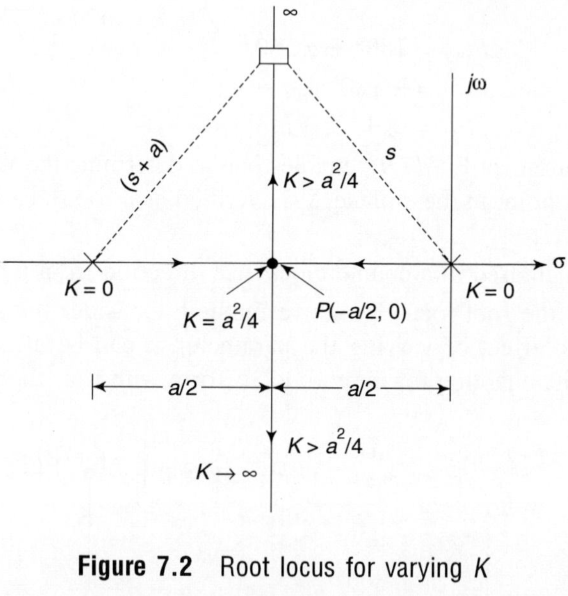

Figure 7.2 Root locus for varying K

For multi-loop closed loop or third or higher order system, we found in section 2.3.3 that by using Mason's signal flow gain formula, we had

$$\Delta(s) = 1 - \sum_j L_{j1} + \sum_j L_{j2} - \sum_j L_{j3} \quad ...(7.5)$$

where L_{jr} = gain product j^{th} possible combination of r non-touching loop.

Hence, we have a characteristic equation which may be written as

$$q(s) = \Delta(s) = 1 + F(s) \quad ...(7.6)$$

To find the roots of the characteristic equation Eq. (7.6) is equal to zero

$$1 + F(s) = 0 \quad ...(7.7)$$

$$F(s) = 1 + j0 \quad ...(7.8)$$

The roots of the characteristic equation must also satisfy this relation.

In general, the function $F(s)$ may be written

$$F(s) = K\frac{N(s)}{D(s)} = \frac{K\,(s+Z_1)(s+Z_2)(s+Z_3)\ldots(s+Z_m)}{(s+P_1)(s+P_3)\ldots(s+P_n)}$$

The required angle and magnitude for the root locus

$$|F(s)| = \frac{K\,|s+Z_1|\,|(s+Z_2|\ldots|(s+Z_m)|}{|(s+P_1|\,|s+P_2|\ldots|s+P_n|}$$

$$|F(s)| = 1 \quad ...(7.9)$$

and $\angle F(s) = \angle(s + Z_1) + \angle(s + Z_2) + \ldots\ [\angle(s + P_1) + \angle(s + P_2) + \angle(s + P_3 + \ldots]$

$$\angle F(s) = 180° \pm q\ 360°$$

$$\angle F(s) = \pm\ 180°\ (2q + 1);$$

$$q = 0, 1, 2, 3 \ldots \quad ...(7.10)$$

The magnitude requirement Eq. (7.9), enables one to determine the value of K for a given root location, s_1. A test point in the s-plane, s_1 is verified as a root location, when Eq. (7.10) is satisfied.

The angles are all increased in a counterclockwise direction from a horizontal line.

To further illustrate the root locus by above method. Consider the second order system shown in Fig. 7.3. The effect of varying the parameter, a can be effectively portrayed by writing the characterisitc eqaution for the root locus form with K as the multiplying factor in the numerator.

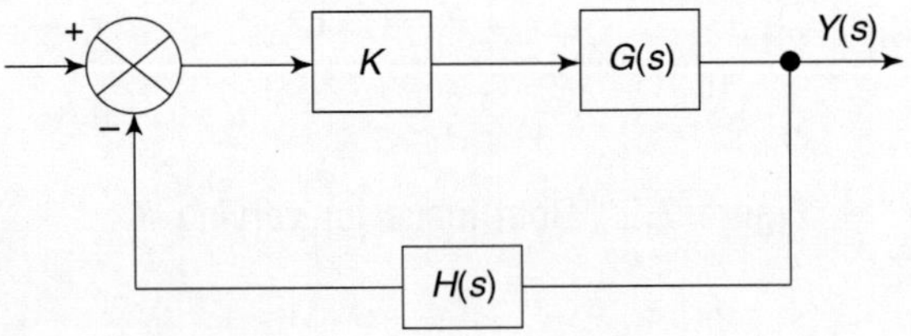

Figure 7.3 Single loop system

The characteristic equation is

$$1 + K\,G(s)\,H(s) = 0 \qquad \text{...(7.11)}$$

Let

$$G(s) = \frac{1}{s\,(s+a)}$$

$$H(s) = 1$$

$$1 + \frac{K}{s\,(s+a)} = 0$$

$$s^2 + as + K = 0 \qquad \text{...(7.12)}$$

or we can write $1 + \dfrac{as}{s^2 + K} = 0$ dividing by $(s^2 + K)$

Then the magnitude criterion is satisfied when

$$\frac{a\,|s_1|}{|s_1^2 + K|} = 1 \qquad \text{...(7.13)}$$

at root s_1.

The angle criterion is

$$\angle s_1 - [\angle(s_1 + j\sqrt{K}) + \angle(s_1 - j\sqrt{K})] = \pm\,180°, \pm\,540° \;....$$

The root locus for the characterisitic equation in Eq. (7.12) is ploted in Fig. 7.4.

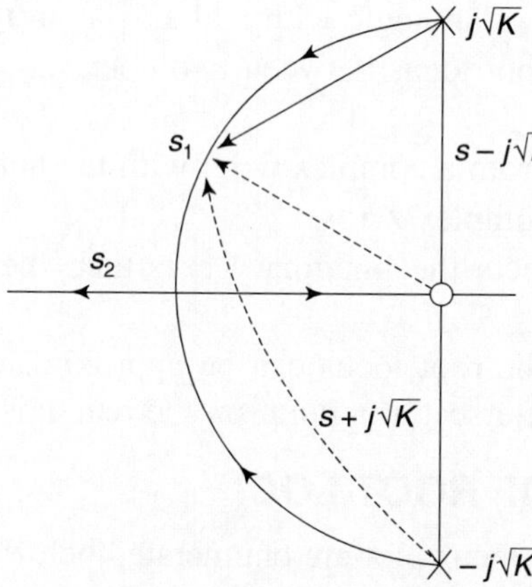

Figure 7.4 Root locus as a function of parameter

Specifically at the root s_1, the magnitude of the parameter, a is found from Eq. (7.13) as

$$a = \frac{|s_1 - j\sqrt{K}|\,|s_1 + j\sqrt{K}|}{|s_1|} \qquad \text{...(7.14)}$$

The roots of the system merge on the real axis at the point s_2 and provide a critically damped response to a step input. The parameters a, has a magnitude at the critically damped roots $s_2 = \sigma_2$ equal to

$$a = \frac{\left|\sigma_1 - j\sqrt{K}\right|\left|\sigma_1 + j\sqrt{K}\right|}{\sigma_2}$$

$$a = \frac{1}{\sigma_2}(\sigma_2^2 + K) \qquad \text{...(7.15)}$$

where σ_2 is evaluated from the s-plane vector lengths as $\sigma_2 = \sqrt{K}$. As a increases beyond the critical value, the roots are both real and distinct; one is larger than σ_2, and another one is smaller.

7.3 FEATURES OF ROOT LOCUS

The salient features of root locus are as follows:

1. The root locus is symmetric about the real axis.
2. The number of branches of the root locus is equal either the number of open loop poles or number of open loop zeros whichever is greater.
3. The locus starts (when $K = 0$) at poles of the loop gain, and ends (when $K \to \infty$) at the zeros.
4. The locus exists on real axis to the left of an odd number of poles and zeros on the axis.
5. The asymptotic lines make an angle with real axis at an intersection point.
6. At some points on the root locus between two open loop poles the root locus break away or break in point.
7. The angle of departure from a complex pole, with the horizontal line.
8. The angle of arrival at complex zeros.
9. If the root locus intersects the imaginary axis then the points of intersection are conjugate.
10. For higher value of K the root locus can be approximated by asymptotic lines and these asymptotic lines intersect at a point on the real axis.

7.4 CONSTRUCTION OF ROOT LOCI

The steps used to sketch a root locus plot are enumerated below:

⇒ Gather background information from transfer function

⇒ Rules

1. Symmetry
2. Number of branches
3. Starting and ending point
4. Locus on real axis
5. Asymptotes as $|s| \to \infty$

6. Breakaway points
7. Angle of departure
8. Angle of arrival
9. Locus crosses imaginary axis

⇒ Given gain "K", determine location of poles

⇒ Given pole location, determine value of "K"

Background information from transfer function

The closed loop transfer function is $T_o(s)$ for the given system in Fig. 7.5.

$$T_o(s) = \frac{KG(s)}{1 + KG(s)\,H(s)}$$

The characteristic equation is

$$1 + KG(s)\,H(s) = 0$$

$$1 + K\frac{N(s)}{D(s)} = 0$$

$$1 + KF(s) = 0 \qquad \text{...(7.16)}$$

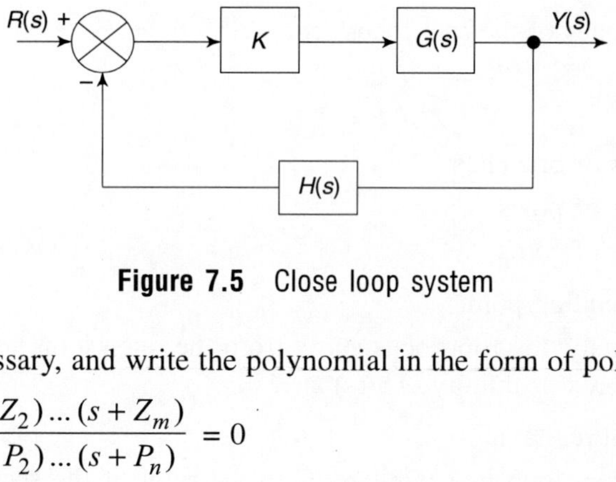

Figure 7.5 Close loop system

Factor $F(s)$ if necessary, and write the polynomial in the form of poles and zeros follows:

$$1 + K\frac{(s + Z_1)(s + Z_2)\ldots(s + Z_m)}{(s + P_1)(s + P_2)\ldots(s + P_n)} = 0$$

$$1 + K\frac{\sum_{i=1}^{m}(s + Z_i)}{\sum_{j=1}^{n}(s + P_j)} \qquad \text{...(7.17)}$$

Locate the poles and zeros on the complex plane (s-plane) with selected symbols. We are usually interested in determining the locus of roots as K varies as

$$0 \le K \le \infty$$

Equation (7.17) is

$$\sum_{j=1}^{n} (s+P_j)+K \sum_{i=1}^{m} (s+Z_i) = 0 \qquad ...(7.18)$$

When we see by the Eq. (7.18)

When $K = 0$

Then Eq. (7.18) has the roots poles only. That means the locus of the roots of the characterisitc equation $1 + KF(s) = 0$ starts at the poles of $F(s)$.

When K approaches infinity.

Then Eq. (7.18) has the roots zeros only. That means the locus of the roots of the characteristic equation at the zeros of $F(s)$.

Now locate the segment of the real axis that are root loci. The root locus on the real axis always lies in a section of the real axis to the left of an odd number of poles and zeros.

Rules

Rule 1. Since the characteristic equation has real coefficients, any complex root must occur in complex conjugate pair (which are symmetric about the real axis). Since the root locus is just a diagram of the roots of the characteristic equation as K varies, it must be symmetrical with respect to the horizontal real axis.

Rule 2. Number of branches of root locus

The number of branches of the root locus are

$$N = P \text{ if } P > Z$$

$$N = Z \text{ if } P < Z$$

where N = number of branches

P = number of poles

Z = number of zeros

Rule 3. Starting and ending point

As discussed previously, the root locus starts from the open loop poles of $F(s)$ at $K = 0$ and terminates at the zeros or infinity of $F(s)$ at $K = \infty$

Rule 4. Root locus on real axis.

In the sum of the open loop poles and zeros to the right of the section is odd, then the locus exists on real axis. While constructing the root-loci on the real axis choose a test point on it. If the sum of P and Z to the right of this point is odd, then this point is a part of the root loci. If it is even, then it is not.

Rule 5. Asymptotes as $|S| \to \infty$

The asymptotes, intersecting at a point x on the real axis are given by

$$x = \frac{\alpha}{q} \qquad ...(7.19)$$

where $\alpha = \sum_{j=1}^{n} P_j - \sum_{i=1}^{m} Z_i$

and $q = n - m$

The angle of asymptotes with real axis is given by

$$\theta = \frac{(2K+1)\times 180^\circ}{q}$$

where $q = n - m$

$$\theta = \frac{(2K+1)\times 180^\circ}{n-m}$$

where $K = 0, 1, 2 \ldots (n - m) - 1$

Rule 6. Breakaway points

To find where the locus breaks away from the axis (or converges on the axis), we note that this always occurs when two (or more) roots intersect. It is a well known fact, that when a polynomial has multiple roots, not only is the value of the polynomial zero, but its derivative is zero also.

Let $$1 + K\frac{N(s)}{D(s)} = 0$$

At the breakaway points, the derivative is

$$\frac{d}{ds}\left[1 + K\frac{N(s)}{D(s)}\right] = 0$$

$$K\left[\frac{N'(s)\,D(s) - D'(s)\,N(s)}{D^2(s)}\right] = 0$$

$$N'(s)\,D(s) - D'(s)\,N(s) = 0 \qquad \text{...(7.20)}$$

(a) break-in point (between to zero)

(b) break-away point (between two poles)

Rule 7. Angle of departure from complex pole. If the loop gain $G(s)$ $H(s)$ has a simple pole on the real axis, we know that the locus will leave from the pole, as $K \to 0$, along the axis. However, if the pole is complex it can leave at any angle. To find the angle at which the locus leaves a complex pole.

$$\phi_d = 180^\circ + (\phi_Z - \phi_P) \qquad \text{...(7.21)}$$

when ϕ_d = angle of departure

ϕ_Z = sum of all the angles subtended by zeros

ϕ_P = sum of all the angles subtended by remaining poles

- Many times, especially for simple root loci, there are no break-away points.

Rule 8. Angle of arrival at complex zero.

If the loop gain, $G(s)\ H(s)$, has a simple zero on the real axis, we know that the locus will arrive at the zero, as $K \to \infty$, along the axis. However, if the zero is complex it can arrive at any angle. To find the angle at which the locus arrives at a complex zero.

$$\phi_a = 180° + (\phi_P - \phi_Z) \qquad ...(7.22)$$

where ϕ_a = angle of arrival

ϕ_Z = sum of all the angles subtended by remaining zeros

ϕ_P = sum of all the angles subtended by poles.

Rule 9. Find where locus crosses imaginary axis.

By using Routh–Hurwitz criterion to determine where the locus crosses the imaginary axis.

Example 7.4.1 $G(s)\ H(s) = \dfrac{K}{s\,(s+4)\,(s^2+4s+5)}$

Solution = Total 3 break point

= 2 break-away point

= 1 break-in point

= zeros lies on infinity. We can calulate only break-away point remaining points are break-in point.

Example 7.4.2 Sketch the root locus for the open loop transfer function of a unity feedback control system given below and determine:

(a) the value of K for $\varepsilon = 0.5$

(b) the value of K of marginal stability

(c) the value of K at $s = -4$

$$G(s) = \frac{K}{s\,(s+1)\,(s+3)}$$

Solution The characteristic equation is

$$1 + G(s)\ H(s) = 0$$

$$1 + \frac{K}{s\,(s+1)\,(s+3)} = 0$$

$$s\ (s+1)\ (s+3) + K = 0$$

The number of branches of the root locus by Rule 2

$$N = 3$$

$\because$ $P > Z$

i.e., $P = 3$

and $Z = 0$

By Rule 3. The root locus start ($K = 0$) at $s = 0$, $s = -1$ and $s = -3$ and terminate at infinity because there is no open loop zero.

By Rule 4. The root locus on the real axis exists between 0 and – 1 and beyond $s = -3$

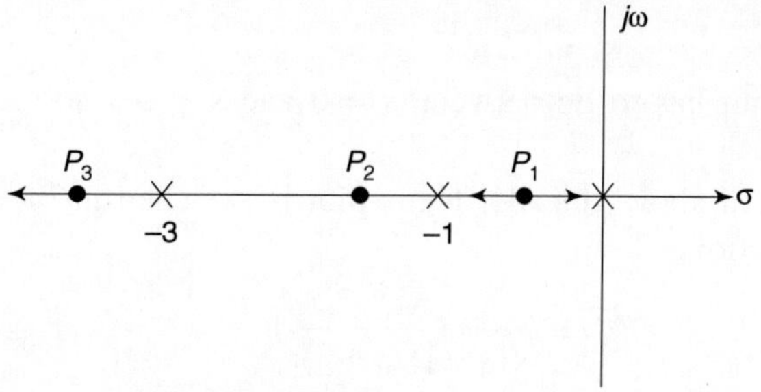

Figure Ex. 7.4.2

By Rule 5. The asymptotes intersect at a point x on the real axis is

$$x = \frac{\sum_{j=1}^{n} P_j - \sum_{i=1}^{m} Z_i}{n - m} = \frac{(0 - 1 - 3) - (0)}{3 - 0} = \frac{-4}{3}$$

$$x = -1.33$$

The angle of asymptotes is given by

$$\theta = \frac{(2K+1)\,180°}{n-m} = \frac{(2K+1)\,180°}{(3-0)}$$

(i) $\theta_1 = \dfrac{[(2 \times 0) + 1] \times 180°}{3 - 0} = 60°$ at $K = 0$

(ii) $\theta_2 = \dfrac{[(2 \times 1) + 1] \times 180°}{3} = 180°$ at $K = 1$

(iii) $\theta_3 = \dfrac{[(2 \times 2) + 1] \times 180°}{3} = 300°$ at $K = 2$

By Rule 6. The characteristic equation is

$$s\,(s + 1)\,(s + 2) + K = 0$$

$$K = -\,(s^2 + 4s^2 + 3s)$$

Differentiating the above equation

$$\frac{dK}{ds} = -\,(3s^2 + 8s + 3)$$

$$\because \quad \frac{dK}{ds} = 0$$

$$3s^2 + 8s + 3 = 0$$

$$s = \frac{-8 \pm \sqrt{(8)^2 - (4 \times 3 \times 3)}}{2 \times 3}$$

$$s = -1.33 \pm 0.88$$

i.e.,
$$s_1 = -0.45$$
$$s_2 = -2.21$$

$\because$ Break-away point has to lie between $s = 0$ and $s = -1$ so the break-away point is -0.45.

As the data calculated above the root locus plot is sketched in Fig. Ex. 7.4.2 (a) for the open loop transfer function.

$$G(s) = \frac{K}{s(s+1)(s+3)}$$

Because the loop gain $G(s)\,H(s)$, has simple pole on the real axis so by Rule 7, the locus will leave from the pole as $K = 0$.

Rule 8. is not lie because there are no zeros.

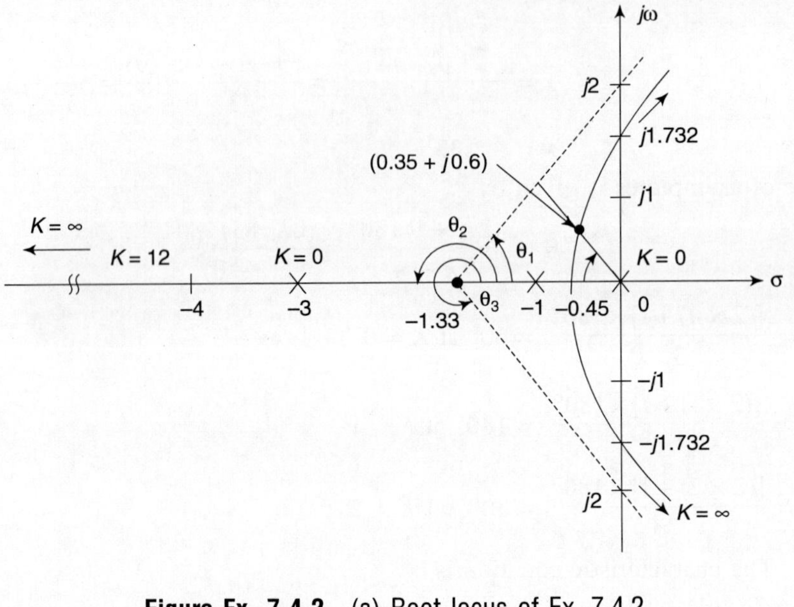

Figure Ex. 7.4.2 (a) Root locus of Ex. 7.4.2

By Rule 9. By Routh-Hurwitz criterion

s^3	1	3
s^2	4	K
s^1	$\frac{12-K}{4}$	0
s^0	K	0

The value of K at which the root locus plot crosses the imaginary axis is determined by first term in s^1 row is zero.

$$\frac{12 - K}{4} = 0$$

$$\therefore \qquad K = 12$$

For $K > 12$ the root lies in the R.H.S. of S plane so the $K = 12$ is value for marginally stable.

By Rule 9 The root locus plot intersects the imaginary axis and the value of s at the intersection point is determined by solving auxiliary equation formed from the s^2 term in Routh array.

$$4s^2 + K = 0$$

$$4s^2 + 12 = 0$$

$$\therefore \qquad s = \pm j\,1.73$$

$$\therefore \qquad \varepsilon = 0.5$$

$$\cos\theta = \varepsilon$$

$$\cos\theta = 0.5$$

$$\theta = 60°$$

So from the origin a line at an angle $\theta = 60°$ is drawn which intersects the root locus plot at $s = (-0.35 + j\,0.6)$.

(b) As the point s lies on the root locus the following equation is satisfied

$$|G(s)\,H(s)| = 1$$

$$\left|\frac{K}{s\,(s+1)\,(s+3)}\right| = 1$$

$$\because \qquad s = (-0.35 + 0.6\,j)$$

$$\left|\frac{K}{(-0.35+0.6\,j)\,(-0.35+0.6\,j+1)\,[-0.35+0.6\,j+3]}\right| = 1$$

$$\frac{K}{1.6} = 1$$

$$K = 1.6 \quad \textbf{Ans.}$$

(c) The value of K as $s = -4$

$$\left|\frac{K}{s\,(s+1)\,(s+3)}\right| = 1$$

$$\frac{K}{-4\,(-4+1)\,(-4+3)} = 1$$

$$\frac{K}{12} = 1$$

$$K = 12 \quad \textbf{Ans.}$$

EXAMPLES

Root Locus Technique

Example 7.1 Plot the root locus pattern of a system whose forward path transfer function

$$G(s) = \frac{K}{s\,(s+2)\,(s+3)}$$

Solution The characteristic equation is

$$1 + G(s)\,H(s) = 0$$

$$1+\frac{K}{s\,(s+2)\,(s+3)}\times 1 = 0$$

$$s\,(s + 2)\,(s + 3) + K = 0$$

$$s^3 + 5s^2 + 6 + K = 0 \qquad ...(1)$$

Rule 2. The number of branches of the root locus

$$N = P \text{ if } P > z$$

$$N = 3$$

$$\because \quad P = 3$$

and $\quad Z = 0$

Rule 3. The root locus starts ($K = 0$) as $s = 0, -2$, and -3 terminates at ($K = \infty$) infinity because there are no zeros.

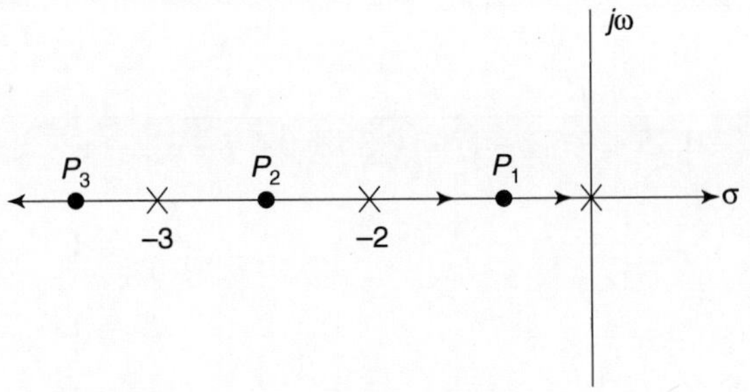

Figure Ex. 7.1 (a) Root locus along real axis

Rule 4. The root locus on the real axis exists between 0 and -2 and beyond -3 as in Fig. Ex. 7.1 (a).

Rule 5: The asymptotes intersect at a point x on the real axis

$$x = \frac{\sum_{j=1}^{n} P_j - \sum_{i=1}^{m} Z_i}{n-m}$$

$$x = \frac{(0-1-3)-0}{3-0} = \frac{-5}{3}$$

$$x = -1.66$$

The angle of asymptotes is given by

$$\theta = \frac{(2K+1)\times 180°}{n-m}$$

$$K = 0, 1, 2$$

(i) $\theta_1 = \dfrac{[(2\times 0)+1]\times 180°}{3-0} = 60°$ at $K = 0$

(ii) $\theta_2 = \dfrac{[(2\times 1)+1]\times 180°}{3} = 180°$ at $K = 1$

(iii) $\theta_3 = \dfrac{[(2\times 2)+1]\times 180°}{3} = 300°$ at $K = 2$

Rule 6. Break-away point

The characteristic equation is

$$s^3 + 5s^2 + 6s + K = 0$$

$$K = -(s^3 + 5s^2 + 6s)$$

Differentiating the above equation

$$\frac{dK}{ds} = -(3s^2 + 10s + 6)$$

$$\frac{dK}{ds} = 0$$

$$3s^2 + 10s + 6 = 0$$

$$s = \frac{-b\pm\sqrt{b^2-4ac}}{2a} = \frac{-10\pm\sqrt{(10)^2-4\times 3\times 6}}{2\times 3}$$

$$s = \frac{-10\pm\sqrt{100-72}}{6} = \frac{-10\pm\sqrt{28}}{6}$$

$$s = -0.784 \text{ and } -2.549$$

Break-away point has to lie between $s = 0$ and $s - 2$ so the point is – 0.784

Rule 7 and Rule 8. There is no complex pole and zero so these points do not exist.

Rule 9. Locus crosses imaginary axis.

The point on imaginary axis can be obtained by Routh array

The characterisitc equation is

$$s^2 + 5s^2 + 6s + K = 0$$

s^3	1	6
s^2	5	K
s^1	$\frac{30-K}{5}$	0
s^0	K	0

For stable system

$$K > 0$$

and
$$\frac{30-K}{5} > 0$$

$$30 - K > 0$$

$$K < 30$$

By solving the auxiliary equation for s^2 term

$$5s^2 + K = 0$$

$$5s^2 = -K$$

$$s^2 = -\frac{K}{5} = -\frac{30}{5} = -6$$

$$s = \pm 2.45\,j$$

As the data calculated above the root locus plot is sketched in Fig. Ex. 7.1 (b).

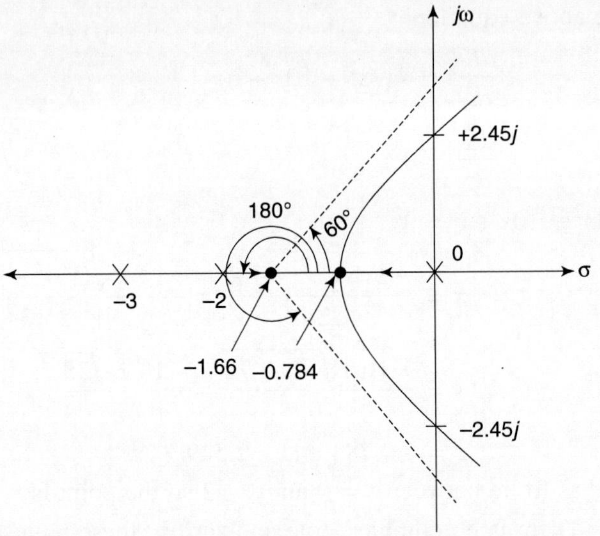

Figure Ex. 7.1 (b) Root locus for the open loop transfer function of Ex. 7.1

Example 7.2 Draw root locus of the following system $G(s)\,H(s) = \dfrac{K\,(s+2)}{(s+3)\,(s^2+2s+2)}$.

(U.P.T.U., 2004–05, CS)

Solution The characteristic equation is

$$1 + G(s)\,H(s) = 0$$

$$1+\frac{K(s+2)}{(s+3)\,(s^2+2s+2)} = 0$$

$$(s + 3)\,(s^2 + 2s + 2) + K\,(s + 2) = 0$$

Rule 2. The number of branches of the root locus $N = P$ if $P > Z$

$\because$ $\qquad P = 3$ and $Z = 1$

so $\qquad N = 3$

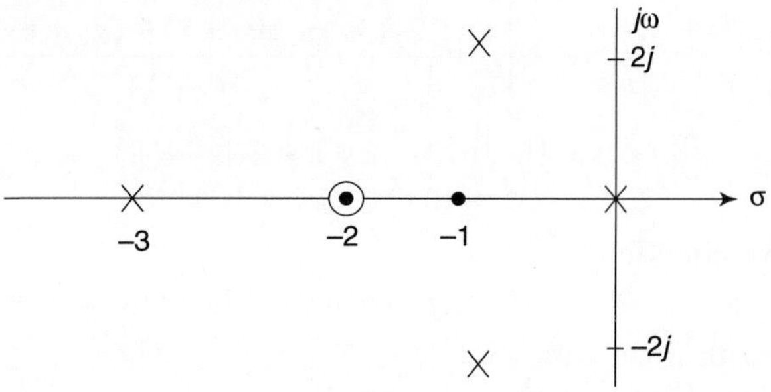

Figure Ex. 7.2 (a) Root locus along real axis

Rule 4. The root locus on the real axis exists between –2 and – 3

Rule 5. The number of asymptotes = $P - Z$

$$= 3 - 1 = 2$$

The asymptotes intersect at a point x on the real axis

$$x = \frac{\sum_{j=1}^{n} P_j - \sum_{i=1}^{m} Z_i}{n-m}$$

$$= \frac{-3-1+2j-1-2j-(-2)}{3-1} = \frac{-5+2}{2}$$

$$x = \frac{-3}{2} == -1.5$$

$$x = -1.5$$

The angle of asymptotes is given by

$$\theta_1 = \frac{(2K+1)\times 180°}{n-1} = \frac{[(2\times 0)+1]\times 180°}{3-1} = 90° \text{ at } K = 0$$

$$\theta_2 = \frac{[(2\times 1)+1]\times 180°}{3-1} = 270° \text{ at } K = 1$$

Rule 6. Break-away point

The characteristic equation is

$(s + 3)\ (s^2 + 2s + 2) + K\ (s + 2) = 0$

$$K = -\frac{(s+3)(s^2+2s+2)}{s+2}$$

$$K = \frac{s^3+5s^2+8s+6}{s+2}$$

No differentiating above equation w.r.t. s

$$\frac{dK}{ds} = -\left[\frac{(s+2)(3s^2+10s+8)-(s^3+5s^2+8s+6)}{(s+2)^2}\right]$$

$$0 = [(s + 2)\ (3s^2 + 10s + 8) - (s^3 + 5s^2 + 8s + 6)]$$

$$2s^3 + 11s^2 + 20s + 10 = 0$$

By solving the above equation

$$s = -\ 2.1 \text{ approx. (Hit and Trail Method)}$$

Rule 9. Intersection with imaginary axis

$$(s + 3)\ (s^2 + 2s + 2) + K\ (s + 2) = 0$$

$$s^3 + 5s^2 + 8s + 6 + Ks + 2K = 0$$

$$s^3 + 5s^2 + (8 + K)\ s + 2K + 6 = 0$$

By the Routh's array

s^3	1	$8 + K$
s^2	5	$2K + 6$
s^1	$\frac{3K+34}{5}$	0
s^0	$6 + 2K$	0

For sustained oscillation

$$\frac{3K+34}{5} = 0$$

$$3K = -\ 34$$

$$K = -\frac{34}{3} = -\ 11.3$$

The auxiliary equation is

$$5s^2 + 6 + 2K = 0$$

$$5s^2 + 6 + 2\ (-\ 11.3) = 0$$

$$5s^2 = 22.6 - 6$$

$$s^2 = \frac{16.6}{5} = 3.33$$

$$s = \pm\ 1.83$$

The value of K is –ve so value of s is real so it does not cut the imaginary axis.

The root locus is shown in Fig. Ex. 7.2 (b).

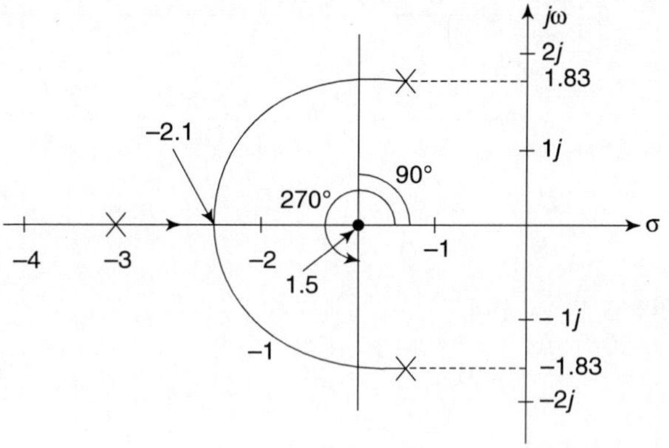

Figure Ex. 7.2 (b) Root locus of Ex. 7.2

Example 7.3 The open loop transfer function of a control system is given by $G(s)\ H(s) = \dfrac{K\,(s+2)^2}{(s+3)^2}$.

Show that the root locus lies on a circle.

Solution **Rule 1** The characteristic equation is

$$1 + G(s)\ H(s) = 0$$

$$1 + \frac{K\,(S+2)^2}{(S+3)^2} = 0$$

$$(s + 3)^2 + K\,(s + 2)^2 = 0$$

Rule 2. The number of branches of the root locus

$$N = P \text{ or } Z \text{ if } P = Z$$

$$N = Z$$

Rule 3. The root locus starts (K = 0) from open loop poles at s = – 3 and terminates ($K = \infty$) and open loop zeros at s = – 2.

Rule 4. Root locus does not lie on real axis because the sum of the open loop poles and zeros to right of the section is even.

Rule 5. The number of asymptotes= $n - m = 2 - 2 = 0$,

i.e., there is no asymptotes.

Rule 6. Break-away point

The characteristic equation is

$$(s + 3)^2 + K\,(s + 2)^2 = 0$$

$$K = -\frac{(s+3)^2}{(s+2)^2}$$

$$\frac{dK}{ds} = -\frac{(s+2)^2 \times 2\,(s+3) \times 1 - (s+3)^2 \times 2\,(s+2) \times 1}{(s+2)^4}$$

$$= -\frac{2\,(s+2)\,(s+3)\,[s+2-s-3]}{(s+2)^4}$$

$$\frac{dK}{ds} = \frac{2\,(s+2)\,(s+3)}{(s+2)^4}$$

To determine breakaway point put

$$\frac{dK}{ds} = 0$$

$$\frac{2\,(s+2)\,(s+3)}{(s+2)^4} = 0$$

$$2\,(s+2)\,(s+3) = 0$$

Hence, the two break-away points are located at

$$s = -2 \text{ and } s = -3$$

which are open loop zeros and poles respectively.

The root locus plot is shown in Fig. Ex. 7.3.

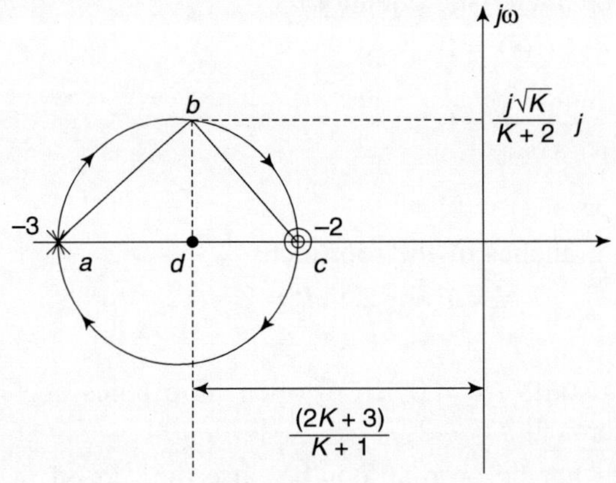

Figure Ex. 7.3 Root locus of Ex. 7.3

The characteristic equation is

$$1 + \frac{K\,(s+2)^2}{(s+3)^2} = 0$$

$$
\begin{aligned}
(s + 3)^2 + K(s + 2)^2 &= 0\\
s^2 + 6s + 9 + K(s^2 + 4s + 4) &= 0\\
s^2 + 6s + 9 + Ks^2 + 4Ks + 4K &= 0\\
(K + 1)s^2 + (4K + 6)s + (9 + 4K) &= 0
\end{aligned}
$$

The roots of above equations are

$$
\begin{aligned}
s_1, s_2 &= \frac{-b \pm \sqrt{b^2 - 4ac}}{2a}\\
&= \frac{-(4K+6) \pm \sqrt{(4K+6)^2 - 4(K+1)(9+4K)}}{2 \times (K+1)}\\
&= \frac{-2(2K+3) \pm \sqrt{(16K^2 + 48K + 36 - 4[9K + 4K^2 + 9 + 4K)}}{2(K+1)}\\
&= \frac{-2(2K+3) \pm \sqrt{16K^2 + 48K + 36 - 36K - 16K^2 - 36 - 16K}}{2(K+1)}\\
&= -\frac{(2K+3)}{(K+1)} \mp \frac{\sqrt{-4K}}{2(K+1)}\\
&= -\frac{(2K+3)}{(K+1)} \mp j\frac{\sqrt{K}}{(K+1)}
\end{aligned}
$$

Roots locus plot in Fig. Ex. 7.3. Let

$$
\begin{aligned}
b &= -\frac{(2K+3)}{K+1} \pm j\frac{\sqrt{K}}{(K+1)}\\
ad &= -3 - \left[-\frac{2K+3}{K+1}\right] = \frac{-3K - 3 + 2K + 3}{K+1} = \frac{-K}{K+1}\\
dc &= -\frac{(2K+3)}{K+1} - (-2) = \frac{-2K - 3 + 2K + 2}{K+1} = -\frac{1}{K+1}
\end{aligned}
$$

$$
\begin{aligned}
ab^2 &= (ad)^2 + (db)^2\\
&= \left(\frac{-K}{K+1}\right)^2 + \left(\frac{\sqrt{K}}{K+1}\right)^2\\
&= \frac{K}{K+1}
\end{aligned}
$$

$$
\begin{aligned}
bc^2 &= dc^2 + bd^2\\
&= \left(-\frac{1}{K+1}\right)^2 + \left(\frac{\sqrt{K}}{K+1}\right)^2\\
&= \frac{1}{K+1}
\end{aligned}
$$

$$
\begin{aligned}
ac &= -3 -- (-2) = -1\\
ac^2 &= 1\\
ac^2 &= ab^2 + bc^2
\end{aligned}
$$

$$ac^2 = \frac{K}{K+1} + \frac{1}{K+1} = \frac{K+1}{K+1} = 1$$

$$ac^2 = ab^2 + bc^2$$

Hence, point b lies on a semicircle.

Example 7.4 Draw the root locus diagram for a unity feedback control system having the forward path transfer function given as $G(s) = \frac{K}{s\,(s+10)\,(s+20)}$. *(U.P.T.U., 2003–04, CS)*

Solution **Rule 1.** The characteristic equation is

$$1 + G(s)\,H(s) = 0$$

$$1 + \frac{K}{s\,(s+10)\,(s+20)} = 0$$

$$s\,(s + 10)\,(s + 20) + K = 0$$

Rule 2. The number of branches of the root locus

$$N = P \text{ if } P > Z$$

$$N = 3 \quad \because P = 3$$

Rule 3. The root locus starts ($K = 0$) at $s = 0$, – 10 and – 20 and terminates ($K = \infty$) on infinity because there are no zeros.

Rule 4. The root locus on the real axis exist between 0 and – 10 and beyond – 20 as shown in Fig. Ex. 7.4 (a).

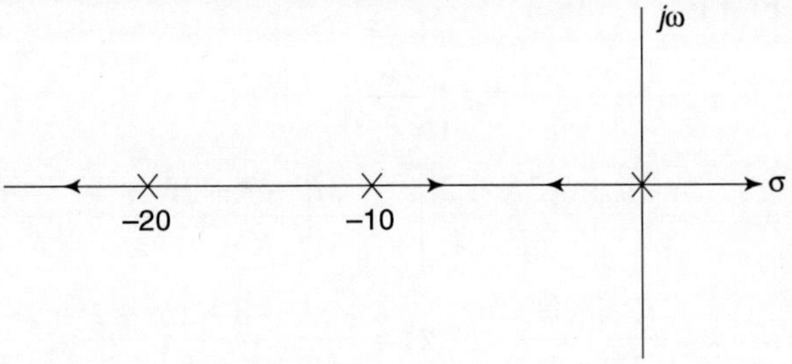

Figure Ex. 7.4 (a) Root locus along real axis

Rule 5. The number of asymptotes = $n - m$

$$= 3 - 0 = 3$$

The asymptotes intersect at point x on real axis

$$x = \frac{\sum_{j=1}^{n} P_j - \sum_{i=1}^{m} Z_i}{n - m}$$

$$x = \frac{(0 - 10 - 20) - 0}{3 - 0} = \frac{-30}{3} = -10$$

The angle of asymptotes for K is given by

$$\theta = \frac{(2K + 1) \times 180°}{n - m}$$

$$\theta_1 = \frac{[(2 \times 0) + 1] \times 180°}{3 - 0} = 60° \text{ at } K = 0$$

$$\theta_2 = \frac{[(2 \times 1) + 1] \times 180°}{3 - 0} = 180° \text{ at } K = 1$$

$$\theta_3 = \frac{[(2 \times 2) + 1] \times 180°}{3 - 0} = 300° \text{ at } K = 2$$

Rule 6. Break-away point

The characteristic equation is

$$s\,(s + 10)\,(s + 20) + K = 0$$

$$K = -\,s\,(s + 10)\,(s + 20)$$

$$= -\,(s^3 + 30s^2 + 200s)$$

$$\frac{dK}{ds} = -\,(3s^2 + 60s + 200)$$

At breakaway point $\qquad \dfrac{dK}{ds} = 0 = -\,(3s^2 + 60s + 200)$

$$s = \frac{-60 \pm \sqrt{(60)^2 - 4 \times 3 \times 200}}{2 \times 3}$$

$$s = \frac{-60 \pm \sqrt{3600 - 2400}}{6}$$

$$s = -4.23 \text{ and } s = -15.77$$

The point 15.77 does not lie on root locus on the real axis so the break-away point is $s = -4.23$.

Rule 9. Point of intesection at imaginary axis

$$s^3 + 30s^2 + 200s + K = 0$$

s^3	1	200
s^2	30	K
s^1	$\dfrac{6000 - K}{30}$	0
s^0	K	

At the intersection of imaginary axis, the system shows sustained oscillation so far that

$$\frac{6000 - K}{30} = 0$$

$$K = 6000$$

The auxiliary equations is given by

$$30s^2 + K = 0$$

$$30s^2\ 6000 = 0$$

$$s^2 = \frac{-6000}{30}$$

$$s = \pm j\sqrt{200}$$

$$s = \pm 14.14\,j$$

$$j\omega = \pm 14.14\,j$$

$$\omega = \pm 14.14 \text{ rad/sec.}$$

As the data calculated above the root locus plot is sketched in Fig. Ex. 7.4 (b).

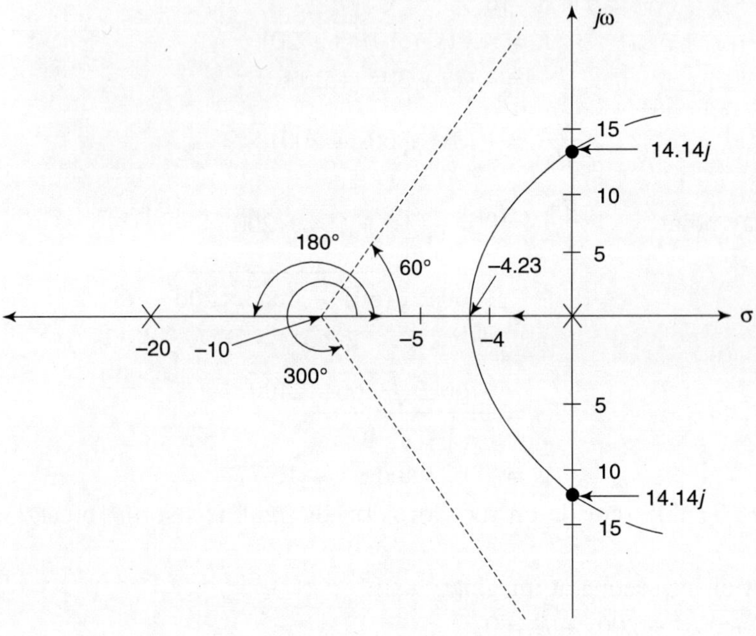

Figure Ex. 7.4 (b) Root locus of Ex. 7.4

Example 7.5 Sketch the root locus plot for the system when open loop transfer function is given by

$$G(s)\,H(s) = \frac{K}{s\,(s+3)\,(s^3+4s+13)}$$

Solution The characteristic equation is

$$1 + G(s)\,H(s) = 0$$

$$1+\frac{K}{s\,(s+3)\,(s^2+4s+13)} = 0$$

$$s\,(s+3)\,(s^2+4s+13)+K = 0$$

Rule 2. The number of branches of the root locus

$$N = P \text{ if } P > Z$$

$$N = 4 \quad \because \; P = 4$$

Rule 3. The root locus starts ($K = 0$) from open loop poles at $s = 0, -3, \pm(-2+j3)$ and terminates on open loop zeros ($K = \infty$) on infinity because there is no zero.

Rule 4. The root locus on the real axis exists between 0 and – 3 as shown in Fig. Ex. 7.5 (a).

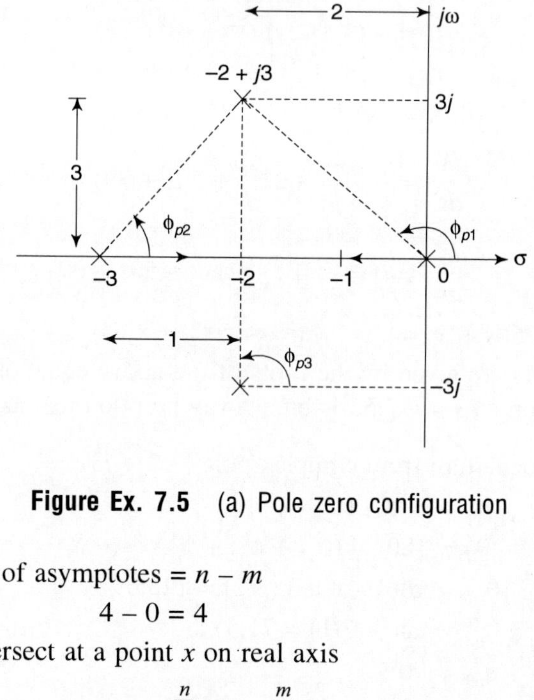

Figure Ex. 7.5 (a) Pole zero configuration

Rule 5. The number of asymptotes = $n - m$

$$4 - 0 = 4$$

The asymptotes intersect at a point x on real axis

$$x = \frac{\sum_{j=1}^{n} P_j - \sum_{i=1}^{m} Z_i}{n-m} = \frac{(0-3-2+3j-2-3j)-(0)}{4-0}$$

$$x = -\frac{7}{4} = -1.75$$

The angle of asymptotes is given by

$$\theta = \frac{(2\quad +1)\times 180^\circ}{n \quad m}$$

$$\theta_1 = \frac{[(2\times 0)+1]\times 180°}{4-0} = 45° \text{ at } K = 0$$

$$\theta_2 = \frac{[(2\times 1)+1]\times 180°}{4-0} = 135° \text{ at } K = 1$$

$$\theta_3 = \frac{[(2\times 2)+1]\times 180°}{4-0} = 225° \text{ at } K = 2$$

$$\theta_4 = \frac{[(2\times 3)+1]\times 180°}{4-0} = 315° \text{ at } K = 3$$

Rule 6. The characteristic equation is

$s\,(s + 3)\,(s^2 + 4s + 13) + K = 0$

$$K = -\,(s^2 + 3s)\,(s^2 + 4s + 13)$$
$$K = -\,(s^4 + 4s^3 + 13s^2 + 3s^3 + 12s^2 + 39s)$$
$$K = -\,(s^4 + 7s^3 + 25s^2 + 39s)$$

$$\frac{dK}{ds} = -\,(4s^3 + 21s^2 + 50s + 39)$$

$$\frac{dK}{ds} = 0 = -\,(4s^3 + 21s^2 + 50s + 39)$$

$4s^2 + 21s^2 + 50s + 39 = 0$

The break-away points are given by the roots of the above equation $s = -\,2.55$ (approx.) satisfy the above equation so $s = -\,2.55$ is breakaway point on real axis.

Rule 7. The angle of departure from complex pole $(-\,2 + j3)$

$$\theta_d = 180 - (\phi_p - \phi_z)$$
$$\theta_d = 180° - (\phi_{p1} + \phi_{p2} + \phi_{p3})$$
$$\theta_{p1} = 180° - \tan^{-1}(3/2) = 123.69°$$
$$\theta_{p2} = \tan^{-1}(3/1) = 71.57$$
$$\theta_{p3} = 90°$$
$$\theta_d = 180 - 123.69° - 71.57° - 90° = -\,105.26° \text{ from } (-\,2 + 3j)$$

The angle of departure from complex role $(-\,2 - 3j)$

$$\theta_d = +\,105.26° \text{ from } (-\,2 - 3j)$$

As the data calculated above the complete root locus is sketched in Fig. Ex. 7.5 (b).

Rule 9: Locus crosses imaginary axis

The characteristic equation is

$s^4 + 7s^3 + 25s^2 + 39s + K = 0$

By Routh array

s^4	1	25	K
s^3	7	39	0
s^2	19.4	K	0
s^1	$\dfrac{756.6 - 7K}{19.4}$	0	
s^0	K		

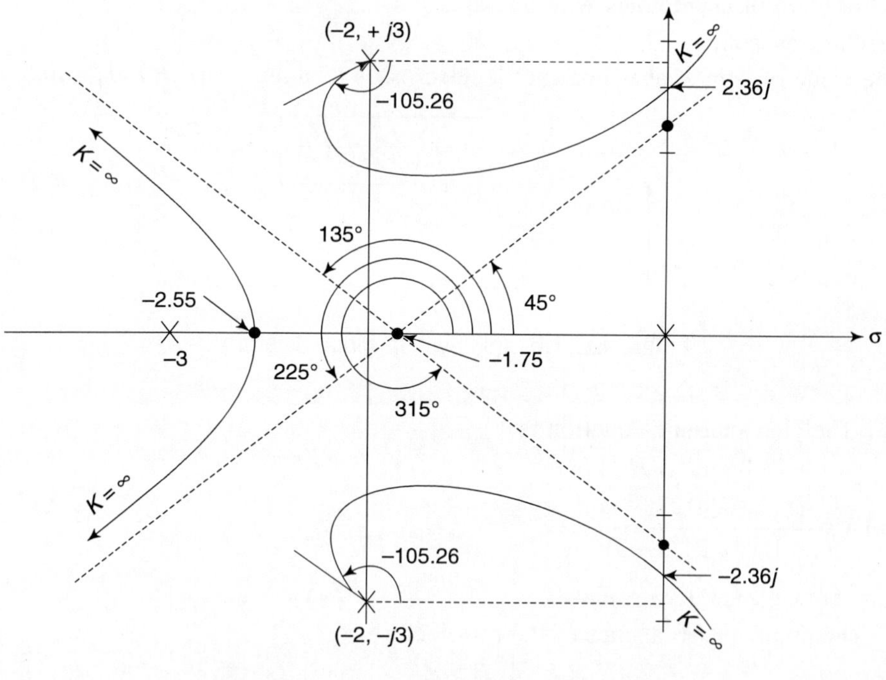

Figure Ex. 7.5 (b) Root locus of Ex. 7.5

For sustained oscillation

$$\frac{756.6 - 7K}{19.4} = 0$$

$$7K = 756.6$$

$$K = 108.1$$

By solving auxiliary equation

$$19.4\, s^2 + K = 0$$

$$19.4\, s^2 = -\,108.1$$

$$s^2 = \frac{-\,108.1}{19.4}$$

$$s^2 = -5.57$$

$$s = \pm j\sqrt{5.57}$$

$$s = \pm 2.36\, j$$

The root locus is shown in Fig. Ex. 7.5 (b).

Example 7.6 Sketch the root locus for the system shown in Fig. Ex. 7.6 (a) determine the following:

(i) Angle of asymptotes

(ii) Intersection of asymptotes with real axis

(iii) Break-away point

(iv) The value of gain K at which root locus crosses $j\omega$ axis. *(U.P.T.U., 2002–03, CS)*

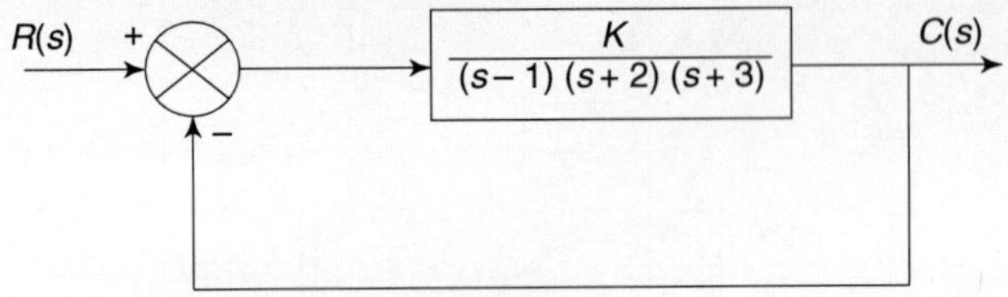

Figure Ex. 7.6 (a) System block diagram

Solution The characteristic equation is

$$1 + G(s)\,H(s) = 0$$

$$1 + \frac{K}{(s-1)(s+2)(s+3)} = 0$$

$$(s-1)(s+2)(s+3) + K = 0$$

Rule 2. The number of branches of the root loci

$$N = P \quad \text{if } P > Z$$

$$N = 3 \quad \because P = 3$$

Rule 3. The root locus starts ($K = 0$) from open loop poles at $s = +1, -2$ and -3 and terminates ($K = \infty$) on infinity.

Rule 4. The root locus on the real axis exists between + 1 and – 2 and beyond – 3 as shown in Fig. Ex. 7.6 (b).

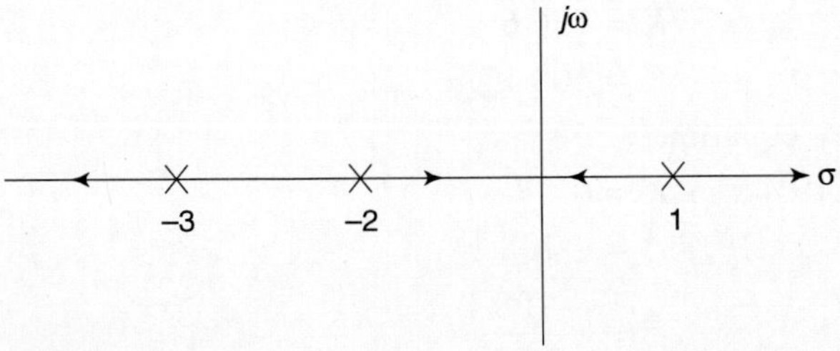

Figure Ex. 7.6 (b) Root locus along real axis

Rule 5. The number of asymptotes $= n - m$

$$= 3 - 0 = 3$$

(i) Angle of asymptotes $\theta = \dfrac{(2K+1)\times 180°}{n-m}$

$$\theta_1 = \frac{[(2\times 0)+1]\times 180°}{3-0} = 60° \text{ at } K = 0$$

$$\theta_2 = \frac{[(2\times 1)+1]\times 180°}{3-0} = 180° \text{ at } K = 1$$

$$\theta_3 = \frac{[(3\times 1)+1]\times 180°}{3-0} = 300° \text{ at } K = 2$$

(ii) Intersection of asymptotes with real axis at

$$x = \frac{\sum_{j=1}^{n} P_j - \sum_{i=1}^{m} Z_i}{n-m} = \frac{(1-2-3)-0}{3-0}$$

$$x = -4/3 = -1.33$$

Rule 6. Break-away point

(iii) The characteristic equation is

$$(s-1)(s+2)(s+3)+K = 0$$

$$K = -(s^2 + 4s^2 + s - 6)$$

$$\frac{dK}{ds} = -(3s^2 + 8s + 1) = 0$$

$$3s^2 + 8s + 1 = 0$$

$$s = \frac{-8 \pm \sqrt{8^2 - 4\times 3\times 1}}{2\times 3} = \frac{-8 \pm \sqrt{64-12}}{6}$$

$$s = \frac{-8 \pm 7.2}{6}$$

$$s = -2.53, -0.133$$

The point -2.53 does not lie on root loci. So the breakaway point is $s = -0.133$

Rules 7 and 8. Do not exist.

Rule 9. Root locus crosses imaginary axis

The characteristic equation is

$$s^3 + 4s^2 + s - 6 + K = 0$$

By Routh array

$$\begin{array}{lcc} s^3 & 1 & 1 \\ s^2 & 4 & -6+K \\ s^1 & \dfrac{4-(-6+K)}{4} & 0 \\ s^0 & -6+K & \end{array}$$

For sustained oscillation

$$\frac{4-(-6+K)}{4} = 0$$

and

$$-6 + K = 0$$
$$K = 6$$
$$4 + 6 - K = 0$$
$$K = 10$$

So auxiliary equation

$$4s^2 - 6 + K = 0$$
$$4s^2 - 6 + 10 = 0$$
$$4s^2 = -4$$
$$s^2 = -1$$
$$s = \pm\sqrt{-1} = \pm j$$
$$j\omega = \pm j$$
$$\omega = 1 \text{ rad/sec.}$$

As the data calculated above the complete root locus plot is sketched in Fig. Ex. 7.6 (c).

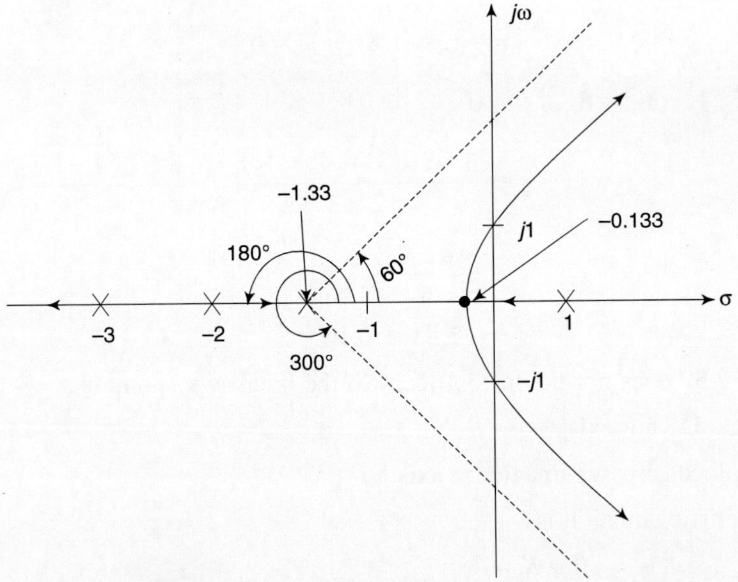

Figure Ex. 7.6 (c) Root locus of Ex. 7.6

Example 7.7 For a control system shown in Fig. Ex. 7.7 (a) plot the root locus of the close loop system. *(U.P.T.U., 2007–08, ACS)*

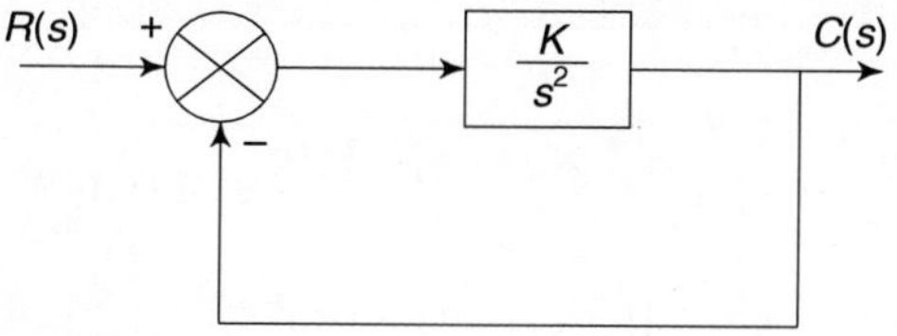

Figure Ex. 7.7 (a) Block diagram of system

Solution The $G(s) = K/s^2$

The characteristic equation is

$$1 + G(s)\,H(s) = 0$$

$$1 + \frac{K}{s^2} = 0$$

$$s^2 + K = 0$$

The transfer function of the closed loop system

$$\frac{C(s)}{R(s)} = \frac{G(s)}{1 + G(s)\,H(s)} = \frac{K/s^2}{1 + K/s^2}$$

$$Y(s) = \frac{K}{s^2 + K}$$

Rule 2. The number of branches of root locus

$$N = P \quad \text{if } P > Z$$

$$N = 2 \quad \because P = Z$$

Rule 3. The root locus starts ($K = 0$) at $s = -j\sqrt{K}, j\sqrt{K}$ and terminates on infinity.

Rule 4. Root locus does not lie on real axis

Rule 5. The number of asymptotes

$$= n - m$$

$$= 2 - 0$$

$$= 2$$

The asymptotes intersect at x on real axis

$$x = \frac{\sum_{j=1}^{n} P_j - \sum_{i=1}^{m} Z_i}{n - m} = \frac{-j\sqrt{K} + j\sqrt{K} - 0}{2 - 0}$$

$$x = 0$$

Angle of asymptotes

$$\theta = \frac{(2K+1)\times 180^\circ}{n-m}$$

$$\theta_1 = \frac{[(2\times)) + 1]\times 180}{2-0} = 90^\circ \text{ at } K = 0$$

$$\theta_2 = \frac{[(2\times 1) + 1]\times 180^\circ}{2-0} = 270^\circ \text{ at } K = 1$$

Rule 6 Calculation of break-away point

The characteristic equation is

$$1 + G(s)\,H(s) = 0$$

$$s^2 + K = 0$$

$$K = s^2$$

$$\frac{dK}{ds} = -\,2s = 0$$

$$s = 0$$

Rules 7, 8 and 9 Do not exist.

As the data calculated above the root locus plot is sketched in Fig. Ex. 7.7 (b).

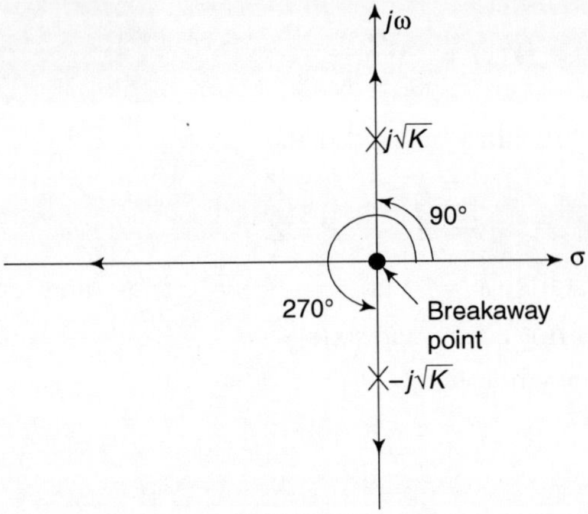

Figure Ex. 7.7 (b) Root locus of Ex. 7.7

Example 7.8 Sketch the root locus plot for the system when open loop transfer function is given by

$$G(s)\,H(s) = \frac{K}{s\,(s+4)\,(s^2+4s+13)}$$

Solution The characteristic equation is

$$1 + G(s)\,H(s) = 0$$

$$1+\frac{K}{s\,(s+4)(s^2+4s+13)} = 0$$

$$s\,(s + 4)\,(s^2 + 4s + 13) + K = 0$$

Rule 2. The number of branches of the root locus

$$N = P \quad \text{if } P > Z$$

$$N = 4 \quad \because P = 4$$

Rule 3. The root locus starts at open loop poles ($K = 0$) at $s = 0, -4$ and $-2 \pm 3j$ and terminates at infinity.

Rule 4. The root locus on the real axis exists between 0 and – 4

Rule 5. The number of asymptotes

$$= n - m$$

$$= 4 - 0 = 4$$

The asymptotes intereset at point x on real axis

$$x = \frac{\sum_{j=1}^{n} P_j - \sum_{i=1}^{m} Z_i}{n-m} = \frac{(0-4-2+3j-2-3j)}{4-0}$$

$$x = -8/4 = -2$$

The pole zero configuration is shown in Fig. Ex. 7.8 (a).

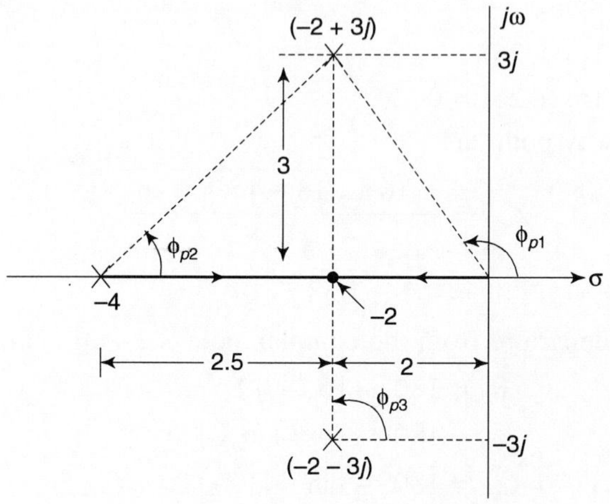

Figure Ex. 7.8 (a) Pole zero configuration

The angle of asymptotes

$$\theta_1 = \frac{[(2\times 0)+1]\times 180^\circ}{4-0} = 45^\circ \text{ at } K = 0$$

$$\theta_2 = \frac{[(2\times 1)+1]\times 180^\circ}{4-0} = 135^\circ \text{ at } K = 1$$

$$\theta_3 = \frac{[(2\times 2)+1]\times 180^\circ}{4-0} = 225^\circ \text{ at } K = 2$$

$$\theta_4 = \frac{[(2\times 3)+1]\times 180^\circ}{4-0} = 315^\circ \text{ at } K = 3$$

Rule 6. Break-away point

The characteristic equation is

$$s\,(s+4)\,(s^2+4s+13)+K=0$$

$$K = -\,(s^2+4s)\,(s^2+4s+13)$$

$$K = -\,[s^4+4s^3+13s^2+4s^3+16s^2+52s]$$

$$K = -\,[s^4+8s^3+29s^2+52s]$$

$$\frac{dK}{ds} = -\,[4s^3+24s^2+58s+52] = 0$$

$$4s^3+24s^2+58s+52=0$$

The break-away points are given by the roots of the above equation, one of the roots is identified (by Hit and Trail method) as = – 2.0 (approx.)

$$s = -\,2 \text{ is breakaway point}$$

and $$\frac{4s^2+24s^2\;58s+52}{s+2} = 0$$

$$\Rightarrow \qquad (4s^2+16s+26)$$

$$\Rightarrow \qquad (s+2)\,(4s^2+16s+26)=0$$

The complex breakaway point are

$$s = \frac{-16 \pm \sqrt{16-4\times 4\times 26}}{2\times 4}$$

$$= -\,2 \pm 1.58\,j$$

Rule 7. The angle of departure from the complex pole $-2+3j$ is given by

$$\theta_d = 180^\circ - (\phi_p - \phi_z)$$

$$\phi_d = 180^\circ - (\phi_{p1} + \phi_{p2} + \phi_{p3})$$

$$\phi_{p1} = 180^\circ - \tan^{-1}(3/2)$$

$$\phi_{p2} = \tan^{-1}(3/2)$$

$$\phi_{p3} = 90^\circ$$

$$\phi_d = 180^\circ - [180^\circ - \tan^{-1}(2/3) + \tan^{-1}(2/3) + 90^\circ]$$

$$\phi_d = 180^\circ - 270^\circ = -\,90^\circ \text{ at } -2+3j$$

and the angle of departure at $-2-3j$ in $\phi_d = 90^\circ$

Rule 9. Root locus crosses the imaginary axis

The characteristic equation is

$$s^4 + 8s^3 + 29s^2 + 52s + K = 0$$

The Routh's array is

s^4	1	29	K
s^3	8	52	
s^2	22.5	K	
s^1	$52.5 - 0.35\,K$		
s^0	K		

For stability

$$52.5 - 0.35\,K \geq 0$$
$$K \leq 148.6$$

The auxiliary equation is

$$22.5\,s^2 + K = 0$$
$$22.5\,s^2 + 148.6 = 0$$
$$s = \pm j\sqrt{\frac{148.6}{22.5}}$$
$$s = \pm j\,2.56 \quad \textbf{Ans.}$$

As the data calculated above the root locus plot is sketch in Fig. Ex. 7.8 (b).

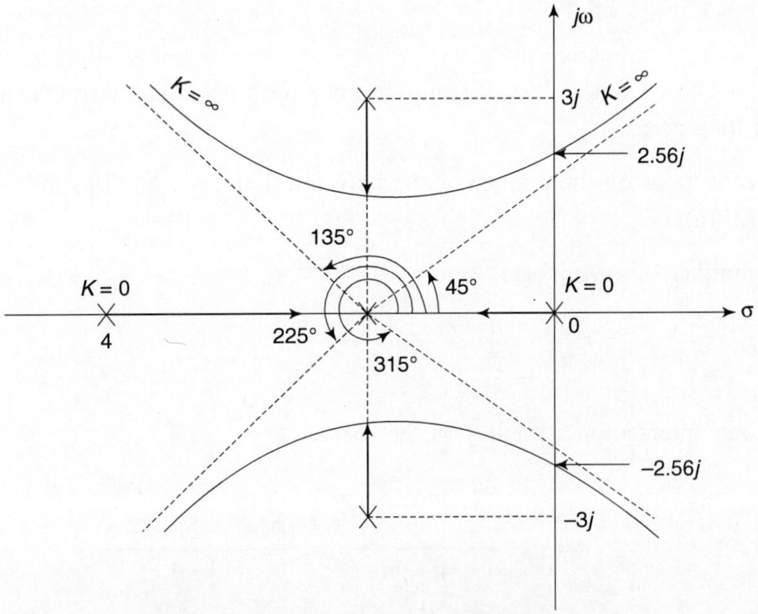

Figure Ex. 7.8 (b) Root locus of Ex. 7.8

Example 7.9 Draw the root locus plot for a unity feedback system whose characteristic equation is given by $s^3 + 3s^2 + (K + 2)\, s + 5K = 0$. *(U.P.T.U., 2006–07, ACS)*

Solution The characteristic equation is

$$s^3 + 3s^2 + (K + 2)\, s + 5K = 0$$

i.e.,
$$1 + G(s)\, H(s) = 0$$

$$1 + G(s)\, H(s)$$

$$\Rightarrow \quad s^3 + 3s^2 + (K + 2)\, s + 5K = 0$$

$$\Rightarrow \quad s^3 + 3s^2 + 2s + Ks + 5K = 0$$

$$\Rightarrow \quad s[s^2 + 3s + 2] + (s + 5)\, K = 0$$

$$\Rightarrow \quad s\,(s + 1)\,(s + 2) + K\,(s + 5) = 0$$

$$\Rightarrow \quad 1 + \frac{K\,(s+5)}{s\,(s+1)\,(s+2)} = 0$$

So
$$G(s) = \frac{K\,(s+5)}{s\,(s+1)\,(s+2)} \text{ and } H(s) = 1$$

Rule 2. The number of branches of the root locus

$$N = P \quad \text{if } P > Z$$

$$N = 3 \quad \because P = 3$$

Rule 3. The root locus starts ($K = 0$) from the open loop poles and terminates ($K = \infty$) on open loop zeros.

Rule 4. The root locus on the real axis exist between 0 and – 1 and beyond – 2 as in Fig. Ex. 7.9 (a).

Rule 5. The number of asymptotes

$$= n - m$$

$$= 3 - 1 = 2$$

The asymptotes intersect at a point x on the real axis

$$x = \frac{\sum_{j=1}^{n} P_j - \sum_{i=1}^{m} Z_i}{n - m} = \frac{(0 - 1 - 2) - (-5)}{3 - 1}$$

$$x = \frac{-3 + 5}{2} = 1$$

Angle of asymptotes

$$\theta = \frac{(2K+1)\times 180°}{n-m}$$

(i) $\theta_1 = \dfrac{[(2\times 0)+1]\times 180°}{3-1} = 90°$ at $K = 0$

(ii) $\theta_2 = \dfrac{[(2\times 1)+1]\times 180°}{3-1} = 270°$ at $K = 1$

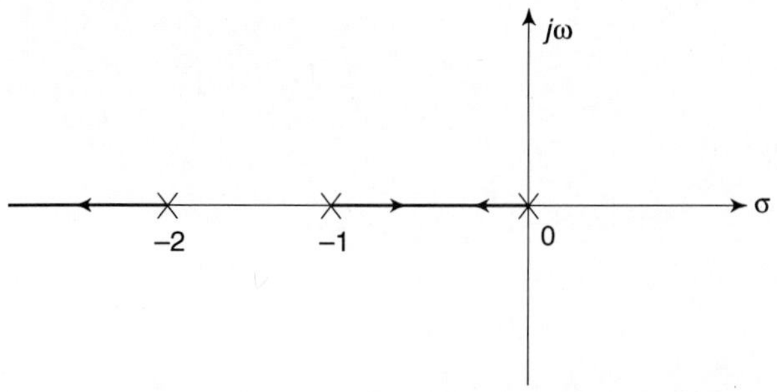

Figure Ex. 7.9 (a) Root locus along real axis

Rule 6. Break-away point

The characteristic equation is

$$1+\frac{K\,(s+5)}{s\,(s+1)\,(s+2)} = 0$$

$$s\,(s+1)\,(s+2) + K\,(s+5) = 0$$

$$K = -\left[\frac{s\,(s+1)\,(s+2)}{(s+5)}\right] = -\frac{(s^3+3s^2+2s)}{s+5}$$

$$K = -\frac{[s\,(s^2+3s+2)]}{(s+5)}$$

$$\frac{dK}{ds} = -\left[\frac{(s^3+3s^2+2s)\times 1-(s+5)\,(3s^2+6s+2)}{(s+5)^2}\right]$$

$$= -\left[\frac{s^3+3s^2+2s-3s^3+21s^2+32s+10}{(s+5)^2}\right]$$

$\Rightarrow \quad 2s^3 - 19s^2 - 30s - 10 = 0$

$\Rightarrow \quad 2s^3 - 19s^2 - 30s + 10 = 0$

By solving the above equation (Hit and Trail method) the root of equation is approximate $s = -\,0.81$.

Rules 7 and 8. Because there is no complex poles and zeros so these rules do not exist.

Rule 9. Root locus intersect on imaginary axis.

The characteristic equation is

$$s^3 + 3s^2 + (K + 2)\, s + 5\, K = 0$$

By Routh's array

$$\begin{array}{lcc} s^3 & 1 & K+2 \\ s^2 & 3 & 5K \\ s^1 & \dfrac{3\,(K+2)-5K}{3} & 0 \\ s^0 & 5K & \end{array}$$

For stable system $K > 0$

$$\frac{3\,(K+2)-5K}{3} > 0$$

$$3\,(K + 2) - 5K > 0$$

$$3\,K + 6 - 5K > 0$$

$$2K < -6$$

$$K < 3$$

For sustained oscillation $K = 3$

The auxiliary equation is $3s^2 + 5K = 0$

$$3s^2 + 5 \times 3 = 0$$

$$3s^2 = -15$$

$$s^2 = \frac{-15}{3} = -5$$

$$s = \pm j\sqrt{5}$$

$$s = \pm\, 2.23\, j$$

As the data calculated above the root locus plot is sketched in Fig. Ex. 7.9 (b).

Example 7.10 Sketch the root locus from the open loop transfer function given below and determine the stability $G(s)\,H(s) = \dfrac{K\,(s+0.1)}{s\,(s-0.2)\,(s^2+s+0.6)}$.

Solution The characteristic equation is

$$1 + G(s)\,H(s) = 0$$

$$1 + \frac{K\,(s+0.1)}{s\,(s-0.2)\,(s^2+s+0.6)} = 0$$

$s\,(s-0.2)\,(s^2+s+0.6)+K\,(s+0.1)=0$

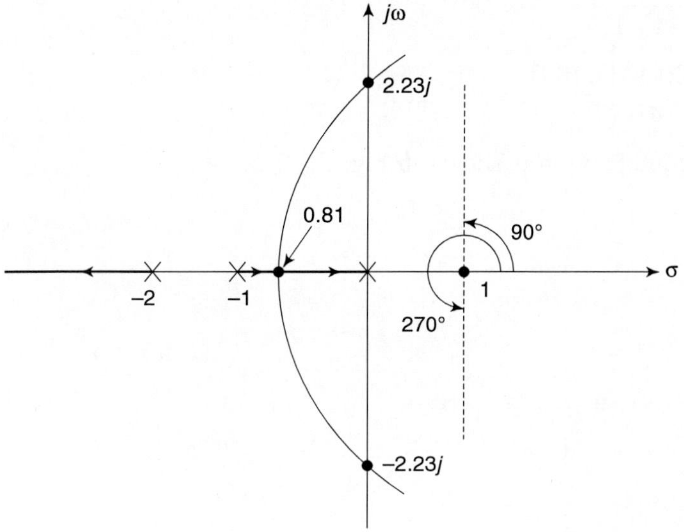

Figure Ex. 7.9 (b) Root locus of Ex. 7.9

Rule 2. The number of root locus branches

$\therefore \quad N = P \quad \text{if } P > Z$

and $\quad N = 4 \quad \because P = 4$

Rule 3. The root locus starts ($K = 0$) from open loop poles and terminates ($K = \infty$) on open loop zeros.

Rule 4. The root locus on the real axis exists between $-\infty$ and -0.1 (zero) and 0 and 0.2 (pole).

Rule 5. The number of asymptotes

$$= n - m = 4 - 1 = 3$$

The asymptotes intersect at a point x on real axis

$$x = \frac{\sum_{j=1}^{n} P_j - \sum_{i=1}^{m} Z_i}{n-m}$$

$$= \frac{(0+0.2-0.50+j\,0.59-0.5-0.59\,j)-(-0.1)}{4-1}$$

$$x = \frac{-0.7}{3} = -0.23$$

The angle of asymptotes

(i) $\theta_1 = \dfrac{[(2\times 0)+1]\times 180^\circ}{4-1} = 60^\circ$ at $K = 0$

(ii) $\theta_2 = \dfrac{[(2\times 1)+1]\times 180^\circ}{4-1} = 180^\circ$ at $K = 1$

(iii) $\theta_3 = \dfrac{[(2\times 2)+1]\times 180^\circ}{4-1} = 300^\circ$ at $K = 2$

The pole zero configuration is shown in Fig. Ex. 7.10 (a).

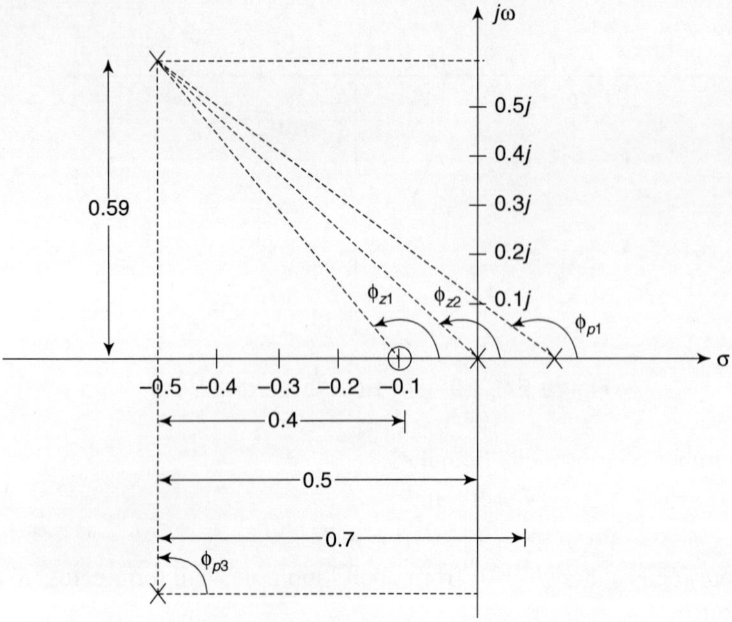

Figure Ex. 7.10 (a) Pole zero configuration

Rule 6. Break-away point

The characteristic equation is

$$[s\,(s - 0.2)\,(s^2 + s + 0.6) + K\,(s + 0.1) = 0$$

$$K = \frac{-[s\,(s-0.2)\,(s^2+s+0.6)}{(s+0.1)}$$

$$\frac{dK}{ds} = \frac{(3s^4+2s^3+0.64s+0.08\,s-0.12)}{(s+0.1)^2}$$

Now put $$\frac{dK}{ds} = 0$$

$$3s^4 + 2s^3 + 0.64s + 0.08s + 0.12 = 0$$

The above equation is nearly satisfied at $s = 0.08$ and $s = 0.37$ so the break-away points can be considered to be located respectively.

$$s = 0.37 \quad \text{and} \quad s = 0.08$$

Rule 7. The angle of departure from the complex pole (– 0.5 + 0.59 *j*) is [Refer Fig. Ex. 7.10 (a)]

$$\theta_d = 180° - (\phi_p - \phi_z)$$

$$\phi_{p1} = 180° - \tan^{-1}\left(\frac{0.59}{0.7}\right) = 139.8°$$

$$\phi_{p2} = 180° - \tan^{-1}\left(\frac{0.59}{0.5}\right) = 130.2°$$

$$\phi_{p3} = 90°$$

$$\phi_{z1} = 180° - \tan^{-1}\left(\frac{0.59}{0.4}\right) = 124.1°$$

So the angle of departure from the pole (– 0.5 + *j* 0.59)

$$\phi_d = 180° - [139.8° + 130.2° + 90° - 124.1°]$$

$$\phi_d = 55.9° \text{ from } -0.5 + 0.59\,j$$

Similarly, ϕ_d from – 0.5 – 0.59 *j*

$$\phi_d = +\,55.9°$$

Rule 9. Intersection of root locus plot with imaginary axis

The characteristic equation is

$$s\,(s - 0.2)\,(s^2 + s + 0.6) + K\,(s + 0.1) = 0$$

$$s^4 + 0.8\,s^3 + 0.4\,s^2 + (K - 0.12)\,s + 0.1\,K = 0$$

The Routh's array is

$$\begin{array}{llll} s^4 & 1 & 0.4 & 0.1\,K \\ s^3 & 0.8 & K - 0.12 & \\ s^2 & 0.55 - 1.25\,K & 0.1\,K & \\ s^1 & \dfrac{-1.25\,K^2 + 0.63\,K - 0.066}{(0.55 - 1.25\,K)} & & \\ s^0 & 0.1K & & \end{array}$$

For sustained oscillation

$$\frac{-1.25\,K^2 + 0.63\,K - 0.066}{(0.55 - 1.25\,K)} = 0$$

$$K = \frac{-0.63 \pm \sqrt{0.63^2 - 4\,(-1.25)\,(-0.066)}}{2 \times (-1.25)}$$

$$K = \frac{-0.372}{-2.5}, \frac{-0.888}{-2.5}$$

$$K = 0.148, \quad 0.352$$

The root locus crosses the imaginary axis at $K = 0.148$ and $K = 0.352$

The auxiliary equation is

$$(0.55 - 1.25\ K)\ s^2 + 0.1\ K = 0$$

Put $K = 0.148$

$$0.55 - (1.25 \times 0.148)\ s^2 + 0.1 \times 0.148 = 0$$

$$s = \pm j\ 0.2$$

Now put $K = 0.352$

$$[0.55 - (1.25 \times 0.352)\ s^2 + 0.1 \times 0.352 = 0$$

$$s = \pm j\ 0.56$$

As the data calculated above the root locus plot is sketched is shown in Fig. Ex. 7.10 (b).

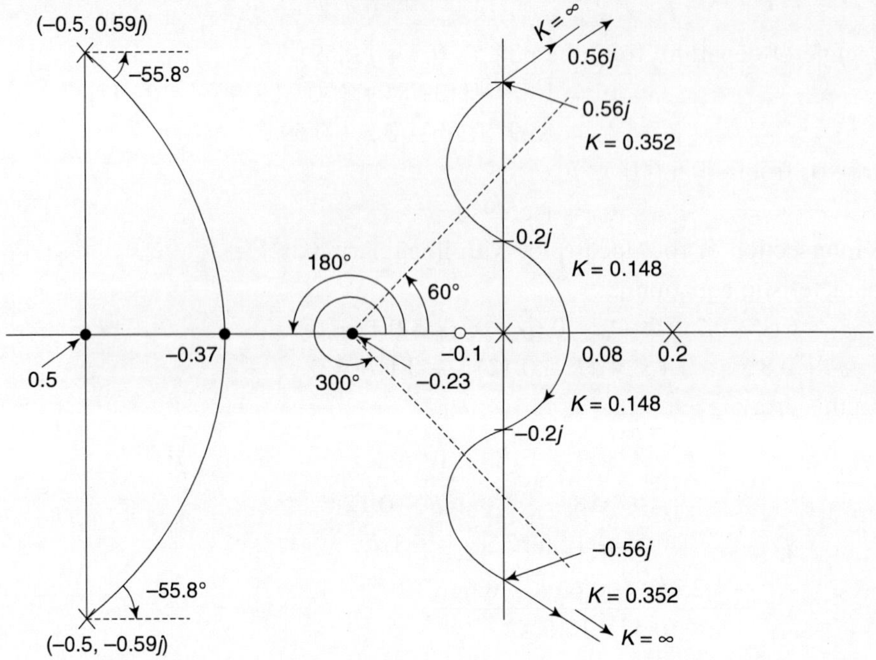

Figure Ex. 7.10 (b) *R* of locus of Ex. 7.10

Example 7.11 Consider the loop transfer function $G(s)\ H(s) = \dfrac{K\ (s+2)}{s^2 + 2s + 2}$.

Construct the root locus of the above function and comment on stability.

(U.P.T.U., 2004–05, ACS)

Solution The characterisitc equation is

$$1 + G(s)\ H(s) = 0$$

$$1+\frac{K(s+2)}{s^2+2s+2}=0$$

$$s^2 + 2s + 2 + K(s + 2) = 0$$

Rule 2. The number of branches of root locus

$$N = P \quad \text{if } P > Z$$

$$N = 2 \quad \because P = 2$$

Rule 3. The root locus starts ($K = \infty$) from open loop poles $s = -1 + j$ and $-1 - j$ and terminates ($K = \infty$) on open loop zero $s = 2$.

Rule 4. The root locus on the real axis exist beyond -2.

Rule 5. The number of asymptotes

$$= n - m$$

$$= 2 - 1 = 1$$

The asymptotes intersect at a point x on real axis

$$x = \frac{\sum_{j=1}^{n} P_j - \sum_{i=1}^{m} Z_i}{n-m} = \frac{(-1+j-1-j)-(-2)}{2-1}$$

$$x = 0$$

The angle of asymptotes

$$\theta = \frac{(2K+1)\times 180°}{n-m} = 60° \text{ at K = 0.}$$

$$\theta_1 = \frac{[(2\times 0)+1]\times 180°}{2-1} = 180° \text{ at } K = 0$$

Rule 6. Break-away point

The characteristic equation is

$$s^2 + 2s + 2 + K(s + 2) = 0$$

$$K = -\frac{(s^2+2s+2)}{s+2}$$

$$\frac{dK}{ds} = -\left[\frac{(s^2+2s+2)\times 1-(s+2)(2s+2)}{(s+2)^2}\right]$$

$$\frac{dK}{ds} = -\left[\frac{s^2+2s+2-2s^2-6s-4}{(s+2)^2}\right]$$

$$\frac{dK}{ds} = -\left[\frac{-s^2-4s-2}{(s+2)^2}\right]$$

Put
$$\frac{dK}{ds} = 0$$
$$s^2 + 4s + 2 = 0$$
$$s = \frac{-4 \pm \sqrt{16 - 4\times1\times2}}{2} = \frac{-4 \pm \sqrt{8}}{2}$$
$$s = -0.586 \text{ and } -3.414$$

Rules 7, 8 Do not exist.

Rule 9. Root locus intersect imaginary axis
$$s^2 + 2s + 2 + (s + 2)\,K = 0$$
$$s^2 + (2 + K)\,s + 2\,K + 2 = 0$$

By Routh's array

s^2	1	$2\,K + 2$
s^1	$2 + K$	0
s^0	$2K + 2$	

For stable system
$$2 + K > 0$$
$$\Rightarrow \quad K < 2$$

for sustained oscillation
$$2 + K = 0$$
$$K = -2$$

The auxiliary equation

$(2 + K)\,s = 0 \Rightarrow s = 0$, i.e., root locus does not intersect imaginary axis.

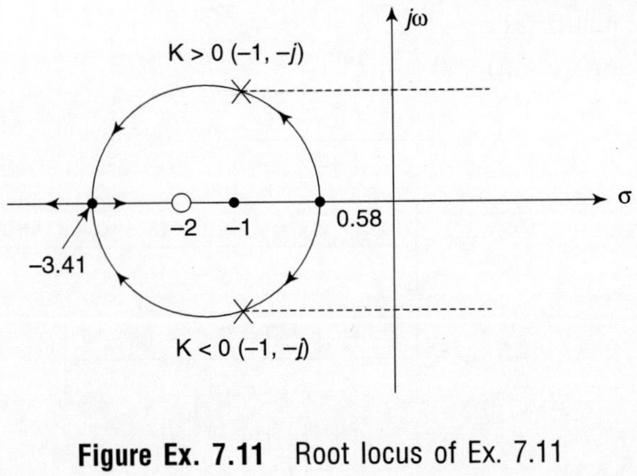

Figure Ex. 7.11 Root locus of Ex. 7.11

Example 7.12 Construct the root locus for $K > 0$ for the transfer function

$$G(s)\,H(s) = \frac{K\,(s+1)}{s\,(s+2)\,(s+4)} \qquad \textit{(U.P.T.U., 2003–04, ACS)}$$

Solution The characteristic equation is

$$1 + G(s)\,H(s) = 0$$

$$1 + \frac{K\,(s+1)}{s\,(s+2)\,(s+4)} = 0$$

$$s\,(s + 2)\,(s + 4) + K\,(s + 1) = 0$$

Rule 2. The number of branches of root locus

$$N = P \quad \text{if } P > Z$$

$$N = 3$$

$\therefore$ $\quad P = 3$

and $\quad Z = 1$

Rule 3. The root locus starts ($K = 0$) from open loop poles $s = 0, -2, -4$ and terminates ($K = \infty$) on open loop zeros $s = -1$.

Rule 4. The root locus on the real axis exist between 0 and -1 (zero) and between -2 and -4.

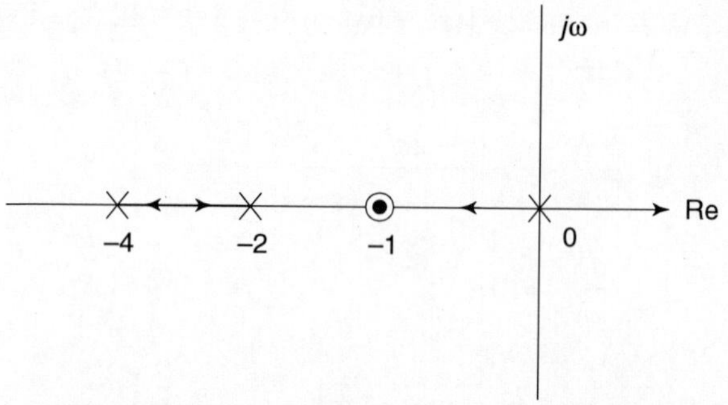

Figure Ex. 7.12 (a) Root locus along real axis

Rule 5. The number of asymptotes

$$= n - m$$

$$= 3 - 1 = 2$$

The angle of asymptotes

$$\theta = \frac{(2K+1)\times 180^\circ}{n-m}$$

(i) $\theta_1 = \dfrac{[(2\times 0)+1]\times 180^\circ}{3-1} = 90^\circ$ at $K = 0$

(ii) $\theta_2 = \dfrac{[(2\times 1)+1]\times 180^\circ}{3-1} = 270^\circ$ at $K = 1$

The asymptotes intersect on real axis at point x

$$x = \frac{\Sigma P_j - \Sigma Z_i}{n-m} = \frac{(0-2-4)-(-1)}{3-1}$$

$$x = \frac{-5}{2} = -2.5$$

$$x = -2.5$$

Rule 6. Break-away point

The characteristic equation is

$$s\,(s+2)\,(s+4) + K\,(s+1) = 0$$

$$K = \frac{-[s\,(s+2)\,(s+4)}{(s+1)}$$

$$K = -\left[\frac{s^3+6s^2+8s}{s+1}\right]$$

$$\frac{dK}{ds} = \frac{-[(s^3+6s^2+8s)\times 1-(s+1)\,(3s^2+12s+8)]}{(s+1)^2}$$

$$\frac{dK}{ds} = \frac{-[(s^3+6s^2+8s-3s^2-12s^2-8s-3s^2-12s-8]}{(\quad 1)}$$

$$\frac{dK}{ds} = -\left[\frac{-2s^3-9s^2-12s-8}{(s+1)^2}\right]$$

Put $$\frac{dK}{ds} = 0$$

$$2s^3 + 9s^2 - 12s - 8 = 0$$

By solving the above equation (By Hit and Trial method) the root $(s + 2.9)$ satisfies the equation so $s = -2.9$ is one root which will represent the break-away point.

Rules 7 and 8. Do not exist (because there is no complex roots)

Rule 9. Root locus intersect imaginary axis

The characteristic equation is

$$s\,(s+2)\,(s+4) + K\,(s+1) = 0$$

$$s^3 + 6s^2 + 8s + K\,(s+1) = 0$$

$$s^3 + 6s^2 + (8+K)\,s + K = 0$$

By Routh's array

s^3	1	$8 + K$
s^2	6	K

$$s^1 \qquad \frac{48 + 6K - K}{6} \qquad 0$$

$$s^0 \qquad K$$

For stable system $K > 0$

$$\frac{48 + 5K}{6} > 0$$

$$48 + 5K > 0$$

$$5K < 48$$

$$K < \frac{48}{5} = 9.6$$

For sustained oscillation

$$\frac{48 + 5K}{6} = 0$$

$$5K = -48$$

$$K = \frac{-48}{5}$$

$$K = -9.6$$

The auxiliary equation is

$$6s^2 + K = 0$$

$$6s^2 - 9.6 = 0$$

$$6s^2 = 9.6$$

$$s^2 = 9.6/6$$

$$s^2 = 1.6$$

$$s = \pm 1.26$$

i.e., it does not interest imaginary axis.

As data calculated above the root locus plot is sketched in Fig. Ex. 7.12 (b).

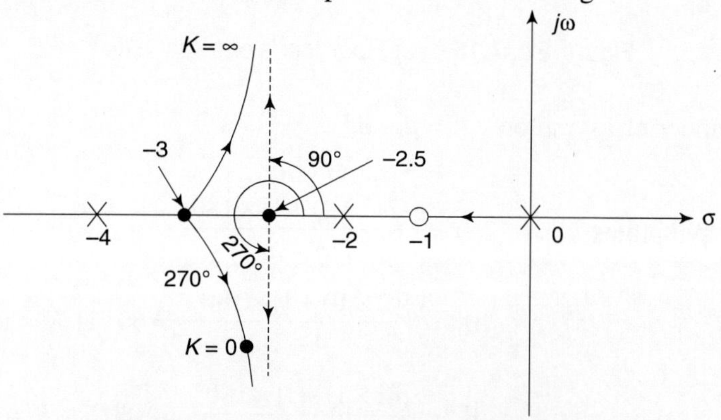

Figure Ex. 7.12 (b) Root locus of Ex. 7.12

Example 7.13 Plot the root locus pattern for a system whose forward path transfer function is $\dfrac{K(s+1)}{s(s+2)(s^2+2s+5)}$.

Solution The characterisitc equation is

$$1 + G(s)\,H(s) = 0$$

$$1+\frac{K(s+1)}{s(s+2)(s^2+2s+5)} = 0$$

$$s\,(s + 2)\,(s^2 + 2s + 5) + K\,(s + 1) = 0$$

Rule 2. The number of branches of the root locus

$$N = P \quad \text{if } P > Z$$

$\therefore$ $\quad P = 4$

and $\quad Z = 1$

So $\quad N = 4$

Rule 3. The root locus starts ($K = 0$) from open loop poles, $s = 0, -2, -1 + 2j, -1 - 2j$ and terminates ($K = \infty$) on open loop zeros at $s = -1$.

Rule 4. The root locus on the real axis exists between 0 and – 1 (zero) and beyond – 2 as shown in Fig. Ex. 7.13 (a).

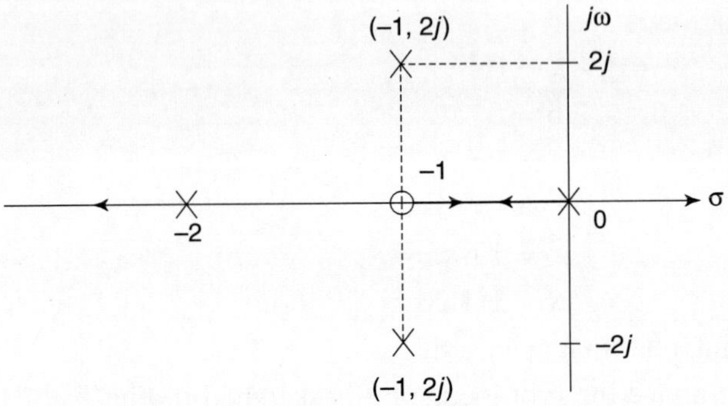

Figure Ex. 7.13 (a) Root loci along real axis

Rule 5. The number of asymptotes $= n - m$

$$= 4 - 1 = 3$$

The angle of asymptotes

$$\theta = \frac{(2K+1)+1]\times 180^\circ}{n-m}$$

$$\theta_1 = \frac{[(2\times 0)+1]\times 180^\circ}{4-1} = 60^\circ \text{ at } K = 0$$

$$\theta_2 = \frac{[(2\times 1)+1]\times 180^\circ}{4-1} = 180^\circ \text{ at } K = 1$$

$$\theta_3 = \frac{[(2 \times 2) + 1] \times 180°}{4 - 1} = 300° \text{ at } K = 2$$

Intersection of asymptotes with real axis

$$x = \frac{\sum_{j=1}^{n} P_j - \sum_{i=1}^{m} Z_i}{n - m} = \frac{(-2 - 1 + 2j - 1 - 2j) - (-1)}{4 - 1}$$

$$x = \frac{-3}{3} = -1$$

Rule 6. Break-away point

The characteristic equation

$s\,(s + 2)\,(s^2 + 2s + 5) + K\,(s + 1) = 0$

$$K = \frac{-[s\,(s + 2)\,(s^2 + 2s + 5)]}{(s + 1)}$$

$$K = \frac{-[(s^2 + 2s)\,(s^2 + 2s + 5)]}{(s + 1)}$$

$$K = \frac{-[s^4 + 2s^3 + 5s^2 + 2s^3 + 4s^2 + 10s)]}{(s + 1)}$$

$$K = \frac{-[s^4 + 4s^3 + 9s^2 + 10s)]}{s + 1}$$

$$\frac{dK}{ds} = \frac{-[(s^4 + 4s^3 + 9s^2 + 10s) \times 1 - (s + 1)\,(4s^3 + 12s^2 + 18s + 10)]}{(s + 1)^2}$$

$$\frac{dK}{ds} = -\left[\frac{s^4 + 4s^3 + 9s^2 + 10s - 4s^4 - 12s^3 - 18s^2 - 10s - 4s^3 - 12s^2 - 18s - 10)}{(s + 1)^2}\right]$$

$$\frac{dK}{ds} = -\frac{[3s^4 - 12s^3 - 21s^2 - 18s - 10]}{(s + 1)^2}$$

Put $\frac{dK}{ds} = 0$

$3s^4 + 12s^3 - 21\,s^2 - 18s - 10 = 0$

By the solution of the above equation, there is no root on real axis so there is no break-away point on real axis.

Rule 7. Angle of departure from complex pole

$$\theta_d = 180° - (\phi_p - \phi_Z)$$

The angle of departure from the pole at $(-1 + 2j)$ refer Fig. Ex. 7.13 (b).

$$\theta_{p1} = 180^\circ - \tan^{-1}\left(\frac{2}{1}\right) = 45^\circ$$

$$= 180^\circ - 63.43^\circ = 116.5^\circ$$

$$\theta_{p2} = \tan^{-1}\left(\frac{2}{2}\right)$$

$$\theta_{p3} = 90^\circ$$

$$\theta_{z1} = 90^\circ$$

$$\theta_d = 180^\circ - (116.5^\circ + 90^\circ + 45^\circ - 90^\circ) = 180^\circ - 161.5^\circ$$

$$\theta_d = 18.5^\circ \text{ from } (-1 + 2j)$$

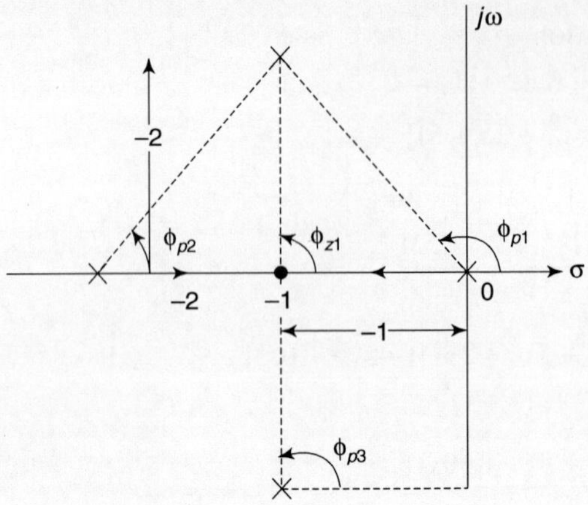

Figure Ex. 7.13 (b) Pole zero configuration

The angle of departure from $(-1 - 2j)$

$$\theta_d = -18.5^\circ$$

Rule 8. Does not exist because there is no complex zero

Rule 9. Intersection of root locus on imaginary axis

The characterisitc equation is

$$s(s+2)(s^2+2s+5) + K(s+1) = 0$$

$$s^4 + 4s^3 + 9s^2 + 10s + K(s+1) = 0$$

$$s^4 + 4s^3 + 9s^2 + (10+K)s + K = 0$$

By Routh's array

s^4	1	9	K
s^3	4	$10 + K$	0
s^2	$\dfrac{26-K}{4}$	K	0

$$s^1 \quad \frac{\left[\frac{(26-K)(10+K)}{4}\right]-4K}{\frac{26-K}{4}} \quad 0$$

$$s^0 \quad K$$

For stable system

$$K > 0$$

and

$$\frac{26-K}{4} > 0$$

$$26 - K > 0$$

$$K < 26$$

For sustained oscillation

$$\frac{\left[\frac{(26-K)(10+K)}{4}\right]-4K}{\frac{26-K}{4}} = 0$$

$$(26 - K)\,(10 + K) - 16\,K = 0$$

$$260 + 26\,K - 10\,K - K^2 - 16\,K = 0$$

$$K^2 = 260$$

$$K = \pm\,16.1$$

The auxiliary equation is

$$\left(\frac{26-K}{4}\right)s^2 + K = 0$$

$$(26 - 16.1)\,s^2 + (4 \times 16.1) = 0$$

$$9.9\,s^2 + 64.4 = 0$$

$$9.9\,s^2 = -\,64.4$$

$$s^2 = \frac{-\,64.4}{9} = -\,6.5$$

$$s = \pm\,j\,\sqrt{6.5}$$

$$s = \pm\,j\,2.55$$

As the data calculated above the root locus plot is sketched in Fig. Ex. 7.13 (c).

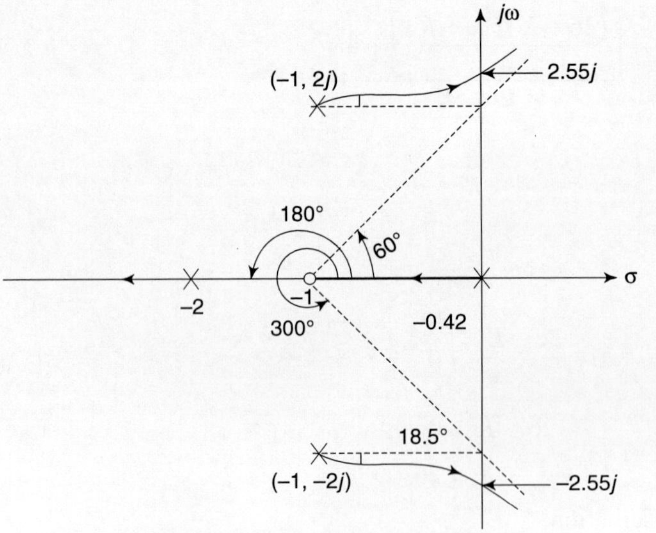

Figure Ex. 7.13 (c) Root locus of Ex. 7.13

Example 7.14 Sketch the root locus for $G(s)\,H(s) = \dfrac{K}{s\,(s+1)\,(s+2)}$.

(U.P.T.U., 2003 (C.O), ACS)

Solution The characteristic equation is

$$1 + G(s)\,H(s) = 0$$

$$1 + \frac{K}{s\,(s+1)\,(s+2)} = 0$$

$$s\,(s+1)\,(s+2) + K = 0$$

Rule 2. The number of branches of root locus

$$N = P \quad \text{if } P > Z$$

$\therefore$ $P = 3$

and $Z = 0$

So $N = 3$

Rule 3. The root locus starts ($K = 0$) from open loop poles $s = 0, -1, -2$ terminates ($K = \infty$) on infinity.

Rule 4. The root locus on the real axis exist between 0 and − 1 and beyond − 2 as shown in Fig. Ex. 7.14 (a).

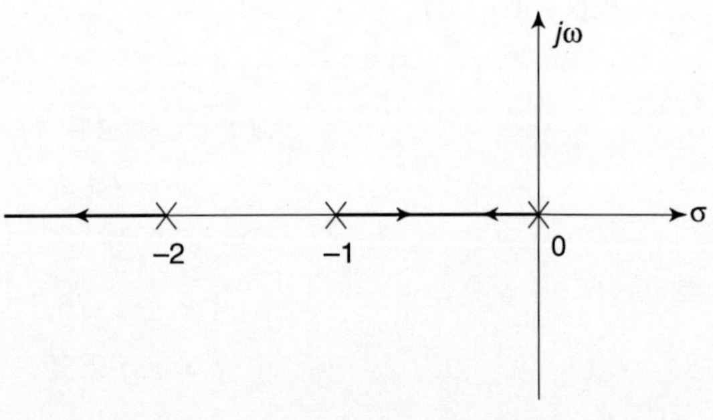

Figure Ex. 7.14 (a) Root loci along real axis

Rule 5. The number of asymptotes

$$= n - m$$

$$= 3 - 0 = 3$$

The angle of asymptotes $\theta = \dfrac{(2K+1)\times 180^\circ}{n-m}$

(i) $\theta_1 = \dfrac{[(2\times 0)+1]\times 180^\circ}{3-0} = 60^\circ$ at $K = 0$

(ii) $\theta_2 = \dfrac{[(2\times 1)+1]\times 180^\circ}{3-0} = 180^\circ$ at $K = 1$

(iii) $\theta_3 = \dfrac{[(2\times 2)+1]\times 180^\circ}{3-0} = 300^\circ$ at $K = 2$

The asymptotes intersect on the real axis at point x

$$x = \frac{\sum_{j=1}^{n} P_j - \sum_{i=1}^{m} Z_i}{n-m} = \frac{(0-1-2)-0}{3-0}$$

$$x = \frac{-3}{3} = -1$$

Rule 6. Breakaway point on real axis

The characteristic equation is

$$s\,(s + 1)\,(s + 2) + K = 0$$

$$K = -\,[(s^2 + s)\,(s + 2)]$$

$$K = -\,[s^3 + 3s^2 + 2s]$$

$$\frac{dK}{ds} = -\,[3s^2 + 6s + 2]$$

Put $\dfrac{dK}{ds} = 0$

$$3s^2 + 6s + 2 = 0$$

$$s = \frac{-6 \pm \sqrt{36-24}}{6} = -\,0.423,\ -\,1.577$$

Since the root locus on real axis lies between 0 and – 1 so the breakaway point is $s = -\,0.423$.

Rules 7 and 8. Do not exist because there is no complex pole and zero.

Rule 9. The intersection of root locus on $j\omega$ axis.

The characteristics equation is

$$s^3 + 3s^2 + 2s + K = 0$$

By Routh's array

s^3	1	2
s^2	3	K
s^1	$\dfrac{6-K}{3}$	0
s^0	K	

For stable system

$$K > 0$$

and
$$\frac{6-K}{3} > 0$$

$$6 - K > 0$$

$$K < 6$$

For sustained oscillation

$$\frac{6-K}{3} = 0$$

$$K = 6$$

The auxiliary equation is

$$3s^2 + K = 0$$

$$3s^2 + 6 = 0$$

$$s^2 = \frac{-6}{3} = -2$$

$$s = \pm j\,1.41$$

As the data calculated above the root locus plot is sketched in Fig. Ex. 7.14 (b).

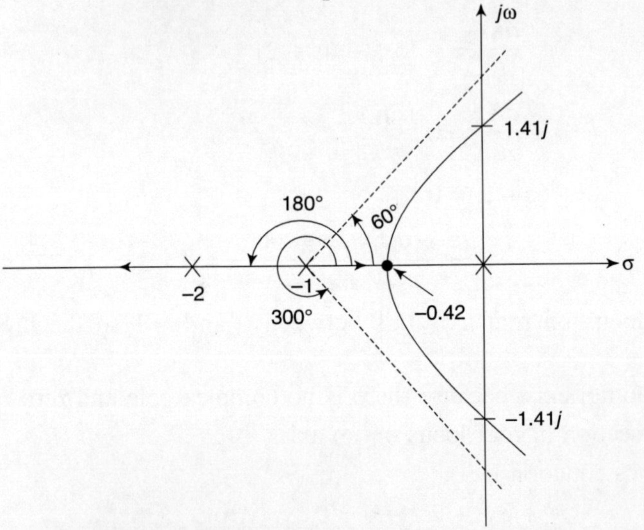

Figure Ex. 7.14 (b) Root locus of Ex. 7.14

Example 7.15 State the rules for construction of root loci of $G(s)\,H(s)$. Find the breakaway points of $G(s)\,H(s) = \dfrac{K}{s\,(s+4)\,(s^2+4s+20)} = 0.$ *(U.P.T.U., 2002–03, ACS)*

Solution For rules of construction of root loci see the section 7.4.

The break-away point

The characteristic equation is

$$1 + G(s)\,H(s) = 0$$

$$1 + \frac{K}{s\,(s+4)\,(s^2+4s+20)} = 0$$

$$s\,(s + 4)\,(s^2 + 4s + 20) + K = 0$$

$$K = -\,s\,(s + 4)\,(s^2 + 4s + 20)$$

$$K = -\,(s^2 + 4s)\,(s^2 + 4s + 20)$$

$$K = -\,[s^4 + 4s^3 + 20s^2 + 4s^3 + 16s^2 + 80s]$$

$$K = -\,[s^4 + 8s^3 + 36s^2 + 80s]$$

$$\frac{dK}{ds} = -\,[4s^3 + 24s^2 + 72s + 80]$$

Put $$\frac{dK}{ds} = 0$$

$$4s^3 + 24s^2 + 27s + 80 = 0$$

The $s = -\,2$ is satisfied in the above equaton, so $s = -\,2$ is one of the root of above equation.

Now

$$\Rightarrow \quad \frac{4s^2 + 24s^2 + 72s + 80}{s+2}$$

$$\Rightarrow \quad (s^2 + 4s + 10)$$

Now solving the equation $s^2 + 4s + 10 = 0$

$$s^2 + 4s + 10 = 0$$

$$s = \frac{-\,4 \pm \sqrt{16 - 40}}{2}$$

$$s = \frac{-\,4 \pm \sqrt{-\,24}}{2}$$

$$s = -\,2 \pm 2.45\,j$$

Thus breakaway points are

$s = -\,2,\ -\,2 + 2.45\,j$ and $-\,2 - 2.45\,j$ **Ans.**

MATLAB PROGRAMS

Program 7.1

```
% Program for finding root locus of the system
```

```
% G(s)H(s) = k(s^2-4s+25)/(s+2)(s+3) & Let k = 1
clear all;
clc;
clf                         % clear the graph on screen
numgh = input('Enter the coefficient of polynomial of numerator:- ')
                            % Define numerator
                            % like [1 -4 25]
dengh = input('Enter the coefficient of polynomial of denominator:- ');
                            % Define denominator
'G(s)H(s)'                  % Display string
GH = tf(numgh,dengh)        % Create transfer function G(s)H(s)
rlocus(GH)                  % Plot root locus
```

Input

```
Enter the coefficient of polynomial of numerator:- [1 -4 25]
Enter the coefficient of polynomial of denominator:- [1 5 6]
```

Output

```
G(s)H(s)
                        s^2 - 4s + 25
Transfer function:      -------------
                        s^2 + 5s + 6
```

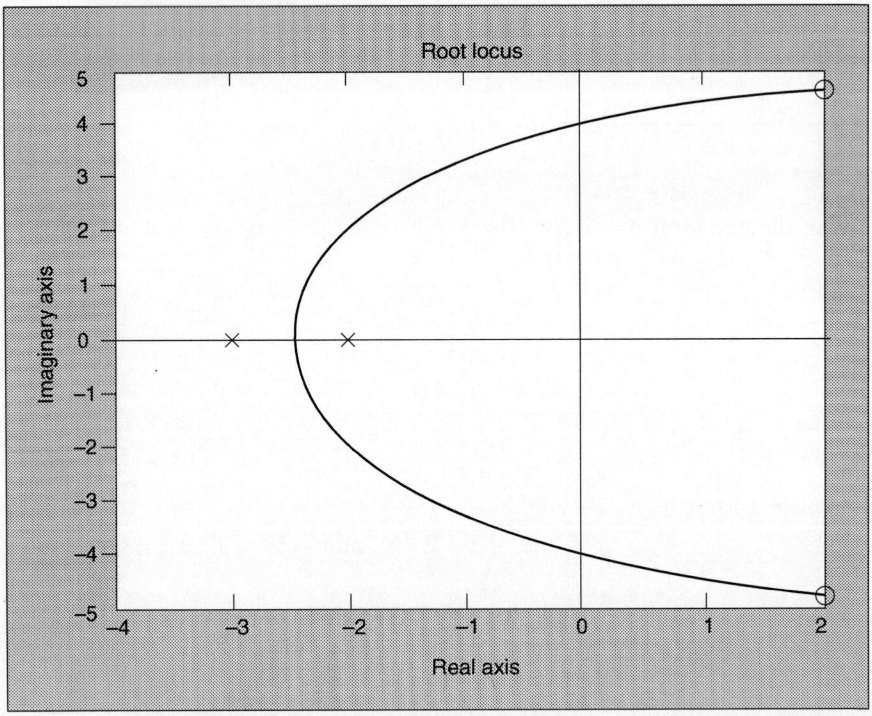

Figure P. 7.1 Root locus of $G(s)\ H(s)$

Program 7.2

```
%Program for finding root locus for given function
% G(s)H(s) = k(s+1.5)/s(s+6)(s+15)
clear all;
clc;                    % clear the figure
clf;                    % define numerator
numgh = [1 1.5];        % Define denominator
'G(s)H(s)'              % Display string
GH = tf(numgh,dengh)    % Create Transfer function
rlocus(GH)              % Plot root locus
hold on
title('Root Locus')     % Add title
K = 0:0.5:50;
rlocus(GH.K)            % Draw smoothed root locus
title('Smoothed Root Locus')
hold on
Mp = input('Enter Percent overshoot:-') % input desired overshoot
zeta = -(log(Mp/100))/sqrt(piA2+[log(Mp/100)]A2)
sgrid(zeta.O)           % overlay desired damping ratio line
                          on root locus
title('[Root locus with % overshoot line]')
[k,p] = rlocfind(GH);   % Generate gain k and close loop poles P,
                        % for point selected intercatively on
                          the root locus
'T(s)'
T = feedback(k*GH,1)    % find close loop transfer function with selected k
```

Input

```
Enter Percent overshoot:- 15
```

Output

```
                                      s + 1.5
Transfer function: G(s)H(s)   ------------------
                              s^3 + 21s^2 + 90s

zeta = 0.5169
selected_point =  -8.8329 + 6.5217i
                               71.41s + 107.1
Transfer function: T(s)   ----------------------------
                          s^3 + 21s^2 + 161.4s + 107.1
```

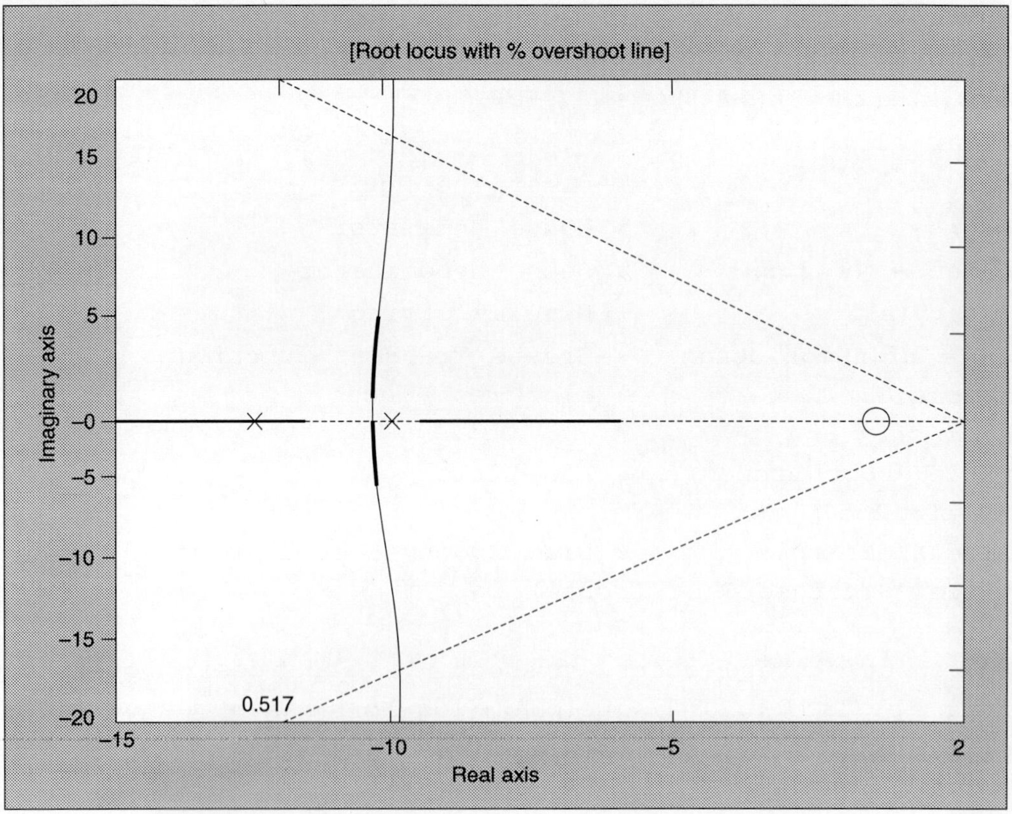

Figure P 7.2 Root locus with percent overshoot line

Program 7.3

```
% Program for Root locus for range of natural frequency and damping ratio
% G(s)H(s) = (s^-4s+20)/(s^2+4s+3)
clear all;
clc;
numgh = input('Enter the coefficient of polynomial of numerator:- ');
                        % Input coefficient of numerator
dengh = input('Enter the coefficient of polynomial of denominator:- ');
                        % Input coefficient of denominator
'G(s)H(s)'
GH = tf(numgh,dengh)  % Create transfer function
rlocus(GH);             % Find root locus
zeta = input('Enter the range of damping ratio:- ');
                        % input range of damping ratio like 0.1:0.5:0.5
```

```
Wn = input('Enter the range of natural frequency:- ')
                         % input range of natural frequency like 0:1:10
sgrid(zeta,Wn);          % Generate Wn and damping ratio grid
                           lines of root locus
title('Root Locus for range of Wn & zeta')
```

Input

```
Enter the coefficient of polynomial of numerator:- [1 -4 20]
Enter the coefficient of polynomial of denominator:- [1 4 3]
Enter the range of damping ratio:- 0.10:0:05:0.5
Enter the range of natural frequency:- 1:1:10
```

Output

```
                        s^2 - 4s + 20
Transfer function:     ---------------
                        s^2 + 4s + 3
```

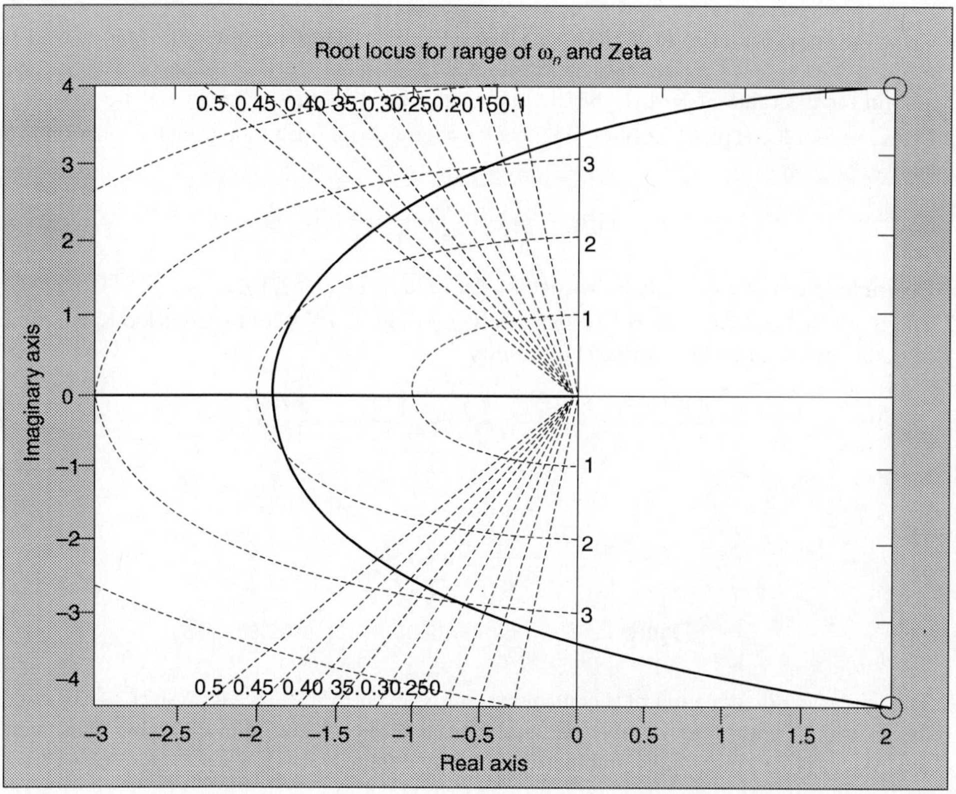

Figure P. 7.3 Root locus for range of Wn and damping ratio

EXCERCISE

7.1 Sketch the root locus plot for each of the following open loop transfer functions by determining

(i) Number of root locus branches

(ii) Angle of asymptotes

(iii) Intersection points of asymptotes on real axis

(a) $G(s)\,H(s) = \dfrac{K}{s\,(s+2.5)}$ (b) $G(s)\,H(s) = \dfrac{K}{s^2\,(s+4)}$

(c) $G(s)\,H(s) = \dfrac{K\,(s+1)}{s^2\,(s+4)}$ (d) $G(s)\,H(s) = \dfrac{K\,(s+4)}{s^2\,(s+1)}$

7.2 Explain the concepts of root locus.

7.3 Write the procedure of constructing root locus.

7.4 Explain how to determine gain margin and phase margin of a closed loop control system from its root loci. Illustrate with the help of an example. *(U.P.T.U., 2007–08, CS)*

7.5 Draw the root locus for $G(s)\,H(s) = \dfrac{K}{s\,(s+1)\,(s+4)}$. Determine the gain margin and phase margin for the value of K such that damping ratio $\xi = 0.07.7$.

7.6 Draw the root locus plot for a unity feedback system whose open loop transfer function is given below

$$G(s) = \frac{K}{s\,(s^2+5s+12.5)}$$

Determine the value of K such that one of the roots is $(-1 + 2.5\,j)$.

7.7 The block diagram of a control system is given in Fig. E.7.7. Plot the root locus as a function of parameter K and comment on the stability.

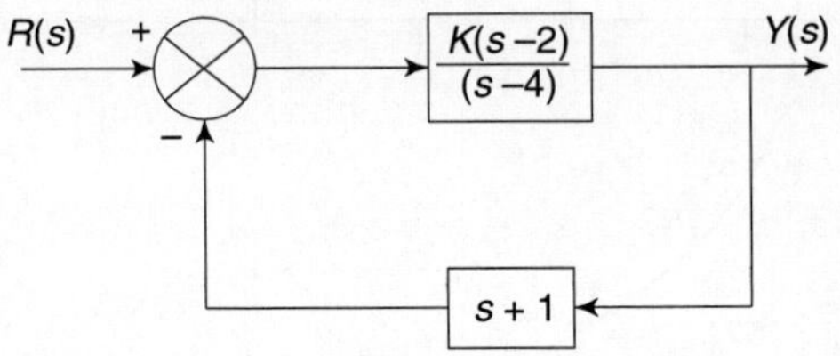

Figure E. 7.7 Block diagram for problem 7.7

7.8 The characteristic equation of a control system is given by $s^3 + 3s^2 + (K+2)\,s + 3K = 0$. Plot the root locus with K as variable parameter from 0 to ∞. Determine the value of K such that the system is (i) marginally stable (ii) critically damped

7.9 Write a short note on root locus.

ANSWERS

7.1 (a) (i) $N = 2$;

(ii) Angle of asymptotes 90°, 270°

(iii) Intersection of asymptotes on real axis – 1.25

(b) (i) $N = 3$

(ii) 60°, 180°, 300°

(iii) $x = -1.33$

(c) (i) $N = 3$

(ii) 90°, 270°

(iii) $x = -1.5$

(d) (i) $N = 3$

(ii) 90°, 270°

(iii) $x = 1.5$

7.5 $K = 2.6$, G.M. = 17 db, P.M. = 53°

7.6 $K = 21$

7.8 (i) It's asymptotic and never crosses the $j\omega$ axis, therefore, the system is stable.

(ii) Break-away point is at – 0.465 and at this point $K = 0.15$ make the system critically damped.

CHAPTER

8 Frequency Response Analysis

8.1 INTRODUCTION

The frequency response of a system is defined as the steady state response of the system to a sinusoidal input signal. The sinusoid is a unique input signal, and the resulting output signal for a linear system as well as the signal throughout the system, is sinusoidal in the steady state. The sinusoidal output differs from the input signal only in amplitude and phase angle.

Consider a linear system with sinusoidal input

$$r(t) = A \sin \omega t \qquad ...(8.1)$$

The system output in steady state is also a sinusodial, and only differs in phase and magnitude so it may be written as

$$y(t) = B \sin (\omega t + \phi) \qquad ...(8.2)$$

where B = Amplitude of the output signal

Let the transfer function of a linear single input, single output system be $M(s)$; then the Laplace transform of the input and the output are related through

$$Y(s) = M(s)\, R(s) \qquad ...(8.3)$$

For sinusoidal steady state analysis, replacing s by $j\omega$ Eq. (8.3) becomes

$$Y(j\omega) = M(j\omega)\, R(j\omega) \qquad ...(8.4)$$

By writing the function $Y(j\omega)$ as

$$Y(j\omega) = |Y(j\omega)| \angle Y(j\omega) \qquad ...(8.5)$$

By the above definitions for $M(j\omega)$ and $R(j\omega)$ Eq. (8.4) becomes

$$|Y(j\omega)| = |M(j\omega)|\, |R(j\omega)| \qquad ...(8.6)$$

and the phase relation

$$\angle Y(j\omega) = \angle M(j\omega) + \angle R(j\omega) \qquad ...(8.7)$$

Thus, for the input and output signal. The amplitude for the output sinusoid is

$$B = A\ |M(j\omega)| \qquad ...(8.8)$$

and the phase of the output is

$$\phi = \angle M(j\omega) \qquad ...(8.9)$$

$$y(t) = A\ |M(j\omega)| \sin(\omega t + \phi) \qquad ...(8.10)$$

Thus, the steady state output signal depends only on the magnitude and phase of $M(j\omega)$ at a specific frequency, ω. Notice that the steady state response as described in Eq. (8.10) is true only for stable systems.

8.2 ADVANTAGES OF THE FREQUENCY RESPONSE ANALYSIS

1. The availability of sinusoid test signals for various ranges of frequencies and amplitudes. Thus, the experimental determination of the frequency response of a system is easily accomplished and is the most reliable and uncomplicated method for the experimental analysis of a system.
2. The design of a system is the frequency domain which provides the designer with control of the bandwidth of a system, and some measure of the response of the system to undesired noise.
3. The transfer function describing the sinusoidal steady state behaviour of a system can be obtained by replacing s with j_w in the system transfer function, $T_0(s)$
4. The effects of noise disturbance and parameter variations are relatively easy to visualize and assess through frequency response.

Disadvantage

The basic disadvantage of the frequency response analysis and design is the indirect link between the frequency and the time domain. Direct correlations between the frequency response and the corresponding transient response characteristics are somewhat tedious.

8.3 CORRELATION BETWEEN TIME AND FREQUENCY RESPONSE

The correlation between frequency response and time response of a second order system is employed as an approximation for higher order system as well. So we will try to find out the correlation between frequency response and second order time response because the correlation is quite tedious to compute for higher order systems.

The second order system of the form shown in Fig. 8.1 has the transfer function

$$\frac{Y(s)}{R(s)} = \frac{\omega_n^2}{s^2 + 2\xi\,\omega_n\,s + \omega_n^2} \qquad ...(8.11)$$

where ξ = damping factor

ω_n = undamped natural frequency

The sinusoidal transfer function of the system is, by putting $s = j\omega$ in Eq. (8.11)

$$\frac{Y(j\omega)}{R(j\omega)} = T_o(j\omega) = \frac{\omega_n^2}{(j\omega)^2 + 2\xi\,\omega_n(j\omega) + \omega_n^2}$$

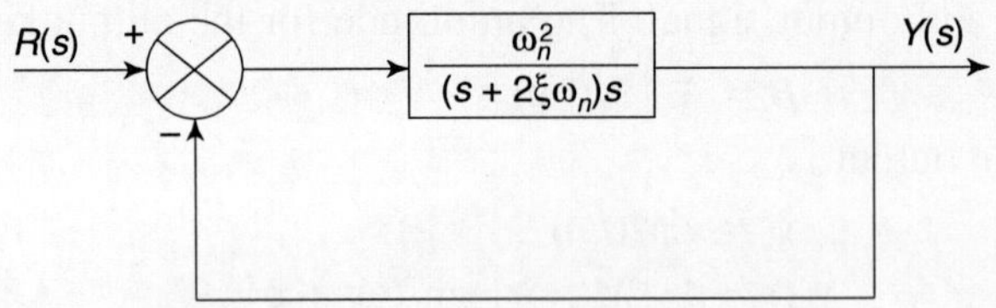

Figure 8.1 Second order system

$$T_o(j\omega) = \frac{\omega_n^2}{\omega_n^2\left[\left(j\frac{\omega}{\omega_n}\right)^2 + 2\xi\left(\frac{j\omega}{\omega_n}\right)+1\right]}$$

$$T_o(j\omega) = \frac{1}{\left(j\frac{\omega}{\omega_n}\right)^2 + 2\xi\left(j\frac{\omega}{\omega_n}\right)+1} \quad ...(8.12)$$

Let $u = \frac{\omega}{\omega_n}$; normalized driving signal frequency substituting u in Eq. (8.12)

$$T_o(j\omega) = \frac{1}{1-u^2+2\xi\, ju} \quad ...(8.13)$$

By Eq. (8.13), we get

$$M = \left|T_o(j\omega)\right| = \frac{1}{\sqrt{(1-u^2)^2+(2\xi u)^2}} \quad ...(8.14)$$

$$\angle T_o(j\omega) = \phi = -\tan^{-1}\frac{2\xi u}{(1-u^2)} \quad ...(8.15)$$

The steady state output for the sinusodial input of unit magnitude and variable frequency (ω) is given by

$$y(t) = \frac{1}{\sqrt{(1-u)^2+(2\xi u)^2}}\sin\left[\omega t - \tan^{-1}\frac{2\xi u}{1-u^2}\right]$$

as Eq. (8.2)

From Eqs. (8.14) and (8.15), it is seen that when

$$u = 0, \quad M = 1 \quad \text{and } \phi = 0$$

$$u = 1, \quad M = \frac{1}{2\xi} \quad \text{and } \phi = -\frac{\pi}{2}$$

$$u \to \infty, \quad M \to 0 \quad \text{and } \phi \to -\pi$$

The characteristic of magnitude and phase angle for normalized frequency (u) for certain value of ξ are shown in Fig. 8.2, respectively.

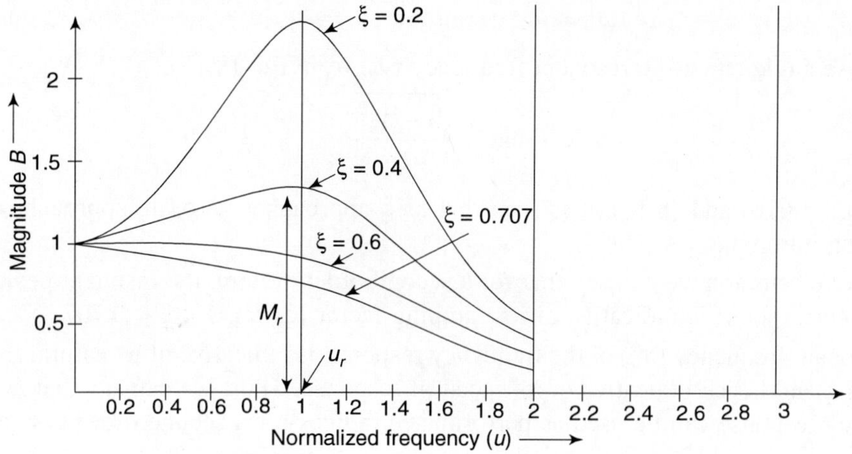

Figure 8.2 Magnitude versus normalized frequency of second order system

In Fig. 8.2 we see that the frequency where M has a peak value is known as the resonant frequency, ω_r.

$$u_r = \frac{\omega_r}{\omega_n}\text{; normalized resonant frequency}$$

Then differentiating Eq. (8.14) w.r.t. u and equal to zero because the slope of the curve at resonant frequency is zero.

$$\left.\frac{dM}{du}\right|_{u=u_r} = -\frac{1}{2}\frac{[-4(1-u_r^2)u_r + 8\xi^2 u_r]}{[(1-u_r^2)^2 + (2\xi u_r)^2]^{3/2}} = 0$$

$$-4u_r + 4u_r^3 + 8\xi^2 u_r = 0$$

$$4u_r^3 - 4u_r + 8\xi^2 + u_r = 0$$

$$4u_r^3 = 4u_r[1 - 2\xi^2]$$

$$u_r^2 = 1 - 2\xi^2$$

$$u_r = \sqrt{1-2\xi^2}$$

$$\omega_r = \omega_n\sqrt{1-2\xi^2} \qquad \because \frac{\omega_r}{\omega_n} = u_r$$

i.e.,

$$\omega_r = \omega_n\sqrt{1-2\xi^2} \qquad \text{...(8.16)}$$

From Eq. (8.14), the maximum value of the magnitude, known as the resonant peak, is given by

$$M_r = \omega_n \sqrt{1 - 2\xi^2} \qquad ...(8.17)$$

The phase angle (ϕ) at the resonant frequency (u_r) from Eq. (8.15).

$$\phi_r = -\tan^{-1} \frac{\sqrt{1 - 2\xi^2}}{\xi}$$

From Eqs. (8.16) and (8.17), it is seen that as ξ approaches zero, ω_r approaches ω_n and M_r approach infinity.

By above discussion we can say that for a second order system, the resonant peak (M_r) of its frequency response is indicative of its damping factor (ξ) for $0 < \xi \leq 0.707$.

The resonant frequency (ω_r) of the frequency response is indicative of its natural frequency for a given ξ and hence indicative of its speed of response. Thus, we can say that M_r and ω_r of frequency response can be used as performance indices for a second order system.

The range of frequency over which $M \geq 0.707$ is defined as bandwith (ω_b) of feedback control system. When we use control system as a low-pass fitter the band width (ω_b) is equal to cut off frequency, ω_c.

The magnitude versus frequency curve is shown in Fig. 8.3 in which the range from point A to B, the magnitude is equal or greater than 0.707 so this range of frequency is called bandwidth.

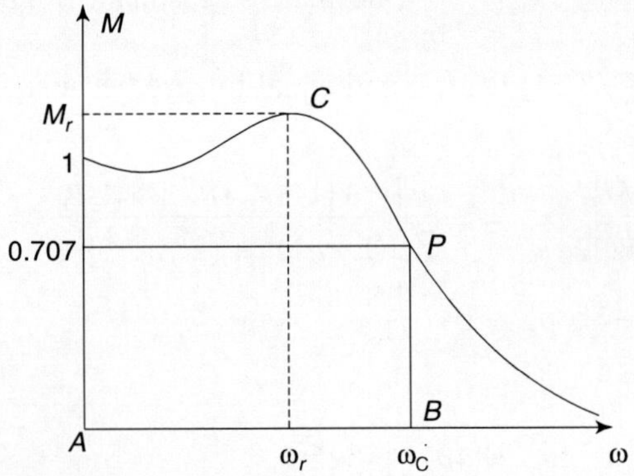

Figure 8.3 Magnitude versus frequency curve

The point B at which the magnitude is equal to 0.707, the frequency at point B is called cut off frequency.

The bandwidth of a control system indicates the following characteristics.

1. Noise filtering characteristics
2. It is measure of the transient response properties

Let the normalized bandwidth be

$$u_b = \omega_b/\omega_n$$

$$M|_{u=u_b} = \frac{1}{\sqrt{[(1-u_b^2)^2 + (2\xi u_b)^2]}} = \frac{1}{\sqrt{2}}$$

$$u_b^4 - 2(1-2\xi^2)u_b^4 - 1 = 0 \quad ...(8.18)$$

The solution of Eq. (8.18) is

$$u_b = \left[1 - 2\xi^2 + \sqrt{(2 - 4\xi^2 + 4\xi^4)}\right]^{1/2} \quad ...(8.19a)$$

It can be approximated in linear form

$$u_b = -1.19\,\xi + 1.85 \quad ...(8.19b)$$

Because bandwidth must be a positive real quantity so the negative sign in quadratic solution and square root of negative sign have been discarded.

By Eq. (8.19) we can say that the normalized bandwidth is a function of damping ξ only.

The plot damping versus normalized bandwidth is shown in Fig. 8.4.

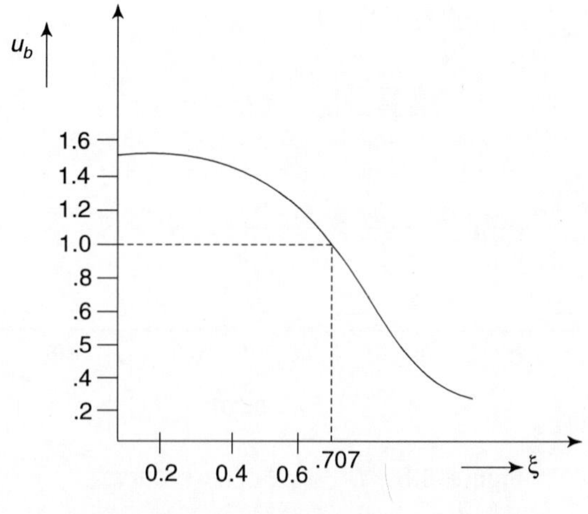

Figure 8.4 Bandwidth versus damping

Now denormalizing the bandwidth from Eq. (8.19)

$$u_b = \left[1 - 2\xi^2 + \sqrt{(2 - 4\xi^2 + 4\xi^4)}\right]^{1/2}$$

$$\therefore \quad u_b = \omega_b/\omega_n$$

$$\frac{\omega_b}{\omega_n} = \left[1 - 2\xi^2 + \sqrt{(2 - 4\xi^2 + 4\xi^4)}\right]^{1/2}$$

$$\omega_b = \omega_n\left[1 - 2\xi^2 + \sqrt{(2 - 4\xi^2 + 4\xi^4)}\right]^{1/2} \quad ...(8.20)$$

In chapter 3 we have studied step response of second order system. So we have

$$\omega_d = \omega_{n1}\sqrt{1-\xi^2} \qquad ...(8.21)$$

$$M_P = \exp\left[-\frac{\pi\xi}{\sqrt{1-\xi^2}}\right] \qquad ...(8.22)$$

where M_P = peak overshoot of step response

ω_d = damped frequency

The comparison of M_r and M_P with ξ is shown in Fig. 8.5. When we see in Fig. 8.5 then we can say that

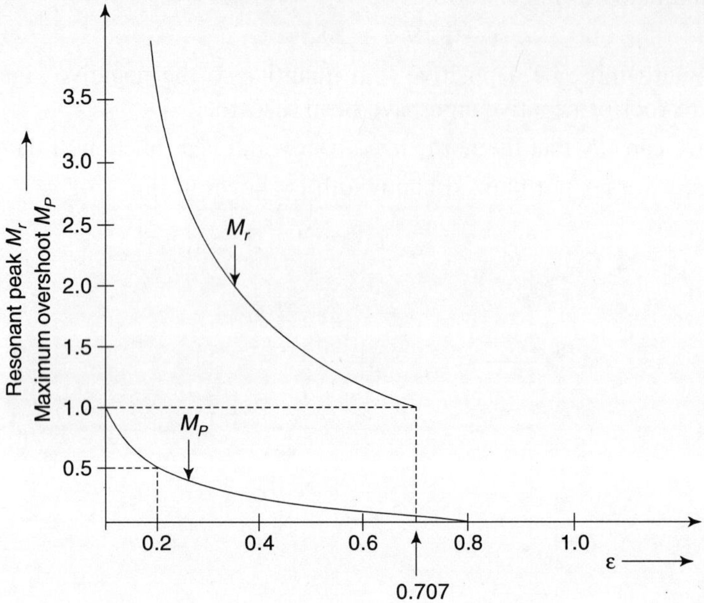

Figure 8.5 M_r M_P versus ξ curve

1. Both the curves M_r as well as M_p are functions of ξ during the range of damping from 0 to 0.707. So we can say that a system with a given value of M_r of its frequency response, must exhibit a corresponding value of M_p if subjected to a step input.
2. For $\xi > 0.707$, the curve for M_r does not exist and the correlation between M_r and M breakes down. But it does not affect the performance of the system because at this range the oscillations are well damped.
3. By Eqs. (8.16) and (8.21) we can say that there is a correlation between resonant frequency (ω_r) of frequency response and damped frequency (ω_d) of the step response.

The ratio of these two frequencies is also

$$\frac{\omega_r}{\omega_d} = \sqrt{\frac{1-2\xi^2}{1-\xi^2}} \qquad ...(8.23)$$

a function of ξ. Figure 8.6 shows the correlation between ω_r and ω_d.

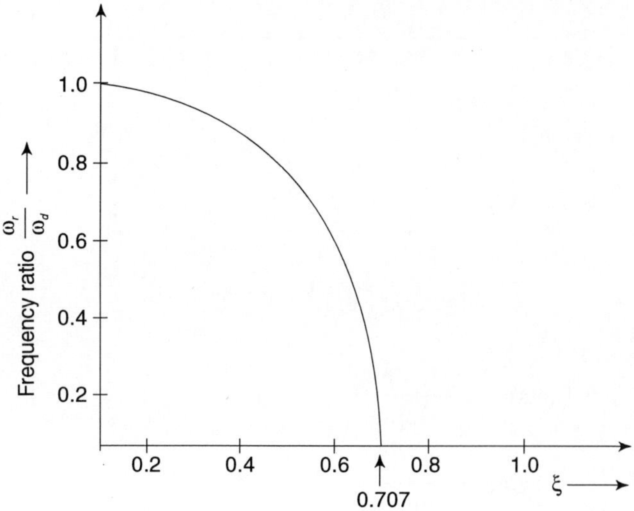

Figure 8.6 Correlation between ω_r and ω_d

So by Eq. (8.20) we can say that bandwidth is a concept for frequency domain for undamped natural frequency, whereas the speed of the response is the time domain concept.

8.4 POLAR PLOTS

The transfer function of a system $G(s)$ can be described in the frequency domain by the relation

$$G(s)\big|_{s=j\omega} = G(j\omega)$$

$$G(j\omega) = R(\omega) + j\,X(\omega) \qquad \text{...(8.24)}$$

where $R(\omega) = \text{Re}\,[G(j\omega)]$

and $x(\omega) = I_m\,[G(j\omega)$

The transfer function can be represented by a phase and magnitude as

$$\left.\begin{aligned} G(j\omega) &= |G(j\omega)|\,e^{j\phi(j\omega)} \\ G(j\omega) &= |G(j\omega)|\,\angle\phi(\omega) \\ G(j\omega) &= M\,\angle\phi \end{aligned}\right\} \qquad \text{...(8.25)}$$

where

$$\text{Magnitude } |G(j\omega)| = \sqrt{[R(\omega)]^2 + [X(\omega)]^2}$$

and phase $$\phi(\omega) = \tan^{-1}\frac{X(\omega)}{R(\omega)}$$

"When the input frequency (ω) is varied from 0 to ∞, the magnitude $|G(j\omega)|$ and phase $\phi(\omega)$ changes so the $G(j\omega)$ traces a locus in the complete plane. This locus is known as polar plot."

Example 8.4.1 Consider a simple *RC* filter shown in Fig. Ex. 8.4.1.

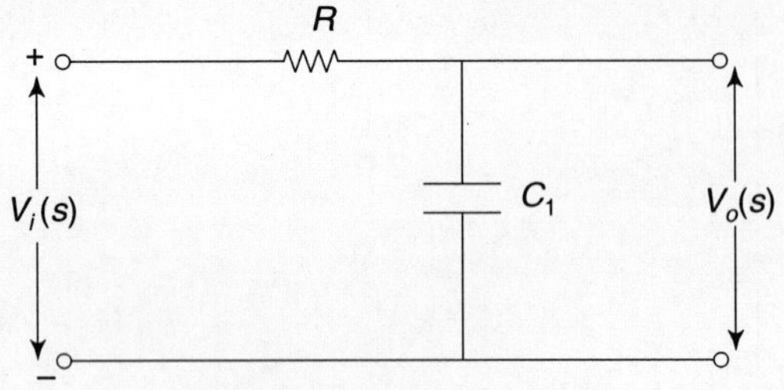

Figure Ex. 8.4.1 RC filter

$$\text{Transfer function} = \frac{\text{Output}}{\text{Input}}$$

$$G(s) = \frac{V_o(s)}{V_i(s)}$$

$\therefore$
$$L[R] = R$$
$$L[C] = 1/Cs$$

so
$$G(s) = \frac{1/Cs}{R + 1/Cs}$$

$$G(s) = \frac{1/Cs}{\frac{RCs+1}{Cs}} = \frac{1}{RCs+1} \quad \text{...(8.26)}$$

Let
$$RC = T$$
$$\omega_1 = \frac{1}{RC} = \frac{1}{T}$$

Now
$$G(j\omega) = \frac{1}{\frac{j\omega}{\omega_1}+1} = \frac{1}{j\left(\frac{\omega}{\omega_1}\right)+1} \quad \text{...(8.27)}$$

$$G(j\omega) = \frac{1 - j(\omega/\omega_1)}{[1 + j(\omega/\omega_1)][1 - j(\omega/\omega_1)]}$$

$$G(j\omega) = \frac{1 - j(\omega/\omega_1)}{1 + (\omega/\omega_1)^2}$$

$$G(j\omega) = \frac{1}{1 + (\omega/\omega_1)^2} - \frac{j(\omega/\omega_1)}{1 + (\omega/\omega_1)^2} \quad \text{...(8.28)}$$

$$G(j\omega) = |G(j\omega)| \angle\phi(\omega)$$

where $|G(j\omega)| = \dfrac{1}{1+(\omega/\omega_1)^2}$

$$\phi(\omega) = \tan^{-1}(\omega/\omega_1)$$

By Eq. (8.28)

When $\omega = 0$

$R(\omega) = 1$

and $X(\omega) = 0$

When $\omega = \infty$

$R(\omega) = 0$

and $X(\omega) = 0$

and when $\omega = \omega_1$

then $R(\omega) = X(\omega)$

$$\phi(\omega) = \tan^{-1}\frac{X(\omega)}{R(\omega)} = 45°$$

i.e., $|G(j\omega_1)| = \dfrac{1}{\sqrt{2}}$

$$\phi(\omega_1) = -45°$$

Now the polar plot is shown in Fig. Ex. 8.1.

In Fig. Ex. 8.4.1 (a) we can say that when ω approaches $+\infty$, we have $|G(j\omega)| \to 0$ anf $\phi(\omega) = -90°$. Similarly, when $\omega = 0$ then $G(j\omega) = 1$ and $\phi(\omega) = 0$.

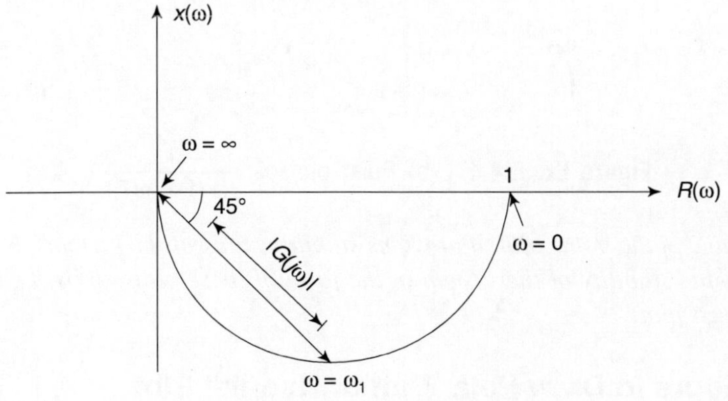

Figure Ex. 8.4.1 (a) Polar plot for RC filter

Now considering the transfer function

$$G(j\omega) = \frac{1}{j\omega(1+j\omega T)} \qquad \text{...(8.29)}$$

Rearranging Eq. (8.29)

$$G(j\omega) = \frac{-T}{1+\omega^2 T^2} - j\frac{1}{\omega(1+\omega^2 T^2)} \quad ...(8.30)$$

From Eq. (8.30), we get

$$\lim_{\omega \to 0} G(j\omega) = -T - j\infty = \infty \angle -90°$$

$$\lim_{\omega \to 0} G(j\omega) = -0 - 0j = 0 \angle 180°$$

The polar plot of transfer function $\frac{1}{j\omega(1+j\omega T)}$ is shown in Fig. Ex. 8.4.1(b)

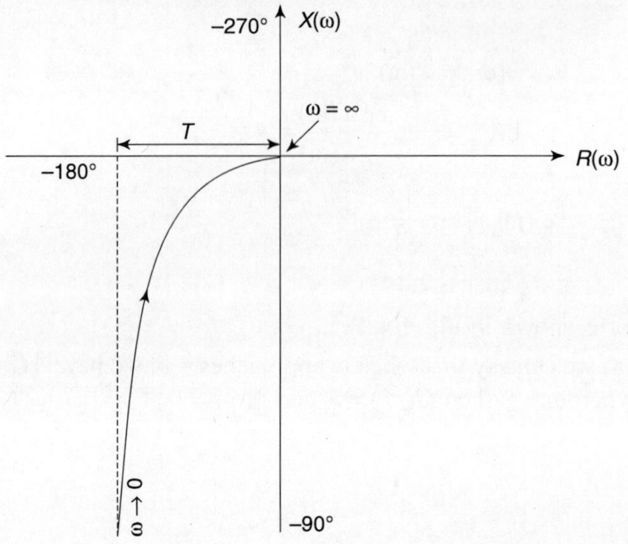

Figure Ex. 8.4.1 (b) Polar plot of $\frac{1}{j\omega(1+j\omega T)}$

- *The advantage of the polar plot, it provides stability study of the systems. N. Nyquist given study about the stability of the system in the form of these plots so by his name it is also called Nyquist plots.*

8.4.1 Procedure to Draw Polar Plot or Nyquist Plot

1. For given function $G(s)$ $H(s)$ substitute $s = j\omega$ into the function.
2. Separate $G(j\omega)$ $H(j\omega)$ is to real and imaginary part by multiplying complex conjugate on both side.
3. Obtain the frequency at which real part is zero that means equated real part is equal to zero.

4. Obtain the frequency at which imaginary part is zero that means equated imaginary part is equal to zero.
5. Now make a table of Re $G(j\omega)$ and Im $G(j\omega)$ at varying frequency (ω) from 0 to ∞.
6. Now plot the polar plot with the help of Table 8.4.
7. This point is only applicable for Nyquist plot apply the Nyquist stability criterion by $N = P - Z$.

where N = Number of encirclement of $(-1 + j0)$ point
= + 've' for anticlockwise direction
= – 've' for clockwise direction
P = Number of poles in the right hand s-plane
Z = Number of zeros is to the right hand s-plane

If $Z = 0$ system is stable
$Z > 0$ system is unstable

Alternative procedure

1. For a given function $G(s)\,H(s)$, substitute $s = j\omega$ in to the function.
2. Obtain the magnitude of the function $G(j\omega)\,H(j\omega)$, i.e., $|G(j\omega)\,H(j\omega)|$.
3. Obtain the angle of the function $G(j\omega)\,H(j\omega)$, i.e., $\angle G(j\omega)\,H(j\omega)$.
4. Obtain the value of magnitude and angle when ω approaches to zero.
5. Obtain the value of magnitude and angle when ω approaches to infinity.
6. Now draw the polar plot.

Table 8.1 Typical polar plots

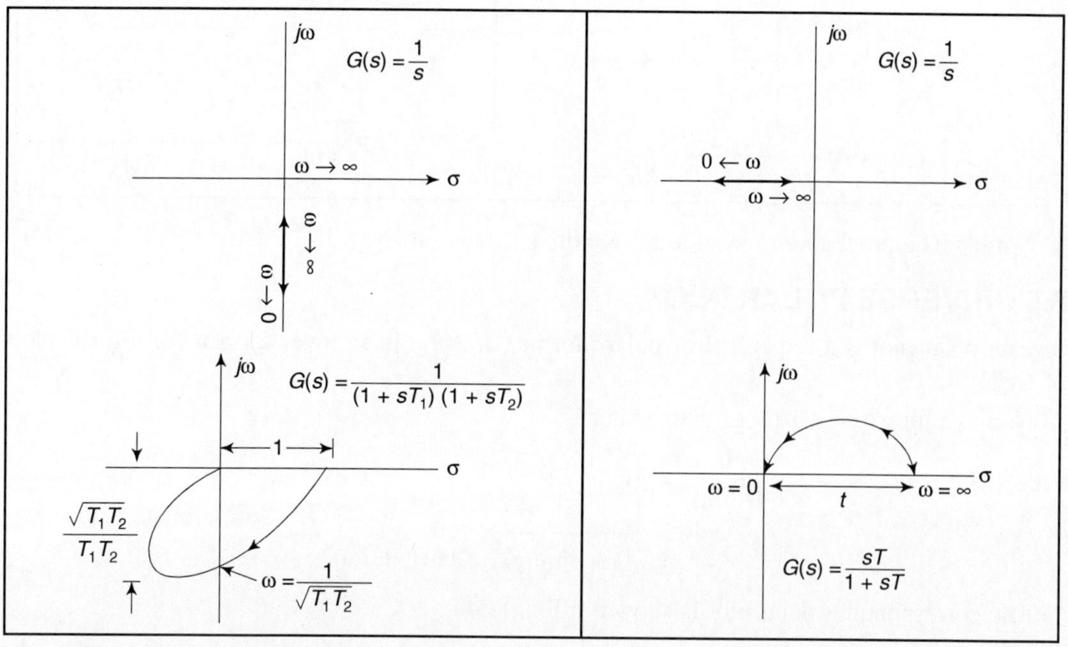

Contd.

Table 8.1 Contd.

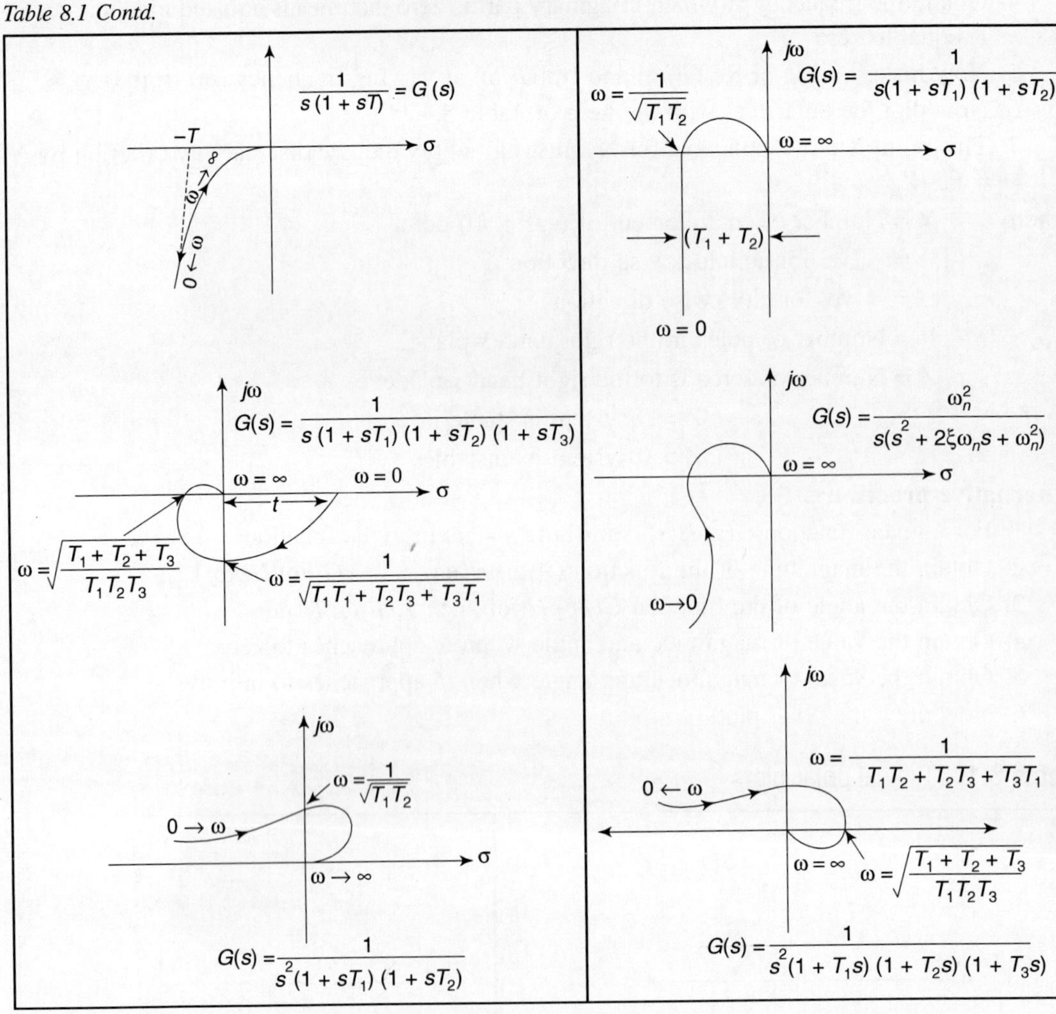

Note: The procedure to sketch these plots is shown in Ex. 8.11 to 8.13

8.5 INVERSE POLAR PLOT

Inverse polar plot is a reciprocal of polar plot i.e., $1/G(j\omega)$ is an inverse polar for a polar plot of $G(j\omega)$.

Now the inverse polar plot for RC filter is

$$\frac{1}{G(j\omega)} = j(\omega/\omega_1) + 1$$

$$= \sqrt{1 + (\omega/\omega_1)^2} \angle \tan^{-1} \omega/\omega_1$$

The corresponding polar plot is shown in Fig. 8.5.

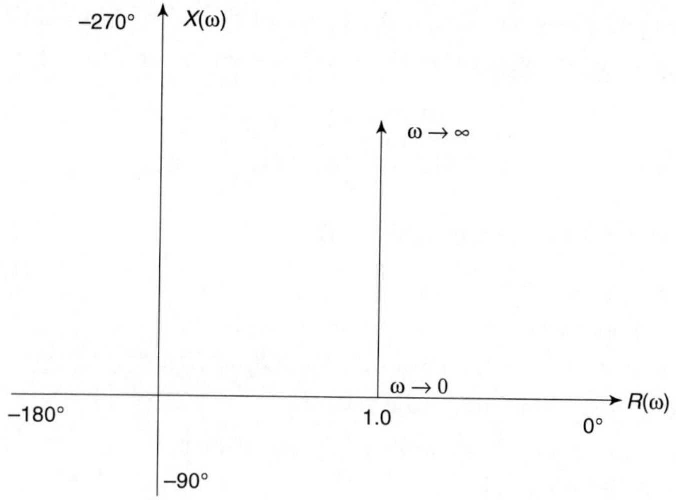

Figure 8.7 Inverse polar plot of $\dfrac{1}{1+j(\omega/\omega_1)}$

The use of inverse polar plot is the stability study for non-unity feedback systems. It is also useful in applying *M*-criterion.

8.6 BODE PLOTS

The scientist H.W. Bode suggested a graphical procedure for determining the stability and frequency response of a control system, based on sinusoidal frequency response.

In general, Bode plot consists of two plots.

1. Magnitude plot in which logarithmic values of M_r are plotted aganist logarithmic values of ω.
2. Phase angle plot in which ϕ_r in degrees are plotted against logarithmic value of ω.

The open loop transfer function of a system is $G(s)\ H(s)$.

The sinusoidal open loop transfer function can be obtained by replacing s by $j\omega$, i.e., $G(j\omega)\ H(j\omega)$ can be expressed in the form of magnitude and phase angle.

8.6.1 Standard Form of $G(j\omega)\ H(j\omega)$

The open loop transfer function in frequency domain may be written as the ratio in polynomial

$$G(j\omega)\ H(j\omega) = \frac{K\,(1+T_1\ j\omega)\,(1+T_2\ j\omega)}{(j\omega)^{\pm P}\ (1+T_a\ j\omega)\,(1+T_b\ j\omega)}$$

The above transfer function shows that there are four basic types of factors

(i) Resultant system gain, K

(ii) Poles or zeros at the origin $(j\omega)^{\pm P}$, $P = 1, 2, 3$ ...

(iii) Simple poles and zeros, i.e., first order factor of the form $(1 + Tj\omega)^{\pm 1}$

(iv) Quadratic factor which cannot be factored into real factor, of the form

$$\left[1+\frac{2\xi}{\omega_n}j\omega+\frac{1}{\omega_n^2}(j\omega)^2\right]^{\pm 1}$$

8.6.2 Procedure to obtain Bode Plot

There are following steps to obtain bode plot

1. Replace $s = j\omega$ to convert it in frequency domain.
2. Find its magnitude as a function of ω, i.e., $|G(j\omega)\,H(j\omega)|$
3. Express the magnitude in dB by the relation

$$|G(j\omega)\,H(j\omega) \text{ in dB} = 20\log_{10}|G(j\omega)\,H(j\omega)|$$

4. Find phase angle by the relation

$$\phi = \tan^{-1}\left(\frac{\text{Imaginary part}}{\text{Real part}}\right) \text{ in degree}$$

5. With required approximation, plot magnitude in dB and phase angle in degree against $\log_{10}\omega$ by varying ω from 0 to ∞.

8.6.3 Plot for the Gain Term, *K*

The magnitude is $G(j\omega)\,H(j\omega) = K$

$$= K + j0$$

$$|G(j\omega)\,H(j\omega)| = \sqrt{K^2+0^2} + K$$

$$\angle G(j\omega)\,H(j\omega) = \tan^{-1}\frac{0}{K} = 0° = 20\log_{10}K$$

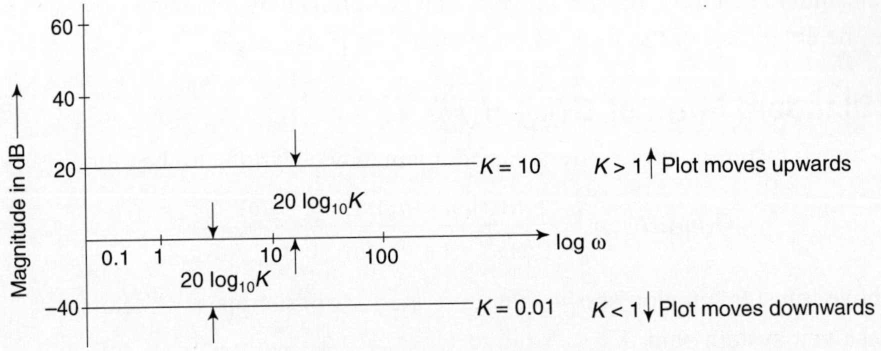

Figure 8.8 (a) Magnitude plot

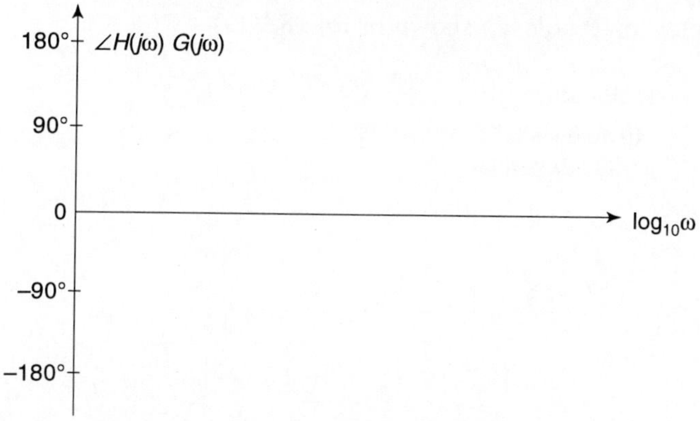

Figure 8.8 (b) Phase angle plot for gain, *K*

8.6.4 Plot for the Poles or Zeros at the Origin $(j\omega)^{\pm P}$

Let us consider a single pole at the origin for simplicity

$$H(j\omega)\ G(j\omega) = \frac{1}{j\omega}$$

$$G(j\omega)\ H(j\omega) = \frac{1}{j\omega + 0}$$

$$|G(j\omega)\ H(j\omega)| = \frac{1}{\sqrt{\omega^2 + 0^2}} = \frac{1}{\omega}$$

$$\begin{aligned}\text{Magnitude in dB} &= 20 \log_{10} 1/\omega = -20 \log \omega \\ &= 0 \text{ dB; } \omega = 1 \\ &= -20 \text{ dB; } \omega = 10 \\ &= -40 \text{ dB; } \omega = 100 \\ &= 20 \text{ dB; } \omega = 0.1\end{aligned}$$

* **10 times change in frequency is called 1 decade**

For two poles

$$G(j\omega)\ H(j\omega) = \frac{1}{(j\omega)^2}$$

$$|G(j\omega)\ H(j\omega)| = \frac{1}{\omega^2}$$

$$\text{Magnitude in dB} = 20 \log_{10} 1/\omega^2 = -40 \log_{10} \omega$$

So it is a straight line of slope – 40 dB/decade

$$|G(j\omega)\ H(j\omega)| \text{ at } \omega = 1$$
$$= 0 \text{ dB}$$

The magnitude plot for P poles is shown in Fig. 8.9 (a).

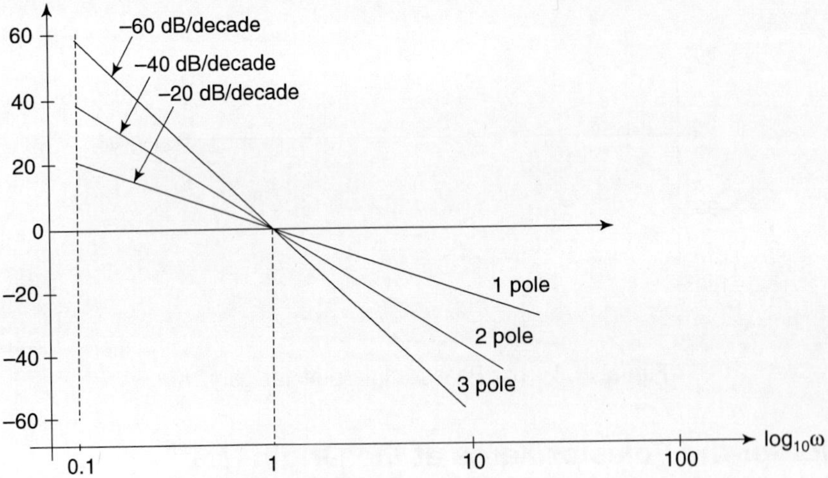

Figure 8.9 (a) Plot for the poles at origin

In general, for P number of zeros at the origin

$$G(j\omega)\,H(j\omega) = (j\omega)^P$$

$$\text{Magnitude in dB} = 20\log_{10}\omega^P$$

$$= 20 \times P\log_{10}\omega$$

i.e., $\text{Slope} = 20 \times P\ \text{dB/decade}$

where P = number of zero at origin

It gives family of lines with slope + 20 dB + 40 dB .../decade as shown in Fig. 8.9 (b).

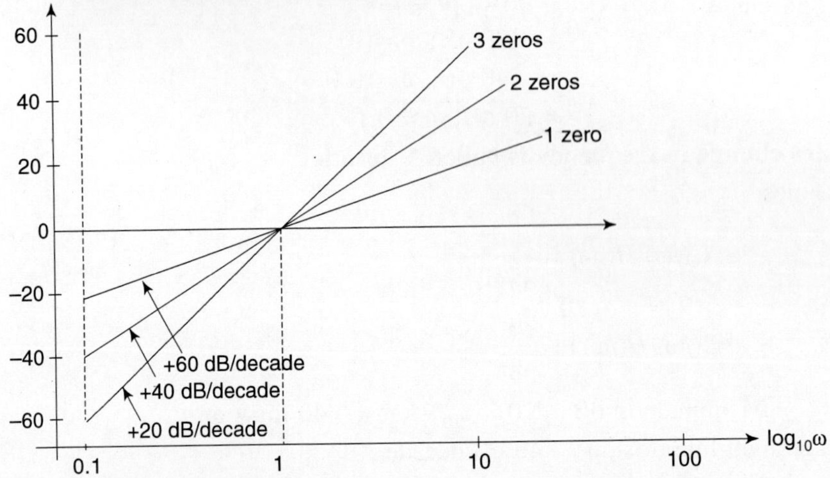

Figure 8.9 (b) Plot for the zeros at origin

Phase angle plot

Consider single pole at the origin

$$G(j\omega)\, H(j\omega) = \frac{1}{j\omega}$$

$$\angle G(j\omega)\, H(j\omega) = \angle \frac{1}{j\omega} = -\angle j\omega = -90^\circ$$

for 2 poles

$$G(j\omega)\, H(j\omega) = \frac{1}{(j\omega)^2}$$

$$\angle G(j\omega)\, H(j\omega) = \angle \frac{1}{(j\omega)^2} = -2\angle j\omega$$

$$= -180^\circ$$

$$\text{Overall phase} = -90^\circ \times P$$

Similarly, for zeros.

$$\text{The total angle} = 90^\circ \times P$$

where P = number of zeros at origin

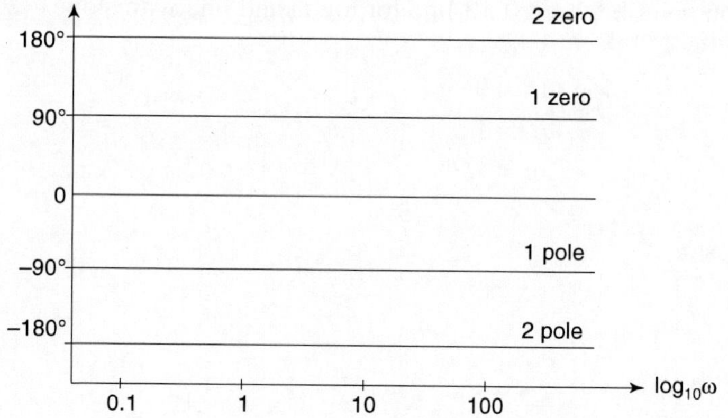

Figure 8.9 (c) Phase angle plot

8.6.5 Plot for the Simple Poles or Zero $(1 + Tj\omega)^{\pm 1}$

Consider a simple pole

$$G(j\omega)\, H(j\omega) = (1 + Tj\omega)^{-1} = \frac{1}{1 + Tj\omega}$$

$$= \frac{1}{1 + Tj\omega}$$

$$|G(j\omega)\, H(j\omega)| = \frac{1}{\sqrt{1+(\omega T)^2}}$$

$$\text{Magnitude in dB} = 20 \log_{10} \left[\sqrt{1+(\omega T)^2}\,\right]^{-1}$$

$$= 20 \log_{10} \left[\sqrt{1+(\omega T)^2}\,\right]$$

We can approximate this into two regions and can draw stright line approximated magnitude plot. The approximation is

(i) For low frequency range

$\omega <<$ —, i.e., $\omega^2\, T^2 << 1$ hence can be neglected

$\therefore$ Magnitude in dB = $-\ 20 \log_{10} 1 = 0$

(ii) For high frequency range

$\omega >> \frac{1}{T}$ i.e., $\omega^2\, T^2 >> 1$

$\therefore$ Magnitude in dB = $-\ 20 \log_{10} \omega T$

i.e., it is a straight line of slope – 20 dB/decade

The intersection point between 0 dB line for log ω and line with slope – 20 dB/decade for high ω is

$$-\log \omega T = 0$$

$$\omega T = 1$$

$$\omega = 1/T$$

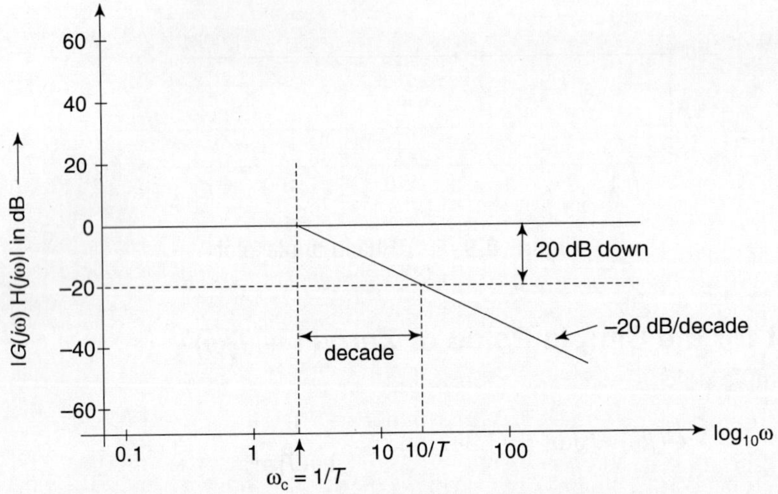

Figure 8.10 (a) Magnitude plot for simple pole

The frequency at which change of slope from 0 dB to – 20 dB/decade occurs is called corner frequency and is denoted by ω_c.

i.e., $$\omega_c = \frac{1}{T}$$

Similarly, for zero

$$G(j\omega)\ H(j\omega) = (1 + Tj\omega)$$

$$|G(j\omega)\ H(j\omega)| = \sqrt{[1+(\omega T)^2]}$$

$$\text{Magnitude in dB} = 20 \log_{10} \sqrt{1+(\omega T)^2}$$

Therefore, the magnitude plot for a simple zero is a straight line of 0 dB upto $\omega_c = \dfrac{1}{T}$ as shown in Fig. 8.10 (b).

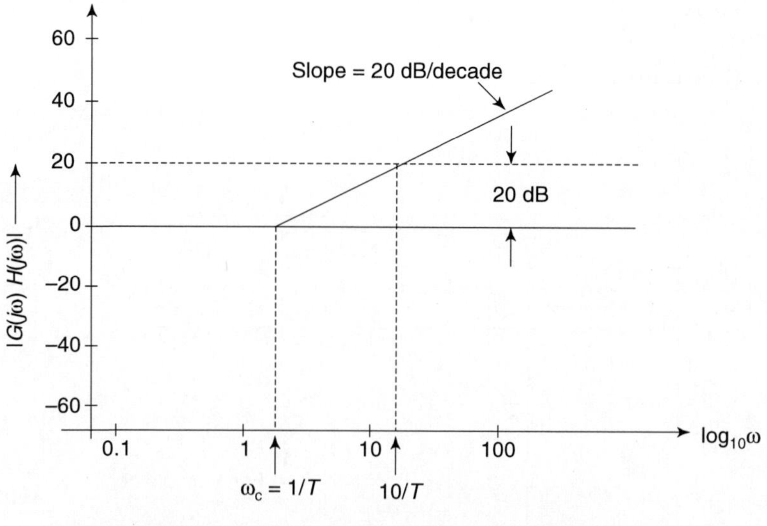

Figure 8.10 (b) Magnitude plot for simple zero

Phase angle plot

For simple pole

$$G(j\omega)\ H(j\omega) = (1 + Tj\omega)^{-1} = \frac{1}{(1+Tj\omega)}$$

$$\angle G(j\omega)\ H(j\omega) = 0 - \tan^{-1} \frac{\omega T}{1} = - \tan^{-1} \omega T$$

For a single zero

$$G(j\omega)\ H(j\omega) = (1 + Tj\omega)$$

$$\angle G(j\omega)\ H(j\omega) = \tan^{-1} \omega T$$

So the shape remains same. Only sign of the angle changes from negative to positive when factor changes from pole to zero. Such plot is to draw by actually calculating angles for different frequency.

ω	$\pm \tan^{-1} \omega T$	(+ for zero, – for pole)
$0.1\omega_c = \dfrac{1}{10T}$	± 5.71	
$0.5\ \omega_c = \dfrac{1}{2T}$	± 26.5	
$\omega_c = \dfrac{1}{T}$	$\pm 45°$	
$2\ \omega_c = \dfrac{2}{T}$	$\pm 63.4°$	
$10\ \omega_c = \dfrac{10}{T}$	$\pm 84.3°$	

These can be shown in Fig. 8.11

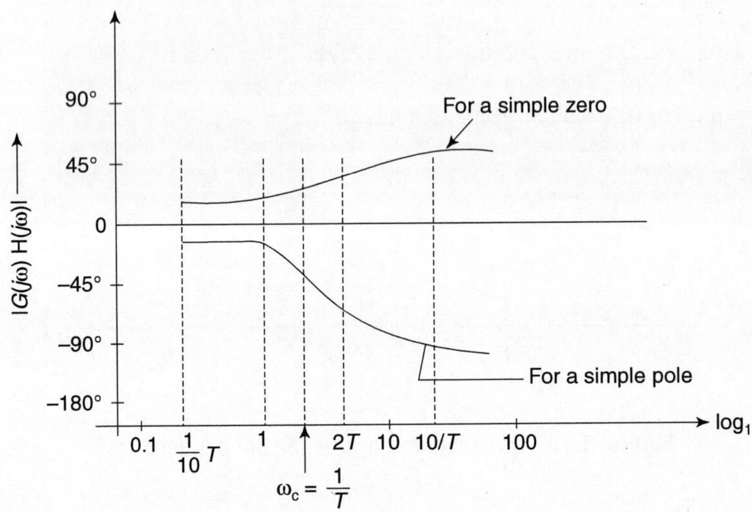

Figure 8.11 Angle contribution by simple pole and zero

8.6.6 Plot for the Quadratic Factor

Generally, the quadratic factor is in the denominator of the transfer function, i.e., *e*

$$G(j\omega)\,H(j\omega) = \frac{1}{1 + \dfrac{2\xi}{\omega_n} j\omega + \dfrac{(j\omega)^2}{\omega_n^2}}$$

$$= \frac{\omega_n^2}{(j\omega)^2 + 2\xi\omega_n\, j\omega + \omega_n^2}$$

$$G(j\omega)\, H(j\omega) = \frac{1}{1 + 2j\xi\dfrac{\omega}{\omega_n} + \left(\dfrac{j\omega}{\omega_n}\right)^2}$$

where ω_n is constant and ω is variable

$$|\, G(j\omega)\, H(j\omega)\,| = \frac{1}{\sqrt{\left[1 - \left(\dfrac{\omega}{\omega_n}\right)^2\right]^2 + 4\xi^2\left(\dfrac{\omega}{\omega_n}\right)^2}}$$

$$\text{Magnitude in dB} = -\,20\log_{10}\sqrt{\left[1 - \left(\frac{\omega}{\omega_n}\right)^2\right]^2 + 4\xi^2\left(\frac{\omega}{\omega_n}\right)^2}$$

(i) For low frequency, i.e., $\omega << \omega_n$

hence $\left(\dfrac{\omega}{\omega_n}\right)^2 << 1$ so neglected

$\therefore$ Magnitude in dB $= -\,20\log_{10} 1 = 0$

(ii) For high frequency, i.e., $\omega >> \omega_n$

$\therefore\quad 4\xi^2\left(\dfrac{\omega}{\omega_n}\right)^2 << \left(\dfrac{\omega}{\omega_n}\right)^4$ because ξ is very low.

$$\therefore\quad \text{Magnitude in dB} = -\,20\log_{10}\sqrt{\left[\left(\frac{\omega}{\omega_n}\right)^2\right]^2}$$

$$= -\,40\log_{10}\frac{\omega}{\omega_n}$$

$$= -\,40\log_{10}\omega + 40\log_{10}\omega_n$$

This is an equation of straight line of slope – 40 dB/decade

$$-\,40\log_{10}\frac{\omega}{\omega_n} = 0$$

$$\frac{\omega}{\omega_n} = 1$$

i.e., $\omega = \omega_n$

$\omega_c = \omega_n$ is corner frequency

The plot for magnitude is shown in Fig. 8.12 (a)

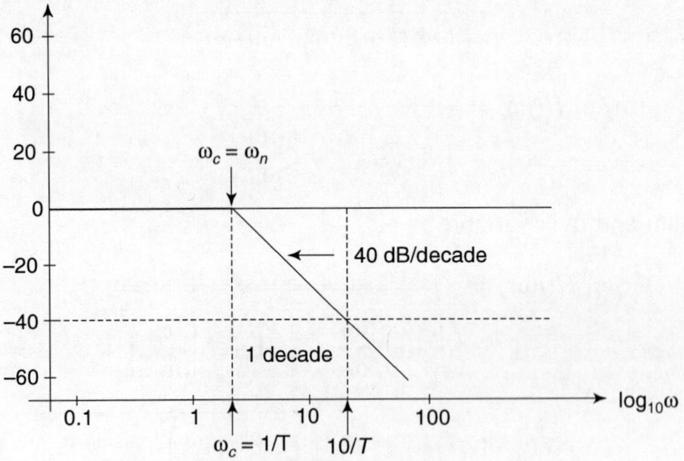

Figure 8.12 (a) Plot for quadratic equation

Phase angle plot

$$G(j\omega)\,H(j\omega) = \tan^{-1}\frac{2\xi\,(\omega/\omega_n)}{1-(\omega/\omega_n)}$$

ω/ω_n	ϕ
0.1	– 3.46°
0.5	– 21.80°
1	– 90°
2	– 158.2°
4	– 170.9°
60	—
∞	– 180°

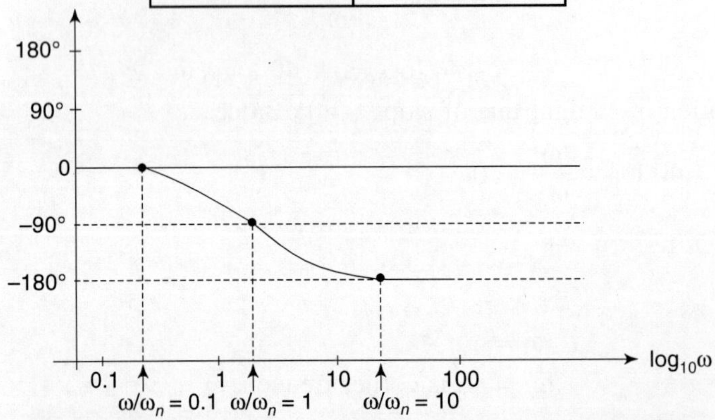

Figure 8.12 (b) Angle contribution by the quadratic poles

For quadratic zeros, sign of the angle should be made positive.

Example 8.6.1 Draw the Bode plot for the system whose open loop transfer function is given below and determine.

(a) The gain margin, (b) the phase margin, and (c) close loop stability

$$G(s)\,H(s) = \frac{4}{s\,(1+0.5\,s)\,(1+0.08\,s)}$$

Solution Putting $s = j\omega$

$$G(j\omega)\,H(j\omega) = \frac{4}{j\omega\,(1+0.5\,j\omega)\,(1+0.08\,j\omega)}$$

Here $K = 4$

$T_1 = 0.5$

and $T_2 = 0.08$

The corner frequencies are

$$\omega_c = \frac{1}{T}$$

$$\omega_{c1} = \frac{1}{T_1} = \frac{1}{0.5} = 2.0 \text{ rad. sec}$$

$$\omega_{c2} = \frac{1}{T_2} = \frac{1}{0.08} = 12.5 \text{ rad. sec}$$

The lower corner frequency is 2.0 rad/sec so the starting frequency of the bode plot is lower than ω_{c1}.

There is one pole at origin so the initial part of Bode plot is – 20 dB/decade.

The intersection point of initial part is at $\omega = K$, i.e., $\omega = 4$.

The slope after one corner frequency i.e., 2 rad/sec is {– 20 + (– 20)}[dB/decade and this slope will continue up to the second corner frequency 12.5 rad/sec.

The slope of the Bode plot after second corner frequency is [– 40 + (– 20)] dB/decade i.e., – 60 dB/decade.

The magnitude plot is shown in Fig. Ex. 8.6.1.

The phase angle plot $\angle G(j\omega)\,H(j\omega)$ for pole at origin is – 90°

Angle due to simple pole $(1 + 0.5\,j\omega)$ is $-\tan^{-1}(0.5\omega)$

Angle due to simple pole $(1 + 0.08\,j\omega)$ is $-\tan^{-1}(0.08\omega)$

The total angle of $G(j\omega)\,H(j\omega)$ is

$$\angle G(j\omega)\,H(j\omega) = \underset{\substack{\uparrow \\ \text{due to pole at origin} \\ (j\omega)^1}}{-90^\circ} \quad \underset{\substack{\uparrow \\ \text{due to simple pole} \\ (1+0.5\,j\omega)}}{-\tan^{-1} 0.5(\omega)} \quad \underset{\substack{\uparrow \\ \text{due to simple pole} \\ (1+0.08\,j\omega)}}{-\tan^{-1}(0.08\omega)}$$

For varying ω

ω rad/sec	1	2	8	10	20	50
$\angle G(j\omega)\, H(j\omega)$	– 121	– 144	– 198	– 207	– 234	– 252

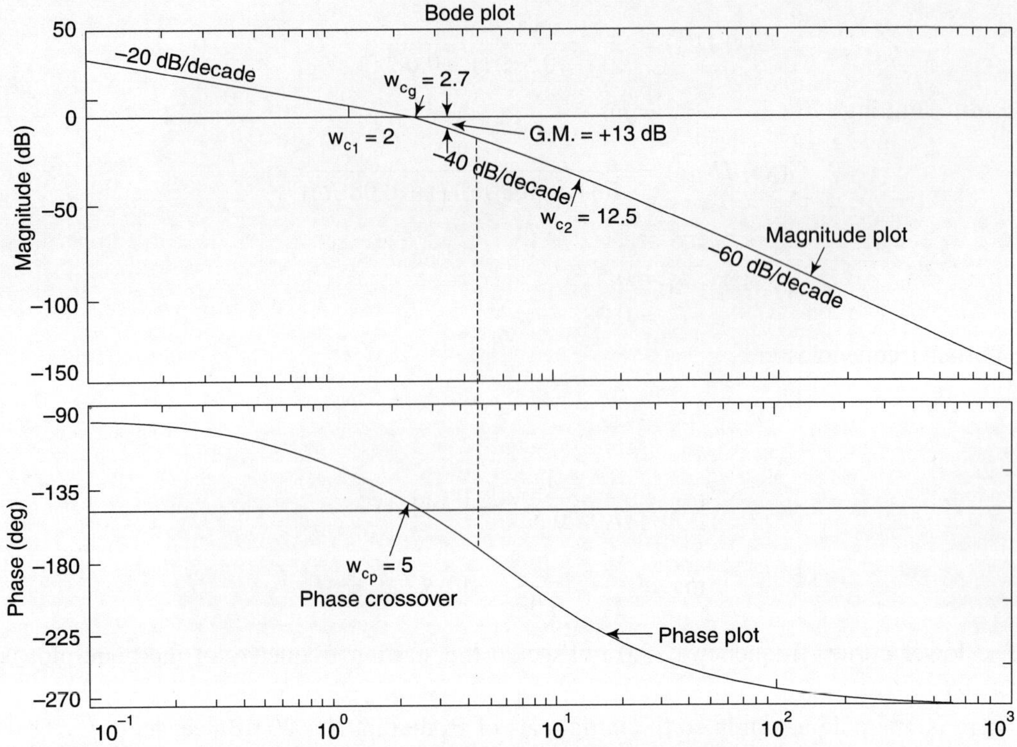

Figure Ex. 8.6.1 Magnitude and phase plot of Ex. 8.6.1

(a) The gain in dB at phase crossover frequency is the gain margin

$$G(j\omega)\, H(j\omega) = -\ 13 \text{ dB}$$

Hence, the gain margin = 13 dB

(b) The phase margin is given by

$$= 180^\circ + \angle G(j\omega)\, H(j\omega) \text{ at gain crossover frequency}$$

$$\text{P.M.} = 180 + (-\ 150^\circ) = 30^\circ$$

(c) The G.M. and P.M. both are positive therefore the close loop system is stable.

Example 8.6.2 Draw the bode plot for the transfer function given below.

$$G(s)\, H(s) = \frac{2\,(s + 0.25)}{s^2\,(s + 1)\,(s + 0.5)}$$

and determine

(i) The gain crossover frequency

(ii) The phase crossover frequency
(iii) The phase margin
(iv) The gain margin
(v) System stability

Solution The transfer function is rewritten as

$$G(s)\ H(s) = \frac{2 \times 0.25\,(1 + s/0.25)}{0.5\,s^2\,(s+1)\left(1 + \dfrac{s}{0.5}\right)}$$

$$G(s)\ H(s) = \frac{(4s+1)}{s^2\,(s+1)\,(2\,s+1)}$$

Comparing the above function with the standard form

$$G(s)\ H(s) = \frac{1\,[1 + T_a\,(s)]}{s^P\,(1 + T_1\,s)\,(1 + T_2\,s)}, \text{ we get}$$

Here $T_a = 4;\quad T_1 = 1;\quad T_2 = 2;\quad P = 2;\quad K = 1$

where P = number of poles at origin

Now putting $s = j\omega$

$$G(j\omega)\ H(j\omega) = \frac{1\,(1 + j\omega)}{(j\omega)^2\,(1 + j\omega)\,(1 + 2j\omega)}$$

Because there are two poles at the origin so the initial part of the bode plot is – 40 dB/decade and intersect 0 dB at $\omega = \sqrt{K} = \sqrt{1} = 1$ rad/sec.

The corner frequency due to zero, i.e., $(1 + j4\omega)$

$$\omega = \frac{1}{4} = 0.25 \text{ rad/sec}$$

$$\omega = \frac{1}{2} = 0.5 \text{ rad/sec due to denominator term}$$

$$\omega = 1 \text{ rad/sec due to denominator term.}$$

After the first corner frequency 0.25 the slope is (– 40 + 20) i.e., – 20 dB/decade

After 0.5 rad/sec, slope is (– 20 – 20) i.e., – 40 dB/decade

After 1 rad/sec. the slope is (– 40 – 20) dB/decde i.e., – 60 dB/decade

The phase angle is given by

$$\angle G(j\omega)\ H(j\omega) = \underset{\substack{\text{due to zero} \\ (1 + 4j\omega)}}{\tan^{-1}(4\omega)} \quad \underset{\substack{\text{due to pole at origin} \\ (j\omega)^2}}{-\,180^\circ} \quad \underset{\substack{\text{due to simple pole} \\ (1 + j\omega)}}{-\,\tan^{-1}(\omega)} \quad \underset{\substack{\text{due to simple pole} \\ (1 + 2\,j\omega)}}{-\,\tan^{-1}(2\omega)}$$

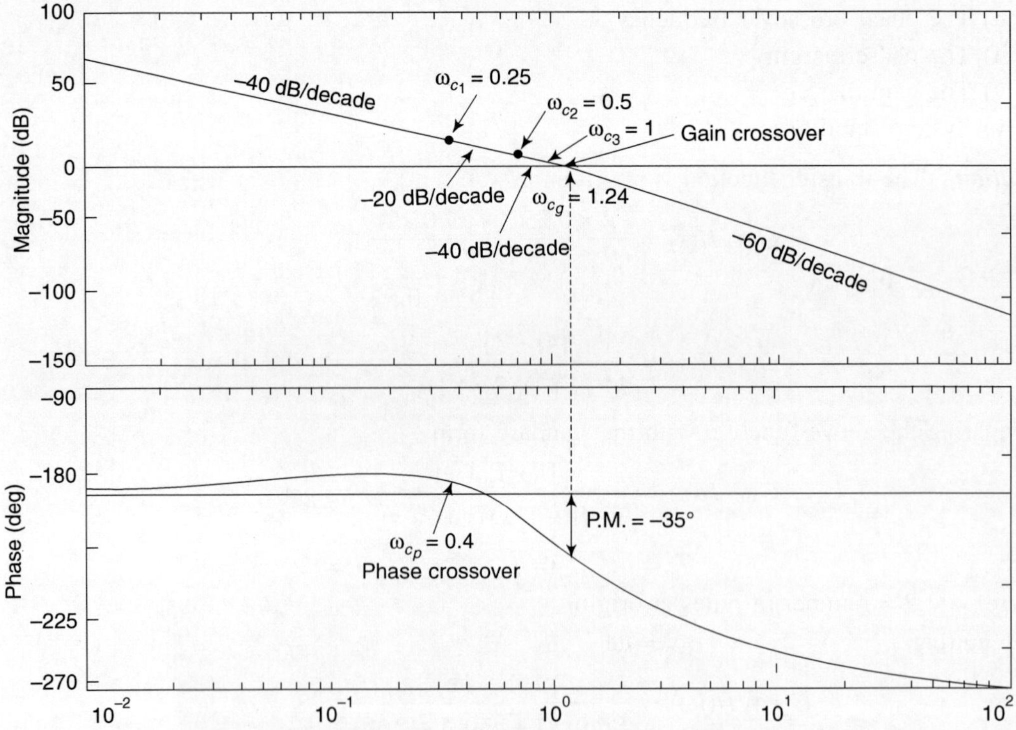

Figure Ex. 8.6.2 Bode plot of Ex. 8.6.2

The phase angle for the frequency range from 0.1 to 5 rad/sec.

ω rad/sec	0.1	0.25	0.5	1	5
$\angle G(j\omega)\, H(j\omega)$	– 175.2	– 175.6	– 188	– 212.4	– 255.8

(i) The gain crosses 0 dB axis at ω = 1.24 rad/sec so the gain crossover frequency is

$$\omega = 1.24 \text{ rad/sec}$$

(ii) The phase crosses – 180° line at ω = 0.4 rad/sec so the phase crossover frequency is

$$\omega = 0.4 \text{ rad/sec}$$

(iii) At gain crossover the phase angle is – 215°, therefore

$$\text{The phase margin} = 180° + (-215°)$$

$$\text{P.M.} = -35°$$ **Ans.**

(iv) At phase crossover the gain is 20 dB so the gain margin is

$$\text{G.M.} = -20 \text{ dB}$$

(v) As both the G.M. and P.M. are negative the system is unstable.

- *The Bode plot is drawn on semilog paper, i.e., lograithmic scale is used. In this paper X-axis is divided into a logarithmic scale which is non-linear. While Y-axis is divided into linear scale, that's why it is called semilog paper. The distance between 1 and 2 is greter than 2 and 3 and so on. But this distance between 1 and 10 is equal to the distance between 10 and 100 or between 100 and 1000 and so on. This distance is called 1 decade. The advantage using the semilog paper is that the multiplication and division of magnitudes get replaced by addition and subtraction, respectively. Another advantage is wide range of frequency which can be accommodated on a single paper.*

- *If the gain crossover frequency < phase crossover frequency, the system is stable or the G.M. and P.M. both are positive.*

  If the gain crossover frequency > phase crossover frequency, the system is unstable or the G.M. and P.M. both are negative.

  If the gain crossover frequency = phase crossover frequency, the system is marginally stable.

EXAMPLES

Bode Plot

Example 8.1 Sketch the Bode plot for the transfer function $G(s) = \dfrac{Ks^{2.}}{(1+0.2s)(1+0.02s)}$.

Determine the system gain K for the gain crossover frequency to be s rad/sec.

Solution Putting $s = j\omega$

$$G(j\omega) = \frac{K(j\omega)^2}{(1+0.2\,j\omega)(1+0.02\,j\omega)}$$

Let $K = 1$

Then $$G(j\omega) = \frac{(j\omega)^2}{(1+0.2\,j\omega)(1+0.02\,j\omega)}$$

Here $T_1 = 0.2$

and $T_2 = 0.02$

The corner frequencies are

$$\omega_c = \frac{1}{T}$$

$$\omega_{c1} = \frac{1}{T_1} = \frac{1}{0.2} = 5 \text{ rad/sec.}$$

$$\omega_{c2} = \frac{1}{T_2} = \frac{1}{0.02} = 50 \text{ rad/sec.}$$

The lower corner frequency is 5 rad/sec so the starting frequency of the Bode plot is lower, ω_{c1}.

There are two zeros at origin so the initial slope of bode plot is

$$\text{Magnitude in dB} = 20 \log_{10} \omega^P = 20\, P \log \omega$$

i.e., $\quad$ Slope = 20 P = 20 × 2 = 40 dB/decade

where $\quad P$ = number of zeros at origin

40 dB/decade the intersection point of initial part is

at $\quad \omega = K$, i.e., $\omega = 1$

The plot for the zeros at origin is shown in Fig. Ex. 8.1 (a)

The slope after one corner frequency, i.e., 5 rad/sec is due to the pole (1 + 0.2 $j\omega$) is

$$= [40 - 20] \text{ dB/decade}$$
$$= 20 \text{ dB/decade}$$

This slope will continue up to the second corner frequency.

The slope after second corner frequency, i.e., 50 rad/sec is due to the pole (1 + 0.02 $j\omega$) is

$$= [20 - 20] \text{ dB/decade}$$
$$= 0 \text{ dB/decade}$$

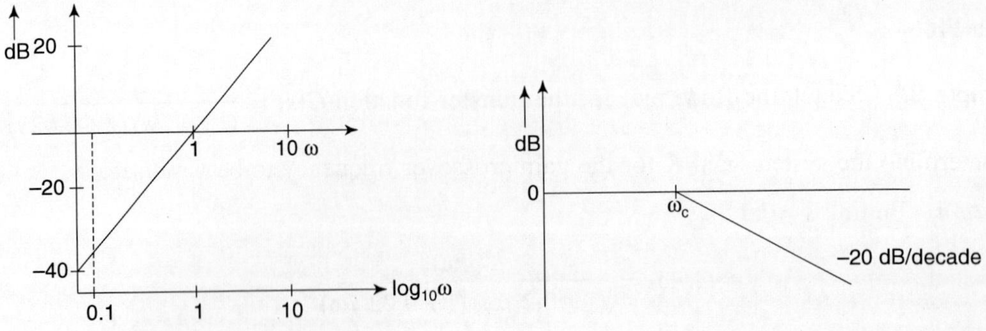

Figure Ex. 8.1 (a) Plot for the zeros at origin, (b) Plot for the real pole

The magnitude plot is shown in Fig. Ex. 8.1 (c).

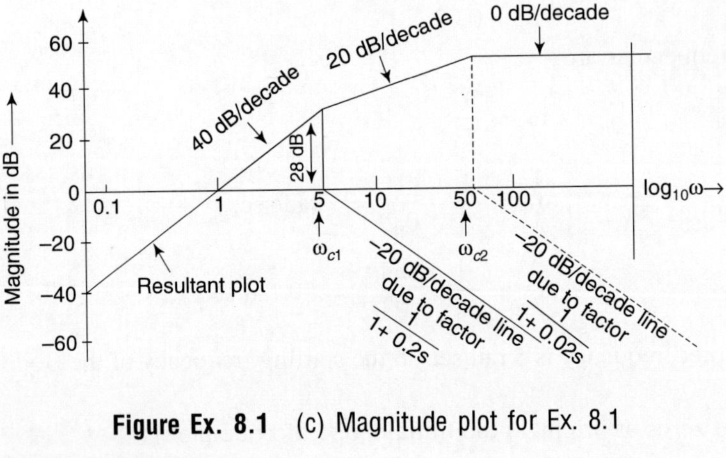

Figure Ex. 8.1 (c) Magnitude plot for Ex. 8.1

The magnitude plot is shown in Fig. Ex. 8.1 (c).

If the gain corner over frequency is required to be 5 rad/sec then the magnitude plot must cross the 0 dB line at 5 rad/sec. For this the plot has to be brought down by 28 dB.

Hence, 20 log $K = -28$.

$$K = 0.04 \quad \textbf{Ans.}$$

Example 8.2 Sketch the bode plot and determine the gain crossover and phase crossover frequencies.

$$G(s) = \frac{10}{s\,(1 + 0.5\,s)\,(1 + 0.1\,s)}$$ (*Pune University*)

Solution Putting $s = j\omega$

$$G(j\omega) = \frac{10}{j\omega\,(1 + 0.5\,j\omega)\,(1 + 0.1\,j\omega)}$$

Here $K = 10$

$T_1 = 0.5$

and $T_2 = 0.1$

The corner frequencies are

$$\omega_c = \frac{1}{T}$$

$$\omega_{c1} = \frac{1}{T_1} = \frac{1}{0.5} = 2 \text{ rad/sec}$$

and $$\omega_{c2} = \frac{1}{T_2} = \frac{1}{0.1} = 10 \text{ rad/sec}$$

The lower frequency is 2 rad/sec so the starting frequency of the bode plot is less than ω_{c1}. There is one pole at the origin so the initial slope of bode plot is

$$M \text{ (in dB)} = +\,20 \log_{10} \frac{1}{\omega^P} = -\,20\,P \log \frac{1}{\omega}$$

i.e., Slope $= 20 \times 1 = -\,20$ dB/decade

The plot for the pole at the origin is shown in Fig. Ex. 8.2 (a).

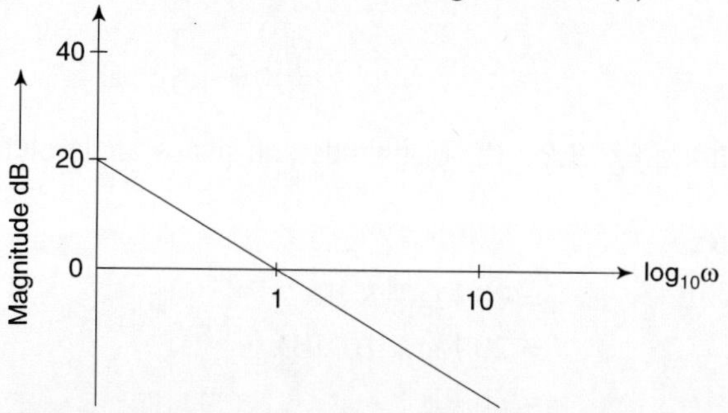

Figure Ex. 8.2 (a) Plot for the pole at the origin

The slope after one corner frequency, i.e., 2 rad/sec is due to the factor $\frac{1}{(1+0.5\,s)}$ is

$$= [-20 + (-20)] \text{ dB/decade}$$
$$= -40 \text{ dB/decade}$$

This slope will continue up to the second corner frequency 10 rad/sec.

The slope after second corner frequency, i.e., 10 rad/sec is due to factor $\frac{1}{(1+0.1\,s)}$ is

$$= [-40 + (-20)] \text{ dB/decade}$$
$$= -60 \text{ dB/decade}$$

The magnitude and phase angle plot for Ex. 8.2 is shown in Fig. Ex. 8.2 (b).

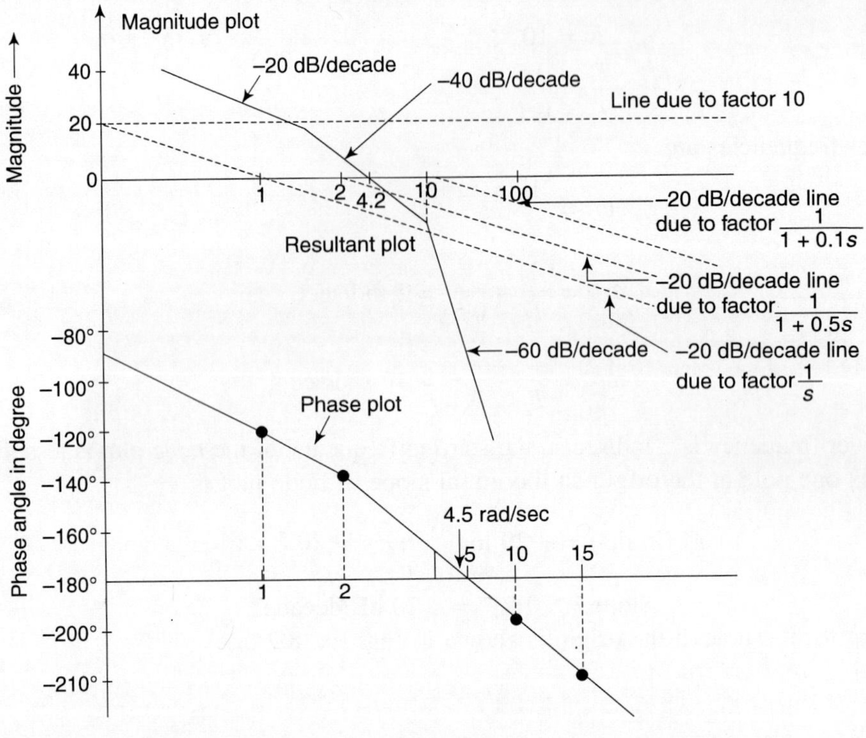

Figure Ex. 8.2 (b) Magnitude and phase angle plot

The slope for term K

$$= 20 \log_{10} K \text{ dB}$$
$$= 20 \log_{10} 10 \text{ dB}$$
$$= 20 \text{ dB}$$

The phase angle plot

$\angle G(j\omega)\ H(j\omega)$ for pole at origin is $-90°$

Angle due to simple pole $(1 + 0.5\ j\omega)$ is $-\tan^{-1}(0.5\omega)$

Angle due to simple pole $(1 + 0.1\ j\omega)$ is $-\tan^{-1}(0.1\ j\omega)$

The total angle of $\angle G(j\omega)\ H(j\omega)$ is

$$\angle G(j\omega)\ H(j\omega) = -90° - \tan^{-1} 0.5\omega - \tan^{-1} 0.1\omega$$

ω rad/sec	0	0.1	1	2	5	10	15
$\angle G(j\omega)\ H(j\omega)$	– 90°	– 93.43	– 122.3	– 146.31	– 184.76	– 213.7	– 228.7

The phase plots are also shown in Fig. Ex. 8.2 (b).

From Fig. Ex. 8.2 (b).

The gain crossover frequency = 42 rad/sec **Ans.**

The phase crossover frequency = 4.5 rad/sec **Ans.**

Example 8.3 Draw Bode plot of the transfer function $G(s) = \dfrac{K}{s\,(s+1)\,(s+5)}$ having unity feedback comment on stability for $K = 10$. *(U.P.T.U., 2004–05, CS)*

Solution The function is

$$G(s) = \frac{K}{s\,(s+1)\,(s+5)}$$

Putting $K = 10$

$$G(s) = \frac{K}{s\,(s+1)\,(s+5)} = \frac{2}{s\,(s+1)\,(0.2s+1)}$$

Putting $s = j\omega$

$$G(j\omega) = \frac{2}{j\omega\,(1+j\omega)\,(1+0.2\,j\omega)}$$

Here $K = 2$

$T_1 = 1$

and $T_2 = 0.2$

The corner frequencies are

$$\omega_{c1} = \frac{1}{} \ \frac{1}{} = 1 \text{ rad/sec}$$

$$\omega_{c2} = \frac{1}{T_2} = \frac{1}{0.2} = 5 \text{ rad/sec}$$

The lower corner frequency is 1 rad/sec so the starting frequency of the Bode is lower than ω_{c1}. There is one pole at origin so the initial slope of the bode plot is

$$\text{Magnitude in dB} = 20 \log \frac{1}{\omega^P}$$

i.e., $$\text{Slope} = +20 \times -1 = -20 \text{ dB/decade}$$

The plot for the pole at origin is shown in Fig. Ex. 8.3 (a).

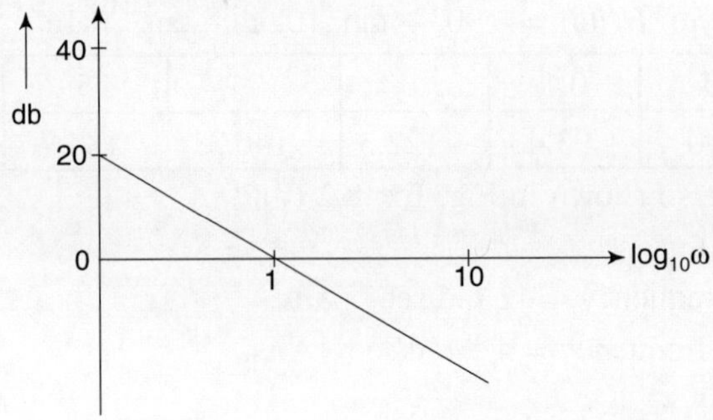

Figure Ex. 8.3 (a) Plot for the pole at origin

The slope for constant $K = 2 = 20 \log_{10} K$

$$= 20 \log_{10} 2$$
$$= 6.02 \text{ dB}$$

The slope after one corner frequency, i.e., 1 rad/sec is due to real pole $(1 + j\omega)$ is

$$= [-20 + (-20)] \text{ dB/decade}$$
$$= -40 \text{ dB/decade}$$

This slope will continue up to the second corner frequency 5 rad/sec.

The slope after second corner frequency, i.e., 5 rad/sec is due to the real pole $(1 + 0.2\,\omega)$ is

$$= [-40 + (-20)] \text{ dB/decade}$$
$$= -60 \text{ dB/decade}$$

The phase angle plot

$\angle G(j\omega)\, H(j\omega)$ for pole at origin is $-90°$

$\angle G(j\omega)\, H(j\omega)$ due to simple pole $(1 + j\omega) = -\tan^{-1} \omega$

$\angle G(j\omega)\, H(j\omega)$ due to simple pole $(1 + 0.2\, j\omega) = -\tan^{-1} 0.2\omega$

The total angle is

$$\angle G(j\omega)\, H(j\omega) = -90° - \tan^{-1}\omega - \tan^{-1} 0.2\omega$$

For varying ω

ω rad/sec	0	0.1	0.2	0.5	1	5	10
$\angle G(j\omega)\, H(j\omega)$	– 90°	– 96.9°	– 103.6	– 122.3	– 146.3	– 213.6	– 237.7

The magnitude and phase plot is shown in Fig. Ex. 8.3 (b).

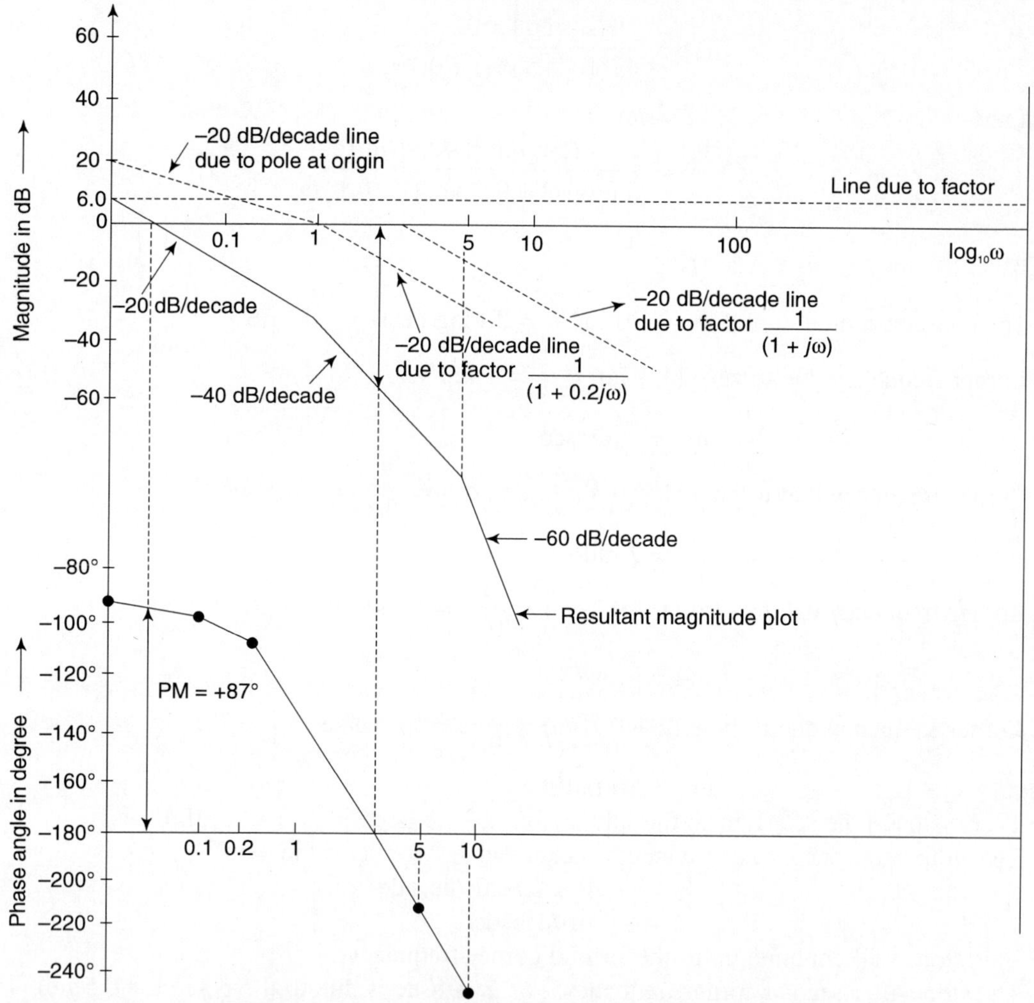

Figure Ex. 8.3 (b) Magnitude and phase plot for Ex. 8.3

By the inspection of bode plot in Fig. Ex. 8.3 (b).

$$\text{Gain margin} = 50 \text{ dB}$$

$$\text{Phase margin} = +\,87^\circ$$

The gain margin and phase margin both are positive so the system is stable.

Example 8.4 Sketch the Bode plot for unity feedback system having the following transfer function $G(s)\,H(s) = \dfrac{(s+1)\,(s+2)}{s\,(s+2)\,(s+10)}$. *(U.P.T.U., 2003–04, CS)*

Solution The given function is

$$G(s) = \frac{(s+1)\,(s+5)}{s\,(s+2)\,(s+10)} = \frac{5\,(s+1)\,(1+0.2\,s)}{2\times 10\,(s)\,(1+0.5\,s\,(1+0.1\,s)}$$

$$G(s) = \frac{(s+1)(1+0.2\,s)}{4s\,(1+0.5\,s)(1+0.1\,s)}$$

Putting $s = j\omega$

$$G(j\omega) = \frac{(1+j\omega)(1+0.2\,j\omega)}{4\,(j\omega)(1+0.5\,j\omega)(1+0.1\,j\omega)}$$

Here $K = \frac{1}{4}$

The constant plot $= 20 \log K = 20 \log (1/4) = -12$ dB

Corner frequency due to zero $(1 + j\omega) = \frac{1}{1} = 1$ rad/sec

i.e., $\omega_{c1} = 1$ rad/sec

Corner frequency due to pole $(1 + 0.5\,j\omega)$, i.e., $\omega_{c2} = \frac{1}{0.5} = 2$ rad/sec

$\omega_{c2} = 2$ rad/sec

Corner frequency due to zero $(1 + 0.2\,j\omega) = \frac{1}{0.2} = 5$ rad/sec

i.e., $\omega_{c3} = 5$ rad/sec

Corner frequency due to pole $(1 + 0.1\,j\omega) = \frac{1}{0.1} = 10$ rad/sec

i.e., $\omega_{c4} = 10$ rad/sec.

There is one pole at origin so the initial slope of the bode plot is – 20 dB/decade.

The slope after one corner frequency, i.e., 1 rad/sec is due to the zero $(1 + j\omega)$ is

$= [-20 + 20]$ dB/decade

$= -0$ dB/decade

This slope will continue up to the second corner frequency

The slope after second corner frequency, i.e., 2 rad/sec is due to the pole $(1 + 0.5\,j\omega)$ is

$= [0 + (-20)]$ dB/decade

$= -20$ dB/decade

This slope will continue up to third corner frequency is 5 rad/sec

The slope after thrid corner frequency, i.e., 5 rad/sec is due to the zero $(1 + 0.2\,j\omega)$ is

$= [-20 + (20)]$ dB/decade

$= 0$ dB/decade

This slope will continue up to fourth corner frequency

The slope after fourth corner frequency, i.e., 10 rad/sec is due to pole $(1 + 0.1\,j\omega)$ is

$= [0 + (-20)]$ dB/decade

$= -20$ dB/decade

The phase angle plot

$\angle G(j\omega)\,H(j\omega)$ for pole at origin is $= -90°$

$\angle G(j\omega)\,H(j\omega)$ due to pole $(1 + 0.5\,j\omega)$ is $= -\tan^{-1} 0.5\,\omega$

$\angle G(j\omega)\ H(j\omega)$ due to pole $(1 + 0.1\ j\omega)$ is $= -\tan^{-1} 0.1\omega$

$\angle G(j\omega)\ H(j\omega)$ due to zero $(1 + j\omega)$ is $= \tan^{-1}\omega$

$\angle G(j\omega)\ H(j\omega)$ due to zero $(1 + 0.2\ j\omega)$ is $= \tan^{-1} 0.2\omega$

The total angle is $\angle G(j\omega)\ H(j\omega)$

$$\angle G(j\omega)\ H(j\omega) = \tan^{-1}\omega + \tan^{-1} 0.2\omega - 90^\circ - \tan^{-1} 0.5\ \omega - \tan^{-1} 0.1\ \omega$$

The angle for varying ω

ω rad/sec	0.1	1	2	5	10	50
$\angle G(j\omega)\ H(j\omega)$	– 86.6°	– 66°	– 61°	– 61°	– 65°	– 83°

The magnitude and phase plot is shown in Fig. Ex. 8.4.

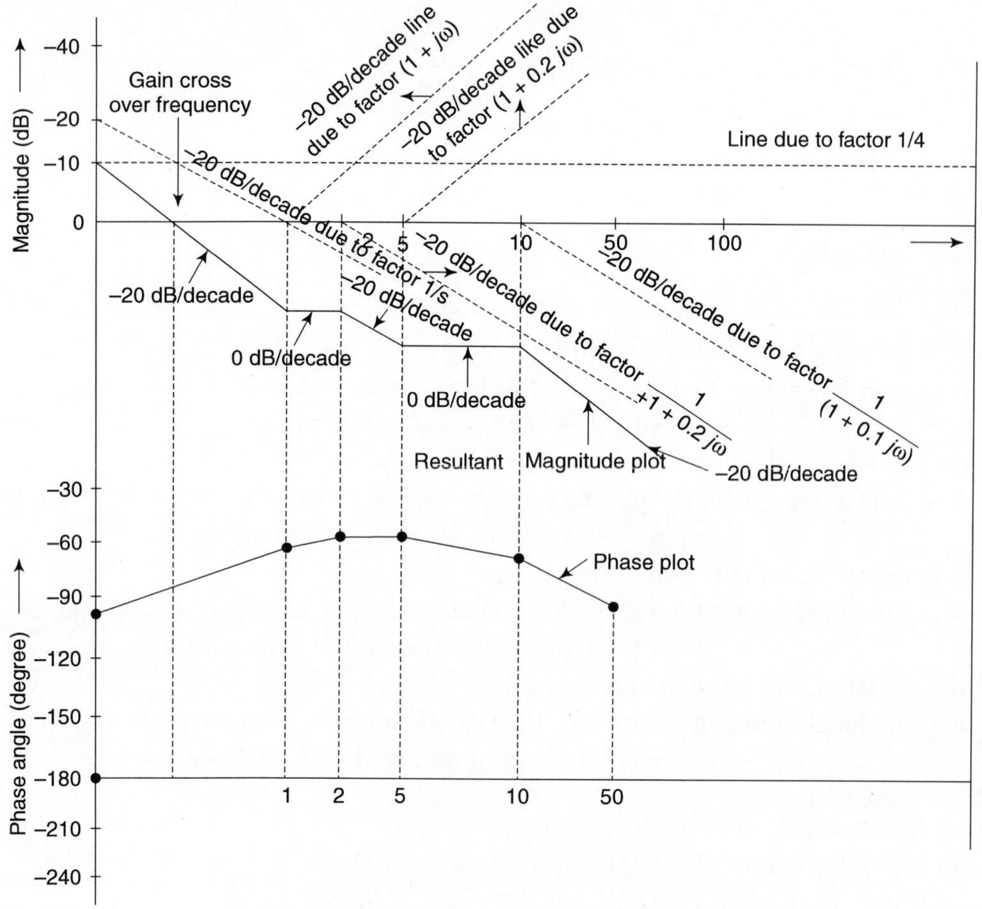

Figure Ex. 8.4 Magnitude and phase plot for Ex. 8.4

Example 8.5 Sketch the Bode plot for a unity feedback system characterised by the loop transfer function $G(s) = \dfrac{(1\ 0.2\ s)(1 + 0.025\ s)}{s^3\ (1 + 0.001\ s)(1 + 0.005\ s)}$.

What is the gain margin of the system. *(U.P.T.U., 2002–03, CS)*

Solution Putting $s = j\omega$

Corner frequency due to zero $(1 + 0.2\,j\omega) = \dfrac{1}{0.2} = 5$

$$\omega_{c1} = 5 \text{ rad/sec}$$

Corner frequency due to zero $(1 + 0.025\,j\omega) = \dfrac{1}{0.025}$

$$\omega_{c2} = 40 \text{ rad/sec}$$

Corner frequency due to pole $(1 + 0.005\,j\omega) = \dfrac{1}{0.005}$

$$\omega_{c3} = 200 \text{ rad/sec}$$

Corner frequency due to pole $(1 + 0.001\,j\omega) = \dfrac{1}{0.001}$

$$\omega_{c4} = 1000 \text{ rad/sec}$$

The system is in type 3, i.e., there are three poles at origin so the initial slope of the Bode plot is

$$\text{Slope} = 20 \times -3 = -60 \text{ dB/decade}$$

Because magnitude in dB $= -20 \log \dfrac{1}{\omega^3} = -60 \log \omega$ dB

$\because \quad K = 1$

So, the slope starts from 20 log 1 = 0 dB

The slope after one corner frequency, i.e., 5 rad/sec is due to the zero $(1 + 0.2\,j\omega)$ is

$$= [-60 + (20)] = -40 \text{ dB/decade}$$

and will continue up to second corner frequency

The slope of after second corner frequency, i.e., 40 rad/sec is due to zero $(1 + 0.025\,j\omega)$ is

$$= [-40 + (20)] \text{ dB/decade} = -20 \text{ dB/decade}$$

and will continue up to third corner frequency

The slope after third corner frequency, i.e., 200 rad/sec is due to pole $(1 + 0.005\,j\omega)$ is

$$= [-20 + (-20)] \text{ dB/decade} = -40 \text{ dB/decade}$$

and will continue up to fourth corner frequency

The slop after fourth corner frequency, i.e., 1000 rad/sec is due to the pole $(1 + 0.001\,j\omega)$ is

$$= [-40 + (-20)] \text{ dB/decade} = -60 \text{ dB/decade}$$

The phase angle plot

$\angle G(j\omega)\,H(j\omega)$ for pole at origin is $= -90° \times 3 = -270°$

$\angle G(j\omega)\,H(j\omega)$ due to pole $(1 + 0.005\,j\omega)$ is $= -\tan^{-1} 0.005\,\omega$

$\angle G(j\omega)\,H(j\omega)$ due to pole $(1 + 0.001\,j\omega)$ is $= -\tan^{-1} 0.001\omega$

$\angle G(j\omega)\,H(j\omega)$ due to zero $(1 + 0.2\,j\omega) = \tan^{-1} 0.2\,\omega$

$\angle G(j\omega)\,H(j\omega)$ due to zero $(1 + 0.025\,j\omega)$ is $= \tan^{-1} 0.025\,\omega$

The total angle is

$\angle G(j\omega)\,H(j\omega) = \tan^{-1}(0.2\omega) + \tan^{-1}(0.025\omega) - 270° - \tan^{-1}(0.005\,\omega) - \tan^{-1}(0.001\omega)$

Angle for varying ω

ω rad/sec	1	5	10	40	100
$\angle G(j\omega)\,H(j\omega)$	$-257.6°$	$-219.6°$	$-196°$	$-155.7°$	$-131.68°$
				200	1000
				$-159°$	$-216°$

From Fig. Ex. 8.5 the gain margin = 43 dB system is stable.

∴ Gain crossover frequency < phase cross over frequency

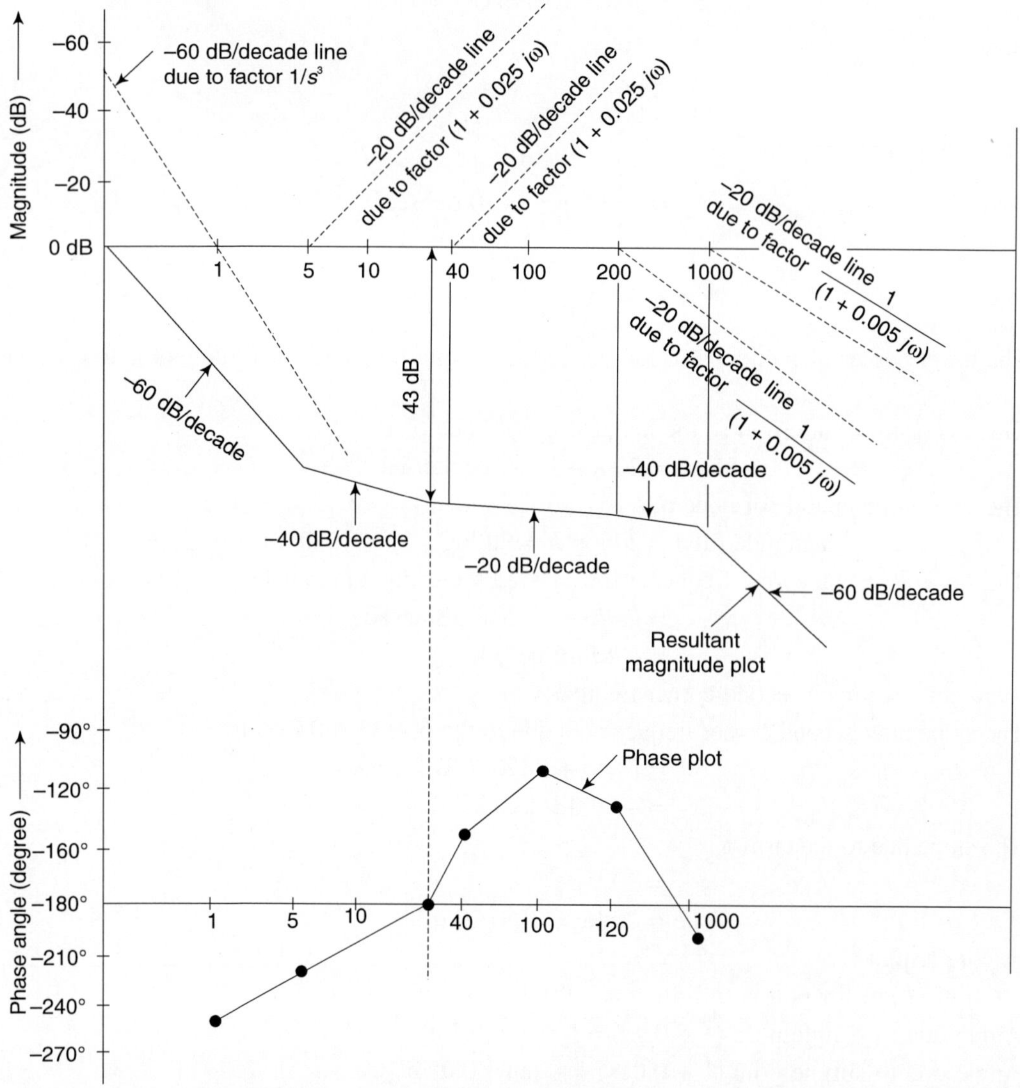

Figure Ex. 8.5 Magnitude and phase plot for Ex. 8.5

Example 8.6 Draw Bode plot for the system having $G(s) = \dfrac{10}{s\,(1 + 0.01\,s)\,(1 + 0.1\,s)}$; $H(s) = 1$

Determine

(a) The gain crossover frequency and corresponding phase margin

(b) The phase crossover frequency and corresponding gain margin (*Pune University*)

Solution The system function is given

$$G(s) = \frac{10}{s\,(1 + 0.01\,s)\,(1 + 0.1\,s)}$$

Here $K = 10$

$T_2 = 0.01$

and $T_1 = 0.1$

The corner frequency are

$$\omega_{c1} = \frac{1}{T_1} = \frac{1}{0.1} = 10 \text{ rad/sec}$$

$$\omega_{c2} = \frac{1}{T_2} = \frac{1}{0.01} = 100 \text{ rad/sec}$$

The lower corner frequency is 10 rad/sec so the starting frequency of Bode plot is lower than ω_{c1}.

There is one pole at origin so the initial slope of bode plot is

$$\text{Slope} = 20 \times -1 = -20 \text{ rad/sec}$$

The starting magnitude of bode plot is

$$\text{Magnitude (dB)} = 20 \log K = 20 \log_{10} 10 = 20 \text{ dB}$$

The slope after one corner frequency, i.e., 10 rad/sec is due to the pole $(1 + 0.1s)$ is

$$= [-20 + (-20)] \text{ dB/decade}$$
$$= -40 \text{ dB/decade}$$

and will continue upon second corner frequency

The slope after second corner frequency is due to the pole $(1 + 0.01s)$ is

$$= [-40 + (-20)] \text{ dB/decade}$$
$$= -60 \text{ dB/decade}$$

The slope due to the term K

$$= 20 \log K$$

$$\text{Slope} = 20 \log_{10} 10 = 20 \text{ dB}$$

Phase angle plot

$\angle G(j\omega)\,H(j\omega)$ for pole at origin is $= -90°$

Angle due to simple pole $(1 + 0.1s) = -\tan^{-1}(0.1\omega)$

Angle due to simple pole $(1 + 0.01s) = -\tan^{-1} 0.01\omega$

The total angle

$\angle G(j\omega)\,H(j\omega) = -90° - \tan^{-1} 0.1\omega - \tan^{-1} 0.01\,\omega$

The angle for varying ω

ω = 0	.1	0.5	1	5	10	50	100
∠G(jω) H(jω) = – 90°	– 90.6°	– 93.15°	– 96.3°	– 119°	141°	– 195°	– 214.3°

The Bode plot is shown in Fig. Ex. 8.6.

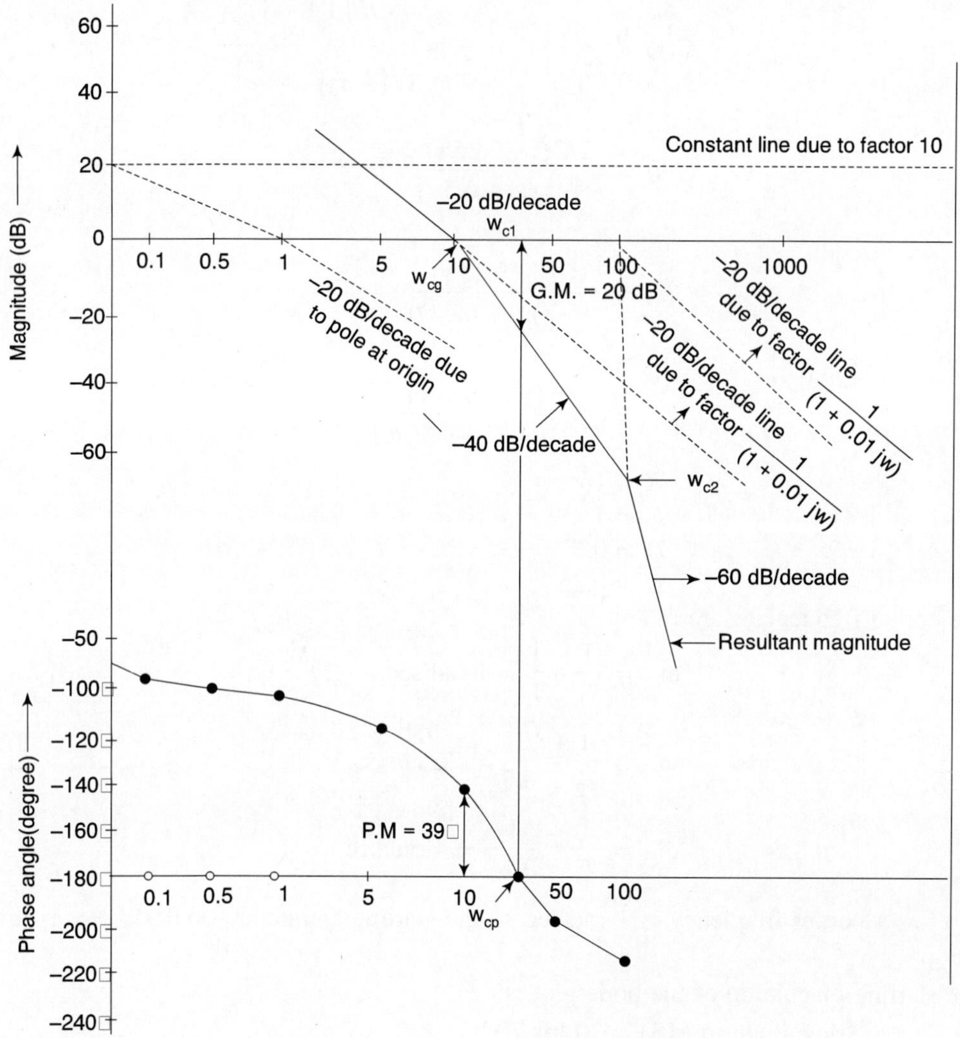

Figure Ex. 8.6 Bode plot for Ex. 8.6

The gain crossover frequency

$$\omega_{cP} = 31.5 \text{ rad/sec}$$

$$\text{Gain margin} = 20 \text{ dB}$$

$$\text{Phase margin} = 39° \quad \textbf{Ans.}$$

Example 8.7 A unity feedback system has a transfer function of $G(s) = \dfrac{K\,(s+4)}{(s-1)\,(s-2)}$. For $K = 8$ draw the bode plot. *(U.P.T.U., 2006–07, ACS)*

Solution The function is

$$G(s) = \frac{K\,(s+4)}{(s-1)\,(s-2)} = \frac{4K\left(1+\frac{s}{4}\right)}{2\,(1-s)\left(1-\frac{s}{2}\right)}$$

$$G(s) = \frac{2K\,(1+0.25\,s)}{(1-s)\,(1-0.5\,s)}$$

for $K = 8$

$$G(s) = \frac{16\,(1+0.25\,s)}{(1-s)\,(1-0.5s)}$$

Putting $s = j\omega$

$$G(j\omega) = \frac{16\,(1+0.25\,j\omega)}{(1-j\omega)\,(1-0.5\,j\omega)}$$

Here $K = 16$

and $T_1 = 1$

$T_2 = 0.5$

and $T_3 = 0.25$

The corner frequencies are

$$\omega_{c1} = \frac{1}{T_1} = \frac{1}{1} = 1 \text{ rad/sec}$$

$$\omega_{c2} = \frac{1}{T_2} = \frac{1}{0.5} = 2 \text{ rad/sec}$$

$$\omega_{c3} = \frac{1}{T_3} = \frac{1}{0.25} = 4 \text{ rad/sec}$$

The lower corner frequency is 1 rad/sec so the starting frequency of Bode plot is lower than ω_{c1}.

The starting magnitude of the bode plot is

$$\text{Magnitude in (dB)} = 20 \text{ lok } K$$
$$= 20 \log_{10} 16$$
$$= 24 \text{ dB (approx.)}$$

The slope after one corner frequency, i.e., 1 rad/sec is due to the pole $(1 - j\omega)$ is

$$= [0 + (-20 \text{ dB})]$$
$$= -20 \text{ dB/decade}$$

and will continue up to second corner frequency

The slope after second corner frequency, i.e., 2 rad/sec is due to pole (1 – 0.5 $j\omega$) is

$$= [-20 + (-20)] \text{ dB/decade}$$
$$= -40 \text{ dB/decade}$$

The slope after third corner frequency, i.e., 4 rad/sec is due to zero (1 + 0.25 $j\omega$) is

$$= [-40 + (20)] \text{ dB/decade}$$
$$= -20 \text{ dB/decade}$$

The phase angle plot

$\angle G(j\omega)\ H(j\omega)$ due to simple pole $(1 - 0.05s) = -\tan^{-1}(-0.05\ \omega) = \tan^{-1} 0.5\ \omega$

$\angle G(j\omega)\ H(j\omega)$ due to simple pole $(1 + s) = -\tan^{-1} - \omega = \tan^{-1}\omega$

$\angle G(j\omega)\ H(j\omega)$ due to simple zero $(1 + 0.25\ s) = \tan^{-1} 0.25\ \omega$

$\angle G(j\omega)\ H(j\omega)$ due to zero $(1 + 0.025\ j\omega)$ is $= \tan^{-1} 0.025\omega$

The total angle is

$\angle G(j\omega)\ H(j\omega) = \tan^{-1} 0.25\omega + \tan^{-1} 0.5\omega + \tan^{-1}\omega$

The angle for varying ω

$\omega =$	0	0.1	0.5	1	2	5	10
$\angle G(j\omega)\ H(j\omega)\ =$	0°	10°	47.7°	85.6°	135°	198.2°	231.2°

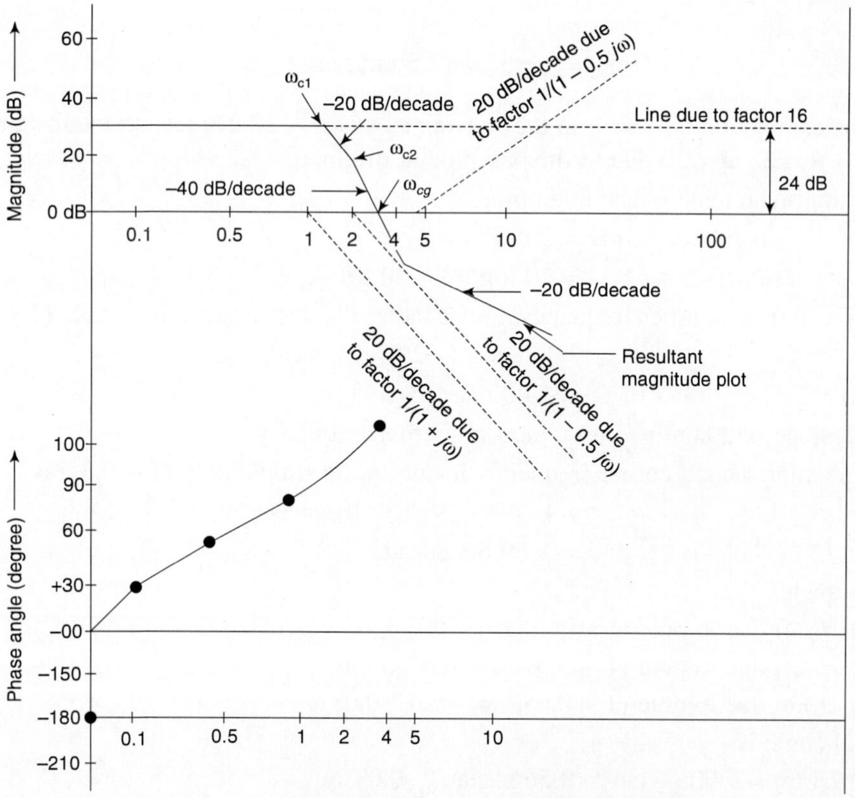

Figure Ex. 8.7 Bode plot for Ex. 8.7

Exmple 8.8 Construct asymptotic plot for the transfer function

$$G(s)\,H(s) = \frac{30}{s\,(1+0.5\,s)\,(1+0.08\,s)}$$

Determine

(a) Gain margin

(b) Phase margin

(c) Closed loop stability *(U.P.T.U., 2003–04, ACS)*

Solution The function is

$$G(s) = \frac{30}{s\,(1+0.5\,s)\,(1+0.08\,s)};\quad H(s) = 1$$

Here $K = 30$

$T_1 = 0.5$

and $T_2 = 0.08$

The corner frequencies are

$$\omega_{c1} = \frac{1}{0.5} = 2 \text{ rad/sec}$$

$$\omega_{c2} = \frac{1}{0.08} = 12.5 \text{ rad/sec}$$

The transfer function is type 1 so the starting slopes is – 20 dB/decade intersecting 0 dB basis at $\omega = K$, i.e., $\omega = 30$ due to the one ploe at origin.

The starting magnitude of the bode plot is

$$\text{Magnitude in dB} = 20 \log K$$
$$= 20 \log 30 = 30 \text{ dB}$$

The slope after one corner frequency, i.e., 2 rad/sec is due to the simple pole $(1 + 0.5s)$ is

$$= [-20 + (-20)] \text{ dB/decade}$$
$$= -40 \text{ dB/decade}$$

And this slope will continue up to second corner frequency.

The slope after second corner frequency is due to the simple pole $(1 + 0.8s)$ is

$$= [-40 + (-20)] \text{ dB/decade}$$
$$= -60 \text{ dB/decade}$$

The phase angle

$\angle G(j\omega)\,H(j\omega)$ due to pole at origin is $= -90°$

$\angle G(j\omega)\,H(j\omega)$ due to pole $(1 + 0.5s) = -\tan^{-1} 0.5\,\omega$

$\angle G(j\omega)\,H(j\omega)$ due to pole $(1 + 0.08s) = -\tan^{-1} 0.08\,\omega$

The total angle is

$\angle G(j\omega)\,H(j\omega) = -90° - \tan^{-1}(0.5\omega) - \tan^{-1}(0.08\,\omega)$

The angle for varying ω

$\omega =$	1	2	8	10	20	50
$\angle G(j\omega)\ H(j\omega) =$	– 121°	– 144°	– 198°	– 207°	– 234°	– 252°

By the inspection of Bode plot in Fig. Ex. 8.8 the gain corssover frequency ω_{cg} = 7.3 rad/sec.

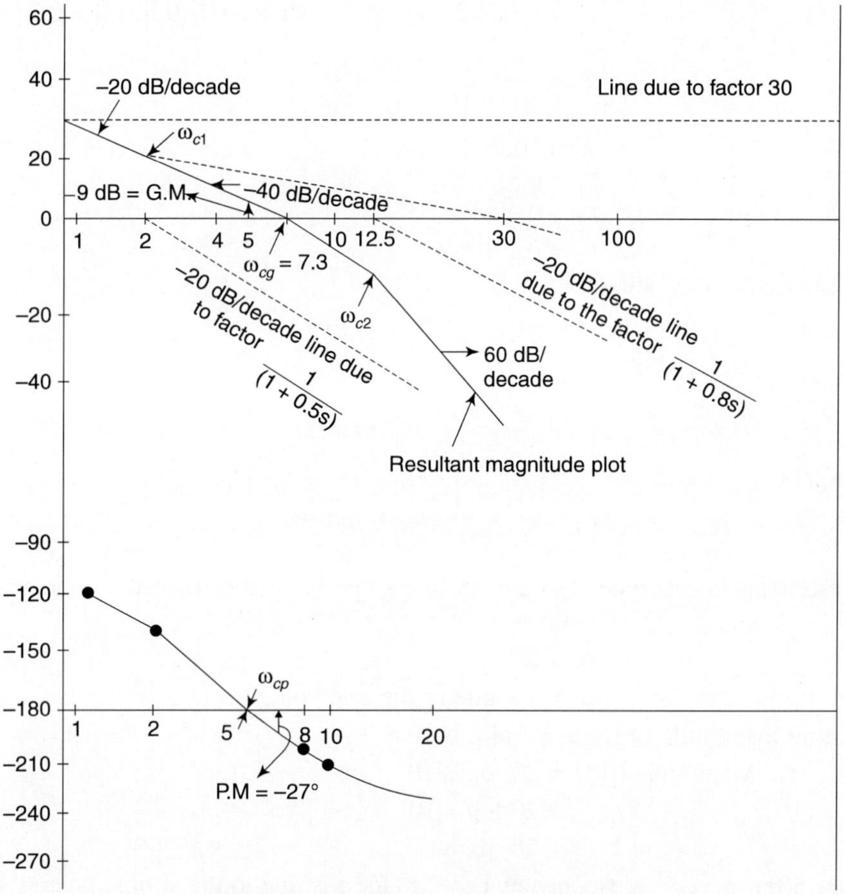

Figure Ex. 8.8 Bode plot for Ex. 8.8

Phase crossover frequency, ω_{cg} = 5 rad/sec

(a) The gain margin at $\omega_{cg} = G\ (j5)\ H(j5) = 9$ dB

Hence, G.M. = – 9 dB

(b) The phase margin is given by

$$\text{P.M.} = 180 = \angle G(j\omega)\ H(j\omega)\Big|_{\text{at } \omega_{cg}}$$

$$= 180° + (-\ 207°) = -\ 27°$$

(c) Since the gain margin and phase margin both are negative so closed loop system is unstable.

Example 8.9 Construct the Bode plot for a unity feeback control system having $G(s) = \dfrac{10\,(s+10)}{s\,(s+2)\,(s+5)}$. From the plot obtain the G.M. and P.M. Comment on the stability.

(*Pune University*)

Solution

$$G(s) = \frac{10\,(s+10)}{s\,(s+2)\,(s+5)} = \frac{10\times 10\,(1+0.1s)}{2\times 5\times s\,(1+0.5\,s)\,(1+0.2\,s)}$$

$$G(s) = \frac{10\,(1+0.1\,s)}{s\,(1+0.5\,s)\,(1+0.2\,s)}$$

Here

$$K = 10$$
$$T_1 = 0.5$$
$$T_2 = 0.2$$
$$T_3 = 0.1$$

The corner frequencies are

$$\omega_{c1} = \frac{1}{T_1} = \frac{1}{0.5} = 2 \text{ rad/sec}$$

$$\omega_{c2} = \frac{1}{T_2} = \frac{1}{0.2} = 5 \text{ rad/sec}$$

$$\omega_{c3} = \frac{1}{T_3} = \frac{1}{0.1} = 10 \text{ rad/sec}$$

The transfer function is type 1 so the starting slope is – 20 dB/decade, intersecting 0 dB axis at

$$\omega = K$$

i.e., $\omega = 10$, due to the one pole at origin

The starting magnitude of the bode plot is

$$\text{Magnitude (dB)} = 20 \log K \text{ dB}$$
$$= 20 \log_{10} 10$$
$$= 20 \text{ dB}$$

The slope after one corner frequency, i.e., 2 rad/sec is due to the simple pole $(1 + 0.5s)$ is

$$= [-20 + (-20)] = -40 \text{ dB/decade}$$

And this slope will continue upto second corner frequency.

The slope after second corner frequency, i.e., 5 rad/sec is due to the simple pole $(1 + 0.2s)$ is

$$= [-40 + (-20)] \text{ dB/decade}$$
$$= -60 \text{ dB/decade}$$

And will continue up to third corner frequency.

The slope after third corner frequency is 10 rad/sec is due to the zero $(1 + 0.1\,s)$ is

$$= [-60 + (20)] \text{ dB/decade}$$
$$= -40 \text{ dB/decade}$$

The phase angle

$\angle G(j\omega)\ H(j\omega)$ for pole at origin is = – 90°

$\angle G(j\omega)\ H(j\omega)$ due to pole $(1 + 0.5s) = -\tan^{-1}(0.5\omega)$

$\angle G(j\omega)\ H(j\omega)$ due to pole $(1 + 0.2s) = -\tan^{-1}(0.2\omega)$

$\angle G(j\omega)\ H(j\omega)$ due to zero $(1 + 0.1s) = -\tan^{-1}(0.1\omega)$

The total angle is

$\angle G(j\omega)\ H(j\omega) = \tan^{-1}(0.1\ \omega) - 90° - \tan^{-1}(0.5\omega) - \tan^{-1}(0.2\omega)$

The angle for varying ω

ω :	0	0.1	0.5	1	2	5	10	20
$\angle G(j\omega)\ H(j\omega)$:	– 90°	– 93°	– 107°	– 122°	– 145°	– 176.6°	– 187°	– 205°

The Bode plot is shown Fig. Ex. 8.9

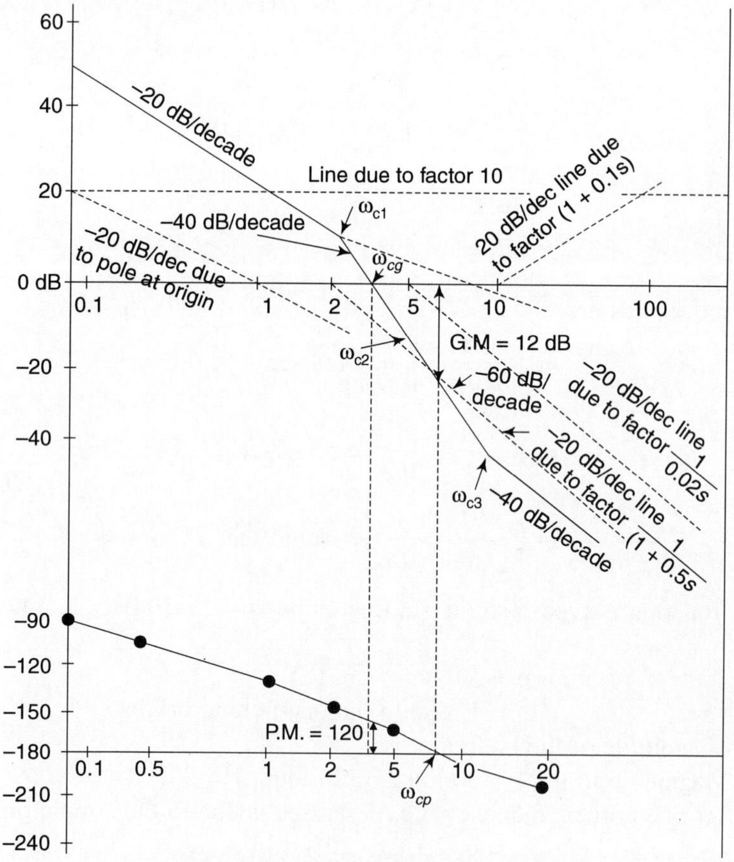

Figure Ex. 8.9 Bode plot for Ex. 8.9

The gain margin at crossover frequency = 12 dB

The phase margin at gain crossover frequency = 120°

Since gain and phase margin are positive system is absolutely stable.

Example 8.10 The open loop transfer function of a unity feedback system is

$$G(s) = \frac{50}{s\,(s+10)\,(s+5)\,(s+1)}$$

(a) Gain margin and phase margin

(b) The value of steady state error coefficient for a gain of 10 dB and the value which will make the closed loop system marginally stable.

Solution The system function is

$$G(s) = \frac{50}{s\,(s+10)\,(s+5)\,(s+1)} = \frac{50}{5 \times 10 \times s\,(s+1)\,(s+0.2\,s)\,(s+0.1\,s)}$$

$$G(s) = \frac{1}{s\,(1+s)\,(1+0.2\,s)\,(1+0.1\,s)}$$

Putting $s = j\omega$

$$G(j\omega) = \frac{}{j\omega\,(1+j\omega)\,(1+0.2\,j\omega)\,(1+0.1\,j\omega)}$$

Here

$$K = 1$$
$$T_1 = 1$$
$$T_2 = 0.2$$

and

$$T_3 = 0.1$$

The corner frequencies are

$$\omega_{c1} = \frac{1}{T_1} = \frac{1}{1} = 1 \text{ rad/sec}$$

$$\omega_{c2} = \frac{1}{T_2} = \frac{1}{0.2} = 5 \text{ rad/sec}$$

$$\omega_{c3} = \frac{1}{T_3} = \frac{1}{0.1} = 10 \text{ rad/sec}$$

The transfer function is type 1 so the starting slope is – 20 dB/decade, intersecting 0 dB axis at

$$\omega = K$$

i.e., $\omega = 1$, due to the one pole at origin

The starting magnitude of the bode plot is

$$\text{Magnitude in (dB)} = 20 \log K = 20 \log_{10} 1 = 0 \text{ dB}$$

The slope after one corner frequency, i.e., 1 rad/sec is due to the simple pole $(1 + j\omega)$ is

$$= [-20 + (-20)] \text{ dB/decade} = -40 \text{ dB/decade}$$

And will continue up to second corner frequency.

The slope after second corner frequency, i.e., 5 rad/sec is due to the simple pole $(1 + 0.2j\omega)$ is

$$= [-40 + (-20)] \text{ dB/decade} = -60 \text{ dB/decade}$$

And will continue up to third corner frequency.

The slope after third corner frequency is due to the simple pole $(1 + 0.1\ \omega)$

$$= [-60 + (-20)]\ \text{dB/decade} = -80\ \text{dB/decade}$$

The phase angle

$\angle G(j\omega)\ H(j\omega)$ for pole at origin is $= -90°$

$\angle G(j\omega)\ H(j\omega)$ due to pole $(1 + s) = -\tan^{-1}(\omega)$

$\angle G(j\omega)\ H(j\omega)$ due to pole $(1 + 0.2s) = -\tan^{-1}(0.2\omega)$

$\angle G(j\omega)\ H(j\omega)$ due to pole $(1 + 0.1s) = -\tan^{-1}(0.1\omega)$

The total angle is

$\angle G(j\omega)\ H(j\omega) = -90° - \tan^{-1}(\omega) - \tan^{-1}(0.2\omega) - \tan^{-1}(0.1\omega)$

$\angle G(j\omega)\ H(j\omega)$ for varying ω

ω :	0	0.1	0.2	0.5	1	2	3	4
$\angle G(j\omega)\ H(j\omega)$:	– 90°	– 97°	– 105°	– 125°	– 152°	– 187°	– 209°	– 227°

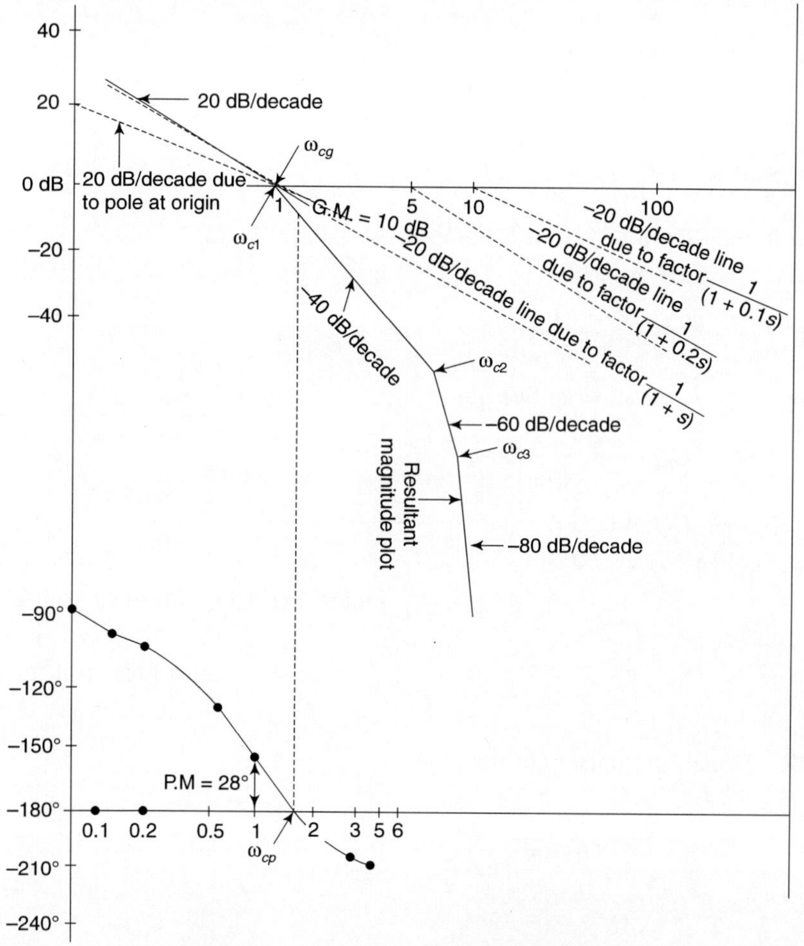

Figure Ex. 8.10 Bode plot for Ex. 8.10

(a) (i) Gain margin = 10 dB

(ii) Phase margin = 28°

(b) Initial slope of the line is – 20 dB/decade and intersects 0 dB line at $\omega = 1$ rad/sec therefore $K_v = 1$

Steady state error coefficient for the system to be marginally stable is found by lifting the magnitude curve by 10 dB (GM). The new magnitude curve is drawn and initial line having a slope of – 20 dB/decade is extended till it intersects 0 dB line. Intersect occurs at $\omega = 3.2$ rad/sec

$\therefore \quad K_v = 3.2$

Polar plot

Example 8.11 Sketch the polar plots of $1/s$ and $1/s^2$.

Solution

(i) When $G(s) = \dfrac{1}{s}$

Putting $s = j\omega$

$$G(s) = \frac{1}{j\omega}$$

If $\omega = 0$, then $G(j\omega) = \infty$

If $\omega = \infty$, then $G(j\omega) = 0$

(ii) When $G(s) = \dfrac{1}{s^2}$

Putting $s = j\omega$, we get

$$G(j\omega) = \frac{1}{(j\omega)^2} = -\frac{1}{\omega^2}$$

If $\omega = 0$, then $G(j\omega) = \infty$

If $\omega = \infty$, then $G(j\omega) = 0$

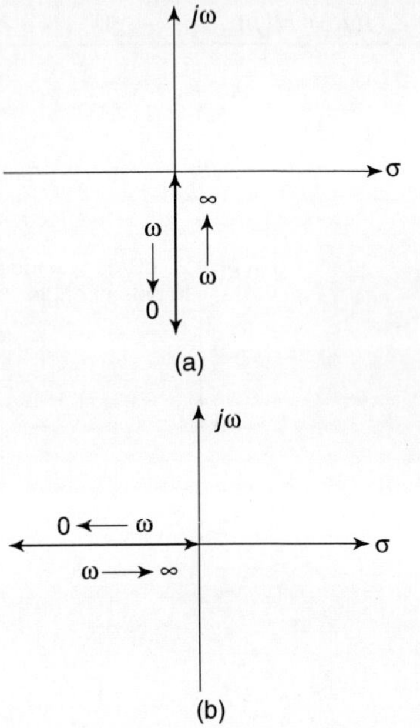

Figure Ex. 8.11 (a) Polar plot for $G(s) = \dfrac{1}{s}$

(b) Polar plot for $G(s) = \dfrac{1}{s^2}$

Example 8.12 Sketch the polar plot for $G(s) = \dfrac{K}{(1+Ts)}$.

Solution

$$G(s) = \frac{K}{(1+Ts)}$$

Putting $s = j\omega$

$$G(j\omega) = \frac{K}{1+j\omega T}$$

Angle $$\phi = \angle G(j\omega) = -\tan^{-1}\omega t$$

$$M = |G(j\omega)| = \frac{K}{\sqrt{\omega^2 T^2 + 1}}$$

When ω approaches infinity

$$M = \lim_{\omega \to \infty} |G(j\omega)| = \lim_{\omega \to \infty} \frac{K}{\sqrt{1 + \omega^2 T^2}}$$

$$M = K$$

$$\phi = \lim_{\omega \to \infty} \angle G(j\omega) = \lim_{\omega \to \infty} (-\tan^{-1}\omega T)$$

$$= -90°$$

When ω approaches zero

$$M = \lim_{\omega \to 0} |G(j\omega)| = \lim_{\omega \to 0} \frac{K}{\sqrt{1 + \omega^2 T^2}} = K$$

$$\phi = \lim_{\omega \to 0} \angle G(j\omega) = \lim_{\omega \to 0} (-\tan^{-1}\omega T) = 0°$$

The polar plot is

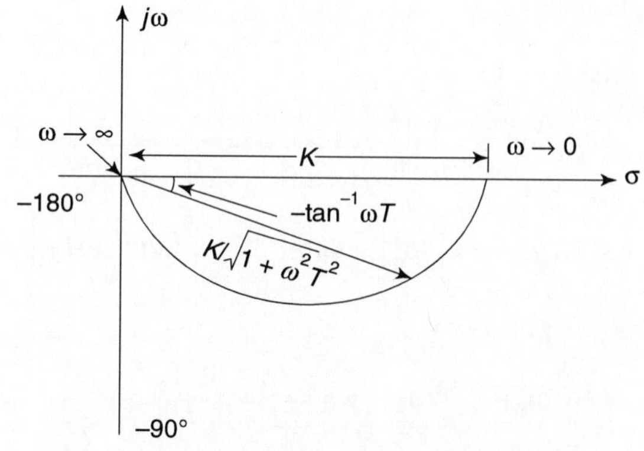

Figure Ex. 8.12 Polar plot for Ex. 8.12

Example 8.13 Sketch the polar plot for $G(s) = \dfrac{1}{(1 + T_1 s)(1 + T_2 s)}$

Solution Putting $$s = j\omega$$

$$G(j\omega) = \frac{1}{(1 + j\omega T_1)(1 + j\omega T_2)}$$

The magnitude is

$$|G(j\omega)| = \frac{1}{\sqrt{1+\omega^2 T_1^2}\,\sqrt{1+\omega^2 T_2^2}} \quad \text{...(i)}$$

The angle is

$$\angle G(j\omega) = -\tan^{-1}\omega T_1 - \tan^{-1}\omega T_2 \quad \text{...(ii)}$$

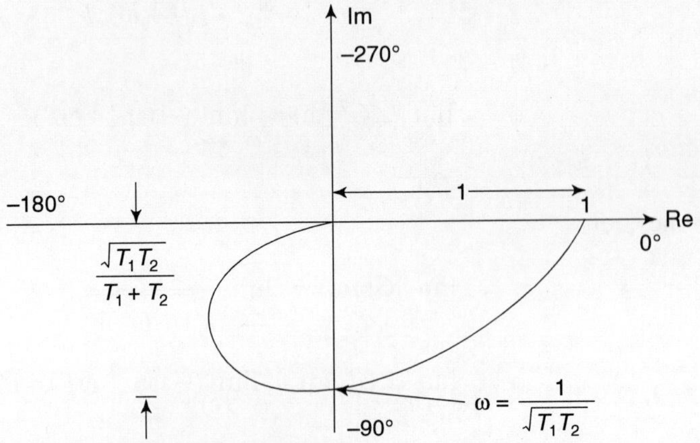

Figure Ex. 8.13 Polar plot for Ex. 8.13

When ω approaches zero, i.e., $\omega \to 0$

$$|G(j\omega)| = \lim_{\omega\to 0}\frac{1}{\sqrt{1+\omega^2 T_1^2}\,\sqrt{1+\omega^2 T_2^2}} = 1$$

and

$$\angle G(j\omega)\,H(j\omega) = \lim_{\omega\to 0}\left[-\tan^{-1}\omega T_1 - \tan^{-1}\omega T_2\right] = 0°$$

When ω approaches ∞, i.e., $\omega \to \infty$

$$|G(j\omega)| = \lim_{\omega\to \infty}\frac{1}{\sqrt{1+\omega^2 T_1^2}\,\sqrt{1+\omega^2 T_2^2}} = 0$$

$$\angle G(j\omega) = \lim_{\omega\to\infty}\angle -\tan^{-1}\omega T_1 - \tan^{-1}\omega T_2$$

$$= -90° - 90° = -180°$$

$$G(j\omega) = \frac{1}{(1+j\omega T_1)(1+j\omega T_2)}$$

Separating $G(j\omega)$ into real and imaginary part by multiplying by complex conjugate

$$G(j\omega) = \frac{(1-j\omega T_1)(1-j\omega T_2)}{(1+j\omega T_1)(1+j\omega T_2)} \times \frac{1}{(1-j\omega T_1)(1-j\omega T_2)}$$

$$G(j\omega) = \frac{(1 - j\omega T_2 - j\omega T_1 + j^2\,\omega^2\,T_1\,T_2)}{(1 + \omega^2 T_1^2)\,(1 + \omega^2\,T_2^2)}$$

$$G(j\omega) = \frac{(1 - \omega^2\,T_1\,T_2) - j(\omega\,T_1 + \omega\,T_2)}{(1 + \omega^2\,T_1^2)\,(1 + \omega^2\,T_2^2)}$$

$$G(j\omega) = \frac{1 - \omega^2\,T_1\,T_2}{(1 + \omega^2\,T_1^2)\,(1 + \omega^2\,T_2^2)} - j\frac{(\omega\,T_1 + \omega\,T_2)}{(1 + \omega^2\,T_1^2)\,(1 + \omega^2\,T_2^2)}$$

Now the real part is equal to zero, so

$$\frac{1 - \omega^2\,T_1\,T_2}{(1 + \omega^2\,T_1^2)\,(1 + \omega^2\,T_2^2)} = 0$$

$$\omega^2\,T_1\,T_2 = 1$$

$$\omega^2 = \frac{1}{T_1\,T_2} \Rightarrow \omega = \frac{1}{\sqrt{T_1\,T_2}}$$

This is the point at which the polar plot crosses the imaginary axis putting the value of ω in Eq. (i), we get

$$|G(j\omega)| = \frac{1}{\sqrt{1 + \dfrac{T_1^2}{T_1\,T_2}}\,\sqrt{1 + \dfrac{T_2^2}{T_1\,T_2}}}$$

$$|G(j\omega)| = \frac{\sqrt{T_1\,T_2}}{T_1 + T_2}$$

Example 8.14 Sketch the polar plots for

(i) Type 0 system $G(s) = \dfrac{3}{(s+1)\,(s+2)}$

(ii) Type 1 $G(s) = \dfrac{12}{s\,(s+1)\,(s+2)}$ *(U.P.T.U., 2006–07, CS)*

Solution (i) $$G(s) = \frac{3}{(s+1)\,(s+2)}$$

$$G(s) = \frac{3}{2\,(1+s)\,(1+0.5\,s)} = \frac{1.5}{(1+s)\,(1+0.5\,s)}$$

Here $T_1 = 1$

and $T_2 = 0.5$

and $K = 1.5$

Solving the problem as Ex. 8.13.

Putting $s = j\omega$

$$G(j\omega) = \frac{1.5}{(1 + j\omega)(1 + 0.5\,\omega j)}$$

The magnitude $$|G(j\omega)| = \frac{K}{\sqrt{1 + (T_1\,\omega^2)^2}\,\sqrt{1 + (T_2\,\omega^2)}} \text{ (from Ex. 8.13)}$$

Putting the value of K, T_1 and T_2

$$|G(j\omega)| = \frac{1.5}{\sqrt{1 + \omega^2}\,\sqrt{1 + (0.5\,\omega)^2}}$$

The angle is $$\angle G(j\omega)\,H(j\omega) = \angle -\tan^{-1}\omega T_1 - \tan^{-1}\omega T_2 \text{ (from Ex. 8.13)}$$

$$\angle G(j\omega)\,H(j\omega) = \angle -\tan^{-1}\omega - \tan^{-1} 0.5\,\omega$$

When ω approaches to zero, i.e., $\omega \to 0$

$$\lim_{\omega \to 0} |G(j\omega)| = \lim_{\omega \to 0} \frac{1.5}{\sqrt{1 + \omega^2}\,\sqrt{1 + (0.5\,\omega)^2}} = 1.5$$

$$\lim_{\omega \to 0} \angle G(j\omega) = \lim_{\omega \to 0} \left[-\tan^{-1}\omega - \tan^{-1} 0.5\,\omega = 0°\right.$$

Now ω approaches to infinity, i.e., $\omega \to \infty$

$$\lim_{\omega \to 0} |G(j\omega)| = \lim_{\omega \to 0} \frac{1.5}{\sqrt{1 + \omega^2}\,\sqrt{1 + (0.5\,\omega)^2}}$$

$$= 0$$

$$\lim_{\omega \to 0} \angle G(j\omega) = \lim_{\omega \to 0} \angle -\tan^{-1}\omega - \tan^{-1} 0.5\,\omega$$

$$= -90° - 90°$$

$$= -180°$$

Separating the real and imaginary part of $G(s)$

$$G(j\omega) = \frac{K(1 - \omega^2\,T_1\,T_2)}{(1 + \omega^2\,T_1^2)(1 + \omega^2\,T_2^2)} - j\,\frac{K(\omega T_1 + \omega T_2)}{(1 + \omega^2\,T_1^2)(1 + \omega^2 T_2^2)}$$

Putting the value of T_1 and T_2

$$G(j\omega) = \frac{1.5(1 - \omega^2 \times 0.5)}{(1 + \omega^2)[1 + (0.5\,w)^2]} = j\,\frac{1.5\,\omega(1 + 0.5)}{(1 + \omega^2)[1 + (0.5\,\omega)^2]}$$

Equating real part of zero

$$\frac{1.5(1 - 0.5\,\omega^2)}{(1 + \omega^2)[1 + (0.5\,\omega)^2]} = 0$$

$$1 - 0.5\,\omega^2 = 0$$

$$\omega^2 = \frac{1}{0.5}$$

$$\omega = 1.414$$

and $$\omega = \pm \infty$$

On equating imaginary part to zero

$$\frac{1.5\,\omega\,(1.5)}{(1+\omega^2)\,[1+(0.5\,\omega)]} = 0$$

$$\omega = 0$$

When $$\omega = 0 \quad |G(j\omega)| = 1.5$$

$$\omega = \infty \quad |G(j\omega)| = 0$$

The polar plot is shown in Fig. Ex. 8.14 (a) by referring Table 8.4.

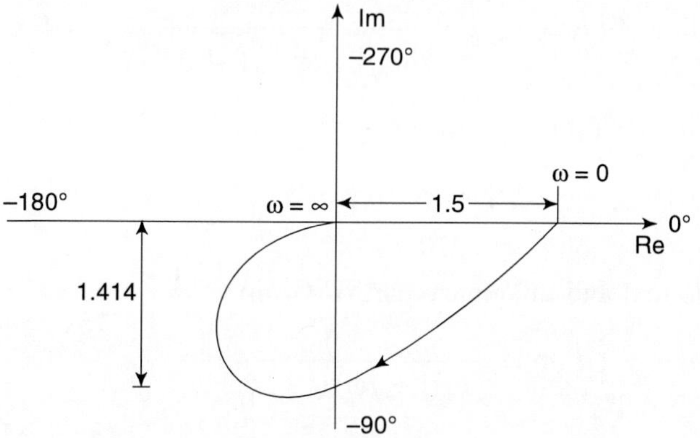

Figure Ex. 8.14 (a) Polar plot of Ex. 8.14 (i)

(ii) $$G(s) = \frac{12}{s\,(s+1)\,(s+2)}$$

Putting $$s = j\omega$$

$$G(j\omega) = \frac{12}{j\omega\,(j\omega+1)\,(j\omega+2)} = \frac{12}{2\,j\omega\,(1+j\omega)\,(1+0.5\,j\omega)}$$

$$G(j\omega) = \frac{6}{j\omega\,(1+j\omega)\,(1+0.5\,j\omega)}$$

Here $$K = 6$$

$$T_1 = 1$$

$$T_2 = 0.5$$

The magnitude is

$$|G(j\omega)| = \frac{K}{\omega\sqrt{(1+\omega^2\,T_1^2)}\,\sqrt{(1+\omega^2\,T_2^2)}}$$

$$|G(j\omega)| = \frac{6}{\omega\sqrt{1+\omega^2}\sqrt{1+(0.5\,\omega)^2}}$$

The angle is

$$\angle G(j\omega)\,H(j\omega) = -\,90^\circ - \tan^{-1}\omega - \tan^{-1}(0.5\omega)$$

When ω approaches to zero, i.e., $\omega \to 0$

$$\lim_{\omega\to 0}|G(j\omega)| = \lim_{\omega\to 0}\frac{6}{\omega\sqrt{1+\omega^2}\sqrt{1+(0.5\,\omega)^2}} = \infty$$

$$\lim_{\omega\to 0}\angle G(j\omega) = \lim_{\omega\to 0}\angle -90^\circ - \tan^{-1}\omega - \tan^{-1}(0.j\omega) = -\,90^\circ$$

When ω approaches to infinity, i.e., $\omega \to \infty$

$$\lim_{\omega\to\infty}|G(j\omega)| = \lim_{\omega\to\infty}\frac{6}{\omega\sqrt{1+\omega^2}\sqrt{1+(0.5\,\omega)^2}} = 0$$

$$\lim_{\omega\to\infty}\angle G(j\omega) = \lim_{\omega\to\infty}\angle -90^\circ - \tan^{-1}\omega - \tan^{-1}(0.5\omega)$$

$$= -\,90^\circ - 90^\circ - 90^\circ$$

$$= -\,270^\circ$$

Now separating the real and imaginary part of $G(j\omega)$

$$G(j\omega) = \frac{6}{j\omega\,(1+j\omega)\,(1+j\,0.5\,\omega)}$$

$$G(j\omega) = \frac{-\,36\,\omega^2}{\omega^2(1+\omega^2)(4+\omega^2)} + j\,\frac{12\cdot\omega\,(2-\omega^2)}{\omega^2\,(1+\omega^2)(4+\omega^2)}$$

Equating the imaginary part to zero

$$\frac{12\cdot\omega^2\,(2-\omega^2)}{\omega^2\,(1+\omega^2)(4+\omega^2)} = 0$$

$$\omega^2 = 2$$

$$\omega = \pm\sqrt{2}$$

and

$$\omega = \pm\,\infty$$

Equating the real part to zero

$$\frac{-\,36\,\omega^2}{\omega^2\,(1+\omega^2)(4+\omega^2)} = 0$$

$$\omega = \infty$$

The polar plot is shown in Fig. Ex. 8.14 (b) by refering Table 8.4 for

$$G(s) = \frac{1}{s\,(1+T_1)(1+T_2\,s)}$$

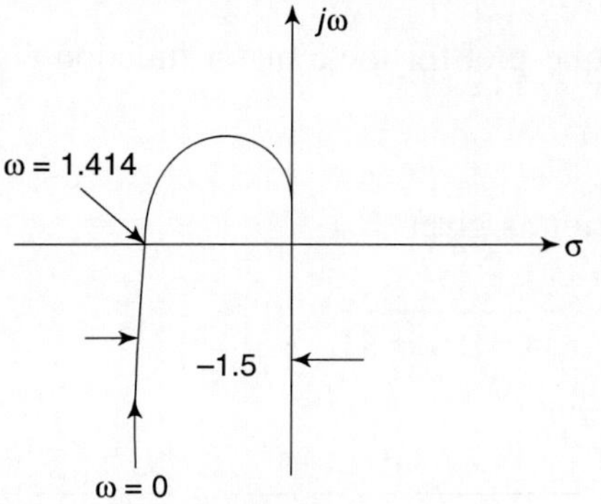

Figure Ex. 8.14 (b) Polar plot for Ex. 8.14 (ii)

$$\omega = \frac{1}{\sqrt{T_1\, T_2}}$$

$$\omega = \frac{1}{\sqrt{1 \times 0.5}}$$

$$\omega = 1.414$$

Example 8.15 Draw the polar plot of the following system

$$G(s) = \frac{1}{s^2\,(1+T_1\, s)\,(1+T_2\, s)\,(1+T_3\, s)}.$$

Solution Putting

$$s = j\omega$$

$$G(j\omega) = \frac{1}{(j\omega)^2\,(1+j\omega\, T_1)\,(1+j\omega\, T_2)\,(1+j\omega\, T_3)}$$

By referring Table 8.4 the polar plot for that function is shown in Fig. Ex. 8.15.

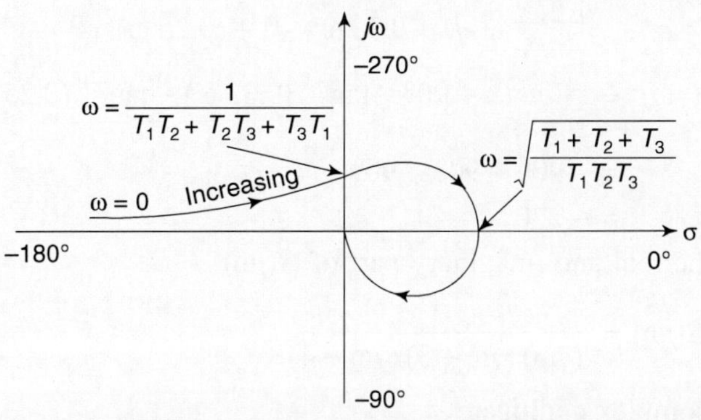

Figure Ex. 8.15 Polar plot of Ex. 8.15

Example 8.16 Sketch the polar plot for the transfer function $G(s) = \dfrac{1}{s\,(s+3)\,(s+4)}$.

(U.P.T.U., 2003–04, CS)

Solution The transfer function is given

$$G(s) = \frac{1}{s\,(s+3)\,(s+4)}$$

Putting $s = j\omega$

$$G(j\omega) = \frac{1}{(j\omega)(j\omega+3)\,(j\omega+4)} = \frac{1}{4\times(j\omega)\,(1+0.33\,\omega j)\,(1+0.25\,\omega j)}$$

$$G(j\omega) = \frac{0.083}{(j\omega)\,(1+0.33\,\omega j)\,(1+0.25\,\omega j)}$$

$$|G(j\omega)| = \frac{0.083}{\omega\sqrt{1+(0.33\,\omega)^2}\,\sqrt{1+(0.25\omega)^2}}$$

The angle is

$$\angle G(j\omega) = -\,90^\circ - \tan^{-1}(0.33\omega) - \tan^{-1} 0.25\omega$$

When ω approaches to zero, i.e., $\omega \to 0$

$$\lim_{\omega\to 0} |G(j\omega)| = \lim_{\omega\to 0} \frac{0.83}{\omega\sqrt{1+(0.33\,\omega)^2}\,\sqrt{1+(0.25\,\omega)^2}} = \infty$$

$$\lim_{\omega\to 0} \angle G(j\omega) = \lim_{\omega\to 0} \angle -90^\circ - \tan^{-1}(0.3\omega) - \tan^{-1}(0.25\,\omega)$$

$$= -\,90^\circ$$

When ω approaches to infinity

$$\lim_{\omega\to\infty} |G(j\omega)| = \lim_{\omega\to\infty} \frac{0.83}{\omega\sqrt{1+0.33\,\omega^2}\,\sqrt{1+(0.25\,\omega)^2}}$$

$$\lim_{\omega\to\infty} \angle G(j\omega) = \lim_{\omega\to\infty} \angle -90^\circ - \tan^{-1}(0.33\,\omega) - \tan^{-1}(0.25\,\omega)$$

$$= -\,90^\circ - 90^\circ - 90^\circ$$

$$= -\,270^\circ$$

Now seaparating the real and imaginary part of $G(j\omega)$

$$G(j\omega) = \frac{1}{(j\omega)\,(j\omega+3)\,(j\omega+4)}$$

Multiplying with complex conjugate

$$G(j\omega) = \frac{1}{(j\omega)\,(j\omega+3)(j\omega+4)} \times \frac{(-\,j\omega)\,(3-j\omega)\,(4-j\omega)}{(-\,j\omega)\,(3-j\omega)\,(4-j\omega)}$$

$$= \frac{(-j\omega)(3-j\omega)(4-j\omega)}{\omega^2(9+\omega^2)(16+\omega^2)} = \frac{-7\omega^2 - j\omega(12-\omega^2)}{\omega^2(9+\omega^2)(16+\omega^2)}$$

$$= \frac{-7\,\omega^2}{\omega^2(9+\omega^2)(16+\omega^2)} - j\frac{\omega(12-\omega^2)}{\omega^2(9+\omega^2)(16+\omega^2)}$$

$$G(j\omega) = -\frac{7}{(9+\omega^2)(16+\omega)^2} - j\frac{12-\omega^2}{\omega(9+\omega^2)(16+\omega^2)}$$

The frequency at which real part is zero

$$\frac{7}{(9+\omega^2)(16+\omega^2)} = 0$$

$$\omega = \infty$$

The frequency at which imaginary part is zero

$$\frac{-12-\omega^2}{\omega(9+\omega^2)(1+\omega^2)} = 0$$

$$12 - \omega^2 = 0$$

$$\omega^2 = 12$$

$$\omega = \sqrt{12} = 3.46$$

The polar plot is shown in Fig. Ex. 8.16.

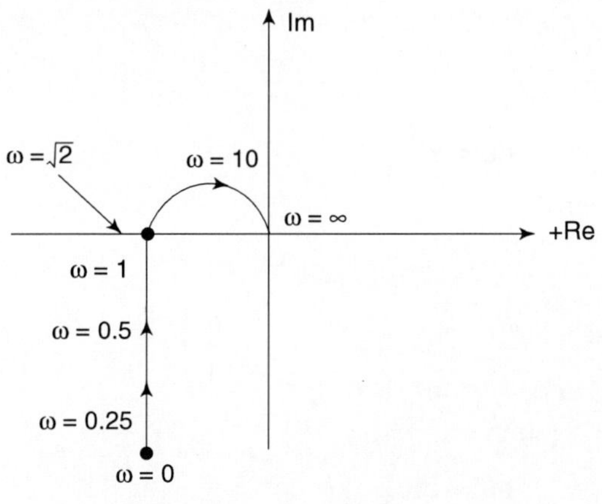

Figure Ex. 8.16 Polar plot of Ex. 8.16

For varying ω the real and imaginary parts

ω	$\omega = 0$	0.25	0.5	1	$\sqrt{12}$	10	∞
Re of $G(j\omega)$	0.0486	– 0.0480	– 0.0466	– 0.0412	– 0.012	0.006	0
Impart of $G(j\omega)$	$-\infty$	1.312	– 0.313	– 0.0647	0	0.0078	0

Example 8.17 Sketch the polar plot for the transfer function $G(s) = \dfrac{10}{s(s+1)(s+2)}$.

(U.P.T.U., 2002–03, CS)

Solution Putting

$$s = j\omega$$

$$G(j\omega) = \frac{10}{j\omega(j\omega+1)(j\omega+2)}$$

Separating real and imaginary part, so multiplying with complex conjugate, we get

$$G(j\omega) = \frac{10}{j\omega(j\omega+1)(j\omega+2)} \times \frac{(-j\omega)(1-j\omega)(2-j\omega)}{(-j\omega)(1-j\omega)(2-j\omega)}$$

$$= \frac{10(-j\omega)(2-3j\omega-\omega^2)}{\omega^2(1+\omega^2)(4+\omega^2)} = \frac{-20j\omega+30(j\omega)^2 = 10j\omega^3}{\omega^2(1+\omega^2)(4+\omega^2)}$$

$$= \frac{-30\omega^2}{\omega^2(1+\omega^2)(4+\omega^2)} - j\frac{20\omega-10\omega^3}{\omega^2(1+\omega^2)(4+\omega^2)}$$

$$G(j\omega) = \frac{-30}{(1+\omega^2)(4+\omega^2)} - j\frac{20-10\omega^2}{\omega(1+\omega^2)(4+\omega^2)}$$

Equating the real part zero

$$\frac{-30}{(1+\omega^2)(4+\omega^2)} = 0$$

$$\omega = \infty$$

The imaginary part is zero

$$\frac{20-10\omega^2}{\omega(1+\omega^2)(4+\omega^2)} = 0$$

$$10\omega^2 = 20$$

$$\omega^2 = 2$$

$$\omega = \pm\sqrt{2}$$

For the real and imaginary part for varying ω, we get

ω	0	1	$\sqrt{2}$	5	50	ω
Re part of $G(j\omega)$	– 7.5	– 3	– 1.67	– 0.84	– 0.000005	0
Impart of $G(j\omega)$	– ∞	1	0	0.86	0.00008	0

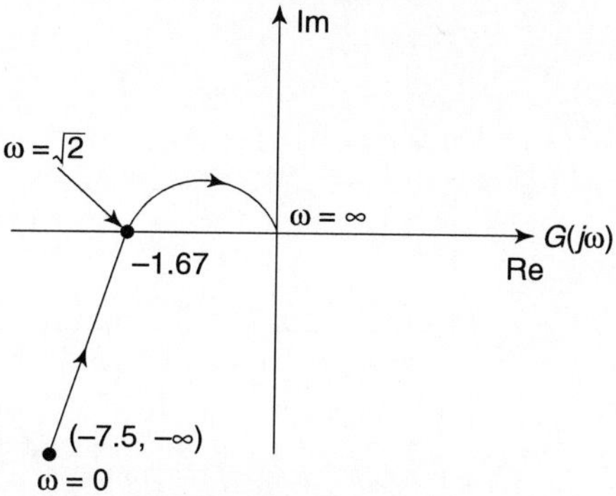

Figure Ex. 8.17 Polar plot of Ex. 8.17

Example 8.18 Draw the polar plot for the given function $G(s) = \dfrac{1}{s\,(1+s\,T_1)\,(1+T_2\,s)}$.

(U.P.T.U., 2005–06, ACS)

Solution For the given function see the Table 8.4 typical Nyquist plot.

MATLAB PROGRAMS

Program 8.1

```
% Program for finding a bode plot for the given function
%G(s) = (s+3)/(s+5)(s^2+3s+20)
clear all
clc
clf                    % clear the graph
numg = input('Enter the coefficient of polynomial of numerator:- ');
                       % Input coefficient of numerator
deng1 = input('Enter the coefficient of First polynomial of denominator:- ');
                       % Input coefficient of denominator
deng2 = input('Enter the coefficient of Second polynomial of denominator:- ')
                       % Input coefficient of denominator
deng = conv(deng1,deng2);% Multiplying two polynomial
'G(s)'                 % Display String
G = tf(numg,deng)      % Create transfer function
bode(G)                % Plot Bode plot
grid on
title('Open loop Frequency response')
```

Input

```
Enter the coefficient of polynomial of numerator:- [1 3]
Enter the coefficient of First polynomial of denominator:- [1 5]
Enter the coefficient of Second polynomial of denominator:- [1 3 20]
```

Output

```
                                    s + 3
Transfer function: G(s)  ----------------------------
                         s^3 + 8s^2 + 35s + 100
```

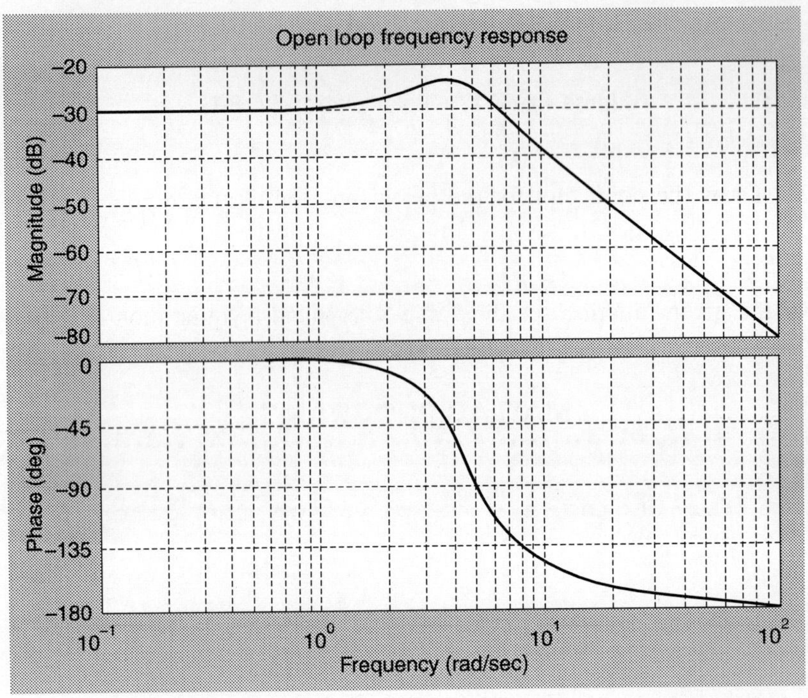

Figure P. 8.1 Open loop frequency response of $G(s)$

Program 8.2

```
% Program for finding bode plot and calculate margins and margins frequency
% G(s)  = 20/(s^2+3s+2)
clear all
clc
clf                        % clear the graph
numg = input('Enter the coefficient of polynomial of numerator:- ');
                           % Input coefficient of numerator
deng = input('Enter the coefficient of First polynomial of denominator:- ');
                           % Input coefficient of denominator
```

```
'G(s)'
G = tf(numg,deng)       % Create transfer function
bode(G)                 % draw bode plot
grid on                 % Show the magnitude and phase line on plot
[Gm, Pm, Wcg, Wcp] = margin(G) % Calculate margin and margin frequency
Gm_db = 20*log10(Gm) % Gain margin in dB
```

Input

```
Enter the coefficient of polynomial of numerator:- [1 3]
Enter the coefficient of First polynomial of denominator:- [1 5]
Enter the coefficient of Second polynomial of denominator :- [1 3 20]
```

Output

```
                                    s + 3
Transfer function: G(s)  -----------------------------
                          s^3 + 8s^2 + 35s + 100

Gm = inf
Pm = 38.9397
Wcg = inf
Wcp = 4.1900
Gm_dB = inf
```

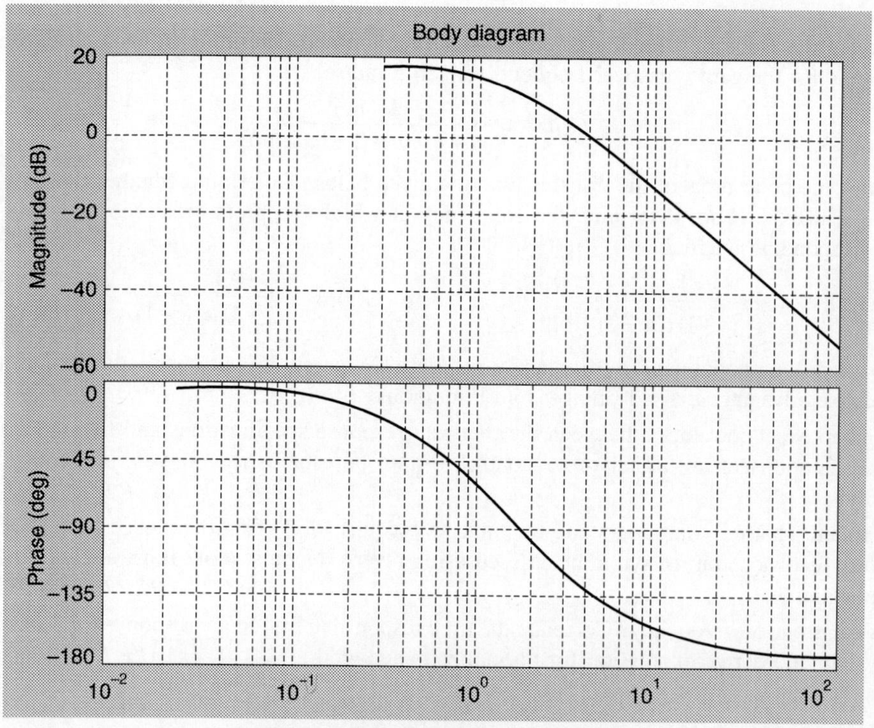

Figure P. 8.2 Bode plot of $G(s)$

EXERCISE

8.1 Explain frequency response analysis in control system.

8.2 Explain polar and inverse polar plot. What is the role of these plots in control system?

8.3 Explain correlation between time and frequency response.

8.4 Sketch the polar plot and inverse polar plot for a unity feedback system with open loop transfer function $G(s) = \dfrac{1}{s(s+1)(s+1)}$ and find the frequency at which $|G(j\omega)| = 1$ and the corresponding phase angle $\angle G(j\omega)$.

8.5 Draw the Bode plot of the following transfer function having unity feedback $G(s) = \dfrac{1}{s(1+0.5s)(1+0.05s)}$.

8.6 Explain the relationship betwwen pole positions an stability.

8.7 Draw Bode plot for $G(s) = \dfrac{20}{s(s+2)(s+10)}$ also determine

(i) Phase crossover frequency ω_{cp}

(ii) Gain crossover frequency ω_{cg}

(iii) Gain margin

(iv) Phase margin

(v) Stability

8.8 Draw the Bode plot for the following system $G(s) = \dfrac{(s+3)}{s(s+1)(s+5)(s+2)}$.

8.9 Sketch the Bode plot for the given transfer function and determine in each case, the system gain K for the gain crossover frequency ω_{cg} to 5 rad/sec.

$$G(j\omega) = \frac{K(j\omega)}{(1+0.2j\omega)(1+0.02j\omega)}$$

8.10 Draw the polar plots of the transfer function given below. Determine whether these plots cross the real axis. If so, determine the frequency at which the plots cross the real axis and the corresponding magnitude $|G(j\omega)|$.

(a) $G(s) = \dfrac{1}{s^2(s+1)(2s+1)}$

(b) $G(s) = \dfrac{1}{s(s+1)(2s+1)}$

8.11 What do you understand by frequency response? What are its various methods? Discuss their relative advantages, disadvantages and limitations. *(U.P.T.U., 2007–08, CS)*

8.12 Can an exact correlation between the transient response specifications and frequency response specifications of an higher order control system be established? If yes how and if no then why? *(U.P.T.U., 2007–08, CS)*

8.13 Explain what do you understand by "inverse Nyquist Plot". Can the closed loop stability of a unity feedback control sytem be assesed using inverse Nyquist plot? Illustrate with the help of an example. *(U.P.T.U., 2007–08, CS)*

8.14 What is cut-off frequency? Correlate the time response to frequency response for a second order system and derive expression for normalized bandwidth. *(U.P.T.U., 2006–07, CS)*

8.15 In connection with Bode plot

(i) Define bandwidth

(ii) Define cut-off frequency

(iii) Define corner frequency

(iv) Explain correction to be made at corner frequency and discuss the importance of bandwidth *(U.P.T.U., 2005–06, CS)*

8.16 Explain polar plot and inverse polar plot with examples. *(U.P.T.U., 2005–06, CS)*

8.17 Define gain margin, and phase margin. Explain their significance regarding stability with an example. *(U.P.T.U., 2004–05, CS)*

8.18 Write a note on Bode plot, how does it help in system design. *(U.P.T.U., 2007–08, ACS)*

8.19 Discuss the time domain analysis and frequency domain analysis of control systems and what information can be obtained through root locus? *(U.P.T.U., 2006–07, ACS)*

8.20 Describe the various frequency domain specifications with proper expression and plots. *(U.P.T.U., 2005–06, ACS)*

8.21 What are *M* and *N* circles ? Explain the physical significance of these circles in stability study. *(U.P.T.U., 2004–05, ACS)*

8.22 Explain the steps for constructing the polar plots. Draw the inverse polar plot of the circuit shown in Fig. E. 8.22. *(U.P.T.U., 2004–05, ACS)*

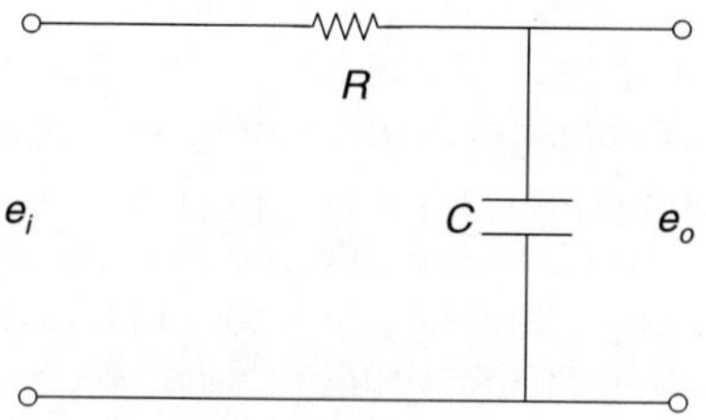

Figure E. 8.22

8.23 Determine the correlation between time and frequency response for second order system and also show impact and damping on bandwidth. *(U.P.T.U., 2004–05, ACS)*

8.24 State the salient features of magnitude curve of Bode plot of a type 0 system. *(U.P.T.U., 2003, CO., ACS)*

ANSWERS

8.4 $|G(j\omega)| = 0.683$ rad/sec

$\angle G(j\omega) = -159°$

8.7 (i) 4.7 rad/sec; (ii) 1 rad/sec; (iii) G.M = 20 dB; (iv) P.M. = 57°; (v) Stable

8.9 0.056

8.10 (a) Does not cross (b) 0.71 rad/sec. 0.67

CHAPTER 9

Stability in Frequency Domain

9.1 INTRODUCTION

In previous chapter, we have already discussed about stability and developed a number of tools to determine stability. We have studied Rouths–Hurwitz criterion which provides absolute stability. The root locus technique was discussed in chapter 7. It is an approach to determine the location of roots in the s-plane, given the open loop pole zero configuration. We have also studied Bode plot which also provides relative stability in frequency domain. So we can say that the stability of a system depends upon the location of the roots of its characteristic equation in complex plane. For a system to be stable, the roots of the characteristic equation should not lie in the right half of the complex plane or on the imaginary axis, for such roots lead to instability or sustained oscillations of the system. Now we continue that discussion in this chapter by showing how frequency response method can be used to investigate stability. The important concepts of phase margin (P.M.), gain margin (G.M.) and bandwidth are developed in the content of Nyquist diagram and Bode plot. Here we are using the technique, the Nyquist stability criterion, which relates the location of the roots of the characteristic equation for the open-loop frequency response of a system.

9.2 MAPPING CONTOURS IN THE COMPLEX PLANE

To determine stability, we must ascertain that all the zeros of function $F(s)$ lie in the left-hand s-plane. Nyquist thus proposed a mapping of the right-hand s-plane. Nyquist thus proposed a mapping of the right hand s-plane into the $F(s)$ plane. So to utilize and understand Nyquist's criterion, we shall first consider briefly the mapping of contourism the complex plane. Basically, a contour map is a counter or trajectory in one plane mapped or translated into another plane by a function $F(s)$.

Now considering a function $F(s)$, which is a ratio of two polynomials. It is assumed that both the polynomials are in the form of product of linear factor as given below:

$$F(s) = \frac{(s - Z_1)(s - Z_2)(s - Z_n)}{(s - P_1)(s - P_2)(s - P_m)} \qquad ...(9.1)$$

Since s is a complex vriable, represented by $s = \sigma + j\omega$ on the complex s-plane.

So the function $F(s)$ is also complex and it can be defined as $F(s) = u + jv$.

In Eq. (9.1) the $Z_1, Z_2, Z_3, \ldots Z_n$ are the zeros of the function $F(s)$ and $P_1, P_2, \ldots P_m$ are the poles of $F(s)$.

Equation 9.1 indicates that for every point s in the s plane at that point $F(s)$ is analytic that means a function $F(s)$ is analytic in the s-plane provided the function and all its derivative exist. Therefore, it can be stated that the function $F(s)$ maps the points in the s-plane into the $F(s)$ plane, if a contour in the s-plane which does not go through any singular point. (If function or its derivatives does not exist in the s-plane are called singular points.)

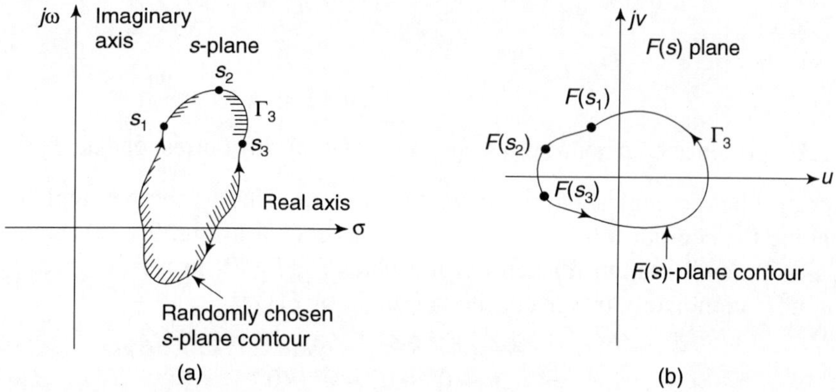

Figure 9.1 (a) s-plane contour, (b) $F(s)$ plane contour

Let us consider a contour in s-plane (Γ_s) having three points s_1, s_2, s_3 and contour is traversed in the colockwise direction, this is generally used in control system theory but opposite to the complex variable theory. The region to the right of a closed contour is considered enclosed by the contour. So the shaded areas in Fig. 9.1 (a) is enclosed by the closed contour.

Figure 9.1 (b) is showing the correspond $F(s)$ plane (Γ_s). We are not interested in the accurate shape of the $F(s)$ plane contour. We are only interested in the encirclement of the origin by the $F(s)$ plane contour. We will be see this in the next section.

Let us consider an s-plane contour which encloses only one of the zeros of $F(s)$, like $s = Z_1$, while all the poles and reamining zeros are distributed in the s-plane outside the contour as shown in Fig. 9.2. (a) and (b).

Equation 9.1 can be written as

$$F(s) = |F(s)| \angle F(s)$$

$$|F(s)| = \frac{|s - Z_1||s - Z_2|}{|s - P_1||s - P_2|} \quad \text{...(9.2)}$$

$$\angle F(s) = \angle s - Z_1 + \angle s - Z_2 + \ldots \angle s - P_1 - \angle s - P_2$$

$$j\omega = \angle\phi_{Z1} + \angle\phi_{Z2} + \angle\phi_{P1}\ \angle\phi_{P2}$$

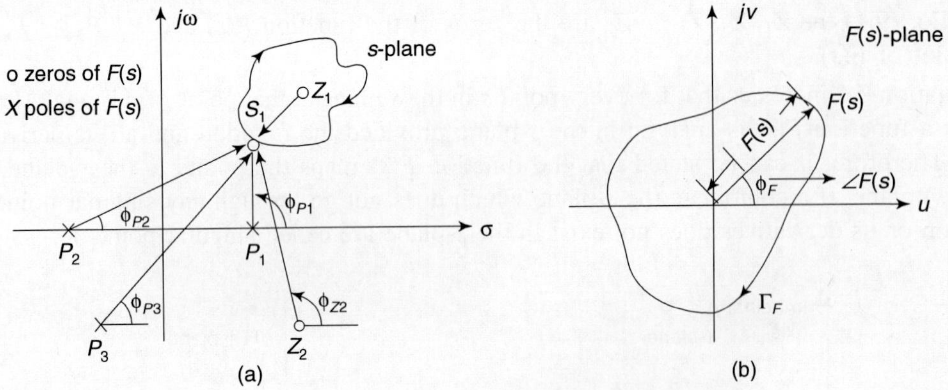

Figure 9.2 (a) *s*-plane contour enclosing a zero of *F*(*s*), (b) Corresponding *F*(*s*) plane

We can determine the angles as *s* traverses the contour. Clearly, the net angle changes as *s* traverses along the contour a full rotation of – 2π as shown in Fig. 9.2 (a) the net rotation for zero Z_1 is – 2π and rotation for zero Z_2 and poles P_1, P_2, P_3 in zero degree. So we can say the contour is completely traversed, the net angle of *F*(*s*) is

$$\begin{aligned}\angle F(s) &= \angle\phi Z_1 + \angle\phi Z_2 - \angle\phi_{P1} - \angle\phi_{p2} - \angle\phi_{p3}\\ &= -2\pi + 0^\circ - 0^\circ - 0^\circ - 0^\circ\\ &= -2\pi, \text{ i.e., } 360^\circ \text{ clockwise}\end{aligned}$$

Since only one zero is enclosed so the *F*(*s*) plane contour encircles the origin once.

Now consider a contour which encloses two zeros, like $s = Z_1$ and $s = Z_2$, the *F*(*s*) plane contour encircles the origin twice is the clockwise direction as shown in Fig. 9.3.

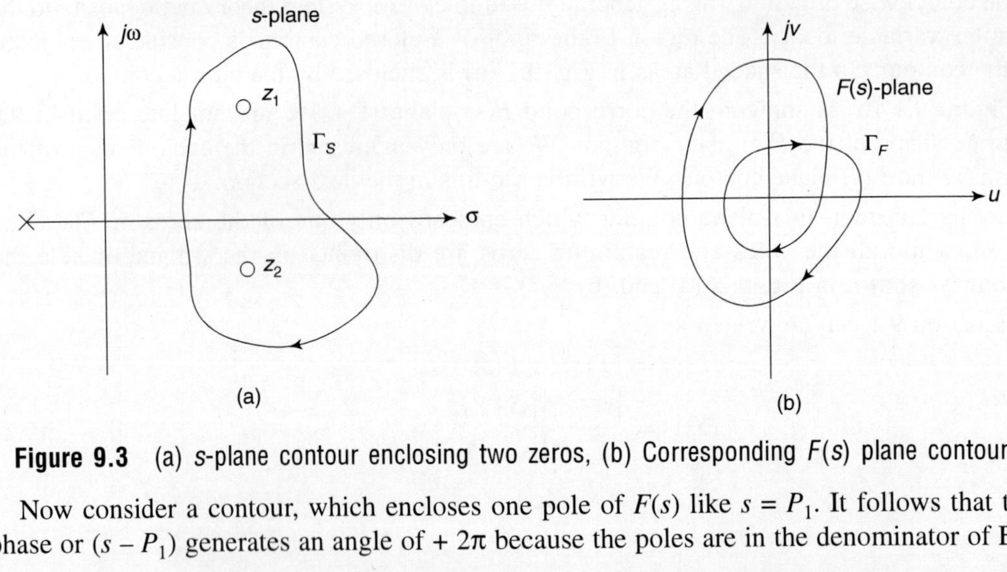

Figure 9.3 (a) *s*-plane contour enclosing two zeros, (b) Corresponding *F*(*s*) plane contour

Now consider a contour, which encloses one pole of *F*(*s*) like $s = P_1$. It follows that the phase or $(s - P_1)$ generates an angle of + 2π because the poles are in the denominator of Eq.

9.1, the $F(s)$ plane contour experiences an angle change of $+ 2\pi$, which encircles the origin in contour clockwise direction as shown in Fig. 9.4 (a) and (b).

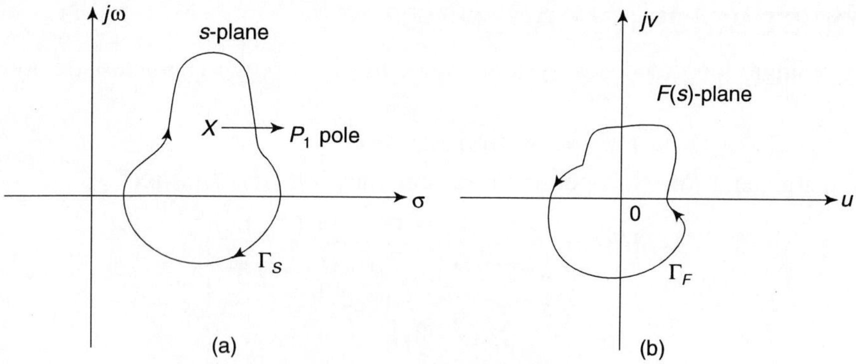

Figure 9.4 (a) s-plane contour enclosing a pole of $F(s)$, (b) Corresponding $F(s)$ plane contour

Thus, if there are P poles and zeros of $F(s)$ enclosed by the s-plane contour, then the corresponding $F(s)$ plane contour must encircle the origin Z times in the clockwise direction and P times in the counterclockwise direction, resulting in a net encirclement of the origin, $(P - Z)$ times in the counterclockwise direction.

$$N = P - Z$$

where N = Number of encirclement of origin

Let us consider a s-plane contour encloses three zeros and one pole so the number of encirclement (N) is

$$N = P - Z$$
$$= 1 - 3 = - 2$$

because N is negative so the direction of $F(s)$ plane contour is clockwise as shown in Fig. 9.5 (a) and (b).

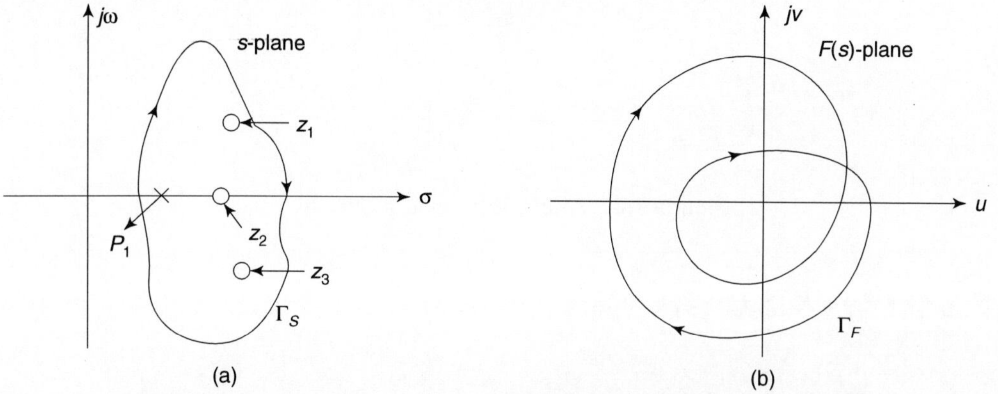

Figure 9.5 (a) s-plane contour enclosing three zeros and one pole, (b) Corresponding $F(s)$ plane counter

- *If the contour in the s-plane is so chosen that it does not enclose any zero or pole, the corresponding contour in the F(s) plane then will not encircle the origin.*

9.3 NYQUIST STABILITY CRITERION

Considering a single loop feedback system shown in Fig. 9.6, the characteristic equation of the system is

$$F(s) = 1 + G(s)\,H(s) = 0 \qquad \text{...(9.3)}$$

We have the general form of open loop transfer function $G(s)\,H(s)$ is

$$\left.\begin{aligned} G(s)\,H(s) &= \frac{(s+Z_1)(s+z_2)\ldots(s+Z_n)}{(s+P_1)(s+P_2)\ldots(s+P_m)} \\ G(s)\,H(s) &= K\frac{\sum\limits_{i=1}^{n}(s+Z_i)}{\sum\limits_{j=1}^{m}(s+P_j)} \end{aligned}\right\} \qquad \text{...(9.4)}$$

Now
$$F(s) = 1 + K\frac{K\sum\limits_{i=1}^{n}(s+Z_i)}{\sum\limits_{j=1}^{m}(s+P_j)}$$

$$F(s) = \frac{\sum\limits_{j=1}^{m}(s+P_j) + K\sum\limits_{i=1}^{n}(s+Z_i)}{\sum\limits_{j=1}^{m}(s+P_j)}$$

$\because$
$$F(s) = 0 = 1 + G(s)\,H(s)$$

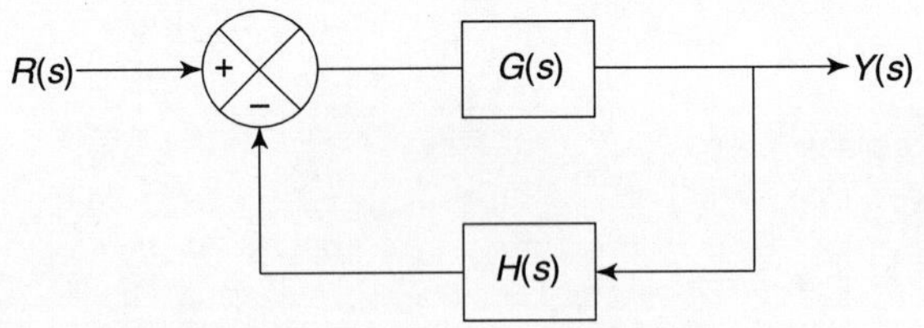

Figure 9.6 A feedback control system

So
$$\sum_{j=1}^{m}(s+P_j) + K\sum_{i=1}^{n}(s+Z_i) = 0 \qquad \text{...(9.5)}$$

$\Rightarrow$
$$\sum_{i=1}^{n}(s+Z_i) = 0$$

From Eq. (9.5) it is seen that the zeros of $F(s)$ at Z_i are the roots of the characteristic equation where $i = 1, 2, 3 \ldots n$ and the poles of $F(s)$ at P_j are same as the open loop poles of the system where $j = 1, 2, 3 \ldots m$.

For the stable system, the roots of the characteristic equation, i.e., zeros of $F(s)$ must lie in the left half of the s-plane. There may be some poles lie in the right half of s-plane but all the zeros of $F(s)$, i.e., the close loop poles, may lie in the left half s-plane, meaning thereby that an open loop unstable system may lead to a closed loop stable operation.

9.3.1 Nyquist Contour

A custom which completely encloses right half of the s-plane, called the Nyquist contour, is represented by c as shown in Fig. 9.7 it is directed clockwise.

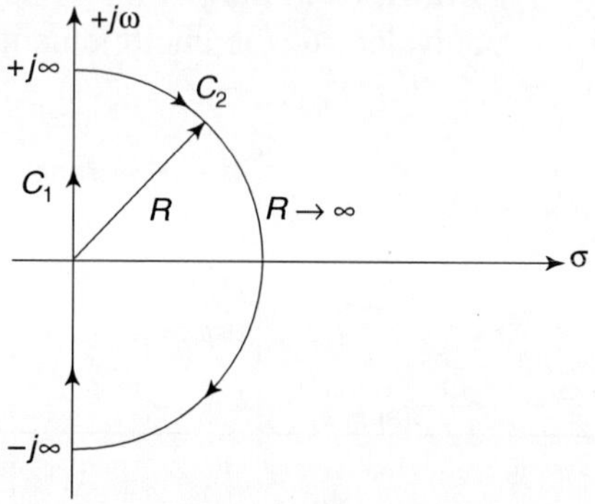

Figure 9.7 The Nyquist contour

For examining the presence of any zero of $F(s) = 1 + G(s) H(s)$ in the right half s-plane, let us consider Nyquist contour, this contour has two segment C_1 and C_2.

The segment C_1 is along imaginary ($j\omega$) axis and C_2 is an arc of infinite radius.

For C_1 $\quad s = j\omega$ with s varying from $-j\omega$ to $+j\infty$

and for C_2 $\quad s = Re^{j\theta}$ with θ varying from $\pi/2$ to 0 to $-\pi/2$ with $R \to \infty$

So we can say that the Nyquist contour encloses all the right half s-plane poles and zeros of

$$F(s) = 1 + G(s) H(s)$$

Let P poles and Z zeros of $F(s)$ exist in the right half s-plane. The $F(s)$ plane contour of s-plane contour encircles the origin of N times when

$$N = P - Z \qquad \ldots(9.6)$$

in the counterclockwise direction

1. If zeros of $F(s) = 1 + G(s) H(s)$ do not exist in the right half s-plane, i.e., $Z = 0$

$$N = P \qquad \ldots(9.7)$$

If Eq. (9.7) is satisfied then the system is stable. This equation is satisfied only when the number of counterclockwise encirclements of the origin of the $F(s)$ plane by the countour Γ_F should equal the number of the right half s-plane poles of $F(s)$ which are the poles of open loop transfer function, $G(s) H(s)$.

2. When $P = 0$ for open loop stable system, the closed loop system is stable if

$$N = P = 0 \quad \text{...(9.8)}$$

By Eq. (9.8) we can say that net encirclements, of the origin of the $F(s)$ plane by the Γ_F contour should be zero.

It is easily observed that

$$F_o(s) = F(s) - 1 \quad \text{...(9.9)}$$

where $F_o(s)$ = open loop transfer function

$F_o(s) = G(s)\, H(s)$

$F_o(s) = 1 + G(s)\, H(s)$

So we can say that the contour of $F_o(s)$ corresponding to the Nyquist contour in the s-plane in the same as contour Γ_F of $F(s)$ is drawn from the point $(-1 + j0)$. Thus, the encirclement of the origin by the contour Γ_F is equivalent to the encirclement of the point $(-1 + j0)$ by the contour Γ_F as shown in Fig. 9.8.

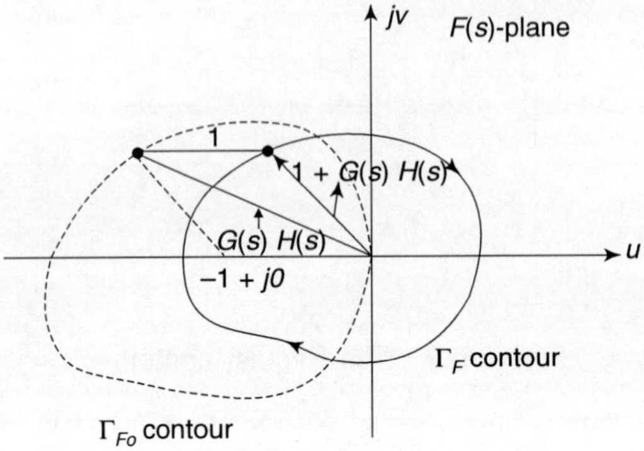

Figure 9.8

The Nyquist stability criterion can be stated as follows:

1. A feedback system is stable if and only if the contour Γ_F in the $F_o(s)$ plane, i.e., open loop transfer function $G(s)\, H(s)$ corresponding to the Nyquist contour in the s-plane does not encircle the $(-1, 0)$ point when the number of poles of $F_o(s)$ in the right hand s-plane is zero ($P = 0$).
2. A feedback system is stable if and only if for the contour Γ_F of the open loop transfer function $G(s)\, H(s)$ corresponding to the Nyquist contour in the s-plane, the number of counterclockwise encirclements of the $(-1, 0)$ point is equal to the number of poles $F_o(s)$ with positive real parts.

The mapping of the Nyquist contour into the contour Γ_F is carried out as follows:

(i) For mapping a imaginary axis substitute $s = j\omega$, this converts the mapping function into a frequency function, i.e., $G(s)\, H(s) = G(j\omega)\, H(j\omega)$.

(ii) In physical system the number of zeros is less than equal to number of poles

i.e., $\lim_{\substack{R \to \infty \\ s = Re^{j\theta}}} G(s)\, H(s) = \text{real constant.}$

i.e., $G(s)\,H(s)$ = real constant

$$s = Re^{j\theta}$$

$$R = \infty \quad = 0 \text{ if } (n < m)$$

Example 9.3.1 Consider a feedback system whose open loop transfer function is given by

$$G(s)\,H(s) = \frac{100}{(s+1)(s/10+1)}$$

Comparing the function with the general form

$$G(s)\,H(s) = \frac{K}{(T_1\,s+1)(T_2\,s+1)} \qquad ...(9.10)$$

We get $K = 100$

$$T_1 = 1$$

and $T_2 = 1/10 = 0.1$

The magnitude of $G(j\omega)\,H(j\omega)$ and phase of $G(j\omega)\,H(j\omega)$ are shown in Table 9.1.

Table 9.1

ω	0	0.1	0.76	1	2	10	20	100	∞
$\lvert G(j\omega)\,H(j\omega)\rvert$	100	96	79.6	70.7	50.2	6.8	2.24	0.10	0
$\angle G(j\omega)\,H(j\omega)$	0	– 5.7°	– 41.5°	– 50.7°	– 74.7°	– 129.3°	– 150.5°	– 173.7°	– 180°

Figure Ex. 9.3.1 (a) and (b) is showing the Nyquist contour and mapping for $G(s)\,H(s)$ function.

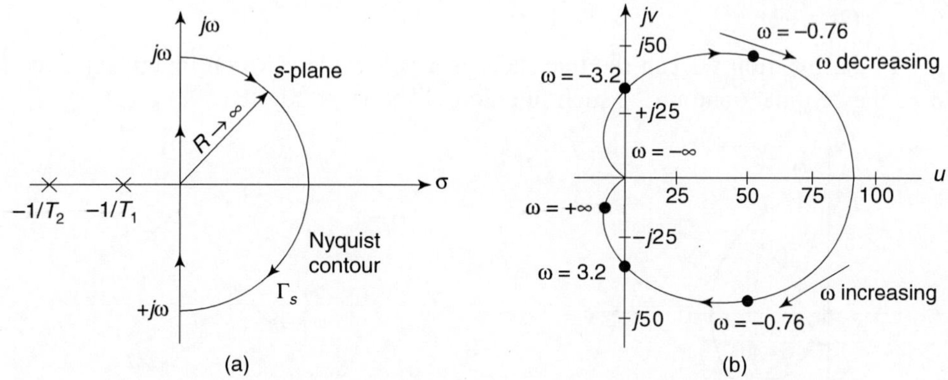

Figure Ex. 9.3.1 (i) (a) Nyquist contour, (b) Polar plot of $G(s)\,H(s)$

Here we note that the number of poles of $G(s)\,H(s)$ in the right hand s-plane is zero, and thus $P = 0$, so for this system to be stable, we require $N = Z = 0$ and the contour must not encircle the $-1 + j0$ point in the $G(s)\,H(s)$ plane. As we investigate the Fig. Ex. 9.3.1 (b) and Eq. 9.10, we find that, the contour does not encircle the $1 + j0$ point, and the system is stable for all $K \geq 0$.

Open loop poles on the $j\omega$ (imaginary) axis

If there are any poles of $G(s)\ H(s)$ exist on imaginary axis in such case the earlier Nyquist contour in s-plane cannot be used as such it needs some modification. If there is any pole on imaginary axis then the Nyquist contour in s-plane is effected by a small semicircle of radius ε, around the pole at imaginary axis. When $\varepsilon \to 0$, the modified Nyquist contour is shown in Fig 9.3.1 (ii).

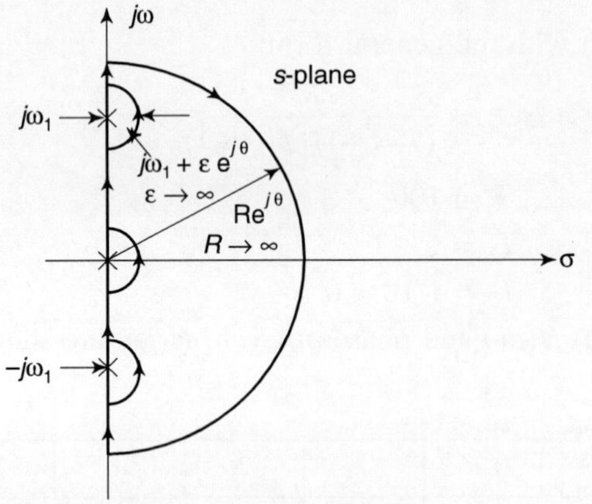

Figure Ex. 9.3.1 (ii) Nyquist contour for poles lies on imaginary axis

Example 9.3.2 Consider a feedback system whose open loop transfer function is given by

$$G(s)\ H(s) = \frac{K}{s\,(T_1\,s+1)}.$$

Solution In the function we can see that there is a pole which lies on imaginary axis, i.e., at origin so the Nyquist contour for such function is shown in Fig. Ex. 9.3.2 (a).

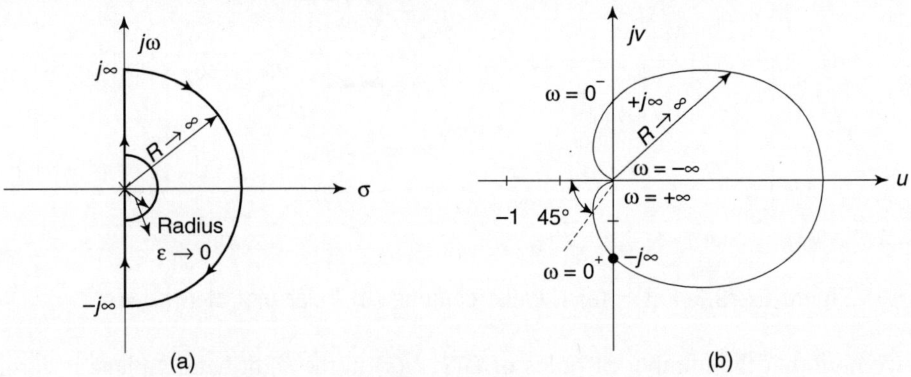

Figure Ex. 9.3.2 (a) Nyquist contour for the given function,
(b) Corresponding $F(s)$-plane of Nyquist contour

The mapping of Nyquist contour is as follows:

1. The small semicircular indent around the pole at origin where $s = \varepsilon e^{j\theta}$ and θ is varying from $-90°$ to 0 to 90°.
2. The portion from $\omega = 0^+$ to $\omega = +\infty$. The portion of the contour Γ_s is mapped by the function $G(s)\,H(s)$ as the real frequency polar plot because $s = j\omega$, and for this part of the contour. This results in the real freqeuncy polar plot shown in Fig. Ex. 9.3.2 (b), we have

$$\lim_{\omega\to+\infty} G(j\omega)\,H(j\omega) = \lim_{\omega\to+\infty} \frac{K}{+j\omega\,(j\omega T+1)}$$

$$= \lim_{\omega\to\infty} \left|\frac{K}{T\,\omega^2}\right| \angle -\pi/2 - \tan^{-1}\omega T \qquad ...(9.11)$$

So the magnitude approaches zero at an angle of $-180°$.

3. The portion from $\omega = +\infty$ to $\omega = -\infty$.

This portion is mapped into the point zero at the origin of the $G(s)\,H(s)$ plane. It is represented by

$$\lim_{R\to\infty} G(s)\,H(s)\big|_{s=\mathrm{R}e^{j\theta}} = \lim_{R\to\infty} \left|\frac{K}{R^2}\right| e^{-2j\phi} \qquad ...(9.12)$$

as ϕ varies from $+90°$ at $\omega = +\infty$ to $\phi = -90°$ at $\omega = -\infty$. So the contour moves from an angle of $-180°$ at $\omega = +\infty$ to an angle of $+180°$ at $\omega = -\infty$. The magnitude of the $G(s)\,H(s)$ contour when $R \Rightarrow \infty$ is always zero or a constant.

4. The portion from $\omega = -\infty$ to $\omega = 0^-$

This portion is mapped by the function $G(s)\,H(s)$ as

$$G(s)\,H(s)\big|_{s=-j\omega} = G(-j\omega)\,H(-j\omega) \qquad ...(9.13)$$

When we investigate Fig. Ex. 9.3.2 (b), we find that irrespective of the value of the gain K and the time constant T_1 the contour does not encircle the -1 point, and the system is always stable.

Example 9.3.3 Consider a system with an open loop transfer function

$$G(s)\,H(s) = \frac{K}{s\,(T_1 s+1)\,(T_2 s+1)} \qquad ...(9.13a)$$

Solution Now investigating the $G(j\omega)\,H(j\omega)$ – lcous. The origin of the s-plane maps into a semicircle of infinite radius, because there is one pole at origin or imaginary axis so there is a small semicircle near this pole with a radius ε, where $\varepsilon \to 0$.

To investigate the stability of the system, it is sufficient to plot the portion of the contour ($\Gamma_{F\omega}$) that is the real freqeuncy polar plot $G(j\omega)\,H(j\omega)$ for $0^+ < \omega < +\infty$.

Put $s = j\omega$ is Eq. (9.13)

$$G(j\omega)\,H(j\omega) = \frac{K}{j\omega\,(j\omega T_1+1)\,(j\omega T_2+1)}$$

$$= \frac{-K(T_1 + T_2) - jK(1/\omega)(1-\omega^2 T_1 T_2)}{j\omega(j\omega T_1 + 1)(j\omega T_2 + 1)}$$

$$\left.\begin{aligned} |G(j\omega)H(j\omega)| &= \frac{K}{\sqrt{\omega^4 (T_1 + T_2)^2 + \omega^2 (1 - \omega^2 T_1 T_2)^2}} \\ \angle G(j\omega)H(j\omega) &= -\tan^{-1} T_1 - \tan^{-1} T_2\,\omega - \pi/2 \end{aligned}\right\} \quad \text{...(9.14)}$$

When $\omega = 0^+$, the magnitude of locus is infinite at an angle of – 90° in $G(s)\,H(s)$ plane when w approaches + ∞, we have

$$\lim_{\omega \to \infty} G(j\omega)H(j\omega) = \lim_{\omega \to \infty} \left| \frac{1}{\omega^3 T_1 T_2} \right| \angle -\pi_2 - \tan^{-1} \omega_1 T - \tan^{-1} \omega_2 T_2$$

$$= \lim_{\omega \to \infty} \left| \frac{1}{\omega^3 T_1 T_2} \right| \angle -3\pi/2$$

So the $G(j\omega)\,H(j\omega)$ approaches a magnitude of zero at an angle of – 270°. To approach at an angle of – 270°, the locus must cross the u-axis in the $G(s)\,H(s)$ – plane, as shown in Fig. Ex. 9.3.3 (b). So it encircles the – 1 point twice so the system is unstable with two roots in the right hand s-plane.

The point at which $G(s)\,H(s)$ locus intersects the real axis can be found when imaginary part of $G(j\omega)\,H(j\omega)$ is zero.

i.e., $jv = 0$

$$v = \frac{-K(1/\omega)(1-\omega^2 T_1 T_2)}{1 + \omega^2 (T_1^2 + T_2^2) + \omega^4 T_1^2 T_2^2}$$

Thus,

When $1 - \omega^2 T_1 T_2 = 0$ or then $v = 0$

i.e., $\omega = \dfrac{1}{\sqrt{T_1 T_2}}$

The magnitude of the real part at this frequency, i.e., u at $\omega = \dfrac{1}{\sqrt{T_1 T_2}}$

$$K = \frac{-K(T_1 T_2)}{1 + \omega^2 (T_1^2 + T_2^2) + \omega^4 T_1^2 T_2^2}$$

Putting $\omega = \dfrac{1}{\sqrt{T_1 T_2}}$

$$u = -\frac{K T_1 T_2}{T_1 + T_2}$$

The open loop transfer function does not have any open loop pole on the R.H.S. So for stability the point (– 1, $j0$) should not be encircled. This is obtained from basic equation

$$N = P - Z \text{ since } P = 0$$

$$N = -Z$$

For closed loop stability, $t = 0$. For this condition to be satisfied

This can only happen if $N = 0$

$$\frac{K(T_1 T_2)}{T_1 + T_2} < 1$$

$$\Rightarrow \qquad K < \frac{T_1 T_2}{T_1 + T_2}$$

$$\Rightarrow \qquad 0 < K < \frac{T_1 T_2}{T_1 + T_2}$$

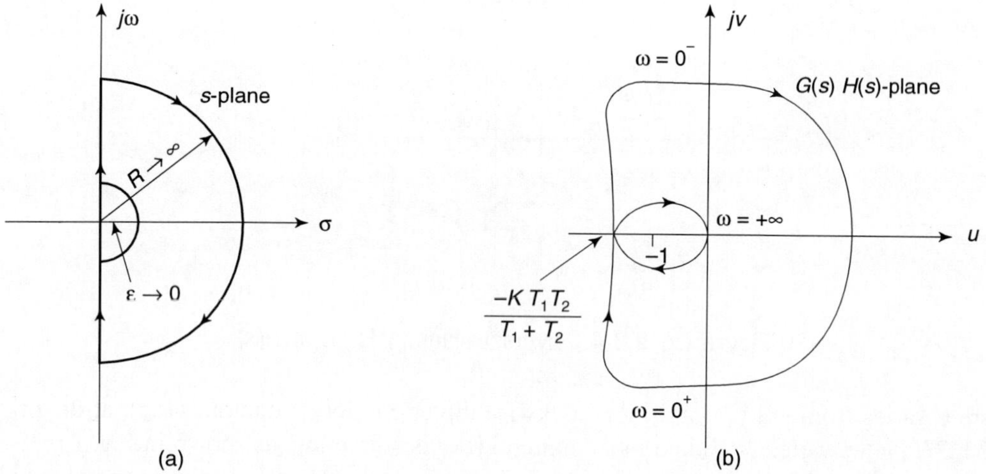

Figure Ex. 9.3.3 (a) Nyquist contour in *s*-plane, (b) Nyquist for the given function

Example 9.3.4 Consider a system with a pole in the right hand *s*-plane, the function is given as

$$G(s)\,H(s) = \frac{K}{s\,(s-1)} \qquad \text{...(9.15)}$$

Solution The open loop transfer function has one pole in the right hand *s*-plane, so $P = 1$. For system to be stable, we require $N = -P = -1$, there is one counterclockwise encirclement of the – 1 point. At the semicircular detour at the origin of the *s*-plane.

We let $s = \varepsilon\, e^{j\phi}$ when $-\pi/2 \le \phi \le \pi/2$. Then we have

$$\lim_{\varepsilon \to 0} \frac{K}{-\varepsilon e^{j\theta}} = \lim_{\varepsilon \to 0} \left|\frac{K}{\varepsilon}\right| \angle -180 - \phi \qquad ...(9.16)$$

So this portion of the contour (Γ_{Fv}) is a semicircle of infinite magnitude in the left-hand $G(s)\ H(s)$ plane shown in Fig. Ex. 9.3.4 when $s = j\omega$

$$G(j\omega)\ H(j\omega) = \frac{K}{j\omega\,(j\omega - 1)} = \frac{K}{(\omega^2 + \omega^4)^{1/2}} = \angle -\pi/2 - \tan^{-1} - \omega$$

$$G(j\omega)\ H(j\omega) = \frac{K}{(\omega^2 + \omega^4)^{1/2}} \angle -\pi/2 + \tan^{-1}\omega \qquad ...(9.17)$$

Finally for the semicircle of radius R and $R \to \infty$

We have,

$$\lim_{R \to \infty} G(s)\ H(s)\Big|_{s = Re^{j\phi}} = \left(\lim_{R\infty} \left|\frac{K}{R^2}\right|\right) e^{-2j\phi} \qquad ...(9.18)$$

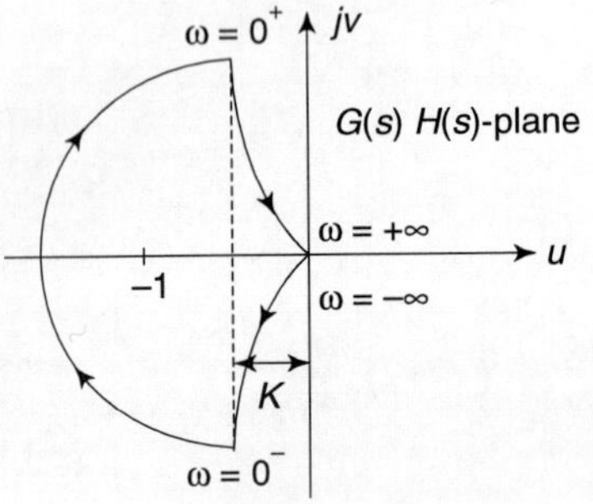

Figure Ex. 9.3.4 Nyquist diagram for $G(s)\ H(s)$

where ϕ varies from $\pi/2$ to $-\pi/2$ in a clockwise direction. So the contour (Γ_{Fv}), at the origin of the GH plane, varies 2π radian in a counterclockwise direction, as shown in Fig. Ex. 9.3.4. The contour (Γ_{Fv}) in the $G(s)\ H(s)$-plane encircles the $-1 + j0$ point once in the clockwise direction, and $N = +1$ so

$$Z = N + P = 2$$

and the system is unstable because two zeros of the characteristic equation, irrespective of the value of gain K, lies in the right of the s-plane.

9.4 ASSESSMENT OF RELATIVE STABILITY

In some previous chapters, we have studied about stability. Some techniques we have seen for the stability information like Routh–Hurwitz criterion, pole zero plot, root locus plot and Bode plot. Out of these methods the Routh array and pole zero plot give information about absolute stability, i.e., whether the system is stable or not but does not give any information about degree of stability, i.e., relative stability. And other methods like Bode plot and root locus plot is a technique by which we can get qualitative information about stability, i.e., relative stability.

In the previous section we have seen the method as Nyquist plot which provides absolute stability, but in this section the use of Nyquist plot shall be extended to extract information regarding relative stability, thus eliminating the need to determine the location of the roots of characteristic equation. Thus, Nyquist plot becomes a powerful tool for designers which provides information regarding both absolute stability and relative stability of a feedback system.

9.4.1 Measure of Relative Stability

Here we would like to determine a similar measure of relative stability useful for the frequency response method. The Nyquist plot provides us with suitable information about relative stability by inspection of the polar plot of the open loop function, $G(s)\ H(s)$.

We have seen in previous section that the stability criterion is merely the non-encirclement of (– 1, 0) points. So we can say that as the polar plot gets closer to (– 1, 0) point, the system tends towards instability.

Now we can consider two different systems the dominant closed loop poles are shown in the s-plane in Fig. 9.9 (a). By the inspection of Fig. 9.9 (a) we can say that system x is more stable than system y because the dominant closed loop poles are located comparatively away to the left from the imaginary axis. The polar plots for both the systems are shown in Fig. 9.9 (b) by the inspection of polar plots we can say that as a polar plot moves closer to (– 1, 0) point the system closed loop poles move closer to the imaginary axis and hence the system becomes relatively less stable and vice versa.

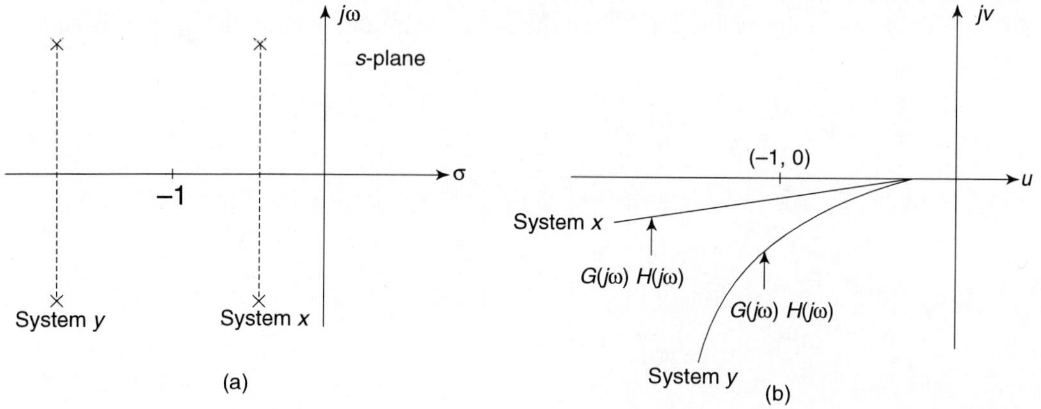

Figure 9.9 (a) Closed-loop poles in s-plane, (b) Polar plot for system x and y

Figure 9.10 shows a typical $G(j\omega)\ H(j\omega)$ plane which crosses the negative real axis at a frequency $\omega = \omega_1$, ω_2 and ω_3 with an intercept of a_1, a_2, a_3. Let the phasor $G(j\omega)\ H(j\omega)$ make an angle of ϕ_1, ϕ_2, ϕ_3 with negative real axis measured positively in counterclockwise direction. It is observed that as $G(j\omega)\ H(j\omega)$ locus approaches (– 1, 0) point, the relative stability reduces.

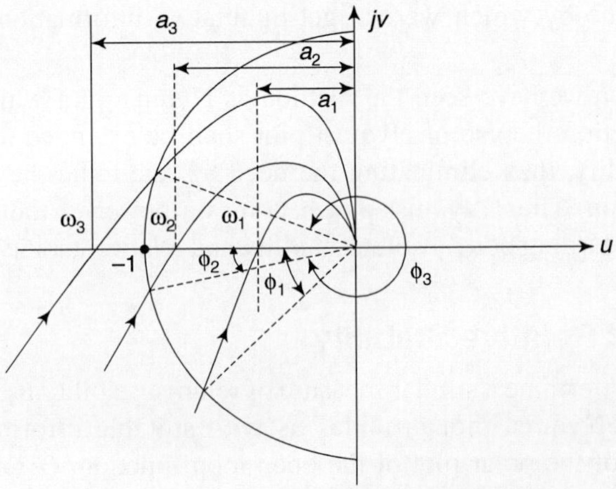

Figure 9.10 Polar plot for $G(j\omega)\ H(j\omega)$

Simultaneously, the value of a_1, a_2, a_3 approaches unity and that of ϕ_1, ϕ_2, ϕ_3 tends to zero. Thus, the relative stability could be measured in terms of the intercept (a_1) or the angle, ϕ. These measures are used to define gain margin and phase margin as practical measures of relative stability.

9.4.2 Gain Margin and Phase Margin

The Nyquist plot shown in Fig. 9.11 for $G(j\omega)\ H(j\omega)$ for the value of K which brings the closed loop system on the verge of instability. For the value of K, the plot passes through the point (– 1, 0). For smaller values of K then the above mentioned value the system is stable.

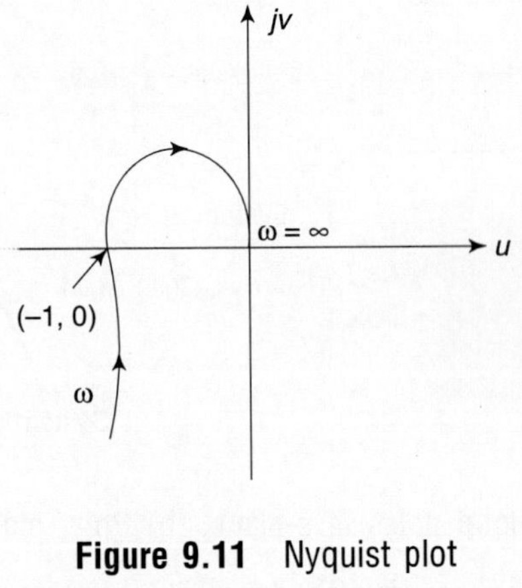

Figure 9.11 Nyquist plot

Let the system open loop transfer function be

$$G(j\omega)\ H(j\omega) = \frac{K}{j\omega\,(j\omega\, T_1 + 1)\,(j\omega\, T_2 + 1)} \qquad ...(9.19)$$

The locus intersects the u axis at a point

$$u = \frac{-K T_1 T_2}{T_1 + T_2} \quad \text{...(9.20)}$$

because the system has the roots on the $j\omega$ axis

when $$u = -1$$

i.e., $$K = \left(\frac{T_1 + T_2}{T_1 T_2}\right)$$

when $$K < \left(\frac{T_1 + T_2}{T_1 T_2}\right) \text{ the system is stable}$$

$$K = \left(\frac{T_1 + T_2}{T_1 T_2}\right) \text{ the system is marginally stable}$$

$$K > \left(\frac{T_1 + T_2}{T_1 T_2}\right) \text{ the system is unstable}$$

When the Nyquist polar is closer to the (– 1, 0) point the system tends towards more oscillation in its output. This closeness of the Nyquist plot to the (– 1. 0) point determines the relative stability of a control system expressed in terms of gain margin and phase margin.

The Nyquist plot for a stable system is shown in Fig. 9.12. The plot crosses the negative real axis at $\omega = \omega_2$ and the corresponding magnitude $|G(j\omega)\, H(j\omega)|$ is less than 1.

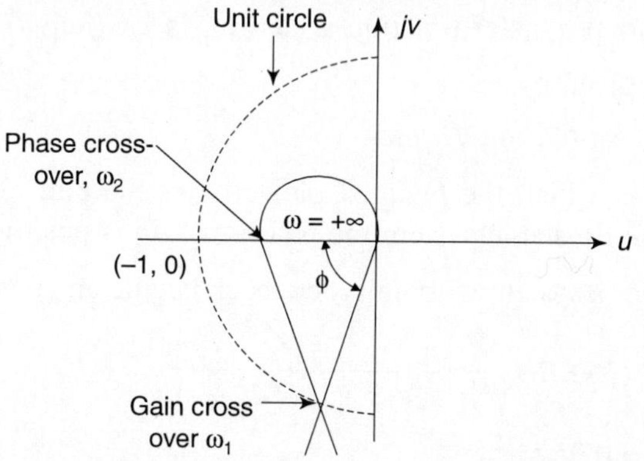

Figure 9.12 G.M. and P.M. from Nyquist plot

If the gain K of the system is further increased so that the plot passes through the point (– 1, 0), the system starts tending towards instability. The magnitude of $|G(j\omega)\, H(j\omega)|$ is equal to 1 at this point. Thus, it is concluded that if the gain is increased by a factor $\left|\frac{1}{G(j\omega)\, H(j\omega)}\right|$ the closed loop system becomes unstable. Thus, there exists a margin for increasing the gain of the system, the frequency (ω_2) at which the Nyquist plot crosses the negative real axis is termed phase crossover frequency.

"The gain margin is a factor by which the gain of stable system is allowed to increase before the system reaches instability."

It is given by

$$\text{Gain margin} = 20 \log_{10} \frac{1}{|G(j\omega)\, H(j\omega)|} \text{ dB}$$

when

$$= 20 \log_{10} \frac{1}{a} \text{ dB} \qquad \text{...(9.20a)}$$

$|G(j\omega_2)\, H(j\omega_2)| < 1$ = G.M. is positive, so system is stable

$|G(j\omega_2)\, H(j\omega_2)| = 1$, G.M. is zero, so system is marginally stable

$|G(j\omega_2)\, H(j\omega_2)| = 1$, G.M. is negative, so the system is unstable.

From Fig. 9.12 for phase margin, it is observed that the unit circle crosses the Nyquist plot at $\omega = \omega_1$. If the Nyquist plot is rotated through an angle in the clockwise direction then the plot passes through the point (– 1, 0) and the closed loop system becomes unstable. The rotation of the plot mentioned above is achieved by changing the phase of $G(j\omega)\, H(j\omega)$. The angle ϕ is known as the phase margin of the system.

"The phase margin of a stable system is the amount of additional phase lag required to bring the system to the point of instability."

Figure 9.12 gives the Nyquist plot for a stable system and the phase margin (ϕ) is positive. The phase of the open loop transfer function at $\omega = \omega_1$ is $\angle G(j\omega_1)\, H(j\omega_1)$.

The phase margin is given by

$$\text{Phase margin} = 180° + \angle G(j\omega_1)\, H(j\omega_1) \qquad \text{...(9.21)}$$

The frequency (ω_1) at which the Nyquist plot crosses the unit circle is known as gain cross- over frequency. For a stable system the phase margin is positive.

Example 9.4.1 Consider a system with unity feedback having an open loop transfer function

$$G(s)\, H(s) = \frac{K}{s\,(T_1\, s + 1)\,(T_2\, s + 1)} \qquad \text{...(9.22)}$$

where $T_1 = 0.2$ and $T_2 = 0.05$

(i) Calculate G.M. and P.M.

(ii) What should be the adjustment into the gain when G.M. is 20 dB and P.M. is 40°.

Solution Substituting $s = j\omega$ in Eq. (9.22) and substituting the value of T_1 T_2

We have
$$G(j\omega)\, H(j\omega) = \frac{K}{j\omega\,(0.2\, j\omega + 1)\,(0.05\, j\omega + 1)}$$

Let $\qquad K = 1$

The magnitude and phase of the function of the Eq. (9.19)

$$|G(j\omega)\,H(j\omega)| = \frac{1}{\sqrt{\omega^2\,(0.2+0.05)^2 + \omega^2\,(1-\omega^2\times 0.2\times 0.05)^2}}$$

$$|G(j\omega)\,H(j\omega)| = \frac{1}{\omega^2\,\sqrt{(1+0.2\,\omega)^2}\,\sqrt{1+0.05\,\omega)^2}}$$

$$\angle G(j\omega)\,H(j\omega) = -\tan^{-1} T_1\,\omega - \tan^{-1} T_2\,\omega - \pi/2$$

$$\angle G(j\omega)\,H(j\omega) = -\tan^{-1} 0.2\,\omega - \tan^{-1} 0.05\,\omega - \pi/2$$

The Nyquist plot for $G(j\omega)\,H(j\omega)$ is shown in Fig. Ex. 9.4.1.

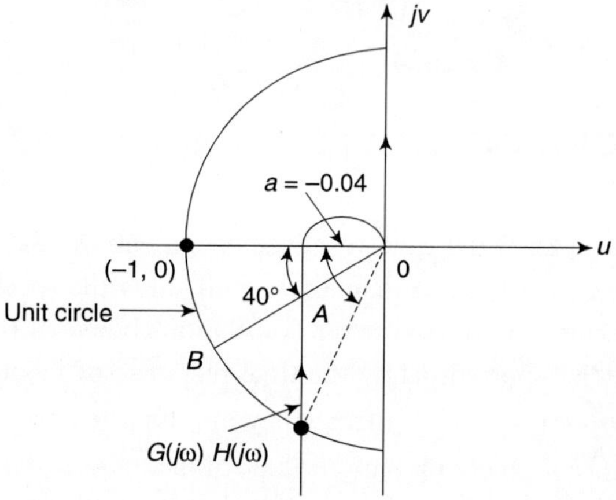

Figure Ex. 9.4.1 Nyquist plot for given function

From the plot $a = -0.04$

(i) By Eq. (9.20). the Gain margin

$$\text{G.M.} = 20\log_{10}\frac{1}{a}\text{ dB}$$

$$= 20\log_{10}\frac{1}{0.04}$$

$$= 20\log_{10} 25 = 20\times 1.40$$

$$= 27.96\text{ dB}$$

Phase Margin by Eq. (9.21) is

$$\text{P.M.} = 180^\circ + \angle G(j\omega_1)\,H(j\omega_1)$$

$$\text{P.M.} = 180^\circ - 256^\circ = -76^\circ$$

$$\text{P.M.} = 76^\circ$$

(ii) If G.M. is 20 dB, we have

$$\text{G.M.} = 20 \log_{10} \frac{1}{a}$$

$$20 = 20 \log_{10} \frac{1}{a}$$

So $\qquad a = 0.1$

This intercept is achieved if the system gain is increased by a factor of 0.1/0.04 = 2.5.

So $\qquad K = 2.5$

If P.M. is 40°, it is obtained if the system gain is increased such that point A is shified to B in Fig. Ex. 9.4.1. This can be achieved if the system gain is increased by a factor of OA/OB = 1/0.191 = 5.24.

Thus $\qquad K = 5.24$

9.5 SPECIFICATIONS OF CLOSED LOOP SYSTEM IN FREQUENCY DOMAIN

As we have studied in chapter 3 the time response specifications like maximum overshoot, settling time, rise time, etc., in the same manner we will study the specifications in frequency domain. It will help us to use the second order correlations between frequency response and transient response to predict approximately the time response of feedback systems.

1. **Resonant frequency (ω_r)** It is a frequency at which resonance peak (M_r) occurs.
2. **Resonance peak (M_r)** It is a maximum value of magnitude of the close loop frequency response.
3. **Cut-off rate** It is the slope of the log-magnitude curve near the cut-off frequency. It indicates the ability of the system to separate the signal from noise.
4. **Bandwidth** It is the range of frequencies for which the system gain is more than 3 dB.
5. **G.M and P.M.** These already defined in previous section.

9.6 CONSTANT *M* AND *N* CIRCLES

Constant M and N circles are graphical techniques, which help in determining the frequency response specification like M_r, ω_r, etc., directly from the open loop frequency response. Constant M contour and constant N-contour is drawn on the complex plane, when these contours are used in conjuction with the open loop frequency response, the value of M_r and ω_r are easily obtained.

9.6.1 Constant *M* Circles

The open loop transfer function $G(j\omega)\ H(j\omega)$ is a complex quantity and can be expressed as

$$G(j\omega)\ H(j\ \ ) = x + jy \qquad ...(9.23)$$

For unity feedback system $H(j\omega) = 1$

So
$$G(j\omega)\,.\,1 = x + jy$$
$$G(j\omega) = x + jy \qquad ...(9.24)$$

The close loop frequency response is
$$T(j\omega) = \frac{Y(j\omega)}{R(j\omega)} = \frac{G(j\omega)}{1+G(j\omega)} \qquad ...(9.25)$$

Substituting the vaule of $G(j\omega)$ in Eq. (9.25)
$$T(j\omega) = \frac{x+jy}{1+x+jy} \qquad ...(9.26)$$

From Eq. (9.26) the magnitude is given by
$$M = |T(j\omega)| = \frac{\sqrt{x^2+y^2}}{\sqrt{(1+x)^2+y^2}}$$
$$M^2 = \frac{x^2+y^2}{(1+x)^2+y^2}$$
$$M^2\,[(1+x)^2+y^2] = x^2+y^2$$
$$(1-M^2)\,x^2 - 2M^2\,x + (1-M^2)\,y^2 = M^2$$
$$x^2 \; \frac{2M^2}{(1-M)^2}\,x + y^2 = \frac{M^2}{1-M^2} \qquad ...(9.27)$$

Adding $\left(\frac{M^2}{1-M^2}\right)^2$ on both sides
$$x^2 - \frac{2M^2}{(1-M)^2}\,x + \left(\frac{M^2}{1-M^2}\right)^2 + y^2 = \frac{M^2}{1-M^2} + \left(\frac{M^2}{1-M^2}\right)^2$$
$$\left(x - \frac{M^2}{1-M^2}\right)^2 + y^2 = \left(\frac{M}{1-M^2}\right)^2 \qquad ...(9.28)$$

Equation (9.28) is the equation of a circle with centre at
$$x_o = \frac{M^2}{1-M^2} \qquad ...(9.29)$$

and
$$y_o = 0$$

and with radius
$$r_o = \frac{M}{1-M^2} \qquad ...(9.30)$$

For different values of M Eq. (9.28) represents a family of circles.

- *On a particular circle the value of M is constant that is why these circles are called constant M circles.*

The centres and radii of M circles for different values of M are given in Table 9.2.

Table 9.2

M	Centre (x, y)	Radius $r = \dfrac{M}{1-M^2}$
0.5	0.33	0.67
1.0	∞	∞
1.2	– 3.27	2.73
1.6	– 1.64	1.03
2.0	– 1.33	0.67
3.0	– 1.13	0.38

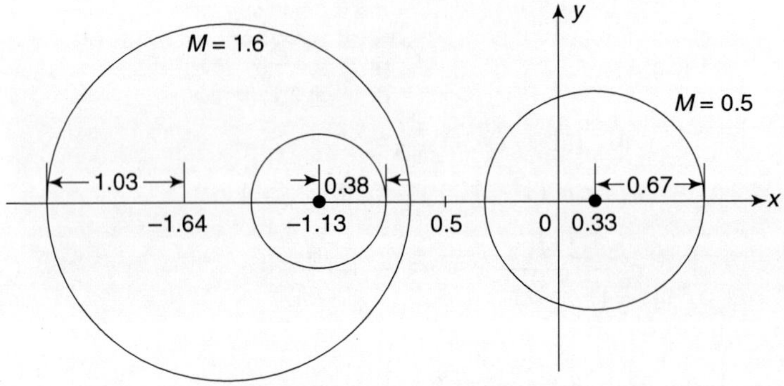

Figure 9.13 Constant M circles

The Nyquist plot is superimposed on M circle and the points of intersection give the magnitude of $T(j\omega)$ at different values of ω.

9.6.2 Constant–*N* Circles

By Eq. (9.26) the phase angle of $T(j\omega)$ is given by

$$\angle T(j\omega) = \angle \frac{x + jy}{1 + x + jy} \qquad ...(9.31)$$

The phase angle is denoted by ϕ, so

$$\phi = \tan^{-1}\frac{y}{x} - \tan^{-1}\frac{y}{1+x}$$

$$\tan\phi = \tan\left[\tan^{-1}\left(\frac{y}{x}\right) - \tan^{-1}\left(\frac{y}{1+x}\right)\right]$$

$$= \frac{\tan\left(\tan^{-1}\left(\frac{y}{x}\right) - \tan\left[\tan^{-1}\frac{y}{1+x}\right]\right.}{1+\tan\left(\tan^{-1}\frac{y}{x}\right)\cdot\tan\left[\tan^{-1}\frac{y}{x}\right]}$$

$$\tan\phi = \frac{\frac{y}{x} - \frac{y}{1+x}}{1+\left(\frac{y}{x}\right)\left(\frac{y}{1+x}\right)}$$

$$\tan\phi = \frac{y}{x^2 + x + y^2} \qquad \text{...(9.32)}$$

Substituting $\tan\phi = N$ in Eq. (9.32), we get

$$N = \frac{y}{x^2 + y^2 + x}$$

$$x^2 + x + y^2 - \frac{y}{N} = 0 \qquad \text{...(9.33)}$$

Adding $\left(\frac{1}{4} + \frac{1}{4N^2}\right)$ on both sides Eq. (9.33)

We get

$$\left(x^2 + x + \frac{1}{4}\right) + \left(y^2 - \frac{y}{N} + \frac{1}{4N^2}\right) = \frac{1}{4} + \frac{1}{4N^2}$$

$$\left(x + \frac{1}{2}\right)^2 + \left(y - \frac{1}{2N}\right)^2 = \frac{1}{4} + \frac{1}{4N^2} \qquad \text{...(9.34)}$$

For different values of N Eq. (9.34) represents a family of circles.

The centre at

$$\left.\begin{aligned} x_o &= \frac{-1}{2} \text{ and } y_o = \frac{1}{2N} \\ \text{and with radius} \quad r_o &= \frac{1}{2N}(N^2+1)^{1/2} \end{aligned}\right\} \qquad \text{...(9.35)}$$

- *On a particular circle the value of N or the value of phase angle of the closed loop transfer function is constant, so these circles are called constant N circles. Since Eq. (9.34) is satisfied for x = 0 and y = 0 and for x = – 1, y = 0, all the N circles pass through the origin and (– 1, 0) point regardless of the value of N.*

The centre and the radius for N circles are given in Table 9.3.

Table 9.3

ϕ	$N = \tan \phi$	Centre (x_o, y_o)	Radius r_o
– 90°	∞	0	0.5
– 60°	– 1.732	– 0.289	0.577
– 50°	– 1.19	– 0.42	0.656
– 30°	– 0.577	0.866	1.0
– 10°	– 0.176	– 2.84	2.88
0°	0	∞	∞
+ 10°	0.176	2.84	2.88
+ 30°	0.577	0.866	1.0
+ 50°	0.19	0.42	0.656
+ 60°	1.732	0.289	0.577
+ 90°	∞	0	0.05

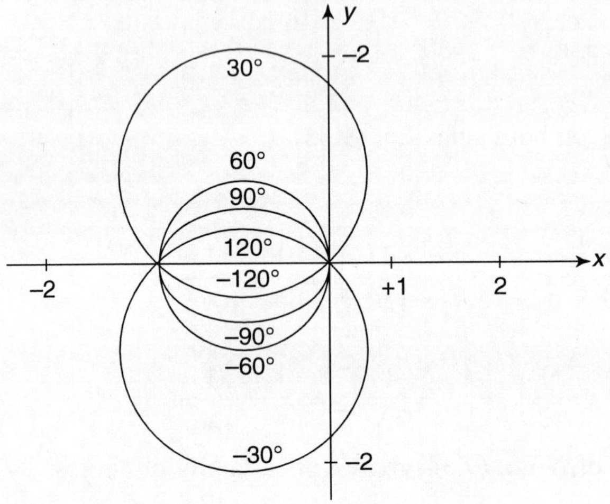

Figure 9.14 Constant *N* circles

Example 9.6.1 Consider a system as in Ex. 9.6.1 whose transfer function is given. By use of constant *M* and constant *N* circles calculate ω_r and M_r.

Now $G(j\omega)$ $H(j\omega)$ – plot superimposed on constant *M* and constant *N* circles. Plot is shown in Fig. Ex. 9.6.1.

In Fig. Ex. 9.6.1 at various values of frequency the *M* and ϕ are read off. With the help of data, the close loop frequency response curves are plotted in Fig. Ex. 9.6.1.

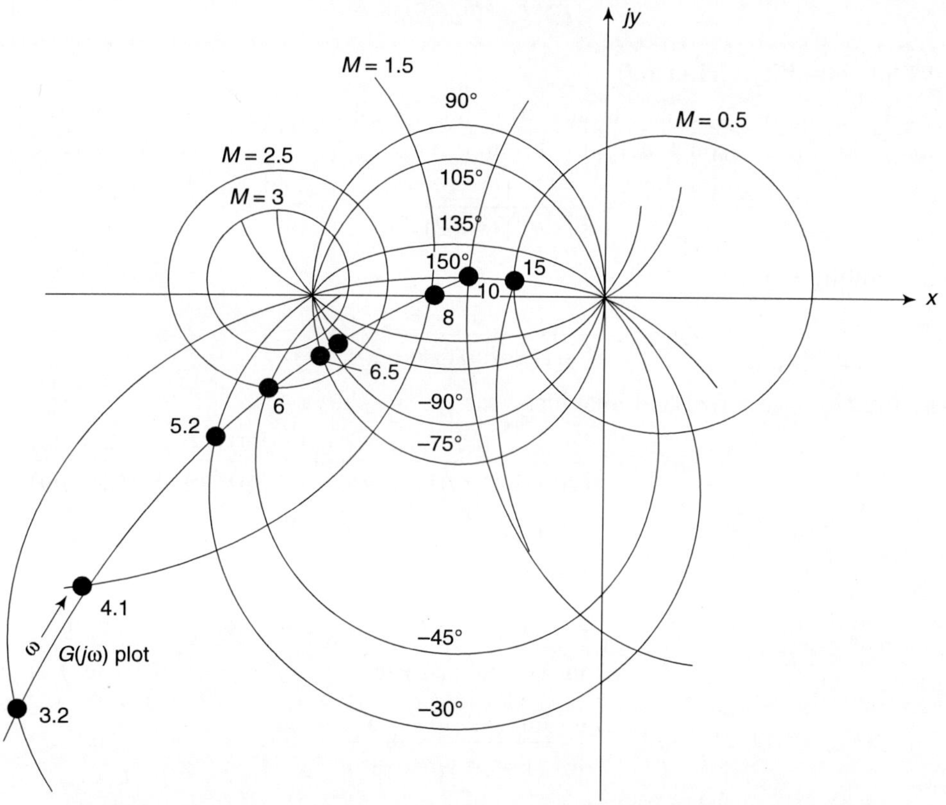

Figure Ex. 9.6.1 (a) Constant M and N circle for units feedback system

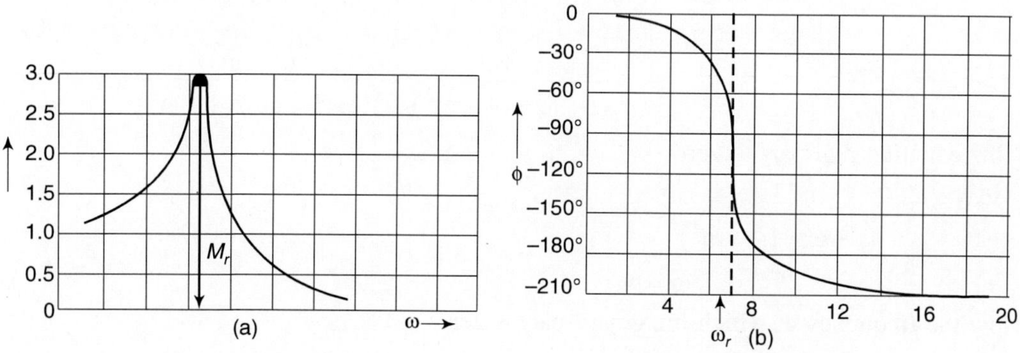

Figure Ex. 9.6.1 (b) Closed loop frequency response

By the curve the specifications are as follows:

1. Resonance peak, $M_r = 3.0$
2. Resonance frequency, $\omega_r = 6.7$ rad/sec.
3. G.M. and P.M. have already been calculated in Ex. 9.4.1.

EXAMPLES

Nyquist Plot (Stability Criterion)

Example 9.1 Using the Nyquist stability criterion, find, if the unity feedback control system having the following transfer function is stable or not. *(U.P.T.U., 2003–04, CS)*

$$G(s) = \frac{1}{s\,(s+1)\,(s+2)}$$

Solution Putting $s = j\omega$

$$G(j\omega) = \frac{1}{j\omega\,(1+j\omega)\,(j\omega+2)}$$

Separating $G(j\omega)$ into real and imaginary part

$$G(j\omega) = \frac{1}{(j\omega)\,(1+j\omega)\,(2+j\omega)} \times \frac{(-j\omega)\,(1-j\omega)\,(2-j\omega)}{(-j\omega)\,(1-j\omega)\,(2-j\omega)}$$

$$= \frac{-(j\omega)\,(1-j\omega)\,(2-j\omega)}{\omega^2\,(1+\omega^2)\,(4+\omega^2)}$$

$$= \frac{(-j\omega-\omega^2)\,(2-j\omega)}{\omega^2\,(1+\omega^2)\,(4+\omega^2)} = \frac{2j\omega-\omega^2-2\omega^2+j\omega^2}{\omega^2\,(1+\omega^2)\,(4+\omega^2)}$$

$$= \frac{-(3\omega^2+2j\omega-j\omega^3)}{\omega^2\,(1+\omega^2)\,(4+\omega^2)}$$

$$= \frac{-3\omega^2}{\omega^2\,(1+\omega^2)\,(4+\omega^2)} - \frac{j\,(2\omega-\omega^3)}{\omega^2\,(1+\omega^2)\,(4+\omega^2)}$$

$$= -\frac{3}{(1+\omega^2)\,(4+\omega^2)} - j\,\frac{(2-\omega^2)}{\omega\,(1+\omega^2)\,(4+\omega^2)}$$

Now equating real part to zero

$$-\frac{3}{(1+\omega^2)\,(4+\omega^2)} = 0$$

$$\omega = \infty$$

Now the frequency at which imaginary part is zero.

$$-\frac{(2-\omega^2)}{\omega\,(1+\omega^2)\,(4+\omega^2)} = 0$$

$$2-\omega^2 = 0$$

$$\omega^2 = 2$$

$$\omega = \pm\sqrt{2}$$

Now for varying ω

$\omega \rightarrow$	0	0.5	1	$\sqrt{2}$	5	10	∞
Re $G(j\omega)$	– 0.75	– 0.567	– 0.3	– 0.23	– 0.044	0.0003	0
Im $G(j\omega)$	– ∞	1.32	– 0.1	0	0.006	0.001	6

Now from Fig. Ex. 9.1, we have seen that

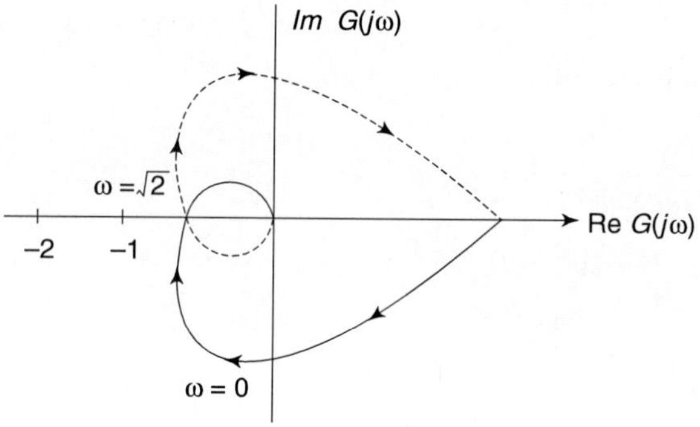

Figure Ex. 9.1 Nyquist plot

Number of encirclements of $(-1 + j0)$ point $N = 0$

Number of poles in the right hand s-plane

$$P = 0$$

So from
$$N = P - Z$$
$$Z = 0 - 0 = 0$$

No zeros in the right hand of s-plane, so the system is absolutely stable.

Example 9.2 Comment on the stability of the system whose open loop transfer function is

$$G(s)\,H(s) = \frac{1}{s\,(1+2s)\,(1+s)}$$

Also find the gain and phase margins. *(AMIE)*

Solution Putting
$$s = j\omega$$

$$G(j\omega)\,H(j\omega) = \frac{1}{j\omega\,(1+2\,j\omega)\,(1+j\omega)}$$

Separating real and imaginary parts of $G(j\omega)\,H(j\omega)$

$$= \frac{1}{(j\omega)\,(1+2j\omega)\,(1+j\omega)} \times \frac{(-j\omega)\,(1-2j\omega)\,(1-j\omega)}{(-j\omega)\,(1-2\,j\omega)\,(1-j\omega)}$$

$$= \frac{(0\,j\omega - 2\omega^2)\,(1-j\omega)}{\omega^2\,(1+4\omega^2)\,(1+\omega^2)}$$

$$= \frac{-j\omega - \omega^2 - 2\omega^2 + 2j\omega^3}{\omega^2(1+4\omega^2)(1+\omega^2)} = \frac{-3\omega^2 - j(\omega - 2\omega^3)}{\omega^2(1+4\omega^2)(1+\omega^2)}$$

$$= \frac{-3\omega^2}{\omega^2(1+4\omega^2)(1+\omega^2)} - \frac{j\omega(1-2\omega^2)}{\omega^2(1+4\omega^2)(1+\omega^2)}$$

$$= \frac{-3}{(1+4\omega^2)(1+\omega^2)} - \frac{j(1-2\omega^2)}{\omega(1+4\omega^2)(1+\omega^2)}$$

The frequency at which imaginary part is zero

$$\frac{1-2\omega^2}{\omega(1+4\omega^2)(1+\omega^2)} = 0$$

$$1 - 2\omega^2 = 0$$

$$\omega^2 = \frac{1}{2}$$

$$\Rightarrow \qquad \omega = \frac{1}{\sqrt{2}} = 0.707$$

$$\omega = 0.707$$

The frequency at which real part is zero

$$\frac{-3}{(1+\omega^2)(1+4\omega^2)} = 0$$

$$\omega = \infty$$

The Nyquist plot is shown in Fig. Ex. 9.2.

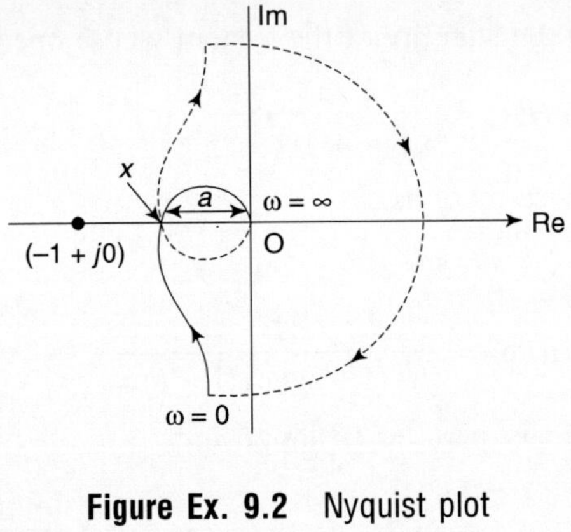

Figure Ex. 9.2 Nyquist plot

The magnitude is

$$|G(j\omega)| = \frac{1}{\omega\sqrt{1+\omega^2}\sqrt{1+4\omega^2}}$$

$$|G(j\omega)|_{\text{at } \omega = 0.707} = \frac{1}{0.707\sqrt{1+(0.707)^2}\sqrt{1+(4\times 0.707^2)}}$$

$$a = 0.66$$

Since $Ox = 0.66$

So point $-1 + j0$ not encircled

$\therefore$ $N = 0$

Also ploes on the R.H.S. plane, i.e., $P = 0$s

$$N = P - Z$$

$$0 = 0 - Z$$

$$Z = 0$$

$Z = 0$, implies closed loop stability and $P = 0$ implies open loop stability

$$\text{Gain margin} = 20 \log_{10} \frac{1}{a} = 20 \log_{10} \frac{1}{0.66} = 3.61 \text{ dB}$$

Frequency at which magnitude becomes unity is given by

$$\omega\sqrt{1+\omega^2}\sqrt{1+4\omega^2} = 1$$

$$\omega^2 (1 + \omega^2)(1 + 4\omega^2) = 1$$

$$\omega^2 = 0.33 \text{ by Hit and Trail method}$$

$\therefore$ $\omega = 0.574$ rad/sec.

Phase margin

$$\phi = -90° - \tan^{-1}\omega - \tan^{-1}\omega + 180°$$

$$= -90° - \tan^{-1} 0.57 - \tan^{-1}(2 \times 0.52) + 180°$$

$$= 11.57° \quad \textbf{Ans.}$$

Example 9.3 Apply Nyquist stabilty criterion to the system with loop transfer function,

$$G(s)\,H(s) = \frac{(4s+1)}{s^2(1+s)(1+2s)}$$

and as certain its stability. *(U.P.T.U., 2002–03, CS)*

Solution Putting $s = j\omega$

$$G(j\omega)\,H(j\omega) = \frac{(1+4j\omega)}{(j\omega)^2(1+j\omega)(1+2j\omega)}$$

Separating real and imaginary parts

$$G(j\omega)\,H(j\omega) = \frac{1+4j\omega}{(j\omega)^2(1+j\omega)(1+2j\omega)} \times \frac{(-j\omega)(1-2j\omega)}{(-j\omega)(1-2j\omega)}$$

$$= \frac{(1+4\,j\omega)(1-3\,j\omega-2\omega^2)}{-\,\omega^2\,(1+\omega^2)(1+4\omega^2)}$$

$$= \frac{1-3j\omega-2\,\omega^2+4j\omega+12\,\omega^2-8\,j\omega^3}{-\,\omega^2\,(1+\omega^2)(1+4\,\omega^2)}$$

$$= \frac{1+10\omega^2}{-\,\omega^2(1+\omega^2)(1+4\omega^2)} - j\,\frac{8\,\omega^3-\omega}{-\,\omega^2\,(1+\omega^2)(1+4\omega^2)}$$

The fequency at which real part is zero

$$\frac{1+10\,\omega^2}{\omega^2\,(1+\omega^2)(1+4\,\omega^2)} = 0$$

$$1 + 10\omega^2 = 0$$

$$\omega^2 = \frac{1}{10}$$

$$\Rightarrow \qquad \omega^2 = -\,0.1$$

$$\omega = \pm\, j\sqrt{0.1}$$

So, real part is never been zero, $\omega = \infty$

The frequency at which imaginary part is zero.

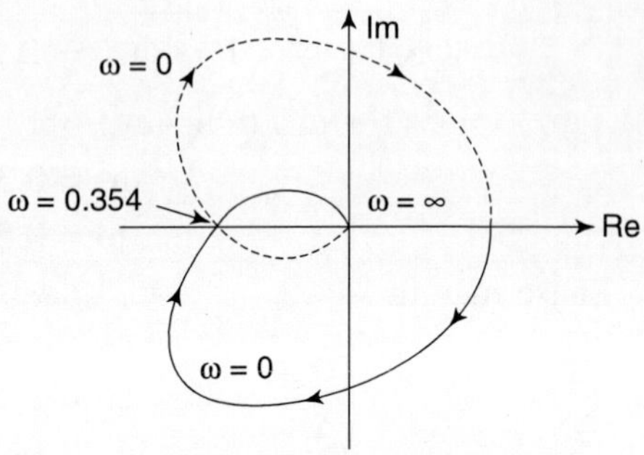

Figure Ex. 9.3 Nyquist plot

$$\frac{8\omega^2-1}{\omega\,(1+\omega^2)(1+4\omega^2)} = 0$$

$$8\omega^2 = 1$$

$$\omega^2 = \frac{1}{8}$$

$$\omega = \pm\frac{1}{\sqrt{8}} = \pm\, 0.354 \text{ rad/sec.}$$

The real and imaginary $G(j\omega)$ at varying frequency

$\omega \rightarrow$	0	0.1	0.354	0.5	1	5	10	∞
Re $G(j\omega)$	$-\infty$	– 104.7	– 10.65	5.6	– 1.1	0.004	0.0003	0
Im $G(j\omega)$	$-\infty$	– 1.9	0	0.8	0.7	0.15	0.002	0

Total number of encirclements

$$N = -\,2$$

Number of open loop poles in right half of s-plane

$$P = 0$$

So
$$N = P - Z$$
$$-2 = 0 - Z$$
$$Z = 2$$

So, there are two closed loop zeros in the right half of s-plane, the system is unstable.

Example 9.4 The open loop transfer function of a unity feedback system is

$$G(s)\,H(s) = \frac{K\,(s+2)}{S\,(s-1)}$$

Solution Let
$$K = 1$$

Putting
$$s = j\omega$$

$$G(j\omega)\,H(j\omega) = \frac{(3+j\omega)}{j\omega\,(-1+j\omega)}$$

Separating real and imaginary parts

$$G(j\omega)\,H(j\omega) = \frac{(3+j\omega)}{(j\omega)\,(-1+j\omega)} \times \frac{(-1-j\omega)\,(-j\omega)}{(-j\omega)\,(-1-j\omega)}$$

$$G(j\omega)\,H(j\omega) = \frac{3[1.33\,\omega - j\,(1-0.33\,\omega^2)]}{\omega\,(1+\omega^2)}$$

$$= \frac{-3.99\,\omega}{\omega\,(1+\omega^2)} - j\,\frac{(1-0.33\,\omega^2)}{\omega\,(1+\omega^2)}$$

The frequency at which imaginary part is zero

$$1 - 0.33\omega^2 = 0$$
$$\omega = 1.74 \text{ rad/sec.}$$

The magnitude of $G(j\omega)\,H(j\omega)$

$$|G(j\omega)\,H(j\omega)| = \frac{3\sqrt{1+(0.33\,\omega)^2}}{\omega\sqrt{1+\omega^2}}$$

The phase of $G(j\omega)\,H(j\omega)$ is

$$\phi = \tan^{-1} 0.33\,\omega - 90^\circ - \tan^{-1}\frac{\omega}{-1}$$

when
$$\omega = 0 \text{ then } \phi = 90^\circ$$
$$\omega = \infty \text{ then } \phi = 270^\circ$$

The Nyquist plot is shown in Fig. Ex. 9.4

when
$$K = 1$$

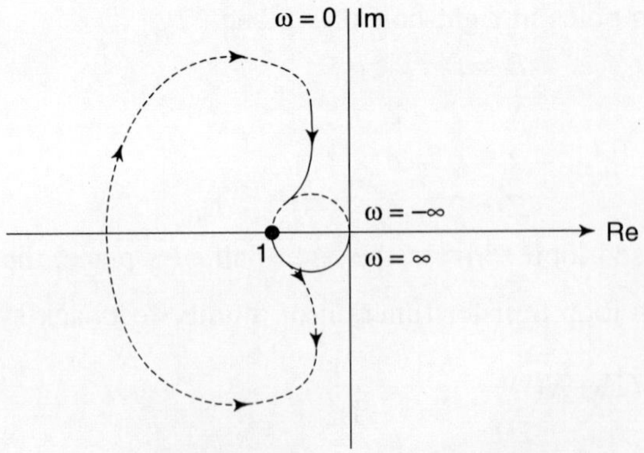

Figure Ex. 9.4 Nyquist plot

The point $-1 + j0$ is lying on the intersection of the curve with negative real axis. As the value of K is increased, point $(-1, j0)$ will move inside the loop.

$$\therefore \quad N = +1 \text{ (since one anticlockwise encirclement)}$$

$$P = 1 \text{ (one pole on R.H.S. of the } s\text{-splane)}$$

$$N = P - Z$$

$$1 = 1 - Z$$

$$\therefore \quad Z = 0$$

Which implies that open loop is unstable as $P = 1$ and, closed loop system is stable as $Z = 0$. If $K = 1$, the system is on the verge of instability.

Example 9.5 For a unity feedback system with a forward path transfer function

$$G(s) = \frac{K}{s(s+3)(s+5)}$$

find the range of gain K for stability, instability and marginal stability, use the Nyquist criterion.

(U.P.T.U., 2002–03, Cs; 2002–03, ACS)

Solution

$$G(s) = \frac{K}{s(s+3)(s+5)}$$

Putting

$$s = j\omega$$

$$G(j\omega) = \frac{K}{(j\omega)(3+j\omega)(5+j\omega)}$$

Separating real and imaginary parts of $G(j\omega)$

$$G(j\omega) = \frac{K}{(j\omega)(3+j\omega)(5+j\omega)} \times \frac{(-j\omega)(3-j\omega)(5-j\omega)}{(-j\omega)(3-j\omega)(5-j\omega)}$$

$$= \frac{K(-j\omega)(15 - 8j\omega - \omega^2)}{\omega^2(9+\omega^2)(25+\omega^2)}$$

$$G(j\omega) = \frac{K(-15\,j\omega - 8\omega^2 + j\omega^3)}{\omega^2\,(9+\omega^2)(25+\omega^2)}$$

$$G(j\omega) = \frac{K(-8\,\omega^2)}{\omega^2\,(9+\omega^2)(25+\omega^2)} - j\,\frac{15\,\omega - \omega^3}{\omega^2(9+\omega^2)(25+\omega^2)}$$

$$= \frac{-8\,K}{(9+\omega^2)(25+\omega^2)} - j\,\frac{15-\omega^2}{\omega\,(9+\omega^2)(25+\omega^2)}$$

The fequency at which real part is zero

$$\frac{-8\,K}{(9+\omega^2)(25+\omega^2)} = 0$$

$$\Rightarrow \qquad \omega = 0$$

The frequency at which imaginary part is zero

$$\frac{15-\omega^2}{\omega\,(9+\omega^2)(25+\omega^2)} = 0$$

$$\omega^2 = 15 \quad \Rightarrow \omega = \pm\,3.88 \text{ rad/sec.}$$

The real and imaginary parts of varying frequencies

ω	0	1	$\sqrt{15}$	5	10	∞
Re $G(j\omega)$	– 0.845 K	– 0.03 K	– 0.008 K	– 0.005 K	– 0.0006 K	0
Im $G(j\omega)$	– ∞	– 0.056	0	0.001	0.011	0

The Nyquist plot is shown in Fig. Ex. 9.5.

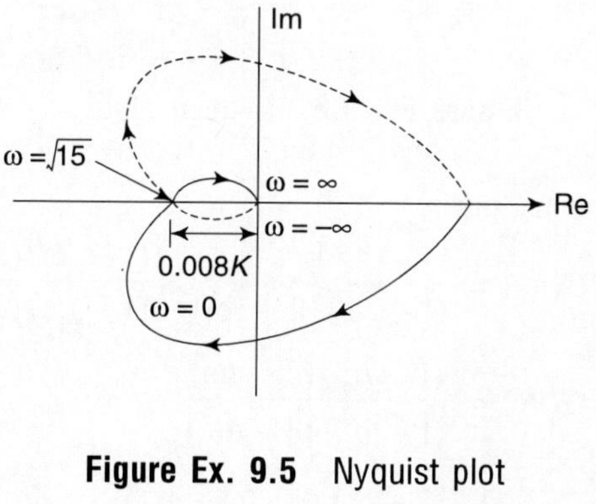

Figure Ex. 9.5 Nyquist plot

Now if the plot crosses real axis at $(-1 + j0)$ point then $-\,0.008\,K = -\,1$

$$K = \frac{-1}{0.008} = 125$$

At this value of $K = 125$ the system is marginally stable if $-0.008\,K < -1$ then the system is absolutely stable.

$$-0.008 < -1$$

$$K < -\frac{1}{0.008} \Rightarrow K < 125$$

if $0.008\,K > -1$, i.e., $K > 125$ the system is unstable.

Example 9.6 An unity negative feedback system has $G(s) = \dfrac{1}{(s+1)(s+2)}$ sketch the Nyquist plot for it. Mark on it the important points with their values and corresponding ω values. *(U.P.T.U., 2007–08, ACS)*

Solution Putting $s = j\omega$

$$G(j\omega) = \frac{1}{(1+j\omega)(2+j\omega)}$$

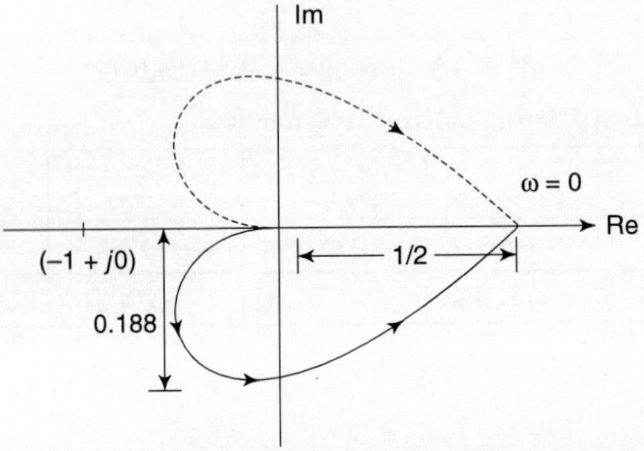

Figure Ex. 9.6 Nyquist plot

Separating real and imaginary parts

$$G(j\omega) = \frac{1}{(1+j\omega)(2+j\omega)} \times \frac{(1-j\omega)(2-j\omega)}{(1-j\omega)(2-j\omega)}$$

$$G(j\omega) = \frac{(1-j\omega)(2-j\omega)}{(1+\omega^2)(4+\omega^2)}$$

$$= \frac{(2-j\omega-2\,j\omega-\omega^2)}{(1+\omega^2)(4+\omega^2)}$$

$$= \frac{(2-\omega^2)-3\,j\omega}{(1+\omega^2)(4+\omega^2)}$$

$$= \frac{2-\omega^2}{(1+\omega^2)(4+\omega^2)} - j\frac{3\,\omega}{(1+\omega^2)(4+\omega^2)}$$

The frequency at which real part is zero

$$\frac{2-\omega^2}{(1+\omega^2)(4+\omega^2)} = 0$$

$$2 - \omega^2 = 0$$

$$\omega^2 = 2 \quad \Rightarrow \omega = \sqrt{2}$$

The frequency at which imaginary part is zero

$$\frac{3\,\omega}{(1+\omega^2)(4+\omega^2)} = 0$$

$$\omega = 0$$

The magnitude is

$$|G(j\omega)| = \frac{\frac{1}{2}}{\sqrt{(1+\omega^2)}\sqrt{1+(\omega/2)^2}} = \frac{0.5}{\sqrt{1+\omega^2}\sqrt{1+(0.5\,\omega)^2}}$$

$$|G(j\omega)|_{\text{at }\omega=\sqrt{2}} = 0.17$$

$$\angle G(j\omega) = -\tan^{-1}\omega - \tan^{-1} 0.5\,\omega$$

$$\lim_{\omega\to 0} |G(j\omega)| = \frac{1}{2}$$

$$\lim_{\omega\to\infty} |G(j\omega)| = 0$$

$$\lim_{\omega\to 0} \angle G(j\omega) = 0°$$

$$\lim_{\omega\to 0} \angle G(j\omega) = -90° - 90°$$

$$= -180°$$

The plot of $G(j\omega)$ the encirclement of point $(-1 + j0)$ is zero, i.e., $N = 0$

The number of ploes on R.H.S. plane is zero

i.e., $P = Z$

$$N = P - Z$$

$$0 = 0 - Z$$

$$Z = 0$$

Hence, system is stable.

Example 9.7 Consider a feedback system with the following open loop transfer fucntion

$$G(s)\,H(s) = \frac{K}{s\,(1+T_1\,s)\,(1+T_2\,s)}$$

investigate the stability of this system for various value of K. *(U.P.T.U., 2005–06, ACS)*

Solution Please refer Ex. 9.3 for this function.

Example 9.8 Consider the feedback control system with the loop transfer function given as $G(s)\,H(s) = \dfrac{5}{s\,(1-s)}$. State about system stability by Nyquist criterion.

(U.P.T.U., 2004–05, ACS)

Solution Putting $s = j\omega$

$$G(j\omega)\,H(j\omega) = \frac{5}{j\omega\,(1-j\omega)}$$

Magnitude is $$|G(j\omega)\,H(j\omega)| = \frac{5}{\omega\,\sqrt{1+\omega^2}}$$

Separating real and imaginary parts of $G(j\omega)\,H(j\omega)$

$$G(j\omega)\,H(j\omega) = \frac{5}{j\omega\,(1-j\omega)} \times \frac{(-j\omega)\,(1+j\omega)}{(-j\omega)\,(1+j\omega)}$$

$$= \frac{5\,(-j\omega+\omega^2)}{\omega^2\,(1+\omega^2)} = \frac{5\,\omega^2 - 5\,\omega j}{\omega^2(1+\omega^2)} = \frac{5\,\omega^2}{\omega^2\,(1+\omega^2)} - j\,\frac{5\,\omega}{\omega^2\,(1+\omega^2)}$$

$$G(j\omega)\,H(j\omega) = \frac{5}{1+\omega^2} - j\,\frac{5}{\omega\,(1+\omega^2)}$$

The freqeuncy at which real part is zero

$$\frac{5}{1+\omega^2} = 0$$

$$\omega = \infty$$

The fequency at which imaginary part is zero

$$\frac{5\,\omega}{\omega^2\,(1+\omega^2)} = 0$$

$$\omega = 0$$

The Nyquist plot is shown in Fig. Ex. 9.8.

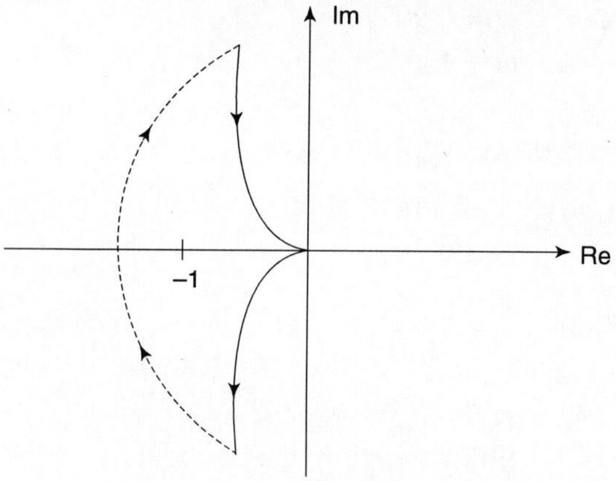

Figure Ex. 9.8 Nyquist plot

Since the locus encloses (– 1, *j*0) point once, $N = 1$ and given

$$P = 1$$

$$N = P - Z$$

$$-1 = 1 - Z$$

$$Z = -2$$

So, the system is unstable.

Example 9.9 The open loop transfer function of a feedback system is $G(s)\,H(s) = \dfrac{K}{s\,(s+1)\,(s+2)}$. Find the range of *K* for stability. Find the value of gain *K* for obtaining a gain margin of 3 dB. For this gain margin, obtain the value and phase margin and phase crossover frequency.

Solution

$$G(s)\,H(s) = \frac{K}{S\,(s+1)\,(s+2)}$$

Putting

$$s = j\omega$$

$$G(j\omega)\,H(j\omega) = \frac{K}{j\omega\,(1+j\omega)\,(2+j\omega)}$$

The angle is

$$\angle G(j\omega)\,H(j\omega) = -90^\circ - \tan^{-1}\omega - \tan^{-1}\omega/2$$

When

$$\omega = 0$$

$$\lim_{\omega \to 0} |G(j\omega)\,H(j\omega)| = \lim_{\omega \to 0} \frac{1}{\omega\sqrt{1+\omega^2}\,\sqrt{4+\omega^2}}$$

$$\lim_{\omega \to 0} \angle G(j\omega)\,H(j\omega) = \lim_{\omega \to 0} \angle -90^\circ - \tan^{-1}\omega - \tan^{-1}\omega/2$$

$$= -90^\circ$$

When $\omega = 1$

$$|G(j\omega)\,H(j\omega)|_{\omega=1} = \frac{K}{\sqrt{10}}$$

$$\angle G(j\omega)\,H(j\omega)|_{\omega=1} = -162^\circ$$

When $\omega = 2$

$$|G(j\omega)\,H(j\omega)|_{\omega=2} = \frac{K}{4\sqrt{10}}$$

$$\angle G(j\omega)\,H(j\omega)|_{\omega=2} = -98^\circ$$

When $\omega = \infty$

$$|G(j\omega)\,H(j\omega)|_{\omega=\infty} = 0$$

$$\angle G(j\omega)\,H(j\omega)|_{\omega=\infty} = -270^\circ$$

The polar plot is shown in Fig. Ex. 9.9.

$$\omega = \frac{1}{\sqrt{T_1 T_2}} = \frac{1}{\sqrt{1 \times 0.5}} = \sqrt{2}$$

$$a = \frac{K}{2}\left(\frac{T_1\,T_2}{T_1 + T_2}\right) = \frac{K}{2}\left(\frac{1 \times 0.5}{1 + 0.5}\right) = 0.167\,K$$

For stability $N = P - Z$

Since $P = 0$, $Z = -N$. But for stability Z should be zero, i.e., N should be zero which means that the point $(-1 + j0)$ should be encircled. This will only happen if

$$a < 1 \quad \text{or} \quad \frac{K}{6} < 1 \quad \text{or} \quad K < 6$$

$$\text{Gain margin} = 20 \log \frac{1}{a} = 20 \log \frac{6}{K}$$

$$3 = 20 \log \frac{6}{K}$$

$$\frac{6}{K} = 1.41$$

$$\Rightarrow \quad K = 4.25$$

Phase crossover freqeuncy

$$|G(j\omega)\,H(j\omega)| = 1 \text{ when gain is 0 dB}$$

$$|G(j\omega)\,H(j\omega)| = \frac{K}{\omega\sqrt{1+\omega^2}\,\sqrt{4+\omega^2}} = 1$$

$$(42.5)^2 = \omega^2\,(1 + \omega^2)\,(4 + \omega^2)$$

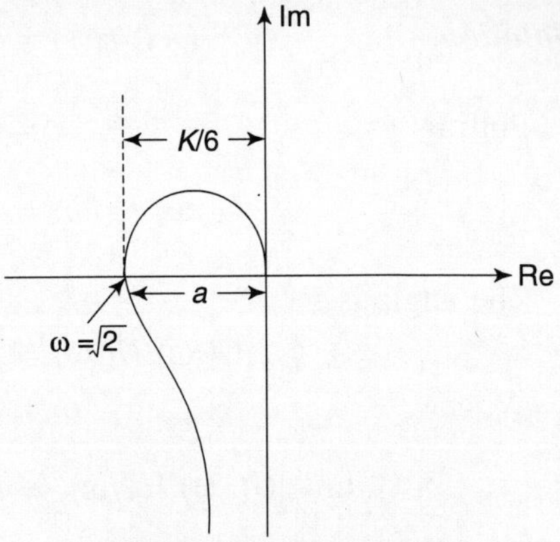

Figure Ex. 9.9 Polar plot

By Hit and Trial method

$$\omega = 1.183 \text{ is the value which satisfies the eqaution.}$$

Phase margin

$$\text{P.M.} = \angle G(j\omega)\, H(j\omega)\,\big|_{\omega = 1.183} + 180°$$

$$= \left\{\left(-90° - \tan^{-1}\frac{\omega}{1} - \tan^{-1}\frac{\omega}{2}\right)\bigg|_{\omega = 1.183}\right\} + 180°$$

$$= -90° - \tan^{-1}\frac{1.183}{1} - \tan^{-1}\frac{1.183}{2} + 180°$$

$$= 9.6° \quad \textbf{Ans.}$$

Example 9.10 Sketch the Nyquist plot and discuss stability of a unity feedback system with

$$G(s) = \frac{K}{s\,(1 + T_1\, s)\,(1 + T_2\, s)}. \qquad (U.P.T.U., 2003, C.O., ACS)$$

Solution Putting $s = j\omega$

$$G(j\omega)\, H(j\omega) = \frac{K}{j\omega\,(j\omega\, T_1 + 1)\,(j\omega\, T_2 + 1)}$$

Separating real and imaginary part

$$G(j\omega)\, H(j\omega) = \frac{K}{j\omega\,(1 + j\omega\, T_1)\,(1 + j\omega\, T_2)} \times \frac{(-j\omega)\,(1 - j\omega\, T_1)\,(1 - j\omega\, T_2)}{(-j\omega)\,(1 - j\omega\, T_1)\,(1 - j\omega\, T_2)}$$

$$= \frac{K\,(-j\omega)\,(1 - j\omega\, T_1)\,(1 - j\omega\, T_2)}{\omega^2\,(1 + \omega^2\, T_1^2)\,(1 + \omega^2\, T_2^2)}$$

$$= \frac{K\omega^2\,(T_1 + T_2) - jK\omega\,(1 - \omega^2\, T_1\, T_2)}{\omega^2\,(1 + \omega^2\, T_1^2)\,(1 + \omega^2\, T_2^2)}$$

$$= \frac{-K\,(T_1 + T_2)}{(1 + \omega^2\, T_1^2)\,(1 + \omega^2\, T_2^2)} - j\,\frac{K\,(1 - \omega^2\, T_1\, T_2)}{\omega\,(1 + \omega^2\, T_1^2)\,(1 + \omega^2\, T_2^2)}$$

The frequency at which imaginary part is zero

$$\frac{K\,(1 - \omega^2\, T_1\, T_2)}{\omega^2\,(1 + \omega^2\, T_1^2)\,(1 + \omega^2\, T_2^2)} = 0$$

$$1 - \omega^2\, T_1\, T_2 = 0$$

$$\omega^2 = \frac{1}{T_1\, T_2} \quad \Rightarrow \omega = \pm\frac{1}{\sqrt{T_1\, T_2}}$$

$$|G(j\omega)\,H(j\omega)|_{\omega = \pm\frac{1}{\sqrt{T_1 T_2}}} = \frac{K\,(T_1 + T_2)}{\left(\frac{1}{T_1\,T_2}\,T_1^2 + 1\right)\left(\frac{1}{T_1\,T_2}\,T_2^2 + 1\right)} = \frac{K\,T_1\,T_2}{(T_1 + T_2)} \qquad ...(1)$$

at $\omega = 0$

$$G(j\omega)\,H(j\omega) = -\,K\,(T_1 + T_2) - j\,\infty$$

$$= \infty\;\angle\pi/2 \text{ at } \omega = \infty \qquad ...(2)$$

$$G(j\omega)\,H(j\omega) = 0\;\angle\pi/2 \qquad ...(3)$$

By Eqs. 1, 2 and 3 the Nyquist plot is shown in Fig. Ex. 9.10.

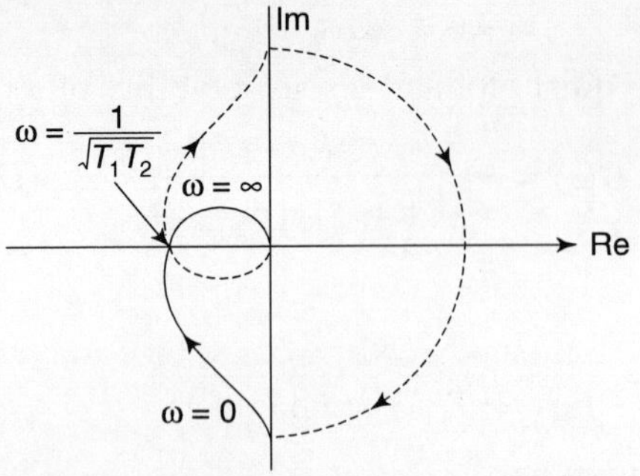

Figure Ex. 9.10

If $\frac{K\,T_1\,T_2}{T_1 + T_2} < 1$, the point $(-\,1 + j0)$ is not encircled by the Nyquist plot and the closed loop system is found to be stable. However, if $\frac{K\,T_1\,T_2}{T_1 + T_2} > 1$ the point $(-\,1, +\,j0)$ is encircled twice in clockwise direction indicating that there are two roots and characteristic equation with positive real part and the closed loop system is unstable.

$$\frac{K\,T_1\,T_2}{T_1 + T_2} < 1$$

$$K \le \left(\frac{1}{T_1} + \frac{1}{T_2}\right) \qquad \textbf{Ans.}$$

Example 9.11 The open loop transfer function of a feedback control system is $G(s)\,H(s) = \frac{K\,(1 + 2s)}{s\,(1 + s)\,(1 + s + s^2)}$. Find the range of K for stability. Find the value of K for the system have a gain margin of 5 dB. With this value of K, find the phase crossover freqeuncy and phase margin.

Solution

$$G(s)\,H(s) = \frac{K\,(1+2s)}{s\,(1+s)\,(1+s+s^2)}$$

Putting $s = j\omega$

$$G(j\omega)\,H(j\omega) = \frac{K\,(1+2j\omega)}{j\omega\,(1+j\omega)\,1+j\omega+(j\omega)^2}$$

The magnitude is

$$|\,G(j\omega)\,H(j\omega)\,| = \frac{\sqrt{1+4\omega}}{\omega\,\sqrt{1+\omega^2}\,\sqrt{(1-\omega^2)^2+\omega^2}}$$

The angle is

$$\angle G(j\omega)\,H(j\omega) = -\,90° - \tan^{-1} 2\,\omega - \tan^{-1}\omega - \tan^{-1}\frac{\omega}{1-\omega^2}$$

When $\omega = 0$

$$|\,G(j\omega)\,H(j\omega)\,| = \infty$$

$$\angle G(j\omega)\,H(j\omega) = -\,90°$$

When $\omega = 1$

$$|\,G(j\omega)\,H(j\omega)\,| = \sqrt{\frac{5}{2}}$$

$$\angle G(j\omega)\,H(j\omega) = 161.5°$$

When $\omega = 0.5$

$$|\,G(j\omega)\,H(j\omega)\,| = 2.8$$

$$\angle G(j\omega)\,H(j\omega) = 10.5.5°$$

When $\omega = \infty$

$$|\,G(j\omega)\,H(j\omega)\,| = 0$$

$$\angle G(j\omega)\,H(j\omega) = 270°$$

Separating into real and imaginary parts and equating the imaginary part to zero, i.e.,

$$1 + 2\omega^2 - 2\omega^4 = 0$$

by Hit and Trial method

$$\omega = 1.17$$

$$|\,G(j\omega)\,H(j\omega)\,|_{\omega\,=\,1.17} = \frac{K\,\sqrt{1+(1.17\times 2)^2}}{1.17\,\sqrt{1+(1.17)^2}\,\sqrt{1-1.17^2)^2+1.17}}$$

$$a = 1.153\,K$$

$$|\,G(j\omega)\,H(j\omega)\,|_{\omega\,=\,1.17} = \tan^{-1}(2\times 1.17) - 90° - \tan^{-1} 1.17 - \tan^{-1}\frac{1.17}{1-1.17^2}$$

$$= 180°$$

$|G(j\omega)\, H(j\omega)| \angle G(j\omega)\, H(j\omega) = 1.153\, K \angle -180°$

The polar plot is shown in Fig. Ex. 9.11.

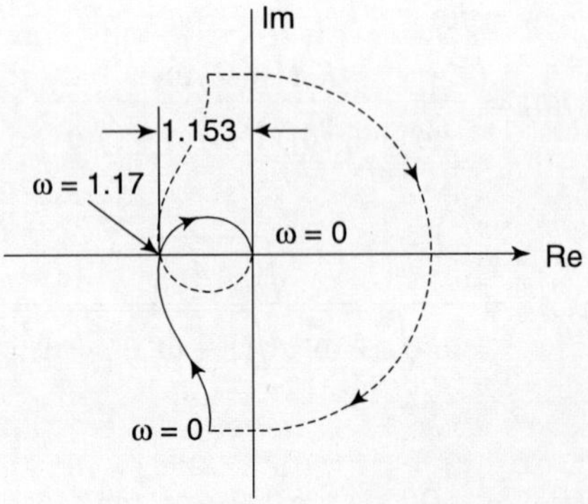

Figure Ex. 9.11 Nyquist plot

If $K = 1$, then point $(-1 + j0)$ will be encircled and the system will be unstable.

If $$K < \frac{1}{1.153} \Rightarrow K < 0.87$$

The system will be stable.

Phase crossover frequency

$$\frac{K\sqrt{1+4\omega^2}}{\omega\sqrt{1+\omega^2}\sqrt{(1-\omega^2)^2+\omega^2}} = 1 \qquad ...(A)$$

$$\text{Gain Margin} = 20 \log \frac{1}{a}$$

$$5 = 20 \log \frac{1}{1.153\, K}$$

$$\frac{1}{1.153\, K} = 1.78$$

$$K = 0.487$$

Putting the value of K in Eq. (A)

$$\frac{0.487\,(1+4\omega^2)}{\omega^2\,(1+\omega^2)\,(1-\omega^2)^2+\omega^2} = 1$$

By Hit and Trial method $\omega = 0.87$ is the phase crossover frequency

Phase margin $= \tan^{-1}(2 \times 0.87) - 90° - \tan^{-1} 0.87 - \tan^{-1}\dfrac{0.87}{1-0.87^2} + 180° = 34.7°$ **Ans.**

MATLAB PROGRAMS

Program 9.1

```
% Program for Nyquist
%G(s)  = 5/(s+1)(s^2+3s+20)
clear all
clc
clf                         %clear the graph
numg = input('Enter the coefficient of polynomial of numerator:-');
                            % Input coefficient of numerator
deng1 = input('Enter the coefficient of first polynomial of denominator:- ');
                            % Input coefficient of denominator
deng2 = input('Enter the coefficient of second polynomial of denominator:- ')
                            % Input coefficient of denominator
deng  = conv(deng1,deng2); % Multiplying two term
'G(s)'                      % Display string
G  = tf(numg,deng)          % Create transfer function
nyquist(G)                  % Make Nyquist diagram
```

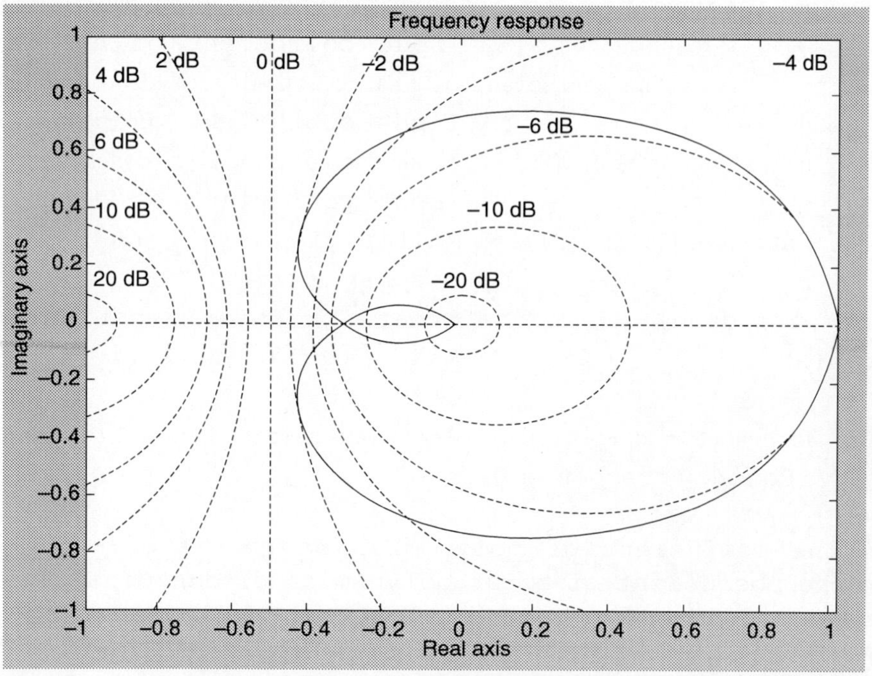

Figure P. 9.1 Nyquist plot of $G(s)$

```
grid on
title('Frequency response')
```

Input

```
Enter the coefficient of polynomial of numerator:- 5
Enter the coefficient of first polynomial of denominator:- [1 1]
Enter the coefficient of second polynomial of denominator :- [1 2 5]
```

Output

Transfer function: $G(s) \quad \dfrac{5}{s\text{^}3 + 3s\text{^}2 + 7s + 5}$

The Nyquist plot is shown in Fig. P. 9.1.

Program 9.2

```
% Program for finding gain margin, phase margin and margin
  frequency of Nyquist plot
% G(s) = 10/(s^2+2)(s^3+2s^2+2s)
clear all
clc
clf                               % clear the figure
numg = input('Enter the coefficient of polynomial of numerator:- ');
                                  % Input coefficient of numerator
deng1 = input('Enter the coefficient of first polynomial of denominator:- ');
                                  % Input coefficient of denominator
deng2 = input('Enter the coefficient of second polynomial of denominator:- ');
                                  % Input coefficient of denominator
deng = conv(deng1,deng2);         % Multiplying two term
'G(s)'                            % Display string
G = tf(numg,deng)                 % Create transfer function
nyquist(G)                        % Make Nyquist diagram
grid on
[Gm, Pm, Wcg, Wcp] = margin(G)    % Calculate margin and margin frequency
title('Frequency response')
```

Input

```
Enter the coefficient of polynomial of numerator:- 30
Enter the coefficient of sirst polynomial of denominator:- [1 0]
Enter the coefficient of second polynomial of denominator:- [1 5 6]
```

Output

Transfer function: $G(s) \quad \dfrac{30}{s\text{^}3 + 5s\text{^}2 + 6s}$

```
Gm = 1.0000
Pm = 4.6373e-007
Wcg = 2.4495
Wcp = 2.4495
The Nyquist plot is shown in Fig. P. 9.2.
```

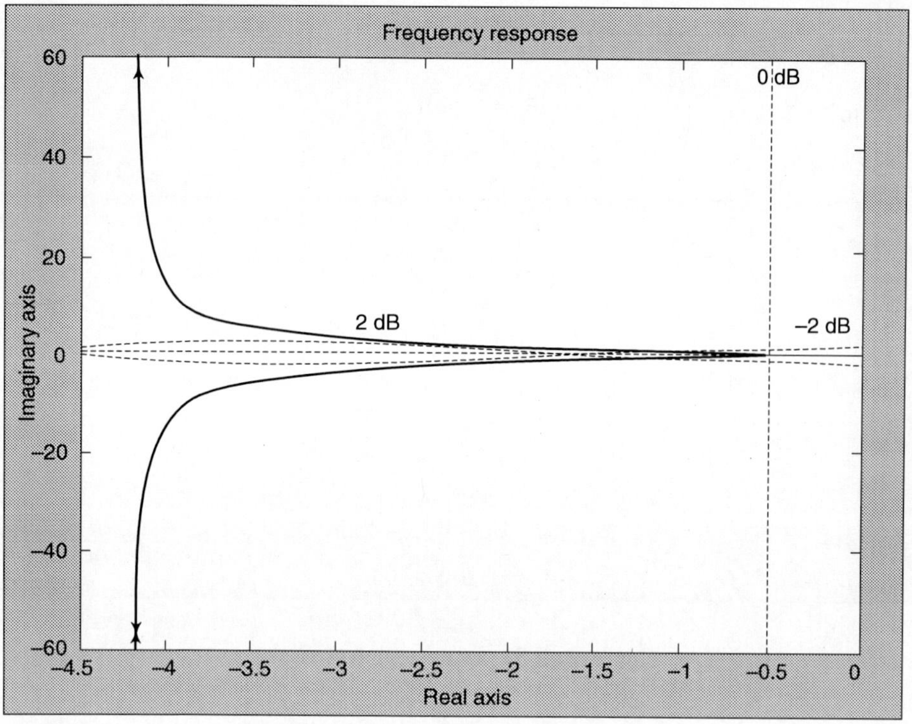

Figure P. 9.2 Nyquist plot for G(s)

Program 9.3

```
% Program to determine the range of k for stability using
  frequency response method
%G(s) = 10/(s+1)(s+2)(s+3)
clear all
clc
clf                       % clear the figure
numg = input('Enter the coefficient of polynomial of numerator:-');
                          % Input coefficient of numerator
deng1 = input('Enter the poles of denominator:- ');
                          % Input coefficient of denominator
```

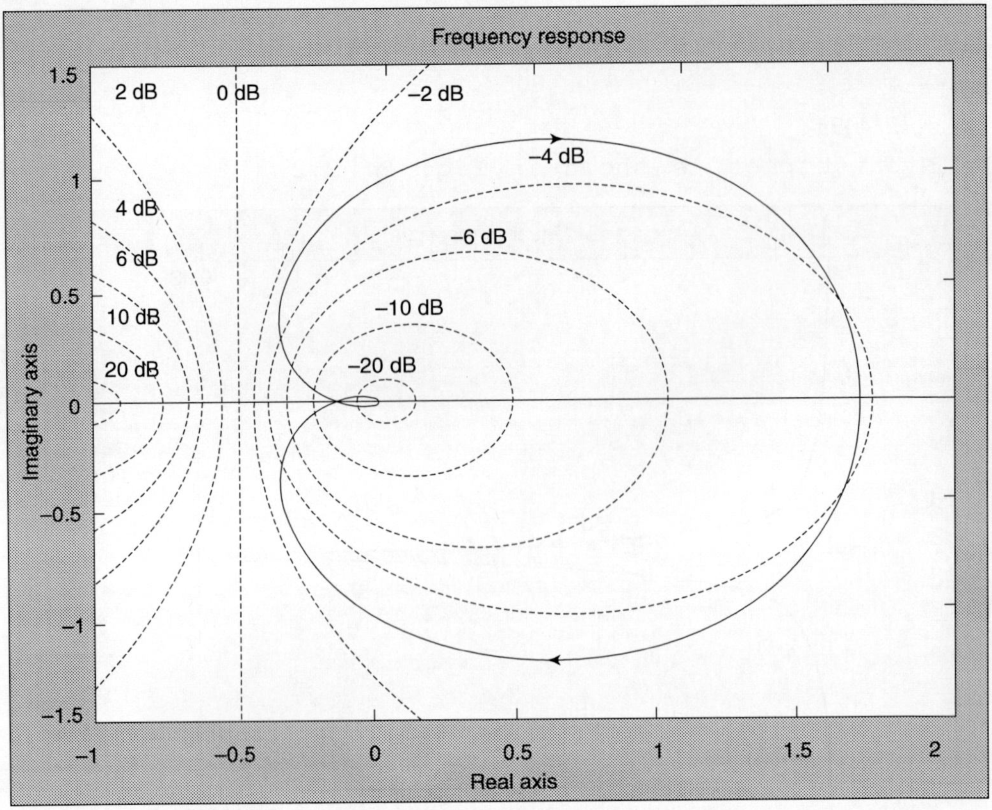

Figure P. 9.3 Nyquist plot of $G(s)$

```
deng = poly(deng1);            % Define denominator
'G(s)'                         % Display String
G = tf(numg,deng)              % Create transfer function
nyquist(G)                     % Make Nyquist diagram
grid on
[Gm, Pm, Wcg, Wcp] = margin(G) % Calculate margin and margin frequency
k = Gm                         % Display k for stability
title('Frequency response')
```

Input

```
Enter the coefficient of polynomial of numerator:- 10
Enter the poles of denominator:- [-1 -2 -3]
```

Output

Transfer function: $G(s) \quad \dfrac{10}{s\text{^}3 + 6s\text{^}2 + 11s + 6}$

```
Gm  =  6.0000
Pm  =  90.0000
Wcg  =  3.3166
Wcp  =  1.0000
k  =  6.0000
The Nyquist plot is shown in Fig. P. 9.3.
```

EXERCISE

9.1 Explain Nyquist stability criterion. *(U.P.T.U., 2003–04, CS)*

9.2 Explain constant M and N circles and their use in control system.

9.3 Draw the Nyquist plot for a system with the open loop transfer function,

$$G(s)\,H(s) = \frac{K\,(1+0.5\,s)\,(s+1)}{(1+10\,s)\,(s-1)}$$

Determine the range of values of K for which the system is stable.

9.4 Check the stability of the system where Nyquist plots are shown in Fig. Ex. 9.4 (a) and (b)

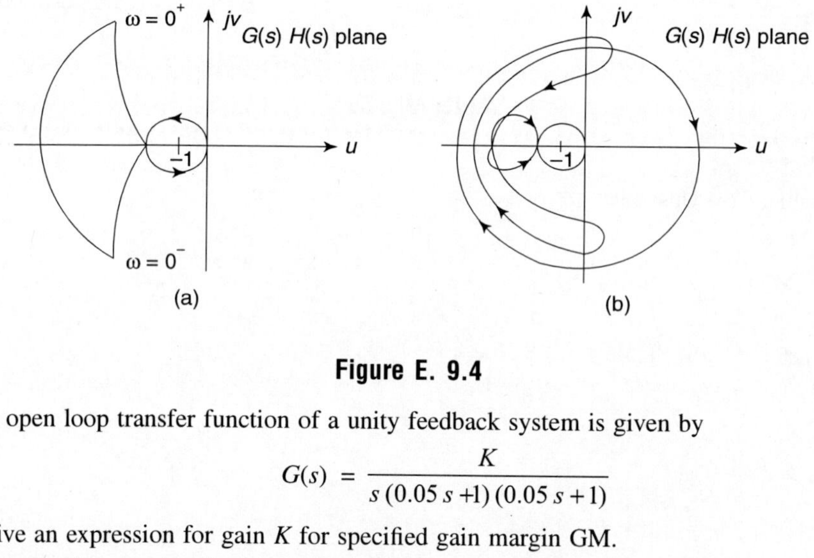

Figure E. 9.4

9.5 The open loop transfer function of a unity feedback system is given by

$$G(s) = \frac{K}{s\,(0.05\,s+1)\,(0.05\,s+1)}$$

Derive an expression for gain K for specified gain margin GM.

9.6 Draw the Nyquist plot for the given function

$G(s)\,H(s) = \dfrac{Ks^2}{s^3+4s+4}$ and examime closed loop stability in terms of K.

9.7 The open loop transfer function of a feedback control system is given by

$$G(s)\,H(s) = \frac{K}{s^2+2s-2}$$

Plot the Nyquist plot and show the closed loop system is stable if $K \geq 2$.

9.8 Draw the Nyquist plot for the control system whose open loop transfer functions are given below and determine the stability.

(a) $G(s)\,H(s) = \dfrac{10}{s\,(s+1)\,(s^2+4)}$

(b) $G(s)\,H(s) = \dfrac{1.5}{(s+1)\,(0.54\,s+1)\,(0.25\,s+1)}$

(c) $G(s)\,H(s) = \dfrac{1}{s\,(s+1)\,(s^2+2s+2)}$

9.9 What is gain margin and phase margin? For the open loop transfer function,

$G(j\omega)\ H(j\omega) = \dfrac{K}{j\omega\,(j\omega\,T_1+1)\,(j\omega\,T_2+1)}$ investigate the stability of the system for various values of K. *(U.P.T.U., 2006–07, CS)*

9.10 Explain Nyquist criterion with an example. *(U.P.T.U., 2005–06, CS)*

9.11 Discuss the basic features of Nyquist criterion and what do you understand by the term stability of the system, explain relative and asymptotic stability. *(U.P.T.U., 2006–07, ACS)*

9.12 Discuss the general procedure of constructing Nyquist plot for stability consideration. *(U.P.T.U., 2006–07, CS)*

ANSWERS

9.3 $\infty > K > 6$

9.4 (a) Stable, (b) Unstable

9.5 $K = \dfrac{1}{G_M}$ (22)

9.6 K 71

9.8 (a) Unstable, (b) Stable, (c) Stable

CHAPTER 10

Introduction to Design

10.1 INTRODUCTION

The important purpose for control engineers is to design an automatic control system. The performance of an automatic control system, we have already discussed in previous chapter in which we have seen the quantitative measures of performance. So we can say that the performance of a control system is its primary importance. By the analysis of previous chapters we can say that a suitable control system is stable, and the result of that particular system is acceptable to input commands. It should be less sensitive to system parameter changes, due to this less steady states error for input commands. So we can say that it is able to reduce the effect of unwanted disturbances.

Every control system which is designed for particular task has to meet certain performance specifications. There are two methods for specifying the performance of control system, these methods we have studied in chapters 3 and 8. One is time response specification or frequency response specification and second one is optimum performance.

An automatic control system that provides an optimum performance without any necessary adjustment is rare. So we need some adjustment which is necessary to compromise among the many demanding specifications, which is make a system suitable and acceptable performance.

At many places in the previous chapters, we have considered the question of design and adjustment of the system parameters to get desirable response and performance. In chapter 3 we have studied the performance indices and measure of performance. Then in chapters 6, 7 and 9 we determined methods of investigating the stability of a control system because we know that a system which is unstable is not acceptable unless it is stable. In chapter 8 we developed suitable measures of performance in terms of the frequency variable, ω. So this frequency response is utilized to design many suitable control systems.

- *The design of a control system deals with the plan or the arrangement of the system structure and then select suitable components and parameters for that system.*

10.2 APPROACHES TO SYSTEM DESIGN

There are two methods of which we can design a system.

1. First we select the configuration of the system with compensators then select the parameters of the compensators to meet the requirements or performance specification.
2. In second method first we find the system that meets the requirements or performance specification after that compute the necessary compensators.

In previous chapters we have used first approach for the system design like the root locus technique which is used to design system parameters. In chapters 8 and 9 we have studied the Bode plot and Nyquist plot for designing a control system to achieve the desired performance specification in terms of phase margin, gain margin, resonant peak, etc.

For adjusting the system response up to desired limit, an additional component is inserted with the structure of the automatic control system. This additional component or device is called compensator. The compensating device may be electric, mechanical, hydraulic, pneumatic or some other type of device.

10.3 PRELIMINARY CONSIDERATIONS

If the performance of a control system is not as per desired specifications, then we require some additional device, known as compensatory network, to obtain the desired performance. The device can be inserted in forward path gain or feedback path gain. When the compensation network is inserted in cascade with forward transfer function it is called cascade compensation. The transfer function of a compensator is $G_c(s)$ as $E_o(s)/E_{in}(s)$.

i.e., $$G_c(s) = \frac{E_o(s)}{E_{in}(s)} \quad ...(10.1)$$

According to the location of compensation network, it is classified in four types.

1. **Cascade compensation.** In cascaded compensation the compensation network is inserted in cascade with the system forward transfer function as shown in Fig. 10.1.

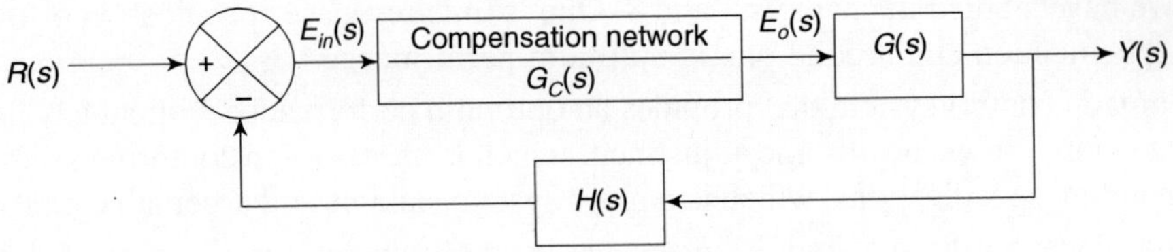

Figure 10.1 Cascade compensation

2. **Feedback compensation.** In feedback compensation, the compensation network can be introduced in feedback path as shown in Fig. 10.2 (a) and (b).

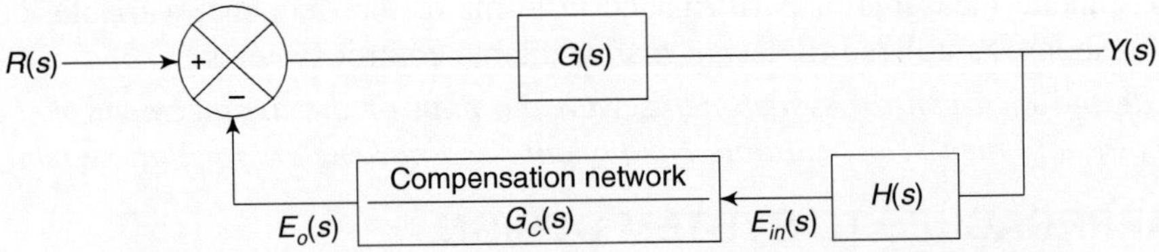

(a)

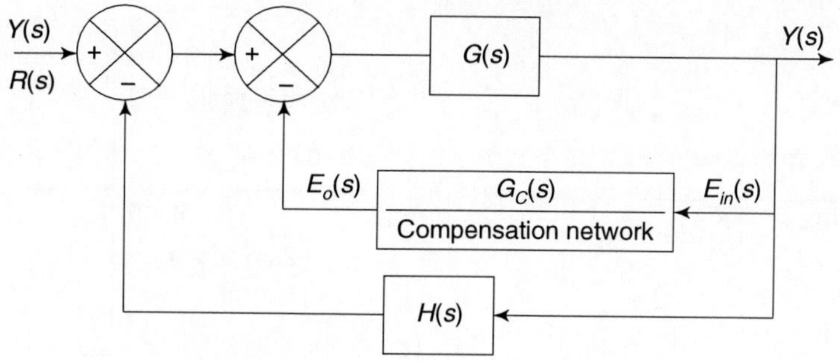

(b)

Figure 10.2 (a) Cascade feedback compensation, (b) Parallel feedback compensation

3. **Output or load compensation.** In this type of compensation, the compensation network is inserted at the output of the system, as shown in Fig. 10.3.

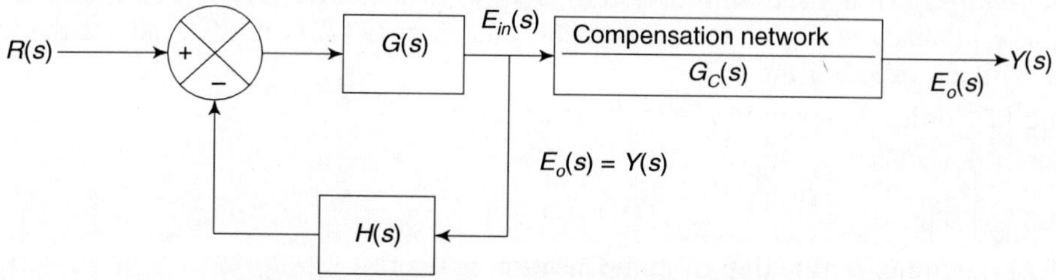

Figure 10.3 Output or load compensation

4. **Input compensation.** In this compensation, the compensation network is inserted at the input side as shown in Fig. 10.4.

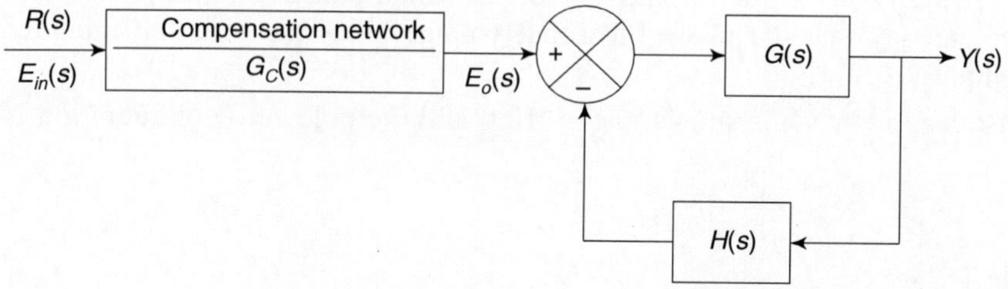

Figure 10.4 Input compensation

The compensation network introduces additional poles/zeros in the actual system due to this its actual transfer function is altered. The effect of adding zero to an unstable system makes it stable as shown in Fig. 10.5 (a) and (b).

Generally, the performance specification of a control system is given in frequency domain like gain margin, phase margin, so the use of Bode plot method is convenient in designing compensation networks for control systems.

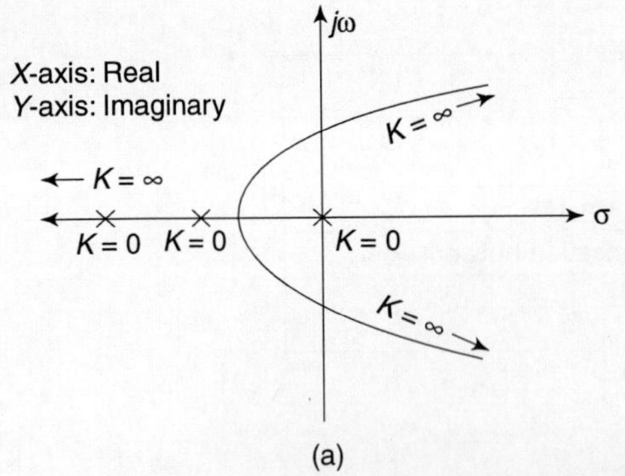

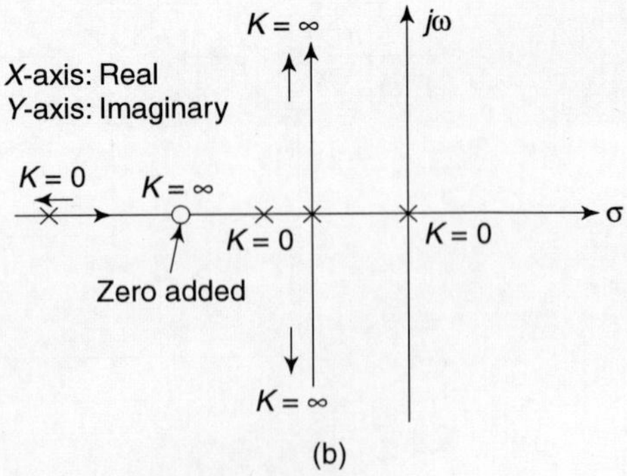

Figure 10.5 (a) Unstable system, (b) Stable system

10.4 PHASE LEAD NETWORK

A compensator $G_c(s)$, is used with a system $G(s)$ so that the overall loop gain can be set to satisfy the e_{ss} (steady state error) requirement, and then $G_c(s)$ is used to adjust the system dynamic without affecting the e_{ss}.

Let the first order compensator with transfer function is

$$G_c(s) = \frac{K\,(s + Z_c)}{(s + P_c)} \qquad ...(10.2)$$

where $G_c(s)$ = transfer function of compensation network

Z_c = zeros inserted for compensation

P_c = poles inserted for compensation

So, the problem of designing is, the selection of Z_c, P_c and K in order to provide a suitable performance.

"When $|Z_c| < |P_c|$, i.e., $\alpha < 1$, the network is called phase lead network". In phase lead network we can say that the phase of output voltage leads the phase of input voltage for sinusoidal input.

The phase lead network is shown in Fig. 10.6 and the pole zero configuration is given in Fig. 10.7.

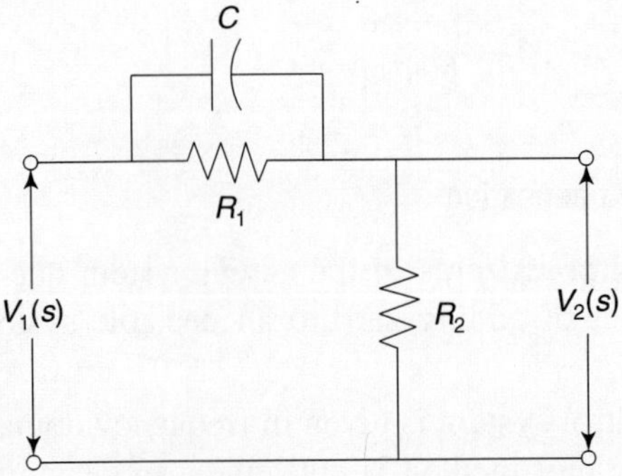

Figure 10.6 Phase lead network

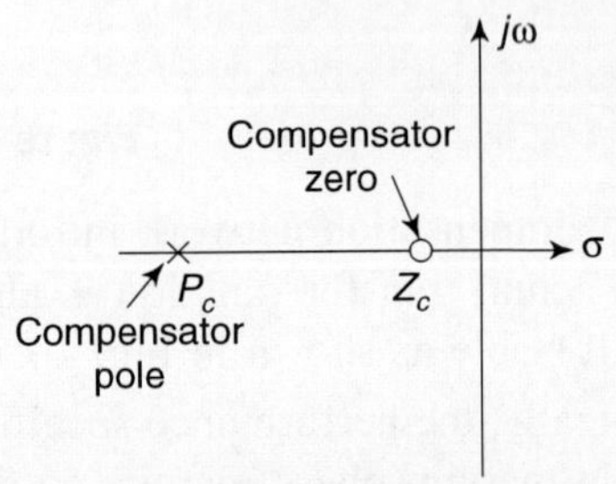

Figure 10.7 Pole zero configuration of phase lead network

The transfer function of the network

$$G_c(s) = \frac{V_2(s)}{V_1(s)} = \frac{(s+Z_c)}{(s+P_c)} \quad ...(10.3)$$

Let $$Z_c = \frac{1}{T}, P_c = \frac{1}{\alpha T}$$

and $$\alpha = \frac{Z_c}{P_c} < 1 \qquad \because Z_c < P_c \text{ so } T > 0$$

$$G_c(s) = \frac{\left(s+\frac{1}{T}\right)}{\left(s+\frac{1}{\alpha T}\right)} = \frac{(T_s+1)/T}{(\alpha sT+1)/\alpha T}$$

$$G_c(s) = \frac{\alpha\,(1+TS)}{(1+\alpha ST)} \quad ...(10.4)$$

By the Fig. 10.6

$$\alpha = \frac{R_2}{R_1+R_2} \quad \text{and } T = R_1\,C$$

Substituting $s = j\omega$ in Eq. 10.4

$$G_c(j\omega) = \frac{\alpha\,(1+j\omega\,T)}{(1+j\omega\alpha\,T)} \quad ...(10.5)$$

The Bode plot for the transfer of Eq. (10.5) is shown in Fig. 10.8.

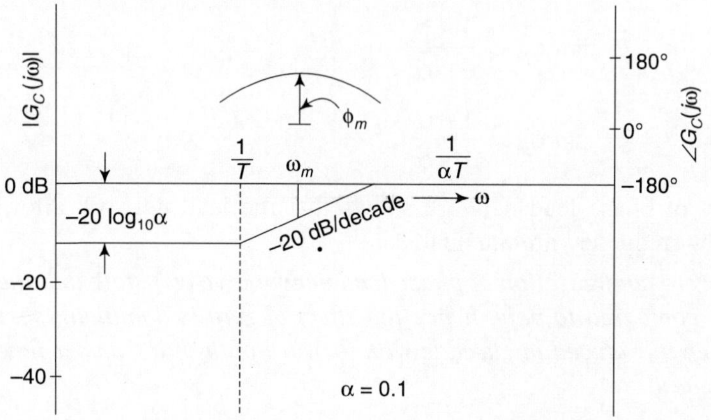

Figure 10.8 Bode plot for phase lead network

The two corner frequencies are

$$\omega_{c1} = \frac{1}{T} = \text{lower corner frequency}$$

$$\omega_{c2} = \frac{1}{\alpha T} \text{ upper corner frequency}$$

The maximum phase lead (ϕ_m) occurs at frequency between upper and lower corner frequencies is called mid-frequency ω_m

$$\log_{10} \omega_m = \frac{1}{2}\left[\log_{10}\left(\frac{1}{T}\right) + \log_{10}\left(\frac{1}{\alpha T}\right)\right]$$

$$\omega_m = \frac{1}{\sqrt{\alpha}\, T} \quad \text{...(10.6)}$$

The phase angle, $\angle G_c(j\omega)$

$$\angle G_c(j\omega) = \tan^{-1}(\omega T) - \tan^{-1}(\omega\, \alpha\, T) \quad \text{...(10.7)}$$

at mid-frequency phase is maximum, so $\omega = \omega_m$ is (10.7)

$$G_c(j\omega) = \phi_m = \tan^{-1}(\omega_m T) - \tan^{-1}(\omega\, \alpha\, T)$$

$$= \tan^{-1}\left(\frac{1}{\sqrt{\alpha}\, T} \times T\right) - \tan^{-1}\left(\frac{1}{\sqrt{\alpha}\, T}\, \alpha\, T\right)$$

$$= \tan^{-1}\left(\frac{1}{\sqrt{2}}\right) - \tan^{-1}\left(\sqrt{\alpha}\right)$$

$$\tan \phi_m = \frac{\frac{1}{\sqrt{\alpha}} - \sqrt{\alpha}}{1 + \frac{1}{\sqrt{\alpha}}\sqrt{\alpha}} = \frac{1-\alpha}{2\sqrt{\alpha}}$$

$$\tan \phi_m = \frac{1-\alpha}{2\sqrt{\alpha}} \quad \text{...(10.8)}$$

$$\sin \phi_m = \frac{1-\alpha}{1+\alpha} \quad \text{...(10.9)}$$

The Bode plot of phase lead network shows that the lead network allows to pass high frequency and low frequency are attenuated.

- *By the pole zero configuration of phase lead network we can state that the zero is nearer to origin as compared to pole, hence the effect of zero is dominant, so, the phase lead network when introduced in cascade with forward path of a transfer function the phase shift is increased.*

10.5 PHASE LAG NETWORK

We have seen the transfer function of the compensation network $G_c(s)$ in previous section is in continuation with this.

"When $|Z_c| > |P_c|$, i.e., $\alpha > 1$, the network is called phase lag network". In phase lag network we can say that the phase of output voltage lags the phase of input voltage for sinusoidal inputs.

The phase lag network is shown in Fig. 10.9 and the pole zero configuration in Fig. 10.10.

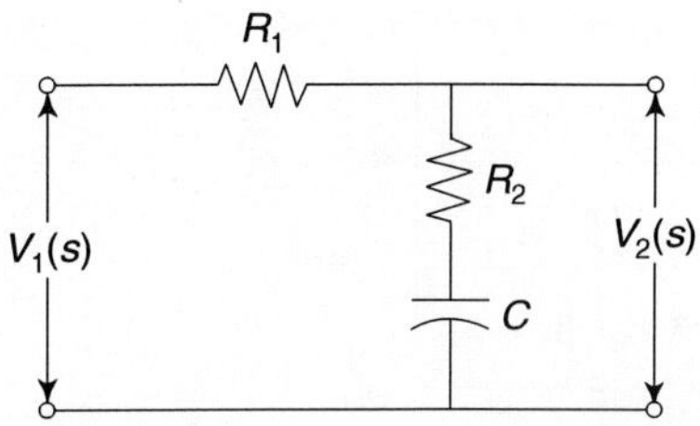

Figure 10.9 Phase lag network

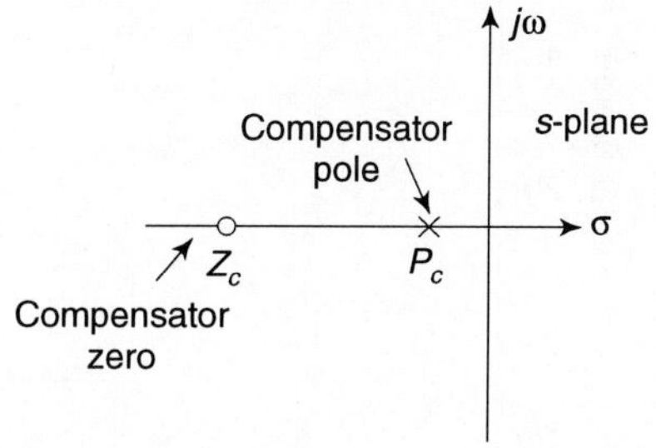

Figure 10.10 Pole zero configuration of phase lag network

The transfer function of the network,

$$G_c(s) = \frac{V_2(s)}{V_1(s)} = \frac{(s+Z_c)}{(s+P_c)} \quad \text{...(10.10)}$$

Let $$Z_c = \frac{1}{T},\ P_c = \frac{1}{\beta T}$$

and $$\beta = \frac{Z_c}{P_c} > 1 \qquad \because Z_c > P_c$$

$$G_c(s) = \frac{\left(s+\frac{1}{T}\right)}{\left(1+\frac{1}{\beta T}\right)} = \frac{(1+sT)}{\beta(1+\beta sT)} \quad \text{...(10.11)}$$

When $\beta > 1$ $$\beta = \frac{R_1+R_2}{R_2} \text{ and } T = R_2 C$$

Putting $s = j\omega$ in Eq. (10.11)

$$G_c(j\omega) = \frac{(1+j\omega T)}{\beta(1+j\omega\beta T)} \quad \text{...(10.12)}$$

The Bode plot for the transfer function of Eq. (10.12) is shown in Fig. 10.11.

The two corner frequencies are

$$\omega_{c1} = \frac{1}{\beta T}; \text{ lower corner frequency}$$

$$\omega_{c2} = \frac{1}{T}; \text{ upper corner frequency}$$

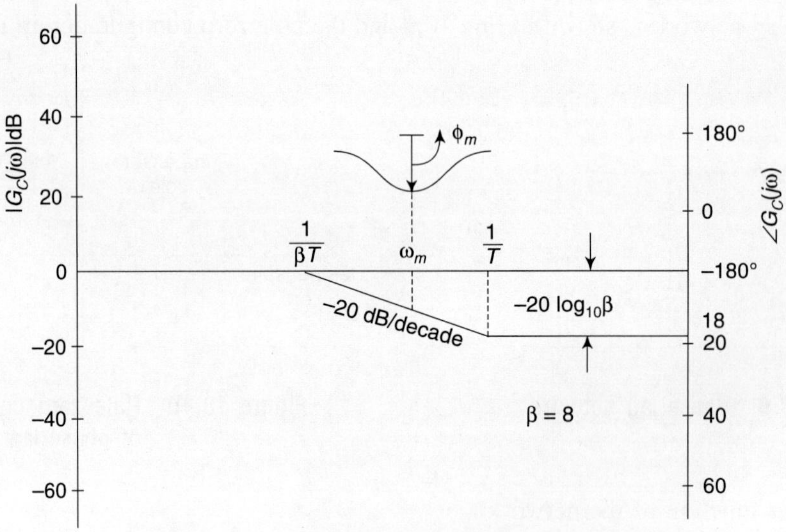

Figure 10.11 Bode plot for phase lag network

The maximum phase lag (ϕ_m) occurs at mid-frequency, ω_m.

$$\log_{10} \omega_m = \frac{1}{2}\left[\log_{10}\left(\frac{1}{\beta T}\right) + \log_{10}\left(\frac{1}{T}\right)\right]$$

$$\omega_m = \frac{1}{\sqrt{\beta}\, T} \quad \text{...(10.13)}$$

The phase angle $\angle G_c(j\omega)$,

$$\angle G_c(j\omega) = \tan^{-1}(\omega T) - \tan^{-1}(\omega\, \beta T)$$

Put

$$\omega = \omega_m = \frac{1}{\sqrt{\beta}\, T}$$

$$\phi_m = \tan^{-1}(\omega_m T) - \tan^{-1}(\omega_m\, \beta\, T)$$

$$\phi_m = \tan^{-1}\left(\frac{1}{\sqrt{\beta}\, t} \times T\right) - \tan^{-1}\left(\frac{1}{\sqrt{\beta}\, T}\beta T\right)$$

$$\phi_m = \tan^{-1}\left(\frac{1}{\sqrt{\beta}}\right) - \tan^{-1}\left(\sqrt{\beta}\right)$$

$$\tan \phi_m = \frac{1-\beta}{2\sqrt{\beta}} \quad \text{...(10.14)}$$

$$\sin \phi_m = \frac{1-\beta}{1+\beta} \quad \text{...(10.15)}$$

By Eq. (10.15) if $\beta > 1$ then $\sin \phi_m$ is negative. It indicates that ϕ_m is negative, i.e., the phase angle is lagging.

The Bode plot of phase lag network shows that the lag network allows to pass low frequencies and high frequencies are attenuated.

- *By the pole zero configuration we can state that the pole is nearer to the origin as compared to zero, hence the effect of pole is dominant, so, the phase lag network when introduced in cascade with forward path of a transfer function, the phase shift is reduced.*

10.6 PHASE LEAD LAG NETWORK

The advantage of phase lead network is that it increases the gain crossover frequency so due to this, the bandwidth is increased, and it improves the speed of response and reduces the overshoot but steady state error is not much improved.

The advantage of phase lag network is that it reduces the gain crossover frequency due to this bandwidth is decreased, so the steady state error (e_{ss}) is improved but the speed and response is reduced.

In both the previous networks we have seen that one is providing improved speed and response but not steady state error and second one is providing improved steady state error but no speed and response. So taking the advantage of both the networks simultaneously we move forward and make another network which is a combination of phase lead and phase lag networks. It will provide both, i.e., improved speed, response and steady state error. And this network is called phase lead lag network as shown in Fig. 10.12 and pole zero configuration of it is shown in Fig. 10.13.

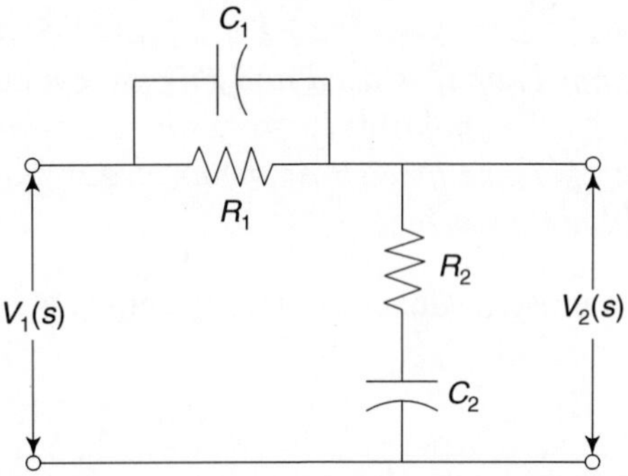

Figure 10.12 Phase lead lag netowrk

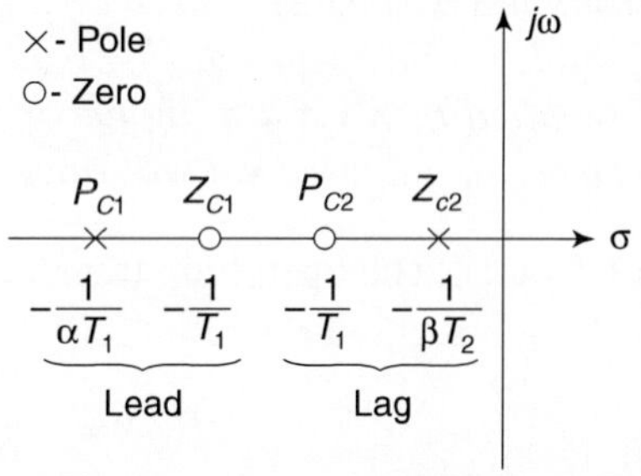

Figure 10.13 Pole zero configuration of phase lead lag network

The transfer function for the phase lead lag network is $G_c(s)$

$$G_c(s) = \frac{V_2(s)}{V_1(s)} = \frac{\alpha(1+sT_1)}{(1+\alpha sT_1)} \cdot \frac{(1+sT_2)}{(1+s\beta T_2)} \qquad ...(10.16)$$

where $T_1 = R_1 C_1$

$T_2 = R_2 C_2$

$\alpha < 1$

$\beta > 1$

α is attenuation factor due to phase lead network is considered.

Put $s = j\omega$ in Eq. (10.16)

$$G_c(j\omega) = \frac{\alpha\,(1 + j\,\omega\,T_1)}{(1 + j\,\omega\,\alpha\,T_1)} \cdot \frac{(1 + j\,\omega\,T_2)}{(1 + j\,\omega\,\beta\,T_2)} \quad ...(10.17)$$

The Bode plot for the Eq. (10.17) is shown in Fig. 10.14.

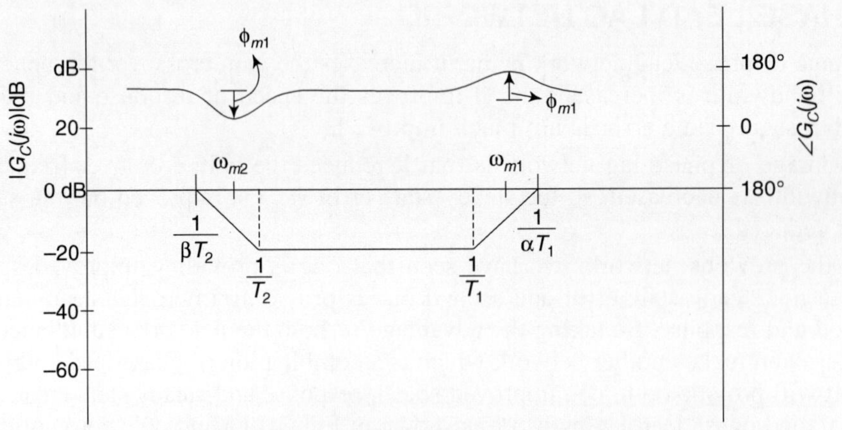

Figure 10.14 Bode plot for phase lead lag network

- *Unstable systems are type 2 or higher, lead compensation is used and this system can be made stable by introducing lead compensation. When improved steady state accuracy is required then we use lag network. When fast response as well as better steady state accuracy is required we will prefer phase lead lag network.*

Example 10.6.1 The open loop transfer function of a unity feedback control system is given by

$$G(s) = \frac{K}{s\left(1 + \frac{1}{2}s\right)}$$

Design a suitable compensation such that the system will have $K_v = 10$ and P.M. = 45°.

Solution

$$K_v = \lim_{s \to 0} s\,G(s) = \lim_{s \to 0} s \cdot \frac{K}{s\left(1 + \frac{1}{2}s\right)}$$

$$K_v = K$$

i.e.,

$$K = 10$$

Thus, the open loop transfer function,

$$G(s) = \frac{10}{s\,(1 + 0.5\,s)}$$

The Bode plot for the above function is plotted in Fig. Ex. 10.6.1. The frequency range selected is ω = 1 rad/sec to ω = 10 rad/sec. The phase angle is given by

$$\angle G(j\omega) = -90° - \tan^{-1}(0.5\,\omega)$$

ω (rad/sec)	1	5	10	50
$\angle G(j\omega)°$	– 116.6°	– 158.2°	– 168.7°	– 177.7°

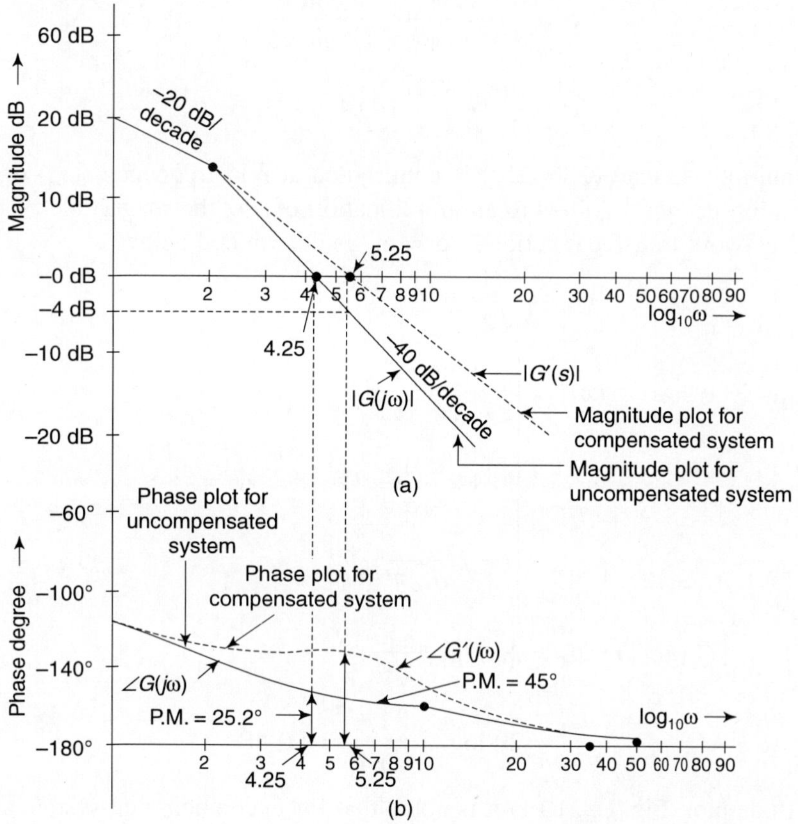

Figure Ex. 10.6.1 (a) Bode plot for gain $|G(j\omega)|$, (b) Bode plot for phase $\angle G(j\omega)$

The corner frequency is

$$\omega_{c1} = \frac{1}{0.5} = 2 \text{ rad/sec.}$$

From the Bode plot in Fig. Ex. 10.6.1 (a) and (b) it is noted that uncompensated system having a phase margin 25.2° and the gain crossover occurs at ω = 4.25 rad/sec.

To increase the phase margin to 45° the gain cross point is to be shifted to a higher value of frequency and this is possible by introducing a phase lead network. The required phase lead is

$$\phi_L = \phi_m - \phi_s + \text{tolerance}$$
$$\phi_L = 45° - 25.2° + \text{tolerance}$$

Let tolerance be 5.2° for simplicity

$$\phi_L = 45° - 25.2° + 5.2°$$
$$\phi_L = 25°$$

The value of α parameter of the phase lead network is given by

$$\alpha = \frac{1-\sin\phi_L}{1+\sin\phi_L} = \frac{1-\sin 25°}{1+\sin 25°}$$

$$\alpha = \frac{1-0.423}{1+0.423} = 0.4$$

The maximum phase lead (ϕ_L = 25°) is contributed at mid frequency (ω_m) of the phase lead compensation network. Allowing an amplification of $1/\alpha$ the magnitude of phase lead compensation network transfer function at $\omega = \omega_m$ is determined below.

$$\because \quad \omega_m = \frac{1}{\sqrt{\alpha}\,T}$$

$$\therefore \quad |G_c(j\omega)| = \left|\frac{1+j\omega_m T}{1+j\omega_m \alpha T}\right|$$

$$= \left|\frac{1+j\dfrac{1}{\sqrt{\alpha}\,T}T}{1+j\dfrac{1}{\sqrt{\alpha}\,T}\alpha T}\right|$$

$$|G_c(j\omega)| \text{ in dB} = 20\log_{10}\frac{1}{\sqrt{\alpha}}$$

$$= 20\log_{10}\frac{1}{\sqrt{0.4}} = 4.0 \text{ dB}$$

From the Bode plot (Fig. Ex. 10.1) it is noted that for uncompensated system at a gain of – 4.0 dB, the frequency is 5.25 rad/sec. The new gain crossover frequency for the compensated system is selected at $\omega = \omega_m$ = 5.25 rad/sec.

The two corner frequencies for the phase lead network are

The lower corner frequency

$$\omega_{c1} = \frac{1}{T} = \sqrt{\alpha}\,\omega_m = \sqrt{0.4}\times 5.25$$

$$\omega_{c1} = 3.3 \text{ rad/sec.}$$

The upper corner frequency

$$\omega_{c2} = \frac{1}{\alpha T} = \frac{\sqrt{\alpha}\,\omega_m}{\alpha} = 8.3 \text{ rad/sec.}$$

The transfer function of the phase lead compensation network with amplifier gain $(1/\alpha)$ is 2.5

$$G_c(j\omega) = \frac{1 + j\omega T}{1 + j\omega \alpha T}$$

$$= \frac{1 + j\omega \dfrac{1}{3.3}}{1 + j\omega \times 0.4 \times \dfrac{1}{3.3}} = \frac{(1 + j\,0.30\,\omega)}{(1 + j\,0.12\,\omega)} \quad \textbf{Ans.}$$

The open loop transfer function of the compensated system is

$$G'(s) = G_c(s)\,G(s)$$

$$= \frac{(1 + j\,0.30\,\omega)}{(1 + j\,0.12\,\omega)} \cdot \frac{10}{s\,(1 + 0.5\,s)} \quad \textbf{Ans.}$$

The plot for compensated system is shown if Fig. Ex. 10.1.

10.7 DESIGN OF CLOSED LOOP SYSTEMS USING COMPENSATION TECHNIQUES IN TIME DOMAIN

The design of closed loop systems in time domain is carried out by root locus tenchique. In this method the original design specifications, when the response is dynamic, are converted into ξ and ω_n of a pair of desired complex conjugate closed loop poles. We assume that the system will be dominated by these complex poles. Therefore, its dynamic behaviour can be approximated by second order system.

Let us consider the location of roots (i.e., complex pole pair) as shown in Fig. 10.15 (a) is desired root location and (b) is uncompensated root locus.

Now we design a compensator like that the least damped complex poles of the resulting transfer function correspond to the desired dominant poles and all other closed loop poles are either located very near to the open loop zeros or are relatively far away from the imaginary axis. It indicates that only dominant poles make contribution to the system dynamic.

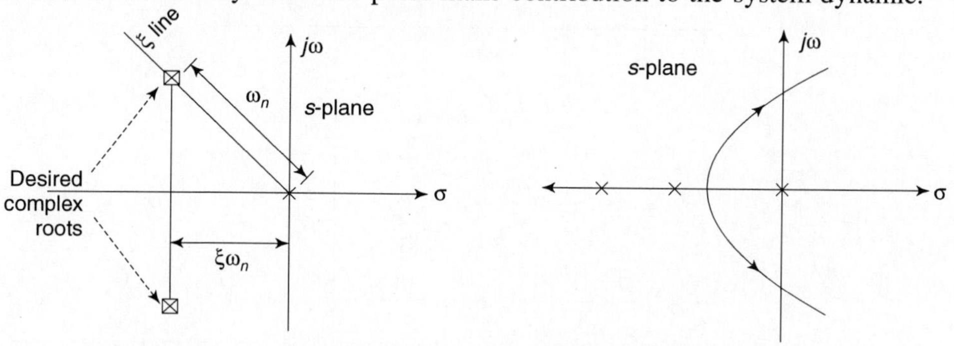

Figure 10.15 (a) Desired root location, (b) Uncompensated root locus

The method of root locus for compensating network is as follows:

Step 1: First list the system specifications then convert them into a desired roots for the dominant roots.

Step 2: Draw the root locus for uncompensated network and determine whether the desired roots can be realized with an uncompensated system.

Step 3: If a compensator is necessary, place the zero of the phase lead network below the desired root location (or to the left of the first two real poles) and for the phase lag network place the poles below the desired root location (or to the right of the first two real zeros).

Step 4: Determine the root location so that the total angle at the desired root location is 180° and therefore is on the compensated root locus.

Step 5: Calculate total system gain at the desired root location then calculate error constant.

Step 6: If error constant is not satisfactory repeat the process.

10.7.1 Lead Compensation in Time Domain

We have seen in Fig 10.15. l(b), the root locus of uncompensated network. The system is unstable so to make it stable we need some compensator network suppose we need lead compensation network, therefore we first locate our desired dominant complex closed loop poles so that the dominant poles satisfy the specification in terms of ξ and ω_n and then zero is added to provide a phase lead by placing it to the left of their first two real poles. But it must not alter the dominance of the desired poles, i.e., zero should not be placed near the origin, because it will result and will dominate the system response, as shown in Fig. 10.16(b) for the system shown in Fig. 10.16(a).

The lead compensator, $G_c(s)$, has to be so designed that the compensated root locus passes through desired location.

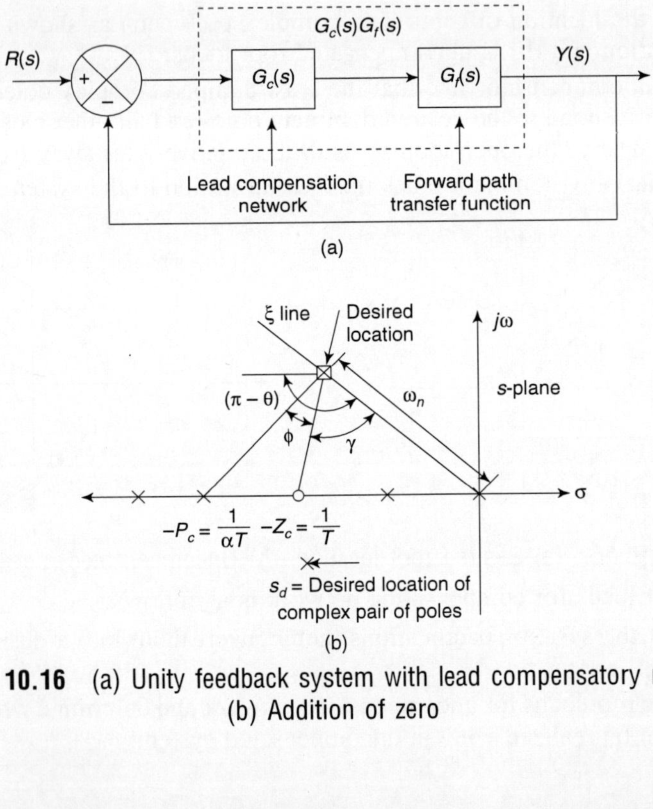

Figure 10.16 (a) Unity feedback system with lead compensatory network, (b) Addition of zero

The angle criterion must be followed, i.e.,

$$\angle G_c(s)\, G_f(s)\Big|_{s=s_d} = \angle G_c(s) + \angle G_f(s)\Big|_{s=s_d} = \pm 180°$$

i.e., $$\angle G_c(s_d) = \pm\ 180° - \angle G_f(s_d) = \phi \qquad ...(10.18)$$

Thus, for the root locus of the lead compensated network is shown in Fig. 10.17. The lead compensator pole zero pair must contribute at an angle (ϕ) given by the Eq. (10.18).

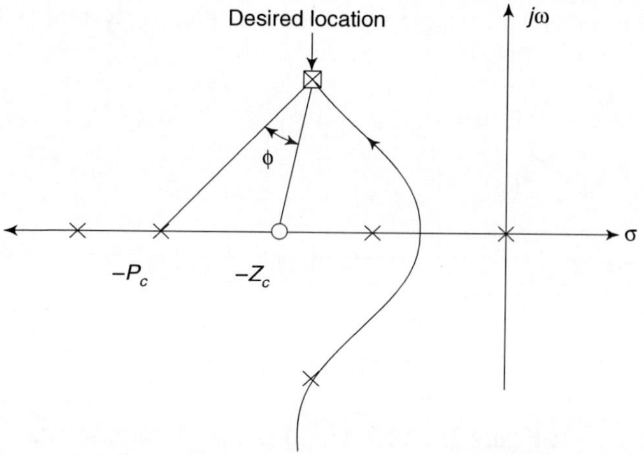

Figure 10.17 Root locus of compensated network

The location of compensatory pole and zero, i.e., P_c and Z_c is not fixed it may vary. So the best location of compensating pole and zero is one which gives largest value of $\alpha = Z_{c,}\ P_c.\ Z_c$ is located by drawing an angle (γ) by making a line between s_d and 0 location, as shown in Fig. 10.16 (b). Similarly, the pole is located by making an angle (ϕ) as shown in Fig. 10.16 (b).

By figure, we have

$$Z_c = \omega_c\left[\frac{\sin\gamma}{\sin(\pi-\theta-\gamma)}\right] \qquad ...(10.19)$$

and $$P_c = \omega_n\left[\frac{\sin(\gamma+\phi)}{\sin(\pi-\theta-\gamma-\phi)}\right] \qquad ...(10.20)$$

We have $$\alpha = \frac{Z_c}{P_c} = \frac{\omega_n\left[\dfrac{\sin\gamma}{\sin(\pi-\theta-\gamma)}\right]}{\omega_n\left[\dfrac{\sin(\gamma+\phi)}{\sin(\pi-\theta-\gamma-\phi)}\right]}$$

$$\alpha = \frac{\sin\gamma\,\sin(\pi-\theta-\gamma-\phi)}{\sin(\pi-\theta-\gamma)\,\sin(\gamma+\phi)} \qquad ...(10.21)$$

For largest α, the condition is

$$\frac{d\alpha}{d\gamma} = 0$$

Equation (10.21) gives

$$\gamma = \frac{1}{2}(\pi - \theta - \phi) \qquad ...(10.22)$$

The above method provides the best location of P_c and Z_c but it is not necessary that the dominant condition is satisfied. If it does not satisfy then readjust the location of compensatory pole and zero.

Example 10.7.1 Let us consider a single loop feedback control system as shown in Fig. Ex. 10.7.1 (a) where

$$G(s) = \frac{K}{s^2} \qquad ...(10.23)$$

When $$H(s) = 1$$

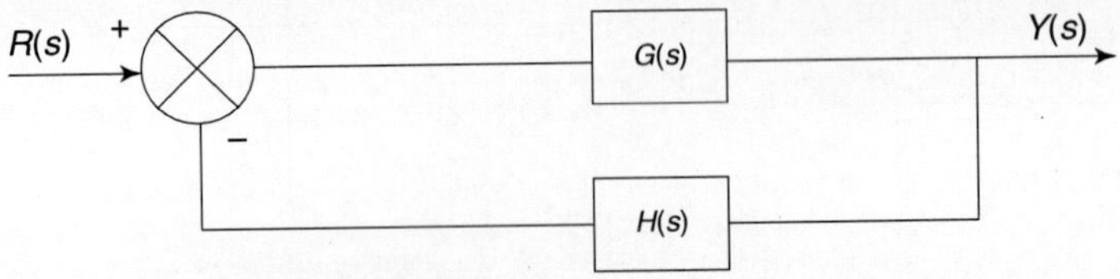

Figure Ex. 10.7.1 (a) Unity feedback system

The characteristic equation for the network

$$1 + G(s)\,H(s) = 0$$

$$1 + \frac{K}{s^2} = 0 \qquad ...(10.24)$$

The root locus of such system shows that the close loop poles always lie on the imaginary axis. Therefore, we desire to compensate this system with a lead compensation network $G_c(s)$, where

$$G_c(s) = \frac{E_o(s)}{E_i(s)} = \frac{(s + Z_c)}{(s + P_c)} \qquad ...(10.25)$$

$$|Z_c| < |P_c|$$

The specification for the system is settling time, $T_s \leq 4$ sec

The maximum overshoot for a step input $\leq 35\%$

$\because$ We have $$\%M_P = e^{-\frac{\xi\pi}{\sqrt{1-\xi^2}}} \times 100 \qquad ...(10.26)$$

So, we get the damping ratio from Eq. (10.26)

$$\xi \geq 0.32$$

The settling time required is

$$T_s = \frac{4}{\xi\,\omega_n} = 4$$

So $$\xi \omega_n = 1$$

Now we choose the desired location for dominant poles as

$$\alpha + j\omega = -1 \pm 2j \qquad ...(10.27)$$

As shown in Fig. Ex. 10.7.1 (b)

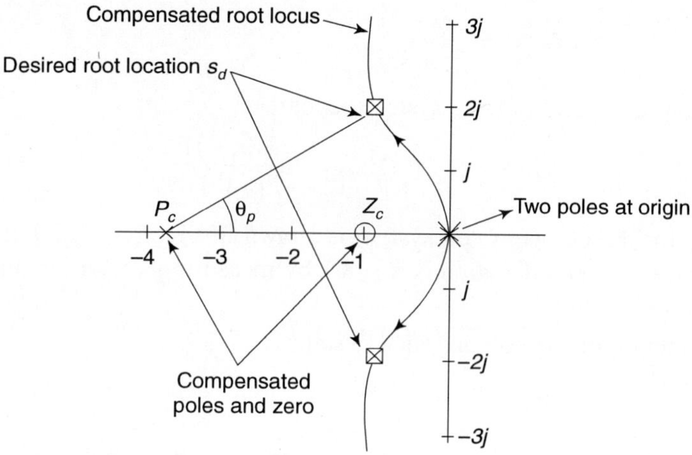

Figure Ex.10.7.1 (b) Root locus for compensated network

Now we insert the zero below the desired location as $s = -1$ as shown in Fig. Ex. 10.7.1 (b).

The angle at s_d

$$\angle G_f(s) = -2 \times 116° = -232°$$

So the angle contribution at s_d required of the lead compensator pole zero pair is

$$\phi = 180° - G_f(s_d) = -180° + 232°$$

$$\phi = 52°$$

For the largest value of α, we have

$$\gamma = -(\pi - \theta - \phi)$$

$$= \frac{1}{2}(180° - \cos^{-1} 0.32 - 52°)$$

$$\gamma = 28.3°$$

Using γ and ϕ $$Z_c = 1 = 1/R_1 C$$

$$P_c = -3.6 = \frac{1}{\left(\dfrac{R_1}{R_1 + R_2}\right) R_1 C}$$

$$\alpha = \frac{Z_c}{P_c} = \frac{1}{3.6} = 0.278 \text{ (Satisfactory)}$$

Let $c = 1\ \mu F$, we obtain

$$R_1 = 1000\ \text{K}\Omega$$

$$R_2 = 2.6\ \text{M}\Omega$$

So, the transfer function of the compensator is

$$G_c(s) = \frac{(s+1)}{(s+3.6)}$$

The overall transfer function of the system becomes

$$G(s) = G_c(s)\ G_f(s) = \frac{K\,(s+1)}{s^2\,(s+3.6)} \qquad ...(10.28)$$

The root locus for the compensated system is shown in Fig. Ex. 10.7.1 (b) by which the gain at desired close loop pole location is $K = 8.1$ by measuring vector length from the poles and zeros to the root location.

The transfer function of the compensated system is

$$G(s) = \frac{8.1\,(s+1)}{s^2\,(s+3.6)} \qquad ...(10.29)$$

The system block diagram is shown in Fig. Ex. 10.7.1 (c)

The acceleration constant is given by

$$K_a = \lim_{s \to 0} s^2\ G(s) = \frac{8.1 \times 1}{3.6}$$

$$K_a = 2.25$$

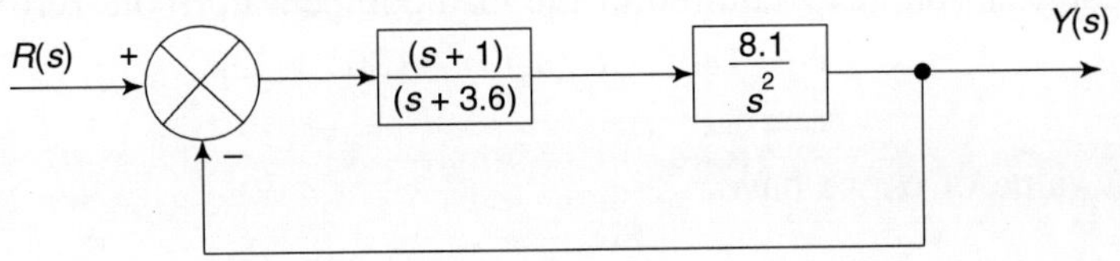

Figure Ex. 10.7.1 (c) Block diagram of compensated system

The steady state performance of this system is quite satisfactory.

Lag Compensation in Time Domain

As explained earlier the lag compensation is used to improve the steady state behaviour of a system. The steps necessary for the design of phase lag network on the *s*-plane as follows:

Step 1: Draw the root locus of uncompensated system.

Step 2: Obtain the system transient performance specifications and place the dominant roots location on the root locus of uncompensated system.

Step 3: Obtain the loop gain at the desired root location and thus the system error constant.

Step 4: Compare this uncompensated error constant with the desired error constant, and calculate the necessary increase.

Step 5: With the known α of the compensator, obtain suitable location of the pole and zero of the compensator so that the compensated root locus will pass through the desired root location.

We have seen in the previous section that the transfer function of the R_c phase lag network is of the form.

$$G_c(s) = \frac{1}{\alpha}\frac{(s+Z_c)}{(s+P_c)} \qquad ...(10.30)$$

As we have

$$Z_c = \frac{1}{T} = \frac{1}{R_2 C}$$

and

$$\alpha = \frac{R_1 + R_2}{R_2}$$

$$P_c = \frac{1}{\alpha T} = \frac{1}{\frac{(R_1+R_2)}{R_2} T}$$

The steady state error of the uncompensated system is

$$\lim_{t\to\infty} e_{ss}(t) = \lim_{s\to 0} s\,G(s)$$

$$e_{ss} = \lim_{s\to 0} s\,G(s) \qquad ...(10.31)$$

The transfer function of forward path

$$G_f(s) = \frac{K \sum_{i=1}^{m} (s+Z_i)}{S^{\gamma} \sum_{j=\gamma+1}^{n} (s+P_j)} \qquad ...(10.32)$$

At certain value of K, this system has a satisfactory transient response.

The pole zero location for lag network is shown in Fig. Ex. 10.7.1 (d).

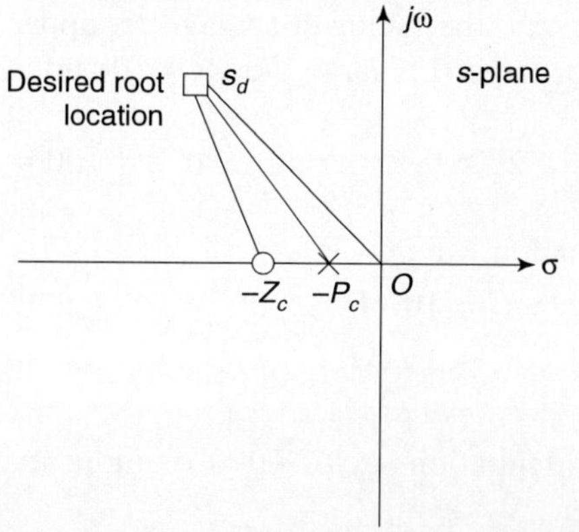

Figure Ex. 10.7.1 (d) Compensator pole zero plot

The velcoity constant

$$K_v = \frac{K \sum_{i=1}^{m} Z_i}{\sum_{j=1}^{n} P_j}$$

Now we will add lag compensator network to the uncompensated system and obtain the compensated velocity constant.

$$K_{vc} = \lim_{s \to 0} s\,\{G_c(s)\,G_f(s)\} = \lim_{s \to 0} \{G_c(s)\}\,K_{vuc}$$

$$= \left(\frac{Z_c}{P_c}\right)\left(\frac{1}{\alpha}\right) K_{vuc} = \left(\frac{Z_c}{P_c}\right)\left(\frac{K}{\alpha}\right)\frac{\sum Z_i}{\sum P_j} \quad \text{...(10.33)}$$

where K_{vc} = velocity constant for compensated network

K_{vuc} = velocity constant for uncompensated network

$\left(\frac{K}{\alpha}\right)$ is a gain on the compensated root locus at the desired root location, s_d. If Z_c and P_c are chosen so that $|Z_c| = \alpha\ |P_c| < 1$ due to this the K_v will be increased at the s_d by the ratio of $Z_c/P_c = \alpha$.

Example 10.7.2 If $Z_c = 0.1$ and $P_c = 0.01$ so the velocity constant of the desired root location (s_d) will be increased by a factor of α, i.e.,

$$\alpha = \text{—} = \frac{0.1}{0.01} = 10$$

If the compensator pole and zero, i.e., P_c and Z_c appear relatively near together so their effect on the location of desired root (s_d) will be small, i.e., negligible. Therefore, the compensator pole-zero combination near the origin compared to ω_n can be used to increase error constant of a feedback system by the factor α.

While altering the s_d slightly, the factor does have an upper limit, typically about 100, because required R and C for large α is large. For $\alpha = 100$ and $Z_c = 0.1$

$$Z_c = 0.1 = \frac{1}{R_2 C} \text{ and } \alpha = 100 = \frac{R_1 + R_2}{R_2}$$

By solving the above equation, we get

$$C = 10\ \mu F$$
$$R_1 = 1\ M\Omega$$
$$R_2 = 99\ M\Omega$$

So we can say that an attenuation, α, of 1000 or more may be obtained by utilizing pneumatic process controller.

Example 10.7.3 Consider the uncompensated system with an open loop transfer function of

$$G_f(s) = \frac{K}{s(s+2)}$$

The system is to be compensated to meet the following specifications

$$\text{Damping ratio } \xi = 0.45$$

$$\text{Velocity error constant} = 20 \text{ sec}^{-1}$$

The root locus of uncompensated system is shown in Fig. Ex. 10.7.3 (a).

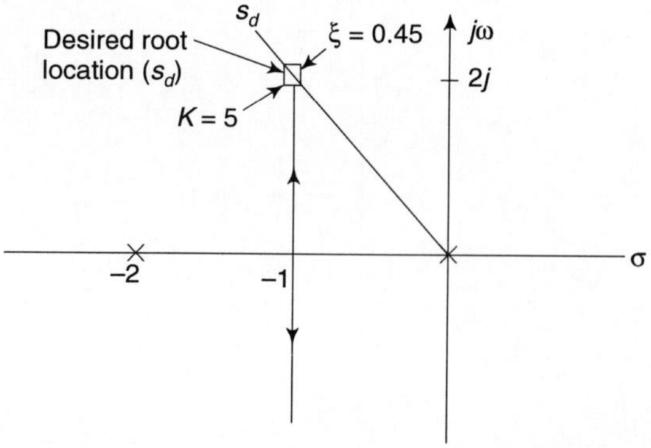

Figure Ex. 10.7.3 (a) Root locus of uncompensated system

The uncompensated root locus is a vertical line at $s = -1$ as shown in Fig. Ex. 10.7.3 (a). The desired root location (s_d) is $-1 \pm 2j$ at which $\xi = 0.45$ and the gain at this root, we have

$$K = (2.24)^2 = 5$$

The velocity constant for uncompensated system is

$$K_v = \lim_{s \to 0} s\,G(s)$$

$$= \lim_{s \to 0} s \cdot \frac{K}{s(s+2)} = \frac{K}{2}$$

$$K_v = \frac{5}{2} = 2.5$$

The required ratio α of the compensator is

$$\left|\frac{Z_c}{P_c}\right| = \alpha = \frac{K_{vc}}{K_{vuc}} = \frac{20}{2.5} = 8$$

By the root locus of Fig. Ex. 10.7.3 (b), we find that we might set $Z_c = -0.1$

$$P_c = \frac{1}{\alpha} = \frac{-0.1}{8} = 0.0125$$

The angle from P_c and Z_c at s_d is approximately 1° so the $s = -1 \pm 2j$ is still the location of the dominant roots.

The compensated system transfer function is

$$G_s(s)\,G(s) = \frac{s\,(s+0.1)}{s\,(s+2)\,(s+0.0125)}$$

where $\left(\dfrac{K}{\alpha}\right) = 5$

or $K = 40$ in order to account for the attenuation of the lag network.

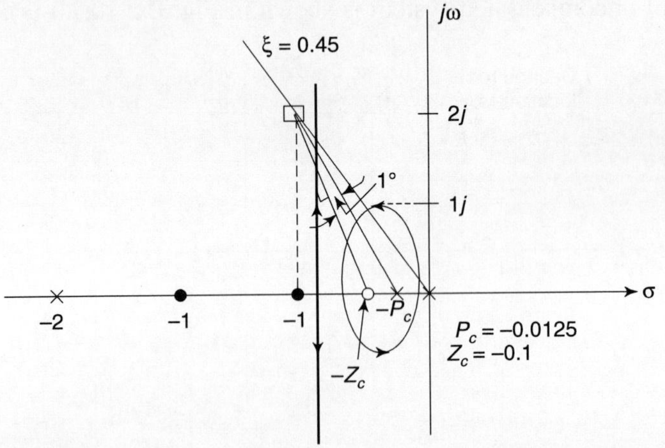

Figure Ex. 10.7.3 (b) Root locus of compensated system

Lag lead Compensation

We have already seen the advantages or use of lead and lag compensation network in which we have seen that lead network is only useful for the improving transient response of the system but limited improvement in steady state error whereas the lag network will provide improved steady state error so we can say that when we have satisfactory steady state error and only we need to improve transient response then we will prefer lead compensation network and when we have satisfactory transient response and we need only improved steady state then we will prefer lag compensation network.

But when we have both steady state error and transient response, unsatisfactory then we will use both lag and lead network simultaneously. It is known as lag-lead compensation network and will improve both transient response and steady state error of the system.

The general form of the function of the lead-lag compensator is pole zero form is

$$G_s(s) = \frac{(s+Z_{c1})\,(s+Z_{c2})}{\underbrace{(s+P_{c1})}_{\substack{\text{Lag compensation}\\ G_{c1}(s)}}\;\underbrace{(s+P_{c2})}_{\substack{\text{Lead compensation}\\ G_{c2}(s)}}} \qquad \text{...(10.34)}$$

$$\beta = \frac{Z_{c1}}{P_{c1}} = \frac{P_{c2}}{Z_{c2}} > 1$$

So $\quad G_c(s) = G_{c1}(s)\,G_{c2}(s)$

In this, generally, we will design first the lead compensation network to meet the specification of transient response. This design needs a suitable value of β. Now we will obtain the error constant for lead compensation network. If this is satisfactory then the design is complete. If it is not so and the error constant is much higher than obtained by lead compensation network then we will design lag compensation network such that the overall system meets the specifications on steady state performance. The lag section can increase the error constant by a factor β. If this increase is sufficient, we proceed to complete lag section design as usual. Otherwise, we must redesign the lead compensation section for a larger value of β.

10.8 DESIGN OF CLOSED LOOP SYSTEMS USING COMPENSATION TECHNIQUES IN FREQUENCY DOMAIN

The design of closed loop system also can be done with compensation network in frequency domain. The Bode diagram is used to design a suitable compensation network in preference to other frequency response plots.

The frequency response of the uncompensated system is added to the response of cascade compensation network.

The frequency domain specification is in the following form:

(a) Phase margin (P.M.) or resonant peak, M_r, i.e., indicates the relative stability.

(b) Bandwidth (ω_b) or resonant frequency, ω_r, i.e., indicates rise time and settling time.

(c) Error constant, i.e., indicates steady state error.

The compensation is carried out using Nyquist plot or Nichols chart. Use of Bode plot is preferred because it is easier to draw and modification in plot is comparatively easy.

When the specification is given in time domain then firstly we will convert it into frequency domain by using correlation between frequency and time domain which is already explain in chapter 3.

Now the correlations are as follows:

$$M_r = \frac{1}{2\xi\sqrt{1-\xi^2}} \quad \text{...(10.35)}$$

$$M_r = \omega_n\sqrt{1-2\xi^2} \quad \text{...(10.36)}$$

$$\phi_{PM} = \tan^{-1}\frac{2\xi}{\left[\sqrt{(1+4\xi^4)}-2\xi^2\right]^{1/2}} \quad \text{...(10.37)}$$

$$\omega_b = \omega_n\left[1-2\xi^2+\sqrt{2-4\xi^2+4\xi^2}\right]^{1/2} \quad \text{...(10.38)}$$

10.8.1 Phase Lead Compensation

The lead compensation network is used when the phase margin of the uncompensated system is unsatisfactory. Then the lead compensation is designed to meet the specified phase margin.

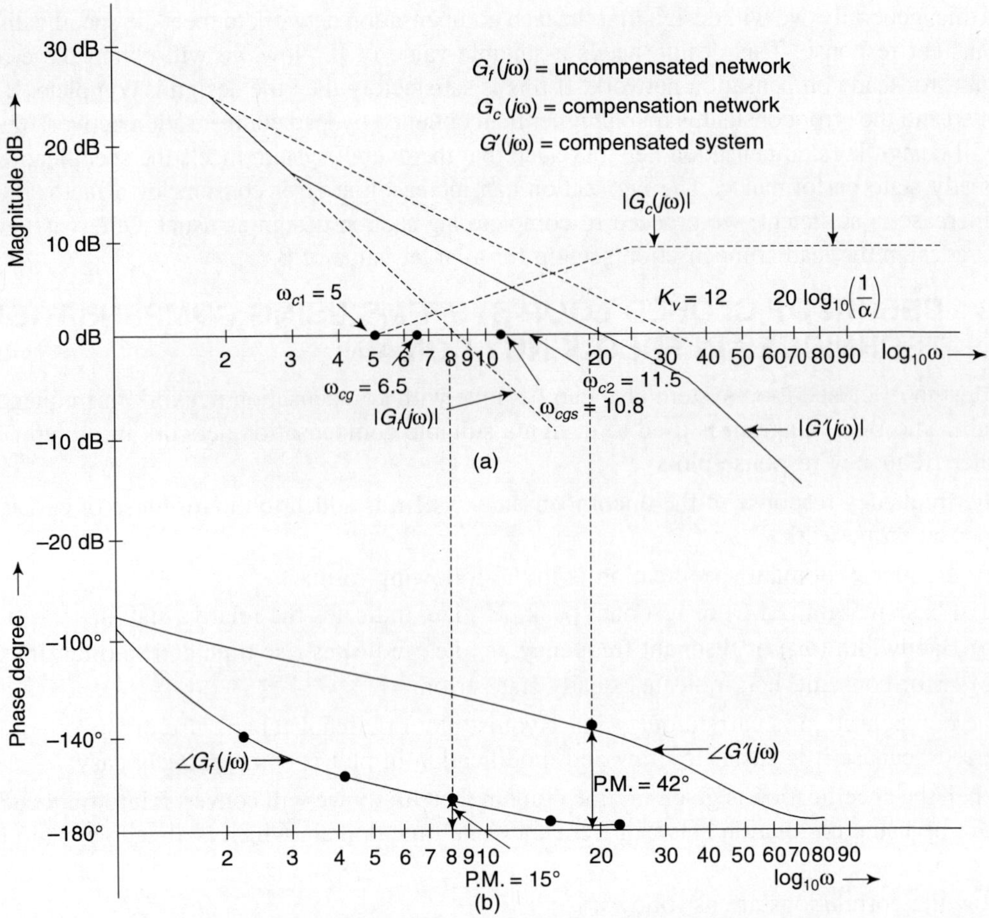

Figure 10.18 (a) Bode plot for G.M., (b) Bode plot for P.M.

Figure 10.18 shows the bode plot of unity feedback system with an open loop transfer function, $G_f(s) = \dfrac{K_v}{s\,(Ts+1)}$.

The velocity constant (K_v) is adjusted to the specified value. Consider a gain crossover frequency of uncompensated system is ω_{cg} and the phase margin of the uncompensated system is ϕ_{uc}. Let ϕ_{uc} fall short of the specified phase margin, ϕ_s. So the additional phase margin is provided by the lead compensation network, so placed that its two corner frequencies on either side of ω_{cg}. This location of the network's Bode plot ensure that the lead is provided over the desired region. As shown in Fig. 10.18 due to the addition of phase lead network, the crossover frequency (ω_{cg}) shifts to the right to some unknown value, ω_{cgs} (shifted gain crossover frequency). Thus, the required phase lead at the new crossover frequency is

$$\phi_L = \phi_s - \phi_{uc} + \varepsilon \qquad \text{...(10.39)}$$

where ϕ_L = required phase lead

ϕ_s = specified phase

ϕ_{uc} = phase margin of uncompensated system

ε = tolerance

The tolerance, generally we will consider 5° – 10° but it may be high for higher slope like 60 dB/decade, etc. We get the required phase lead ϕ_c, ϕ_L at shifted gain crossover frequency, ω_{cgs}. The frequency of maximum phase lead ω_m of the network must be made to coincide with ω_{cgs}.

Thus, we get

$$\omega_{cgs} = \omega_m$$

$$\therefore \qquad \phi_m = \phi_L$$

The parameter (α) can be obtained

$$\alpha = \frac{1-\sin\phi_L}{1+\sin\phi_L}$$

The corner frequencies of the lead network can be calculated as

The lower corner frequency

$$\omega_{c1} = \frac{1}{T} = \omega_m\sqrt{\alpha}$$

The upper corner frequency

$$\omega_{c2} = \frac{1}{\alpha T} = \omega_m\sqrt{\alpha}$$

Procedure for deisgn a system

1. Obtain the loop gain (K) to satisfy error constant.
2. Obtain the phase margin (ϕ_{uc}) for uncompensated system.
3. Determine the phase lead required with allowing a small amount of safety (tolerance)
$$\phi_L = \phi_s - \phi_{uc} + \varepsilon$$
4. Consider $\phi_L = \phi_m$ and obtain parameter α
$$\alpha = \frac{1-\sin\phi_m}{1+\sin\phi_m}$$
5. Evaluate 10 log (1/α) at ω_m. Locate that frequency at which the uncompensated system has a gain – 10 log (1/α). This is $\omega_{cgs} = \omega_m$.
6. Calculate the corner frequencies
$$\omega_{c1} = \frac{1}{T} = \omega_m\sqrt{\alpha}$$
$$\omega_{c2} = \frac{1}{\alpha T} = \omega_m/\sqrt{\alpha}$$

7. Draw the Bode plot for magnitude and phase of the compensated system and check resulting P.M., if P.M. is low then increase the value of ε and repeat from step 3.

Example 10.8.1 The example based on phase lead network is already done in Ex. 10.7.2.

Phase Lag Compensation

The design of phase lag network which is suitable for compensating a feedback system can be readily accomplished on the Bode diagram. In Bode diagram we will simply add the frequency response of the compensator to the Bode diagram of the uncompensated system to obtain satisfactory system frequency response.

The transfer function of phase lag compensation network is $G_c(j\omega)$.

$$G_c(j\omega) = \frac{(1 + j\omega T)}{(1 + j\omega \alpha T)} \qquad ...(10.40)$$

The Bode diagram of the phase lag network is shown in Fig. Ex. 10.8.1 for two values of α. On the Bode diagram, the magnitude of the pole and zero of the compensator is much smaller than the smallest pole of the uncompensated system. So we can say that the phase lag is not the useful effect of the compensator; it is the attenuation – 20 log α that is the useful effect for compensation. The phase lag network is used to provide an attenuation and therefore to lower the crossover frequency of the system. At lower crossover frequency, we usually find increased phase margin of the system.

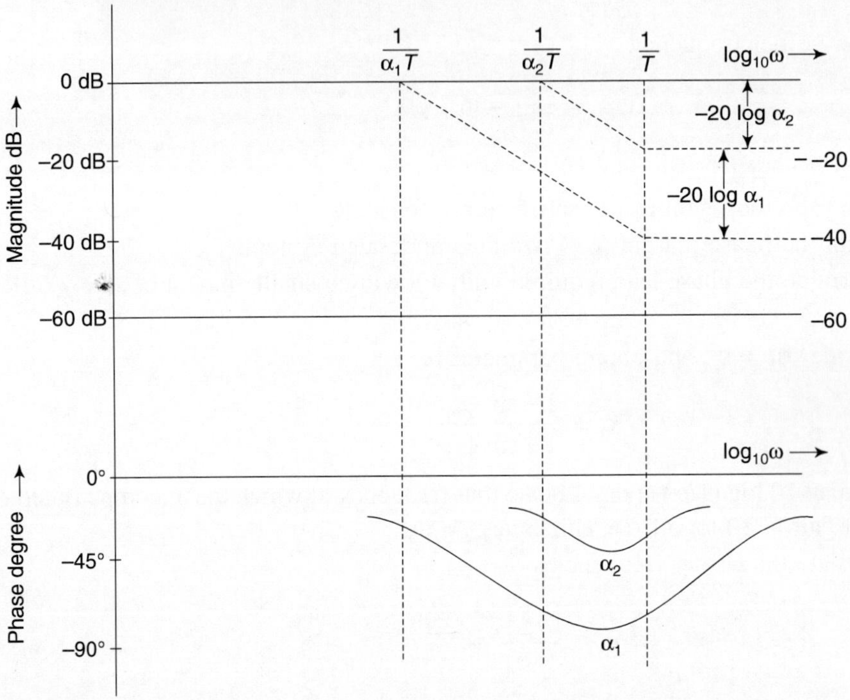

Figure Ex. 10.8.1 Bode plot for phase lag network at two values of α

The design procedure for phase lag network.

1. Obtain the Bode plot for uncompensated system and determine the phase margin of the system and compare it with desired P.M.
2. Obtain the gain crossover frequency (ω_{cg}) where the phase margin is satisfied (consider 5° tolerance for phase lag network).
3. Place the zero of the compensator one decade below the new gain crossover frequency ω_{cgs}, i.e., shifted gain crossover frequency.
4. Measure the necessary attenuation at ω_{cgs} to ensure that the magnitude curve crosses at this frequency.
5. Calculate α by the attenuation – 20 log α at ω_{cgs}.
6. Calculate the pole, $\omega_p = \dfrac{1}{\alpha T} = \dfrac{\omega_2}{\alpha}$

Example 10.8.2 Let us consider the system, whose open loop transfer function is

$$G_f(s) = \frac{K}{s\,(s+1)\,(s+2)}.$$

The system is to be compensated to meet the following specifications.

Damping ratio $\xi = 0.5$

Settling time $t_s = 10$ sec

Velocity error constant, $K_v \geq 5\ \text{sec}^{-1}$

With the help of Eqs. 10.37 and 10.38

$$t_s = \frac{3}{\xi\,\omega_n} \Rightarrow \omega_n = \frac{3}{\xi\,t_s}$$

$$\omega_n = \frac{3}{0.5 \times 10} = 0.6 \text{ rad/sec}$$

$$\omega_b = \omega_n\left[1 - 2\xi^2 + \sqrt{(2 - 4\xi^2 + 4\xi^2}\,\right]^{1/2}$$

$$= 0.6\left[1 - 2 \times 0.5^2 + \sqrt{2}\right]^{1/2}$$

$$= 0.6\ [1 - 0.5 + 1.41]^{1/2}$$

$$= 0.6\ [1.91]^{1/2}$$

$$\omega_b = 0.83 \text{ rad/sec}$$

$$\phi_m = \tan^{-1}\left\{\frac{2\,\xi}{\left[\sqrt{(1 + 4\,\xi^4)} - 2\,\xi^2\right]^{1/2}}\right\}$$

$$= \tan^{-1}\left[\frac{2 \times 0.5}{\left[\sqrt{(1 + 0.5^4 \times 4)} - 2 \times 0.5^2\right]^{1/2}}\right]$$

$$= \tan^{-1}\left\{\frac{1}{\left[\sqrt{1.25}-0.5\right]^{1/2}}\right\} = \tan^{-1}\left\{\frac{1}{0.786}\right\} = \tan^{-1}[1.27]$$

$$= 51.8°$$

For 5% tolerance band, the specified phase margin is

$$\phi_m = 47°$$

Now the open loop transfer function of the uncompensated system is

$$G_f(s) = \frac{K/2}{s(s+1)(0.5s+1)}$$

The specification on K_v is met by choosing $K = 20$

Thus, $$G_f(j\omega) = \frac{10}{j\omega(j\omega+1)(j0.5\omega+1)}$$

The Bode plot of $G_f(j\omega)$ is shown in Fig. Ex. 10.6 from which it is found that the crossover frequency, $\omega_{cg} = 2.42$ rad/sec, and the phase margin, $(\phi_{uc}) = -28°$ so the uncompensated system is unstable.

It is further seen from the Bode plot that neglecting the phase lag contribution of the network, the specified phase margin of 47° is obtained, if the crossover frequency is 0.527 rad/sec. Since this is fairly low, the upper network corner frequency cannot be taken to its left in order to avoid large time constants. This indicates that phase lag contribution of the network at the new crossover frequency will be considerable and may be guessed at 12°. The uncompensated system must, therefore, make a phase margin contribution of

$$\phi_{La} = \phi_m + \varepsilon$$

$$\phi_{La} = 47° + 12° = 59°$$

at ω_{cs} which is found to be 0.37 rad/sec placing the upper corner frequency of the compensator two octaves below ω_{cgs}, we have

$$\omega_{c2} = \frac{1}{T} = \frac{\omega_{cgs}}{(2)^2} = \frac{0.37}{4}$$

$$\omega_{c2} = 0.09 \text{ rad/sec}$$

To bring the log-magnitude curve down to 0 dB at ω_{c2}, the lag network must provide an attenuation of 20 dB, therefore

$$20 \log \alpha = 20 \qquad \alpha = 10$$

The lower corner frequency of the network is

$$\omega_{c1} = \frac{1}{\alpha T} = \frac{0.09}{10\alpha} = 0.09 \text{ rad/sec}$$

The transfer function of the lag network is

$$G_c(s) = \frac{1}{10}\left(\frac{s+0.09}{s+0.009}\right)$$

$$G_c(s) = \frac{(11\,s+1)}{(111\,s+1)}$$

Phase lag introduced by the lag network at ω_{c2} is

$$\text{Phase lag} = \tan^{-1}(11\,\omega_{c2}) - \tan(111\,\omega_{c1})$$
$$= \tan^{-1}(11 \times 0.37) - \tan^{-1} \times (111 \times 0.37)$$
$$= 76.2° - 88.6° = 12°$$

Therefore safety margin of $\varepsilon = 12°$ is satisfied. The open loop transfer function of the compensated system becomes

$$G(s) = \frac{10\,(11\,s+1)}{s\,(s+1)\,(0.5\,s+1)\,(111\,s+1)}$$

The bode plot of $G(j\omega)$ is shown in Fig. Ex. 10.8.2 from where the phase margin of compensated system is 47°.

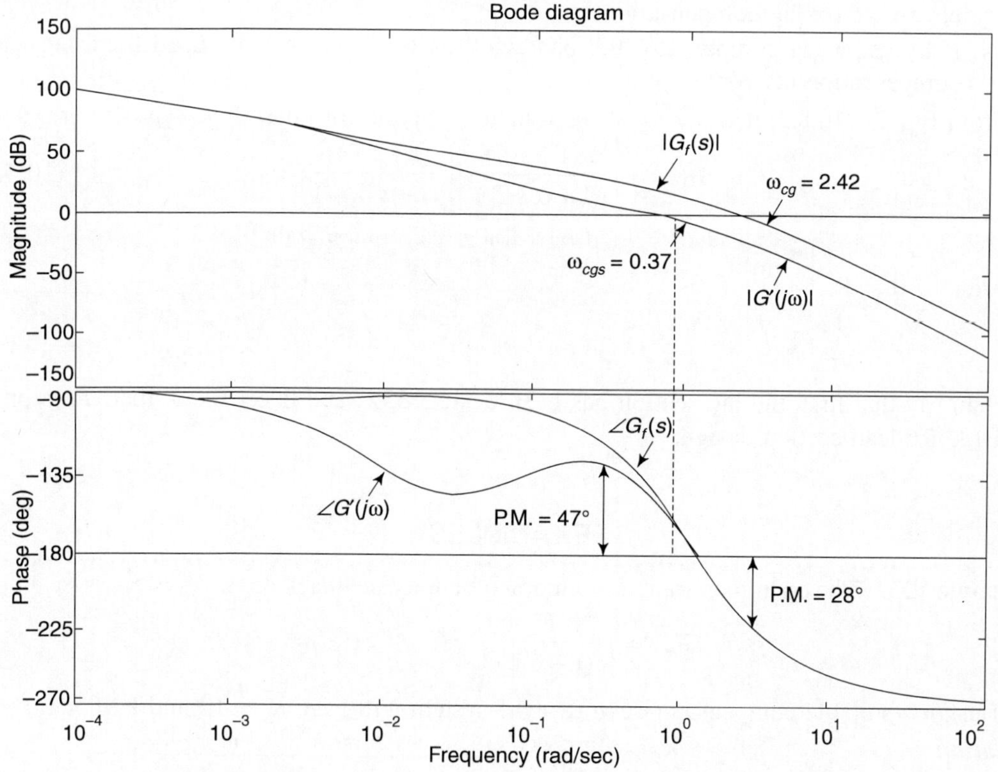

Figure Ex. 10.8.2 Design of lag compensator for type 1 system

Phase Lead Lag Compensation

As we have discussed earlier that lead compensation network is used with the uncompensated system to improve the phase margin and it also speeds up the response and increases the bandwidth. For higher order systems and those have large error constants, large leads are required for compensation due to this it increases excessively large bandwidth which is not desirable for the purpose of noise transmission. For such type of systems those have large error constants, lag compensation is preferred provided the uncompensated system is not absolutely unstable.

The use of large compensation reduces the system's bandwidth and speeds down the response. So we can say that for large error constant and moderately large bandwidth, it may not be possible to meet the specification through either lag or lead compensation only. In this condition we can use a lag-lead compensator in which lag section is used to deal with error constant, and lead section is used to deal with phase margin and give desired bandwidth. Procedure to designed lead-lag compensation:

1. First check the P.M. and bandwidth of uncompensated system for specified error constant.
2. Compare the bandwidth of uncompensated system with specified bandwidth.
3. If obtained bandwidth < specified bandwidth, lead compensation may be used. If it is not so we try lag compensation.
4. If by using lag compensator the bandwidth is too low. Then its need for a lag-lead compensation network.

From Fig Ex. 10.8.2, it is seen that the transfer function for phase lead lag network is

$$G_c(s) = \left[\underbrace{\frac{\alpha(1+sT)_1}{(1+\alpha s T_1)}}_{\text{Lead section}}\right]\left[\underbrace{\frac{(1+sT)_2}{(1+s\beta T_2)}}_{\text{Lag section}}\right] \quad \text{...(10.41)}$$

where $\alpha < 1$

$$\beta = \frac{1}{\alpha} > 1$$

We can say that first, the lag section has been designed, T_2 and β get fixed, then T_1 is only variable for lead section design.

EXAMPLES

Example 10.1 The open loop transfer function of a unity feedback control system is given by

$$G(s) = \frac{K}{s(1+0.2s)}$$

Design a suitable compensator such that the system will have $K_v = 10$ and P.M. = 60°.

Solution

$$K_v = \lim_{s\to 0} s\,G(s)$$

$$= \lim_{s\to 0} s\cdot\frac{K}{s(1+0.2s)} = K$$

$$\therefore \quad K_v = K = 10$$

The Bode plot for the above transfer function is plotted in Fig. Ex. 10.1.

The corner frequency is

$$\omega_{c1} = \frac{1}{T_1} = \frac{1}{0.2} = 5 \text{ rad/sec}$$

As the system is type 1, the inifital slope of the Bode plot is 20 dB/decade and the initial ordinate is

$$20 \log_{10} (10) = 20.0 \text{ dB}$$

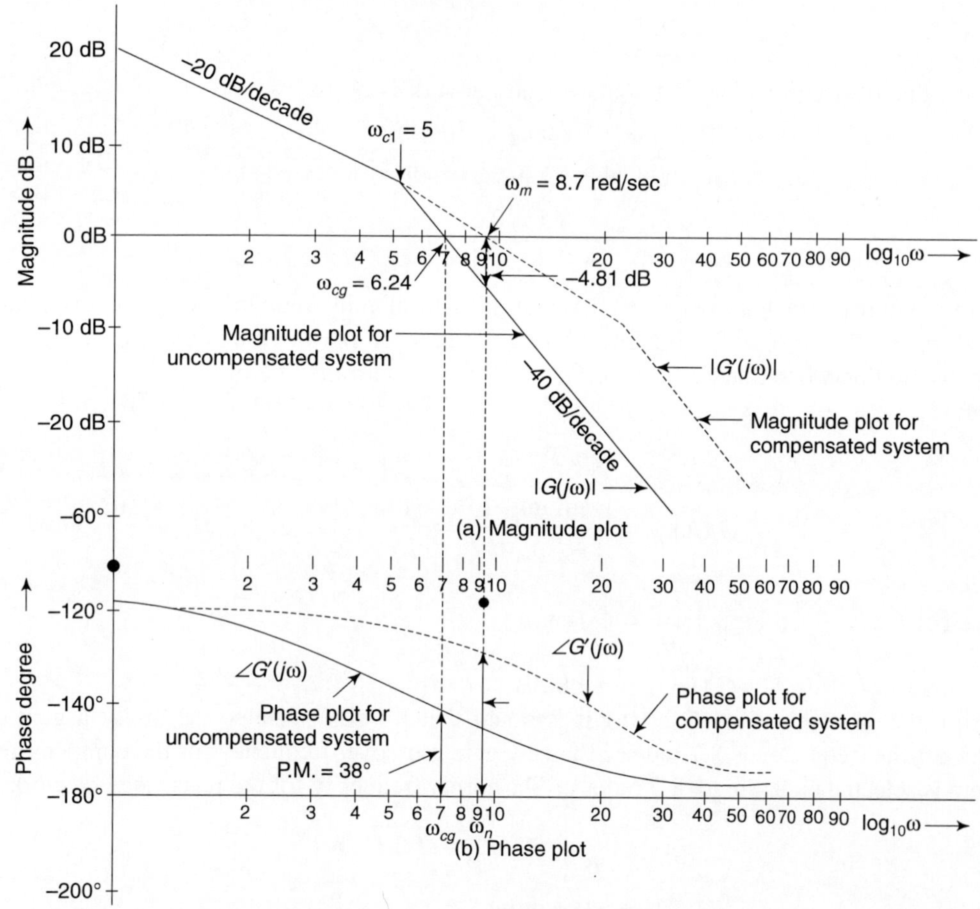

Figure Ex. 10.1 Bode plot for Ex. 10.1

The slope of the maganitude plot after ω = 5 rad/sec is [– 20 + (– 20)] dB/decade

$$= -40 \text{ dB/decade}$$

The frequency range selected is ω = 1 rad/sec to ω = 10 rad/sec. The phase angle is

$$\angle G(j\omega) = -90° - \tan^{-1} (0.2\,\omega)$$

ω	1	5	10	50
$\angle G(j\omega)$	– 101.3	– 135	– 153.3	– 177.3

From the Bode plot it is noted that uncompensated system is stable having a phase margin of 38° and the gain crossover occurs at $\omega = 6.24$ rad/sec.

To increase the phase margin to 60° the gain crossover point is to be shifted to a higher value of frequency and this is possible by introducing a phase lead compensation network. The required phase lead is

$$\phi_L = \phi_m - \phi_s + \text{tolerance}$$
$$\phi_L = 60^\circ - 38^\circ + 8^\circ$$
$$\phi_L = 30^\circ$$

The value of α parameter of the phase lead network is given by

$$\alpha = \frac{1-\sin\phi_L}{1+\sin\phi_L} = \frac{1-\sin 30^\circ}{1+\sin 30^\circ}$$
$$= \frac{1-0.5}{1+0.5} = \frac{0.5}{1.5} = 0.33$$

The maximum phase lead ($\phi_L = 30°$) is contributed at mid-frequency (ω_m) of the phase lead compensation netowrk. Allowing an amplification of $1/\alpha$ the magnitude of phase lead compensation network transfer function at $\omega = \omega_m$ is determined below.

$$\omega_m = \frac{1}{\sqrt{\alpha}\,T}$$

$$G_c(j\omega_m) = \left|\frac{1+j\omega_m T}{1+j\omega_m\,\alpha T}\right| = \frac{1}{\sqrt{\alpha}}$$

in decibel $\qquad 20\log_{10}\frac{1}{\sqrt{\alpha}} = G_c(j\omega_m)$

$$G_c(j\omega_m) = -4.8 \text{ dB}$$

From the Bode plot Fig. Ex. 10.1 it is noted that for compensated system at a gain of – 4.81 dB the frequency is 8.7 rad/sec the new gain crossover frequency for the compensated system is selected at $\omega = \omega_m = 8.7$ rad/sec. The corner frequency for the phase lead network is

$$\omega_{c1} = \frac{1}{T} = \sqrt{\alpha}\,\omega_m = \sqrt{0.33} \times 8.7$$
$$= 5.0 \text{ rad/sec}$$

The transfer function of the phase lead compensation network with amlifier gain $(1/\alpha) = 3.0$ is

$$G_c(j\omega) = \frac{1+j\omega T}{1+j\,\omega\,\alpha T} = \frac{1+j\omega\frac{1}{5}}{1+j\omega\times 0.33\times\frac{1}{5}}$$

$$= \frac{(1+j\,0.2\,\omega)}{(1+j\,0.07\,\omega)} \quad \textbf{Ans.}$$

The open loop transfer function of the compensated system is

$$G'(S) = G_c(s)\, G(s)$$

$$G'(S) = \frac{(1 + 0.2\, s)}{(1 + 0.07\, s)} \times \frac{10}{s\,(1 + 0.2\, s)}$$

Example 10.2 The open loop transfer function, $G(s) = \dfrac{K}{s\,(s+1)\,(0.5\,s+1)}$.

Compensate the system such that, $K_v = 5\ \text{sec}^{-1}$ and phase margin is at least 40° and the gain margin is at least 10 dB with lag compensator. *(U.P.T.U., 2002–03, ACS)*

Solution

$$K_v = \lim_{s \to 0} s\, G(s) = \lim_{s \to 0} \frac{1}{s\,(s+1)\,(0.5\,s+1)}$$

$\therefore$

$$K_v = K = 5\ \text{sec}^{-1}$$

Now

$$G(s) = \frac{5}{s\,(s+1)\,(0.5\,s+1)}$$

Putting

$$s = j\omega$$

$$G(j\omega) = \frac{5}{j\omega\,(j\omega+1)\,(0.5\,j\omega+1)}$$

The corner frequencies are

$$\omega_{c1} = \frac{1}{T_1} = \frac{1}{1} = 1 \text{ rad/sec}$$

$$\omega_{c2} = \frac{1}{T_2} = \frac{1}{0.5} = 2 \text{ rad/sec}$$

As the system is type 1, the initial slope of the bode plot is 20 dB/decade and the initial ordinate is

$$-\,20 \log_{10} (5) = 13.9 \text{ dB}$$

The slope of the magnitude after first corner frequency, i.e., 1 rad/sec is due to the simple pole $(j\omega + 1)$ is [– 20 + (– 20)] dB/decade = – 40 dB/decade.

The slope after second corner frequency, i.e., 2 rad/sec is due to the pole $(0.5\, j\omega + 1)$ is

$$= [-\,40 + (-\,20)] \text{ dB/decade}$$

$$= -\,60 \text{ dB/decade}$$

The phase angle is

$$\angle G(j\omega) = -\,90° - \tan^{-1} \omega - \tan^{-1} (0.5\quad)$$

For varying ω

ω :	0	1	2	5	10
$\angle G(j\omega)$:	– 90°	– 161.6°	– 198.4°	– 236.8°	– 252.9°

The Bode plot is shown in Fig. Ex. 10.2.

From the Bode plot in Fig. Ex. 10.2 the phase margin is found to be – 13°, which means that the system is unstable. With a desired P.M. = 40°, the gain crossover frequency must be shifted to a lower value and this is possible by using a phase lag compensation network.

However, the use of phase lag network will reduce the phase angle to some extent at a new gain crossover frequency, therefore, some tolerance, say 5°, is taken into consideration, thus phase angle of the compensated system at new gain crossover frequency 40° + 5° = 45°.

For the compensated system at ω = 0.56 rad/sec the phase angle is 45°. The new gain crossover frequency is selected as ω = 0.56 rad/sec. At this new gain crossover frequency the magnitude of the uncompensated system is 17 dB, which is to be neutralized by phase lag network, therefore, at real gain crossover.

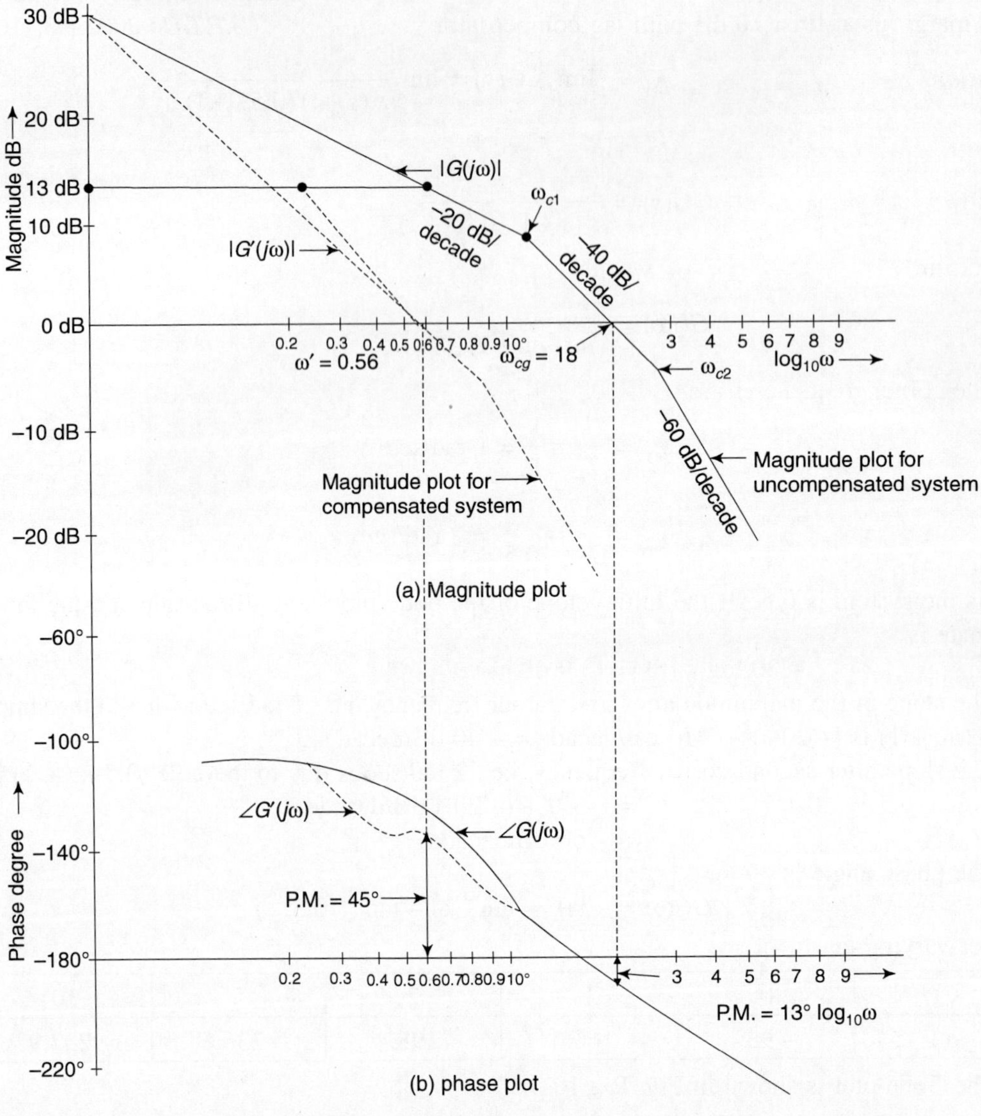

Figure Ex. 10.2 Bode plot of Ex. 10.2

$$- 20 \log_{10} \beta = - 17 \text{ dB}$$
$$\beta = 7.08$$

The upper corner frequency of the phase lag compensation network is selected one decade earlier than the new gain crossover frequency, hence the upper corner frequency for the phase lag network is

$$\omega' = \frac{0.56}{10} = 0.056 \text{ rad/sec}$$

Therefore, the upper corner frequency

$$= \frac{1}{T} = 0.056 \text{ rad/sec}$$

The lower crossover frequency

$$= \frac{1}{\beta T} = 0.048 \text{ rad/sec}$$

$$= \frac{0.056}{7.08} = 0.007 \text{ rad/sec}$$

The transfer function of the phase lag network is

$$G_c(j\omega) = \frac{(1 + j\omega T)}{(1 + j\omega \beta T)} = \frac{\left(1 + j\omega \dfrac{1}{0.056}\right)}{\left(1 + j\omega \times 7.09 \times \dfrac{1}{0.056}\right)}$$

$$= \frac{(1 + j\,17.86\,\omega)}{(1 + j\,126.6\,\omega)} = \frac{(1 + j\,17.86\,\omega)}{(1 + j\,126.6\,\omega)}$$

The open loop transfer function of the compensated system is

$$G'(s) = G_c(s)\, G(s)$$

$$G'(s) = \frac{1 + j\,17.8\,(\omega)}{1 + j\,126.6\,(\omega)} \times \frac{5}{s\,(s + 1)\,(0.5\,s + 1)}$$ **Ans.**

Example 10.3 The open loop transfer function of a unity feedback control system is given by $G(s) = \dfrac{K}{s\,(1 + 0.5\,s)\,(1 + 0.2\,s)}$ it is desired that

(i) for a unit ramp input the steady state error of the output position be less than 0.125°/ degree/sec

(ii) P.M. $\geq 30°$

(iii) G.M. ≥ 10 dB

Design a suitable compensation network.

Solution

$$K_v = \frac{\text{Desired output velocity}}{\text{Steady state error}}$$

$$K_v = \frac{1}{0.125} = 8 \text{ sec}^{-1}$$

$$K_v = \lim_{s \to 0} s\,G(s)$$

$$= \lim_{s \to 0} s \cdot \frac{K}{s\,(1+0.5\,s)\,(1+0.2\,s)}$$

$$K_v = K = 8 \text{ sec}^{-1}$$

So the open loop transfer function for the system is

$$G(s) = \frac{8}{s\,(1+0.5s)\,(1+0.2\,s)}$$

The corner frequencies are

$$\omega_{c1} = \frac{1}{T_1} = \frac{1}{0.5} = 2 \text{ rad/sec}$$

$$\omega_{c2} = \frac{1}{T_2} = \frac{1}{0.2} = 5 \text{ rad/sec}$$

As the system is type 1, the initial slope of the bode plot is – 20 dB/decade and the initial ordinate is

$$20 \log_{10} 8 = 18 \text{ dB}$$

The slope of the magnitude after ω = 2 rad/sec is due to the pole (1 + 0.5 s) is [– 20 + (– 20)] dB/decade.

$$= -40 \text{ dB/decade}$$

The slope after second corner frequency, i.e., 5 rad/sec is due to the pole (1 + 0.2 s) is

$$= [-40 + (-20)] \text{ dB/decade}$$
$$= -60 \text{ dB/decade}$$

The phase angle is given by

$$\angle G(j\omega) = -90° - \tan^{-1}(0.5\,\omega) - \tan^{-1}(0.2\,\omega)$$

ω :	0.1	1	2	5	10
$\angle G(j\omega)$:	– 74°	– 127.1°	– 156.8°	– 203.1°	– 232°

The Bode plot for the above transfer function is shown in Fig. Ex. 10.3.

From the Bode plot it is observed that the gain crossover and phase crossover frequencies are very near to each other as such the system is nearly marginally stable, to make the system relatively more stable, i.e., with a desired P.M. ≥ 30°, the gain crossover frequency must be shifted to a lower value and this is possible by using a phase lag compensation network.

However, the used phase lag network will reduce the phase angle to some extent at new gain crossover frequency, therefore, some tolerance, say 5°, taken into consideration, thus phase angle of the compensated system at new gain crossover frequency is 30° + 5° = 35°.

For the compensated system at ω = 1.6 rad/ sec the phase angle is 35°. The new gain crossover frequency is selected as ω_{cg}^1 = 1.6 rad/sec. At this new gain crossover frequency the magnitude of the uncompensated system is 11 dB which is to be neutralized by phase lag network, therefore, at new gain crossover frequency

$$-20 \log_{10} \beta = -11 \text{ dB}$$

$$\beta = 3.55$$

The upper corner frequency

$$= \frac{0.56}{10} = 0.56 \text{ rad/sec}$$

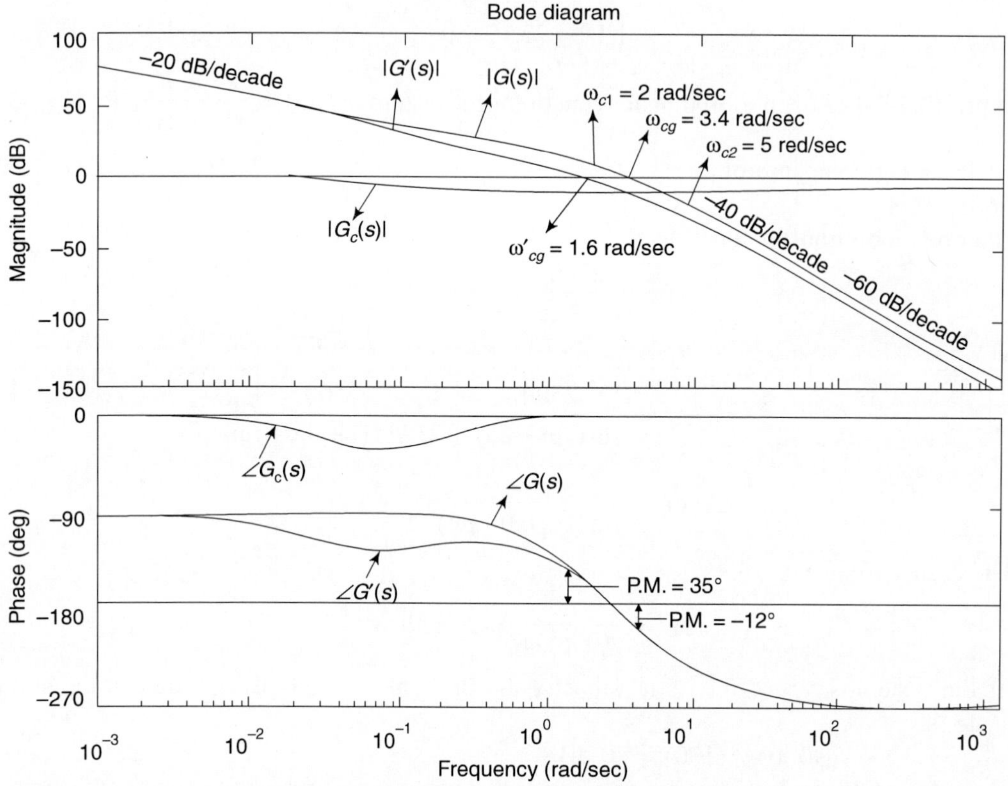

Figure Ex. 10.3 Bode plot for Ex. 10.3

Therefore, the upper corner frequency

$$= \frac{1}{T} = \frac{1.6}{10} = 0.16 \text{ rad/sec}$$

The lower crossover frequency

$$= \frac{1}{\beta T} = \frac{0.16}{3.55} = 0.045 \text{ rad/sec.}$$

The transfer function of the phase lag network is

$$G_c(j\omega) = \frac{(1+j\omega T)}{(1+j\omega \beta T)} = \frac{1+j\omega \dfrac{1}{0.16}}{1+j\omega \times 3.5 \times \dfrac{1}{0.16}}$$

$$= \frac{(1+j\,6.25\,\omega)}{(1+j\,22.19\,\omega)}$$

The open loop transfer function of the compensated system is

$$G'(s) = G_c(s)\,G(s)$$

$$G'(s) = \frac{(1+j\,6.25\,\omega)}{1(1+j\,22.19\,\omega)} \times \frac{8}{s\,(1+0.5s)\,(1+0.2s)}$$

Example 10.4 The open loop transfer function of a system is $G(s) = \dfrac{2500\,K}{s(s+2s)}$, P.M. $\geq 45°$.

Design a lag compensator. *(U.P.T.U., CO, ACS)*

Solution Let $K = 1$

The open loop transfer function is

$$G(s) = \frac{2500}{s\,(s+25)}$$

Putting $s = j\omega$

$$G(j\omega) = \frac{2500}{j\omega\,(j\omega+25)} = \frac{2500}{25\,j\omega\,(1+0.04\,j\omega)}$$

$$G(j\omega) = \frac{100}{j\omega\,(1+0.04\,j\omega)}$$

The corner frequency is

$$\omega_c = \frac{1}{T} = \frac{1}{0.04} = 25 \text{ rad/sec}$$

As the system is type 1, the initial slope of the Bode plot is – 20 dB/decade and the initial ordinate is

$$-\,20\log_{10}(100) = 40 \text{ dB}$$

The slope of the magnitude after first corner frequency, i.e., 25 rad/sec is due to the simple pole (1 + 0.04 s) is [– 20 + (– 20)] dB/decade

The phase angle is

$$\angle G(j\omega) = -\,90° - \tan^{-1}(0.04\,\omega)$$

ω :	1	5	10	25	50
$\angle G(j\omega)$:	– 92.2°	– 101.3°	– 111.8°	– 135°	– 153.4°

The Bode plot is shown in Fig. Ex. 10.4 . From the Bode plot the phase margin is found to be 28° at gain crossover frequency is 46.5 rad/sec with a desired P.M. = 45°, the gain crossover frequency must be shifted to a lower value and this is possible by using a phase lag compensation network.

For the compensated system at ω = 25 rad/sec the phase angle is 45°. So the new gain crossover frequency is selected as ω_{cg}^{1} = 25 rad/sec. At this new gain crossover frequency the magnitude of the uncompensated system is 9 dB, which is to be neutralized by phase lag network, therefore at new gain crossover.

$$-20 \log_{10} \beta = -9$$

$$\beta = 2.82$$

The upper corner frequency

$$= \frac{0.56}{10} = 0.056 \text{ rad/sec}$$

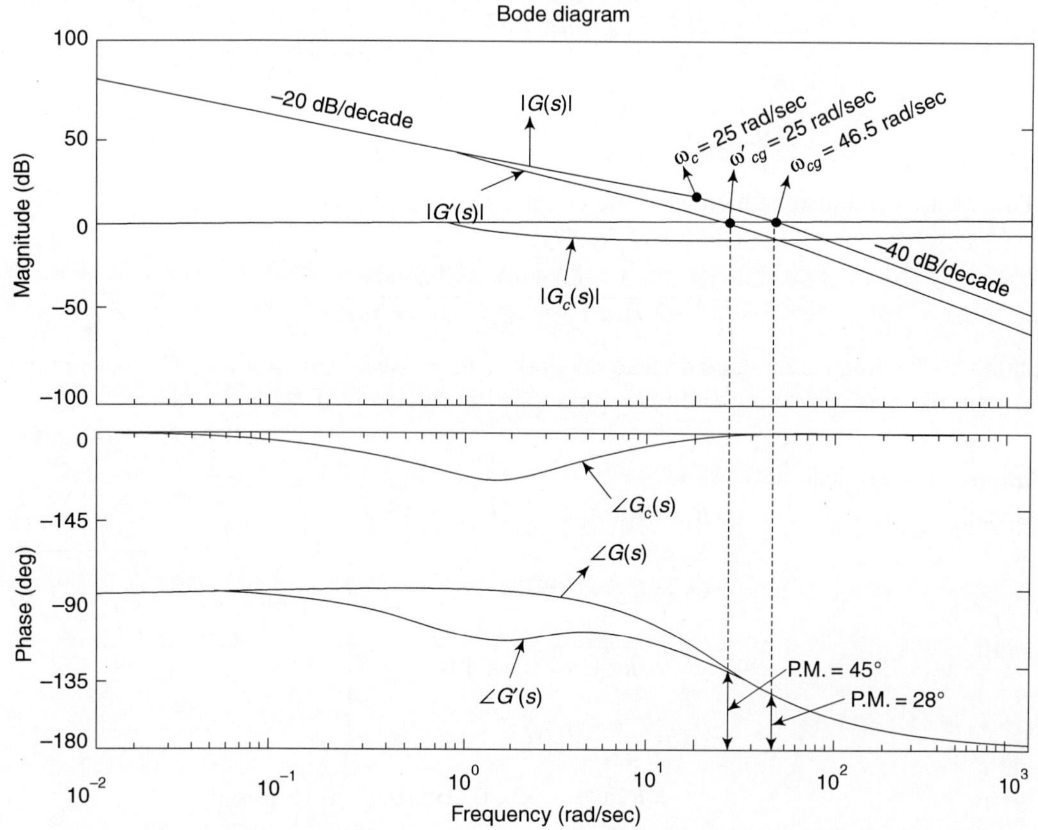

Figure Ex. 10.4 Bode plot of Ex. 10.4

The upper corner frequency of the phase lag compensation network is selected one decade earlier than the new gain crossover frequency, hence the upper corner frequency for the phase lag network is

$$\omega' = \frac{25}{10} = 2.5 \text{ rad/sec } 10$$

So the upper corner frequency

$$= \frac{1}{T} = 2.5 \text{ rad/sec}$$

The lower corner frequency

$$= \frac{1}{\beta T} = \frac{2.5}{2.82} = 0.89 \text{ rad/sec}$$

The transfer function of the phase lag network is

$$G_c(j\omega) = \frac{(1 + j\omega T)}{(1 + j\omega \beta T)} = \frac{\left(1 + j\omega \frac{1}{2.5}\right)}{\left(1 + j\omega \times 2.82 \times \frac{1}{2.5}\right)}$$

$$G_c(j\omega) = \frac{(1 + j\,0.4\,\omega)}{(1 + j\,1.13\,\omega)}$$

The open loop transfer function of the compensated system is

$$G'(s) = G_c(s)\; G(s)$$

$$G'(s) = \frac{(1 + j\,0.4\,\omega)}{(1 + j\,1.13\,\omega)} \times \frac{2500}{S\,(s + 25)}$$

Example 10.5 Realise the lag and lead network if $R_1 = 1.8\ K\Omega$, $R_2 = 3\ K\Omega$, $C = 0.1\ \mu F$ and determine the max. phase lag and max. phase lead respectively if $\omega_m = 0.2$ rad/sec.

(U.P.T.U., 2003–04, ACS)

Solution The lag lead network is

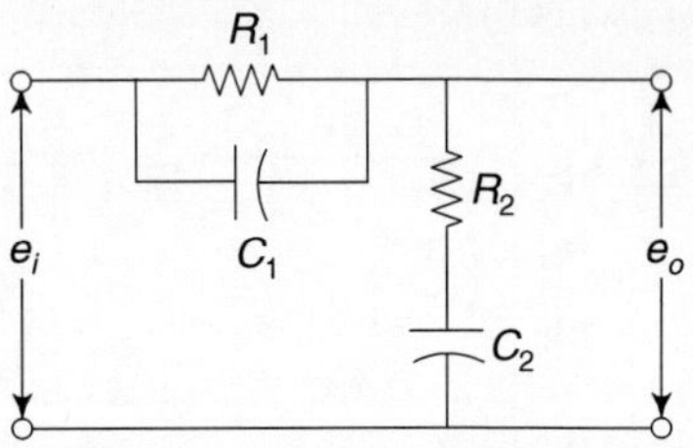

Given

$$R_1 = 1.8\ K\Omega$$

$$R_2 = 3\ K\Omega$$

$$C_1 = C_2 = C = 0.1\ \mu F$$

and

$$\alpha = \frac{R_2}{R_1 + R_2} = \frac{3}{3 + 1.8}$$

$$\alpha = \frac{3}{4.8} = 0.625$$

$$T = R_1\, C = 1.8 \times 0.1 \times 10^{-3} = 0.18 \text{ m sec}$$

Maximum phase lead

$$\sin \phi_m = \frac{1 - \alpha}{1 + \alpha} = \frac{1 - 0.625}{1 + 0.625}$$

$$\sin \phi_m = 0.23$$

$$\phi_m = \sin^{-1}(0.23)$$

$$\phi_m = 13.3° \quad \textbf{Ans.}$$

Now $$\beta = \frac{R_1 + R_2}{R_2} = \frac{1.8+3}{3} = \frac{4.8}{3} = 1.6$$

Maximum phase lag

$$\sin \phi_m = \frac{1-\beta}{1+\beta} = \frac{1-1.6}{1+1.6}$$

$$\sin \phi_m = \frac{0.6}{2.6} = -0.23$$

$$\phi_m = -13.3° \quad \textbf{Ans.}$$

The negative sign indicates that ϕ_m is negative which means that the phase angle is lagging.

Example 10.6 Design a phase lead compensation network for a system having open loop transfer function given below such that the system will have acceleration error coefficient $K_a = 100 \text{ sec}^{-1}$ and suitable phase margin for stable operation.

$$G(s)\,H(s) = \frac{K}{s^2\,(1+0.05\,s)}$$ *(U.P.T.U., 2003–04, ACS)*

Solution Assume P.M. = 20°

$$K_a = \lim_{s\to 0} s^2\,G(s)\,H(s) = \lim_{s\to 0} s^2 \times \frac{K}{s^2\,(1+0.05\,s)}$$

$$K_a = K = 100$$

Thus, the open loop transfer function is

$$G(s)\,H(s) = \frac{100}{s^2\,(0.05\,s+1)}$$

The corner frequency for uncompensated system is

$$\omega_{c1} = \frac{1}{T} = \frac{1}{0.05} = 20 \text{ rad/sec}$$

As the system is type 2, the initial slope of the Bode plot is – 40 dB/decade and the initial ordinate is

$$20 \log_{10} (100) = 40 \text{ dB}$$

The slope of the magnitude after first corner frequency, i.e., 20 rad/sec is due to the simple pole (1 + 0.5 s)

$$= [-40 + (-20)] \text{ dB/decade}$$
$$= -60 \text{ dB/decade}$$

The phase angle is

$$\angle G(j\omega) = -180° - \tan^{-1} (0.05\,\omega)$$

ω :	1	10	20	50	100
$\angle G(j\omega)\,H(j\omega)$:	– 182.8°	– 206.5°	– 225°	– 248.2°	– 258.6°

The Bode plot is shown in Fig. Ex. 10.6. From the Bode plot, it is noted that the phase margin is – 26° at gain crossover frequency is 9.5 rad/sec and the system is unstable.

To increase the phase margin to 20° the gain crossover point is to be shifted to a higher value of frequency and this is possible by introducing a phase lead compensation network.

The required phase lead is

$$\phi_L = \phi_m - \phi_s + \text{tolerance}$$
$$\phi_L = 20 - (-26°) + 10°$$
$$\phi_L = 56°$$

The value of α parameter of the phase lead network is given by

$$\alpha = \frac{1-\sin\phi_L}{1+\sin\phi_L} = \frac{1-\sin 56°}{1+\sin 56°}$$

$$\alpha = \frac{1-0.829}{1+0.829} = 0.093$$

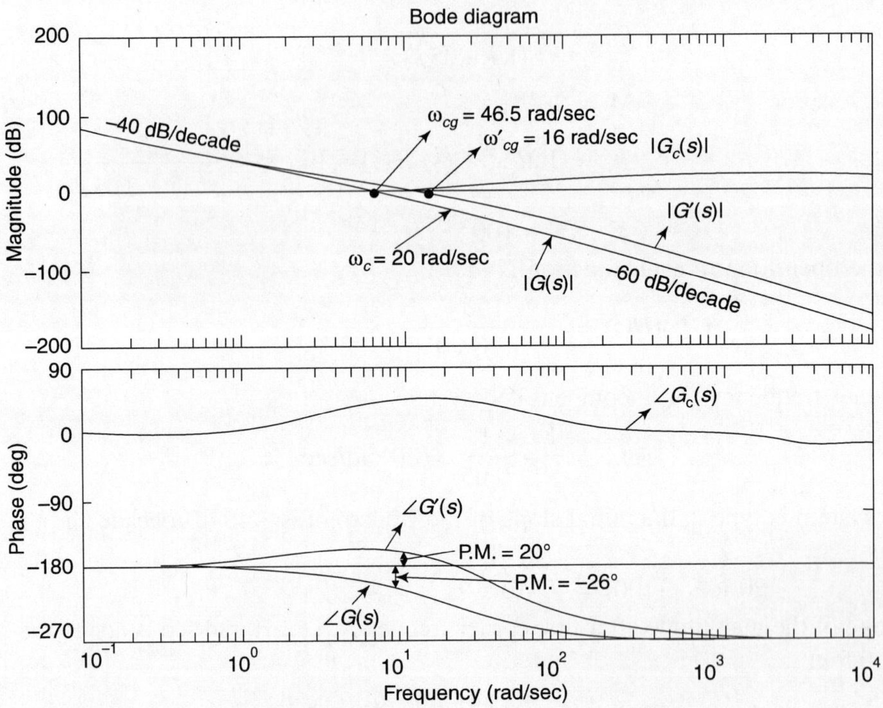

Figure Ex. 10.6 Bode plot for Ex. 10.6

The maximum phase lead (ϕ_L = 56°) is contributed at mid-frequency (ω_m) of the phase lead compensation network. Allowing an amplification of 1/α the magnitude of phase lead compensation network function at $\omega = \omega_m$ is determined below.

$$\omega_m = \frac{1}{\sqrt{\alpha}\,T}$$

$$G_c(j\omega) = \frac{(1+j\omega_m T)}{(1+j\omega_m \alpha T)} = \frac{1}{\sqrt{\alpha}}$$

in decibel $\quad 20 \log_{10} \frac{1}{\sqrt{\alpha}} = G_c(j\omega)$

$$20 \log_{10} \frac{1}{\sqrt{0.093}} = G_c(j\omega)$$

$$G_c(j\omega) = 10.3 \text{ dB}$$

From the Bode plot it is noted that for compensated system at a gain of – 10.3 dB the frequency is 16 rad/sec. The new gain crossover frequency for the compensated system is selected at $\omega_{cg}^1 = \omega_m = 16$ rad/sec.

The two corner frequencies for the phase lead network are

$$\text{Lower corner frequency} = \frac{1}{T} = \sqrt{\alpha}\, \omega_n$$

$$= \sqrt{0.093} \times 16 = 4.9 \text{ rad/sec}$$

$$\text{Upper corner frequency} = \frac{1}{\alpha T} = \frac{\sqrt{\alpha}\, \omega_m}{\alpha} = \frac{\omega_m}{\sqrt{\alpha}} = 52.4 \text{ rad/sec.}$$

The transfer function of the phase lead network with amplifier gain $\frac{1}{\alpha} = \frac{1}{0.093} = 15.75$ is

$$G_c(j\omega) = \frac{1+j\omega T}{1+j\omega \alpha T} = \frac{(1+j\,0.2\,\omega)}{(1+j\,0.0186\,\omega)}$$

The open loop transfer function of the compensated system is

$$G'(s) = G_c(s)\, G(s)\, H(s)$$

$$G'(s) = \frac{(1+0.2\,s)}{(1+0.0186\,s)} \times \frac{100}{s^2\,(1+0.05\,s)} \qquad \textbf{Ans.}$$

Example 10.7 A unity feedback control system has open loop transfer function

$$G(s) = \frac{K}{s\,(0.1\,s+1)\,(0.2\,s+1)}$$

The system is required to satisfy the following performance specification: $K_v = 30$, P.M. ≥ 40, bandwidth $\omega_m = 5$ rad/sec. Design a lag compensator. *(U.P.T.U., 2003–04, ACS)*

Solution $\quad K_v = \lim_{s \to 0} s\, G(s) = \lim_{s \to 0} \frac{K}{s\,(0.1\,s+1)\,(0.2\,s+1)}$

$$K_v = K = 30 \text{ sec}^{-1}$$

The open loop transfer function is

$$G(s) = \frac{30}{s\,(0.1\,s+1)\,(0.2\,s+1)}$$

Put

$$s = j\omega$$

$$G(j\omega) = \frac{30}{j\omega\,(0.1\,j\omega + 1)\,(0.2\,j\omega + 1)}$$

The corner frequencies are

$$\omega_{c2} = \frac{1}{T_2} = \frac{1}{0.1} = 10 \text{ rad/sec}$$

$$\omega_{c1} = \frac{1}{T_1} = \frac{1}{0.2} = 5 \text{ rad/sec}$$

As the system is type 1, the initial slope of the bode plot is – 20 dB/decade and the initial ordinate is

$$20 \log_{10} (30) = 29.5 \text{ dB}$$

The slope of the magnitude after first corner frequencies, i.e., 5 rad/sec is due to the simple pole (0.2 s + 1) is

$$= [-20 + (-20)] \text{ dB/decade}$$
$$= -40 \text{ dB/decade}$$

The slope of the magnitude after second corner frequency, i.e., 10 rad/sec is due to the simple pole (0.1 s + 1) is

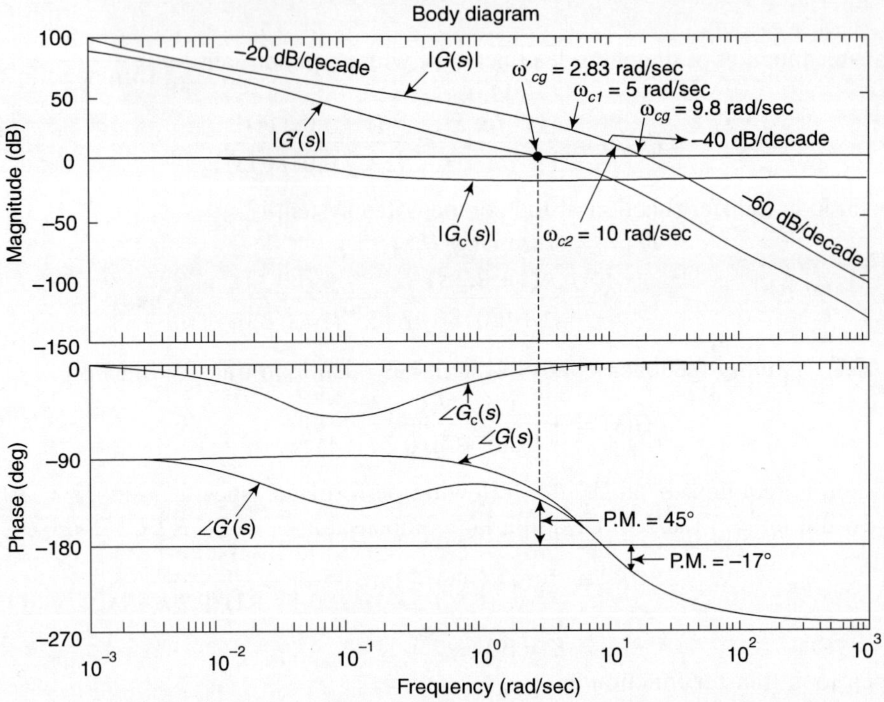

Figure Ex. 10.7 Bode plot of Ex. 10.7

$$= [-40 + (-20)] \text{ dB/decade}$$
$$= -60 \text{ dB/decade}$$

The phase angle is

$$\angle G(j\omega) = -90^\circ - \tan^{-1}(0.2\,\omega) - \tan^{-1}(0.1\,\omega)$$

ω :	0	1	5	10	20	50
$\angle G(j\omega)$:	– 90°	– 107°	– 161.6°	– 198.4°	– 229.3°	– 252.9°

The Bode plot is shown in Fig. Ex. 10.7.

From the Bode plot in Fig. Ex. 10.7, the phase margin is found to be – 17°, which means that the system is unstable. With a desired phase margin = 40°, the gain crossover frequency must be shifted to a lower value and this is possible by using a phase lag compensation.

However, the used phase lag network will reduce the phase angle to some extent at a new gain crossover frequency, therefore, some tolerance say 5° is taken into account, thus phase angle of the compensated system at new gain crossover frequency is 40 + 5° = 45°.

For the compensated system at ω = 2.83 rad/sec the phase angle is 45°. The new gain crossover frequency is selected as ω'_{cg} = 2.83 rad/sec. At this new gain crossover frequency the magnitude of the uncompensated system is 19 dB, which is to be neutralized by phase lag network, therefore at new gain crossover

$$-20 \log_{10} \beta = -19 \text{ dB}$$
$$\beta = 8.91$$

The upper corner frequency of the phase lag compensation network is selected one decade earlier than the new gain crossover frequency, hence the upper corner frequency for the phase lag network is

$$\omega' = \frac{2.83}{10} = 0.283 \text{ rad/sec}$$

Therefore, the upper comer frequency

$$= \frac{1}{T} = 0.283 \text{ rad/sec}$$

The lower corner frequency

$$= \frac{1}{\beta T} = \frac{0.283}{8.91} = 0.032 \text{ rad/sec}$$

The transfer function of the phase lag network is

$$G_c(j\omega) = \frac{(1 + j\omega T)}{(1 + j\omega \beta T)} = \frac{\left(1 + j\omega \dfrac{1}{0.283}\right)}{\left(1 + j\omega \times 8.91 \times \dfrac{1}{0.283}\right)}$$

$$G_c(j\omega) = \frac{(1 + j\,3.5\,\omega)}{(1 + j\,31.48\,\omega)}$$

The open loop transfer function of the compensated system is

$$G'(S) = G_c(s)\, G(s)$$

$$G'(S) = \frac{(1 + j\,3.5\,\omega)}{(1 + j\,31.48\,\omega)} \times \frac{30}{s\,(0.1\,s + 1)\,(0.2\,s + 1)} \quad \textbf{Ans.}$$

Example 10.8 Consider the system shown in Fig. Ex. 10.8. Design lead compensator of this system to meet the following specification.

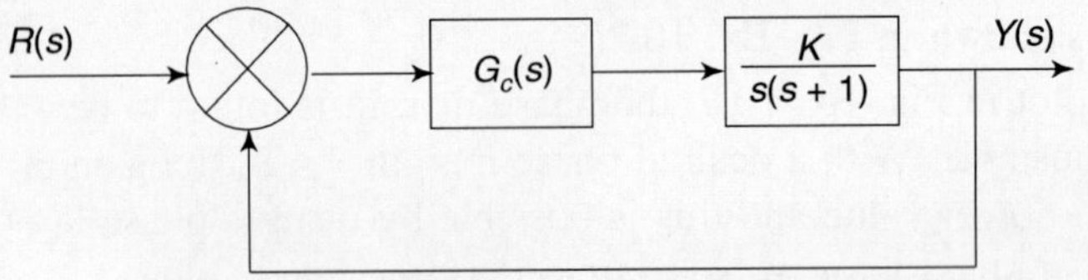

Figure Ex. 10.8 (a) Compensated system

$$\xi = 0.7 \text{ (damping ratio)}$$

$$t_s = 1.4 \text{ sec (settling time)}$$

Velocity error constant $K_v = 2\ \text{sec}^{-1}$

Solution Maximum phase lead

$$\phi_m = \tan^{-1}\left\{\frac{2\,\xi}{\left[(1 + 4\,\xi^4)^{1/2} - 2\,\xi^2\right]^{1/2}}\right\}$$

$$= \tan^{-1}\left\{\frac{2 \times 0.7}{\left[(1 + 4 \times 0.7^4)^{1/2} - 2 \times 0.7^2\right]^{1/2}}\right\} = 65°$$

For 5% tolerance band $t_s = \dfrac{3}{\xi\,\omega_n} \Rightarrow \omega_n = \dfrac{3}{\xi\,t_s}$

$$\omega_n = \frac{3}{0.7 \times 1.4} = 3.06 \text{ rad/sec}$$

$$K_v = 2\ \text{sec}^{-1}$$

Open loop transfer function of an uncompensated system is

$$G(s) = \frac{K}{s\,(s + 1)} = \frac{2}{s\,(s + 1)}$$

The Bode plot of the function is shown in Fig. Ex. 10.8 (b).

The phase angle is

$$\angle G(j\omega) = -\,90° - \tan^{-1}\omega$$

ω :	0.1	0.2	0.5	1	2	5	10
$\angle G(j\omega)$:	– 96°	– 101°	– 166.5°	– 135°	– 153°	– 169°	– 174°

From the Bode plot, the phase margin is 35.5°. But the phase margin required is 65°.

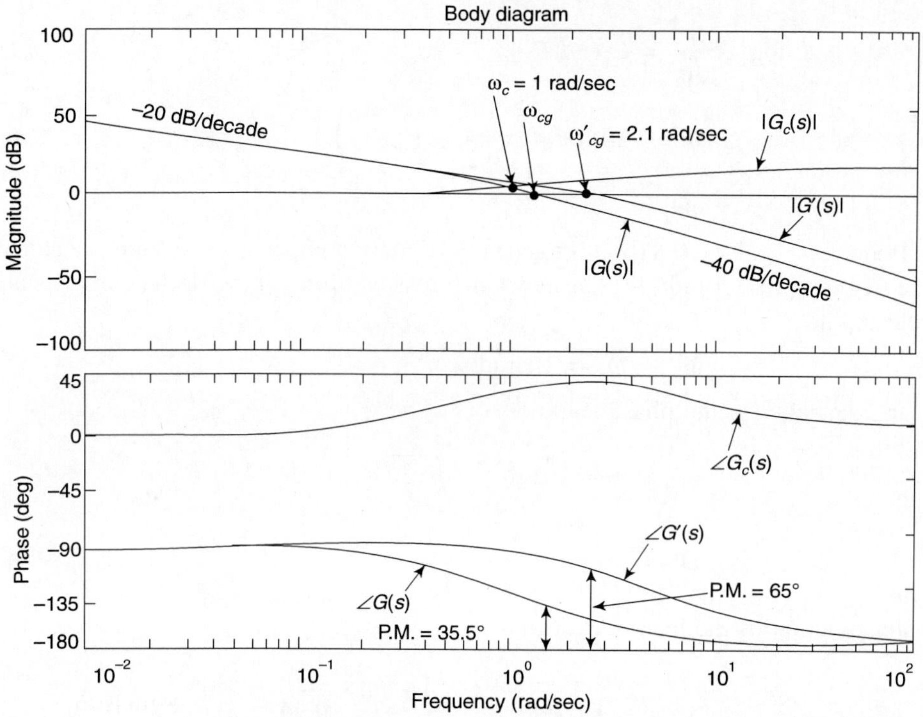

Figure Ex. 10.8 (b) Bode plot of Ex. 10.8

So, the necessary phase lead

$$\phi_L = \phi_m - \phi_s + \text{tolerance}$$

$$\phi_L = 65 - 35.5° + 9.5°$$

$$\phi_L = 40° \text{ (say)}$$

The value of α parameter of the phase lead network is given by

$$\alpha = \frac{1 - \sin \phi_L}{1 + \sin \phi_L} = \frac{1 - \sin 40°}{1 + \sin 40°}$$

$$\alpha = 0.2$$

The maximum phase lead (ϕ_L= 40°) is contributed at mid-frequency (ω_m) of the phase lead compensation network. Allowing on amplification of $1/\alpha$ the magnitude of phase lead compensation network transfer function at $\omega = \omega_m$ is determined below.

$$\omega_m = \frac{1}{\sqrt{\alpha}\, T}$$

$$G_c(j\omega) = \left| \frac{1 + j\,\omega_m T}{1 + j\,\omega_m\, \alpha\, T} \right| = \frac{1}{\sqrt{\alpha}}$$

in decibel $$20\log_{10}\frac{1}{\sqrt{\alpha}} = G_c(j\omega)$$

$$G_c(j\omega) = 20\log_{10}\frac{1}{\sqrt{0.2}}$$

$$|G_c(j\omega)| = -6.98 \text{ dB}$$

From the Bode plot Fig. Ex. 10.8 (b) it is to be noted that for compensated system at a gain of – 6.96 dB the frequency is 2.1 rad/sec. The new gain crossover frequency for the compensated system is selected at

$$\omega'_{cg} = \omega_m = 2.1 \text{ rad/sec.}$$

The corner frequency for the phase lead network is

$$\omega_{c1} = -= \sqrt{\alpha}\,\omega = \sqrt{0.20\times 2.1} = 0.94$$

$$\omega_{c2} = \frac{1}{\alpha T} = \frac{\omega}{\sqrt{\alpha}} = \frac{2.1}{\sqrt{0.2}} = 4.7$$

The transfer function of the lead network is

$$G_c(j\omega) = \frac{1+j\omega T}{1+j\omega\alpha T} = \frac{1+j\omega\times\dfrac{1}{0.94}}{1+j\omega\times 0.2\times\dfrac{1}{0.94}} = \frac{(1+j\omega\,1.06)}{(1+j\omega\,0.21)}$$

The open loop transfer function of the compensated system is

$$G'(s) = G_c(s)\,G(s)$$

$$G'(s) = \frac{(1+j\,1.06\,\omega)}{(1+j\,0.21\,\omega)}\times\frac{2}{S(S+1)} \quad \textbf{Ans.}$$

Example 10.9 For the system shown in Fig. Ex. 10.9 (a), design a lead compensator such that $\omega_m = 4$ rad/sec and $\xi = 0.5$ for the compensated system.

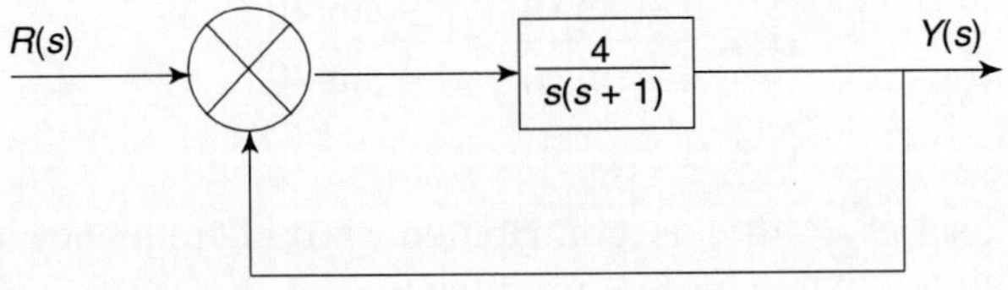

Figure Ex. 10.9 (a) Uncompensated system

Solution

$$\theta = \cos^{-1}(\xi) = \cos^{-1}(0.5)$$

$$\theta = 60°$$

$$\phi = -180° - \angle G_j(s_d)$$

$$\phi = -180° - (-120° - 90°)$$
$$\phi = 30°$$

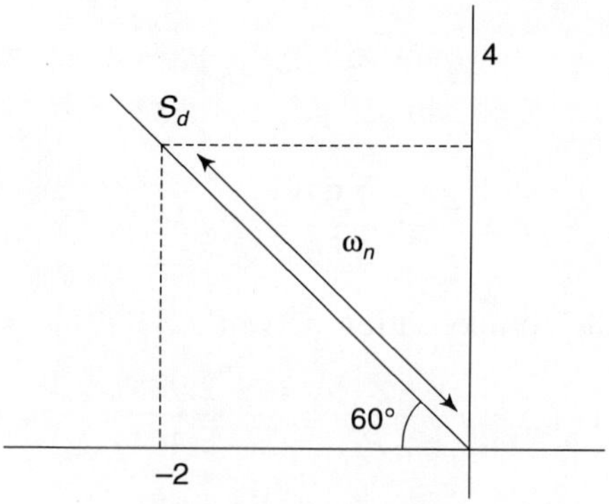

Figure Ex. 10.9 (b)

For maximum value of α

$$r = \frac{1}{2}(180 - \theta - \phi)$$
$$= \frac{1}{2}(180° - 60° - 30°)$$
$$r = 45°$$

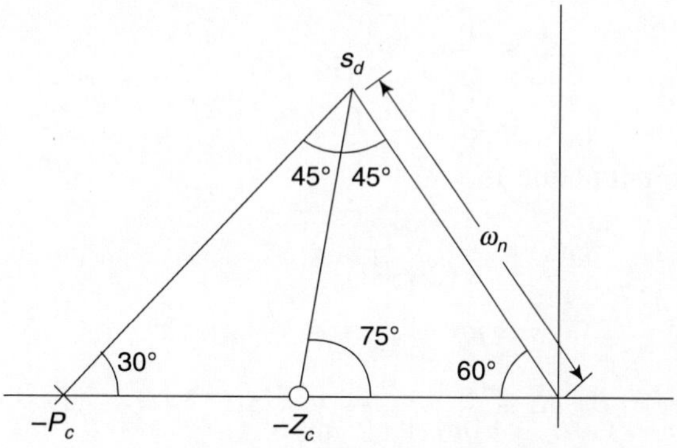

Figure Ex. 10.9 (c)

By sine rule $$\frac{Z_c}{\sin 45°} = \frac{\omega_n}{\sin 75°}$$

$$Z_c = \frac{4 \sin 45°}{\sin 75°}$$

Also $$\frac{P_c}{\sin 75°} = \frac{\omega_n}{\sin 45°}$$

$$P_c = \frac{4 \sin 75°}{\sin 45°} = 5.464$$

$$\alpha = \frac{Z_c}{P_c} = \frac{2.928}{5.464} = 0.536$$

The transfer function of the compensation network

$$G_c(s) = \frac{(s + Z_c)}{(s + P_c)} = \frac{(s + 2.928)}{(s + 5.464)}$$

$$G_c(s) = 0.536 \left(\frac{0.341\, s + 1}{0.183\, s + 1} \right) \quad \textbf{Ans.}$$

Example 10.10 A unity feedback system has an open loop transfer function of $G(s) = \frac{4}{s\,(2s + 1)}$. It is desired to obtain a phase margin of 44°, without sacrificing the K_v of the system. Design a suitable lead network and compute the value of network components assuming any suitable impedance level. *(U.P.T.U., 2006–07, CS)*

Solution $$K_v = \lim_{s \to 0} s\, G(s)$$

$$K_v = \lim_{s \to 0} s \cdot \frac{4}{s\,(1 + 2s)}$$

$$K_v = 4$$

The open loop transfer function is

$$G(s) = \frac{4}{s\,(2\,s + 1)}$$

Putting $$s = j\omega$$

$$G(j\omega) = \frac{4}{j\omega\,(1 + 2\,j\omega)}$$

Here $$K = 4$$

$$T_1 = 2$$

The corner frequency is

$$\omega_{c1} = \frac{1}{T_1} = \frac{1}{2} = 0.5 \text{ rad/sec}$$

There is one pole at the origin so the initial slope of the bode plot is – 20 dB/decade

The initial point of the slope is

$$= 20 \log_{10} K = 20 \log_{10} 4$$
$$= 12.04 \text{ dB}$$

The slope after one corner frequency, i.e., 0.5 rad/sec is due to the real pole $(1 + 2\,j\omega)$ is

$$= [-20 + (-20)] \text{ dB/decade}$$
$$= -40 \text{ dB/decade}$$

The phase angle plot

$\angle G(j\omega)$ for pole at origin is $= -90°$

$\angle G(j\omega)$ due to simple pole $(1 + 2j\omega) = -\tan^{-1} 2\,\omega$

The total angle is

$$\angle G(j\omega) = -90° - \tan^{-1} 2\,\omega$$

For verying ω

ω :	0	0.1	0.2	0.5	1	2	5	10
∠G(jω) :	– 90°	– 101.3°	– 111.8°	– 135	– 153°	– 165.9°	– 174.3°	– 177.1°

The Bode plot for uncompensated system is shown in Fig. Ex. 10.10.

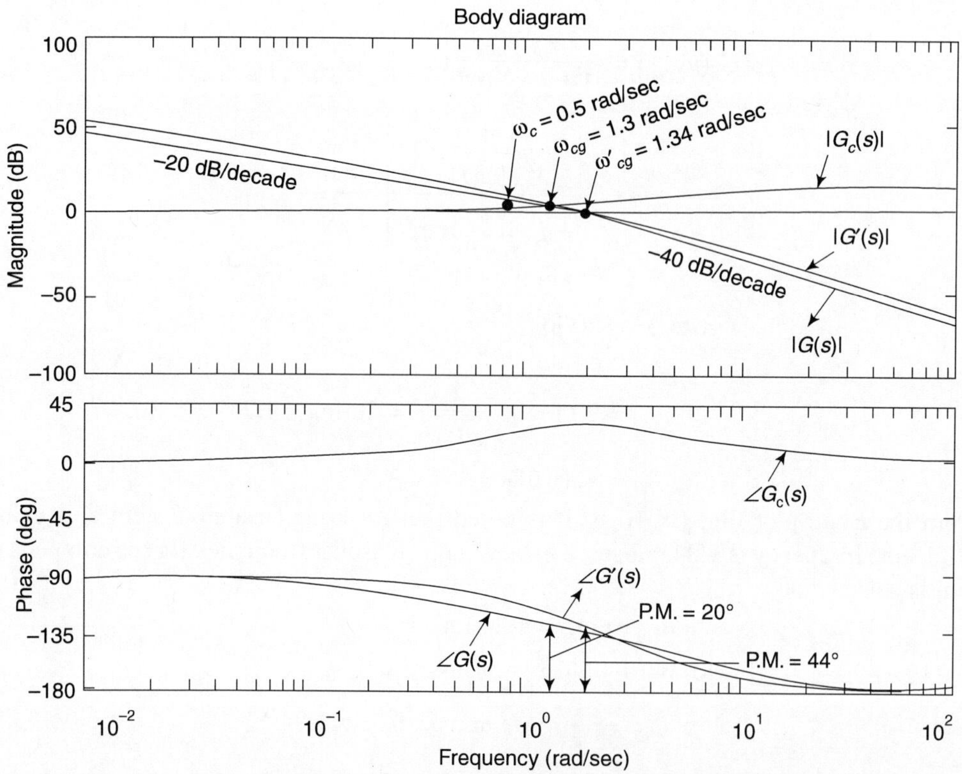

Figure Ex. 10.10 Bode plot for Ex. 10.10

From the bode plot in Fig. Ex. 10.10 it is noted that uncompensated system having a phase margin is 20°, i.e., $\phi = 2°$. To increase the phase margin to 44°, i.e., $\phi_s = 44°$ the gain crossover point is to be shifted to a higher value of frequency and this is possible by introducing a phase lead network. The required phase lead is

$$\phi_L = \phi_m - \phi_s + \text{tolerance}$$

$$\phi_L = 44° - 20° + 6°$$

Let tolerance be 6° for simplicity

$$\phi_L = 30°$$

The value of a parameter of the phase lead network is given by

$$\alpha = \frac{1 - \sin\phi_L}{1 + \sin\phi_L} = \frac{1 - \sin 30°}{1 + \sin 30°}$$

$$\alpha = 0.33$$

The maximum phase lead ($\phi_L = 30°$) is contributed at mid-frequency (ω_m) of the phase lead compensation network. Allowing an amplification of $1/\alpha$ the magnitude of phase lead compensation network transfer function at $\omega = \omega_m$ is determined below.

$$\therefore \quad \omega_m = \frac{1}{\sqrt{\alpha}\, T}$$

$$|G_c(j\omega_m)| = \left|\frac{1 + j\omega_m T}{1 + j\omega_m \alpha T}\right|$$

$$= \left|\frac{1 + j\dfrac{1}{\sqrt{\alpha}\, T} \times T}{1 + j\dfrac{1}{\sqrt{\alpha}\, T} \times \alpha T}\right| = \frac{1}{\sqrt{\alpha}}$$

$$|G_c(j\omega_m)| = 20 \log_{10} \frac{1}{\sqrt{\alpha}}$$

$$= 20 \log_{10} \frac{1}{\sqrt{0.33}} = 20 \log_{10} 1.74$$

$$= 4.81 \text{ dB}$$

From the Bode plot Fig. Ex 10.10 it is noted that for compensated system at a gain of –4.81 dB the frequency is 1.84 rad/sec. The new gain crossover frequency for the compensated system is selected at

$$\omega'_{cg} = \omega_m = 1.84 \text{ rad/sec.}$$

The two corner frequency for the phase lead network are

$$\omega_{c1} = \frac{1}{T} = \sqrt{\alpha}\, \omega_m = \sqrt{0.33} \times 1.84$$

$$\omega_{c2} = 1.06 \text{ rad/sec}$$

The upper corner frequency

$$\omega_{c2} = \frac{1}{\alpha T} = \frac{\sqrt{\alpha}\,\omega_m}{\alpha} = \frac{\omega_m}{\sqrt{\alpha}}$$

$$= \frac{1.84}{\sqrt{0.33}} = 3.20 \text{ rad/sec}$$

The transfer function of the phase lead compensation network with amplifier gain ($1/\alpha$) is 3.

$$G'(s) = \frac{1 + j\omega T}{1 + j\,\omega\,\alpha T} = \frac{1 + j\omega \dfrac{1}{1.06}}{1 + j\omega \times 0.33 \dfrac{1}{1.06}}$$

$$= \frac{(1 + 0.94\,\omega j)}{(1 + 0.31\,\omega j)} \quad \textbf{Ans.}$$

The open loop transfer function of the compensated system is

$$G'(s) = G_c(s)\,G(s)$$

$$= \frac{(1 + 0.95\,\omega j)}{(1 + 0.31\,\omega j)} \times \frac{4}{s\,(2\,s + 1)} \quad \textbf{Ans.}$$

The plot for the compensated system is shown in Fig. Ex. 10.1 (a) and (b).

Example 10.11 The open loop transfer function of a unity feedback system is $G(s) = \dfrac{K}{s\,(s+1)}$.

It is desired to have the velocity error constant K_v = 12 sec^{-1} and phase margin as 40°. Design a lead compensator to meet the above specifications. *(U.P.T.U., 2002–03, CS)*

Solution Velocity error constant, K_v = 12 sec^{-1}

We know, $$K_v = \lim_{s \to 0} s\,G(s) = \lim_{s \to 0} s \cdot \frac{K}{s\,(s+1)} = K$$

$$K_v = K = 12 \text{ sec}^{-1}$$

Now from the Bode plot drawn for the above uncompensated system, i.e., $G(s) = \dfrac{12}{s\,(s+1)}$ is shown in Fig. Ex. 10.11.

$$\angle G(j\omega) = -\,90° - \tan^{-1} 2\,\omega$$

ω :	0.1	0.5	1	2	5	8	10
$\angle G(j\omega)$:	– 95.1°	– 116.6°	– 135°	– 153°	– 168.7°	– 172.9°	– 174.3°

The phase margin, ϕ_s = 18°.

The phase margin desired, ϕ_m = 40°

So, additional phase lag required

$$\phi_L = \phi_m - \phi_s + \text{tolerance}$$

$$\phi_L = 40° - 18° + 5 = 27°$$

Now the value of constant α can be found

$$\alpha = \frac{1 - \sin \phi_L}{1 + \sin \phi_L} = \frac{1 - \sin 27°}{1 + \sin 27°} = \frac{0.546}{1.454}$$

$$\alpha = 0.376$$

The frequency at which ϕ_m is achieved

$$\omega_m = \frac{1}{\sqrt{\alpha}\, T} \Rightarrow T = \frac{1}{\omega_m \sqrt{\alpha}}$$

The phase lead network is given by

$$G_c(j\omega_m) = \left| \frac{1 + j\omega_m T}{1 + j\omega_n \alpha T} \right| = \left| \frac{1 + j\dfrac{1}{T\sqrt{\alpha}} T}{1 + j\dfrac{1}{T\sqrt{\alpha}} \alpha T} \right|$$

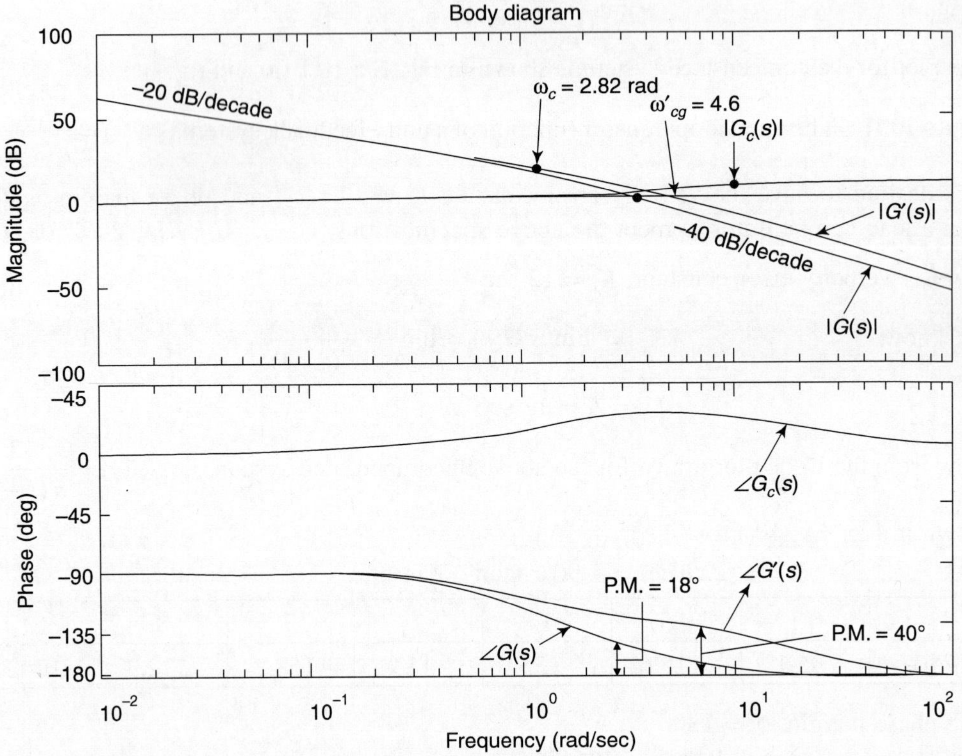

Figure Ex. 10.11 Bode plot for Ex. 10.11

$$= \frac{1}{\sqrt{\alpha}}$$

in decibel, $20 \log |G_c(j\omega)| = 20 \log \frac{1}{\sqrt{0.376}}$

$$= 4.23 \text{ dB}$$

From the Bode plot the frequency at – 4.23 dB is 4.6 rad/sec

So, $\omega'_{cg} = \omega_m = 4.6$ rad/sec.

The lower corner frequency

$$= \frac{1}{T} + \sqrt{\alpha}\ \omega_m$$

$$= \sqrt{0.376} \times 4.6$$

$$= 2.82 \text{ rad/sec}$$

The upper corner frequency, $\frac{1}{\alpha T} = \frac{\omega_m \sqrt{\alpha}}{\alpha} = 7.5$ rad/sec

The transfer function of the phase lead compensation network with amplifier gain

$$a = \frac{1}{\alpha} = \frac{1}{0.376} = 2.86$$

$$G_c(j\omega) = \frac{1 + j\omega T}{1 + j\omega\, \alpha T} = \frac{1 + j\omega \times \frac{1}{2.82}}{1 + j\omega \times 0.376 \times \frac{1}{2.82}}$$

$$= \frac{1 + j\, 0.355\, \omega}{1 + j\, 0.133\, \omega}$$

Now if $R_1 = 1\ K\Omega$

Then $\alpha = \frac{R_2}{R_1 + R_2}$

$$R_2\, (\alpha + 1) = -\, R_1\, \alpha$$

$$R_2 = \frac{R_1\, \alpha}{\alpha - 1}$$

$$= \frac{-1000 \times 0.376}{0.376 - 1} = 602\ \Omega$$

$$T = R_1\, C \ \Rightarrow C = \frac{T}{R_1} \Rightarrow \frac{1/2.82}{1000}$$

$C = 0.354 \times 10^{-3}\ F$

$C = 354\ \mu\text{F}$ **Ans.**

Example 10.12 Consider type 1 system with an open loop transfer function of $G(s) = \dfrac{K}{s\,(s+1)\,(s+4)}$ the system is to be compensated to meet the following specifications

(i) $\xi = 0.5$

(ii) Undamped natural frequency $\omega_m = 2$. *(U.P.T.U., 2005–06, ACS)*

Solution

$$G(s) = \frac{K}{s\,(s+1)\,(s+4)}$$

$\xi = 0.5$ and $\omega_m = 2$ rad/sec for uncompensated system using the transient response specification the desired dominant closed loop poles are found to lie at (by drawing root locus for uncompensated system)

$$s_d = 1 \pm j\,1.73 \text{ (by section 10.7)}$$

The contribution required from the lead compensation pole zero pair is

$$\begin{aligned}\phi &= 180° - \angle G_f(s_d)\\ &= -\,180° - (30° - 90° - 120°)\\ &= -\,180° + 240° = 60°\end{aligned}$$

Further, it is observed that the open loop poles as $s = -\,1$ lies directly below the desired location place the compensator zero close to this pole to its left, i.e., – 1.2.

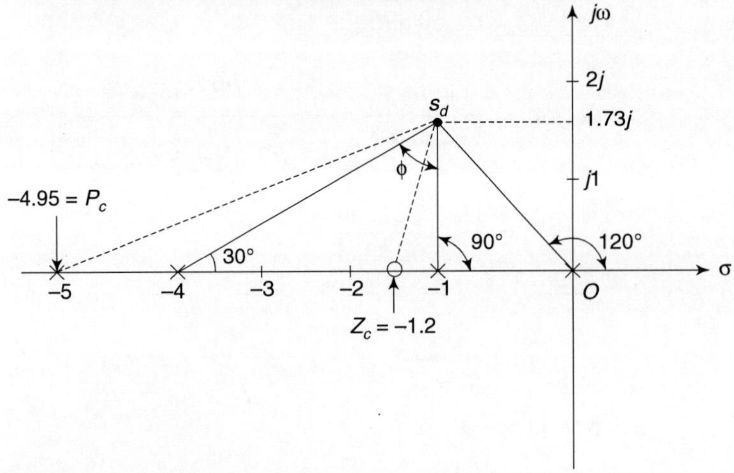

Figure Ex. 10.12 Phase lead compensator

Now join the compensator zero Z_c to s_d and locate the compensator pole by making an angle of $\phi = 60°$, the location of pole is found to be at – 4.95.

Now the open loop transfer function of the compensated system becomes

$$\begin{aligned}G'(s) &= G(s) \times G_c(s)\\ &= \frac{30.4}{s\,(s+1)\,(s+4)} \times \frac{(s+1.2)}{(s+4.95)}\end{aligned}$$

$$G'(s) = \frac{30.4\,(s+1.2)}{s\,(s+1)\,(s+4)\,(s+4.95)}$$

The velocity error constant

$$K_v = \lim_{s\to 0} s\cdot G(s) = \lim_{s\to 0} \frac{30.4\,(s+1.2)}{s\,(s+1)\,(s+4)\,(s+4.95)}$$

$$K_v = 1.84 \quad \textbf{Ans.}$$

Example 10.13 Consider a feedback control system with the open loop transfer function with unity feedback $G(s) = \dfrac{K}{s\,(s+1)}$

The required specifications are: *(U.P.T.U., 2004–05, ACS)*

(i) The phase margin of the system must be greater than 45°.

(ii) When the input is a ramp function, the steady state error of the output is positive should be less than 0.1 deg/sec of the final output velcoity design a phase lead compensation by Bode plot and clearly show specific points.

Solution The procedure for solution of this function is given in previous example (10.11)

Let $$K_v = 12\ \text{sec}^{-1}$$

$$\phi_m = 45°$$

Draw the Bode plot of above system with $K_v = 12$. The phase margin of uncompensated system is found to be $\phi_s = 15°$ [See Fig. Ex. 10.11] if a lead compensation is used, the phase lead required at the new cross over is given by

$$\phi_L = \phi_m - \phi_s + \text{tolerance}$$

$$\phi_L = 45° - 15° + 5° = 35°$$

$$\phi_L = 350°\ (\text{say})$$

The value of constant α can be found

$$\alpha = \frac{1-\sin\phi_L}{1+\sin\phi_L} = \frac{1-\sin 35°}{1+\sin 35°} = 0.271$$

The magnitude contribution of the compensating network at ω_m is

$$20\log_{10}\frac{1}{\sqrt{\alpha}} = 20\log\frac{1}{\sqrt{0.271}}$$

$$= 5.66\ \text{dB}$$

Therefore, the frequency at which the uncompensated system has a magnitude of – 5.66 dB becomes the new crossover frequency $\omega'_{cg} = \omega_m$

$$\omega'_{cg} = \omega_m = 4.7\ \text{rad/sec}$$

The lower corner frequency

$$\omega_{c1} = \frac{1}{T_1} = \omega_m\sqrt{\alpha} = 4.7\times\sqrt{0.271}$$

$$\omega_{c1} = 2.44 \text{ rad/sec.}$$

Upper corner frequency

$$\omega_{c2} = \frac{1}{\alpha T} = \frac{\omega_m}{\sqrt{\alpha}} = 9.02 \text{ rad/sec}$$

The transfer function of the lead network is

$$a = \frac{1}{\alpha} = 3.69$$

$$G_c(j\omega) = \frac{1 + j\omega T}{1 + j\omega \alpha T} = \frac{1 + j\omega \times \frac{1}{3.69}}{1 + j\omega \times 0.271 \times \frac{1}{3.69}}$$

$$= \frac{1 + j\,0.2710\,\omega}{1 + j\,0.0734\,\omega}$$

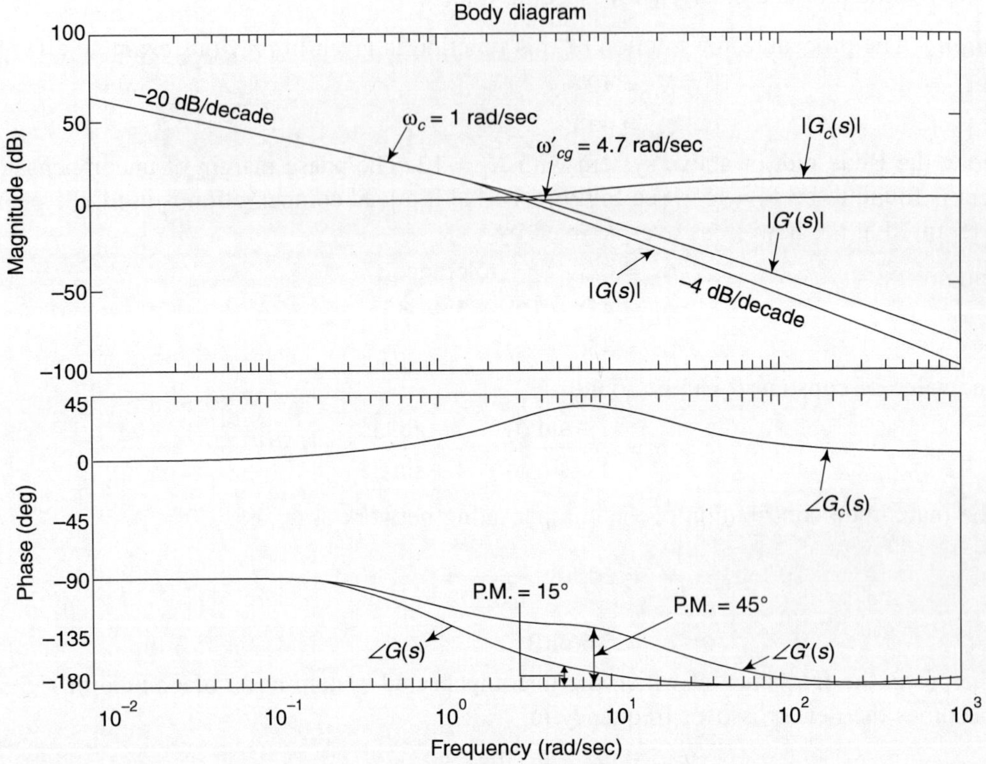

Figure Ex. 10.13 Bode plot for Ex. 10.13

The open loop transfer function of the compensated system is

$$G'(s) = G(s)\, G_c(s)$$

$$= \frac{12}{s\,(s+1)} \times \frac{(1+0.2710\;\omega j)}{(1+0.0734\;\omega j)}$$

$$G'(s) = \frac{12(1+0.2710\;\omega j)}{s\,(s+1)\,(1+0.073\;\omega j)} \quad \textbf{Ans.}$$

The Bode plot for Ex. 10.13 is shown if Fig. Ex. 10.13.

Example 10.14 Design a lag-lead compensator for $G(s) = \dfrac{K}{s\,(1+0.1\,s)\,(1+0.2\,s)}$ such that $K_v = 100$ and phase margin > 45°. *(U.P.T.U., 2003, CO, ACS)*

Solution

$$G(s) = \frac{K}{s\,(1+0.1\,s)\,(1+0.2\,s)}$$

Let us taking

$$K_v = 30\ \text{sec}^{-1}$$

$$\text{P.M.} \geq 50°$$

Putting

$$s = j\omega$$

$$G(j\omega) = \frac{K}{j\omega(1+0.1\,j\omega)\,(1+0.2\,j\omega)}$$

The corner frequencies are

$$\omega_{c1} = \frac{1}{T_1} = \frac{1}{0.2} = 5\ \text{rad/sec}$$

$$\omega_{c2} = \frac{1}{T_2} = \frac{1}{0.1} = 10\ \text{rad/sec}$$

The Bode plot start from

$$20 \log_{10} K = 20 \log 30$$

$$= 29.5\ \text{dB}$$

As the system is type 1, the initial slope of the plot is – 20 dB/decade.

The slope of the magnitude plot after ω = 5 rad/sec is

$$= [-20 + (-20)]\ \text{dB/decade}$$

$$= -40\ \text{dB/decade}$$

The slope of the after second corner frequency, i.e., ω = 10 rad/sec.

$$= [-40 + (-29)]\ \text{dB//decade}$$

$$= -60\ \text{dB/decade}$$

The phase angle is

$$\angle G(j\omega) = -90° - \tan^{-1}(0.2\,\omega) - \tan^{-1} 0.1\,\omega$$

ω :	1	2	5	10	50
$\angle G(j\omega)$:	– 107°	– 123.1°	– 161.6°	– 198.4°	– 252.98°

It is easily follows that K_v = 30 satisfy the specification on K_v. The Bode plot of open loop transfer function with this value of K is shown below, from which it is found that the uncompensated system has a crossover frequency of 9.8 rad/sec and a phase margin of –17°. Thus, the uncompensated system is therefore unstable for the specified K_v.

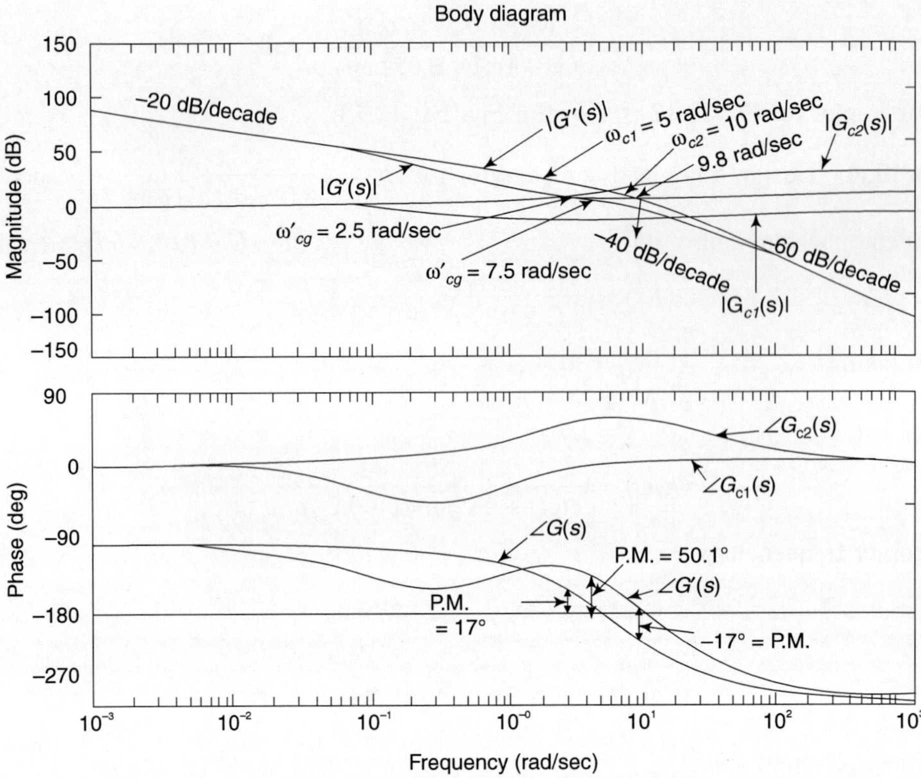

Figure Ex. 10.14 Bode plot of Ex. 10.14

It is seen from the Bode plot Fig. Ex. 10.14 of the uncompensated system that it must be brought down by 20 dB the new crossover frequency to be 2.5 rad/sec, i.e., = ω'_{cg2} 2.5 rad/sec.

This gives the β-parameter of the lag section as

$$20 \log \beta = 20$$

$$\beta = 10$$

Let us choose τ_1 of the lag section to be 1 rad/sec. The lag section design is thus complete and its transfer function is given by

$$G_{c1}(s) = \frac{(s+1)}{(10\,s+1)}$$

It is found that the lag section compensated system has a phase margin of 17°.

We now proceed to design the lead section.

This design is constrained by the fact that $\alpha = \frac{1}{\beta} = \frac{1}{10} = 0.1$ is already fixed. The maximum lead provided by the lead section is

$$\phi_m = \sin^{-1}\left\{\frac{\left(1-\frac{1}{\beta}\right)}{\left(1+\frac{1}{\beta}\right)}\right\} = 56°$$

To fully utilize the lead effect, we choose the compensated crossover frequency to coincide with ω_m, which is then obtained as that frequency where the lag compensated system has a dB of

$$-10 \log \beta = -10$$

This value read off from bode plot is 7.5 rad/sec

i.e., $$\omega'_{cg1} = \omega_m = 7.5 \text{ rad/sec}$$

$$\omega_m = \sqrt{\frac{\beta}{\tau_2}} = 7.5$$

$$\tau_2 = 0.425 \text{ and } \frac{\tau_2}{\beta} = 0.0425$$

The transfer function of the lead section is

$$G_{c2}(s) = \frac{(0.425\, s+1)}{(0.0425\, s+1)} = \frac{(1+\tau_2\, s)}{\left(1+\frac{\tau_2}{\beta}\, s\right)}$$

The open loop transfer function of the lag lead compensated system

$$G'(s) = G_{c1}(s)\, G_{c2}(s)\, G_c(s)$$

$$G'(s) = \frac{30\,(s+1)\,(0.425\, s+1)}{s\,(0.1\, s+1)\,(0.2\, s+1)\,(10\, s+1)\,(0.0425\, s+1)}$$ **Ans.**

Example 10.15 Derive the transfer function of lead and lag compensating network. Discuss the different aspect of selecting a compensating network. *(U.P.T.U., 2006–07, ACS)*

Solution Applying KVL in input loop

We get $$V_i(s) = \left[\frac{R_1 \frac{1}{cS}}{R_1 + \frac{1}{cS}}\right]$$

$$V_i(s) = \frac{[(R_1+R_2)+R_1\, R_2\,(Cs)]}{(R_1 Cs+1)} \quad \text{...(i)}$$

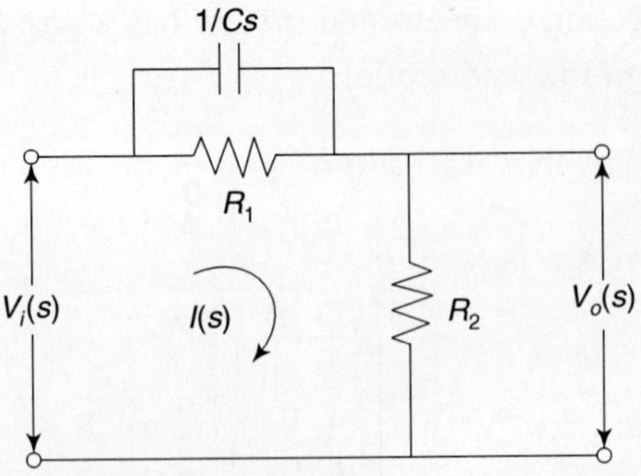

Figure Ex. 10.15 (a) Phase lead network

Applying KVL in output loop

$$V_o(s) = R_2 \, I(s) \quad \text{...(ii)}$$

Putting the value of $I(s)$ from Eq. (ii) to Eq. (i), we get

$$V_i(s) = \frac{[(R_1 + R_2) + R_1 \, R_2 \, (Cs)]}{(R_1 \, Cs + 1)} \left[\frac{V_o(s)}{R_2} \right]$$

$$\frac{V_o(s)}{V_i(s)} = \frac{R_2 \, (R_1 \, Cs + 1)}{(R_1 + R_2) + R_1 \, R_2 \, Cs)}$$

$$\frac{V_o(s)}{V_i(s)} = \frac{R_2}{R_1 + R_2} \left[\frac{(R_1 \, Cs + 1)}{1 + \frac{R_1 \, R_2}{R_1 + R_2} Cs} \right]$$

$$\frac{V_o(s)}{V_i(s)} = \alpha \left[\frac{1 + s \, T_1}{1 + s \, T_2} \right]$$

where $\alpha = \dfrac{R_2}{R_1 + R_2}$

$$T_1 = R_1 \, C$$

and $T_2 = \dfrac{R_1 R_2}{R_1 + R_2} C$

The block diagram representation is given as

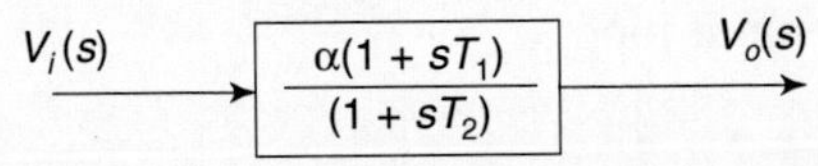

Figure Ex. 10.15 (b) Block diagram of phase lead network

- *It is observed that the zero is nearer to origin as compared to pole hence the effect of zero is dominant, therefore, the phase lead network when introduced in cascade with forward path of a transfer function the phase shift is increased.*

Phase Lag Network

Applying KVL at input side

$$V_i(s) = R_1\, I(s) + R_2\, I(s) + \frac{1}{Cs} I(s)$$

$$V_i(s) = \left[R_1 + R_2 + \frac{1}{Cs} \right] I(s)$$

$$V_i(s) = \left\{ \frac{(R_1 + R_2)\, Cs + 1}{Cs} \right\} I(s) \quad ...(1)$$

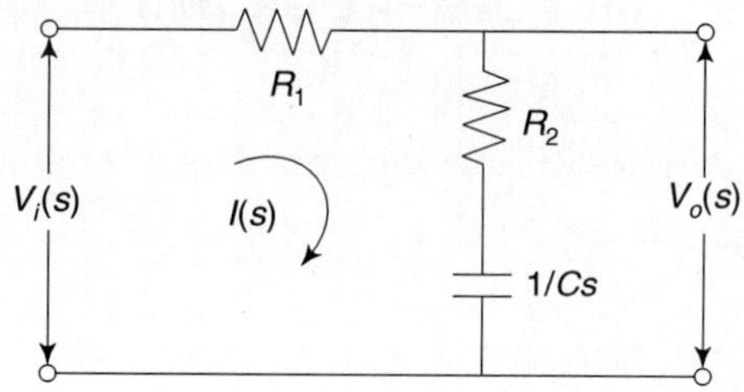

Figure Ex. 10.15 (c) Phase lag network

Applyin KVL at output side

$$V_o(s) = R_2\, I(s) + \frac{1}{Cs} I(s)$$

$$V_o(s) = \left\{ R_2 + \frac{1}{Cs} \right\} I(s)$$

$$V_o(s) = \frac{(R_2\, Cs + 1)}{Cs} I(s) \quad ...(2)$$

Putting the value of $I(s)$ from Eq. (2) to Eq. (1), we get

$$V_i(s) = \left[\frac{(R_1 + R_2)\, Cs + 1}{Cs} \right] \times \frac{V_o(s)}{\frac{(R_2\, Cs + 1)}{Cs}}$$

$$\frac{V_o(s)}{V_i(s)} = \frac{(R_2\, Cs + 1)}{[R_1 + R_2)\, Cs + 1]}$$

$$\frac{V_o(s)}{V_i(s)} = \frac{(1 + s\, T_1)}{(1 + s\, T_2)}$$

where $T_1 = R_2\, C$

and $T_2 = (R_1 + R_2)\, C$

The block diagram representation is given in Fig. Ex. 10.15 (d).

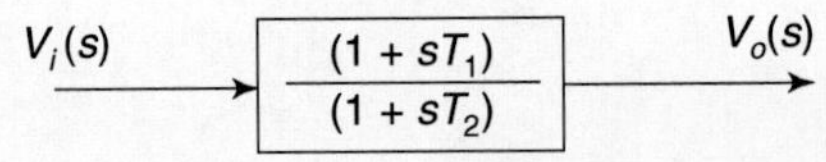

Figure Ex. 10.15 (d) Block diagram of phase lag network

- *It is observed that the pole is nearest to origin compared to zero, hence the pole is dominant, therefore, the phase lag network when introduced in cascade with forward path of a transfer function the phase shift is reduced.*

MATLAB PROGRAMS

Program 10.1

```
% Program for finding maximum phase lead and maximum phase lag.
clear all
clc
R1 = input(' Enter the value of Resistance R1 in Ohm :-')
R2 = input(' Enter the value of Resistance R2 in Ohm :-')
C1 = input(' Enter the value of Capacitance C1 in Farad :-')
C2 = input(' Enter the value of Capacitance C2 in Farad :-')
alpha = R2/(R1+R2);
'Time constant'
T = R1*C1
'Maximum Phase Lead is :-'
K_lead = asin((1-alpha)/(1+alpha));  % K_lead is in radian
Max_Phase_Lead ± (K_lead* 180)/pi    % phase is in degree
'Maximum Phase Lag is :-'
beta = (R1+R2)/R2;
K_lag = asin((1-beta)/(1+beta));     % K_Lag is in radian
Max_Phase_Lead = (K_lag*180)/pi      % phase is in degree
```

INPUT:

```
Enter the value of Resistance R1 in Ohm :- 1800
Enter the value of Resistance R2 in Ohm :- 3000
Enter the value of Capacitance C1 in Farad :- 0.0000001
Enter the value of Capacitance C2 in Farad :- 0.0000001
```

OUTPUT:

```
Time constant
T= 1.8000e-004
Maximum Phase Lead is :-
Max_Phase_Lead = 13.3424
```

```
Maximum Phase Lag is :-
Max_Phase_Lead = -13.3424
```

Program 10.2

```
% Program for designing a system by root locus techniques
% For a given open loop transfer function given below, designe an ideal
% derivative compensator to yield a 26% overshoot, with a threefold
% reduction in settling time
% G(s) = k/s(s+4)(s+6)
clf;                         % Clear graph on screen
'Uncompensated System'       % Display label
numg = 1 ;                   % Define numerator of G(s)
deng = poly([0 -4 -6]);      % Define denominator of G(s)
'G(s)'                       % Display label
G = tf(numg,deng)
Mp = input('Enter desired percent overshoot:-');
                             % input desired overshoot Let 16%
zeta = -log(Mp/100)/sqrt(pi^2+[log(Mp/100)]^2);
                             % Calculate damping ratio
rlocus(G);                   % Root locus of uncompensated system
sgrid(zeta,0)                % desired percent overshoot lines
title('Uncompenstaed root locus with, num2str(Mp),')
[k,p] = rlocfind(G);         % Generate gain k, and closed loop
                               poles p for point selected
'Close-loop poles'
p
f = input('Give pole number that is operating point:-')
                           % chose uncompensated system dominant pole
pole
operating_point = p(f) % Display uncompensated dominant pole
gain = k               % Display gain of uncompensated system
estimated_Ts = 4/abs(real(p(f)))    % Display Ts
Estimated_Tp = pi/abs(imag(p(f)))   % Display Tp
estimated_Mp = Mp      % Display uncompensated percent overshoot
estimated_zeta = zeta  % Display uncompensated damping ratir
estimated_Wn = sqrt(real(p(f))^A2+imag(p(f)A2))
                       % uncompensated natural frequency
num_Kv = conv([1 0], numg); % set numerator to evaluate Kv
den_Kv = deng;         % set denominator to evaluate Kv
```

```
sG = tf(num_Kv,den_Kv);
G = minreal(sG)        % Cancel common poles and zeros
Kv = dcgain(k*sG)      % uncompensated Kv
ess = 1/Kv             % uncompensated steady state error
'T(s)'
T = feedback(k*G,1)    % find uncompensated T(s)
step(T)                % Step response of uncompensated system
title(['uncompensated system step response'])
'Press any key to compensated system'
pause
'Compensated system'
desired_Ts = input('enter desired settling time:- ');  % input desired Ts
Wn = 4/(desired_Ts*zeta);                                % Calculate Wn
desired_pole = (-zeta*Wn)+(Wn*sqrt(1-zeta^2)*pi);
                       % Find desired dominant pole location
angle_desired_pole = (180/pi);
angle(polyval(numg,desired_pole)/polyval(deng,desired_pole));
                       % calculate angular
contribution to desired pole without compensator
PD_angle = 180 - angle_desired_pole;
   % find required angular contribution from PD compensator
Zc = ((imag(desired_pole)/tan(PD_angle*pi/180))-real(desired_pole));
                      % calculate PD zero location
'PD compensator'
numc = [1 Zc];        % define numerator of compensator
dene = [0 1];         % define denominator of compensator
'Gc(s)'
Gc = tf(numc,denc)
'G(s)Gc(s)'
Ge = G*Gc
rlocus(Ge,0:0.5:50)  % Plot PD compensated system root locus
sgrid(zeta,O);
title(['PD compensated root locus'])
[k,p] = rlocfind(Ge) % generate k and poles
'Close loop poles:-'
p
fc = input('Give pole number that is operating point:-');
'Compensated root locus'
operating_point = p(fc);
```

```
gain = k
estimated_Ts = 4/abs(real(p(fc))) % Display PD compensated Ts
Estimated_Tp = pi/abs(imag(p(fc)))% Display PD compensated Tp
estimated_Mp = Mp                 % Display compensated percent overshoot
estimated_zeta = zeta             % Display compensated damping ratir
estimated_Wn = sqrt(real(p(fc))^2+imag(p(fc)^2))
                                  % uncompensated natural frequency
s=tf([1 0],1);
sGe = s*Ge;
sGe =minreal(sGe);
Kv = dcgain(k*sGe);
ess = 1/Kv;
'Tc(s)'
T = feedback(k*Ge,1)
'Press any key to continue'
pause
```

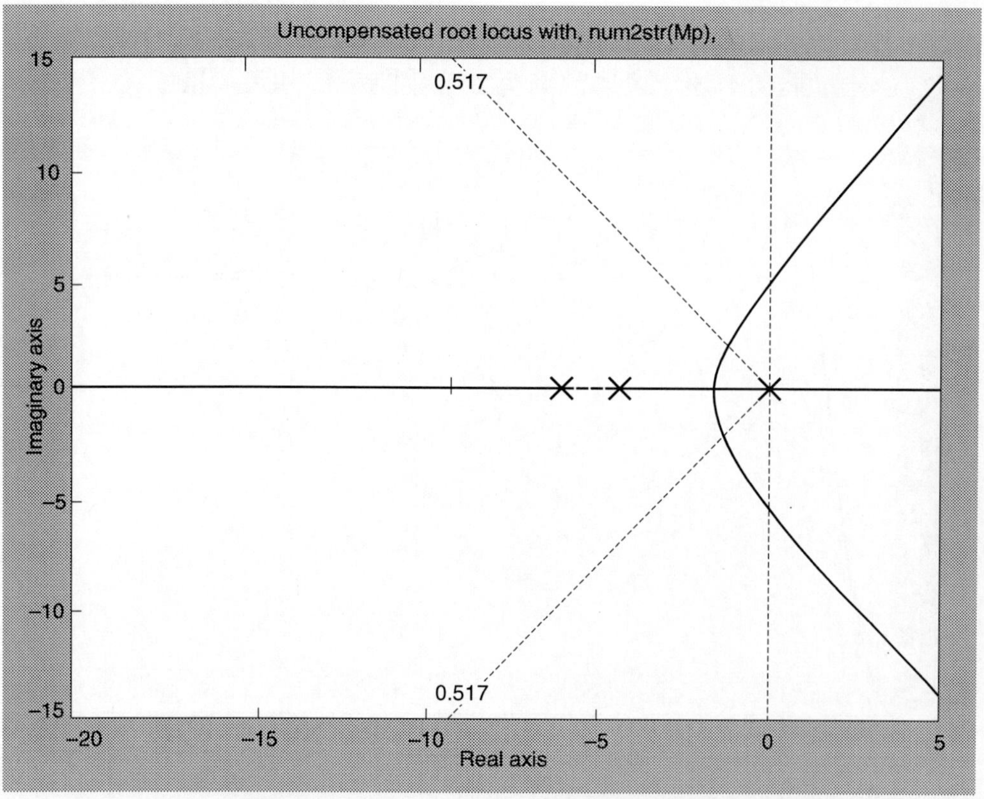

Figure P. 10.2 (a) Root locus of uncompensated system

```
step(T)
title('PD compensated system response')
```

INPUT

```
Uncompensated system
Enter desired percent overshoot:- 15
Give pole number that is operating point:- 2
selected_point = -8.2405 - 0.2329i
```

Compensated system

```
enter desired settling time:- 12
selected_point = -4.2275 - 0.0248i
```

OUTPUT

Unompensated system

```
Close-loop pole
p =   -8.2496
      -0.8752 + 2.9654i
      -0.8752 - 2.9654i
```

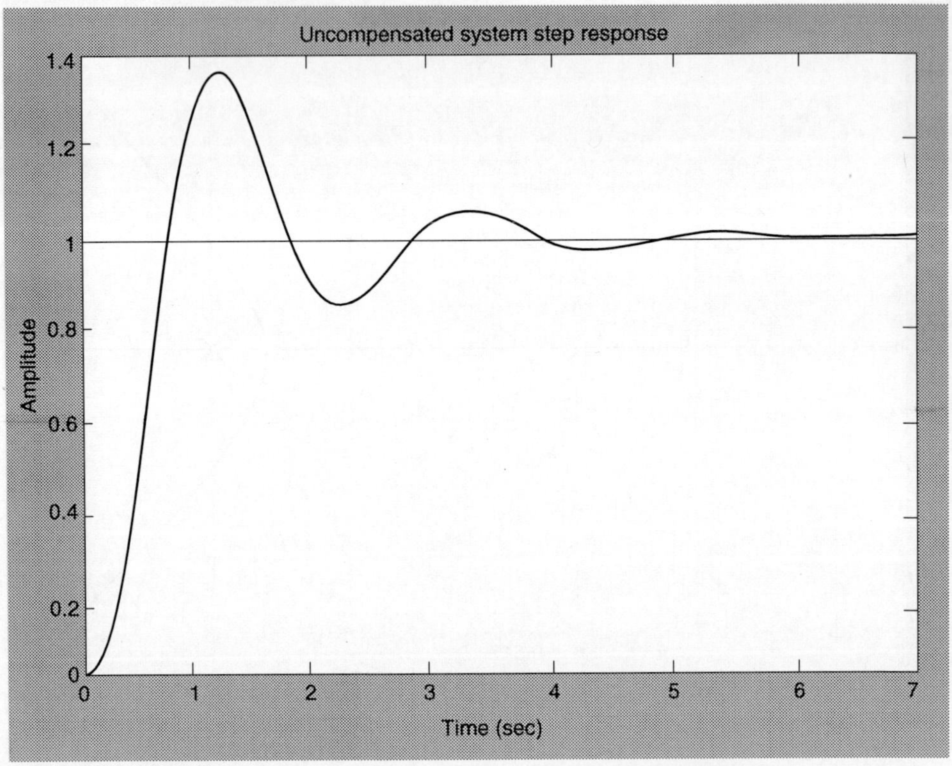

Figure P. 10.2 (b) Step response of uncompensated system

```
operating_point = -0.8752 + 2.9654i
gain = 78.8651
estimated_Ts = 4.5704
Estimated_Tp = 1.0594
estimated_Mp = 15
estimated_zeta = 0.5169
estimated_Wn = 0 + 2.1035i
```

Transfer function: $\dfrac{1}{s^2 + 10s + 24}$

```
Kv = 3.2860
ess = 0.3043
```

Transfer function: $T(s) = \dfrac{78.87}{s^3 + 10s^2 + 24s + 78.87}$

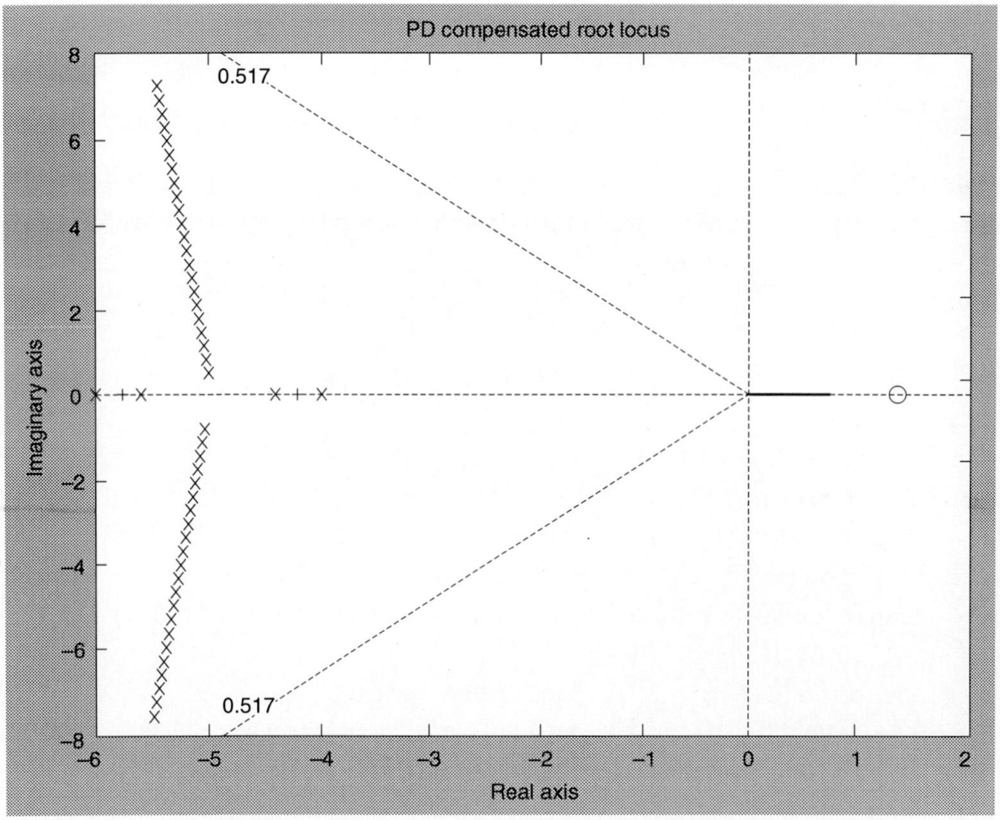

Figure P. 10.2 (c) Root locus of derative compensated system

PD compensateed system

```
Transfer function: Gc(s)
s - 1.401
                                   s - 1.401
Transfer function: G(s) Gc(s) = -------------------
                                 s^3 + 10s^2 + 24s

k = 0.3047
p = -5.7884
  -4.2290
  0.0174
gain = 0.3047
estimated_Ts = 0.9458
Estimated_Tp = inf
estimated_Mp = 15
estimated_zeta = 0.5169
estimated_Wn = 4.2290
                                  0.3047 s - 0.4268
Transfer function: Tc(s) = ------------------------------
                           s^3 + 10s^2 + 24.3s - 0.4268
```

Program 10.3

```
% Program for designing a system using bode plot by gain adjustment
% G(s) = 50/s(s+25)(s+21)
clear all;
clc
clf;                          % Clear the figure
numg = 10;
deng = poly([-1 -4 -9]);
G = tf(numg,deng)
Mp = input('Enter desired percent overshoot:- ')
zeta = (-log(Mp/100))/(sqrt(pi^2+log(Mp/100)^2));
Pm = atan(2*zeta/(sqrt(-2*zeta^2+sqrt(1+4*zeta^4))))*(180/pi);
w = 0.01 :0.01:1000;
[M P] = bode(G,w);            % find bode plot
Ph = -180+Pm;                 % required phase angle
for k = 1:1:length(P);% search bode data for required phase angle
  if P(k)<=0
    M = M(k);
     'Required k'
```

```
    k = 1/M
    break
  end
end
T = feedback(k*G,1);
step(T)
title('Close loop Response')
```

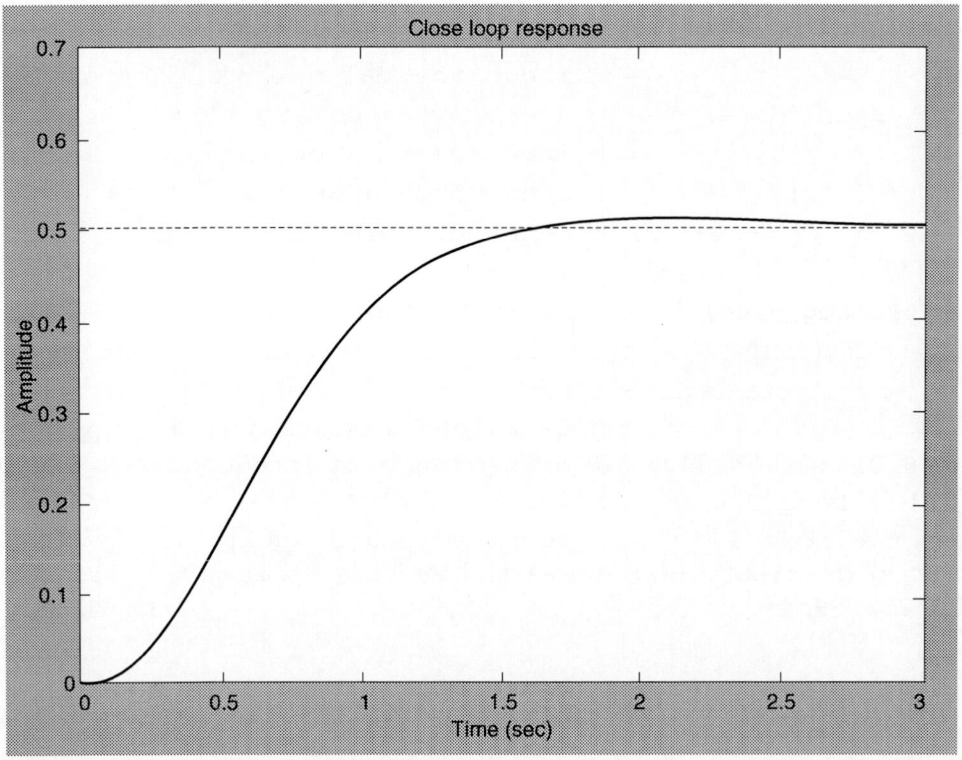

Figure P. 10.3 Close loop response

INPUT

```
Enter desired percent overshoot:- 20
```

Output

Transfer function: $\dfrac{10}{s^3 + 14s^2 + 49s + 36}$

```
Required k
k = 3.6002
```

Program 10.4

```
% Program to design a lag compensator for a given system in P. 8.2, using
% bode plot to design a lag compensator to yield a tenfold improvement in
% steady state error over the gain compensated system and keeping the
% percent overshoot at 10%
clear all;
clc;
clf;                        % clear the flag
k = input('Enter the value of k to meet ess:-');
                            % input the value of k
Mp = input('Enter the percent overshoot Mp:-');
                            % input desired overshoot
numg = 50*k;                % Define numerator
deng = poly([0-36-81]);% define denominator
'G(s)'
G = tf(numg,deng)          % Create transfer function
zeta = (-log(Mp/100))/(sqrt(pi^2+log(Mp/100)^2));% Calculate damping ratio
Pm = atan(2*zeta/(sqrt(-2*zeta^2+sqrt(1+4*zeta^4))))*(180/pi)+10;
                            % Calculate maximum phase
w = 0.01 :0.01:1000;        % define range of frequency
[M P] = bode(G,w);          % find bode plot
Ph = -180+Pm;               % required phase angle
for k = 1:1 :length(P);     % search bode data for required phase angle
  if P(k)-Ph<=0
    M = M(k);
    Wf = w(k);
    'Required k'
    k=1/M
    break
  end
end
Wh = Wf/10;
Wl = (Wh/M);
numc = [1 Wh];
denc = [1 Wl];
Kc = Wl/Wh;
'Lag Compensator'
Kc
```

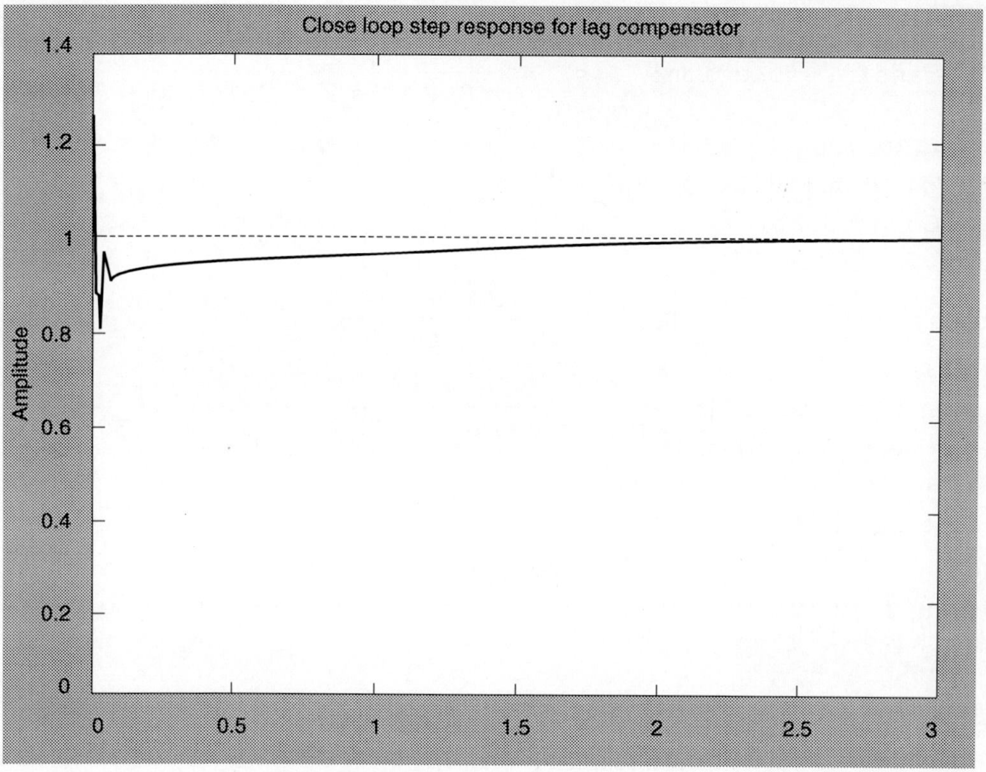

Figure P. 10.4 Close loop response

```
'Gc(s)'
Gc = tf(Kc*numc,denc)
Gc(s)G(s)
GcG = Gc*G;
s = tf ([10], 1);
sGcG = s*GcG
sGcG = minreal(sGcG);
Kv = dcgain(sGcG)
T = feedback(k*GcG,1);
step(T)
title('Close loop step response for lag compensator')
```

INPUT

```
Enter the value of k to meet ess:- 0.2
Enter the percent overshoot Mp:- 15
```

OUTPUT

```
                                         10
10 Transfer function: G(s) = ---------------------
                              s^3 + 117s^2 + 2916s

Required k
k = 3.7253e+003
Lag Compensator
Kc = 3.7253e+003
                          3725s + 4467
Transfer function: Gc(s) = ------------
                            s + 4467
                                3.725e004 s^2 + 4.467e004 s
Transfer function: Gc(s)G(s) = ------------------------------------------------
                        s^4 + 4584^3 + 5.255e005 s^2 + 1.302e007 s

Kv = 0.0034
```

EXERCISE

10.1 The system whose specification such as

(i) $K_v \geq 5$

(ii G.M. $\geq$ 6 dB

(iii) P.M. $\geq$ 25 and the transfer function is $G(s)\, H(s) = \dfrac{K}{s\,(1+0.5\,s)^2}$

Design a phase lead compensation network to satisfy the above specifications.

10.2 Explain the compensation technique in control system.

10.3 A feedback system with unity feedback has open loop transfer function

$$G(s) = \frac{K}{s\,(s+1)\,(0.2\,s+1)}$$

Design phase lag compensation for the system to achieve the following specification

(i) $K_v = 8$

(ii) P.M. = 40° and compute crossover frequency of the compensated and uncompensated system.

10.4 A unity feedback control system has an open loop transfer function of $G(s) = 1/s^2$. Design a compensation network such that the P.M. of 45° is achieved without sacrificing system velocity error constant and sketch the bode plot of compensated system.

10.5 Design a lead compensator such that the system having following open loop transfer function will have a P.M. of 45°

$$G(s)\, H(s) = \frac{10}{(s+1)\,(1+0.2\,s)^2}$$

10.6 Given the system below in Fig. E. 10.6, design an ideal derivative compensator to yield a 16% overshoot, with a threshold reduction in settling time.

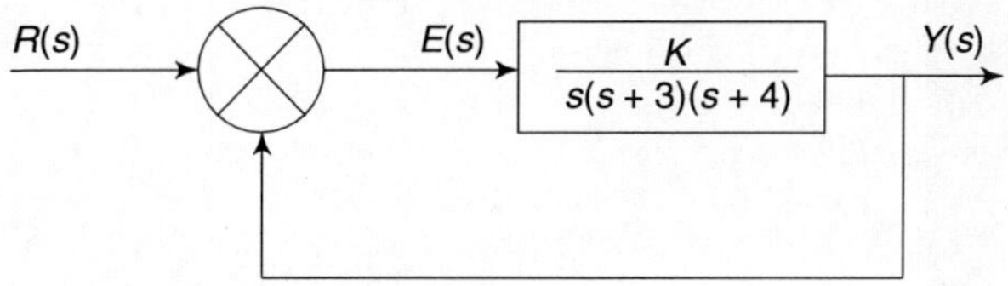

Figure E. 10.6

10.7 For a unity feedback system with a forward transfer function

$$G(s) = \frac{K}{s(s+50)(s+120)}$$

Use frequency response techniques to find the value of gain K, to yield a closed loop step response with 20% overshoot.

10.8 Make a technical comparison between cascade lead and cascade lag compensation used for linear control systems, support with suitable examples. *(U.P.T.U., 2007–08, CS)*

10.9 Explain the effect of adding a pole to a second order system. When if it useful to add a zero to the system. *(U.P.T.U., 2005–06, CS)*; *(U.P.T.U., 2004–05, CS)*

10.10 Explain lag, lead, lead lag compensation and give examples. Where are these used? Take suitable example which have both lag and lead features and justify. *(U.P.T.U., 2004–05, CS)*

10.11 Explain the difference between phase-lead, and phase lag compensators. Draw their Bode plots. *(U.P.T.U., 2003–04, CS)*

10.12 A first order lead compensator has its pole at $-P_c$ and its zero at $-\alpha P_c$. Comment upon the value of α. Sketch the Bode plot for the compensation what will be maximum phase angle contribution of the compensator and at what ω value? *(U.P.T.U., 2007–06, ACS)*

10.13 Comment upon the application area of lag compensator. Give a simple RC circuit to implement it. Derive its transfer function. *(U.P.T.U., 2007–08, ACS)*

10.14 Drive the transfer function of lead and lag compensation network. Discuss the different aspect of selecting a compensating network. *(U.P.T.U., 2006–07, ACS)*

10.15 Explain about phase lead compensation by root locus method with proper expression and plots. *(U.P.T.U., 2005–06, ACS)*

10.16 Discuss about lag lead compensation with suitable diagram and also represent on s-plane clearly. *(U.P.T.U., 2004–05, ACS)*

10.17 Explain about phase lag compensation by root locus method with proper expression and plot. *(U.P.T.U., 2004–05, ACS)*

ANSWERS

10.3 $G_c(s) = \dfrac{6.67\,s+1}{66.67\,s+1}$

ω_c (uncompensated) = 2.6 rad/sec

ω_c (compensated) = 0.64 rad/sec

10.4 $G_c(s) = \dfrac{1.61\,s+1}{0.214\,s+1}$, lead network

10.7 K = 194, 200

CHAPTER

11 Review of State Variable Technique

11.1 INTRODUCTION

In the previous chapters, we have studied several methods to analyse and design of feedback control system like root locus, frequency response methods. In these methods we have seen that the physical system can be modelled in the form of transfer function. With the transfer function we can analyse and design system easily. But there are certain disadvantages using transfer function for system analysis and design as follows:

1. The transfer function model is only applicable to LTI (linear time invariant) systems.
2. It is only applicable to SISO (single input single output) systems.
3. It gives the information only for a given input and does not give any information about internal state of the system.
4. Transfer function of the system is only defined under zero initial conditions. So we can say that the design and analysis of feedback control system using classical method like root locus frequency response, etc., is not suitable for complex systems and may not lead to a control system which yields an optimum performance in some defined manner.

So from the above discussion we can say that the classical method for analysis and design a control system is not sufficient to get the optimum result. We need some more general method which represents the system along with the output and gives information about internal state of the system variables at some predetermined points. That means we need some method which overcomes the previous disadvantages. Such approach is state variable technique. It is a time domain approach which provides basic concepts for modern control theory and optimization of the system response.

11.2 STATE SPACE MODEL USING PHASE VARIABLES

In this section, we discuss a state-space model of control system using phase variables as a state variable. The state model of the control system is easily determined if the system is known in the form of transfer function or differential equation.

The general form of an 4th order linear differential equation for which the output is $y(t)$ and input is $u(t)$ is

$$y^n + a_1\, y^{n-1} + \cdot\, a_{n-1}\, \dot{y} + a_n\, y = b_o\, u^m + b_1\ u^{m-1} + \ldots\ b_{m-1}\, u + b_m\, u \qquad \ldots(11.1)$$

For the LTI system a_i and b_j are constants, m and n are integers with $m \leq n$

$$y^n \underline{\Delta} \frac{d^n y}{dt^n}$$

The transfer function of the system which is obtained from Eq. (11.1) is considering the initial conditions are zero,

$$T(s) = \frac{Y(s)}{U(s)} = \frac{b_o\, s^m + b_1\, s^{m-1} \ldots + b_{n-1}\, s + b_m}{s^n + a_1\, s^{n-1} + \ldots\, a_{n-1}\, s + a_n} \qquad \ldots(11.2)$$

When $m = n$,

$$T(s) = \frac{b_o\, s^n + b_1\, s^{n-1} \ldots + b_{n-1}\, s + b_n}{s^n + a_1\, s^{n-1} + \ldots\, a_{n-1}\, s + a_n} \qquad \ldots(11.3)$$

So we can obtain a state model from the transfer function [$T(s)$] of the system with initial conditions are zero, then relax zero initial conditions to arbitrary initial conditions. First we consider that $Z = 0$ that means transfer function does not have zeros.

$$T(s) = \frac{Y(s)}{U(s)} = \frac{b}{s^n + a_1\, s^{n-1} + \ldots\, a_{n-1}\, s + a_n} \qquad \ldots(11.4)$$

By Eq. (11.4) the differential equation is

$$y^n + a_1\, y^{n-1} + \ldots\, a_{n-1}\, \dot{y} + a_n y = bu \qquad \ldots(11.5)$$

Let us consider

$$\begin{aligned} x_1 &= y \\ x_2 &= \dot{y} \\ &\ldots\ldots\ldots\ldots \\ &\ldots\ldots\ldots\ldots \\ x_n &= y^{n-1} \end{aligned} \qquad \ldots(11.6)$$

By Eq. (11.5) is reduced to a set of n first order differential equations.

$$\begin{aligned} \dot{x}_1 &= x_2 \\ \dot{x}_2 &= x_3 \\ &= x_4 \\ &\ldots\ldots\ldots\ldots \\ &\ldots\ldots\ldots\ldots \\ \dot{x}_{n-1} &= x_n \\ \dot{x}_n &= -\, a_n\, x_1 - a_{n-1}\, x_2 \ldots\, a_1\, x_n + bu \end{aligned} \qquad \ldots(11.7)$$

Equation (11.7) result in the following state equations

$$\underbrace{\begin{bmatrix} \dot{x}_1 \\ \dot{x}_2 \\ \vdots \\ \dot{x}_{n-1} \\ \dot{x}_n \end{bmatrix}}_{x} = \underbrace{\begin{bmatrix} 0 & 1 & 0 & \cdots & 0 \\ 0 & 0 & 1 & \cdots & 0 \\ \vdots & \vdots & \vdots & & \\ 0 & 0 & 0 & \cdots & 1 \\ -a_n & -a_{n-1} & -a_{n-2} & \cdots & -a_1 \end{bmatrix}}_{A} \underbrace{\begin{bmatrix} x_1 \\ x_2 \\ x_3 \\ \vdots \\ x_{n-1} \\ x_n \end{bmatrix}}_{x} + \underbrace{\begin{bmatrix} 0 \\ 0 \\ \vdots \\ 0 \\ b \end{bmatrix}}_{B} u \quad \text{...(11.8)}$$

So $$\dot{X} = AX + Bu \quad \text{...(11.9a)}$$

By Eqs. (11.8) and (11.9) we can say that matrix A is the matrix in which the upper off diagonal element has l's and the last row, has negative of the coefficients of the original differential equation and rest elements are zero. A matrix is known as Bush matrix or companion form.

The block diagram of state model for Eq. (11.9) is shown in Fig. 11.1.

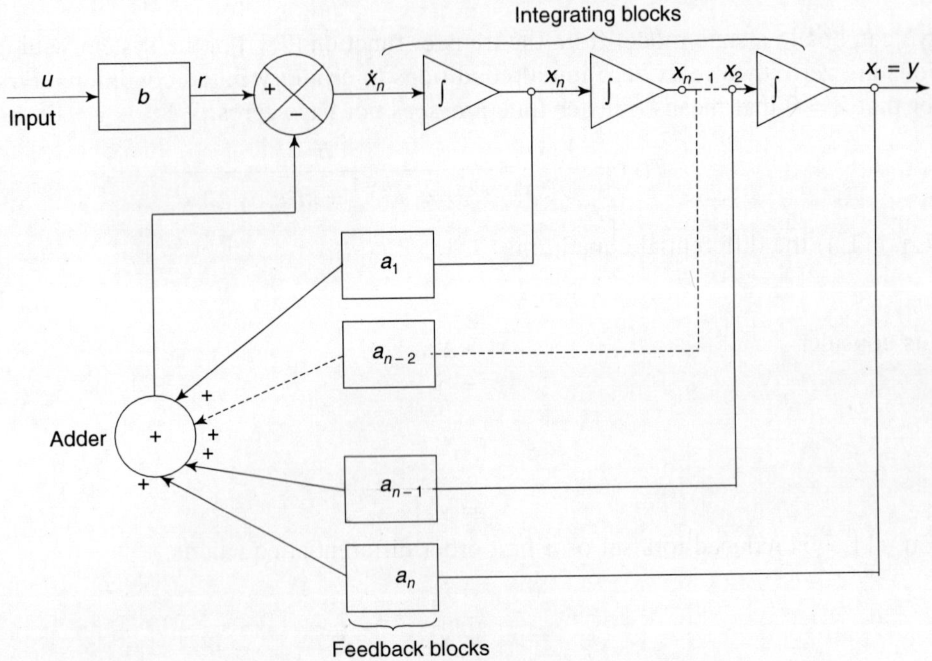

Figure 11.1 Block diagram of state model

The matrix B has only last element is non-zero. The output of the model is

$$y = x_1$$

So the output equation is given by

$$y = CX \quad \text{...(11.9b)}$$

where $c = [1, 0 .... 0]$

By the above discussion we can say that if there is a transfer function of the system which does not have zeros we can drive the state model of such system through the differential equation. But when a transfer function has zeros as well, the resulting differential equation contains terms which are derivatives of the control force or input (u) and the method discussed above cannot be applied as such. For such system has zeros also another method with the signal flow graph is used below.

Phase Variable Formulation for Transfer Function with Poles and Zeros

Let us consider a fourth order transfer function by Eq. (11.3)

$$T(s) = \frac{Y(s)}{U(s)} = \frac{b_o\, s^4 + b_1\, s^3 + b_2\, s^2 + b_3\, s + b_4}{s^4 + a_1\, s^3 + a_2\, s^2 + a_3\, s + a_4} \qquad ...(11.10)$$

We find out four state variables x_1, x_2, x_3 and x_4. The signal flow graph must have at least four integrators Eq. (11.10) may be re-arranged as

$$T(s) = \frac{\left[(b_o + b_1/s + b_2/s^2 + b_3/s^3 + b_4/s^4)\right]s^4}{\left[(1 + a_1/s + a_2/s^2 + a_3/s^3 + a_4/s^4)\right]s^4}$$

$$T(s) = \frac{b_o + b_1/s + b_2/s^2 + b_3/s^3 + b_4/s^4}{1-(-a_1/s - a_2/s^2 - a_3/s^3 - a_4/s^4)} \qquad ...(11.11)$$

In chapter 1, it was shown that the transfer function and signal flow graphs are related by Mason's gain formula, reproduced below:

$$T_o(s) = \frac{1}{\Delta}\left[\sum_{i=1}^{n} P_i\, \Delta_i\right] \qquad ...(11.12)$$

where G_i = path gain of ith forward path

Δ = 1– (sum of loop gains of all individual loops) + (sum of gain products of all possible combinations of two non-touching loops) – (sum of gain products of all possible combinations of three non-touching loops) + ...

$$\Delta = 1 - \sum_j L_{j1} + \sum_j L_{j2} - \sum_j L_{j3} + \ldots$$

Δ_i = it is formed by removing loop gains from A that touch the ith forward path.

Comparing Eq. (11.10) with (11.11) we observe that a signal flow graph of Eq. (11.9) may consist of

(a) Four feedback loops (touching each other) with gains $-\,a_1/s$, $-\,a_2/s^2$, $-\,a_3/s^3$ and a_4/s^4.

(b) Five forward paths which touch the loops and have gains b_o, b_1/s, b_2/s^2, b_3/s^3 and b_4/s^4.

A signal flow graph configuration which satisfies the above requirements is shown in Fig 11.2.

From the figure we have

$$y = x_1 + b_o\, u$$

$$\dot{x}_1 = -\,a_1\,(x_1 + b_o\, x) + x_2 + b_1\, u$$

$$\left.\begin{aligned}\dot{x}_1 &= -\,a_1\,x_1 + x_2 + (b_1 - a_1\,b_o)\,u\\ \dot{x}_2 &= -\,a_2\,x_1 + x_3 + (b_2 - a_2\,b_o)\,u\\ \dot{x}_3 &= -\,a_3\,x_1 + x_4 + (b_3 - a_3\,b_o)\,u\\ \dot{x}_4 &= a_4\,x_1 + (b_4 - a_4\,b_o)\,u\end{aligned}\right\} \qquad ...(11.13)$$

or

$$\begin{bmatrix}\dot{x}_1\\ \dot{x}_2\\ \dot{x}_3\\ \dot{x}_4\end{bmatrix} = \begin{bmatrix}-\,a_1 & 1 & 0 & 0\\ -\,a_2 & 0 & 0 & 1\\ -\,a_3 & 0 & 0 & 1\\ -\,a_4 & 0 & 0 & 0\end{bmatrix}\begin{bmatrix}x_1\\ x_2\\ x_3\\ x_4\end{bmatrix} + \begin{bmatrix}b_1 - a_1\,b_0\\ b_2 - a_2\,b_0\\ b_3 - a_3\,b_0\\ b_4 - a_4\,b_0\end{bmatrix} u \qquad ...(11.14a)$$

$$y = [1\,0\,0\,0]\begin{bmatrix}x_1\\ x_2\\ x_3\\ x_4\end{bmatrix} + b_o\,u \qquad ...(11.14b)$$

The initial conditions, x_1 (0), x_2 (0), x_3 (0), x_4 (0), can be obtained in terms of given initial conditions

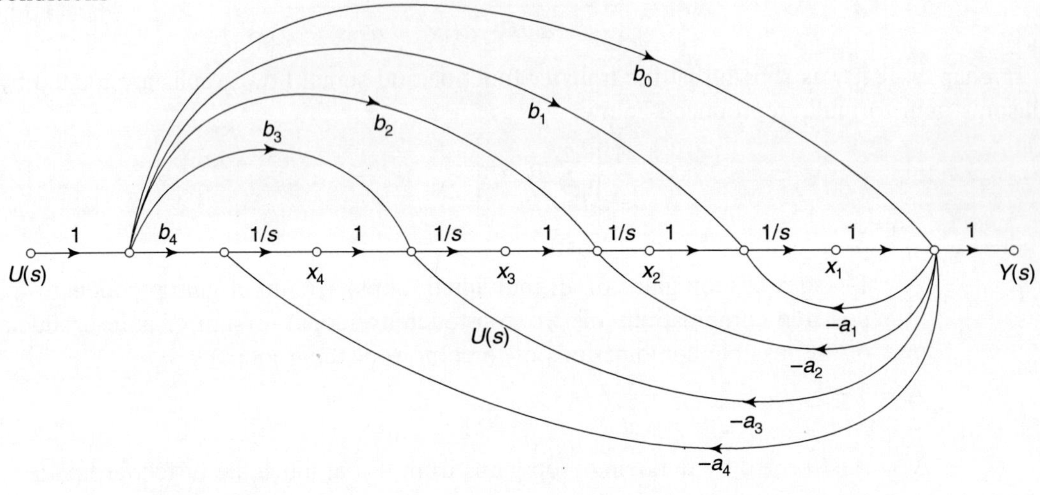

Figure 11.2 Signal flow graph for fourth order system of Eq. 11.10

y (0), $\dot{y}$ (0), $\ddot{y}$ (0), $\dddot{y}$ (0), u (0), $\dot{u}$ (0) $\ddot{u}$ (0) $\dddot{u}$ (0) from Eq. (11.13). The result is given below:

$$\begin{bmatrix}x_1(0)\\ x_2(0)\\ x_3(0)\\ x_4(0)\end{bmatrix} = \begin{bmatrix}1 & 0 & 0 & 0\\ a_1 & 1 & 0 & 0\\ a_2 & a_1 & 1 & 0\\ a_3 & a_2 & a_1 & 1\end{bmatrix}\begin{bmatrix}y(0)\\ \dot{y}(0)\\ \ddot{y}(0)\\ \dddot{y}(0)\end{bmatrix} + \begin{bmatrix}-\,b_0 & 0 & 0 & 0\\ -\,b_1 & -\,b_0 & 0 & 0\\ -\,b_2 & -\,b_1 & -\,b_0 & 0\\ -\,b_3 & -\,b_2 & -\,b_1 & -\,b_0\end{bmatrix}\begin{bmatrix}u(0)\\ \dot{u}(0)\\ \ddot{u}(0)\\ \dddot{u}(0)\end{bmatrix} \qquad ...(11.15)$$

The result given in Eqs. (11.13) and (11.14) can easily be generalized for an nth order differential equation.

$$y^n + a_1\, y^{n-1} + \ldots\, a_{n-1}\, \dot{y} + a_n\, y = b_o\, u^m + \ldots\, b_{m-1}\, \dot{u} + b_m\, u \qquad \ldots(11.16)$$

As alternate phase variable formation for the transfer function of Eq. (11.11) may be obtained as follows:

The transfer function $G(s)$ of Fig. 11.3 may be divided into two parts as shown in Fig. 11.4.

$$G(s) = \frac{Y(s)}{U(s)} = \frac{X_1(s)}{U(s)} \cdot \frac{Y(s)}{X_1(s)} \qquad \ldots(11.17)$$

$$\frac{X_1(s)}{U(s)} = \frac{1}{s^4 + a_1\, s^3 + a_2\, s^2 + a_3\, s + a_4} \qquad \ldots(11.18)$$

$$\frac{Y(s)}{X_1(s)} = b_o\, s^4 + b_1\, s^3 + b_2\, s^2 + b_3\, s + b_4 \qquad \ldots(11.19)$$

The transfer function of Eq. (11.17) is without zeros and therefore its phase variable representation may be obtained from Eq. (11.8).

$$\begin{bmatrix} \dot{x}_1 \\ \dot{x}_2 \\ \dot{x}_3 \\ \dot{x}_4 \end{bmatrix} = \begin{bmatrix} 0 & 1 & 0 & 0 \\ 0 & 0 & 1 & 0 \\ 0 & 0 & 0 & 1 \\ -a_4 & -a_3 & -a_2 & -a_1 \end{bmatrix} \begin{bmatrix} x_1 \\ x_2 \\ x_3 \\ x_4 \end{bmatrix} + \begin{bmatrix} 0 \\ 0 \\ 0 \\ 1 \end{bmatrix} u \qquad \ldots(11.20a)$$

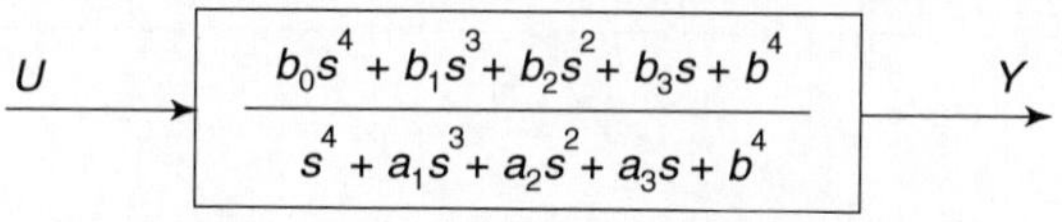

Figure 11.3 Fourth order system

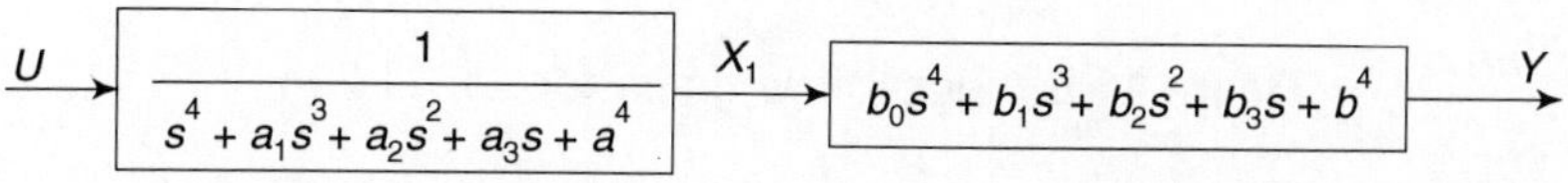

Figure 11.4 Alternate representation of Fig. 11.3

From the transfer function of Eq. (11.8), we get

$$y = b_o\, \ddddot{x}_1 + b_1\, \dddot{x}_1 + b_2\, \ddot{x}_1 + b_3\, \dot{x}_1 + b_4\, x_1$$

$$= b_o\,(-a_4\, x_1 - a_3\, x_2 - a_2\, x_3 - a_1\, x_4 + u) + b_1\, x_4 + b_2\, x_3 + b_3\, x_2 + b_4\, x_1$$

$$y = (b_4 - a_4\, b_o)\, x_1 + (b_3 - a_3\, b_o)\, x_2 + (b_2 - a_2\, b_o)\, x_3 + (b_1 - a_1\, b_o)\, x_4 + b_o\, u \qquad \ldots(11.20b)$$

Output Eq. 11.14 (b) can be written in vector matrix form as

$$y = [(b_4 - a_4\, b_0)\,(b_3 - a_3\, b_0)\,(b_2 - a_2\, b_0)\,(b_1 - a_1\, b_0)] \begin{bmatrix} x_1 \\ x_2 \\ x_3 \\ x_4 \end{bmatrix} + b_0\, u \qquad \text{...(11.20c)}$$

The signal flow graph corresponding to Eq. 11.20 (a) and (c) is shown in Fig. 11.5. Equation 11.20 (a) and c) can be generalized for n^{th} order system.

- *The disadvantage of phase variable formulation is that the phase variables, in general, are not physical variables of the system and therefore are not available for measurement and control purposes.*

We can say that phase variable formulation is only useful for the system having zeros also. With the poles, phase variables are simple to realize mathematically, they are not a practical set a state variables from measurement and control point of view also, phase variables do not provide any advantage.

Canonical valuables are most suitable from analysis point of view.

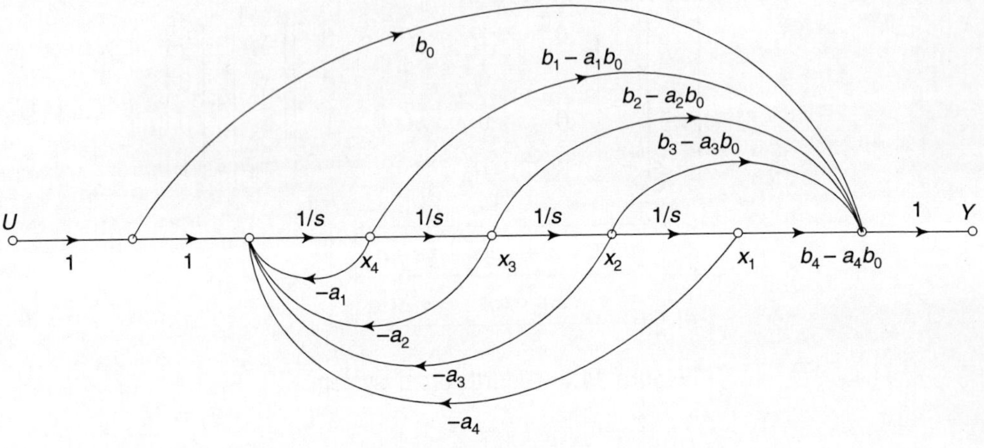

Figure 11.5 Signal flow graph for Eq. 11.20

11.3 STATE SPACE MODEL USING CANONICAL VARIABLES

In canonical variables of a system, the matrix A is a diagonal matrix. The use of state space model with canonical variables is also known as normal form and plays an important role in control theory.

11.3.1 Parallel Decomposition of Transfer Function

We have the transfer function for this model

$$T(s) = \frac{Y(s)}{U(s)} = \frac{b_o s^n + b_1 s^{n-1} + ... b_{n-1} s + b_n}{s^n + a_1 s^{n-1} + ... a_{n-1} s + a_n} \quad ...(11.21)$$

Let us assume that the denominator is known in factored form and the poles of transfer function, located at $P_1, P_2, P_3, ... P_n$ are all distinct.

The partial fraction of Eq. (11.21)

$$T(s) = \frac{Y(s)}{U(s)} = b_o + \sum_{i=1}^{n} \frac{C_i}{(s - P_i)} \quad ...(11.22)$$

where $\quad C_i$ = residues of the poles at $s = P_i$

The block diagram for the transfer function is shown in Fig. 11.6. The state equation is

$$\dot{x}_1 = p_i x_i + uy(t) \quad ...(11.23a)$$

where $\quad i = 1, 2, 3 ... n$

The output, $y(t)$, is given

$$y(t) = C_1 x_1 + C_2 x_2 + ... C_n x_n + b_o u \quad ...(11.23b)$$

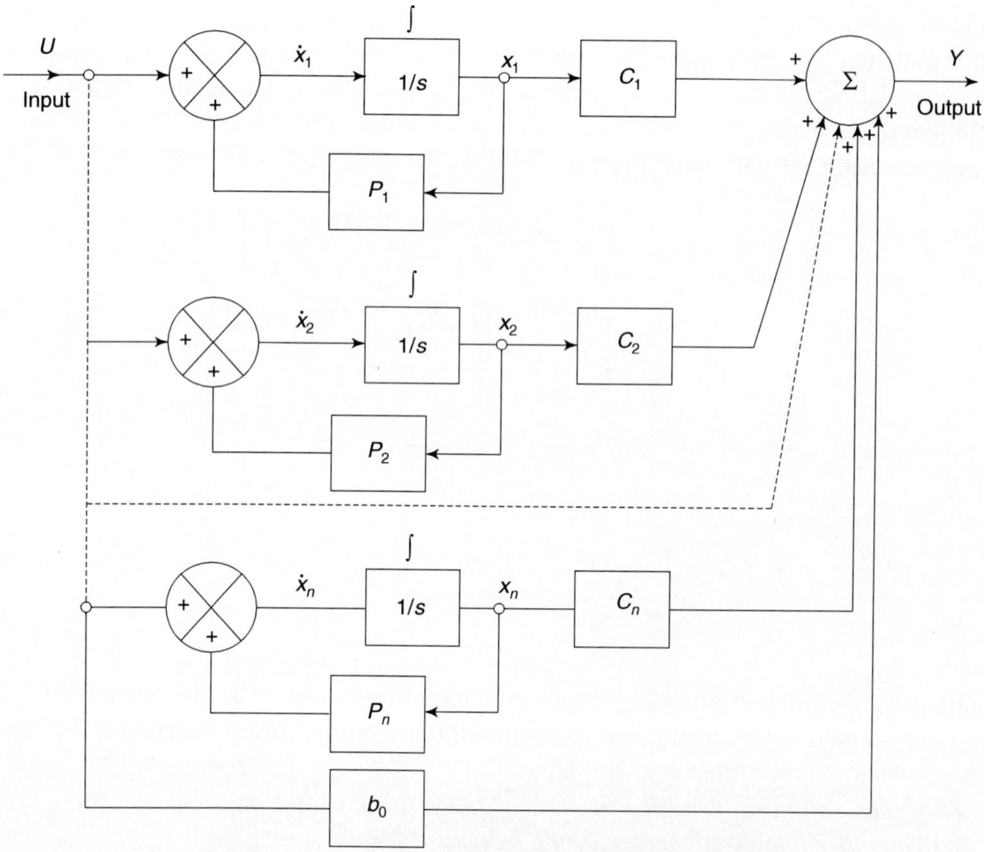

Figure 11.6 Block diagram of canonical state model

Equation (11.23) represnets the canonical state model. By representing this state model in the vector matrix form as

$$\begin{bmatrix} \dot{x}_1 \\ \dot{x}_2 \\ \vdots \\ \dot{x}_n \end{bmatrix} = \underbrace{\begin{bmatrix} P_1 & 0 & 0 & \cdots & 0 \\ 0 & P_2 & 0 & \cdots & 0 \\ 0 & 0 & P_3 & \cdots & 0 \\ \vdots & \vdots & \vdots & \vdots & \vdots \\ 0 & 0 & 0 & \cdots & P_n \end{bmatrix}}_{A} \begin{bmatrix} x_1 \\ x_2 \\ x_3 \\ \vdots \\ x_n \end{bmatrix} + \underbrace{\begin{bmatrix} 1 \\ 1 \\ 1 \\ \vdots \\ 1 \end{bmatrix}}_{B} u \qquad \text{...(11.24a)}$$

$$y = [C_1\ C_2\ \ldots\ C_n] \begin{bmatrix} x_1 \\ x_2 \\ x_3 \\ \vdots \\ x_n \end{bmatrix} + b_0\, u \qquad \text{...(11.24b)}$$

By observing Eq. (11.24), we get that the matrix A is a diagonal matrix and the poles of the transfer function, $T(s)$, represents the diagonal elements. All elements of vector matrix B of the state model are all unity, and the elements of the row vector (C) are the residues of the system poles.

Alternate canonical state model is

$$\begin{bmatrix} \dot{x}_1 \\ \dot{x}_2 \\ \vdots \\ \dot{x}_n \end{bmatrix} = \begin{bmatrix} P_1 & 0 & 0 & 0 & \cdots & 0 \\ 0 & P_2 & 0 & 0 & \cdots & 0 \\ 0 & 0 & P_3 & 0 & \cdots & 0 \\ 0 & \vdots & \vdots & \vdots & \vdots & \vdots \\ 0 & 0 & 0 & 0 & \cdots & P_n \end{bmatrix} \begin{bmatrix} x_1 \\ x_2 \\ x_3 \\ \vdots \\ x_n \end{bmatrix} + \begin{bmatrix} C_1 \\ C_2 \\ C_3 \\ \vdots \\ C_n \end{bmatrix} u \qquad \text{...(11.25a)}$$

$$= [1 1 1 \ldots 1] \begin{bmatrix} x_1 \\ x_2 \\ x_3 \\ \vdots \\ x_n \end{bmatrix} + b_0\, u \qquad \text{...(11.25b)}$$

Equation (11.25) represents the decoupled nature of canonical or normal form model. By the decoupled nature we can say that in normal form, then first order differential equations are independent of each other.

- *The canonical variables, like phase variables, are not real physical variables of the system, this is the disadvantage of the canonical form.*

11.3.2 Parallel Decomposition of a Transfer Function Having Repeated Roots of its Characteristic Equation

The method used earlier for obtaining the state model of a transfer function, considered distinct roots of its characteristic equation. For the case of repeated roots the procedure for obtaining the state model of a given transfer function is given below:

Let the transfer function be

$$\frac{Y(s)}{U(s)} = \sum_{i=1}^{n} \frac{C_i}{(s - P_{i-r})(s - P_r)^m} \qquad ...(11.26)$$

where C_1 = residues of the poles at $s = P_i$ and P_r

P_i = Poles (i = 1, 2, 3 ... n)

P_r = Repeated poles (r = 1, 2, 3, ... m)

In the block diagram for above function let i = 3 and r = 2 (for simplicity)

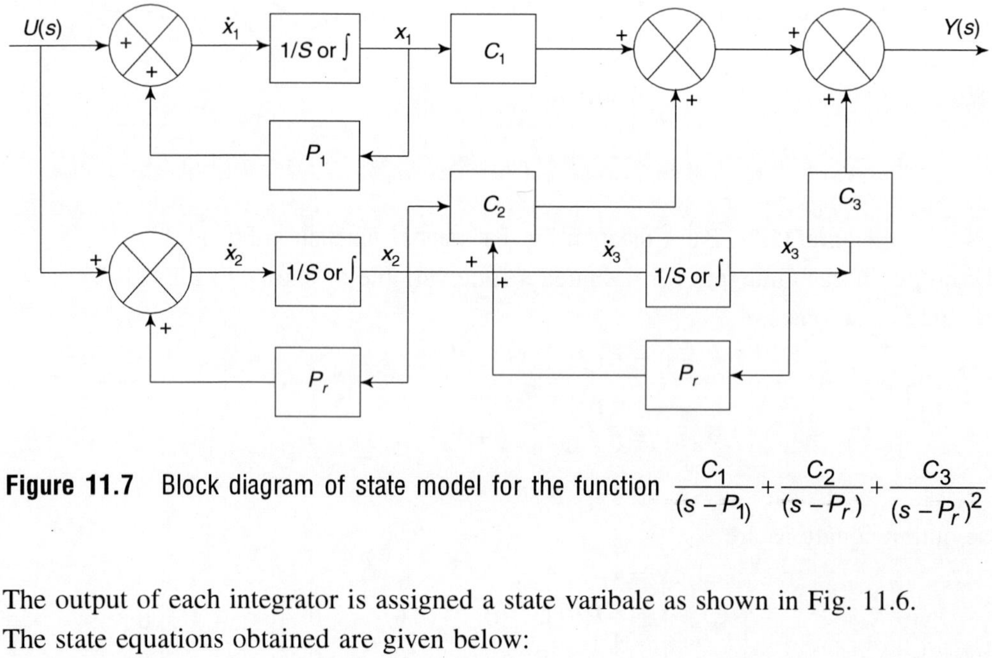

Figure 11.7 Block diagram of state model for the function $\frac{C_1}{(s - P_1)} + \frac{C_2}{(s - P_r)} + \frac{C_3}{(s - P_r)^2}$

The output of each integrator is assigned a state varibale as shown in Fig. 11.6.

The state equations obtained are given below:

$$\dot{x}_1 = P_1 x_1 + u$$

$$\dot{x}_2 = P_r + x_2 + u$$

$$\dot{x}_3 = x_2 + P_r x_3$$

The output equation is

$$y = C_1 x_1 + C_2 x_2 + C_3 x_3$$

The state model is

$$\begin{bmatrix} \dot{x}_1 \\ \dot{x}_2 \\ \dot{x}_3 \end{bmatrix} = \begin{bmatrix} P_1 & 0 & 0 \\ 0 & P_r & 0 \\ 0 & 1 & P_r \end{bmatrix} \begin{bmatrix} x_1 \\ x_2 \\ x_3 \end{bmatrix} + \begin{bmatrix} 1 \\ 1 \\ 0 \end{bmatrix} u$$

The matrix A is called the Jordan's canonical form.

11.3.3 Cascade Decomposition of Transfer Function

The cascade decomposition is used when the given transfer function does not have zeros. The function is given in the form of

$$\frac{Y(s)}{U(s)} = \sum_{i=1}^{n} \frac{1}{(s-p_i)} \qquad \text{...(11.27)}$$

where $i = 1, 2, 3, \ldots n$

The block diagram of that function is given in Fig. 11.8.

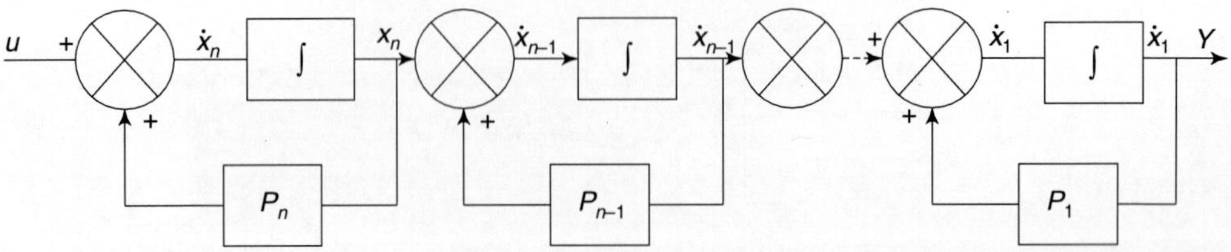

Figure 11.8 Block diagram for the transfer function in Eq. 11.27

The output of each integrator is assigned a state variable as shown in Fig. 11.8.

The state equations are

$$\dot{x}_1 = P_1 x_1 + x_2$$
$$\vdots$$
$$\ddot{x}_{n-1} = P_{n-1} x_{n-1} + x_n$$
$$\dot{x}_n = P_n x_n + u$$

The output equations are

$$y = x_1$$

$$\begin{bmatrix} \dot{x}_1 \\ \dot{x}_2 \\ \vdots \\ \dot{x}_{n-1} \\ \dot{x}_n \end{bmatrix} = \begin{bmatrix} P_1 & 1 & 0 & \cdots & 0 & 0 \\ 0 & P_2 & 1 & \cdots & 0 & 0 \\ \vdots & \vdots & \vdots & \vdots & \vdots & \vdots \\ 0 & 0 & 0 & \cdots & P_{n-1} & 1 \\ 0 & 0 & 0 & \cdots & 0 & P_n \end{bmatrix} \begin{bmatrix} x_1 \\ x_2 \\ \vdots \\ x_{n-1} \\ x_n \end{bmatrix} + \begin{bmatrix} 0 \\ 0 \\ \vdots \\ 0 \\ 1 \end{bmatrix} u \qquad \text{...(11.28)}$$

$$y = x_1 \qquad \text{...(11.29)}$$

The coefficients of matrix A of a state equation, obtained as per cascade decomposition of the transfer function, give a triangular matrix known as pole zero form.

11.3.4 Direct Decomposition of the Transfer Function

The direct decomposition of the transfer function is performed for the system, when the transfer function is given in the following form.

$$\frac{Y(s)}{U(s)} = \frac{1}{s^n + a_1 s^{n-1} + a_2 s^{n-2} + \ldots a_{n-1} s + a_n} \quad \ldots(11.30)$$

where n = order of the system

In the block diagram of such system, n integrators and n summing junctions are used. The block diagram is shown in Fig. 11.9.

The output of each integrator is assigned a state variable and the state model obtained is given below:

$$\begin{aligned}
\dot{x}_1 &= x_2 \\
\dot{x}_2 &= x_3 \\
&\vdots \\
&= x_{n-2} \\
\dot{x}_n &= -a_n x_1 - a_{n-1} x_2 - \ldots a_1 x_{n-1} - a x_n + u
\end{aligned}$$

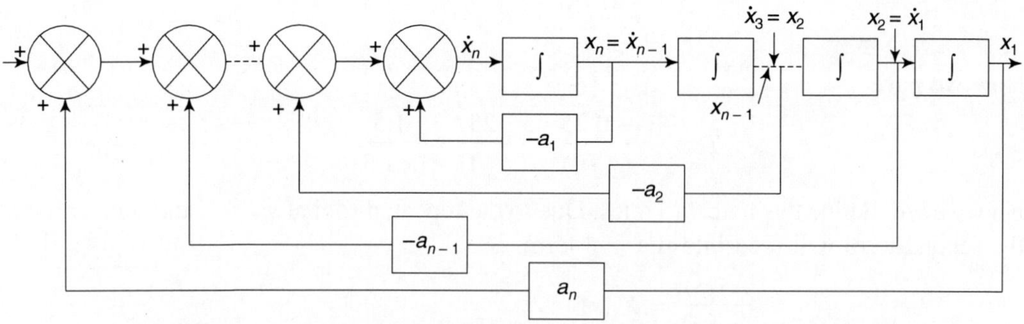

Figure 11.9 Block diagram representation for transfer function Eq. 11.30

The output is $y = x_1$.

The state model

$$\begin{bmatrix} \dot{x}_1 \\ \dot{x}_2 \\ \vdots \\ \dot{x}_{n-1} \\ \dot{x}_n \end{bmatrix} = \begin{bmatrix} 0 & 1 & 0 & \cdots & 0 & 0 & 0 \\ 0 & 0 & 1 & \cdots & 0 & 0 & 0 \\ \vdots & & & & & & \\ 0 & 0 & 0 & \cdots & 1 & 0 & 0 \\ a_n & a_{n-1} & a_{n-2} & \cdots & a_3 & a_2 & a_1 \end{bmatrix} \begin{bmatrix} x_1 \\ x_2 \\ \vdots \\ x_{n-1} \\ x_n \end{bmatrix} + \begin{bmatrix} 0 \\ 0 \\ \vdots \\ 0 \\ 1 \end{bmatrix} u \quad \ldots(11.31)$$

The coefficient matrix (A) of the state equation obtained as per direct decomposition of a transfer function has a unit element above its diagonal and negtive coefficients in the last row. All other elements are zero. Such a matrix is known as Bush's form.

Example 11.3.1 The block diagram of the system is shown in Fig. Ex. 11.3.1. Write down the canonical state variable form for the complete system also draw the block diagram in state variable form.

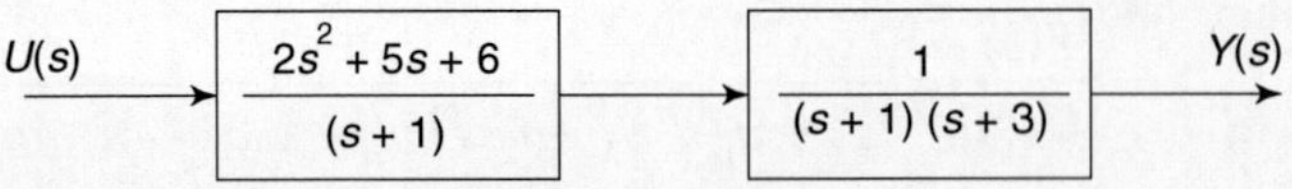

Figure Ex. 11.3.1

The transfer function of the system is

$$T(s) = \frac{Y(s)}{U(s)} = \frac{2s^2 + 5s + 6}{(s+1)^2\,(s+3)}$$

By the partial fraction

$$T(s) = \frac{2s^2 + 5s + 6}{(s+1)^2\,(s+3)} = \frac{A}{(s+1)} + \frac{B}{(s+1)} + \frac{C}{(s+3)}$$

By solving the above equation, we get

$$B = 2.25$$

$$A = -\,0.25$$

$$C = 1.5$$

Now we have

$$T(s) = \frac{-\,0.25}{(s+1)} + \frac{2.25}{(s+1)} + \frac{1.5}{(s+3)}$$

Now we will define the state variables. Due to the repeated root at $s = -\,1$, the state variable of the second term will feed into the first term.

$$\frac{X_2(s)}{(s+1)} = X_1(s)$$

$$\Rightarrow \qquad x_2 = \dot{x}_1 + x_1$$

$$\Rightarrow \qquad \dot{x}_1 = -\,x_1 + x_2 \qquad \text{...(a)}$$

$$\frac{U(s)}{(s+1)} = X_2(s)$$

$$\Rightarrow \qquad u = \dot{x}_2 + x_2$$

$$\Rightarrow \qquad \dot{x}_2 = -\,x_2 + u \qquad \text{...(b)}$$

$$\frac{U(s)}{(s+3)} = X_3(s)$$

$$\Rightarrow \qquad u = \dot{x}_3 + 3x_3$$

$$\Rightarrow \qquad \dot{x}_3 = -\,3x_3 + u \qquad \text{...(c)}$$

The output

$$y = 0.25\, x_1 + 2.25\, x_2 + 1.5\, x_3 \qquad \text{...(d)}$$

Now in the matrix form

$$\begin{bmatrix} \dot{x}_1 \\ \dot{x}_2 \\ \dot{x}_3 \end{bmatrix} = \begin{bmatrix} \overbrace{\begin{array}{cc} -1 & 1 \\ 0 & -1 \end{array}}^{\text{Jordan block}} & \begin{array}{c} 0 \\ 0 \end{array} \\ \begin{array}{cc} 0 & 0 \end{array} & -3 \end{bmatrix} \begin{bmatrix} x_1 \\ x_2 \\ x_3 \end{bmatrix} + \begin{bmatrix} 0 \\ 1 \\ 1 \end{bmatrix} u \qquad \text{...(e)}$$

$$y = [-0.25 \quad 2.25 \quad 1.5] \begin{bmatrix} x_1 \\ x_2 \\ x_3 \end{bmatrix} \qquad \text{...(f)}$$

The state variable diagram is shown in Fig. Ex. 11.3.1.

The dotted block in matrix A of Eq. (e) is not in decoupled form because due to the repeated root at $s = -1$. This dotted block is known as Jordan block.

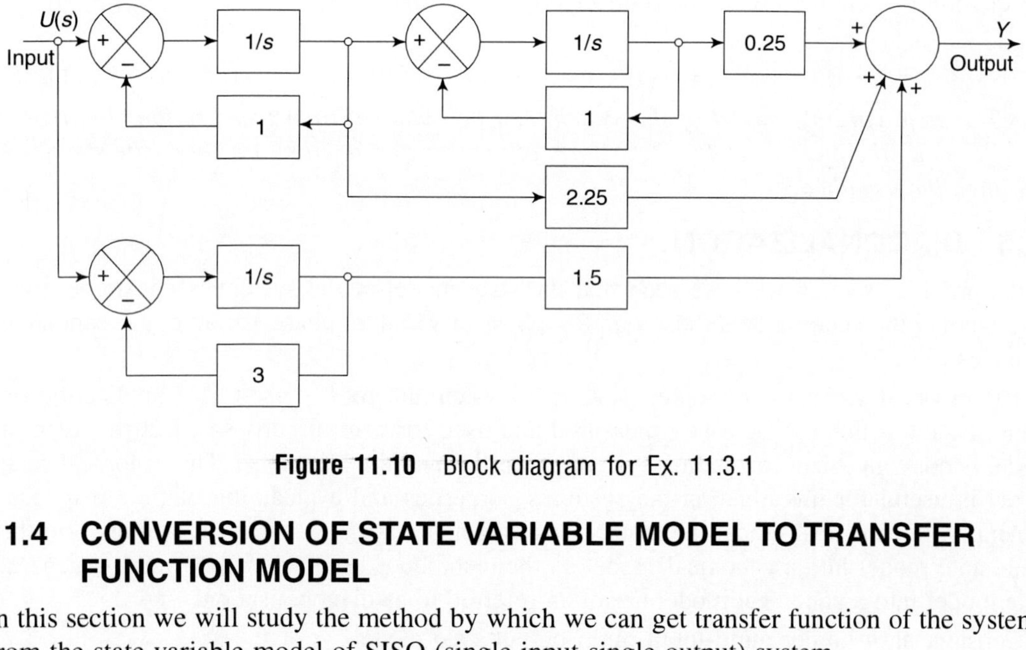

Figure 11.10 Block diagram for Ex. 11.3.1

11.4 CONVERSION OF STATE VARIABLE MODEL TO TRANSFER FUNCTION MODEL

In this section we will study the method by which we can get transfer function of the system from the state variable model of SISO (single input single output) system.

The general equation for state model

$$\dot{x} = Ax + Bu \qquad \text{...(11.32)}$$

$$y = Cx + Du \qquad \text{...(11.33)}$$

The transfer function can be obtained as follows:

Taking the Laplace transform of Eqs. (11.32) and (11.33), we have

$$s\,X(s) = AX(s) + BU(s) \quad \text{...(11.34)}$$

$$Y(s) = CX\,(s) + DU(s) \quad \text{...(11.35)}$$

By Eq. (11.34).

$$s\,X(s) - AX(s) = BU(s)$$

$$X(s)[sI - A) = BU(s)$$

$$X(s) = [sI - A]^{-1}\,BU(s)$$

Substituting the value of $X(s)$ in Eq. (11.35), we get

$$Y(s) = C\,[(sI - A)^{-1}\,BU(s)] + DU(s)$$

Assuming initial conditions are zero

$$Y(s) = [C\,(sI - A)^{-1}\,B + D]\,U(s)$$

$$\frac{Y(s)}{U(s)} = T(s) = C\,(sI - A)^{-1}\,B + d \quad \text{...(11.36)}$$

$$T(s) = C\,\frac{\text{adj of } (sI - A)\,B}{\text{def } (sI - A)} + d \quad \text{...(11.37)}$$

Equation (11.37) represents the transfer function of the model.

The characteristic equation is

$$|\,sI - A\,| = 0 \quad \text{...(11.38)}$$

- *The state variable model of the system can be obtained from transfer function model by taking the inverse Laplace transform of the transfer function, i.e., opposite step from the above method.*

11.5 DIAGONALIZATION

In the previous section we have seen that the state model of the system is not unique. We have studied the state models those employ physical variable, phase variable and canonical variables.

The physical variables for representation of system are mostly useful for application on point of view which can be easily measured and used for control purposes. But this type of model is not convenient for finding system properties and time response. The canonical state model is useful for investigating the system's porperties and evaluataion of time response. Generally, in canonical model the matrix A is diagonal matrix. So we need to transform the other state model into a canonical model so the techniques used for transforming a general state model into a canonical model are often referred to as diagonalisation.

Consider an nth order multi-input-multi-output state model

$$\dot{x} = AX + Bu \quad \text{...(11.39a)}$$

$$y = Cx + Du \quad \text{...(11.39b)}$$

Let us consider Eqs. (11.39a) and (11.39b) representing general state model in which matrix A is non-diagonal matrix.

Let us define a new state vector, $X = PZ$

$$X = PZ$$

where P is a $(n \times n)$ non-singular constant matrix. By this transformation the above model is

$$P\dot{Z} = APZ + Bu$$

$$\dot{Z} = P^{-1} APZ + P^{-1} Bu \quad ...(11.40)$$

$$y = CPZ + Du \quad ...(11.41)$$

If matrix P is selected such that $P^{-1} AP$ is a diagonalized matrix A so the model represented by the Eqs. (11.40) and (11.41) is canonical state model. The matrix M is called diagonilising matrix.

Equations (11.40) and (11.41) may be written as

$$= \tilde{P}z + \tilde{B}u \quad ...(11.42a)$$

$$y = \tilde{c}z + Du \quad ...(11.42b)$$

where $\tilde{P} = P^{-1} A P$ (diagonal matrix)

$\tilde{B} = P^{-1} B$

$\tilde{C} = C P$

Now, an important part of this transformation is the selection of matrix P. The matrix P is selected by the use of eigenvectors. So now we will briefly discuss the eigenvalue and eigenvectors.

Eigenvalues and Eigenvectors

Let us consider the matrix A of state model is

$$A = \begin{bmatrix} A_{11} & A_{12} & A_{13} \\ A_{21} & A_{22} & A_{23} \\ A_{31} & A_{32} & A_{33} \end{bmatrix}_{(3\times 3)} \quad ...(11.43)$$

Corresponding to A the characteristic equation

$$|\lambda I - A| = 0 \quad ...(11.44)$$

By solving Eq. (11.44) we will get the value of $\lambda = \lambda_1, \lambda_2, \lambda_3$ those are called eigenvalues.

Now find the eigenvectors

The eigenvectors (V_i) associated with eigenvalues λ_i, $i = 1, 2, 3 \ldots n$ can be obtained by the co-factor of the matrix

Now

V_1 associated with λ_1

$$(\lambda_1 I - A) = V_1 \begin{bmatrix} C_{11} \\ C_{12} \\ C_{13} \end{bmatrix} \quad ...(11.45)$$

V_2 associated with

$$V_2 = (\lambda_2 I - A) = \begin{bmatrix} C_{21} \\ C_{22} \\ C_{23} \end{bmatrix}$$

Similarly,

$$V_3 = (\lambda_3 I - A) = \begin{bmatrix} C_{31} \\ C_{32} \\ C_{33} \end{bmatrix}$$

The diagonalizing matrix or model matrix P is a vector matrix

$$P = [V_1 \;\vdots\; V_2 \;\vdots\; V_3] \qquad \text{...(11.46a)}$$

$$P = \begin{bmatrix} C_{11} & C_{21} & C_{31} \\ C_{12} & C_{22} & C_{32} \\ C_{13} & C_{23} & C_{33} \end{bmatrix} \qquad \text{...(11.46b)}$$

Now it can be verified that

$$\tilde{P} = P^{-1} AP = \begin{bmatrix} \lambda_1 & 0 & 0 \\ 0 & \lambda_2 & 0 \\ 0 & 0 & \lambda_3 \end{bmatrix}$$

which is a diagonal matrix with eigenvalues of A as its diagonal elements. Basically, $\tilde{P}$ could be written directly without the need to compute $P^{-1} AP$.

When the A and $P^{-1} AP$, i.e., $\tilde{P}$ have the same characteristic equation; therefore the eigenvalues are invariant under this condition.

When A is expressed in the form given below.

$$A = \begin{bmatrix} 0 & 1 & 0 & \cdots & 0 \\ 0 & 0 & 1 & \cdots & 0 \\ \vdots & \vdots & \vdots & & \vdots \\ 0 & 0 & 0 & \cdots & 1 \\ -a_n & -a_{n-1} & -a_{n-2} & \cdots & -a_1 \end{bmatrix} \qquad \text{...(11.47)}$$

Then the model matrix can be shown to be a special matrix called the Vandermonde matrix.

$$P = \begin{bmatrix} 1 & 1 & \cdots & 1 \\ \lambda_1 & \lambda_2 & \cdots & \lambda_n \\ \lambda_1^2 & \lambda_2^2 & \cdots & \lambda_n^2 \\ \vdots & \vdots & & \\ \lambda_1^{n-1} & \lambda_2^{n-1} & \cdots & \lambda_n^{n-1} \end{bmatrix} \qquad \text{...(11.48)}$$

Example 11.5.1 Consider a matrix A given below.

$$A = \begin{bmatrix} -5 & 2 & 0 \\ 0 & -4 & 1 \\ 0 & 0 & -6 \end{bmatrix}$$

Find the model matrix or diagonalizing matrix $\tilde{P}$.

The characteristic equation corresponds to the matrix A.

$$|\lambda I - A| = 0$$

$$\begin{bmatrix} \lambda+5 & -2 & 0 \\ 0 & \lambda+4 & -1 \\ 0 & 0 & \lambda+6 \end{bmatrix} = 0$$

$$(\lambda + 5)(\lambda + 4)(\lambda + 6) = 0$$

Therefore, the eigenvalues of matrix A are

$$\lambda_1 = 5$$
$$\lambda_2 = 4$$
$$\lambda_3 = -6$$

Now the eigenvectors

V_1 associated with λ_1

$$V_1 = \text{co-factor of } (\lambda_1 I - A)$$

$$= \text{co-factor of } \begin{bmatrix} 1 & -2 & 0 \\ 0 & 0 & -1 \\ 0 & 0 & 2 \end{bmatrix} = \begin{bmatrix} 1 \\ 0 \\ 0 \end{bmatrix}$$

Similarly, V_2 is associated with λ_2

$$V_2 = \text{co-factor of } (\lambda_2 I - A)$$

$$= \text{co-factor of } \begin{bmatrix} 0 & -2 & 0 \\ 0 & -1 & -1 \\ 0 & 0 & 1 \end{bmatrix} = \begin{bmatrix} 2 \\ 1 \\ 0 \end{bmatrix}$$

and V_3 is associated with λ_3

$$V_3 = \text{co-factor of } (\lambda_3 I - A)$$

$$= \text{co-factor of } \begin{bmatrix} -1 & -2 & 0 \\ 0 & -2 & -1 \\ 0 & 0 & 0 \end{bmatrix} = \begin{bmatrix} 2 \\ -1 \\ 2 \end{bmatrix}$$

where V_1, V_2, V_3 are given vectors of A

So $$P = \begin{bmatrix} 1 & 2 & 2 \\ 0 & 1 & -1 \\ 0 & 0 & 2 \end{bmatrix}$$ **Ans.**

Now
$$\tilde{P} = P^{-1} A P$$

$$\tilde{P} = \begin{bmatrix} 1 & 2 & 2 \\ 0 & 1 & -1 \\ 0 & 0 & 2 \end{bmatrix}^{-1} \begin{bmatrix} -5 & 2 & 0 \\ 0 & -4 & 1 \\ 0 & 0 & -6 \end{bmatrix} \begin{bmatrix} 1 & 2 & 2 \\ 0 & 1 & -1 \\ 0 & 0 & 2 \end{bmatrix}$$

$$\tilde{P} = \begin{bmatrix} 1 & -2 & -2 \\ 0 & 1 & 0.5 \\ 0 & 0 & 0.5 \end{bmatrix} \begin{bmatrix} -5 & 2 & 0 \\ 0 & -4 & 1 \\ 0 & 0 & -6 \end{bmatrix} \begin{bmatrix} 1 & 2 & 2 \\ 0 & 1 & -1 \\ 0 & 0 & 2 \end{bmatrix}$$

$$\tilde{P} = \begin{bmatrix} -5 & 0 & 0 \\ 0 & -4 & 0 \\ 0 & 0 & -6 \end{bmatrix}$$

By the above matrix it is proved that $\tilde{P}$ is a diagonal matrix in which the diagonal elements are the eigenvalue of matrix A.

Example 11.5.2 Consider the following state space model of a system and transform the model into canonical model.

$$\begin{bmatrix} \dot{x}_1 \\ \dot{x}_2 \\ \dot{x}_3 \end{bmatrix} = \begin{bmatrix} 0 & 1 & 0 \\ 0 & 0 & 1 \\ -6 & -11 & -6 \end{bmatrix} \begin{bmatrix} x_1 \\ x_2 \\ x_3 \end{bmatrix} + \begin{bmatrix} 0 \\ 0 \\ 6 \end{bmatrix} u$$

$$y = [1 \ 0 \ 0] \begin{bmatrix} x_1 \\ x_2 \\ x_3 \end{bmatrix}$$

Solution The general equation of the state space model is

$$\dot{x} = Ax + Bu$$
$$y = Cx + Du$$

$$A = \begin{bmatrix} 0 & 1 & 0 \\ 0 & 0 & 1 \\ -6 & -11 & -6 \end{bmatrix}$$

$$B = \begin{bmatrix} 0 \\ 0 \\ 6 \end{bmatrix}$$

$$C = [1 \ 0 \ 0]$$
$$D = 0$$

The characteristic equation of matrix A is

$$|\lambda I - A| = 0$$

$$\begin{bmatrix} \lambda & -1 & 0 \\ 0 & \lambda & -1 \\ 6 & 11 & \lambda+6 \end{bmatrix} = 0$$

$$(\lambda + 1)(\lambda + 2)(\lambda + 3) = 0$$

$$\lambda = -1, -2, -3 \text{ are the eigenvalues}$$

$$P = \begin{bmatrix} 1 & 1 & 1 \\ \lambda_1 & \lambda_2 & \lambda_3 \\ \lambda_1^2 & \lambda_2 2 & \lambda_3^2 \end{bmatrix} = \begin{bmatrix} 1 & 1 & 1 \\ -1 & -2 & -3 \\ 1 & 4 & 9 \end{bmatrix}$$

We have the canonical equation

$$\dot{Z} = P^{-1} A P Z + P^{-1} Bu$$

$$y = C P Z + Du$$

$$\begin{bmatrix} \dot{Z}_1 \\ \dot{Z}_2 \\ \dot{Z}_3 \end{bmatrix} = \begin{bmatrix} 1 & 1 & 1 \\ -1 & -2 & -3 \\ 1 & 4 & 9 \end{bmatrix}^{-1} \begin{bmatrix} 0 & 1 & 0 \\ 0 & 0 & 1 \\ -6 & -11 & -6 \end{bmatrix} \begin{bmatrix} 1 & 1 & 1 \\ -1 & -2 & -3 \\ 1 & 4 & 9 \end{bmatrix} \begin{bmatrix} Z_1 \\ Z_2 \\ Z_3 \end{bmatrix}$$

$$\begin{bmatrix} \dot{Z}_1 \\ \dot{Z}_2 \\ \dot{Z}_3 \end{bmatrix} = \begin{bmatrix} 3 & 2.5 & 0.5 \\ -3 & -4 & -1 \\ 1 & 1.5 & 0.5 \end{bmatrix} \begin{bmatrix} 0 & 1 & 0 \\ 0 & 0 & 1 \\ -6 & -11 & -6 \end{bmatrix} \begin{bmatrix} 1 & 1 & 1 \\ -1 & -2 & -3 \\ 1 & 4 & 9 \end{bmatrix} \begin{bmatrix} z_1 \\ z_2 \\ z_3 \end{bmatrix} + \begin{bmatrix} 3 & 2.5 & 0.5 \\ -3 & -4 & 1 \\ 1 & 1.5 & 0.5 \end{bmatrix} \begin{bmatrix} 0 \\ 0 \\ 6 \end{bmatrix} u$$

$$\begin{bmatrix} \dot{Z}_1 \\ \dot{Z}_2 \\ \dot{Z}_3 \end{bmatrix} = \begin{bmatrix} -1 & 0 & 0 \\ 0 & -2 & 0 \\ 0 & 0 & -3 \end{bmatrix} \begin{bmatrix} Z_1 \\ Z_2 \\ Z_3 \end{bmatrix} + \begin{bmatrix} 3 \\ -6 \\ 3 \end{bmatrix} u$$

The output

$$y = cPZ$$

$$y = [1 \ 0 \ 0] \begin{bmatrix} 1 & 1 & 1 \\ -1 & -2 & -3 \\ 1 & 4 & 9 \end{bmatrix} \begin{bmatrix} Z_1 \\ Z_2 \\ Z_3 \end{bmatrix}$$

$$y = [1 \ 1 \ 1] \begin{bmatrix} Z_1 \\ Z_2 \\ Z_3 \end{bmatrix} \quad \textbf{Ans.}$$

11.6 CONTROLLABILITY AND OBSERVABILITY

In control system we need to answer two basic questions for deciding whether or not a control solution exists. In this, basically we control the pole location of the closed loop system. We are saying simply that the control signal, *u*, can control the behaviour of each state variable in $x(t_o)$. If any one of the state variables cannot be controlled by the control signal, u, then we cannot place the poles of the system at desired location, the question those arise are:

1. Whether the system can be transferred or not from any initial state to any desired state within finite time by applying suitable control force?
2. If we know the output vector for a finite length of time, whether the initial state of the system can be determined or not?

'Kalman' has given the answer to these basic questions known as controllability and observability.

"If it is possible to transfer the system state from a desired initial state, $x(t_o)$, to a desired final state, $x(t_o)$, in specified finite time by a control vector, $u(t)$, the system is said to be controllable; otherwise, the system is uncontrollable.

"If the initial state vector, $x(t_o)$, can be completely identified by measurements of the output, $y(t)$, over a finite time length, the system is said to be observable, otherwise the system is said to be unobservable.

The Controllability Matrix

As mentioned earlier, the concept of controllability involves, to design state feedback for a plant under any representation or choice of state variables, a matrix can be derived that must have a particular property if all state variables are to be controlled by the plant input u. We now state the requirement for controllability, including the form, and property.

An nth order system whose state equation is

$$\dot{x} = Ax + Bu \quad \text{...(11.49)}$$

where $x = n$ dimensional state vector

$u =$ control force

$A = n \times n$ matrix

$B = n \times 1$ matrix

Equation (11.49) is completely controllable if the matrix

$$C_M = [B \ AB \ A^2B \ A^n \ B] \quad \text{...(11.50)}$$

is of rank n. Where C_M is called controllability matrix. The above method for test of controllability is given by 'Kalman.'

- *Complete controllable means that all state variables are controllable.*

Example 11.6.1 The state equation for the system is given by

$$\begin{bmatrix} \dot{x}_1 \\ \dot{x}_2 \\ \dot{x}_3 \end{bmatrix} = \begin{bmatrix} -1 & 1 & 0 \\ 0 & -1 & 0 \\ 0 & 0 & -2 \end{bmatrix} \begin{bmatrix} x_1 \\ x_2 \\ x_3 \end{bmatrix} + \begin{bmatrix} 0 \\ 1 \\ 1 \end{bmatrix} u$$

Check whether the system is controllable or not.

Solution We have

$$A = \begin{bmatrix} -1 & 1 & 0 \\ 0 & -1 & 0 \\ 0 & 0 & -2 \end{bmatrix}; B = \begin{bmatrix} 0 \\ 1 \\ 1 \end{bmatrix}$$

By the Kalman method

$$C_M = [B \ : AB \ : A^2B]$$

$$AB = \begin{bmatrix} -1 & 1 & 0 \\ 0 & -1 & 0 \\ 0 & 0 & -2 \end{bmatrix}\begin{bmatrix} 0 \\ 1 \\ 1 \end{bmatrix} = \begin{bmatrix} 1 \\ -1 \\ -2 \end{bmatrix}$$

$$A^2\,B = \begin{bmatrix} -1 & 1 & 0 \\ 0 & -1 & 0 \\ 0 & 0 & -2 \end{bmatrix}\begin{bmatrix} -1 & 1 & 0 \\ 0 & -1 & 0 \\ 0 & 0 & -2 \end{bmatrix}\begin{bmatrix} 0 \\ 1 \\ 1 \end{bmatrix}$$

$$A^2\,B = \begin{bmatrix} -1 & -2 & 0 \\ 0 & 1 & 0 \\ 0 & 0 & 4 \end{bmatrix}\begin{bmatrix} 0 \\ 1 \\ 1 \end{bmatrix} = \begin{bmatrix} -2 \\ 1 \\ 4 \end{bmatrix}$$

Now $$C_M = [B \ : \ AB \ : \ A^2\,B]$$

$$C_M = \begin{bmatrix} 0 & 1 & -2 \\ 1 & -1 & 1 \\ 1 & -2 & 4 \end{bmatrix}$$

The rank of the C_M matrix is 3 because $|C_M| = -1$, i.e., not zero. We conclude that the system is controllable.

Example 11.6.2 Determine whether the system is controllable.

$$\dot{x} = Ax + Bu$$

$$\begin{bmatrix} \dot{x}_1 \\ \dot{x}_2 \\ \dot{x}_3 \end{bmatrix} = \begin{bmatrix} -1 & 1 & 2 \\ 0 & -1 & 5 \\ 0 & 3 & -4 \end{bmatrix}\begin{bmatrix} x_1 \\ x_2 \\ x_3 \end{bmatrix} + \begin{bmatrix} 2 \\ 1 \\ 1 \end{bmatrix} u$$

Solution

$$A = \begin{bmatrix} -1 & 1 & 2 \\ 0 & -1 & 5 \\ 0 & 3 & -4 \end{bmatrix}; B = \begin{bmatrix} 2 \\ 1 \\ 1 \end{bmatrix}$$

$$AB = \begin{bmatrix} -1 & 1 & 2 \\ 0 & -1 & 5 \\ 0 & 3 & -4 \end{bmatrix}\begin{bmatrix} 2 \\ 1 \\ 1 \end{bmatrix} = \begin{bmatrix} 1 \\ 4 \\ -1 \end{bmatrix}$$

$$A^2B = \begin{bmatrix} -1 & 1 & 2 \\ 0 & -1 & 5 \\ 0 & 3 & -4 \end{bmatrix}\begin{bmatrix} -1 & 1 & 2 \\ 0 & -1 & 5 \\ 0 & 3 & -4 \end{bmatrix}\begin{bmatrix} 2 \\ 1 \\ 1 \end{bmatrix}$$

$$= \begin{bmatrix} 1 & 4 & -5 \\ 0 & 16 & -25 \\ 0 & -15 & 31 \end{bmatrix} \begin{bmatrix} 2 \\ 1 \\ 1 \end{bmatrix} = \begin{bmatrix} 1 \\ -9 \\ 16 \end{bmatrix}$$

The controllability matrix (C_M) is

$$C_M = [B : AB : A^2B]$$

$$C_M = \begin{bmatrix} 2 & 1 & 1 \\ 1 & 4 & -9 \\ 1 & -1 & 16 \end{bmatrix}$$

The $|C_M| = 80$ is not zero and the rank of the C_M matrix is 3 so the system is controllable.

Observability Matrix

As we have seen controllability. Now we will see observability. This is an another important requirement for system design. In order to determine observability for system under any representation or choice of state variables, a matrix can be desired that must have a particular property if all state variables are to be observed at the output. We now state the requirements for observability.

An nth order system whose state and output equations are, respectively.

$$\dot{x} = Ax + Bu$$

$$y = Cx$$

is completely observable if the matrix

$$O_M = \begin{bmatrix} C \\ CA \\ CA^2 \\ \vdots \\ CA^{n-1} \end{bmatrix} \text{ is of rank } n, \quad \text{...(11.51)}$$

where O_M is called the observability matrix.

Example 11.6.3 The state and output equation is as follows:

$$\begin{bmatrix} \dot{x}_1 \\ \dot{x}_2 \\ \dot{x}_3 \end{bmatrix} = \begin{bmatrix} 0 & 1 & 0 \\ 0 & 0 & 1 \\ -4 & -3 & -2 \end{bmatrix} \begin{bmatrix} x_1 \\ x_2 \\ x_3 \end{bmatrix} + \begin{bmatrix} 0 \\ 0 \\ 1 \end{bmatrix} u$$

$$y = [0 \ 5 \ 1]\, x$$

Determine whether the system is observable or not?

Solution

$$C = [0 \ 5 \ 1],$$

$$A = \begin{bmatrix} 0 & 1 & 0 \\ 0 & 0 & 1 \\ -4 & -3 & -2 \end{bmatrix}$$

$$CA = [0 \; 5 \; 1]\begin{bmatrix} 0 & 1 & 0 \\ 0 & 0 & 1 \\ -4 & -3 & -2 \end{bmatrix}$$

$$CA = [-4 \; -3 \; 3]$$

$$CA^2 = [0 \; 5 \; 1]\begin{bmatrix} 0 & 1 & 0 \\ 0 & 0 & 1 \\ -4 & -3 & -2 \end{bmatrix}\begin{bmatrix} 0 & 1 & 0 \\ 0 & 0 & 1 \\ -4 & -3 & -2 \end{bmatrix}$$

$$CA^2 = [0 \; 5 \; 1]\begin{bmatrix} 0 & 0 & 1 \\ -4 & -3 & -2 \\ 8 & 2 & 1 \end{bmatrix}$$

$$CA^2 = [-12 \; -13 \; -9]$$

Now
$$O_M = \begin{bmatrix} C \\ CA \\ CA^2 \end{bmatrix} = \begin{bmatrix} 0 & 5 & 1 \\ -4 & -3 & 3 \\ -12 & -13 & -9 \end{bmatrix}$$

The rank of $O_M = 3$ because $|O_M| = -344$ is non-zero so the system is observable.

11.7 TIME DOMAIN SOLUTION OF STATE EQUATION

In previous sections, we discussed various state models used to represent a system. Now we can solve the state equations directly in the time domain using a method closely allied to the classical solution of differential equation. We will find that the final solution consists of two parts: one is zero input response or natural response and second one is zero state response or forced response as follows:

The state equation is

$$\dot{x}(t) = Ax(t) + Bu(t) \qquad ...(11.52)$$

$$x(t_o) = x_o \qquad ...(11.53)$$

Multiplying both sides by e^{-At}, we get

$$e^{-At}[\dot{x}(t) - A\,x(t)] = e^{-At}\,Bu(t) \qquad ...(11.54)$$

Integrating on both sides of Eq. (11.54), with respect to t between the limit 0 and t, we get

$$\int_0^t e^{-At}\,[\dot{x}(t) - A\,x(t) = \int_0^t e^{-At}\,B\,u(\tau)\,d\tau$$

$$\int_0^t \cdot \frac{d}{dt} e^{-At}\,x(t) = \int_0^t e^{-At}\,u(\tau)\,d\tau$$

$\because$
$$e^{-At}[\dot{x}(t) - Ax(t)] = \frac{d}{dt} e^{-At}\,u(\tau)\,d\tau$$

Now $$\left[e^{-At}\, x(t)\right]_0^t = \int_0^t e^{-At}\, B\, u(\tau)\, d\tau$$

$$e^{-At}\, x(t) - x\,(0) = \int_0^t e^{A\tau}\, Bu(\tau)\, d\tau$$

Now pre-multiplying both sides by e^{At}, we have

$$e^{At}\,.\, e^{-At}\, x(t) - e^{At}\, x(0) = e^{At}\int_0^t e^{-At}\, Bu(\tau)\, d\tau$$

$$x(t) - e^{At}\, x(0) = \int_0^t e^{A(t-\tau)}\, Bu(\tau)\, d\tau$$

$$x(t) = \underbrace{e^{At}\, x(0)}_{\substack{\text{Zero input response}\\ \text{or natural response}}} + \int_0^t \underbrace{e^{A(t-\tau)}\, Bu(\tau)\, d\tau}_{\substack{\text{Zero state response or}\\ \text{forced response}}} \qquad ...(11.55)$$

Equation (11.55) represents the time domain solution of state equation.

In the above discussion, the state transition matrix has been defined as

$$\phi(t) = e^{At} \qquad ...(11.56)$$

where the initial times is $t = 0$

In general, for a LTI (Linear Time Invarient) system the initial time is $t = t_0$ then the state transition matrix becomes

$$\phi(t - t_0) = e^{A(t-t_0)} \qquad ...(11.57)$$

(a) *When only initial conditions are considered and input (u) is zero, then the response is called zero input response or natural response.*

(b) *When only input is considered and initial conditions are zero, then the response is called zero state response or forced response.*

Properties of state transition matrix

1. $\phi(t) = e^{At}$

   Put $t = 0$

$$\phi(0) = e^{A(0)}$$

$$\therefore \quad \phi(0) = I \qquad ...(11.58)$$

i.e., state transition matrix is identity matrix when $t = 0$

2. $\phi(t) = e^{At}$

   Post-multiplying be e^{At} on both sides

$$\phi(t)\, e^{-At} = e^{At}\, e^{-At} = I$$

   Pre-multiplying both sides by $\phi^{-1}(t)$

$$\phi^{-1}(t)\, \phi(t)\, e^{-At} = \phi^{-1}(t)\, I$$

$$\therefore \quad e^{-At} = \phi(-\, t)$$

$$\phi^{-1}(t)\,\phi(t)\,\phi(-t) = \phi^{-1}(t)$$
$$\phi(t)\,\phi(-t) = 1$$
$$\boxed{\phi(-t) = \phi^{-1}(t)} \quad \text{...(11.59)}$$

3. $$\phi(t_1 + t_2) = e^{A(t_1 + t_2)}$$
$$= e^{At_1}\,e^{At_2}$$
$$\boxed{\phi(t_1 + t_2) = \phi(t_1)\,\phi(t_2)} \quad \text{...(11.60)}$$

4. $$[\phi(t)]^n = (e^{At}\,e^{AT}\,e^{At}\,...\,n \text{ times})$$
$$[\phi(t)]^n = e^{Ant}$$

$\therefore$ $$\boxed{[\phi(t)]^n = \phi(nt)} \quad \text{...(11.61)}$$

5. $$\phi'(t) = \frac{d}{dt}e^{At}$$
$$\phi'(t) = A\,e^{At}$$
$$\boxed{\phi'(t) = A\,\phi(t)} \quad \text{...(11.62)}$$

Various Methods of Evaluation of State Transition Matrix (e^{At})

There are three methods for evaluation of e^{At}

1. Laplace transform method

   We have

$$\dot{x} = Ax + Bu$$
$$e^{At} = L^{-1}\,[(sI - A)^{-1}] \quad \text{...(11.63)}$$

where $$(sI - A)^{-1} = \frac{\text{adj of } (sI - A)}{\text{determinant of } (sI - A)}$$

$$(sI - A)^{-1} = \frac{\text{adj of } (sI - A)}{|sI - A|}$$

2. By Caley Hamilton solution (numerical computation)

$$\phi(t) = e^{At} = I + At + \frac{A^2\,t^2}{2!} + \frac{A^3\,t^3}{3!} + ...$$

i.e., $$e^{At} = I + At + \frac{A^2\,t^2}{2!} + \frac{A^3\,t^3}{3!} + ...$$

where I = identify matrix of order equal to the order of matrix A

A = coefficient matrix of state equation

3. By use of $\tilde{P}$ matrix

   We know that

$$\tilde{P} = P^{-1}\,A\,P$$
$$A = P\,\tilde{P}^m\,P^{-1}$$
$$A^2 = P\,\tilde{P}\,P^{-1}\cdot P\,\tilde{P}\,P^{-1} = P\,\tilde{P}^2\,P^{-1}$$

Similarly, $$A^m = P\,\tilde{P}^m\,P^{-1}$$

$$e^{At} = I + At + \frac{A^2\,t^2}{2!} + ...$$

$$= I + P\,\tilde{P}\,P^{-1}\,t + \frac{P\,\tilde{P}^2\,P^{-1}\,t^2}{2} + ...$$

$$= P\,e^{\tilde{P}\,t}\,P^{-1}$$

Find the eigenvalue. Let $\lambda_{1,}$ λ_2, λ_n be the eigenvalues

The find $\tilde{P}$ matrix

$$\tilde{P} = \begin{bmatrix} \lambda_1 & 0 & \cdots \\ 0 & \lambda_2 & \cdots \\ \vdots & & \\ 0 & 0 & \lambda_n \end{bmatrix}$$

Find e^{At} $$e^{\tilde{P}t} = \begin{bmatrix} e^{\lambda_1 t} & \cdots & \cdots \\ \cdots & e^{\lambda_2 t} & \cdots \\ \cdots & \cdots & e^{\lambda_n t} \end{bmatrix}$$

Find e^{At} $$\boxed{e^{AT} = P\,e^{\tilde{P}t}\,P^{-1}} \qquad ...(11.64)$$

where P = Model matrix.

EXAMPLES

Example 11.1 Determine the state transition matrix for the given system matrix $A = \begin{bmatrix} 1 & 0 \\ 1 & 1 \end{bmatrix}$ by Laplace transform method. *(U.P.T.U., 2006–07, CS)*

Solution Given

$$A = \begin{bmatrix} 1 & 0 \\ 1 & 1 \end{bmatrix}$$

$$(sI - A) = \begin{bmatrix} s & 0 \\ 0 & s \end{bmatrix} - \begin{bmatrix} 1 & 0 \\ 1 & 1 \end{bmatrix}$$

$$(sI - A) = \begin{bmatrix} s-1 & 0 \\ -1 & s-1 \end{bmatrix} \qquad ...(1)$$

$$[sI - A]^{-1} = \frac{\text{adj}\,(sI - A)}{|sI - A|}$$

From Eq. (1), we get

$$|sI - A| = (s-1)^2 - 0$$
$$= (s-1)^2$$

$$\text{adj of } (sI - A) = \begin{bmatrix} s-1 & 0 \\ 1 & s-1 \end{bmatrix}$$

$$\frac{\text{adj of } (sI - A)}{|sI - A|} = \frac{1}{(s-1)^2}\begin{bmatrix} s-1 & 0 \\ 1 & s-1 \end{bmatrix}$$

$$\frac{\text{adj of } (sI - A)}{|sI - A|} = \begin{bmatrix} \dfrac{1}{s-1} & 0 \\ \dfrac{1}{(s-1)^2} & \dfrac{1}{s-1} \end{bmatrix}$$

$$|sI - A| = \frac{\text{adj of } (sI - A)}{|sI - A|} = \begin{bmatrix} \dfrac{1}{s-1} & 0 \\ \dfrac{1}{(s-1)^2} & \dfrac{1}{s-1} \end{bmatrix}$$

We know that

$$\phi(t) = L^{-1}\,(sI - A)^{-1}$$

$$\phi(t) = L^{-1}\begin{bmatrix} \dfrac{1}{s-1} & 0 \\ \dfrac{1}{(s-1)^2} & \dfrac{1}{s-1} \end{bmatrix}$$

By inverse Laplace transform

$$\phi(t) = \begin{bmatrix} t^t & 0 \\ te^t & e^t \end{bmatrix} = e^{At} \quad \textbf{Ans.}$$

Example 11.2 Determine whether the system represented by state equation is controllable or not?

$$\begin{bmatrix} \dot{x}_1 \\ \dot{x}_2 \\ \dot{x}_3 \end{bmatrix} = \begin{bmatrix} 0 & 1 & 0 \\ 0 & 0 & 1 \\ -6 & -11 & 6 \end{bmatrix}\begin{bmatrix} x_1 \\ x_2 \\ x_3 \end{bmatrix} + \begin{bmatrix} 0 \\ 0 \\ 1 \end{bmatrix} u \quad \textit{(U.P.T.U., 2006–07, CS)}$$

Solution We have

$$A = \begin{bmatrix} 0 & 1 & 0 \\ 0 & 0 & 1 \\ -6 & -11 & 6 \end{bmatrix}; B = \begin{bmatrix} 0 \\ 0 \\ 1 \end{bmatrix}$$

By the Kalman method $C_M = [B \quad AB \quad A^2B]$

$$AB = \begin{bmatrix} 0 & 1 & 0 \\ 0 & 0 & 1 \\ -6 & -11 & -6 \end{bmatrix}\begin{bmatrix} 0 \\ 0 \\ 1 \end{bmatrix} = \begin{bmatrix} 0 \\ 1 \\ -6 \end{bmatrix}$$

$$A^2B = \begin{bmatrix} 0 & 1 & 0 \\ 0 & 0 & 1 \\ -6 & -11 & -6 \end{bmatrix}\begin{bmatrix} 0 & 1 & 0 \\ 0 & 0 & 1 \\ -6 & -11 & -6 \end{bmatrix}\begin{bmatrix} 0 \\ 0 \\ 1 \end{bmatrix}$$

$$= \begin{bmatrix} 0 & 0 & 1 \\ -6 & -11 & 6 \\ 36 & 60 & 25 \end{bmatrix}\begin{bmatrix} 0 \\ 0 \\ 1 \end{bmatrix} = \begin{bmatrix} 1 \\ -6 \\ 25 \end{bmatrix}$$

Now $C_M = [B : \ AB : \ A^2B]$

$$C_M = \begin{bmatrix} 0 & 0 & 1 \\ 0 & 1 & -6 \\ 1 & -6 & 25 \end{bmatrix}$$

The rank of the C_M matrix is 3 because the $|C_M| = -1$, i.e., not zero. We conclude that the system is controllable.

Example 11.3 Write the state equations of an electrical network shown in Fig. Ex. 11.3 (a).

(U.P.T.U., 2006–07, ACS; U.P.T.U., 2004–05, ACS)

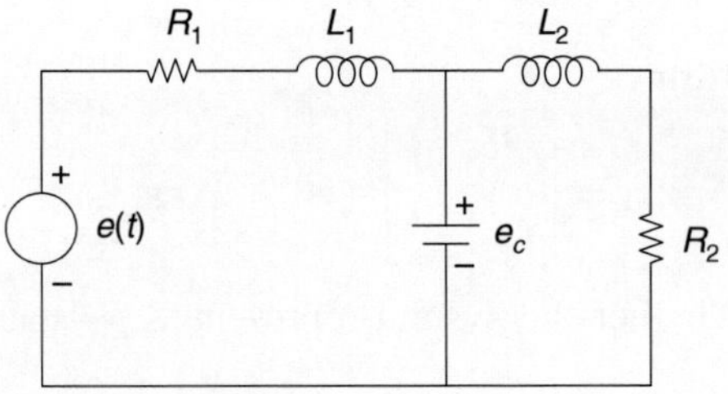

Figure Ex. 11.3 (a)

Solution The differential equations are written as

$$R_1 I_1 + L_1 \frac{dI_1}{dt} + V_c = e(t)$$

$$\frac{dI_1}{dt} = \frac{-R_1}{L_1} I_1 - \frac{1}{L_1} V_c + \frac{e(t)}{L_1}$$

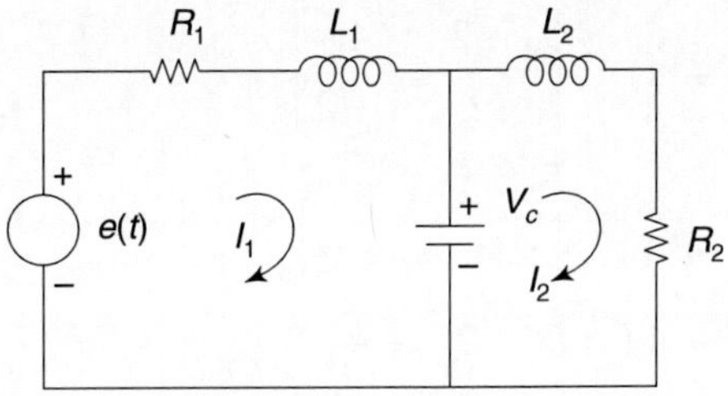

and $$R_2 I_2 + L_2 \frac{dI_2}{dt} - V_c = 0 \quad ...(1)$$

$$\frac{dI_2}{dt} = -\frac{R_2}{L_2} I_2 + \frac{V_c}{L_2} \quad ...(2)$$

and $$\frac{dV_c}{dt} = \frac{I_1}{C} - \frac{I_2}{C} \quad ...(3)$$

The output equation is

$$Y_1 = I_1$$
$$Y_2 = I_2$$

Now putting $I_1 = x_1$, $I_2 = x_2$ and $V_c = x_3$

Now writing Eqs. (1), (2) and (3) in matrix form and substitute the above value

$$\begin{bmatrix} \dot{x}_1 \\ \dot{x}_2 \\ \dot{x}_3 \end{bmatrix} = \begin{bmatrix} -\frac{R_1}{L_1} & 0 & -\frac{1}{L_1} \\ 0 & -\frac{R_2}{L_2} & \frac{1}{L_2} \\ \frac{1}{C} & -\frac{1}{C} & 0 \end{bmatrix} \begin{bmatrix} x_1 \\ x_2 \\ x_3 \end{bmatrix} + \begin{bmatrix} \frac{1}{L_1} \\ 0 \\ 0 \end{bmatrix} \begin{bmatrix} e \\ t \\ 0 \\ 0 \end{bmatrix}$$

$$[Y] = \begin{bmatrix} 1 & 0 \\ 0 & 1 \end{bmatrix} \begin{bmatrix} x_1 \\ x_2 \end{bmatrix} \quad \textbf{Ans.}$$

Example 11.4 Consider the system $\dot{x} = Ax$ with $x_0 = x(0)$, where $A = \begin{bmatrix} -2 & -4 \\ 1 & -2 \end{bmatrix}$. Find $\phi(t)$ and the solution X for $x_0 = [1 \quad 1]^T$. *(U.P.T.U., 2005–06, ACS)*

Solution Given

$$A = \begin{bmatrix} -2 & -4 \\ 1 & -2 \end{bmatrix}$$

$$(sI - A) = \begin{bmatrix} s & 0 \\ 0 & s \end{bmatrix} - \begin{bmatrix} -2 & -4 \\ 1 & -2 \end{bmatrix}$$

$$(sI - A) = \begin{bmatrix} s+2 & 4 \\ -1 & s+2 \end{bmatrix}$$

$$|sI - A| = \begin{bmatrix} s+2 & 4 \\ -1 & s+2 \end{bmatrix}$$

$$|sI - A| = (s + 2)^2 + 4$$

$$\text{adj of } (sI - A) = \begin{bmatrix} s+2 & -4 \\ 1 & s+2 \end{bmatrix}$$

$$[sI - A]^{-1} = \frac{\text{adj of } (sI - A)}{|sI - A|} = \frac{\begin{bmatrix} s+2 & -4 \\ 1 & s+2 \end{bmatrix}}{(s+2)^2 + 4}$$

$$[sI - A]^{-1} = \begin{bmatrix} \dfrac{s+2}{(s+2)^2+4} & \dfrac{-4}{(s+2)^2+4} \\ \dfrac{1}{(s+2)^2+4} & \dfrac{s+2}{(s+2)^2+4} \end{bmatrix}$$

$$L^{-1}\,[sI - A]^{-1} = L^{-1}\begin{bmatrix} \dfrac{s+2}{(s+2)^2+4} & \dfrac{-4}{(s+2)^2+4} \\ \dfrac{1}{(s+2)^2+4} & \dfrac{s+2}{(s+2)^2+4} \end{bmatrix}$$

By inverse Laplace transform

$$\phi(t) = e^{At} = \begin{bmatrix} e^{-2t}\cos 2t & -e^{-2t}\sin 2t \\ \dfrac{1}{4}s^{-2t}\sin 2t & e^{-2t}\cos 2t \end{bmatrix}$$

We have

$$x(t) = \phi(t)\, x_0$$

$$x(t) = \begin{bmatrix} e^{-2t}\cos 2t & -e^{-2t}\sin 2t \\ \dfrac{1}{4}e^{-2t}\sin 2t & e^{-2t}\cos 2t \end{bmatrix}\begin{bmatrix} 1 \\ 1 \end{bmatrix}$$

$$x(t) = \begin{bmatrix} e^{-2t}\cos 2t - e^{-2t}\sin 2t \\ \dfrac{1}{4}e^{-2t}\sin 2t + e^{-2t}\cos 2t \end{bmatrix}$$

$$x(t) = e^{-2t}\begin{bmatrix} \cos 2t - \sin 2t \\ \dfrac{1}{4}\sin 2t + \cos 2t \end{bmatrix} \quad \textbf{Ans.}$$

Example 11.5 Examine the controllability and observability conditions of the system given below using canonical form.

$$\begin{bmatrix} \dot{x}_1 \\ \dot{x}_2 \\ \dot{x}_3 \end{bmatrix} = \begin{bmatrix} 2 & -2 & 3 \\ 1 & 1 & 1 \\ 1 & 3 & -1 \end{bmatrix} \begin{bmatrix} x_1 \\ x_2 \\ x_3 \end{bmatrix} + \begin{bmatrix} 11 \\ 1 \\ -14 \end{bmatrix} u$$

$$Y = [-3 \quad 5 \quad -2] \begin{bmatrix} x_1 \\ x_2 \\ x_3 \end{bmatrix}$$

(*U.P.T.U., 2005–06, ACS*)

Solution Given

$$A = \begin{bmatrix} 2 & -2 & 3 \\ 1 & 1 & 1 \\ 1 & 3 & -1 \end{bmatrix}; B = \begin{bmatrix} 11 \\ 1 \\ -14 \end{bmatrix}$$

$$C = [-3 \quad 5 \quad -2]$$

$$A^2 = \begin{bmatrix} 2 & -2 & 3 \\ 1 & 1 & 1 \\ 1 & 3 & -1 \end{bmatrix} \begin{bmatrix} 2 & -2 & 3 \\ 1 & 1 & 1 \\ 1 & 3 & -1 \end{bmatrix} = \begin{bmatrix} 5 & 3 & 1 \\ 4 & 2 & 3 \\ 4 & -2 & 7 \end{bmatrix}$$

Controllability test

$$C_M = [B : AB : A^2B]$$

$$AB = \begin{bmatrix} 2 & -2 & 3 \\ 1 & 1 & 1 \\ 1 & 3 & -1 \end{bmatrix} \begin{bmatrix} 11 \\ 1 \\ -14 \end{bmatrix} = \begin{bmatrix} 22 \\ -2 \\ 28 \end{bmatrix}$$

$$A^2B = \begin{bmatrix} 5 & 3 & 1 \\ 4 & 2 & 3 \\ 4 & -2 & 7 \end{bmatrix} \begin{bmatrix} 11 \\ 1 \\ -14 \end{bmatrix} = \begin{bmatrix} 44 \\ 4 \\ -56 \end{bmatrix}$$

Now $$C_M = [B : AB : A^2B]$$

$$C_M = \begin{bmatrix} 11 & 22 & 44 \\ 1 & -2 & 4 \\ -14 & 28 & -56 \end{bmatrix}$$

Rank of CM is 2. Hence, $C_M \neq 3$ so the system is uncontrollable.

Observability test

$$O_M = \begin{bmatrix} C \\ CA \\ CA^2 \end{bmatrix}$$

$$CA = [-3 \quad 5 \quad -2]\begin{bmatrix} 2 & -2 & 3 \\ 1 & 1 & 1 \\ 1 & 3 & -1 \end{bmatrix}$$

$$CA^2 = [-3 \quad 5 \quad -2]$$

$$CA^2 = [-3 \quad 5 \quad -2]\begin{bmatrix} 5 & 3 & 1 \\ 4 & 2 & 3 \\ 4 & -2 & 7 \end{bmatrix}$$

$$CA^2 = [-3 \quad 5 \quad -2]$$

$$O_M = \begin{bmatrix} -3 & 5 & -2 \\ -3 & 5 & -2 \\ -3 & 5 & -2 \end{bmatrix}$$

We see that the rank of matrix O_M is not equal to 3, it is 1. Hence, the system is unobservable.

Example 11.6 Consider the third order system.

$$\begin{bmatrix} \dot{x}_1 \\ \dot{x}_2 \\ \dot{x}_3 \end{bmatrix} = \begin{bmatrix} 0 & 1 & 0 \\ 0 & 0 & 1 \\ 0 & -30 & -11 \end{bmatrix}\begin{bmatrix} x_1 \\ x_2 \\ x_3 \end{bmatrix} + \begin{bmatrix} 0 \\ 0 \\ 1 \end{bmatrix} u$$

$$y = [1 \quad 0 \quad 0]\begin{bmatrix} x_1 \\ x_2 \\ x_3 \end{bmatrix}$$

Find the suitable model, matrix such that the poles at $s = -2, -5, -6$.

(*U.P.T.U., 2004–05, ACS*)

Solution The general equation of the state space model is

$$\dot{x} = Ax + Bu$$

$$y = Cx + Du$$

So,

$$A = \begin{bmatrix} 0 & 1 & 0 \\ 0 & 0 & 1 \\ 0 & -30 & -11 \end{bmatrix} \quad B = \begin{bmatrix} 0 \\ 0 \\ 1 \end{bmatrix}$$

and

$$C = [1 \quad 0 \quad 0]$$

When A is expressed in the form given below

$$A = \begin{bmatrix} 0 & 1 & 0 & \cdots & 0 \\ 0 & 0 & 1 & \cdots & 0 \\ \vdots & & & & \\ -a_n & -a_{n-1} & \cdots & \cdots & -a_1 \end{bmatrix}$$

$$P = \begin{bmatrix} 1 & 1 & 1 \\ \lambda_1 & \lambda_2 & \lambda_n \\ \lambda_1^2 & \lambda_2^2 & \lambda_n^2 \\ \vdots & \vdots & \vdots \\ \lambda_1^{n-1} & \lambda_2^{n-1} & \lambda_n^{n-1} \end{bmatrix}$$

So according to above ondition for given pole at $\lambda_1 = -2, \lambda_2 = -5, \lambda_3 = -6$

$$P = \begin{bmatrix} 1 & 1 & 1 \\ -2 & -5 & -6 \\ 4 & 25 & 36 \end{bmatrix} \quad \textbf{Ans.}$$

Example 11.7 Consider a single-input-output system having an overall transfer function $T(s)$ given by $T(s) = \dfrac{s^2 + 4s + 4}{s^3 + 5s^2 + 4s}$.

Represents the system state model in terms of the canonical variables and determine a set of coefficients matrix A, B, C. *(U.P.T.U., 2004–05, ACS)*

Solution The transfer function of the system is

$$T(s) = \frac{Y(s)}{U(s)} = \frac{s^2 + 4s + 4}{s^3 + 5s^2 + 4s} = \frac{s^2 + 4s + 4}{s(s^2 + 5s + 4)}$$

$$\frac{Y(s)}{U(s)} = \frac{s^2 + 4s + 4}{s\,(s + 1)\,(s + 4)} \qquad \text{...(1)}$$

Solving the equation by partial fraction

$$\frac{s^2 + 4s + 4}{s\,(s + 1)\,(s + 4)} = \frac{A}{s} + \frac{B}{(s + 1)} + \frac{C}{(s + 4)} \qquad \text{...(2)}$$

$$\frac{s^2 + 4s + 4}{s\,(s + 1)\,(s + 4)} = \frac{A\,(s + 1)\,(s + 4) + B\,(s + 4)\,s + Cs\,(s + 1)}{s\,(s + 1)\,(s + 4)}$$

$$s^2 + 4\,s + 4 = A\,(s + 1)\,(s + 4) + Bs\,(s + 4) + Cs\,(s + 1) \qquad \text{...(3)}$$

Putting $s = 0$ in Eq. (3), we get

$$4 = A\,(1)\,(4) + B\,(0)\,(4) + C\,(0)$$

$$4 = 4A$$

$$\Rightarrow \quad A = 1$$

Now putting $s + 1 = 0$ in Eq. (3), we get

$$(-1)^2 + 4\,(-1) + 4 = A\,(0)\,(-1 + 4) + B\,(-1 + 4)\,(-1) + C\,(-1)\,(0)$$

$$1 = -3\,B$$

$$\Rightarrow \quad B = -1/3$$

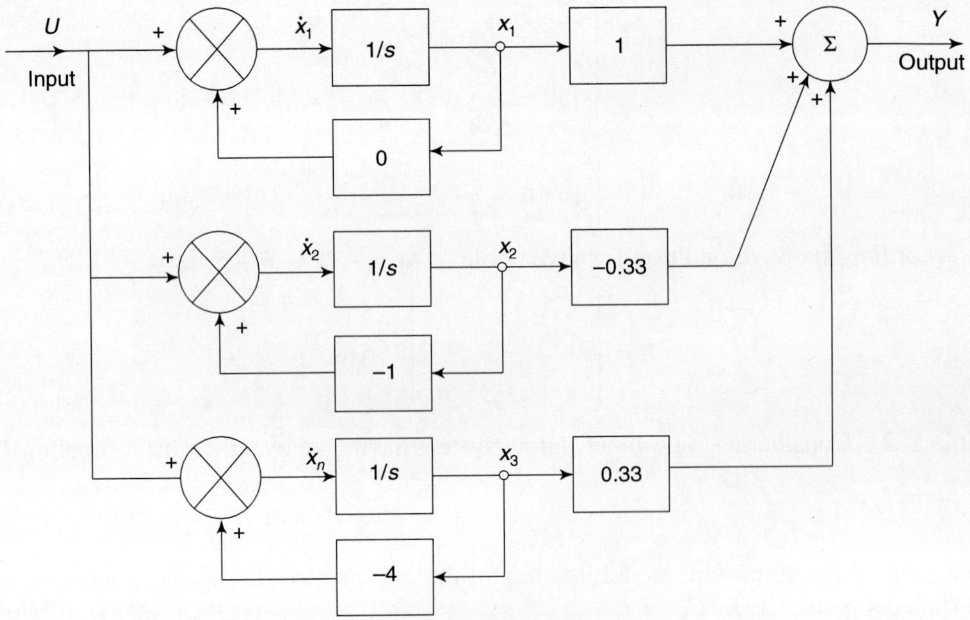

Figure Ex. 11.7 Block diagram of canonical state model of Ex. 11.7

Now putting $s + 4 = 0$ in Eq. (3), we get

$$(-4)^2 + 4(-4) + 4 = A(-4+1)(0) + B(0)(-4) + C(-4)(-4+1)$$

$$16 - 16 + 4 = 12\, C$$

$$C = \frac{4}{12} = \frac{1}{3}$$

Now putting the value of A, B, C in Eq. (2), we get

$$T(s) = \frac{1}{s} + \frac{-\frac{1}{3}}{s+1} + \frac{\frac{1}{3}}{s+4}$$

$$T(s) = \frac{1}{s} - \frac{0.33}{s+1} + \frac{0.33}{s+4} \qquad ...(4)$$

Comparing the Eq. (4) with the standard Eq. (11.20) is

$$T(s) = b_0 + \sum_{i=1}^{n} \frac{C_i}{(s - P_i)}$$

Hence $i = 3$

So, $b_0 = 0$, $C_1 = 1$; $C_2 = -0.33$ and $C_3 = 0.33$ and $P_1 = 0$, $P_2 = -1$, $P_3 = -4$

Now the block diagram for the transfer function is shown in (Fig. 11.6)

The state equations are [Eq. (11.23a)

$$\dot{x}_1 = u$$

$$\dot{x}_2 = x_2 + u$$

$$\dot{x}_3 = 4x_3 + u$$

The output $y(t)$ is given by Eq. (11.23 b)

$$y(t) = C_1 x_1 + C_2 x_2 + \ldots c_n x_n + bu$$

$$y(t) = x_1 - 0.33 x_2 + 0.33 x_3$$

The state model in vector matrix form as

$$\begin{bmatrix} \dot{x}_1 \\ \dot{x}_2 \\ \dot{x}_3 \end{bmatrix} = \begin{bmatrix} 0 & 0 & 0 \\ 0 & -1 & 0 \\ 0 & 0 & -4 \end{bmatrix} \begin{bmatrix} x_1 \\ x_2 \\ x_3 \end{bmatrix} + \begin{bmatrix} 1 \\ 1 \\ 1 \end{bmatrix} u \Bigg\} \quad \ldots(A)$$

$$y = [1 \quad -0.33 \quad 0.33] \begin{bmatrix} x_1 \\ x_2 \\ x_3 \end{bmatrix} \quad \textbf{Ans.}$$

By observing Eq. (*A*), we get that the matrix *A* is a diagonal matrix and the poles of the transfer function $T(s)$ represent the diagonal elements. All elements of vector matrix *B* of the state model are all unity, and the elements of the row vector (*c*) are the residues of the system poles.

Example 11.8 Find e^{At} by the Laplace transform method and Cayley Hamilton theorem for

$$A = \begin{bmatrix} 0 & 1 \\ -12 & -7 \end{bmatrix} \quad (U.P.T.U., 2003, C.O., ACS)$$

Solution Given

$$A = \begin{bmatrix} 0 & 1 \\ -12 & -7 \end{bmatrix}$$

$$(sI - A) = \begin{bmatrix} s & 0 \\ 0 & s \end{bmatrix} - \begin{bmatrix} 0 & 1 \\ -12 & -7 \end{bmatrix} = \begin{bmatrix} s & -1 \\ 12 & s+7 \end{bmatrix}$$

The determinant of $(sI - A)$

$$|sI - A| = s(s + 7) + 12 = s^2 + 7s + 12$$

$$= (s + 4)(s + 3)$$

The adjoint of $(sI - A)$

$$\text{adj of } (sI - A) = \begin{bmatrix} s+7 & +1 \\ -12 & s \end{bmatrix}$$

$$[sI - A]^{-1} = \frac{\text{adj of } (sI - A)}{|sI - A|}$$

$$= \frac{1}{(s+3)(s+4)} \begin{bmatrix} s+7 & +12 \\ -12 & s \end{bmatrix}$$

The state transition matrix e^{At}

$$e^{At} = L^{-1}\,[(sI - A)^{-1}] = \phi(t)$$

$$= L^{-1}\begin{bmatrix} \dfrac{s+7}{(s+3)(s+4)} & \dfrac{1}{(s+3)(s+4)} \\ -\dfrac{12}{(s+3)(s+4)} & \dfrac{s}{(s+3)\,9s+4)} \end{bmatrix}$$

By partial fraction power

$$\frac{(s+7)}{(s+3)(s+4)} = \frac{A}{(s+4)} + \frac{B}{(s+3)} = \frac{-3}{(s+4)} + \frac{4}{(s+3)}$$

Now

$$e^{At} = L^{-1}\begin{bmatrix} \dfrac{-3}{s+4} + \dfrac{4}{s+3} & -\dfrac{1}{(s+4)} + \dfrac{1}{(s+3)} \\ -12\left[\dfrac{1}{(s+4)} + \dfrac{1}{(s+3)}\right] & \dfrac{4}{s+4} + \dfrac{-3}{s+3} \end{bmatrix}$$

By inverse Laplace transform

$$e^{At} = \begin{bmatrix} -3\,e^{-4t} + 4\,e^{-3t} & -e^{-4t} + e^{-3t} \\ 12\,e^{-4t} - 12\,e^{-3t} & 4\,e^{-4t} - 3\,e^{-3t} \end{bmatrix} \quad \textbf{Ans.}$$

Caley Hamilton solution

$$\phi(t) = e^{At} = I + At + \frac{A^2 t^2}{2!} + \frac{A^3 t^3}{3!} + \ldots$$

$$A = \begin{bmatrix} 0 & 1 \\ -12 & -7 \end{bmatrix};$$

$$A^2 = \begin{bmatrix} 0 & 1 \\ -12 & -7 \end{bmatrix}\begin{bmatrix} 0 & 1 \\ -12 & -7 \end{bmatrix} = \begin{bmatrix} -12 & -7 \\ 84 & 37 \end{bmatrix}$$

$$A^3 = \begin{bmatrix} -12 & -7 \\ 84 & 37 \end{bmatrix}\begin{bmatrix} 0 & 1 \\ -12 & -7 \end{bmatrix} = \begin{bmatrix} 84 & 37 \\ -444 & -175 \end{bmatrix}$$

$$\phi(t) = \begin{bmatrix} 1 & 0 \\ 0 & 1 \end{bmatrix} + \begin{bmatrix} 0 & t \\ -12\,t & -7\,t \end{bmatrix} + \begin{bmatrix} \dfrac{-12\,t^2}{2} & -\dfrac{7\,t^2}{2} \\ \dfrac{84\,t^2}{2} & \dfrac{37\,t^2}{2} \end{bmatrix} + \begin{bmatrix} \dfrac{-84\,t^3}{6} & -\dfrac{37\,t^2}{6} \\ -\dfrac{444+3}{6} & -\dfrac{175\,t^2}{6} \end{bmatrix} + \ldots$$

$$= \begin{bmatrix} 1 - 6\,t^2 + 14\,t^3 + \ldots & t - \dfrac{7\,t^2}{2} + \dfrac{37}{6}\,t^3 + \ldots \\ -12\,t + 42\,t^2 - 74\,t^3 + \ldots & 1 + 7\,t + \dfrac{37\,t^2}{2} - \dfrac{175\,t^3}{6} + \ldots \end{bmatrix} \quad \textbf{Ans.}$$

Example 11.9 For the state matrix

$$A = \begin{bmatrix} 0 & 1 \\ -8 & -6 \end{bmatrix} \text{ Find } e^{At} \text{ by any two methods } (\textit{U.P.T.U., 2002–03, ACS})$$

Solution Given

$$A = \begin{bmatrix} 0 & 1 \\ -8 & -6 \end{bmatrix}$$

1. By Caley Hamilton method, we have

$$e^{At} = I + At + \frac{A^2 t^2}{2!} + \frac{A^3 t^3}{3!} + \ldots \qquad \ldots(1)$$

$$A^2 = \begin{bmatrix} 0 & 1 \\ -8 & -6 \end{bmatrix}\begin{bmatrix} 0 & 1 \\ -8 & -6 \end{bmatrix} = \begin{bmatrix} -8 & -6 \\ 48 & 38 \end{bmatrix}$$

$$A^3 = \begin{bmatrix} -8 & -6 \\ 48 & 28 \end{bmatrix}\begin{bmatrix} 0 & 1 \\ -8 & -6 \end{bmatrix} = \begin{bmatrix} 48 & 28 \\ -224 & -120 \end{bmatrix}$$

Putting the value of A, A^2, A^3 in Eq. (1), we get

$$e^{At} = \begin{bmatrix} 1 & 0 \\ 0 & 1 \end{bmatrix} + \begin{bmatrix} 0 & 1 \\ -8 & -6 \end{bmatrix} t + \begin{bmatrix} -8 & -6 \\ 48 & 28 \end{bmatrix}\frac{t^2}{2!} + \begin{bmatrix} 48 & 28 \\ -224 & -120 \end{bmatrix}\frac{t^3}{3!} + \ldots$$

$$e^{At} = \begin{bmatrix} 1 & 0 \\ 0 & 1 \end{bmatrix} + \begin{bmatrix} 0 & t \\ -8t & -6t \end{bmatrix} + \begin{bmatrix} \frac{-8t^2}{2} & \frac{-6t^2}{2} \\ \frac{48t^2}{2} & \frac{28t^2}{2} \end{bmatrix} + \begin{bmatrix} \frac{48t^3}{6} & \frac{28t^3}{6} \\ \frac{-224t^3}{6} & \frac{-120t^3}{6} \end{bmatrix} + \ldots$$

$$e^{At} = \begin{bmatrix} 1 & 0 \\ 0 & 1 \end{bmatrix} + \begin{bmatrix} 0 & t \\ -8t & -6t \end{bmatrix} + \begin{bmatrix} -4t^2 & -3t^2 \\ 24t^2 & 14t^2 \end{bmatrix} + \begin{bmatrix} 8t^3 & \frac{14}{3}t^3 \\ \frac{-112}{3}t^3 & -20t^3 \end{bmatrix} + \ldots$$

$$e^{At} = \begin{bmatrix} 1 - 4t^2 + 8t^3 + \ldots & t - 3t^2 + \frac{14}{3}t^3 + \ldots \\ -8t + 24t^2 - \frac{122}{3}t^3 + \ldots & 1 - 6t + 14t^2 - 20t^3 + \ldots \end{bmatrix} \quad \textbf{Ans.}$$

2. By Laplace transform method, we have

$$e^{At} = L^{-1}\,[(sI - A)^{-1}]$$

$$(sI - A) = \begin{bmatrix} s & 0 \\ 0 & s \end{bmatrix} - \begin{bmatrix} 0 & 1 \\ -8 & -6 \end{bmatrix} = \begin{bmatrix} s & -1 \\ 8 & s+6 \end{bmatrix}$$

The determinant of $(sI - A)$ is

$$|sI - A| = s(s + 6) + 8 = s^2 + 6s + 8$$

The adjoint of $(sI - A)$ is

$$\text{adj of } (sI - A) = \begin{bmatrix} s+6 & 1 \\ -8 & s \end{bmatrix}$$

Now $$(sI - A)^{-1} = \frac{\text{adj of } (sI - A)}{|sI - A|} = \frac{1}{s^2 + 6s + 8}\begin{bmatrix} s+6 & 1 \\ -8 & s \end{bmatrix}$$

$$(sI - A)^{-1} = \begin{bmatrix} \dfrac{s+6}{s^2+6s+8} & \dfrac{1}{s^2+6s+8} \\ \dfrac{-8}{s^2+6s+8} & \dfrac{s}{s^2+6s+8} \end{bmatrix} \qquad ...(2)$$

By the partial fraction of each term

$$\frac{s+6}{s^2+6s+2} = \frac{2}{(s+2)} - \frac{1}{(s+4)} \qquad ...(A)$$

$$\frac{1}{s^2+6s+2} = \frac{1}{2\,(s+2)} - \frac{1}{2\,(s+4)} \qquad ...(B)$$

$$\frac{-8}{s^2+Cs+8} = \frac{-4}{s+2} + \frac{4}{s+4} \qquad ...(C)$$

$$\frac{s}{s^2+6s+8} = \frac{-1}{s+2} + \frac{2}{s+4} \qquad ...(D)$$

Substituting the value from A, B, C, D in Eq. (2), we get

$$(sI - A) = \begin{bmatrix} \dfrac{2}{s+2} - \dfrac{1}{s+4} & \dfrac{1}{2\,(s+2)} - \dfrac{1}{2\,(s+4)} \\ \dfrac{-4}{s+2} + \dfrac{4}{s+4} & \dfrac{-1}{s+2} + \dfrac{2}{s+4} \end{bmatrix}$$

$$e^{At} = L^{-1}\,[(sI - A)^{-1}]$$

$$e^{At} = L^{-1}\begin{bmatrix} \dfrac{2}{s+2} - \dfrac{1}{s+4} & \dfrac{1}{2\,(s+2)} - \dfrac{1}{2\,(s+4)} \\ \dfrac{-4}{(s+2} + \dfrac{4}{s+4} & -\dfrac{1}{(s+2)} + \dfrac{2}{s+4} \end{bmatrix}$$

By inverse Laplace transform method

$$e^{At} = \begin{bmatrix} 2\,e^{-2t} - e^{4t} & \dfrac{1}{2}e^{-2t} - \dfrac{1}{2}e^{-4t} \\ -4\,e^{-2t} + 4\,e^{-4t} & -e^{-2t} + 2\,e^{-4t} \end{bmatrix} \quad \textbf{Ans.}$$

Example 11.10 Check observability of the system given below

$$\dot{x}_1 = 2x_2$$
$$\dot{x}_2 = -2x_1 - 4x_2 + u$$
$$y = x_1 + x_2$$

Solution

$$\begin{bmatrix} \dot{x}_1 \\ \dot{x}_2 \end{bmatrix} = \begin{bmatrix} 0 & 2 \\ -2 & -4 \end{bmatrix} \begin{bmatrix} x_1 \\ x_2 \end{bmatrix} + \begin{bmatrix} 0 \\ 1 \end{bmatrix} u$$

$$y = [1 \quad 1] \begin{bmatrix} x_1 \\ x_2 \end{bmatrix}$$

$$A = \begin{bmatrix} 0 & 2 \\ -2 & -4 \end{bmatrix}, c\ [1 \quad 1]$$

$$CA = [1 \quad 1] \begin{bmatrix} 0 & 2 \\ -2 & -4 \end{bmatrix} = [-2 \quad -2]$$

$$O_M = \begin{bmatrix} C \\ \cdots \\ CA \end{bmatrix} = \begin{bmatrix} 1 & 1 \\ -2 & 2 \end{bmatrix}$$

The rank of $O_M \neq 2$ and the $|O_M| = 0$ so the system is unobservable.

Example 11.11 Check controllability and observability of the system given below.

$$\dot{x}_1 = 4x_1 + x_2 + u$$
$$\dot{x}_2 = 3x_1 - 2x_2 + u$$
$$y = x_1 + x_2$$

Solution The system given is

$$\begin{bmatrix} \dot{x}_1 \\ \dot{x}_2 \end{bmatrix} = \begin{bmatrix} -4 & 1 \\ -3 & -2 \end{bmatrix} \begin{bmatrix} x_1 \\ x_2 \end{bmatrix} + \begin{bmatrix} 1 \\ 1 \end{bmatrix} u$$

$$y = [1 \quad 1] \begin{bmatrix} x_1 \\ x_2 \end{bmatrix}$$

Here

$$A = \begin{bmatrix} -4 & 1 \\ -3 & -2 \end{bmatrix}; B = \begin{bmatrix} 1 \\ 1 \end{bmatrix}$$

$$C = [1 \quad 1]$$

Controllability test

$$B = \begin{bmatrix} 1 \\ 1 \end{bmatrix}$$

$$AB = \begin{bmatrix} -4 & 1 \\ -3 & -2 \end{bmatrix}\begin{bmatrix} 1 \\ 1 \end{bmatrix} = \begin{bmatrix} -3 \\ -5 \end{bmatrix}$$

$$C_M = [B : AB]$$

$$C_M = \begin{bmatrix} 1 & -3 \\ 1 & -5 \end{bmatrix}$$

The determinant of $C_M = -2$ and the rank of C_M is 2. Hence, the system is controllable.

Observability test

$$C = [1 \quad 1]$$

$$CA = [1 \quad 1]\begin{bmatrix} -4 & 1 \\ -3 & -2 \end{bmatrix} = [-7 \quad -1]$$

$$O_M = \begin{bmatrix} C \\ \cdots \\ CA \end{bmatrix} = \begin{bmatrix} 1 & 1 \\ -7 & -1 \end{bmatrix}$$

The determinant of O_M is 6 and the rank of the matrix is 2. Hence, the system is observable.

Example 11.12 Verify the controllability given that

$$\dot{x}_1 = x_2 + u_1$$

$$\dot{x}_2 = 2x_3$$

$$\dot{x}_3 = -3x_2 - 4x_3 + u_1 + u_2$$

Solution The system given

$$\begin{bmatrix} \dot{x}_1 \\ \dot{x}_2 \\ \dot{x}_3 \end{bmatrix} = \begin{bmatrix} 0 & 1 & 0 \\ 0 & 0 & 2 \\ 0 & -3 & -4 \end{bmatrix}\begin{bmatrix} x_1 \\ x_2 \\ x_3 \end{bmatrix} + \begin{bmatrix} 0 & 1 \\ 0 & 0 \\ 1 & 1 \end{bmatrix}\begin{bmatrix} u_1 \\ u_2 \end{bmatrix}$$

Here

$$A = \begin{bmatrix} 0 & 1 & 0 \\ 0 & 0 & 2 \\ 0 & -3 & -4 \end{bmatrix} \text{ and } B \begin{bmatrix} 0 & 1 \\ 0 & 0 \\ 1 & 1 \end{bmatrix}$$

$$AB = \begin{bmatrix} 0 & 1 & 0 \\ 0 & 0 & 2 \\ 0 & -3 & -4 \end{bmatrix}\begin{bmatrix} 0 & 1 \\ 0 & 0 \\ 1 & 1 \end{bmatrix} = \begin{bmatrix} 0 & 0 \\ 2 & 2 \\ -4 & -4 \end{bmatrix}$$

$$A^2B = \begin{bmatrix} 0 & 1 & 0 \\ 0 & 0 & 2 \\ 0 & -3 & -4 \end{bmatrix}\begin{bmatrix} 0 & 1 & 0 \\ 0 & 0 & 2 \\ 0 & -3 & -4 \end{bmatrix}\begin{bmatrix} 0 & 1 \\ 0 & 0 \\ 1 & 1 \end{bmatrix}$$

$$= \begin{bmatrix} 0 & 0 & 2 \\ 0 & -6 & -8 \\ 0 & 12 & 10 \end{bmatrix} \begin{bmatrix} 0 & 1 \\ 0 & 0 \\ 1 & 1 \end{bmatrix} = \begin{bmatrix} 2 & 2 \\ -8 & -8 \\ 10 & 10 \end{bmatrix}$$

The controllability test matrix

$$C_M = [B : \quad AB : \quad A^2B]$$

$$C_M = \begin{bmatrix} 0 & 1 & 0 & 0 & 2 & 2 \\ 0 & 0 & 2 & 2 & -8 & -8 \\ 1 & 1 & -4 & -4 & 10 & 10 \end{bmatrix}$$

Collect the column from C_M to form a square 3 × 3 to form non-zero determinant matrix as given below:

$$\begin{bmatrix} 0 & 1 & 0 \\ 0 & 0 & 2 \\ 1 & 1 & -4 \end{bmatrix} = 0 \begin{vmatrix} 0 & 2 \\ 1 & -4 \end{vmatrix} - (1) \begin{vmatrix} 0 & 2 \\ 1 & -4 \end{vmatrix} + 0 \begin{vmatrix} 0 & 0 \\ 1 & 1 \end{vmatrix} = 2$$

Hence $$\begin{bmatrix} 0 & 1 & 0 \\ 0 & 0 & 2 \\ 1 & 1 & -4 \end{bmatrix} \neq 0$$

So, the rank of C_M is equal to 3 (M = 3), hence the system is completely controllable.

Example 11.13 Check for controllability and observability of a system having the following coefficient matrices.

$$A = \begin{bmatrix} 0 & 2 & 0 \\ 0 & 0 & 1 \\ -5 & -10 & -6 \end{bmatrix} \quad B = \begin{bmatrix} 1 \\ 0 \\ 4 \end{bmatrix}$$

and $$C = [10 \quad 5 \quad 6]$$

Solution **Controllability test**

$$B = \begin{bmatrix} 1 \\ 0 \\ 4 \end{bmatrix} \text{ and } A = \begin{bmatrix} 0 & 2 & 0 \\ 0 & 0 & 1 \\ -5 & -10 & -6 \end{bmatrix}$$

$$AB = \begin{bmatrix} 0 & 2 & 0 \\ 0 & 0 & 1 \\ -5 & -10 & -6 \end{bmatrix} \begin{bmatrix} 1 \\ 0 \\ 4 \end{bmatrix} = \begin{bmatrix} 0 \\ 4 \\ -29 \end{bmatrix}$$

$$A^2B = \begin{bmatrix} 0 & 2 & 0 \\ 0 & 0 & 1 \\ -5 & -10 & -6 \end{bmatrix} \begin{bmatrix} 0 & 2 & 0 \\ 0 & 0 & 1 \\ -5 & -10 & -6 \end{bmatrix} \begin{bmatrix} 1 \\ 0 \\ 4 \end{bmatrix} = \begin{bmatrix} 8 \\ -29 \\ 134 \end{bmatrix}$$

The controllability test matrix is

$$C_M = [B \ : \ AB \ : \ A^2B]$$

$$C_M = \begin{bmatrix} 1 & 0 & 8 \\ 0 & 4 & -29 \\ 4 & -29 & 134 \end{bmatrix}$$

The determinant of C_M is – 433. The rank of C_M is 3 ($n = 3$). Hence, the system is controllable

Observability test

$$C = [10 \quad 5 \quad 6]$$

$$CA = [10 \quad 5 \quad 6]\begin{bmatrix} 0 & 2 & 0 \\ 0 & 0 & 1 \\ -5 & -10 & -6 \end{bmatrix} = [-30 \quad -40 \quad -31]$$

$$CA^2 = [10 \quad 5 \quad 6]\begin{bmatrix} 0 & 2 & 0 \\ 0 & 0 & 1 \\ -5 & -10 & -6 \end{bmatrix} = \begin{bmatrix} 0 & 2 & 0 \\ 0 & 0 & 1 \\ -5 & -10 & -6 \end{bmatrix}$$

$$CA^2 = [10 \quad 5 \quad 6]\begin{bmatrix} 0 & 2 & 0 \\ 0 & 0 & 1 \\ 30 & 50 & 26 \end{bmatrix} = [155 \quad 220 \quad 146]$$

The observability test matrix is

$$O_M = \begin{bmatrix} C \\ \ddot{C}A \\ \ddot{C}A^2 \end{bmatrix} = \begin{bmatrix} 10 & 5 & 6 \\ -30 & -40 & -31 \\ 155 & 250 & 146 \end{bmatrix}$$

The determinant of $O_M \neq 0$ and the rank of O_M is 3. Hence, the system is completely observeable.

Example 11.14 Given that

$$\dot{x}(t) = \begin{bmatrix} 0 & 1 \\ -2 & 3 \end{bmatrix} x(t) \text{ find } \phi(t) \text{ or } e^{At}$$

Solution Given that

$$A = \begin{bmatrix} 0 & 1 \\ -2 & 3 \end{bmatrix}$$

$$A^2 = \begin{bmatrix} 0 & 1 \\ -2 & 3 \end{bmatrix}\begin{bmatrix} 0 & 1 \\ -2 & 3 \end{bmatrix} = \begin{bmatrix} -2 & 3 \\ -6 & 7 \end{bmatrix}$$

$$A^3 = \begin{bmatrix} -2 & 3 \\ -6 & 7 \end{bmatrix}\begin{bmatrix} 0 & 1 \\ -2 & 3 \end{bmatrix} = \begin{bmatrix} -6 & 7 \\ -14 & 15 \end{bmatrix}$$

By Caley Hamilton Method

$$e^{At} = I + At + \frac{A^2 t^2}{2!} + \frac{A^3 t^3}{3!} + ...$$

$$e^{At} = \begin{bmatrix} 1 & 0 \\ 0 & 1 \end{bmatrix} + \begin{bmatrix} 0 & 1 \\ -2 & 3 \end{bmatrix} t + \begin{bmatrix} -2 & 3 \\ -6 & 7 \end{bmatrix} \frac{t^2}{2} + \begin{bmatrix} -6 & 7 \\ -14 & -15 \end{bmatrix} \frac{t^3}{6} + ...$$

$$e^{At} = \begin{bmatrix} 1 & 0 \\ 0 & 1 \end{bmatrix} + \begin{bmatrix} 0 & t \\ -2t & 3t \end{bmatrix} + \begin{bmatrix} -t^2 & \frac{3t^2}{2} \\ -3t^2 & \frac{7}{2}t^2 \end{bmatrix} + \begin{bmatrix} -t^3 & \frac{7}{6}t^3 \\ -\frac{7}{3}t^3 & \frac{5}{2}t^3 \end{bmatrix} + ...$$

$$e^{At} = \begin{bmatrix} 1 - t^2 - t^3 + ... & t + \frac{3}{2}t^2 + \frac{7}{6}t^3 + ... \\ -2t - 3t^2 - \frac{7}{3}t^3 & 1 + 3t + \frac{7}{2}t^2 + \frac{5}{2}t^3 + ... \end{bmatrix}$$ **Ans.**

Example 11.15 Given that $\begin{bmatrix} \dot{x}_1 \\ \dot{x}_2 \end{bmatrix} = \begin{bmatrix} 1 & 0 \\ 1 & 1 \end{bmatrix} \begin{bmatrix} x_1 \\ x_2 \end{bmatrix} + \begin{bmatrix} 1 \\ 1 \end{bmatrix}$

Find $\phi(t)$ by Laplace transform method.

Solution Here

$$A = \begin{bmatrix} 1 & 0 \\ 1 & 1 \end{bmatrix}$$

$$(sI - A) = \begin{bmatrix} s & 0 \\ 0 & s \end{bmatrix} - \begin{bmatrix} 1 & 0 \\ 1 & 1 \end{bmatrix} = \begin{bmatrix} s-1 & 0 \\ -1 & s-1 \end{bmatrix}$$

The determinant of $(sI - A)$

$$|sI - A| = (s-1)^2$$

The adj of $(sI - A)$ is $= \begin{bmatrix} s-1 & 0 \\ 1 & s-1 \end{bmatrix}$

$$(sI - A)^{-1} = \frac{1}{(s-1)^2} \begin{bmatrix} s-1 & 0 \\ 1 & s-1 \end{bmatrix} = \begin{bmatrix} \frac{1}{s-1} & 0 \\ \frac{1}{(s-1)^2} & \frac{1}{s-1} \end{bmatrix}$$

$$e^{At} = L^{-1}\,[sI - A)^{-1}]$$

$$e^{At} = L^{-1} \begin{bmatrix} \frac{1}{s-1} & 0 \\ -\frac{1}{(s-1)^2} & \frac{1}{(s-1)} \end{bmatrix}$$

$$e^{At} = \begin{bmatrix} e^t & 0 \\ te^{-t} & e^t \end{bmatrix} \quad \textbf{Ans.}$$

Example 11.16 Obtain the state model of the system whose transfer function is given by $T(s) = \dfrac{s^2 + 2s + 3}{s^3 + 2s^2 + 3s + 1}$ and hence, find its state transition matrix.

Solution The transfer function given

$$T(s) = \frac{s \ + 3s + 3}{s^3 + 2s^2 + 3s + 1}$$

$$T(s) = \frac{3}{s^3 + 2s^2 + 3s + 1} + \frac{3s}{s^3 + 2s^2 + 3s + 1} + \frac{s^2}{s^3 + 2s^2 + 3s + 1}$$

By direct decomposition (from section 11.2)

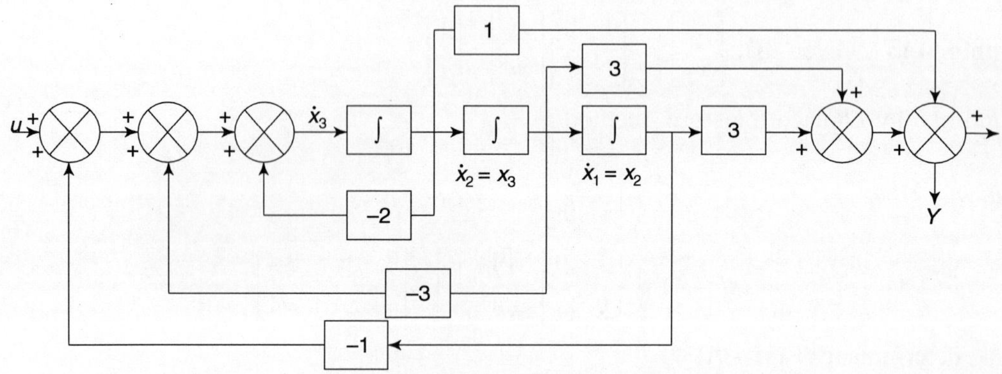

Figure Ex. 11.16 State mode in terms of phase variable

From the above state mode shown in Fig. Ex. 11.16

$$\dot{x}_1 = x_2$$

$$\dot{x}_2 = x_3$$

$$\dot{x}_3 = -x_1 - 3x_2 - 2x_3 + u$$

$$y = 3x_1 + 3x_2 + x_3$$

The state model in phase variable form is

$$\begin{bmatrix} \dot{x}_1 \\ \dot{x}_2 \\ \dot{x}_3 \end{bmatrix} = \begin{bmatrix} 0 & 1 & 0 \\ 0 & 0 & 1 \\ -1 & -3 & -2 \end{bmatrix} \begin{bmatrix} x_1 \\ x_2 \\ x_3 \end{bmatrix} + \begin{bmatrix} 0 \\ 0 \\ 1 \end{bmatrix} u$$

$$y = [3 \quad 3 \quad 1]\begin{bmatrix} x_1 \\ x_2 \\ x_3 \end{bmatrix} \quad \textbf{Ans.}$$

State transition matrix

$$A = \begin{bmatrix} 0 & 1 & 0 \\ 0 & 0 & 1 \\ -1 & -3 & -2 \end{bmatrix}$$

$$(sI - A) = \begin{bmatrix} s & 0 & 0 \\ 0 & s & 0 \\ 0 & 0 & s \end{bmatrix} - \begin{bmatrix} 0 & 1 & 0 \\ 0 & 0 & 1 \\ -1 & -3 & -2 \end{bmatrix} = \begin{bmatrix} s & -1 & 0 \\ 0 & s & -1 \\ 1 & 3 & s+2 \end{bmatrix}$$

The determinant of $(sI - A)$ is

$$|sI - A| = s(s^2 + 2s + 3) + 1 = s^3 + 2s^2 + 3s + 1$$

$$\text{adj of } |sI - A| = \begin{bmatrix} s^2 = 2s + 3 & s+2 & 1 \\ 1 & s(s+2) & s \\ -s & -(3s+1) & s^2 \end{bmatrix}$$

$$(sI - A) = \frac{\text{adj}(sI - A)}{|sI - A|} = \frac{1}{s^3 + 2s^2 + 3s + 1}\begin{bmatrix} s^2 + 2s + 3 & s+2 & 1 \\ 1 & s(s+2) & s \\ -s & -(3s+1) & s^2 \end{bmatrix}$$

$$e^{At} = L^{-1}\{(sI - A)\}$$

$$e^{At} = L^{-1}\left\{\frac{1}{s^3 + 2s^2 + 3s + 1}\begin{bmatrix} s^2 + 2s + 3 & s+2 & 1 \\ 1 & s(s+2) & s \\ -s & -(3s+1) & s^2 \end{bmatrix}\right\} \quad \textbf{Ans.}$$

Example 11.17 Consider a single input output control system having an overall transfer function $T(s)$ given by $T(s) = \dfrac{s^2 + 4s + 4}{s^3 + 5s^2 + 4s}$ represents the system state model in terms of phase variables. *(U.P.T.U., 2004–05, ACS)*

Solution

$$T(s) = \frac{s^2 = 4s + 4}{s^3 + 5s^2 + 4s}$$

Dividing the numerator and denominator by the highest power of s (as in sec 11.2).

$$T(s) = \frac{\frac{1}{s^3}(s^2 + 4s + 4)}{\frac{1}{s^3}[s^3 + 5s^2 + 4s]} = \frac{s^{-1} + s^{-2} + 4s^{-3}}{1 - \left[\frac{5}{s} - \frac{4}{s^2}\right]}$$

$$T(s) = \frac{1/s + 4/s^2 + 4/s^3}{1 - [-5/s - 4/s^2]}$$

Comparing this with Eq. (11.11), we get

$$b_0 = 0, \quad b_1 = 1, \quad b_2 = 4, \quad b_3 = 4$$

and $$a_1 = 5, \quad a_2 = 4$$

The signal flow graph for above function is shown in Fig. Ex. 11.17 (as Fig. 11.2)
The state equations are (see Eq. 11.13)

$$\dot{x}_1 = -5x_1 + u + x_2$$

$$\dot{x}_2 = 4x_1 + x^3 + 4u$$

and $$\dot{x}_3 = 4u$$

$$y = x_1$$

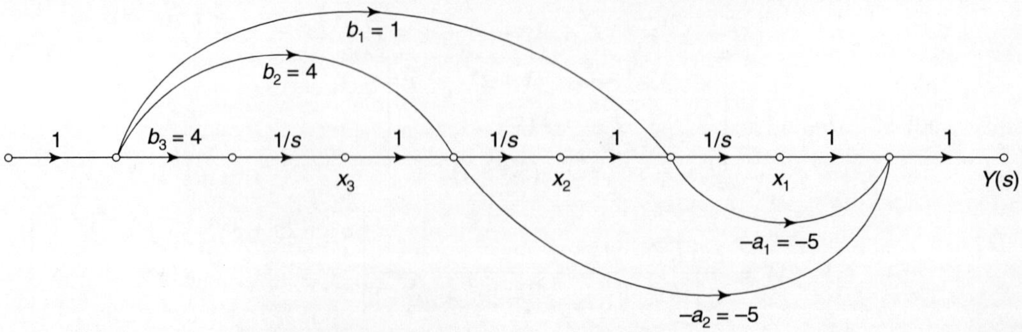

Figure Ex. 11.17 SFG for given system

The state model is

$$\begin{bmatrix} \dot{x}_1 \\ \dot{x}_2 \\ \dot{x}_3 \end{bmatrix} = \begin{bmatrix} -5 & 1 & 0 \\ -4 & 0 & 1 \\ 0 & 0 & 0 \end{bmatrix} \begin{bmatrix} x_1 \\ x_2 \\ x_3 \end{bmatrix} + \begin{bmatrix} 1 \\ 4 \\ 4 \end{bmatrix} u$$

$$y = [1 \quad 0 \quad 0] \begin{bmatrix} x_1 \\ x_2 \\ x_3 \end{bmatrix} \quad \textbf{Ans.}$$

Example 11.18 Obtain phase variable representation for a system whose transfer function is given by

$$T(s) = \frac{3s^3 + 4s^2 + 3s + 30}{s^3 + 8s^2 + 14s + 20}$$

Solution Dividing the numerator and denominator by the highest power of s (as in sec. 11.2)

$$T(s) = \frac{\frac{1}{s^3}\left[3+\frac{4}{s}+\frac{3}{s^2}+\frac{20}{s^3}\right]}{\frac{1}{s^3}\left[1+\frac{8}{s}+\frac{14}{s^2}+\frac{20}{s^3}\right]} = \frac{3+\frac{4}{s}+\frac{3}{s^2}+\frac{20}{s^3}}{1+\frac{8}{s}+\frac{14}{s^2}+\frac{20}{s^3}}$$

Comparing above function with Eq. (11.11)

i.e.,
$$T(s) = \frac{b_0+\frac{b_1}{s}+\frac{b_2}{s^2}+\frac{b_3}{s^3}}{1+\frac{a_1}{s}+\frac{a_2}{s^2}+\frac{a_3}{s^3}}$$

We get

$$b_0 = 3,\quad b_1 = 4,\quad b_2 = 3,\quad b_3 = 20$$
$$a_1 = +\,8,\quad a_2 = +\,14,\quad b_3 = +\,20$$

The signal flow graph is shown in Fig. Ex. 11.18 (as in Fig. 11.2)

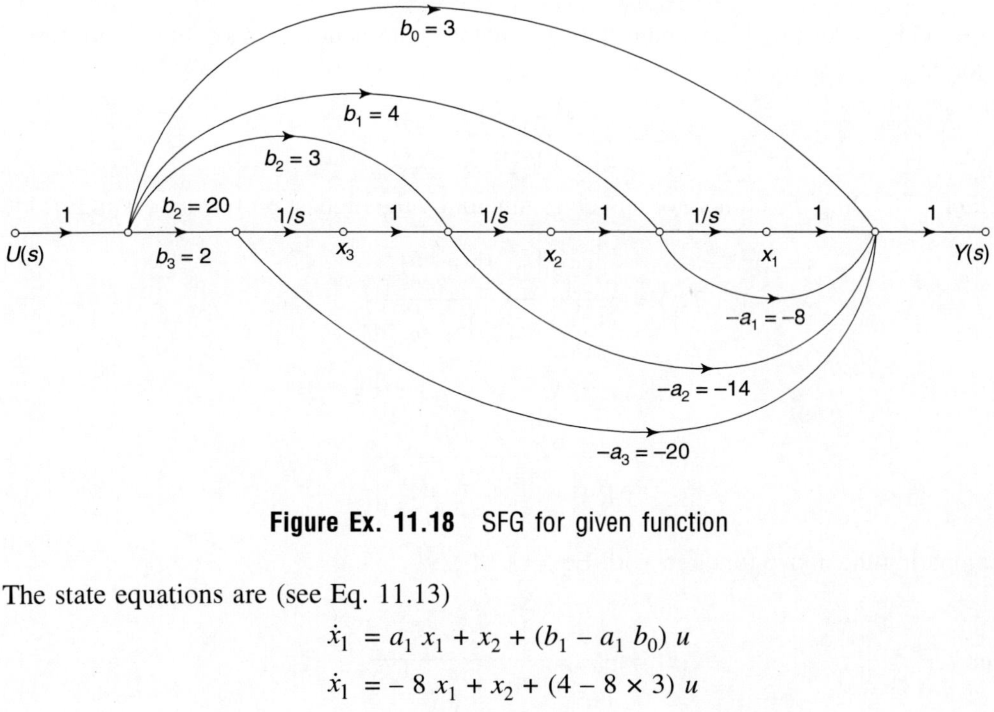

Figure Ex. 11.18 SFG for given function

The state equations are (see Eq. 11.13)

$$\dot{x}_1 = a_1\, x_1 + x_2 + (b_1 - a_1\, b_0)\, u$$

$$\dot{x}_1 = -\,8\, x_1 + x_2 + (4 - 8 \times 3)\, u$$

$$\dot{x}_1 = -\,8\, x_1 + x_2 - 20\, u \qquad ...(1)$$

$$\dot{x}_2 = -\,a_2\, x_1 + x_3 + (b_2 - a_2\, b_0)\, u$$

$$\dot{x}_2 = -\,14\, x_1 + x_3 + (3 - 14 \times 3)\, u$$

$$\dot{x}_2 = -\,14\, x_1 + x_3 - 39\, u \qquad ...(2)$$

$$\dot{x}_3 = - a_3 x_1 + x_4 + (b_3 - a_3 b_0) u$$

$$\dot{x}_3 = - 20 x_1 + (20 - 20 \times 3) u$$

$$\dot{x}_3 = - 20 x_1 - 40 u \quad \text{...(3)}$$

$$y = [1 \quad 0 \quad 0] \begin{bmatrix} x_1 \\ x_2 \\ x_3 \end{bmatrix} + b_0 u$$

$$y = [1 \quad 0 \quad 0] \begin{bmatrix} x_1 \\ x_2 \\ x_3 \end{bmatrix} + 3 u$$

$$\begin{bmatrix} \dot{x}_1 \\ \dot{x}_2 \\ \dot{x}_3 \end{bmatrix} = \begin{bmatrix} -8 & 1 & 0 \\ -14 & 0 & 1 \\ -20 & 0 & 0 \end{bmatrix} \begin{bmatrix} x_1 \\ x_2 \\ x_3 \end{bmatrix} + \begin{bmatrix} -20 \\ -39 \\ -40 \end{bmatrix} u \quad \textbf{Ans.}$$

Example 11.19 Construct the signal flow graph and state model for a system whose transfer function is

$$T(s) = \frac{s^2 + 3s + 6}{s^3 + 12s^2 + 3s + 1}$$

Solution Dividing the numerator and denominator by the highest power of s (as in Eq. 11.2)

$$T(s) = \frac{\frac{1}{s^3}[s^2 + 3s + 6]}{\frac{1}{s^3}[s^3 + 12s^2 + 3s + 1]}$$

$$T(s) = \frac{\frac{1}{s} + \frac{3}{s^2} + \frac{6}{s^3}}{1 - \left(-\frac{12}{s} - \frac{3}{s^2} - \frac{1}{s^3}\right)}$$

Comparing the above function with Eq. (11.12).

i.e.,
$$T(s) = \frac{b_0 + \frac{b_1}{s} + \frac{b_2}{s^2} + \frac{b_3}{s^3}}{1 - \left(-\frac{a_1}{s} - \frac{a_2}{s^2} - \frac{a_3}{s^3}\right)}$$

We get

$$b_o = 0, \quad b_1 = 1, \quad b_2 = 3, \quad b_3 = 6$$
$$a_1 = 12, \quad a_2 = 3, \quad a_3 = 1$$

The signal flow graph for the above function is shown in Fig. Ex. 11.19 (as in Fig. 11.2)

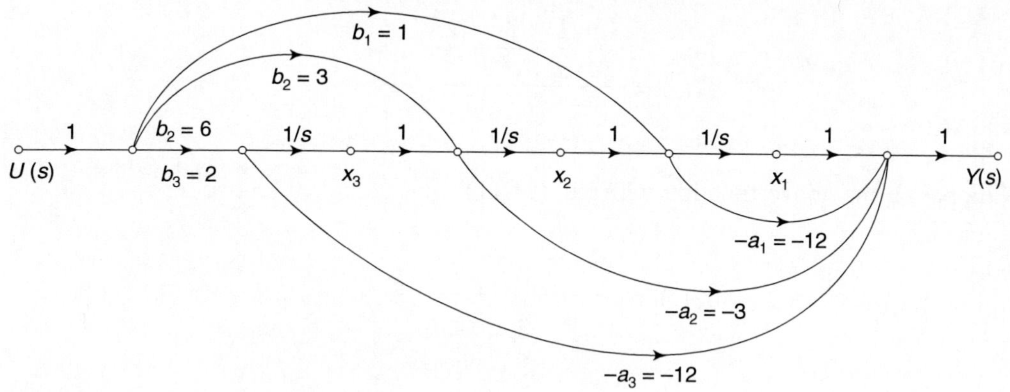

Figure Ex. 11.19 SFG for given function

The state equations are (see Eq. 11.13)

$$\dot{x}_1 = a_1 x_1 + x_2 + (b_1 - a_1 b_0) u$$

$$\dot{x}_1 = 12 x_1 + x_2 + (1) u \quad ...(1)$$

$$\dot{x}_2 = - a_2 x_1 + x_3 + (b_2 - a_2 b_0) u$$

$$\dot{x}_2 = - 3 x_1 + x_3 + 3 u \quad ...(2)$$

$$\dot{x}_3 = - a_3 x_1 + x_4 + (b_3 - a_3 b_0) u$$

$$\dot{x}_3 = - x_1 + 6 u \quad ...(3)$$

i.e., the equation is

$$\begin{bmatrix} \dot{x}_1 \\ \dot{x}_2 \\ \dot{x}_3 \end{bmatrix} = \begin{bmatrix} -12 & 1 & 0 \\ -3 & 0 & 1 \\ -1 & 0 & 0 \end{bmatrix} \begin{bmatrix} x_1 \\ x_2 \\ x_3 \end{bmatrix} + \begin{bmatrix} 1 \\ 3 \\ 6 \end{bmatrix} u$$

$$y = [1 \quad 0 \quad 0] \begin{bmatrix} x_1 \\ x_2 \\ x_3 \end{bmatrix} \quad \textbf{Ans.}$$

Example 11.20 The transfer function of a system is given by $T(s) = \dfrac{s^2 + 3s + 2}{s^3 + 9s^2 + 26s + 24}$.

Determine the state model with phase variable formulation.

Solution Given that $T(s) = \dfrac{s^2 + 3s + 2}{s^3 + 9s^2 + 26s + 24}$

Dividing the numerator and denominator by the highest power of s (as in sec. 11.2)

$$T(s) = \frac{\frac{1}{s} + \frac{3}{s^2} + \frac{2}{s^3}}{1 - \left(-\frac{9}{s} - \frac{26}{s^2} - \frac{24}{s^3}\right)}$$

Comparing the above function with Eq. (11.11)

i.e.,
$$T(s) = \frac{b_0 + \frac{b_1}{s} + \frac{b_2}{s^2} + \frac{b_3}{s^3}}{1 - \left(-\frac{a_1}{s} - \frac{a_2}{s^2} - \frac{a_3}{s^3}\right)}$$

We get

$$b_o = 0, \quad b_1 = 1, \quad b_2 = 3, \quad b_3 = 2$$
$$a_1 = 9, \quad a_2\ 26 \quad a_3 = 24$$

The signal flow graph is shown in Fig. Ex. 11.20 (as in Fig. 11.2)

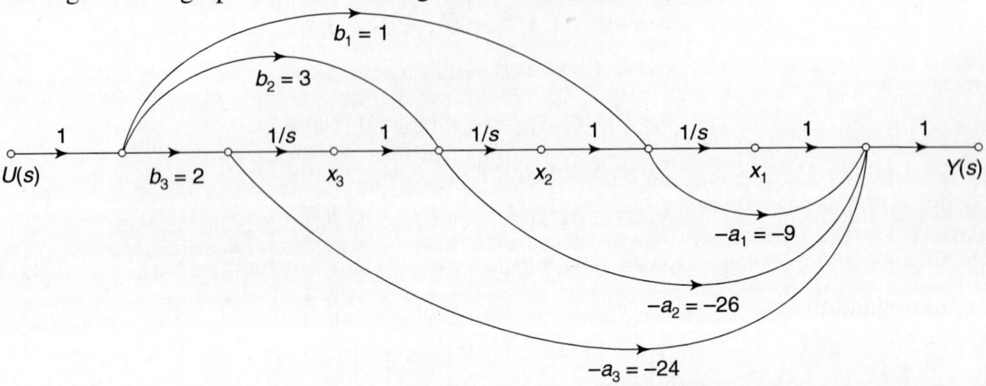

Figure Ex. 11.20 SFG for given function

The state equations are (see Eq. 11.13)

$$\dot{x}_1 = -a_1 x_1 + x_2 + (b_1 - a_1 b_0)\, u$$

$$\dot{x}_1 = -9 x_1 + x_2 + u \qquad \text{...(1)}$$

$$\dot{x}_2 = -a_2 x_1 + x_3 + (b_2 - a_2 b_0)\, u$$

$$\dot{x}_2 = -26 x_1 + x_3 + 3 u \qquad \text{...(2)}$$

$$\dot{x}_3 = -a_3 x_1 + x_3 + (b_3 - a_3 b_0)\, u$$

$$\dot{x}_3 = -24 x_1 + 2 u \qquad \text{...(3)}$$

Now
$$\begin{bmatrix} \dot{x}_1 \\ \dot{x}_2 \\ \dot{x}_3 \end{bmatrix} = \begin{bmatrix} -9 & 1 & 0 \\ -26 & 0 & 1 \\ -24 & 0 & 0 \end{bmatrix} \begin{bmatrix} x_1 \\ x_2 \\ x_3 \end{bmatrix} + \begin{bmatrix} 1 \\ 3 \\ 2 \end{bmatrix} u$$

The output equation is

$$y = [1 \quad 0 \quad 0]\begin{bmatrix} x_1 \\ x_2 \\ x_3 \end{bmatrix} + b_0\, u$$

$$y = [1 \quad 0 \quad 0]\begin{bmatrix} x_1 \\ x_2 \\ x_3 \end{bmatrix} \quad \textbf{Ans.}$$

Example 11.21 Obtain the time response of the system given below

$$\begin{bmatrix} \dot{x}_1 \\ \dot{x}_2 \end{bmatrix} = \begin{bmatrix} 0 & 1 \\ -2 & 0 \end{bmatrix}\begin{bmatrix} x_1 \\ x_2 \end{bmatrix}$$

and
$$y = [1 \quad -1]\begin{bmatrix} x_1 \\ x_2 \end{bmatrix}$$

Given
$$x(0) = \begin{bmatrix} 1 \\ 1 \end{bmatrix}$$

Solution The time response is given by

$$x(t) = e^{At}\, x(0)$$

Evaluation of e^{At}

$$(sI - A) = \begin{bmatrix} s & 0 \\ 0 & s \end{bmatrix} - \begin{bmatrix} 0 & 1 \\ -2 & 0 \end{bmatrix} = \begin{bmatrix} s & -1 \\ 2 & s \end{bmatrix}$$

$$|sI - A| = s^2 + 2$$

$$\text{adj of } (sI - A) = \begin{bmatrix} s & 1 \\ -2 & s \end{bmatrix}$$

$$(sI - A)^{-1} = \frac{\text{adj of } (sI - A)}{|sI - A|} = \frac{1}{(s^2 + 2)}\begin{bmatrix} s & 1 \\ -2 & s \end{bmatrix}$$

$$(sI - A)^{-1} = \begin{bmatrix} \dfrac{s}{s^2 + 2} & \dfrac{1}{s^2 + 2} \\ \dfrac{2}{s^2 + 2} & \dfrac{s}{s^2 + 2} \end{bmatrix}$$

$$e^{At} = L^{-1}\{(sI - A)^{-1}\}$$

$$e^{At} = L^{-1}\begin{bmatrix} \dfrac{s}{s^2 + 2} & \dfrac{1}{s^2 + 2} \\ -\dfrac{2}{s^2 + 2} & \dfrac{s}{s^2 + 2} \end{bmatrix} = \begin{bmatrix} \cos\sqrt{2}\,t & \dfrac{1}{\sqrt{2}}\sin\sqrt{2}\,t \\ -\sqrt{2}\sin\sqrt{2}\,t & \cos\sqrt{2}\,t \end{bmatrix}$$

Now $x(t) = e^{At}\, x(0)$

$$e^{At} = \begin{bmatrix} \cos\sqrt{2}\,t & \frac{1}{\sqrt{2}}\sin\sqrt{2}\,t \\ -\sqrt{2}\sin\sqrt{2}\,t & \cos\sqrt{2}\,t \end{bmatrix}\begin{bmatrix}1\\1\end{bmatrix} = \begin{bmatrix} \cos\sqrt{2}\,t + \frac{1}{\sqrt{2}}\sin\sqrt{2}\,t \\ -\sqrt{2}\sin\sqrt{2}\,t + \cos\sqrt{2} \end{bmatrix}$$

Ans.

$$y(t) = x_1 - x_2$$

$$= \cos\sqrt{2}\,t + \frac{1}{\sqrt{2}}\sin\sqrt{2}\,t + \sqrt{2}\sin\sqrt{2}\,t - \cos\sqrt{2}\,t$$

$$y(t) = \frac{3}{\sqrt{2}}\sin\sqrt{2}\,t \quad \textbf{Ans.}$$

Example 11.22 For the system given below

$$\begin{bmatrix}\dot{x}_1\\ \dot{x}_2\end{bmatrix} = \begin{bmatrix}0 & 1\\ -2 & -1\end{bmatrix}\begin{bmatrix}x_1\\ x_2\end{bmatrix} + \begin{bmatrix}0\\1\end{bmatrix}u$$

$$\begin{bmatrix}(0)\\(0)\end{bmatrix} = \begin{bmatrix}1\\0\end{bmatrix}$$

and $u(t) = 1$

Obtain

1. Zero input response
2. Zero state response
3. Total response

Solution

1. Zero input response is

$$x(t) = e^{At}\, x(0)$$

$$e^{At} = L^{-1}\{(sI - A)^{-1}\}$$

$$|sI - A| = \begin{bmatrix}s & 0\\ 0 & s\end{bmatrix} - \begin{bmatrix}0 & 1\\ -2 & -1\end{bmatrix} = \begin{bmatrix}s & -1\\ 2 & s+1\end{bmatrix}$$

$$|sI - A| = s\,(s+1) + 2$$

$$= s^2 + s + 2$$

$$\text{adj of } (sI - A) = \begin{bmatrix}s+1 & 1\\ -2 & s\end{bmatrix}$$

$$(sI - A)^{-1} = \frac{\text{adj of } (sI - A)}{|sI - A|} = \frac{1}{s^2 + s + 2}\begin{bmatrix}s+1 & 1\\ -2 & s\end{bmatrix}$$

$$(sI - A)^{-1} = \begin{bmatrix} \dfrac{s+1}{s^2+s+2} & \dfrac{1}{s^2+s+2} \\ -\dfrac{2}{s^2+s+2} & \dfrac{s}{s^2+s+2} \end{bmatrix}$$

$$e^{At} = L^{-1}\{(sI - A)^{-1}\}$$

$$e^{At} = L^{-1}\left\{\begin{bmatrix} \dfrac{s+1}{s^2+s+2} & \dfrac{1}{s^2+s+2} \\ -\dfrac{2}{s^2+s+2} & \dfrac{s}{s^2+s+2} \end{bmatrix}\right\}$$

$$x(t) = e^{At}\, x(0)$$

$$x(t) = L^{-1}\begin{bmatrix} \dfrac{s+1}{s^2+s+2} & \dfrac{1}{s^2+s+2} \\ -\dfrac{2}{s^2+s+2} & \dfrac{s}{s^2+s+2} \end{bmatrix}\begin{bmatrix} 1 \\ 0 \end{bmatrix}$$

$$x(t) = L^{-1}\begin{bmatrix} \dfrac{s+1}{s^2+s+2} \\ -\dfrac{2}{s^2+s+2} \end{bmatrix} = L^{-1}\begin{bmatrix} \dfrac{(s+0.5)+0.5}{(s+0.5)^2+(1.32)^2} \\ \dfrac{-2.1.32}{[(s+0.5)^2+(1.32)^2]1.32} \end{bmatrix}$$

$$x(t) = L^{-1}\begin{bmatrix} \dfrac{(s+0.5)}{(s+0.5)^2+(1.32)^2} + \dfrac{0.5}{1.32}\times\dfrac{1.32}{(s+0.5)^2+(1.32)^2} \\ -1.5\cdot\dfrac{1.32}{9s+0.5)^2+(1.32)^2} \end{bmatrix}$$

Taking inverse Laplace transform

$$x(t)\big|_{ZIR} = \begin{bmatrix} e^{-0.5t}\cos 1.32\,t + 0.38\,e^{-0.5t}\sin 1.32\,t \\ -1.5\,e^{-0.5t}\sin 1.32\,t \end{bmatrix} \quad \textbf{Ans.}$$

2. Zero state response is given by

$$x(t)\big|_{ZSR} = e^{At}\,B\,U(s)$$

$$= L^{-1}\begin{bmatrix} \dfrac{s+1}{s^2+s+2} & \dfrac{1}{s^2+s+2} \\ \dfrac{2}{s^2+s+2} & \dfrac{s}{s^2+s+2} \end{bmatrix}\begin{bmatrix} 0 \\ 1 \end{bmatrix}\frac{1}{s}$$

$$= L^{-1}\begin{bmatrix} \dfrac{1}{s(s^2+s+2)} \\ \dfrac{1}{s}\cdot\dfrac{s}{(s^2+s+2)} \end{bmatrix} = L^{-1}\begin{bmatrix} \dfrac{1}{s(s^2+s+2)} \\ \dfrac{1}{s^2+s+2} \end{bmatrix}$$

$$= L^{-1}\begin{bmatrix} \dfrac{1}{2}\cdot\dfrac{1}{s}\dfrac{2}{s^2+s+2} \\ \dfrac{1}{1.32}\dfrac{1.32}{(s+0.5)^2+(1.32)^2} \end{bmatrix}$$

Taking inverse Laplace transform, the zero state response is

$$x(t)\,|_{ZIR} = \begin{bmatrix} \dfrac{1}{3}\left\{1-\dfrac{e^{-0.5t}}{\sqrt{1-0.28^2}}\sin\left[1.32\,t+\tan^{-1}\dfrac{\sqrt{1-0.28^2}}{0.28}\right\}\right. \\ \dfrac{1}{1.32}\,e^{-0.5t}\sin 1.32\,t \end{bmatrix}$$

Ans.

3. The total response

$$x(t) = \text{zero input response} + \text{zero state response}$$

$$\begin{bmatrix} x_1(t) \\ x_2(t) \end{bmatrix} = \begin{bmatrix} e^{-0.5t}\cos 1.32\,t + 0.38\,e^{-0.5t}\sin 1.32\,t \\ -1.5\,e^{-0.5t}\sin 1.32\,t \end{bmatrix}$$

$$+\begin{bmatrix} 0.33\,\{1\,1.04\,e^{-0.5t}\sin(1.32\,t+\tan^{-1}3.42\} \\ 0.8\,e^{-0.5t}\cdot\sin 1.32\,t \end{bmatrix}$$ **Ans.**

Example 11.23 A system is described by state equation given below. Solve for the output $y(t)$:

$$\begin{bmatrix} \dot{x}_1 \\ \dot{x}_2 \end{bmatrix} = \begin{bmatrix} 0 & 1 \\ -12 & -7 \end{bmatrix}\begin{bmatrix} x_1 \\ x_2 \end{bmatrix} + \begin{bmatrix} 1 \\ 1 \end{bmatrix} u(t)$$

$$x(0) = \begin{bmatrix} 10 \\ 0 \end{bmatrix} \text{ and } u(t) = \text{unit step input}$$

$$y(t) = [1 \quad -1]\begin{bmatrix} x_1 \\ x_2 \end{bmatrix}$$ *(U.P.T.U., 2003, CO, ACS)*

Solution Given

$$A = \begin{bmatrix} 0 & 1 \\ -12 & -7 \end{bmatrix}$$

We have already solved for state transition matrix e^{At} as in Ex. 11.8

$$e^{At} = \begin{bmatrix} -3e^{-4t} + 4e^{-3t} & -e^{-4t} + e^{-3t} \\ 12e^{-4t} - 12e^{-3t} & 4e^{-4t} - 3e^{-3t} \end{bmatrix}$$

$$x(s) = e^{At}\, x(0) + \int_0^t e^{A(t-\tau)}\, Bu(\tau)\, dc$$

$$e^{At}\, x(0) = \begin{bmatrix} -3e^{-4t} + 4e^{-3t} & -e^{-4t} + e^{-3t} \\ 12e^{-4t} - 12e^{-3t} & 4e^{-4t} - 3e^{-3t} \end{bmatrix} \begin{bmatrix} 10 \\ 0 \end{bmatrix}$$

$$= \begin{bmatrix} -30e^{-4t} + 40e^{-3t} \\ 120e^{-4t} - 120e^{-3t} \end{bmatrix} \quad ...(1)$$

The term $e^{At}\, B$ is

$$e^{-At}\, B = \begin{bmatrix} -4e^{4\tau} + 5e^{3\tau} \\ 16e^{4\tau} - 15e^{3\tau} \end{bmatrix} \begin{bmatrix} 1 \\ 1 \end{bmatrix} [i]$$

$$= \int_0^t \begin{bmatrix} -4e^{4\tau} + 5e^{3\tau} \\ 16e^{4\tau} - 15e^{3\tau} \end{bmatrix} d\tau = \begin{bmatrix} -e^{4\tau} + \frac{5}{3}e^{3\tau} \\ 4e^{4\tau} - 5e^{3\tau} \end{bmatrix}_0^\tau$$

$$= \begin{bmatrix} -e^{4t} + \frac{5}{3}e^{3t} \\ 4e^{4t} - 5e^{3t} \end{bmatrix} - \begin{bmatrix} -1 + \frac{5}{3} \\ 4 - 5 \end{bmatrix} \quad ...(2)$$

$$= \begin{bmatrix} -e^{4t} + \frac{5}{3}e^{3t} - \frac{2}{3} \\ 4e^{4t} - 5e^{3t} + 1 \end{bmatrix}$$

Collecting the terms (1) and (2), we get

$$x(t) = e^{At}\, x(0) + \int_0^t e^{A)t-\tau)}\, B\, u(\tau)\, d\tau$$

$$x(t) = \begin{bmatrix} -30e^{-4t} + 40e^{-3t} \\ 120e^{-4t} - 120e^{-3t} \end{bmatrix} + e^{At} \int_0^t e^{A(t-\tau)}\, B\, u(\tau)\, d\tau$$

$$x(t) = \begin{bmatrix} -30\,e^{-4t} - 40\,e^{-3t} \\ 120\,e^{-4t} - 12 \cdot e^{-3t} \end{bmatrix} + \begin{bmatrix} -3\,e^{-4t} + 4\,e^{-3t} & -e^{-4t} + e^{-3t} \\ 12\,e^{-4t} - 12\,e^{-3t} & 4\,e^{-4t} - 3\,e^{-3t} \end{bmatrix} \times \begin{bmatrix} -e^{4t} + \frac{5}{3}e^{-3t} - \frac{2}{3} \\ 4\,e^{4t} - 5\,e^{3t} + 1 \end{bmatrix}$$

and $y(t) = x_1(t) - x_2(t)$ **Ans.**

Example 11.24 The transfer function of the system represents the system state model in terms of canonical variables and determine a set of coefficient matrix A, B, C.

$$T(s) = \frac{Y(s)}{U(s)} = \frac{1}{(s+1)} \cdot \frac{1}{(s+2)} \cdot \frac{1}{(s+3)} \quad \text{...(1)}$$

Solution For the decomposition of the transfer function (1), the output of term $\frac{1}{(s+1)}$ is given as input to $\frac{1}{(s+2)}$ and the output of this is given as input to $\frac{1}{(s+3)}$. The output of $\frac{1}{(s+3)}$ is the system output.

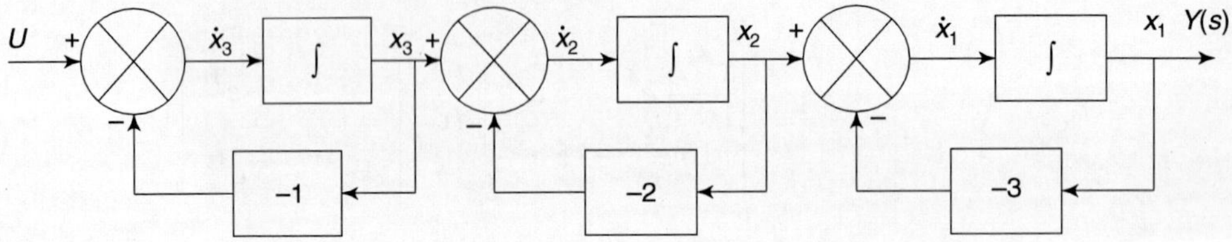

Figure Ex. 10.24 Block diagram for the Ex. 11.24

The state equations based on Fig. Ex. 11.24 are given below

$$\dot{x}_1 = -3\,x_1 + x_2 \quad \text{...(1)}$$

$$\dot{x}_2 = -2\,x_2 + x_3 \quad \text{...(2)}$$

$$\dot{x}_3 = -x_3 + u \quad \text{...(3)}$$

The output equation is

$$y = x_1 \quad \text{...(4)}$$

Now

$$\begin{bmatrix} \dot{x}_1 \\ \dot{x}_2 \\ \dot{x}_3 \end{bmatrix} = \begin{bmatrix} -4 & 1 & 0 \\ 0 & -3 & 1 \\ 0 & 0 & -2 \end{bmatrix} \begin{bmatrix} x_1 \\ x_2 \\ x_3 \end{bmatrix} + \begin{bmatrix} 0 \\ 0 \\ 1 \end{bmatrix} u \quad \text{...(5)}$$

$$y = [1 \quad 0 \quad 0] \begin{bmatrix} x_1 \\ x_2 \\ x_3 \end{bmatrix}$$

By observing Eq. (5), we get

$$A = \begin{bmatrix} -4 & 1 & 0 \\ 0 & -3 & 1 \\ 0 & 0 & -2 \end{bmatrix}, B = \begin{bmatrix} 0 \\ 0 \\ 1 \end{bmatrix}$$

$$C = [1 \quad 0 \quad 0]$$

Example 11.25 Consider a single input, single output control system having an overall transfer function $T(s)$ given by $T(s) = \dfrac{1}{(s+2)(s+4)(s+6)}$ represents the system state model.

Solution The given transfer function is

$$T(s) = \frac{1}{(s+2)(s+4)(s+6)} \qquad ...(1)$$

Solving the equation by partial fraction

$$\frac{1}{(s+2)(s+4)(s+6)} = \frac{A}{(s+2)} + \frac{B}{(s+4)} + \frac{C}{(s+6)} \qquad ...(2)$$

$$\frac{}{(s+2)(s+4)(s+6)} = \frac{A(s+4)(s+6) + B(s+20(s+6) + C(s+2)(s+4)}{(s+2)(s+4)(s+6)}$$

$$1 = A(s+4)(s+6) + B(s+2)(s+6) + C(s+2)(s+4) \qquad ...(3)$$

Putting $s + 2 = 0$ in Eq. (3), we get

$$1 = A(-2+4)(-2+6)$$

$$1 = 8A$$

$\Rightarrow$ $A = 1/8$

Putting $s + 4 = 0$ in Eq. (3), we get

$$1 = B(-4+2)(-4+6)$$

$$1 = -4B$$

$\Rightarrow$ $B = -1/4$

Now putting $s + 6 = 0$ in Eq. (3), we get

$$1 = C(-6+2)(6+4)$$

$$1 = 8C$$

$\Rightarrow$ $C = 1/8$

Putting the value of A, B, C in Eq. (2)

$$\frac{Y(s)}{U(s)} = \frac{\frac{1}{8}}{(s+2)} + \frac{-\frac{1}{4}}{(s+4)} + \frac{\frac{1}{8}}{(s+6)} \qquad ...(4)$$

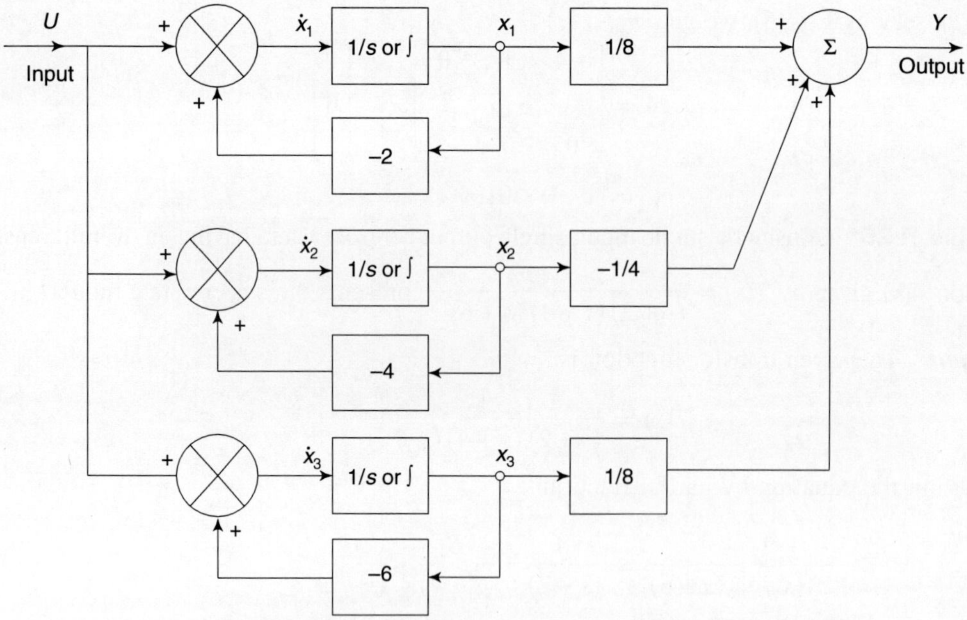

Figure Ex. 11.25 Block diagram for transfer function of Ex. 11.25

Comparing Eq. (4) with Eq. 11.21

i.e.,
$$T(s) = b_o + \sum_{i=1}^{n} \frac{C_i}{(s_i - P_i)}$$

Here
$$i = 3$$
$$b_0 = 0, \quad c_1 = 1/8, \quad c_2 = -1/4, \quad c_3 = 1/8$$
$$P_1 = -2, \quad P_2 = -4, \quad P_3 = -6$$

Now the block diagram for the transfer function is shown below (Fig. 11.6)

The state equation are [by Eq. (11.23 (a)],

$$\dot{x}_1 = -2\,x_1 + u$$
$$\dot{x}_2 = 4\,x_2 + u$$
$$\dot{x}_3 = 6\,x_3 + u$$

The output $y(t)$ is given by Eq. (11.23 b)

$$y(t) = C_1\,x_1 + C_2\,x_2 + C_3\,x_3$$
$$y(t) = \frac{1}{8}x_1 - \frac{1}{4}x_2 + \frac{1}{8}x_3$$

The state model in vector matrix form

$$\begin{bmatrix} \dot{x}_1 \\ \dot{x}_2 \\ \dot{x}_3 \end{bmatrix} = \begin{bmatrix} -2 & 0 & 0 \\ 0 & -4 & 0 \\ 0 & 0 & -6 \end{bmatrix} \begin{bmatrix} x_1 \\ x_2 \\ x_3 \end{bmatrix} + \begin{bmatrix} 1 \\ 1 \\ 1 \end{bmatrix} u$$

$$y = \begin{bmatrix} \frac{1}{8} & -\frac{1}{4} & \frac{1}{8} \end{bmatrix} \begin{bmatrix} x_1 \\ x_2 \\ x_3 \end{bmatrix} \quad \textbf{Ans.}$$

Example 11.26 Consider a single input, single output system having an overall transfer function $T(s)$ given by $T(s) = \dfrac{Y(s)}{U(s)} = \dfrac{2s^2 + 8s + 7}{(s+4)^2 (s+2)}$, represents the system state model.

Solution

$$T(s) = \frac{2s^2 + 8s + 7}{(s+4)^2 (s+2)} \qquad \text{...(1)}$$

Solving the equation by partial fraction

$$\frac{2s^2 + 8s + 7}{(s+4)^2 (s+2)} = \frac{A}{(s+2)} + \frac{B}{(s+4)} + \frac{C}{(s+4)^2} \qquad \text{...(2)}$$

$$2s^2 + 8s + 7 = A\,(s+4)^2 + B\,(s+2)\,(s+4) + C\,(s+2) \qquad \text{...(3)}$$

Putting $s + 2 = 0$ in Eq. (3), we get

$$2\,(-2)^2 + 8\,(-2) + 7 = A\,(-2+4)^2 + 0$$

$$8 - 16 + 7 = 4\,A$$

$$A = -1/4$$

Now putting $s + 4 = 0$ in Eq. (3), we get

$$2\,(-4)^2 + 8\,(-4) + 7 = 0 + C\,(-4+2) + 0$$

$$32 - 32 + 7 = 2\,C$$

$$C = -7/2$$

Putting $s = 0$ in Eq. (3), we get

$$7 = 16\,A + 8\,B + 2\,C \qquad \text{...(4)}$$

Now putting A, C in Eq. (4), we get

$$7 = 16 \times \frac{1}{4} + 8\,B + 2 \times \frac{-7}{2}$$

$$7 = -4 + 8\,B - 7$$

$$8\,B = 18$$

$$B = \frac{18}{8} = \frac{9}{2}$$

Now putting the value of A, B, C in Eq. (3), we get

$$\frac{Y(s)}{R(s)} = \frac{1/4}{(s+2)} + \frac{9/2}{(s+4)} + \frac{-7/2}{(s+4)^2} \qquad \text{...(5)}$$

The block diagram is shown in Fig. Ex. 11.26.

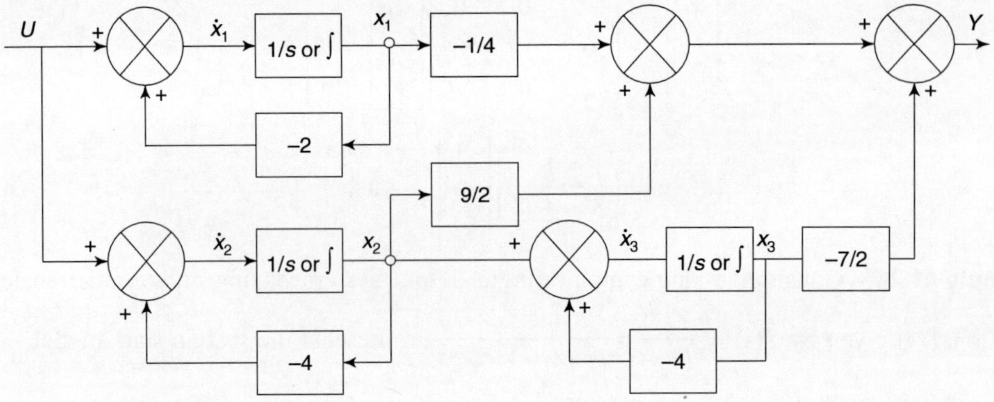

Figure Ex. 11.26 Block diagram for transfer function of Ex. 11.26

The state equations are

$$\dot{x}_1 = -2x_1 + u$$

$$\dot{x}_2 = 4x_2 + u$$

$$\dot{x}_3 = x_2 - 4x_3$$

The output equation is $$y = \frac{1}{4}x_1 + \frac{9}{2}x_2 - \frac{7}{2}x_3$$

The state model in vector matrix form

$$\begin{bmatrix} \dot{x}_1 \\ \dot{x}_2 \\ \dot{x}_3 \end{bmatrix} = \begin{bmatrix} -2 & 0 & 0 \\ 0 & -4 & 0 \\ 0 & 1 & -4 \end{bmatrix} \begin{bmatrix} x_1 \\ x_2 \\ x_3 \end{bmatrix} + \begin{bmatrix} 1 \\ 1 \\ 0 \end{bmatrix} u$$

$$y = \begin{bmatrix} -\frac{1}{4} & \frac{9}{2} & \frac{-7}{2} \end{bmatrix} \begin{bmatrix} x_1 \\ x_2 \\ x_3 \end{bmatrix}$$ **Ans.**

Example 11.27 Consider the transfer function of the system given

$$\frac{Y(s)}{U(s)} = \frac{1}{s^3 + 6s^2 + 24s + 20}$$ represent the system state model.

Solution The function given

$$T(s) = \frac{1}{s^3 + 6s^2 + 24s + 20}$$

The corresponding differential equation for the above transfer function is of third order, therefore, three integrators and three summing points are used to draw the block diagram.

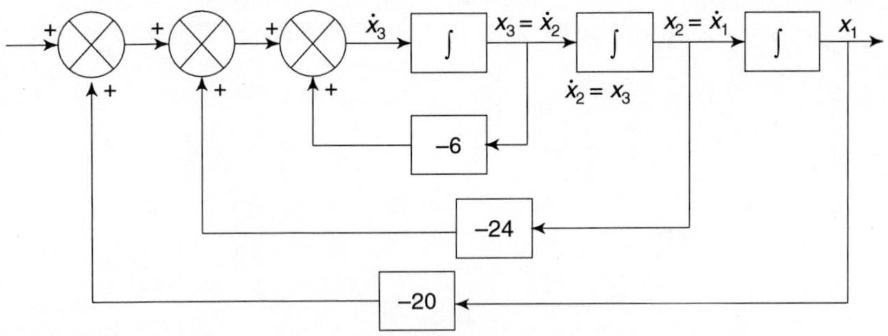

Figure Ex. 11.27 Block diagram for the transfer function of Ex. 11.27

The state equations are

$$\dot{x}_1 = x_2$$

$$\dot{x}_2 = x_3$$

$$\dot{x}_3 = -20\,x_1 - 24\,x_2 - 6\,x_3 + u$$

The output is $y = x_1$

The state model in vector matrix form

$$\begin{bmatrix} \dot{x}_1 \\ \dot{x}_2 \\ \dot{x}_3 \end{bmatrix} = \begin{bmatrix} 0 & 1 & 0 \\ 0 & 0 & 1 \\ -20 & -24 & -6 \end{bmatrix} \begin{bmatrix} x_1 \\ x_2 \\ x_3 \end{bmatrix} + \begin{bmatrix} 0 \\ 0 \\ 1 \end{bmatrix} u$$

$$y = [1 \quad 0 \quad 0] \begin{bmatrix} x_1 \\ x_2 \\ x_3 \end{bmatrix} \qquad \textbf{Ans.}$$

Example 11.28 The transfer function of a system is given by $\dfrac{Y(s)}{U(s)} = \dfrac{s^2 + 6s + 5}{s^3 + 5s^2 + 20s + 35}$.

Determine the state model. Use direct decomposition.

Solution $$\frac{Y(s)}{U(s)} = \frac{5}{s^3 + 5s^2 + 20s + 35} + \frac{6s}{s^3 + 5s^2 + 20s + 35} + \frac{s^2}{s^3 + 5s^2 + 20s + 35}.$$

The state block diagram for $\dfrac{s^2 + 6s + 5}{s^3 + 5s^2 + 20s + 35}$ is shown in Fig. Ex. 11.28.

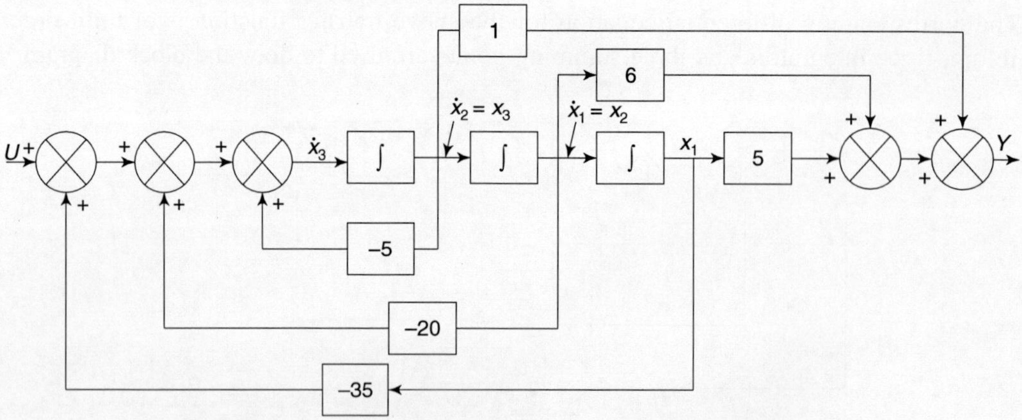

Figure Ex. 11.28 State block diagram for the term $\dfrac{s^2+6s+5}{s^3+5s^2+20s+35}$

The state equations are

$$\dot{x}_1 = x_2$$

$$\dot{x}_2 = x_3$$

and

$$\dot{x}_3 = -35\,x_1 - 20\,x_2 - 5\,x_1 + u$$

$$y = 5\,x_1 + 6\,x_2 + x_3$$

And the state model in vector matrix form

$$\begin{bmatrix} \dot{x}_1 \\ \dot{x}_2 \\ \dot{x}_3 \end{bmatrix} = \begin{bmatrix} 0 & 1 & 0 \\ 0 & 0 & 1 \\ -35 & -29 & -5 \end{bmatrix} \begin{bmatrix} x_1 \\ x_2 \\ x_3 \end{bmatrix} + \begin{bmatrix} 0 \\ 0 \\ 1 \end{bmatrix} u$$

$$y = [5 \quad 6 \quad 1] \begin{bmatrix} x_1 \\ x_2 \\ x_3 \end{bmatrix} \quad \textbf{Ans.}$$

Example 11.29 Using cascade method decomposition the transfer function $\dfrac{Y(s)}{U(s)} = \dfrac{(s+2)}{(s+3)(s+5)}$ and obtain the state model.

Solution The function is given

$$\frac{Y(s)}{U(s)} = \frac{(s+2)}{(s+3)(s+5)} = \frac{(s+2)}{(s+3)} \cdot \frac{1}{(s+5)}.$$

The block diagram for the function is shown in Fig. Ex. 11.29. The output of each integration is assigned a state variable.

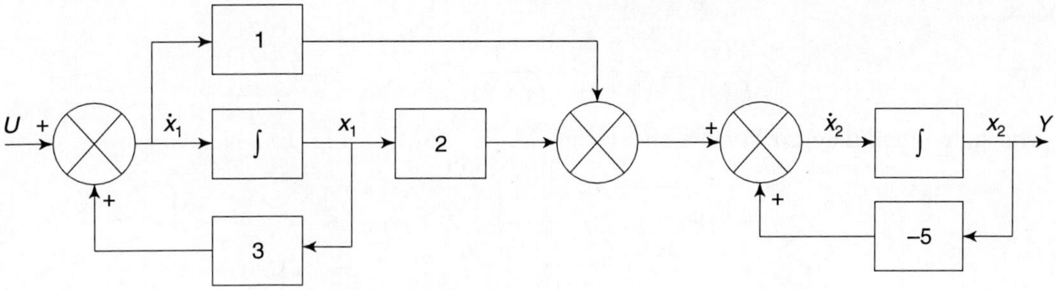

Figure Ex. 11.29 State block diagram for $\dfrac{Y(s)}{U(s)}$

The state equations are

$$\dot{x}_1 = 3\,x_1 + u$$
$$\dot{x}_2 = 2\,x_1 + \dot{x}_1 - 5\,x_2$$

∵
$$\dot{x}_1 = -\,3\,x_1 + u$$

So
$$\dot{x}_2 = 2\,x_1 - 3\,x_1 - 5\,x_2 + u$$
$$\dot{x}_2 = -\,x_1 - 5\,x_2 + u$$

The output is
$$y = x_2$$

The state model is
$$\begin{bmatrix}\dot{x}_1\\ \dot{x}_2\end{bmatrix} = \begin{bmatrix}-1 & 0\\ 2 & -2\end{bmatrix}\begin{bmatrix}x_1\\ x_2\end{bmatrix} + \begin{bmatrix}1\\ 1\end{bmatrix}u$$

$$y = [0 \quad 1]\begin{bmatrix}x_1\\ x_2\end{bmatrix}$$ **Ans.**

Example 11.30 The transfer function of a system is given by $\dfrac{Y(s)}{U(s)} = \dfrac{(s+3)}{(s+1)(s+2)}$. Obtain the state model in canonical form using parallel decomposition method.

Solution The function is given

$$\frac{Y(s)}{U(s)} = \frac{(s+3)}{(s+1)(s+2)} \qquad ...(1)$$

By the partial fraction

$$\frac{(s+3)}{(s+1)(s+2)} = \frac{A}{(s+1)} + \frac{B}{(s+2)}$$

$$(s + 3) = A\,(s + 2) + B\,(s + 1) \qquad ...(2)$$

Solving Eq. (2), we get

$$A = 2$$

and $$B = -1$$

So $$\frac{Y(s)}{U(s)} = \frac{2}{(s+1)} + \frac{-1}{(s+2)}$$

The state block diagram for the above function is shown in Fig. Ex. 11.30.

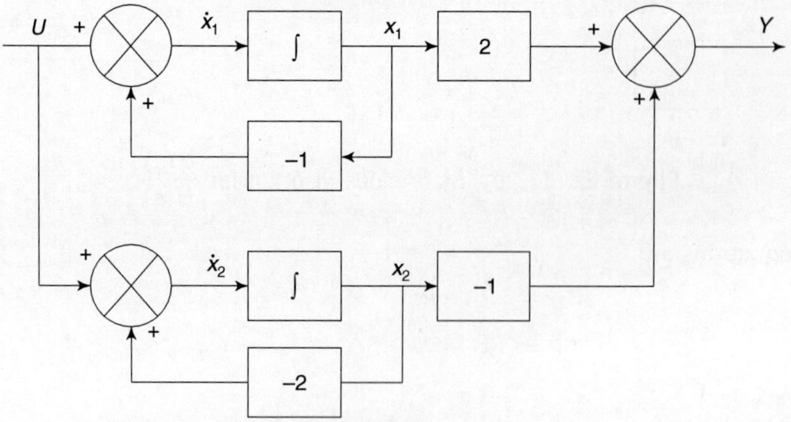

Figure Ex. 11.30 State block diagram for Ex. 11.30

The output of each integrator is assigned a state variable.
The state equations are

$$\dot{x}_1 = -x_1 + u$$

$$\dot{x}_2 = 2x_2 + u$$

$$y = 2x_1 - x_2$$

The state model is $$\begin{bmatrix} \dot{x}_1 \\ \dot{x}_2 \end{bmatrix} = \begin{bmatrix} -1 & 0 \\ 0 & -2 \end{bmatrix} \begin{bmatrix} x_1 \\ x_2 \end{bmatrix} + \begin{bmatrix} 1 \\ 1 \end{bmatrix} u$$

$$y = [2 \quad -1] \begin{bmatrix} x_1 \\ x_2 \end{bmatrix}$$ **Ans.**

Example 11.31 Consider a matrix A given below

$$A = \begin{bmatrix} -1 & 3 & 0 \\ 0 & -3 & 1 \\ 0 & 0 & -5 \end{bmatrix}$$

Find the model matrix or diagonalizing matrix.

Solution $$A = \begin{bmatrix} -1 & 3 & 0 \\ 0 & -3 & 1 \\ 0 & 0 & -5 \end{bmatrix}$$

The characteristic equation is

$$|\lambda I - A| = 0$$

$$\begin{bmatrix} \lambda+1 & -3 & 0 \\ 0 & \lambda+3 & -1 \\ 0 & 0 & \lambda+5 \end{bmatrix} = 0$$

$$(\lambda + 1)(\lambda + 3)(\lambda + 5) = 0$$

Therefore, the eigenvalues of matrix A are

$$\lambda_1 = 1, \quad \lambda_2 = 3, \quad \lambda_3 = -5$$

Then the model matrix can be shown to be a special matrix called the Vander model matrix.

$$P = \begin{bmatrix} 1 & 1 & 1 \\ \lambda_1 & \lambda_2 & \lambda_3 \\ \lambda_1^2 & \lambda_2^2 & \lambda_3^2 \end{bmatrix} = \begin{bmatrix} 1 & 1 & 1 \\ -1 & -3 & -5 \\ 1 & 9 & 25 \end{bmatrix}$$

$$\tilde{P} = P^{-1} A P$$

$$\tilde{P} = \begin{bmatrix} 1 & 1 & 1 \\ -1 & -3 & -5 \\ 1 & 9 & 25 \end{bmatrix}^{-1} \begin{bmatrix} -1 & 3 & 0 \\ 0 & -3 & 1 \\ 0 & 0 & -5 \end{bmatrix} \begin{bmatrix} 1 & 1 & 1 \\ -1 & -3 & -5 \\ 1 & 9 & 25 \end{bmatrix}$$

Observability

$$C = \begin{bmatrix} 0 & 0 & 1 \\ 1 & 1 & 0 \end{bmatrix}$$

$$CA = \begin{bmatrix} 0 & 0 & 1 \\ 1 & 1 & 0 \end{bmatrix} \begin{bmatrix} -3 & 1 & 1 \\ -1 & 0 & 1 \\ 0 & 0 & 1 \end{bmatrix} = \begin{bmatrix} 0 & 0 & 1 \\ -4 & 1 & 2 \end{bmatrix}$$

$$CA^2 = \begin{bmatrix} 0 & 0 & 1 \\ -4 & 1 & 2 \end{bmatrix} \begin{bmatrix} -3 & 1 & 1 \\ -1 & 0 & 1 \\ 0 & 0 & 1 \end{bmatrix} = \begin{bmatrix} 0 & 0 & 1 \\ 11 & -4 & -1 \end{bmatrix}$$

The observable matrix is

$$O_M = \begin{bmatrix} C \\ \cdots \\ CA \\ \cdots \\ CA^2 \end{bmatrix} = \begin{bmatrix} 0 & 0 & 1 \\ 1 & 1 & 0 \\ 0 & 0 & 1 \\ -4 & 1 & 2 \\ 0 & 0 & 1 \\ 11 & -4 & -1 \end{bmatrix}$$

Let us choose three-row matrix

$$\Rightarrow \begin{bmatrix} -4 & 1 & 2 \\ 0 & 0 & 1 \\ 11 & -4 & -1 \end{bmatrix}$$

The determinant of this matrix $\neq 0$, i.e., – 5 and the rank is 3 so the system is controllable.

$$\tilde{P} = \begin{bmatrix} -1 & 0 & 0 \\ 0 & -3 & 0 \\ 0 & 0 & -5 \end{bmatrix} \quad \textbf{Ans.}$$

Example 11.32 Find the state transition matrix for the following unforced system.

$$\begin{bmatrix} \dot{x}_1 \\ \dot{x}_2 \\ \dot{x}_3 \end{bmatrix} = \begin{bmatrix} -2 & 1 & 0 \\ 0 & -2 & 1 \\ 0 & 0 & -2 \end{bmatrix} \begin{bmatrix} x_1 \\ x_2 \\ x_3 \end{bmatrix}$$

(*ME Pune University*)

Solution By Laplace method

$$(sI - A) = \begin{bmatrix} s+2 & 1 & 0 \\ 0 & s+2 & -1 \\ 0 & 0 & s+2 \end{bmatrix}$$

The determinant of $(sI - A)$

$$(sI - A) = (s + 2)\,(s + 2)\,(s + 2) = (s + 2)^3$$

$$\text{The adj. of } (sI - A) = \begin{bmatrix} (s+2)(s+2) & 0 & 0 \\ (s+2) & (s+2)(s+2) & 0 \\ 1 & (s+2) & (s+2)(s+2) \end{bmatrix}$$

$$[(sI - A)]^{-1} = \frac{\text{adj of } (sI - A)}{|sI - A|}$$

$$= \frac{1}{(s+2)^3} \begin{bmatrix} (s+2)(s+2) & 0 & 0 \\ (s+2) & (s+2)(s+2) & 0 \\ 1 & (s+2) & (s+2)(s+2) \end{bmatrix}$$

$$= \begin{bmatrix} \dfrac{1}{(s+2)} & 0 & 0 \\ \dfrac{1}{(s+2)^2} & \dfrac{1}{(s+2)} & 0 \\ \dfrac{1}{(s+2)^3} & \dfrac{1}{(s+2)^2} & \dfrac{1}{(s+2)} \end{bmatrix}$$

$$e^{At} = L^{-1}\,[(sI - A)^{-1}]$$

$$e^{At} = L^{-1}\begin{bmatrix} \dfrac{1}{(s+2)} & 0 & 0 \\ \dfrac{1}{(s+2)^2} & \dfrac{1}{(s+2)} & 0 \\ \dfrac{1}{(s+2)^2} & \dfrac{1}{(s+2)^2} & \dfrac{1}{(s+2)} \end{bmatrix}$$

$$e^{At} = \begin{bmatrix} e^{-2t} & 0 & 0 \\ t\,e^{-2t} & e^{-2t} & 0 \\ \dfrac{1}{2}r^2\,e^{-2t} & t\,e^{-2t} & e^{-2t} \end{bmatrix}$$

$\because$ $$L\{t^n e^{-at}\} = \frac{n!}{(s+a)^{n+1}}$$ **Ans.**

Example 11.33 Determine the state controllability and observability of the system described by

$$\begin{bmatrix} \dot{x}_1 \\ \dot{x}_2 \\ \dot{x}_3 \end{bmatrix} = \begin{bmatrix} -3 & 1 & 1 \\ -1 & 0 & 1 \\ 0 & 0 & 1 \end{bmatrix}\begin{bmatrix} x_1 \\ x_2 \\ x_2 \end{bmatrix} + \begin{bmatrix} 0 & 1 \\ 0 & 0 \\ 2 & 1 \end{bmatrix} u$$

$$y = \begin{bmatrix} 0 & 0 & 1 \\ 1 & 1 & 0 \end{bmatrix} x$$ *(ME Pune University)*

Solution

$$A = \begin{bmatrix} -3 & 1 & 1 \\ -1 & 0 & 1 \\ 0 & 0 & 1 \end{bmatrix} \text{ and } B = \begin{bmatrix} 0 & 1 \\ 0 & 0 \\ 2 & 1 \end{bmatrix}$$

$$C = \begin{bmatrix} 0 & 0 & 1 \\ 1 & 1 & 0 \end{bmatrix}$$

$$AB = \begin{bmatrix} -3 & 1 & 1 \\ -1 & 0 & 1 \\ 0 & 0 & 1 \end{bmatrix}\begin{bmatrix} 0 & 1 \\ 0 & 0 \\ 2 & 1 \end{bmatrix} = \begin{bmatrix} 2 & -2 \\ 2 & 0 \\ 2 & 1 \end{bmatrix}$$

$$A^2 B = \begin{bmatrix} -3 & 1 & 1 \\ -1 & 0 & 1 \\ 0 & 0 & 1 \end{bmatrix}\begin{bmatrix} -3 & 1 & 1 \\ -1 & 0 & 1 \\ 0 & 0 & 1 \end{bmatrix}\begin{bmatrix} 2 & -2 \\ 2 & 0 \\ 2 & 1 \end{bmatrix}$$

$$A^2B = \begin{bmatrix} +8 & -3 & -1 \\ 3 & -1 & 0 \\ 0 & 0 & 1 \end{bmatrix}\begin{bmatrix} 0 & 1 \\ 0 & 0 \\ 2 & 1 \end{bmatrix} = \begin{bmatrix} -2 & -5 \\ 0 & 3 \\ 0 & 1 \end{bmatrix}$$

The controllability matrix

$$C_M = [B \;:\; AB \;:\; A^2B]$$

$$C_M = \begin{bmatrix} 0 & 1 & 2 & -2 & -2 & -5 \\ 0 & 0 & 2 & 0 & 0 & 3 \\ 2 & 1 & 2 & 1 & 0 & 1 \end{bmatrix}$$

The column collect from C_M

Let it be $\begin{bmatrix} 0 & 1 & 2 \\ 0 & 0 & 2 \\ 2 & 1 & 2 \end{bmatrix}$

and determinant of it is 4 and the rank is 3 so the system is controllable.

MATLAB PROGRAMS

Program 11.1

```
% Programe for conversion to state space model from transfer funcion model
% G(s)  =  (s+3)/(s+1)(s+2)(s+3)
numg  =  input('Enter the coefficient of polynomial of numerator:-');
                          % Define numerator
deng1 =input('Enter the poles of denominator:- ');
deng =poly(deng1);      % define denominator
'G(s)'
G  =  tf(numg,deng)      % create transfer function
[Ac Bc Cc Dc]  =  tf2ss(numg,deng)
  % Transform in to canonical state  space from transfer function
r  =  roots([1 2 4])
P  =  10*[r'  10*real(r(1))]
K =acker(Ac,Bc,P)
```

Input

```
Enter the coefficient of polynomial of numerator:- [1 3]
Enter the poles of denominator:- [-1 -2 -3]
```

Output

```
                                   s + 3
Transfer function: G(s)  =  ---------------------
                            s^3 + 6s^2 + 11s + 6
```

```
Ac = - 6 - 11 - 6
       1   0   0
       0   1   0
Bc = 1
     0
     0
Cc = 0 1 3
Dc = 0
```

Program 11.2

```
% Program for conversion state-space to transfer function model
% dX/dt = AX +BU
% Y = Cx
clear all;
clc
A = input('Enter the coefficient matrix A:- '); % input coefficient matrix A
B = input ('Enter the coefficient matrix B:- ') ;% input coefficient matrix B
C = input('Enter the coefficient matrix C:-') ; % input coefficient matrix C
D = input('Enter the coefficient matrix D:- '); % input coefficient matrix D
[num, den] = ss2tf(A,B,C,D);
'T(s)'
T = tf(num,den)
```

Input

```
Enter the coefficient matrix A:- [1 0; -2 1]
Enter the coefficient matrix B:- [1;1]
Enter the coefficient matrix C:- [1 2]
Enter the coefficient matrix D:- [0]
```

Output

Transfer function: $T(s) = \dfrac{3s - 7}{s\text{^}2 - 2s + 1}$

Program 11.3

```
% Program for controllability and observability
clear all;
clc;
A = input('Enter the coefficient matrix A :-');
B = input('Enter the coefficient matrix B :-');
C = input('Enter the coefficient matrix C :- ');
```

```
n = size(A);
if (n == 3)
Cm = [B     A*BA*A*B]
'Determinant is'
D1 = det(Cm)
'Rank is'
R1 = rank(Cm)
if (R1 == n)
  'The System is Controllable'
else
  'The System is Uncontrollable'
end
Om = [C;C*A;C*A*A]
'determinant is'
D2 = det(Cm)
'Rank is'
R2 = rank(Cm)
if (R2 == n)
  'The System is Observable'
else
  'The System is Unobservable'
end
end
if (n ==2)
Cm = [B     A*B]
'Determinant is'
D1 = det(Om)
'Rank is'
R1 = rank(Om)
if (R1 == n)
'The System is Controllable'
else
  'The System is Uncontrollable'
end
Om = [C;C*A]
'Determinant is'
D2 = det(Om)
```

```
'Rank is'
R2 = rank(Om)
if (R2 == n)
'The System is Observable'
else
  'The System is Unobservable'
end
end
```

Input

```
Enter the coefficient matrix A :- [1 0;-2 1]
Enter the coefficient matrix B :- [1 ;0]
Enter the coefficient matrix C :- [1 1]
```

Output

```
Cm =  1     1
     0    -2
Determinant is
D1 = -2
Rank is
R1 =2
The System is Controllable
Om =   1   1
      -1   1
Determinant is
D2 = -2
Rank is
R2 = 2
The System is Observable
```

Program 11.4

```
% Program for finding state Transition matrix
clear all                   % Clear all variable
clc                         % clear screen
A = input('Enter the coefficient matrix A :-');
n = size(A);
Syms s;
Char_Equa = s*eye(n)-A;    % Sl-A
lnv_Char= inv(Char_Equa);  % inverse of Sl-A
```

```
Transition_matrix =ilaplace(lnv_Char)
% Inverse Laplace transform of inverse characterstic equation
```

Input

```
Enter the coefficient matrix A :- [1 0;-2 1]
```

Output

```
Transition_matrix - [exp(t),        0]
             [-2*t*exp(t),      exp(t)]
```

EXERCISE

11.1 Explain controllability and observability of the system.

11.2 A feedback system has a closed loop transfer function:

$$\frac{Y(s)}{U(s)} = \frac{10\,(s+4)}{s\,(s+1)\,(s+3)}$$

Construct the different models for this system and give block diagram representation for each state model.

11.3 Consider a state model

$$\text{Where } A = \begin{bmatrix} 0 & 1 & 0 \\ 0 & 0 & 1 \\ -40 & -34 & -10 \end{bmatrix};\ B = \begin{bmatrix} 0 \\ 0 \\ 1 \end{bmatrix}$$

Suggest a suitable transformation matrix P.

11.4 A feedback system is represented by a signal flow graph shown in Fig. E. 11.4

(a) Construct a state mode

(b) Diagonalize the matrix A of the state model

(c) Determine the stability of the system

(d) Determine $Y(s)/U(s)$

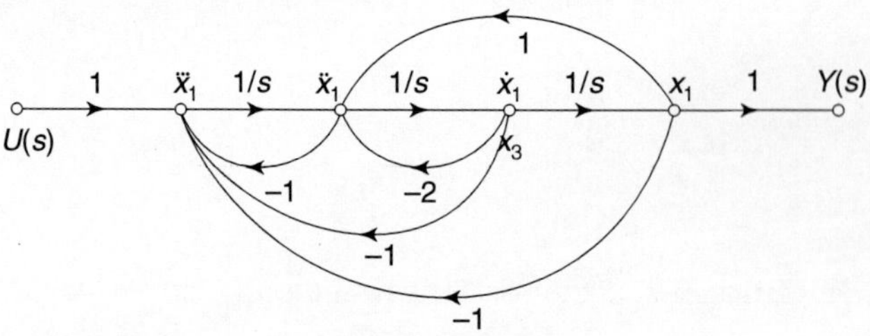

Figure E. 11.4

11.5 For the plant $G(s) = \dfrac{100\,(s+10)}{ss\,(s+3)\,(s+1)}$ represented in the state space in phase variable form by

$$\dot{x} = \begin{bmatrix} 0 & 1 & 0 \\ 0 & 0 & 1 \\ 0 & -36 & -15 \end{bmatrix} x + \begin{bmatrix} 0 \\ 0 \\ 1 \end{bmatrix} u$$

$$y = [\ 1000\ 100\ 0\ ]\ x$$

design for phase variable feedback gains to yield 5% overshoot and a peak time of 0.3 sec.

11.6 Determine whether the system $\dot{x} \begin{bmatrix} -1 & 1 & 2 \\ 0 & -1 & 5 \\ 0 & 3 & -4 \end{bmatrix} x + \begin{bmatrix} 2 \\ 1 \\ 1 \end{bmatrix} =$ is controllable.

11.7 Design a linear state feedback controller to yield 20% overshoot and a settling time of 2 sec for a plant.

$$G(s) = \frac{(s+6)}{(s+9)(s+8)(s+7)}$$

that is represented in state space is cascade form by

$$\dot{x} = \begin{bmatrix} -7 & 1 & 0 \\ 0 & -8 & 1 \\ 0 & 0 & -9 \end{bmatrix} \begin{bmatrix} x_1 \\ x_2 \\ x_3 \end{bmatrix} + \begin{bmatrix} 0 \\ 0 \\ 1 \end{bmatrix} u$$

$$y = [-1 \quad 1 \quad 0]\ x$$

11.8 Determine whether the system

$$\begin{bmatrix} \dot{x}_1 \\ \dot{x}_2 \\ \dot{x}_3 \end{bmatrix} = \begin{bmatrix} 0 & 1 & 0 \\ 0 & 0 & 1 \\ -4 & -3 & -2 \end{bmatrix} \begin{bmatrix} x_1 \\ x_2 \\ x_3 \end{bmatrix} + \begin{bmatrix} 0 \\ 0 \\ 1 \end{bmatrix} u$$

$$y = [0 \quad 5 \quad 1]\ x$$

is observable.

11.9 Determine whether the system

$$\dot{x} = \begin{bmatrix} -2 & 1 & -3 \\ 0 & -2 & 1 \\ -7 & -8 & -9 \end{bmatrix} x + \begin{bmatrix} 2 \\ 1 \\ 2 \end{bmatrix} u$$

$$y = [4 \quad 6 \quad 8]\ x$$

is observable.

11.10 Prove that the eigenvalues of a matrix remain unchanged under linear transformation. Illustrate this fact with the help of an example. *(U.P.T.U., 2007–08, CS)*

11.11 What are controllability and observability of a control system? Give the methods of their testing. What is the importance of controllability and observability in the design of control system? *(U.P.T.U., 2007–08, CS)*

11.12 Define state transition matrix, give its properties. Determine the state transition matrix for the given system matrix.

$A = \begin{bmatrix} 1 & 0 \\ 1 & 1 \end{bmatrix}$ by Laplace transform method. *(U.P.T.U., 2006–07, CS)*

11.13 Write a short note on solving state equations. *(U.P.T.U., 2007–08, ACS)*

11.14 What is state variable? Mention the advantages of state space approach? *(U.P.T.U., 2006–07, ACS)*

11.15 Define the following terms with appropriate expression

(i) State and state variables
(ii) State transition matrix
(iii) Controllability and observability

(U.P.T.U., 2005–06, ACS)

ANSWERS

11.2 (i) Phase variable form

A = [0 1 0 ; 0 0 1 ; 0 -3 -4]

B = [0; 0; 1]

C = [40 10 0]

(ii) Canonical form

A = [0 0 0; 0 -1 0 ; 0 0 -3]

B = [40/3 ; -15 ; 5/3]

C = [1 1 1]

(iii) Canonical form

A = [0 0 0 ; 0 -1 0; 0 0 -3]

S = [1 ; 1; 1]

C = [40/3 ; -15 5/3]

11.4 (a) $\dot{x}_1 = [0\ 1\ 0;\ 0\ 0\ 1;\ -1\ 0\ -3]\ x + [0\ ;\ 0;\ 1]\ u \quad y = x_1$

(b) $\lambda_{1,\,2} = 0.052 \pm j\,0.565,\ \lambda_3 = -3.104 \quad P^{-1}AP = \begin{bmatrix} \lambda_1 & 0 & 0 \\ 0 & \lambda_2 & 0 \\ 0 & 0 & \lambda_3 \end{bmatrix}$

(c) Unstable 1

(d) $\dfrac{1}{s^3 + 35^2 + 1}$

11.5 K = [2094 373.1 14.97]

11.6 Controllable

11.7 K_x = [– 40.23 62.24 –14]

11.8 Observable

11.9 Observable

Appendix—A

LIST OF SYMBOLS

$\% M_p$	Maximum peak overshoot
$T_o(s)$	Overall transfer function
$R(s)$	Laplace transform of input of a systems
$Y(s)$	Laplace transform of output of a system
A	Ampere, unit of electrical current
A	System matrix for state space representation
B	Input matrix for state space representation
C	Output matrix for state space representation
$E(s)$	Laplace transform of the error
C_M	Controllability matrix
$F(s)$	Laplace transform of $F(t)$
G	Electrical conductance
$G(s)$	Forward path transfer function
$G_c(s)$	Transfer function of compensation, network
$G'(s)$	Transfer function of compensated system
G_M $G.M.$	Gain margin
H	Henry, unit of electrical inductance
$H(s)$	Feedback path transfer function
I	Identity matrix
$i(t)$	Electrical current in amperes
K	Controller feedback gain
K_a	Acceleration error constant
Kv	Velocity error constant

L	Electrical inductance in henries
$L\{\ \}$	Laplace transform
$L^{-1}\{\ \}$	Inverse Laplace transform
O_M	Observability matrix
P	Model matrix
$\tilde{P}$	Diagonalization matrix
P_c	Compensator pole
R	Electrical resistance in ohms
$r(t)$	Input to a system
rad/s	Radian/second
sec	Second
s_d	Dominant root location in s-plane
T	Time constant
$T(s)$	Closed loop transfer function
t_p	Peak time in sec
t_r	Rise time in sec
t_s	Settling time in sec
u	Input or control vector for state space representation
$u(t)$	Unit step input
x	State vector for state space representation
$\dot{x}$	Time derivative of state variable
$\dot{X}$	Time derivative of state vector
y	Output vector for state space representation
$y(t)$	Output scalar for state space representation
Z_c	Compensator zero
α	Pole scaling factor for a lead compensator
β	Pole scaling factor for a lag compensator
ξ	Damping ratio
e^{At}	State transition matrix
$\phi(t)$	State transition matrix
$\omega(t)$	Angular velocity in rad/sec
ω_d	Damped frequency of oscillation in rad/sec
ω_n	Natural frequency in rad/sec
W_{cg}	Gain crossover frequency
ω_{cp}	Phase crossover frequency
$\lvert G(j\omega)\rvert$	Magnitude plot (resultant) for bode plot of uncompensated system
$\angle G(j\omega)$	Phase plot for Bode plot of uncompensated system

$\lvert G_c(j\omega) \rvert$	Magnitude plot for compensation network
$\angle G_c(j\omega)$	Phase plot for compensation network
$\lvert G'(j\omega) \rvert$	Magnitude plot for compensated system
$\angle G'(j\omega)$	Phase plot for compensated system
$E_o(s)$	Output of compensation network
$E_i(s)$	Input of compensation network
ϕ_L	Phase lead required
ϕ_m	Maximum phase
ω_m	Mid-frequency
ω_c	Corner frequency
ω_{c1}	Lower corner frequency
ω_{c2}	Upper corner frequency
ϕ_s	Phase specified
ω_b	Bandwidth frequency
ω_r	Rise frequency
$\int$	Integrator
λ	Roots of characteristic equation

Appendix—B

MULTIPLE-CHOICE QUESTIONS

1. The transfer function is defined as
   (a) The ratio of Laplace transform of output to Laplace transform of input initial condition as zero
   (b) The ratio of input to output
   (c) The ratio of output to input
   (d) The ratio of Laplace transform of input to the Laplace transform of output considering initial condition as zero
2. The transfer function relating E and R for the block diagram given below

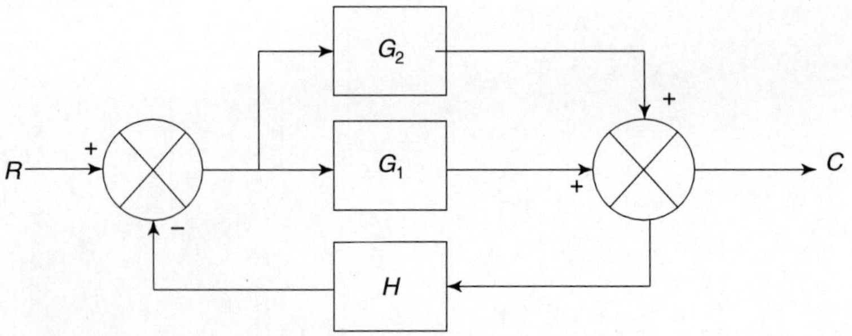

(a) $\dfrac{1}{1+G_1\,G_2\,H}$ (b) $\dfrac{1}{1+(G_1+G_2)\,H}$

(c) $\dfrac{G_1\,G_2}{1+(G_1+G_2)\,H}$ (d) $\dfrac{G_1\,G_2}{1+G_1G_2\,H}$

3. The transfer function for the network given

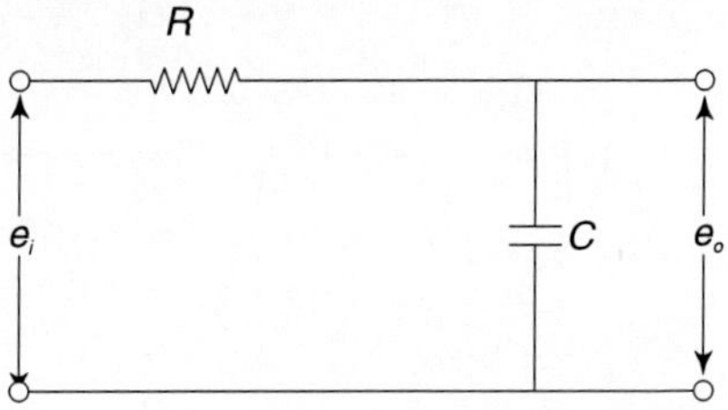

(a) $RCs + 1$

(b) $\dfrac{1}{\frac{R}{C}s+1}$

(c) $\dfrac{R}{C}s+1$

(d) $\dfrac{1}{RCs+1}$

4. The block diagram given below

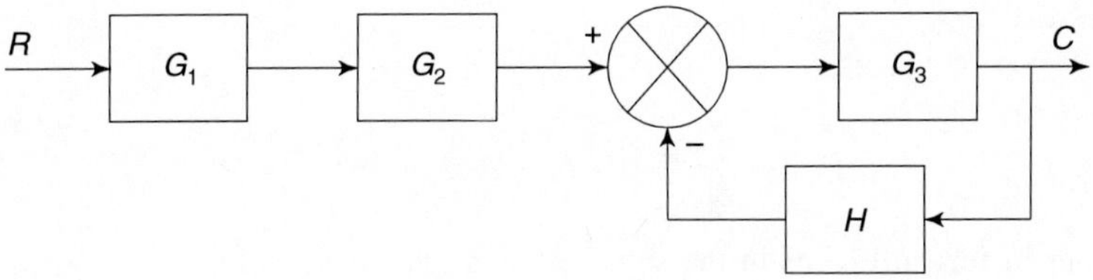

is equivalent to

(a) $\dfrac{G_1\, G_2 G_3}{1+G_3\, H}$

(b) $\dfrac{(G_1+G_2)\, G_2}{1+G_3\, H}$

(c) $\dfrac{G_1\, G_3+G_2\, G_3}{1+G_3\, H}$

(d) $\dfrac{G_1+G_2\, G_3}{1+G_3\, H}$

5. The signal flow graph for the system whose block diagram is shown below will be

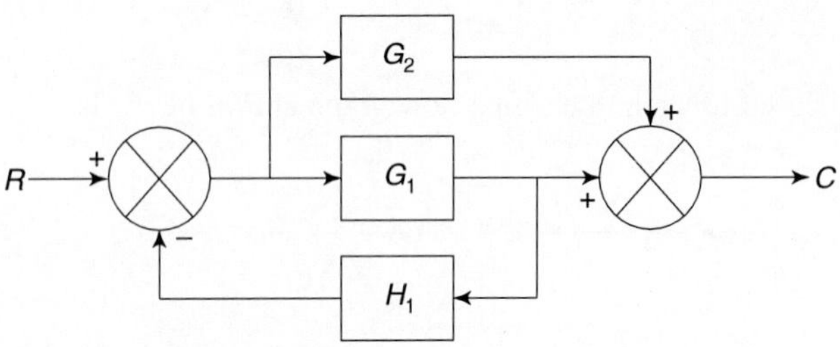

(a)

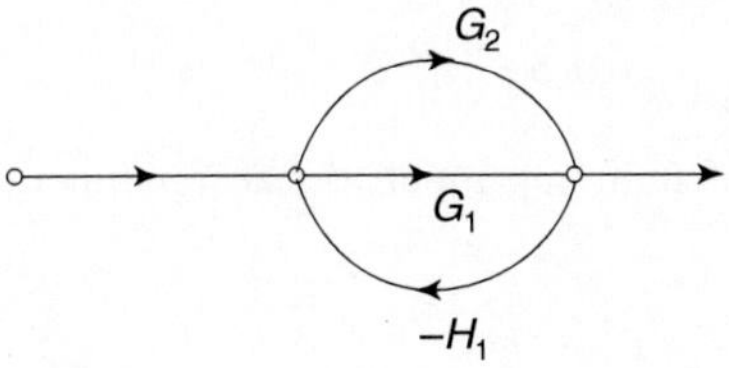

(b)

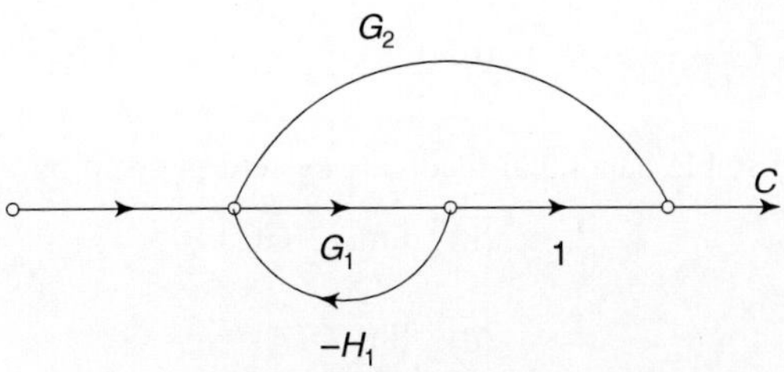

(c) 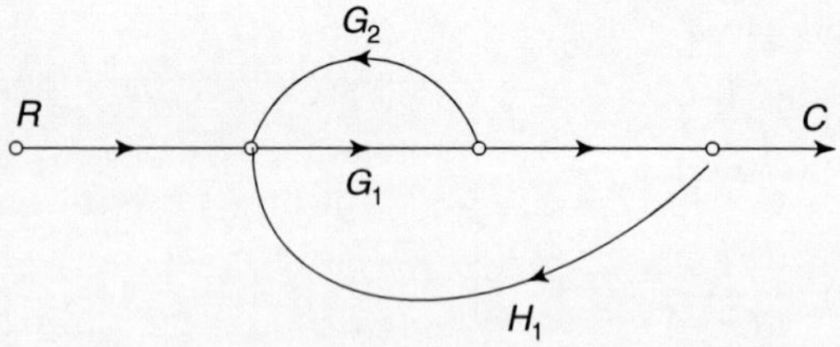

(d) 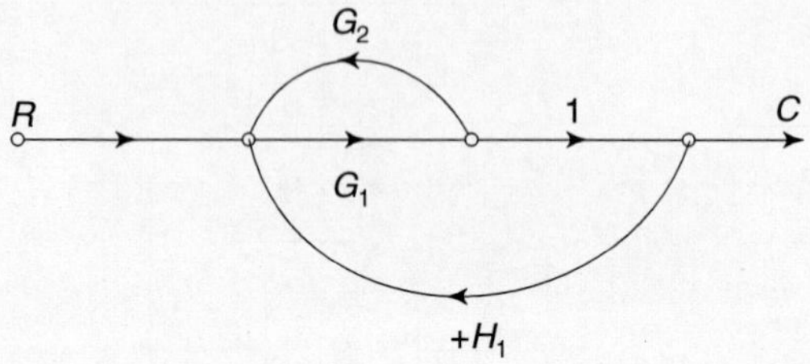

6. The number of forward paths in the signal flow graph shown below is

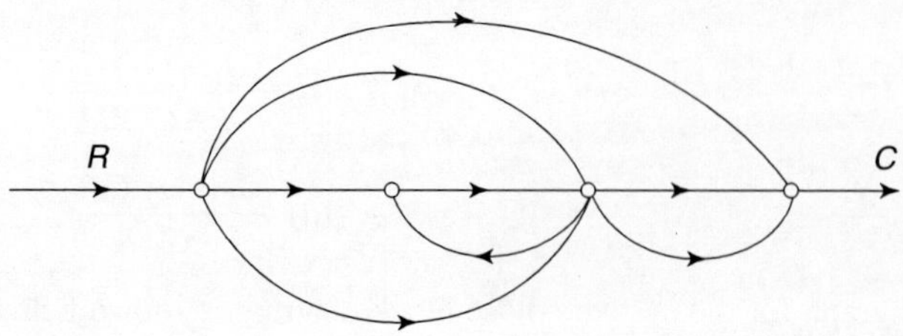

(a) 1 (b) 2
(c) 7 (d) 5

7. The number of closed loops in the signal flow graph shown below is

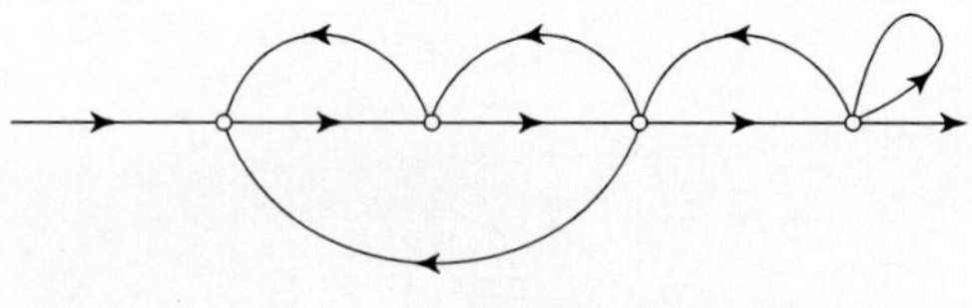

(a) 2 (b) 3
(c) 4 (d) 5

8. The steady value of the output on application of unit step input to transfer function $\frac{C(s)}{R(s)} = \frac{2.5}{s^2 + 2s + 10}$ will be

(a) 0.25 (b) 1
(c) 1/2.5 (d) 2.5

9. The acceleration error constant of a stable canonical feedback system is given by

(a) $\lim_{s \to 0} G(s)$ (b) $\lim_{s \to 0} s^2 G(s)$

(c) $\lim_{s \to 0} s\, G(s)$ (d) $\lim_{s \to 0} G(s)$

10. The velocity error constant of a stable canonical feedback system is given by

(a) $\lim_{s \to 0} s^2 G(s)$ (b) $\lim_{s \to 0} G(s)$

(c) $\lim_{s \to 0} s\,G(s)$ (d) None

11. The first column of Routh array contains the following integers – 6, – 8, – 4, – 3. The system is

(a) Stable (b) Unstable

(c) Marginally stable (d) None of these

12. The first column of Routh array contains the following integers 2, 3, – 6, 8, – 7. The system is

(a) Stable (b) Marginally stable

(c) Unstable (d) None of these

13. Laplace transform of doublet is

(a) $1/s^3$ (b) Unity

(c) $1/s$ (d) s

14. The inductance is not used to fabricate a lag network because

(a) Its size is big (b) It has high reactance

(c) It produces time delay and hysteresis loss (d) None of these

15. For the system $\frac{C(s)}{R(s)} = \frac{35}{s^2 + 6s + 25}$ the damping factor 3 and damped frequency of oscillation will be

(a) $\xi = 0.4,\ \omega_d = 6$ (b) $\xi = 0.6,\ \omega_d = 5$

(c) $\xi = 0.3,\ \omega_d = 5$ (d) $\xi = 0.5,\ \omega_d = 3$

16. Which one of the following transfer functions will have greatest maximum overshoot.

(a) $\frac{3}{s^2 + 2s + 9}$ (b) $\frac{25}{s^2 + 2s + 16}$

(c) $\frac{36}{s^2 + 2s + 36}$ (d) $\frac{49}{s^2 + 2s + 49}$

17. If the poles of a control system lie on the imaginary axis in place the system will be

(a) Stable (b) Unstable

(c) Marginally stable (d) Conditionally stable

18. The steady state error for a type 2 system subjected to unit ramp input is

(a) 1 (b) ∞

(c) $1/K$ (d) 0

19. What is the centre of asymptotes for a open loop transfer function $G(s) = \frac{3(s+2)}{s^2(s+4)}$

(a) – 1 (b) – 3

(c) 0 (d) – 4

20. The steady state error is determined as the difference between the reference input and the system output at
    (a) $t = t_p$ (b) $t = 0$
    (c) $t = \infty$ (d) t = time constant
21. The settling time of a second order system on the basis is given by
    (a) $t_s = \frac{\xi \omega_n}{4}$ (b) $t_s = \frac{4\xi}{\omega_n}$
    (c) $t_s = 4\xi\omega_n$ (d) $t_s = \frac{4}{\xi \omega_n}$
22. Peak time and ω_n are related as
    (a) $\omega_n = \frac{t_p\sqrt{1-\xi^2}}{\pi}$ (b) $\omega_n = \frac{1}{\pi t_p\sqrt{1-\xi^2}}$
    (c) $\omega_n = \frac{\pi}{t_p\sqrt{1-\xi^2}}$ (d) $\omega_n = \frac{t_p}{\pi\sqrt{1-\xi^2}}$
23. In general, which of the following system is preferred?
    (a) Overdamped (b) Critical damped
    (c) Underdamped (d) Undamped
24. The lead-lag compensation will improve
    (a) Transient response (b) Steady state response
    (c) Transient and steady state response (d) None of the above
25. The damping factor of the system is unity, the system is
    (a) Underdamped (b) Critically damped
    (c) Overdamped (d) Unstable
26. An increase in damping ratio
    (a) Increases rise time (b) Decreases rise time
    (c) Does not affect rise time (d) Keeps the time within limit
27. Derivative feedback control
    (a) Increases rise time (b) Increases overshoot
    (c) Does not affect the steady state error (d) Decreases steady state error
28. If M_p = 100%, then damping ratio is
    (a) 0 (b) 0.5
    (c) 1 (d) infinity
29. In open loop system, the control action depends upon the
    (a) Input (b) Desired output
    (c) Input and output both (d) None of these
30. Finite steady state error
    (a) Varies directly with K (b) Is independent of K
    (c) Varies inversely with K (d) None of the above

31. The Routh-Harwitz criterion gives
   (a) Absolute stability
   (b) Relative stability
   (c) Gain margin
   (d) Phase margin
32. For the Routh array given below

s^3	1	s
s^2	3	K
s^1	$\frac{15-K}{3}$	
s^0	K	

   The system is stable if
   (a) $K = 15$
   (b) $K > 3$
   (c) $K < 3$
   (d) $K > 15$
33. Laplace transform of unit step function is
   (a) $\frac{1}{s^2}$
   (b) 1
   (c) $1/s$
   (d) s
34. Laplace transform of unit ramp function is
   (a) s^2
   (b) s
   (c) 7
   (d) $\frac{1}{s^2}$
35. If the bandwidth of a control system is increased, the noise in the system will be
   (a) Zero
   (b) No effect
   (c) Decreased
   (d) Increased
36. Gain margin is a measure of
   (a) Relative stability of system
   (b) Absolute stability of system
   (c) Unstability of system
   (d) None of these
37. The lag compensation in the control system is achieved by
   (a) Adding pole in the transfer function of the system
   (b) Adding zero in the transfer function of the system
   (c) Both (a) and (b)
   (d) None
38. The lead compensation in the control system is achieved by adding
   (a) Pole
   (b) Zero
   (c) Both (a) and (b)
   (d) None
39. In frequency domain analysis frequencies varies with
   (a) Amplitude only
   (b) Phase only
   (c) Both (a) and (b)
   (d) None
40. A capasitance can be used to fabricate a lag network.
   (a) True
   (b) False

41. The open loop transfer function of a control system has 5 poles and 3 zeros. The number of asymptotes in root locus for the system is
(a) 3
(b) 1
(c) 2
(d) 5

42. Roots of the characteristic equation are nothing but
(a) Zeros of the system
(b) Poles of the system
(c) Both (a) and (b)
(d) None of these

43. One of the following methods can be used to determine the absolute stability only of a control system
(a) Routh stability criterion
(b) Bode plot
(c) Root locus
(d) None of these

44. A control system is stable when
(a) Roots of characteristic equation have negative real parts.
(b) Impulse response approaches zero when time tends to infinity.
(c) Roots of characteristic equation have negative real parts.
(d) None of these.

45. Routh tables are separately constructed for two control systems. The first columns obtained (2, 1, 3, – 5) for the system *A* and (– 2, – 3 – 4 – 6) for system *B*. Then
(a) *A* and *B* both are stable
(b) *A* and *B* both are unstable
(c) *A* is stable but *B* is not
(d) *B* is stable but *A* is not

46. Laplace transform is applicable to
(a) Linear system
(b) Nonlinear system
(c) Both (a) and (b)
(d) None of the above

47. Presence of non-linearity in a control system tends to introduce
(a) Instability
(b) Transient error
(c) Steady state error
(d) All of the above

48 The phase crossover frequency at which the phase angle is
(a) – 90°
(b) – 180°
(c) 90°
(d) 180°

49. The gain crossover frequency at which the gain is
(a) 20 dB
(b) – 20 dB
(c) 0 dB
(d) – 10 dB

50. In a overdamped system, the damping factor is of the order of
(a) Unity
(b) Greater than unity
(c) Less than unity
(d) None of the above

51. Damping is a function of
(a) $\frac{1}{\text{Gain}}$
(b) $\frac{1}{\sqrt{\text{Gain}}}$
(c) Gain
(d) $\sqrt{\text{Gain}}$

52. Which of these is not in frequency domain
    (a) Root locus (b) Nyquist plot
    (c) Bode plot (d) None
53. The maximum peak overshoot is given by
    (a) $M_P = e^{-\frac{\xi\pi}{\sqrt{\xi^2 - 1}}}$ (b) $M_P = e^{-\frac{\xi\pi}{\sqrt{1-\xi^2}}}$
    (c) $M_P = e^{-\frac{\xi\pi}{\sqrt{1-\xi^2}}}$ (d) $M_P = e^{-\frac{\xi\pi}{\sqrt{\xi^2 - 1}}}$
54. The peak time is defined by
    (a) $t_p = \dfrac{2\pi}{\omega_n \sqrt{1-\xi^2}}$ (b) $t_p = \dfrac{\pi}{\omega_n \sqrt{\xi^2 - 1}}$
    (c) $t_p = \dfrac{2\pi}{\omega_n \sqrt{1-\xi^2}}$ (d) $t_p = \dfrac{\pi}{\omega_n \sqrt{1-\xi^2}}$
55. The transient response of a feedback system as compared to that without feedback.
    (a) Rises at slow rate (b) Rises at faster rate
    (c) Decays slowly (d) None
56. The lag network increases the bandwidth of the system
    (a) True (b) False
57. The phase shift of $G(s) = \dfrac{1}{s^3}$ is
    (a) – 270° (b) – 90°
    (c) 180° (d) – 180°
58. The phase shift of $G(s) = \dfrac{1}{(1+0.25\,s)}$ is
    (a) $-\tan^{-1} 0.15\,\omega$ (b) $-\tan^{-1} \omega$
    (c) – 90° (d) $-\tan^{-1} 0.25\,\omega$
59. The phase shift of $G(s) = (1 + 10s)$ is
    (a) $\tan^{-1}$ (b) $-\tan^{-1} 10\,\omega$
    (c) $+\tan^{-1} 10\,\omega$ (d) None
60. The slope of $G(s) = \dfrac{1}{s^2}$ is
    (a) – 20 dB/decade (b) – 60 dB/decade
    (c) – 40 dB/decade (d) None
61. The slope of $G(s) = s$ is
    (a) 20 dB/decade (b) – 20 dB/decade
    (c) – 40 dB/decade (d) 40 dB/decade
62. The steady state error of a type one second order system to unit ramp input is
    (a) $\dfrac{2\xi}{\omega_n}$ (b) $2\,\xi\,\omega_n$

(c) $\frac{\omega_n}{2\xi}$ (d) $\frac{2\omega_n}{\xi}$

63. The steady state error of a type one second order system to unit parabolic input is

(a) 0 (b) $2\,\xi\,\omega_n$

(c) ∞ (d) $\frac{2\xi}{\omega_n}$

64. The rise time for type one second order system is given by

(a) $t_r = \frac{\pi - \phi}{\omega_n\sqrt{1-\xi^2}}$ (b) $t_r = \frac{\pi - \sin^{-1}\frac{\sqrt{1-\xi^2}}{\xi}}{\omega_n\sqrt{1-\xi^2}}$

(c) $t_r = \frac{\pi}{\xi\sqrt{1-\xi^2}}$ (d) $t_r = \frac{\cos^{-1}\xi}{\omega_n\sqrt{1-\xi^2}}$

65. The resonant peak is given by

(a) $M_r = \frac{1}{\xi\sqrt{1-\xi^2}}$ (b) $M_r = \frac{1}{2\xi\sqrt{1-\xi^2}}$

(c) $M_r = \frac{1}{\xi\sqrt{1+\xi^2}}$ (d) None

66. The characteristic equation having more number of zeros than the poles, the number of root locus branches will be equal to

(a) Number of zeros (b) Number of poles

(c) Number of zeros – number of poles (d) Number of zeros + number of poles

67. If the characteristic equation of a system is $s^2 + 4s + 100 = 0$ the value of ω_n and ξ will be

(a) 10 rad/sec 0.4 (b) 5 rad/sec 0.2

(c) 10 rad/sec 0.2 (d) 5 ras/ sec 0.4

68. To define the transfer function, all initial conditions should be

(a) Zero (b) Infinity

(c) Finite (d) None

69. Zero initial condition means that the system is

(a) Working with zero reference signal

(b) At rest and no energy is stored in any of its component

(c) Working with zero stored energy

(d) none of these

70. For the transfer function $G(s) = \frac{K}{s\,(s^2 + 5s + 25)}$ the number of asymptotes is

(a) 3 (b) 2

(c) 1 (d) 6

71. The angles of asymptotes in question 70 are
(a) 60° , 180°, 360° (b) 60° , 180°, 240°
(c) 60°, 180°, 300° (d) none of these
72. The terms of Bode plot, the system is stable if
(a) P.M. and G.M. both are positive (b) P.M. and G.M. both are negative
(c) P.M. negative but G.M. positive (d) P.M. = G.M.
73. The characteristic equation of a system is $s^2 + 2s + 8 = 0$. The damping ratio will be
(a) < 1 (b) > 1
(c) 1 (d) None
74 If a closed loop control system operated at a point on imaginary axis, the system is
(a) Stable (b) Overdamped
(c) Underdamped (d) Marginally stable
75. The characteristic equation of a system is $s^6 + s^5 + 5s^4 + 3s^3 + 2s^2 - 4s - 8 = 0$. The number of roots on R.H. plane are
(a) 1 (b) 2
(c) 3 (d) 6
76. Which statement is not correct for open loop control system?
(a) Simple construction and ease of maintenance
(b) Less expensive
(c) Disturbances cause error
(d) To maintain the required quality is the output, recalibration is not required.
77. In a feedback system the transient response
(a) Decays at a constant rate (b) Gets magnified
(c) Decays slowly (d) Decays more quickly
78. Introduction of negative feedback in a system does not lead to reduction is
(a) Bandwidth (b) Distortion
(c) Instability (d) Overall gain
79. The frequency and time domain are related through
(a) Laplace transform (b) Fourier integral
(c) Laplace transform of Fourier integral (d) Laplace transform and Fourier integral
80. The characteristic equation of a second order system is given by $s^2 + 2\xi\omega_o s + \omega_0^2 = 0$. The term ω_o is called
(a) Overdamped natural frequency (b) Undamped critical frequency
(c) Undamped natural frequency (d) None of these
81. In the above system, if $\xi = 1$, the poles of transfer function will be
(a) Real and equal (b) Imaginary and equal
(c) Complex conjugate (d) Equal to – 1
82. In the same system, if $\xi < 1$, the poles are
(a) Real and equal (b) Imaginary and equal
(c) Complex conjugate (d) Real and unequal

83. In the same system if $\xi < 1$, the system will exhibit

(a) Damped response (b) Oscillatory response

(c) An overshoot (d) All of these

84. For the following differential equation $2\frac{d^2y}{dt^2} + 4\frac{dy}{dt} + 8y = 8x,$ the damping ratio is

(a) 1 (b) 0.7

(c) 2 (d) 0.5

85. The time constant for the differential equation in the above question is

(a) 0.5 (b) 1

(c) 0.2 (d) 1.5

86. A servomechanism is represented by the equation $\frac{d^2c_m}{dt^2} + 6.4\frac{dc_m}{dt} = 160\ \xi,\ \xi = (3 - 0.4\ c_m)$

where c_m is motor output shaft position. The damping ratio ξ, will be given

(a) 1.0 (b) 0.80

(c) 0.6 (d) 0.4

87. Static error coefficients are used as a measure of the effectiveness of closed loop system for specified

(a) Position input signal (b) Velocity input signal

(c) Acceleration input signal (d) All of these

88. The closed loop frequency response of a second order system of transfer function $G(s)$ is of the form

(a)

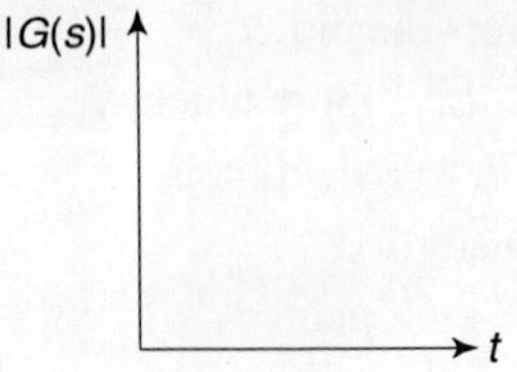

(b)

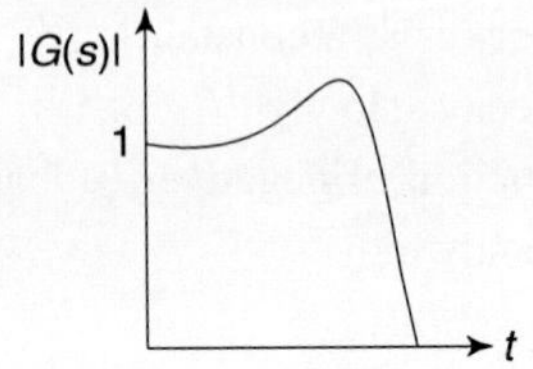

(c)

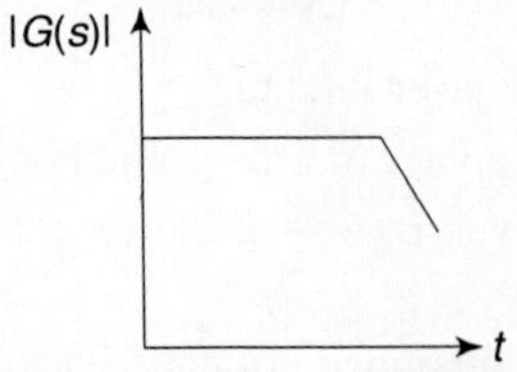

(d) none of these

89. Natural frequency of a unity feedback control system of transfer function $G(s) = 10/s\ (s + 1)$ is

(a) 3.16 rad/sec (b) 0.5 rad/sec

(c) 4.6 rad/sec (d) none of these

90. The damped natural frequency of an underdamped second order system is given by

(a) $\omega_d = \omega_n$ (b) $\omega_d = \xi\ \omega_n$

(c) $\omega_d = \omega_n\ \sqrt{1-\xi^2}$ (d) $\omega_n = \omega_d\ \sqrt{1-\xi^2}$

91. In most systems, an increase in gain leads to
    (a) Larger damping ratio
    (b) Smaller damping ratio
    (c) Constant damping ratio
    (d) None of these
92. If the complex poles of a system have greater real parts, then overshoot is
    (a) Less
    (b) More
    (c) Not affected
    (d) None of these
93. The overshoot of the system, $\frac{16K}{s(s^2+2s+16)}$ for a step input applied would be
    (a) 60%
    (b) 40%
    (c) 20%
    (d) 10%
94. The transfer function of a first order control system is of the type
    (a) $\frac{1}{Ts^2+1}$
    (b) $\frac{1}{Ts+1}$
    (c) Ts
    (d) $\frac{1}{Ts}$
95. The transfer function of an integral controller is of the type
    (a) K_c
    (b) Ts
    (c) $\frac{1}{Ts}$
    (d) $\frac{\quad}{Ts}$
96. The transfer function of a proportional controller is of the type
    (a) Ts
    (b) K_o
    (c) $\frac{1}{Ts+1}$
    (d) $\frac{1}{Ts}$
97. The position error coefficient for a unity feedback system is defined as
    (a) $\lim_{s\to 0} \frac{G(s)}{s}$
    (b) $\lim_{s\to 0} sG(s)$
    (c) $\lim_{s\to 0} G(s)$
    (d) none of these
98. The type – 0 system has a finite non-zero value of
    (a) K_p
    (b) K_v
    (c) K_a
    (d) Either of these
99. The type 1 system has
    (a) No net pole at the origin
    (b) Simple pole at the origin
    (c) Net pole at the origin
    (d) Two poles at the origin
100. Conditionally stable system is one which exhibits poor stability at
    (a) Increased value of open loop gain
    (b) Local frequencies
    (c) Reduced values and open loop gain
    (d) None of these
101. The position and velocity error of a type 2 system are
    (a) Constant, infinity
    (b) Zero, constant
    (c) Zero, zero
    (d) Constant, constant

102. In the s-plane, the unstable region is
 (a) Second and third quadrants including imaginary axis
 (b) First and fourth quadrants including imaginary axis except origin.
 (c) First and second quadrants including real axis.
 (d) Second and fourth quadrants including real axis except the origin.

103. A system has some roots with real parts equal to zero, but none with positive real part is
 (a) Absolutely stable. (b) Relatively stable.
 (c) Marginally stable. (d) Absolutely unstable.

104. When a step function is applied to the input of a system, the output remains below a certain level for all the time, the system is
 (a) Stable. (b) Unstable.
 (c) Not necessarily stable. (d) Always unstable.

105. A system has poles at – 1 and – 5 and zeros at 1 and – 2, the system is
 (a) Stable (b) Unstable
 (c) Marginally stable (d) Highly stable

106. The transfer function of a system is $\dfrac{100\,(1+0.25\,s)}{(1+0.5\,s)}$. The phase shift at $\omega = 0$ and $\omega = \alpha$ will be
 (a) 0° and – 90° (b) 90° and – 90°
 (c) 90° and 180° (d) 0° and 180°

107. The gain of 6 dB is equal to
 (a) – 6 dB/decade (b) 6 dB /decade
 (c) 20 dB /decade (d) 40 dB/decade

108. Phase margin of a system is used to specify
 (a) Relative stability (b) Absolutely stability
 (c) Time response (d) Frequency response

109. Gain margin is the reciprocal of gain at the frequency at which the phase angle becomes
 (a) 0° (b) 90°
 (c) 180° (d) 270°

110. Gain crossover frequency is defined as
 (a) $|G(j\omega)\,H(j\omega)| = 1$ (b) $|G(j\omega)\,H(j\omega)| = 0$
 (c) $|G(j\omega)\,H(j\omega)| = \infty$ (d) $|G(j\omega)\,H(j\omega)| = \dfrac{1}{\sqrt{2}}$

111. The bode plot for the network given below is of the type

(a) 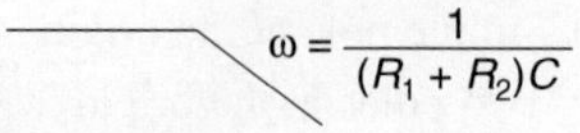

(b) 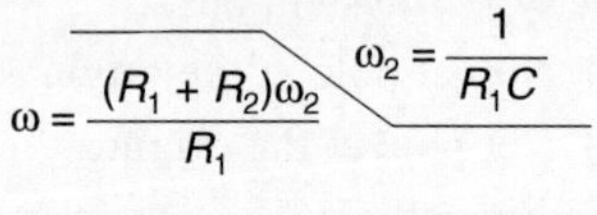

(c) 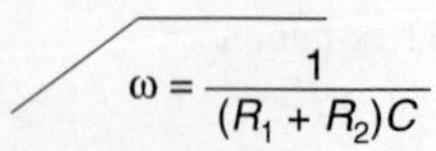

(d) 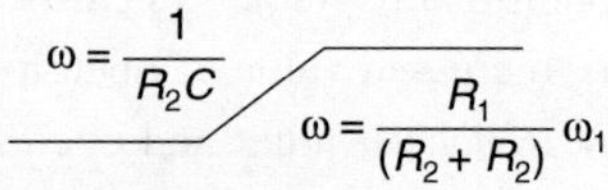

112. Which of the following techniques is not applicable to non-linear system?
 (a) Quasi linearization (b) Functional analysis
 (c) Phase-plane representation (d) Nyquist criterion

113. The open loop transfer function of a system is $G(s) = \dfrac{10}{s(s+2)(s+5)}$ the system is of type

(a) 0 (b) 1
(c) 2 (d) 3

114. A feedback control system has an open loop transfer function given by $G(j\omega) = \dfrac{40}{j\omega(1+j\omega/5)}$.

The phase margin is
(a) 15° (b) 21.8
(c) 31.8 (d) None of these

115. A position control system is a
(a) Automatic regulated system (b) Servomechanism
(c) Process control system (d) Stochastic control system

116. The closed loop gain of the system shown below

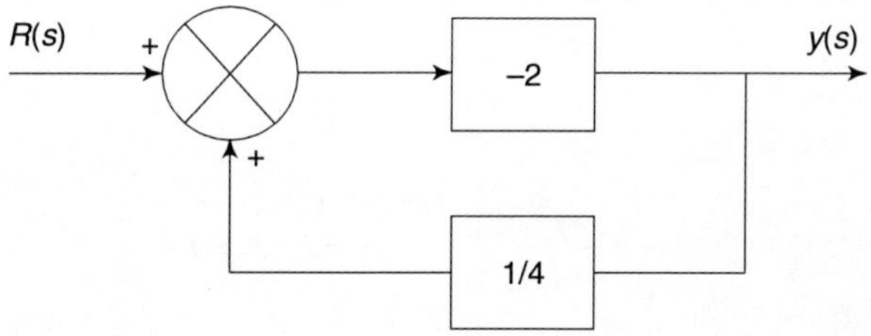

(a) – 4/3 (b) – 4
(c) 4 (d) Non of these

117. Given the transfer function $G(s) = \dfrac{K}{s^3(1+Ts)}$. State the type and order of the system

(a) 2 and 3 (b) 3 and 2
(c) 3 and 4 (d) None of these

118. Transfer function $G(s) = \dfrac{K}{s^2(1+Ts)}$. This open-loop system is

(a) Stable (b) Unstable
(c) Marginally stable (d) Conditionally stable

119. The type number of a transfer function denotes
(a) The number of poles at origin (b) The number of zeros at origin
(c) The number of finite poles (d) The number of poles at infinity

120. The transfer function is $\dfrac{K}{(s+1)(s+2)(s+3)}$ the breakaway point will be between

(a) 0 and – 1 (b) – 1 and – 2
(c) – 2 and – 3 (d) beyond – 3

121. The second order system defined by $\frac{25}{s^2 + 5s + 25}$ is given a step input. The time taken for the output to settle within ± 2% of input is

(a) 1.2*s* (b) I.6*s*

(c) 2s (d) 0.4*s*

122. Which input yields natural response?

(a) Step input (b) Sinusoidal input

(c) Impulse input (d) Ramp input

123. The polar plot of a closed loop system with a transfer function $\frac{G}{1+GH}$ is drawn for

(a) $G(s)$ (b) $1 + GH$

(c) $G(s)\ H(s)$ (d) $\frac{G}{1+GH}$

124. Nyquist plot of a system is given below. The system is

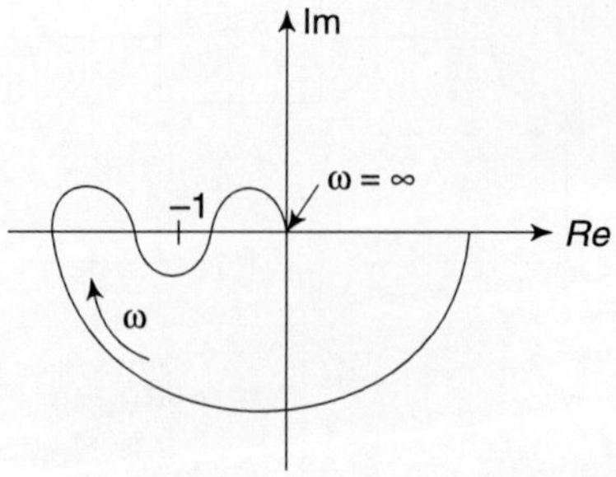

(a) Marginally stable (b) Conditionally stable

(c) Stable (d) Unstable

125. What is the type of system for the Nyquist plot of a system shown below?

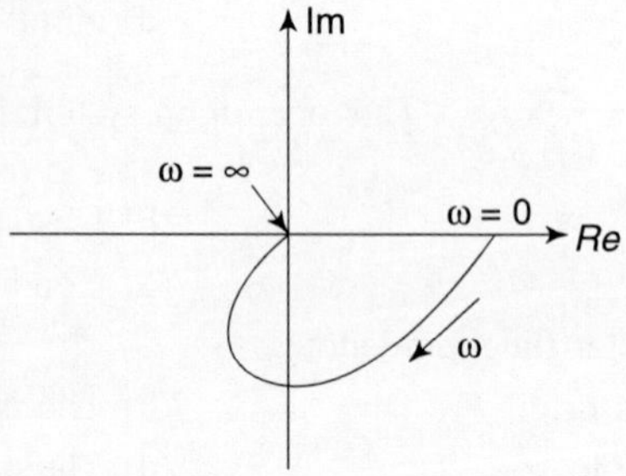

(a) 0 (b) 1

(c) 2 (d) 3

126. The polar plot of a transfer function is shown below. It can be stated that:

(a) The finite zero is closer to the origin than the finite pole

(b) The finite pole is closer to the origin than the zero

(c) The system is unstable

(d) None of these

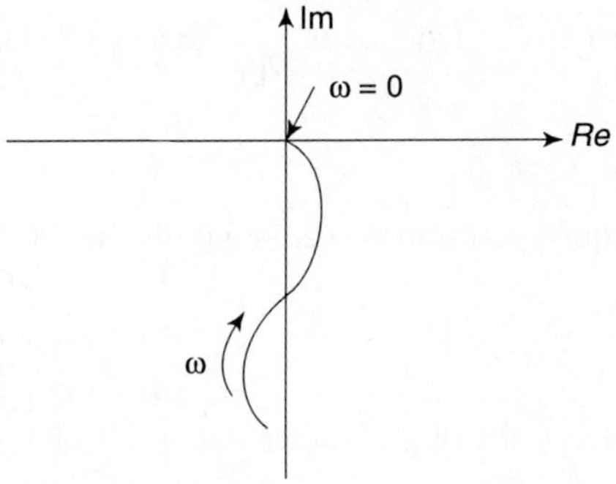

127. A minimum phase system with no zero has a phase angle of – 27° at gain crossover frequency. The system is

(a) Stable
(b) Unstable
(c) Marginally stable
(d) Conditionally stable

128. The transfer function of a closed loop system is $G(s) = \dfrac{K(s+3)}{1+S-s^2+s^3}$. The system is

(a) Stable
(b) Unstable
(c) Marginally stable
(d) Conditionally stable

129. The number of sign changes in the entries in the first column of Routh's array denotes

(a) The number of zeros of the system is the L.H. plane
(b) The number of open loop poles in L. H. plane
(c) The number of open loop zeros in R.H. plane
(d) The number of roots of characteristic polynomial is R.H.P.

130. The gain by a system is 10 at some frequency. In terms of dB, it is

(a) 0 dB
(b) 1 dB
(c) 20 dB
(d) 100 dB

131. 20 dB/decade corresponds to

(a) 3 dB/octane
(b) 6 dB/octane
(c) 10 dB/octane
(d) 20 dB/octane

132. A closed loop system is shown below. When will the system K be very much underdamped?

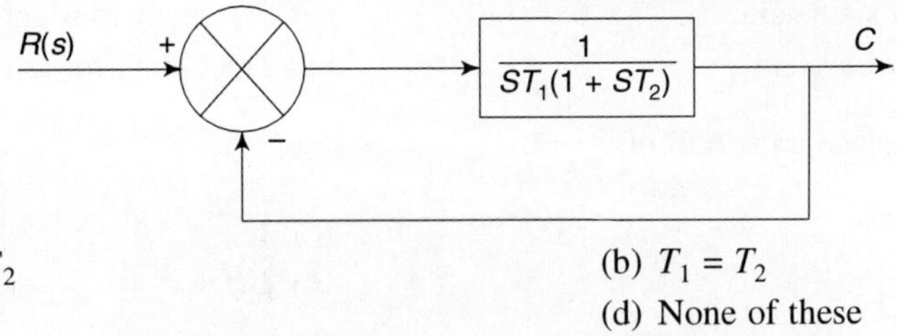

(a) $T_1 << T_2$
(b) $T_1 = T_2$
(c) $T_1 > T_2$
(d) None of these

133. An integral controller is used to improve the transient response of the first order system. If $G(s) = \frac{1}{1+s}$ and the system is operated in closed loop with unity feedback. What is the value of T_i if integral controller transfer function is $\frac{1}{T_i\, s}$ to provide damping ratio of 0.5?

(a) 0.5
(b) 2
(c) 1
(d) 4

134. The unit circle of the Nyquist plot transforms into 0 dB line of the amplitude plot of the bode diagram

(a) At 0 frequency
(b) At low frequency
(c) At high frequency
(d) At any frequency

135. The phase angle versus frequency plot is shown below. It belongs to

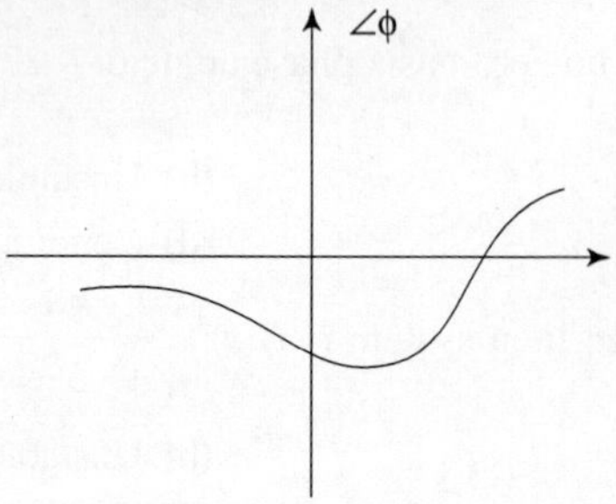

(a) PID controller
(b) Lag-lead network
(c) PD controller
(d) PI controller

136. The transfer function is $\frac{1+0.5\,s}{1+s}$. It represents a

(a) Lead network
(b) Lag network
(c) Lag-lead network
(d) Proportional controller

137. The open loop control system is one in which

(a) Output is dependent on control input
(b) Output is independent of control input
(c) Only system parameters have effect on the control
(d) None of these

138. Transfer function of a system is used to calculate

(a) The steady state gain
(b) The main constant
(c) The order of system
(d) The output for any given input

139. The inverse Laplace transform of $\frac{2}{1+s}$ is

(a) $2\,e^{-1}$
(b) $2e$
(c) $2\,e^{-2t}$
(d) $2\,(t+1)$

140. The poles of $F(s) = \frac{s^2 - 16}{s^5 - 7s^4 - 30s^3}$ are located at

(a) $s = 0$ (tripple pole) , + 3, 10 (b) $s = 0$ tripple pole, – 3 and 10
(c) $s = 4, 4$ (d) $s = 0, 4, 16$

141. For the characteristic equation $s^2 + 4\ 8s + 72 = 0$ the ξ, and ω_n respectively are
(a) 0.212, 8.1 rad/sec (b) 0.299, 8.66 rad/sec
(c) 0.283, 8.48 rad/sec (d) 0.30, 4.21 rad/ sec

142. In the system if the pole lies off the real axis, then the system is
(a) Overdamped (b) Critically damped
(c) Underdamped (d) None of these

143. The open loop transfer function represents
(a) Type 0 system (b) Type I system
(c) Type III system (d) Type II system

144. According to Routh's array the roots of polynomial $F(s) = 0$ all lie in the left half of the s-plane if
(a) All the elements of the first row of the Routh's tabulation are of the same sign.
(b) All the elements of the first column are of the same sign.
(c) Elements of the array of opposite sign alternatively.
(d) Elements of the first column are of opposite sign alternatively.

145. Consider a closed loop system having the following open loop transfer function $G(s)\ H(s) = \dfrac{K}{s\,(T\,s - 1)}$ Nyquist plot for $G(s)\ H(s)$ is shown below. The system is

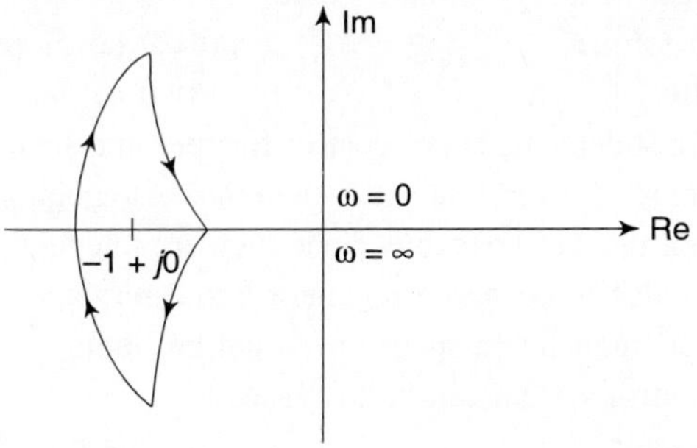

(a) Stable (b) Unstable
(c) Conditionally stable (d) None of these

146. A system has loop gain as $G(s)\ H(s) = \dfrac{K}{s\,(s+1)\,(s+2)\,(s+3)}$. The number of poles and zeros respectively are
(a) 1, 2 (b) 1, 4
(c) 4, 0 (d) 1, 3

147. In the above question, the number of separate root loci is
(a) 1 (b) 2
(c) 3 (d) 4

148. In the above question, the root loci lying on the real axis are between

(a) $s = -1$ and $s = 0$, $s = -1$ and $s = -2$ (b) $s = -1$ and $s = -2$, $s = -2$ and $s = -3$

(c) $s = -1$ and $s = 0$, $s = -1$ and $s = -3$ (d) $s = -1$ and $s = 0$, $s = -2$ and $s = -3$

149. The root loci of a system has three asymptotes. The system can have

(a) Three poles (b) Five poles and two zeros

(c) Four poles and one zero (d) All of these

150. Which of the following statements is not true for root locus technique?

(a) It is used to obtain closed loop pole configuration from open loop poles and zeros

(b) It is most useful for single input single output

(c) It provides the pattern of movement of closed loop poles when open loop gain varies

(d) None of these

151. Consider the pair of equation of a system is $x_2 = t_{12} x_1 + t_{32} x_3$ and $x_3 = t_{13} x_2 + t_{43} x_4$. The signal flow graph of the system will be

(a) 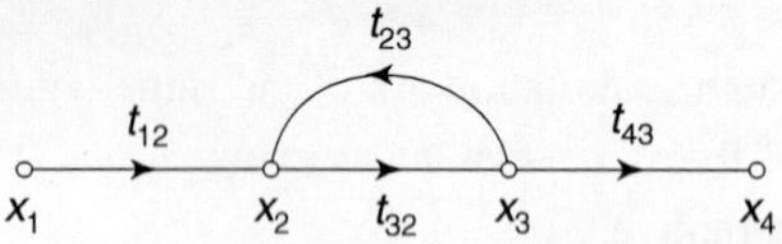

(b) 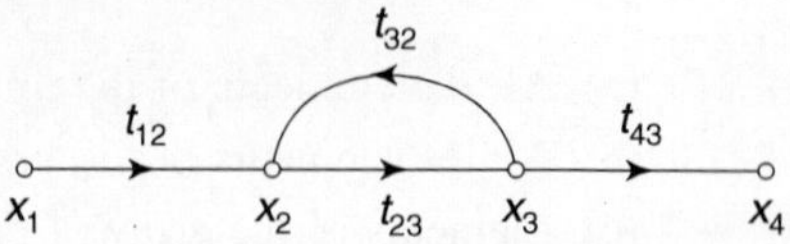

(c) 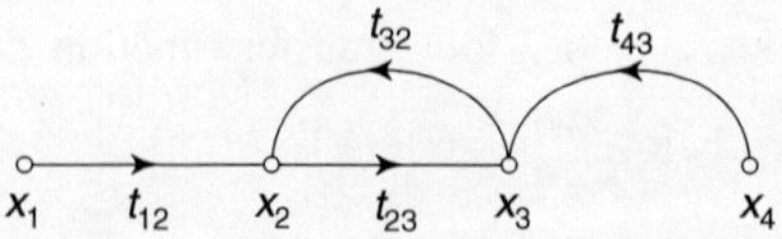

(d) None of these

152. One of the disadvantages of a servomotor is that

(a) It has low starting torque (b) It develops commutation problem

(c) It has low reliability (d) It can handle only light load

153. Which of the following statements is not correct for servomechanism?

(a) Steady state accuracy of a servo is better than that of regulator

(b) Some servos do not need to be stable, since they are intended for use with steady signals

(c) A motor may be added to convert a regulator into a servo.

(d) A servo with better frequency response need not be stable

154. The bandwidth of a control system can be increased by

(a) Phase lag compensator (b) Phase lead compensator

(c) Phase lag lead compensator (d) All of these

155. Adding in the zeros in the transfer function causes

(a) Lag compensation (b) Lead compensation

(c) Lead-lag compensation (d) None of these

156. Adding of poles in the transfer function causes

(a) Lag compensation (b) Lead compensation

(c) Lead-lag compensation (d) None of these

157. In a feedback system, lead compensator

(a) Speeds up the transient response

(b) Increases the margin of stability

(c) Increases the system error constant to some extent

(d) All of these

158. Which compensator will be used to increase the damping of a badly underdamped system?
    (a) Phase lag (b) Phase lead
    (c) Phase lag-lead (d) None of these
159. Excessive noise in control system can cause
    (a) Reduction in bandwidth (b) Reduction in gain
    (c) Saturation in amplifying stages (d) Oscillations
160. The stator of a synchros is made of
    (a) Laminated silicon steel (b) Cast iron
    (c) Stainless steel (d) None of these
161. The servomotor differs from other motors in the sense that it has
    (a) High inertia and high torque (b) Low inertia and low torque
    (c) Low inertia and high torque (d) Entirely different construction
162. Which of the following devices can be used to control the position of very small loads?
    (a) d.c. servomotor (b) Synchro
    (c) a.c. servomotor (d) P.M. M.C. movement
163. The Selsyn servo system provides a
    (a) Overvoltage protection
    (b) Mechanical gearing between two shafts
    (c) Electrical gearing between two shafts
    (d) None of these
164. With the knowledge of state space representation, the transfer function of the system
    (a) Can be determined completely (b) Cannot be determined
    (c) Can be determined partly (d) None of these
165. Which of the following statements is incorrect for state space method?
    (a) It reduce complexity
    (b) It gives compact and convenient form
    (c) Direct programming is not possible on digital computer
    (d) Direct programming is not possible on analog computer
166. Which of the following is a set of state variables for third order system?
    (a) $\ddot{y}(t)$, $\dot{y}(t)$ (b) $\ddot{y}(t)$, $\dot{y}(t)$ $y(t)$
    (c) $\ddot{y}(t)$, $\dddot{y}(t)$, $\dot{y}(t)$ (d) $\ddot{y}(t)$, $y(t)$
167. Which of the following systems provides excellent transient as well as steady state response?
    (a) Proportional action (b) PI action
    (c) PD action (d) P1D action
168. Which of the following statements is a serious disadvantage of electronic control systems?
    (a) Their reliability is low (b) These are temperature sensitive
    (c) These are difficult to operate (d) None of these
169. Which of the following elements is not a final control element?
    (a) Electro-pneumatic convertor (b) Potentiometer
    (c) Servomotor (d) Control value

170. The output of the controller in a control system is given to
(a) Sensor (b) Comparator
(c) Amplifier (d) Final control element

171. The on-off controller is a
(a) Linear system (b) Non-linear system
(c) Digital system (d) Discontinuous system

172. The input to a controller is a
(a) Sensed signal (b) Error signal
(c) Desired variable value (d) Servo signal

173. Essentially a controller is a
(a) Comparator (b) Clipper
(c) Amplifier (d) Sensor

174. When derivative action is included in a proportional controller, the proportional band
(a) Decreases (b) Remains unaffected
(c) Increases (d) Depends upon derivative time constant

175. The phase lag produced by transportation delays
(a) Decreases linearly with frequency (b) Increases linearly with frequency
(c) Inversely proportional to frequency (d) Independent of frequency

ANSWERS

1. (a)	2. (b)	3. (d)	4. (a)	5. (b)	6. (c)
7. (d)	8. (a)	9. (b)	10. (c)	11. (a)	12. (c)
13. (d)	14. (c)	15. (b)	16. (d)	17. (c)	18. (d)
19. (a)	20. (c)	21. (d)	22. (c)	23. (c)	24. (c)
25. (b)	26. (b)	27. (c)	28. (a)	29. (a)	30. (c)
31. (a)	32. (d)	33. (c)	34. (d)	35. (d)	36. (a)
37. (b)	38. (a)	39. (c)	40. (a)	41. (c)	42. (b)
43. (a)	44. (c)	45. (d)	46. (a)	47. (c)	48. (b)
49. (c)	50. (b)	51. (b)	52. (a)	53. (c)	54. (d)
55. (b)	56. (b)	57. (a)	58. (d)	59. (c)	60. (c)
61. (a)	62. (a)	63. (c)	64. (a)	65. (b)	66. (a)
67. (c)	68. (a)	69. (b)	70. (a)	71. (c)	72. (a)
73. (a)	74. (d)	75. (a)	76. (d)	77. (d)	78. (a)
79. (d)	80. (c)	81. (a)	82. (c)	83. (d)	84. (d)
85. (b)	86. (d)	87. (d)	88. (b)	89. (a)	90. (c)
91. (b)	92. (a)	93. (b)	94. (b)	95. (c)	96. (b)
97. (c)	98. (a)	99. (b)	100. (c)	101. (c)	102. (b)
103. (c)	104. (c)	105. (a)	106. (a)	107. (c)	108. (a)

109. (c)	110. (a)	111. (d)	112. (d)	113. (b)	114. (b)
115. (b)	116. (a)	117. (c)	118. (b)	119. (a)	120. (b)
121. (b)	122. (c)	123. (c)	124. (b)	125. (a)	126. (a)
127. (b)	128. (b)	129. (d)	130. (c)	131. (b)	132. (a)
133. (c)	134. (b)	135. (b)	136. (b)	137. (b)	138. (d)
139. (a)	140. (b)	141. (c)	142. (c)	143. (d)	144. (a)
145. (a)	146. (c)	147. (d)	148. (c)	149. (d)	150. (d)
151. (c)	152. (a)	153. (b)	154. (b)	155. (a)	156. (b)
157. (d)	158. (b)	159. (c)	160. (c)	161. (c)	162. (b)
163. (c)	164. (a)	165. (c)	166. (b)	167. (d)	168. (b)
169. (b)	170. (d)	171. (b)	172. (b)	173. (a)	174. (b)
175. (b)					

References

1. Ogata, K., "Modern Control Engineering", PHI Pvt. Ltd. New Delhi, Third Edition, 1997.
2. Nagrath, I.J., and M. Gopal, "Control System Engineering", New Age International Publisher, New Delhi, Fifth Edition, 2007.
3. Kuo, B. C., "Automatic Control System", PHI Pvt. Ltd, New Delhi, Seventh Edition, 1995.
4. Richard C. Dorf and Robert H. Bishop, "Modern Control System", Addison-Wesley, Eighth Edition, 1997.
5. Norman S. Nise, "Control System Engineering", Wiley India, Fourth Edition, 2003.
6. Manke, B.S., "Linear Control System', Khanna Publishers. Third Edition, 2012.
7. Astrom, K.J., and Wittenmark, B. "Computer Controlled Systems" Prentice Hall, Englewood Cliffs, NJ, 1984.
8. Ogata, K., "Discrete-Time Control Systems. Prentice Hall, Englewood Cliffs, NJ, 1987.
9. Phillips, C.L., and Nagle, H.T., Jr. "Digital Control System Analysis and Design". Prentice Hall, Englewood Cliffs, NJ, 1984.
10. Tou, J. "Digital and Sampled-Data Control System McGraw Hill, New York, 1959.

Index

A

B

C

D

E

F